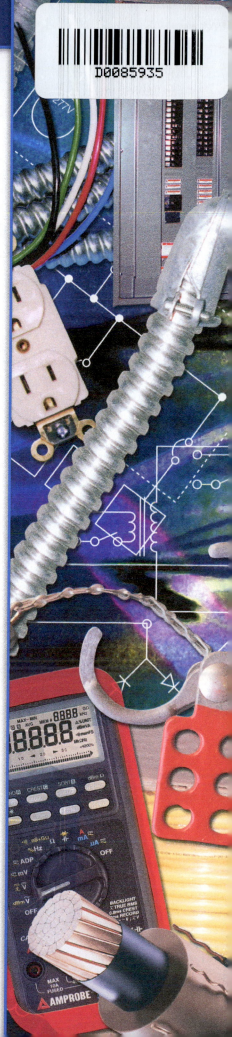

Electrical

Level Four

Trainee Guide
Tenth Edition

Pearson

NCCER

Chief Executive Officer: Don Whyte
President: Boyd Worsham
Chief Operations Officer: Katrina Kersch
Electrical Curriculum Project Manager: Veronica Westfall
Director, Product Development: Tim Davis
Senior Production Manager: Erin O'Nora
Senior Manager of Projects: Chris Wilson
Testing/Assessment Project Manager: Elizabeth Schlaupitz
Project Assistant: Lauren Corley
Lead Technical Writer: Veronica Westfall
Technical Writers: Gary Ferguson, Veronica Westfall, John Mueller, Chris Wilson

Managing Editor: Graham Hack
Desktop Publishing Manager: James McKay
Art Manager: Carrie Pelzer/Karyn Payne
Multimedia Project Manager: Alan Youngblood
Digital Content Coordinator: Rachael Downs
Production Specialists: Gene Page, Eric Caraballoso, Tim Douglas
Production Assistance: Adrienne Payne, David Gregoire, Edward Fortman, Joanne Hart, Olga Trofymenko
Editors: Jordan Hutchinson, Karina Kuchta, Hannah Murray
Product Development Program Specialist: Tim Douglas

Composition: NCCER
Printer/Binder: LSC Communications
Cover Printer: LSC Communications
Text Fonts: Palatino and Univers
Content Technologies: Gnostyx

Credits and acknowledgments for content borrowed from other sources and reproduced, with permission, in this textbook appear at the end of each module.

ISBN-13:   978-0-13-691078-7
ISBN-10:   0-13-691078-5

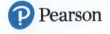

13 2022

Preface

To the Trainee

Electricity powers the applications that make our daily lives more productive and efficient. The demand for electricity has led to vast job opportunities in the electrical field. Electricians constitute one of the largest construction occupations in the United States, and they are among the highest-paid workers in the construction industry. According to the U.S. Bureau of Labor Statistics, the demand for trained electricians is projected to increase, creating more job opportunities for skilled craftspeople in the electrical industry.

Electricians install electrical systems in structures such as homes, office buildings, and factories. These systems include wiring and other electrical components, such as circuit breaker panels, switches, and lighting. Electricians follow blueprints, the *National Electrical Code®*, and state and local codes. They use specialized tools and testing equipment, such as ammeters, ohmmeters, and voltmeters. Electricians learn their trade through craft and apprenticeship programs. These programs provide classroom instruction and on-the-job learning with experienced electricians.

We wish you success as you embark on your first year of training in the electrical craft, and hope that you will continue your training beyond this textbook. There are more than 700,000 people employed in electrical work in the United States, and there are many opportunities awaiting those with the skills and desire to move forward in the construction industry.

New with *Electrical Level Four*

NCCER and Pearson are pleased to present *Electrical Level Four*, which has been updated to meet the 2020 *National Electrical Code®* and includes revisions to the Module Examinations.

In addition to the 2020 *NEC®* changes, this edition of *Electrical Level Four* features several updated modules. *Advanced Controls* (Module ID 26407-20) has been revised to reflect current control and adjustable frequency drive technology. Additionally, the controls presented in *HVAC Controls* (Module ID 26408-20) also feature a great deal of updated technology. The 2020 *NEC®* featured heavy changes to the way loads are calculated and that is reflected in the updated *Load Calculations - Feeders and Services* (Module

ID 26401-20). Additionally, the 2020 *NEC®* had many changes to installations in special locations, such as marinas, which are presented in the revised module *Special Locations* (Module ID 26412-20). Figures and tables throughout the modules have been updated to reflect the NEC and new technology.

Our website, **www.nccer.org**, has information on the latest product releases and training.

Your feedback is welcome. You may email your comments to **curriculum@nccer.org** or send general comments and inquiries to **info@nccer.org**.

NCCER Standardized Curricula

NCCER is a not-for-profit 501(c)(3) education foundation established in 1996 by the world's largest and most progressive construction companies and national construction associations. It was founded to address the severe workforce shortage facing the industry and to develop a standardized training process and curricula. Today, NCCER is supported by hundreds of leading construction and maintenance companies, manufacturers, and national associations. The NCCER Standardized Curricula was developed by NCCER in partnership with Pearson, the world's largest educational publisher.

Some features of the NCCER Standardized Curricula are as follows:

- An industry-proven record of success
- Curricula developed by the industry, for the industry
- National standardization providing portability of learned job skills and educational credits
- Compliance with the Office of Apprenticeship requirements for related classroom training (*CFR 29:29*)
- Well-illustrated, up-to-date, and practical information

NCCER also maintains the NCCER Registry, which provides transcripts, certificates, and wallet cards to individuals who have successfully completed a level of training within a craft in NCCER's Curricula. *Training programs must be delivered by an NCCER Accredited Training Sponsor in order to receive these credentials.*

Special Features

In an effort to provide a comprehensive and user-friendly training resource, this curriculum showcases several informative features. Whether you are a visual or hands-on learner, these features are intended to enhance your knowledge of the construction industry as you progress in your training. Some of the features you may find in the curriculum are explained below.

Introduction

This introductory page, found at the beginning of each module, lists the module Objectives, Performance Tasks, and Trade Terms. The Objectives list the knowledge you will acquire after successfully completing the module. The Performance Tasks give you an opportunity to apply your knowledge to real-world tasks. The Trade Terms are industry-specific vocabulary that you will learn as you study this module.

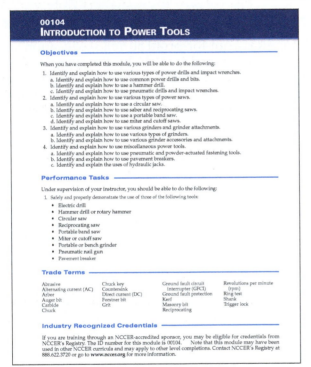

Figures and Tables

Photographs, drawings, diagrams, and tables are used throughout each module to illustrate important concepts and provide clarity for complex instructions. Text references to figures and tables are emphasized with *italic* type.

Notes, Cautions, and Warnings

Safety features are set off from the main text in highlighted boxes and categorized according to the potential danger involved. Notes simply provide additional information. Cautions flag a hazardous issue that could cause damage to materials or equipment. Warnings stress a potentially dangerous situation that could result in injury or death to workers.

Trade Features

Trade features present technical tips and professional practices based on real-life scenarios similar to those you might encounter on the job site.

Bowline Trivia

Some people use this saying to help them remember how to tie a bowline: "The rabbit comes out of his hole, around a tree, and back into the hole."

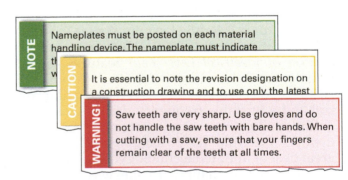

NOTE — Nameplates must be posted on each material handling device. The nameplate must indicate...

CAUTION — It is essential to note the revision designation on a construction drawing and to use only the latest...

WARNING! — Saw teeth are very sharp. Use gloves and do not handle the saw teeth with bare hands. When cutting with a saw, ensure that your fingers remain clear of the teeth at all times.

Case History

Case History features emphasize the importance of safety by citing examples of the costly (and often devastating) consequences of ignoring best practices or OSHA regulations.

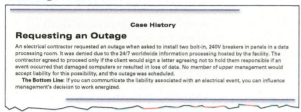

Going Green

Going Green features present steps being taken within the construction industry to protect the environment and save energy, emphasizing choices that can be made on the job to preserve the health of the planet.

Did You Know

Did You Know features introduce historical tidbits or interesting and sometimes surprising facts about the trade.

Step-by-Step Instructions

Step-by-step instructions are used throughout to guide you through technical procedures and tasks from start to finish. These steps show you how to perform a task safely and efficiently.

> Perform the following steps to erect this system area scaffold:
>
> *Step 1* Gather and inspect all scaffold equipment for the scaffold arrangement.
>
> *Step 2* Place appropriate mudsills in their approximate locations.
>
> *Step 3* Attach the screw jacks to the mudsills.

Trade Terms

Each module presents a list of Trade Terms that are discussed within the text and defined in the Glossary at the end of the module. These terms are presented in the text with **bold, blue** type upon their first occurrence. To make searches for key information easier, a comprehensive Glossary of Trade Terms from all modules is located at the back of this book.

> During a rigging operation, the **load** being lifted or moved must be connected to the apparatus, such as a crane, that will provide the power for movement. The connector—the link between the load and the apparatus—is often a sling made of synthetic, chain, or **wire rope** materials. This section focuses on three types of slings:

Section Review

Each section of the module wraps up with a list of Additional Resources for further study and Section Review questions designed to test your knowledge of the Objectives for that section.

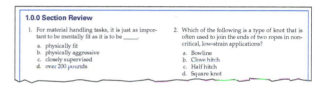

Review Questions

The end-of-module Review Questions can be used to measure and reinforce your knowledge of the module's content.

NCCER Standardized Curricula

NCCER's training programs comprise more than 80 construction, maintenance, pipeline, and utility areas and include skills assessments, safety training, and management education.

Boilermaking
Cabinetmaking
Carpentry
Concrete Finishing
Construction Craft Laborer
Construction Technology
Core Curriculum: Introductory
 Craft Skills
Drywall
Electrical
Electronic Systems Technician
Heating, Ventilating, and Air
 Conditioning
Heavy Equipment Operations
Heavy Highway Construction
Hydroblasting
Industrial Coating and Lining
 Application Specialist
Industrial Maintenance Electrical
 and Instrumentation Technician
Industrial Maintenance Mechanic
Instrumentation
Ironworking
Manufactured Construction
 Technology
Masonry
Mechanical Insulating
Millwright
Mobile Crane Operations
Painting
Painting, Industrial
Pipefitting
Pipelayer
Plumbing
Reinforcing Ironwork
Rigging
Scaffolding
Sheet Metal
Signal Person
Site Layout
Sprinkler Fitting
Tower Crane Operator
Welding

Maritime

Maritime Industry Fundamentals
Maritime Electrical
Maritime Pipefitting
Maritime Structural Fitter
Maritime Welding
Maritime Aluminum Welding

Green/Sustainable Construction

Building Auditor
Fundamentals of Weatherization
Introduction to Weatherization
Sustainable Construction
 Supervisor
Weatherization Crew Chief
Weatherization Technician
Your Role in the Green
 Environment

Energy

Alternative Energy
Introduction to the Power Industry
Introduction to Solar Photovoltaics
Power Generation Maintenance
 Electrician
Power Generation I&C
 Maintenance Technician
Power Generation Maintenance
 Mechanic
Power Line Worker
Power Line Worker: Distribution
Power Line Worker: Substation
Power Line Worker: Transmission
Solar Photovoltaic Systems Installer
Wind Energy
Wind Turbine Maintenance
 Technician

Pipeline

Abnormal Operating Conditions,
 Control Center
Abnormal Operating Conditions,
 Field and Gas
Corrosion Control
Electrical and Instrumentation
Field and Control Center
 Operations
Introduction to the Pipeline
 Industry
Maintenance
Mechanical

Safety

Field Safety
Safety Orientation
Safety Technology

Supplemental Titles

Applied Construction Math
Tools for Success

Management

Construction Workforce
 Development Professional
Fundamentals of Crew Leadership
Mentoring for Craft Professionals
Project Management
Project Supervision

Spanish Titles

Acabado de concreto: nivel uno
 (*Concrete Finishing Level One*)
Aislamiento: nivel uno
 (*Insulating Level One*)
Albañilería: nivel uno
 (*Masonry Level One*)
Andamios (*Scaffolding*)
Carpintería: Formas para
 carpintería, nivel tres
 (*Carpentry: Carpentry Forms, Level
 Three*)
Currículo básico: habilidades
 introductorias del oficio
 (*Core Curriculum: Introductory Craft
 Skills*)
Electricidad: nivel uno
 (*Electrical Level One*)
Herrería: nivel uno
 (*Ironworking Level One*)
Herrería de refuerzo: nivel uno
 (*Reinforcing Ironwork Level One*)
Instalación de rociadores: nivel uno
 (*Sprinkler Fitting Level One*)
Instalación de tuberías: nivel uno
 (*Pipefitting Level One*)
Instrumentación: nivel uno, nivel
 dos, nivel tres, nivel cuatro
 (*Instrumentation Levels One through
 Four*)
Orientación de seguridad
 (*Safety Orientation*)
Paneles de yeso: nivel uno
 (*Drywall Level One*)
Seguridad de campo
 (*Field Safety*)

Acknowledgments

This curriculum was revised as a result of the farsightedness and leadership of the following sponsors:

ABC of Iowa
ABC of Western Pennsylvania
Beacon Electrical Contractors
Cianbro Corporation
Elm Electrical, Inc.
Gaylor Electric, Inc.
Gould Construction Institute

Industrial Management and Training Institute
National Field Services
Madison Comprehensive High School
Putnam Career and Technical Center
Tri-City Electrical Contractors

This curriculum would not exist were it not for the dedication and unselfish energy of those volunteers who served on the Authoring Team. A sincere thanks is extended to the following:

Paul Asselin
David Coelho
Tim Dean
Tim Ely
Mark Kozloski
Dan Lamphear

David Lewis
John Lupacchino
Gerard McDonald
Scott Mitchell
John Mueller
Steve Newton

Ron Otts
Mike Powers
Josiha Schuh
Wayne Stratton
James Westfall

NCCER Partners

American Fire Sprinkler Association
Associated Builders and Contractors, Inc.
Associated General Contractors of America
Association for Career and Technical Education
Construction Industry Institute
Construction Users Roundtable
Gulf States Shipbuilders Consortium
ISN Software Corporation
Manufacturing Institute
Mason Contractors Association of America
Merit Contractors Association of Canada
NACE International
National Association of Women in Construction
National Insulation Association
National Technical Honor Society
NAWIC Education Foundation
North American Crane Bureau
North American Technician Excellence
Pearson
Prov

SkillsUSA®
Steel Erectors Association of America
University of Florida, M. E. Rinker Sr., School of
    Construction Management

Contents

Module One
Load Calculations – Feeders and Services

Using load calculations to accurately size protection devices and circuit conductors ensures that loads are safely supported in an electrical system. These calculations must be made in accordance with *NEC®* requirements, which establish the minimum standards for electrical calculations. This module covers basic calculations for commercial and residential applications, including raceway fill, conductor derating, and voltage drop. (Module ID 26401-20; 20 Hours)

Module Two
Health Care Facilities

In health care facilities, reliable electrical systems and adequate backup power are vital for the safety of patients. For this reason, these systems are strictly regulated in the *NEC®*, as well as in state and local codes. This module covers the installation, alarm system, and backup system requirements of electrical systems in health care facilities, including the requirements for life safety and critical circuits. (Module ID 26402-20; 10 Hours)

Module Three
Standby and Emergency Systems

In places of assembly, power systems provide important functions such as lighting, evacuation routes and others during emergency situations. This module explains the *NEC®* installation requirements for electric generators and storage batteries used during such emergency situations. (Module ID 26403-20; 10 Hours)

Module Four
Basic Electronic Theory

Electronics is a science that deals with the behavior and effect of electron flow in specific substances (which are used to create electronic devices), such as semiconductors. Electronics can be distinguished from electricity in terms of the voltages and currents used. This module explains the function and operation of basic electronic devices, including semiconductors, diodes, rectifiers, and transistors. (Module ID 26404-20; 10 Hours)

Module Five
Fire Alarm Systems

Fire alarms provide an essential service that protects both human life and property from the effects of fire. Fire alarms can be complex systems made up of many different technologies. Numerous codes govern fire alarms to ensure that they operate in useful and predictable ways. This module explores the technologies, codes, and wiring approaches used to assemble a fire alarm system. Installation and troubleshooting techniques are also examined. (Module ID 26405-20; 15 Hours)

Module Six
Specialty Transformers

Transformers are used to increase or decrease voltage coming from a power source. They can be constructed in a variety of configurations for different applications. This module covers various types of transformers, and provides information on selecting, sizing, and installing them. (Module ID 26406-20; 10 Hours)

Module Seven

Advanced Controls

Control systems are what regulate and direct the behavior of devices within an electrical system. They vary in complexity and consist of a variety of components, which provide different types of control. This module discusses applications and operating principles of various control system components, such as solid-state relays, reduced-voltage starters, and adjustable-frequency drives. It also covers basic troubleshooting procedures. (Module ID 26407-20; 20 Hours)

Module Eight

HVAC Controls

Heating, ventilation, and air conditioning (HVAC) systems are among the electrically powered and controlled systems that electricians will encounter, especially in residential and commercial construction. During installation, electricians will be called upon to provide power and control connections to the various components of these systems. For that reason, it is important that electricians develop a basic understanding of HVAC systems and their components. (Module ID 26408-20; 15 Hours)

Module Nine

Heat Tracing and Freeze Protection

Electrical heat-tracing systems keep piping systems at or above a given temperature for the purposes of freeze protection and temperature maintenance. In cold climates, freeze protection is needed because the thermal insulation installed around pipes is not enough to prevent the process fluids from freezing. This module presents heat-tracing and freeze-protection systems along with various applications and installation requirements. (Module ID 26409-20; 10 Hours)

(continued)

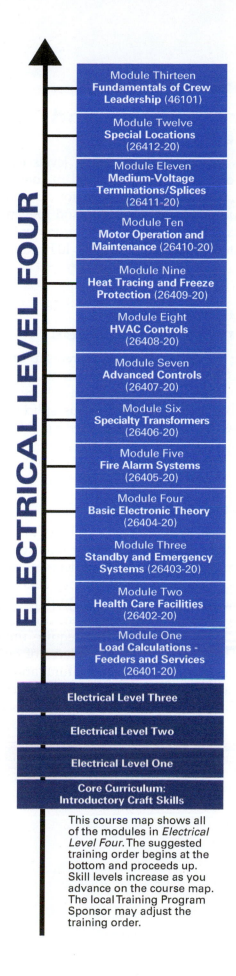

ELECTRICAL LEVEL FOUR

Module Thirteen
Fundamentals of Crew Leadership (46101)

Module Twelve
Special Locations (26412-20)

Module Eleven
Medium-Voltage Terminations/Splices (26411-20)

Module Ten
Motor Operation and Maintenance (26410-20)

Module Nine
Heat Tracing and Freeze Protection (26409-20)

Module Eight
HVAC Controls (26408-20)

Module Seven
Advanced Controls (26407-20)

Module Six
Specialty Transformers (26406-20)

Module Five
Fire Alarm Systems (26405-20)

Module Four
Basic Electronic Theory (26404-20)

Module Three
Standby and Emergency Systems (26403-20)

Module Two
Health Care Facilities (26402-20)

Module One
Load Calculations - Feeders and Services (26401-20)

Electrical Level Three

Electrical Level Two

Electrical Level One

Core Curriculum: Introductory Craft Skills

This course map shows all of the modules in *Electrical Level Four*. The suggested training order begins at the bottom and proceeds up. Skill levels increase as you advance on the course map. The local Training Program Sponsor may adjust the training order.

Module Ten
Motor Operation and Maintenance

For motors to operate at optimum levels over a long span of time, proper care and maintenance is required. This module covers motor care procedures, including cleaning, testing, and preventive maintenance. Basic troubleshooting procedures are also presented. (Module ID 26410-20; 10 Hours)

Module Eleven
Medium-Voltage Terminations/Splices

Both taping systems and manufactured slip-on kits are used to make splices and terminations. This module identifies types of medium-voltage cable and describes how to make various splices and terminations. It also covers hi-pot testing. (Module ID 26411-20; 10 Hours)

Module Twelve
Special Locations

As an electrician, it is your responsibility to familiarize yourself with all of the *NEC®* requirements for any electrical installation you undertake, as well as any local or regional codes that may also apply. This module describes the *NEC®* requirements for selecting and installing equipment, enclosures, and devices for special locations that require unique attention. These locations include places of public assembly, theaters, carnivals, agricultural and livestock facilities, marinas, swimming pools, and temporary facilities. (Module ID 26412-20; 20 Hours)

Module Thirteen
Fundamentals of Crew Leadership

When a crew is assembled to complete a job, one person is appointed the leader. This person is usually an experienced craft professional who has demonstrated leadership qualities. While having natural leadership qualities helps in becoming an effective leader, it is more true that "leaders are made, not born." Whether you are a crew leader or want to become one, this module will help you learn more about the requirements and skills needed to succeed. (Module ID 46101; 22.5 Hours)

Glossary

Load Calculations – Feeders and Services

OVERVIEW

Using load calculations to accurately size protection devices and circuit conductors ensures that loads are safely supported in an electrical system. These calculations must be made in accordance with *NEC®* requirements, which establish the minimum standards for electrical calculations. This module covers basic calculations for commercial and residential applications, including raceway fill, conductor derating, and voltage drop.

Module 26401-20

Trainees with successful module completions may be eligible for credentialing through the NCCER Registry. To learn more, go to **www.nccer.org** or contact us at 1.888.622.3720. Our website, **www.nccer.org**, has information on the latest product releases and training.

Your feedback is welcome. You may email your comments to **curriculum@nccer.org**, send general comments and inquiries to **info@nccer.org**, or fill in the User Update form at the back of this module.

This information is general in nature and intended for training purposes only. Actual performance of activities described in this manual requires compliance with all applicable operating, service, maintenance, and safety procedures under the direction of qualified personnel. References in this manual to patented or proprietary devices do not constitute a recommendation of their use.

26401-20 V10.0

Objectives

When you have completed this module, you will be able to do the following:

1. Perform basic load calculations in accordance with *National Electrical Code®* (*NEC®*) requirements.
   a. Make adjustments in conductor size for various installations.
   b. Calculate feeder ampacity.
   c. Apply tap rules.
   d. Apply demand factors.
2. Make service calculations for residential installations.
   a. Calculate the minimum service size for simple electrical installations.
   b. Make service calculations for single-family dwellings.
   c. Make service calculations for multi-family dwellings.
3. Make service calculations for commercial installations.
   a. Size commercial and industrial lighting loads.
   b. Calculate loads for retail stores.
   c. Calculate loads for office buildings.
   d. Calculate loads for restaurants.
   e. Calculate loads for hotels and motels.
   f. Perform optional calculations for schools.
   g. Size shore power circuits for marinas and boatyards.
   h. Make farm load calculations.
   i. Size motor circuits.

Performance Tasks

This is a knowledge-based module. There are no Performance Tasks.

Trade Terms

Building	Separately derived system	Service lateral
Dwelling unit	Service	Service point
Feeder	Service conductors	Switchboard
Garage	Service drop	Tap conductors
Overhead service conductors	Service equipment	Underground service conductors

Industry Recognized Credentials

If you are training through an NCCER-accredited sponsor, you may be eligible for credentials from NCCER's Registry. The ID number for this module is 26401-20. Note that this module may have been used in other NCCER curricula and may apply to other level completions. Contact NCCER's Registry at 888.622.3720 or go to **www.nccer.org** for more information.

> **NOTE**
> NFPA 70®, *National Electrical Code®* and *NEC®* are registered trademarks of the National Fire Protection Association, Quincy, MA.

Contents

1.0.0 Basic Load Calculations 1

  1.1.0 Conductor Adjustments 2

    1.1.1 Continuous Duty, Raceway Fill, and Ambient Temperature 2

    1.1.2 Voltage Drop 4

    1.1.3 Temperature Limitations for Terminating Conductors 7

  1.2.0 Calculating Feeder Ampacity 7

  1.3.0 Tap Rules 9

    1.3.1 The 10' (3 m) Tap Rule 9

    1.3.2 The 25' (7.5 m) Tap Rule 9

    1.3.3 Taps Supplying a Transformer – Primary and Secondary
          Not Over 25' (7.5 m) 9

    1.3.4 The 100' (30 m) Tap Rule 9

    1.3.5 Outside Taps of Unlimited Length 10

    1.3.6 Transformer Secondary Conductors 10

    1.3.7 Tap Calculations 10

  1.4.0 Demand Factors 11

    1.4.1 Demand Factors for Neutral Conductors 11

    1.4.2 Service Conductor Considerations 11

    1.4.3 Lighting Loads 12

2.0.0 Service Calculations 14

  2.1.0 Calculations for Simple Electrical Installations 15

    2.1.1 Rural Pump House Sample Calculation 15

    2.1.2 Roadside Vegetable Stand 16

  2.2.0 Calculations for Single-Family Dwellings 17

    2.2.1 Sizing the Neutral Conductor 19

    2.2.2 Single-Family Dwelling Standard Calculation 20

    2.2.3 Standard Calculation for Single-Family Dwelling Using
          Electric Heat 21

    2.2.4 Optional Calculation for a Single-Family Dwelling 22

  2.3.0 Calculations for Multi-Family Dwellings 23

    2.3.1 Multi-Family Dwelling Sample Calculation 24

    2.3.2 Multi-Family Dwelling Calculation Using Optional Method 25

3.0.0 Commercial Occupancy Calculations 28

  3.1.0 Commercial and Industrial Lighting Loads 28

    3.1.1 Lighting Loads 29

    3.1.2 General Lighting Loads 29

    3.1.3 Show Window Loads 29

    3.1.4 Track Lighting Loads 29

    3.1.5 Outside Lighting Loads 30

    3.1.6 Sign and Outline Lighting Loads 30

3.2.0	Loads for Retail Stores	30
3.3.0	Loads for Office Buildings	32
3.3.1	Calculations for the Primary Feeder	34
3.3.2	Main Service Calculations	34
3.4.0	Loads for Restaurants	35
3.5.0	Loads for Hotels and Motels	35
3.6.0	Optional Calculations for Schools	35
3.7.0	Shore Power Circuits for Marinas and Boatyards	37
3.8.0	Farm Load Calculations	38
3.9.0	Motor Circuits	38

Figures and Tables

Figure 1	Load center length calculation	5
Figure 2	Basic steps in load calculation	12
Figure 3	Floor plan of a small pump house	15
Figure 4	Roadside vegetable stand	16
Figure 5	Single-family dwelling	17
Figure 6	Completed calculation form	18
Figure 7	*NEC*® requirements for multi-family dwellings	23
Figure 8	Floor plan of a retail store	31
Figure 9	One-line diagram of an office building	33
Figure 10	Restaurant service specifications	36
Figure 11	Summary of farm calculations	39
Table 1	Main Service Calculation Example	34

This page is intentionally left blank.

1.0.0 BASIC LOAD CALCULATIONS

Objective

Perform basic load calculations in accordance with *National Electrical Code®* (*NEC®*) requirements.

a. Make adjustments in conductor size for various installations.
b. Calculate feeder ampacity.
c. Apply tap rules.
d. Apply demand factors.

Trade Terms

Building: A structure that stands alone or that is separated from adjoining structures by firewalls.

Dwelling unit: A single unit providing complete and independent living facilities for one or more persons, including permanent provisions for living, sleeping, cooking, and sanitation. A one-family dwelling consists solely of one dwelling unit. A two-family dwelling consists solely of two dwelling units. A multi-family dwelling is a building containing three or more dwelling units.

Feeder: All circuit conductors between the service equipment, the source of a separately derived system, or other supply source and the final branch circuit overcurrent device.

Garage: A building or portion of a building in which one or more self-propelled vehicles can be kept for use, sale, storage, rental, repair, exhibition, or demonstration purposes.

Overhead service conductors: The overhead conductors between the service point and the first point of connection to the service-entrance conductors at the building or other structure.

Service: The conductors and equipment for delivering electric energy from the serving utility to the wiring system of the premises served.

Service conductors: The conductors from the service point to the service disconnecting means.

Service drop: The overhead conductors between the utility electric supply system and the service point.

Service lateral: The underground conductors between the utility electric supply system and the service point.

Service point: The point of connection between the facilities of the serving utility and the premises wiring.

Switchboard: A large single panel, frame, or assembly of panels on which are mounted on the face, back, or both, switches, overcurrent and other protective devices, buses, and usually instruments. Switchboards are generally accessible from the rear as well as from the front and are not intended to be installed in cabinets.

Tap conductors: As defined in *NEC Article 240*, a tap conductor is a conductor, other than a service conductor, that has overcurrent protection ahead of its point of supply that exceeds the value permitted for similar conductors that are protected as described in *NEC Section 240.4*.

Underground service conductors: The underground conductors between the service point and the first point of connection to the service-entrance conductors in a terminal bus.

The purpose of **feeder** circuit load calculations is to determine the size of feeder circuit overcurrent protection devices and feeder circuit conductors using *NEC®* requirements. These calculations help to ensure that loads within the circuit are adequately supported, resulting in a safe, stable system. Once the feeder circuit load is accurately calculated, feeder circuit components may be sized to serve the load safely. Feeder circuits typically supply transformers, motor control centers, and subpanels.

> **NOTE**
> The load calculations/methods presented in this module use the same concepts and equations that were presented in the "Load Calculations – Branch and Feeder Circuits" module from *Electrical Level Three* (Module ID 26301-20). It is strongly suggested that you review NCCER Module 26301-20 prior to completing this module, as its topics provide a basis for these calculations.

NEC Article 215 covers feeder circuits. *NEC Section 215.2* states that feeders shall have an ampacity that is not less than that required to supply the load, as computed in *NEC Article 220*. The minimum feeder-circuit conductor size, before the application of any adjustment or correction factors, shall have an allowable ampacity of not less than 100% of the noncontinuous load plus 125% of the continuous load.

It is important to remember that the *NEC®* establishes minimum standards for electrical calculations. Since most journeyman and master electrician examinations are based upon the minimum *NEC®* requirements, this module will focus on those requirements. Common practice in the electrical industry is to design and plan electrical systems to make allowance for future expansion. Providing conduit and raceways larger than the minimum required by the *NEC®* may reduce the labor required in pulling conductors, which may offset the higher cost of the larger conduit. A service is likely to be sized larger than the minimum required by the *NEC®* so that future expansion of the electrical system will be possible.

Electrical calculations can be divided into three sections:

- Branch circuits
- Feeders
- Services

The total load connected to the branch circuits determines the load on feeders, and the total feeder load determines the load on the service conductors and equipment. When calculating loads, the logical sequence is to determine the load on branch circuits, calculate the feeder load, and then determine the service load.

1.1.0 Conductor Adjustments

The allowable current-carrying capacity (ampacity) of conductors may require adjustment based on several factors. After complying with *NEC Section 110.14(C)*, the conductors selected may have to be increased in size to accommodate the altered allowable ampacity for that conductor. Feeder conductor size is adjusted in the following situations:

- The load to be supplied is a continuous duty load per *NEC Section 215.2(A)(1)(a)*.
- There are more than three current-carrying conductors in a raceway per *NEC Section 310.15(C)(1)*.
- The ambient temperature that the conductors will pass through exceeds the temperature ratings for conductors listed in *NEC Table 310.15(B)(1)* correction factors.
- Although not mandatory, an increase in conductor size may be made when there is a voltage drop exceeding 3% for feeders or 5% for the combination of the feeder and branch circuit that was caused by excessive voltage drop per *NEC Section 215.2(A)(1)(b), Informational Note No. 2*.

1.1.1 Continuous Duty, Raceway Fill, and Ambient Temperature

Feeder conductors are adjusted when the load (or a part of the load) connected to the feeder is a continuous duty load. A continuous duty load is a load that operates without interruption for more than 3 hours straight.

NEC Table 310.16 lists allowable ampacities for insulated conductors rated up to and including 2,000V. The ampacities listed apply when no more than 3 current-carrying conductors are installed in a raceway or are part of a cable assembly. These ampacities are based upon an ambient temperature of 30°C (86°F).

If the number of current-carrying conductors in the conduit or cable exceeds 3, the ampacity of the conductors must be adjusted using *NEC Table 310.15(C)(1)*. If the ambient temperature exceeds 30°C (86°F), the ampacity of the conductors must be adjusted using the temperature correction factors listed in *NEC Table 310.15(B)(1)*.

When derating conductors, it is permitted to use the allowable ampacity listed for the temperature rating of the conductor insulation instead of the selection ampacity, which is restricted to the 60°C (140°F) or 75°C (167°F) column. For example, the minimum size of THHN copper permitted to supply a load of 130A is No. 1 AWG copper, based on the 75°C (167°F) column of *NEC Table 310.16*. When derating this conductor, the ampacity used will be the 145A, as shown in the 90°C (194°F) column.

NOTE

The selection ampacity for choosing a conductor to carry a load is taken from either the 60°C (140°F) or 75°C (167°F) column of *NEC Table 310.16* as required by *NEC Section 110.14(C)*. When determining the ability of that conductor to carry the load under adverse conditions, the allowable ampacity of the conductor is used based on its actual insulation temperature rating. The size of a conductor may also be increased when the length of the feeder conductor requires a larger circular mil area to ensure a low enough resistance to allow the needed current to flow.

Example 1:

What is the adjusted ampacity for each of six No. 4/0 THHN copper current-carrying conductors in a single conduit?

Solution:

Per *NEC Table 310.16*, the ampacity of No. 4/0 THHN is 260A. Per *NEC Table 310.15(C)(1)*, the ampacity must be adjusted by 80% (four through six current-carrying conductors). To find 80% of the ampacity, multiply by 0.8:

$$80\% \text{ of } 260A =$$
$$260A \times 0.8 = 208A \text{ for each conductor}$$

Example 2:

What is the maximum adjusted ampacity for each of nine No. 3/0 current-carrying THWN copper conductors in a single conduit?

Solution:

Per *NEC Table 310.16*, the ampacity of No. 3/0 THWN is 200A. Per *NEC Table 310.15(C)(1)*, the ampacity must be adjusted by 70% (seven through nine current-carrying conductors), as follows:

$$200A \times 0.7 = 140A \text{ for each conductor}$$

Example 3:

A feeder is being installed to supply a load of 375A. The feeder must be routed through a room with an ambient temperature of 98°F. Will 500 kcmil THHN copper conductors be able to carry the load under these conditions?

Solution:

Per *NEC Table 310.16*, 500 kcmil THHN conductors have an ampacity of 430A. Per the correction factors of *NEC Table 310.15(B)(1)*, the ampacity of a type THHN conductor must be multiplied by a factor of 0.91 to account for the ambient temperature:

$$430A \times 0.91 = 391.3A$$

The 500 kcmil THHN conductors will meet the requirements of the *NEC®*.

Example 4:

What is the minimum size THHN copper conductor required for a feeder supplying a continuous load of 85A and a noncontinuous load of 105A while in an ambient temperature of 34°C and with a total of four current-carrying conductors in the same raceway?

Solution:

To determine the required conductor size, you must calculate the total load to be connected. First, find 125% of the continuous load by multiplying it by 1.25:

$$85A \times 1.25 = 106.25A$$

Then, add the noncontinuous load to determine the total load:

$$106.25A + 105A = 211.25A$$

Per *NEC Table 310.16*, No. 4/0 THHN is selected to carry the load of 211.25A using the 75°C column.

Now, find the total connected load before any adjustments are made (without multiplying the noncontinuous load by 1.25), as follows:

$$85A + 105A = 190A$$

Per *NEC Table 310.15(B)(1)*, the ampacity of a type THHN conductor when in an ambient temperature of 35°C must be multiplied by a factor of 0.96. The allowable ampacity of No. 4/0 THHN copper is 260A.

$$260A \times 0.96 = 249.6A$$

Per *NEC Table 310.15(C)(1)*, the adjustment factor for finding allowable ampacity of a conductor run in a raceway with 4 to 6 current-carrying conductors is 80%. To calculate this, multiply by 0.8:

$$249.6A \times 0.8 = 199.68A$$

The No. 4/0 THHN conductors at 199.68A will be able to carry the load of 190A under these adverse conditions. This conductor satisfies the requirements of *NEC Section 215.2(A)(1)*.

Reciprocals – When derating, reciprocals can be used to quickly find a conductor to serve a load. The reciprocal function on a calculator is typically shown as 1/x.

> **NOTE**
>
> A reciprocal is defined as the value by which you multiply a number in order to obtain a result of 1. The reciprocal of a number is usually written as a fraction, and is equal to 1 over that number. For example, the reciprocal of 3 is ⅓.

Example:

What size THHN conductor would be needed to supply a 150A load when there are 4 to 6 current-carrying conductors in a raceway with an ambient temperature of 110°F?

Solution:

The adjustment factor for 4 to 6 current-carrying conductors is 80%, or 0.8 [*NEC Table 310.15(C)(1)*] The correction factor at this ambient temperature is 0.87 [*NEC Table 310.15(B)(1)*]. First, find the reciprocals of these two factors. Then, find the product of these reciprocals and the load current to size the conductor:

Calculate the reciprocals:

$\frac{1}{0.8} = 1.25$

$\frac{1}{0.87} = 1.1494$

Multiply:

$150A \times 1.25 \times 1.1494 = 215.5A$

In the 90°C column of *NEC Table 310.16*, you will find that the minimum size wire needed for these conditions is No. 3/0 THHN copper.

1.1.2 Voltage Drop

NEC Section 215.2(A)(1)(b), Informational Note No. 2 recommends that the voltage drop for feeders should not exceed 3% to the farthest outlet, and 5% for the combination of feeders and branch circuits to the farthest outlet. When both feeders and branch circuits are combined, the maximum voltage drop allowed for the branch circuits is usually 3%, then the voltage drop allowed for the feeder can be no more than 2%. There are two formulas used to calculate voltage drop for single-phase circuits with resistive loads (power factor = 1.0).

Formula 1:

$$VD = \frac{2 \times L \times R \times I}{1,000}$$

Where:

VD = voltage drop (in volts)
  L = load center or total length in feet
  R = conductor resistance per *NEC Chapter 9, Table 8*
  I = current

The number 1,000 represents 1,000' or less of conductor.

Formula 2:

$$VD = \frac{2 \times L \times K \times I}{CM}$$

Where:

VD = voltage drop
  L = load center or total length in feet
  K = constant (12.9 for copper conductors; 21.2 for aluminum conductors)
  I = current
CM = area of the conductor in circular mils (from *NEC Chapter 9, Table 8*)

To calculate the percentage of voltage drop, divide the voltage drop (VD) by the applied voltage (V) and multiply by 100:

$$\text{Percent voltage drop} = \frac{VD}{V} \times 100$$

Where:

VD = voltage drop
  V = applied voltage

The result is divided by circuit voltage and then multiplied by 100 to determine the percentage voltage drop. If the branch circuit voltage drop exceeds a specified amount, the conductor size is increased to compensate. For a 120V circuit, the maximum voltage drop (in volts) for a permitted 3% drop is:

3% of 120V =
120V × 0.03 = 3.6V

For a 240V, single-phase (1∅) feeder combined with branch circuits, the maximum voltage drop (in volts) would normally be:

5% of 240V =
240V × 0.05 = 12V

In commercial fixed load applications, if the feeder load is not concentrated at the end of the feeder but is spread out along the feeder due to multiple taps, the load center length of the feeder should be calculated and used in the voltage drop formulas. This is because the total current does not flow the complete length of the feeder. If the full

The K Factor

To find the K factor, multiply the resistance per foot of the conductor (found in *NEC Chapter 9, Table 8*) by the circular mils of the conductor and divide by 1,000. For example, No. 6 AWG = 26,240 CM × 0.491 = 12,883.84 ÷ 1,000 = 12.88 (round to 12.9).

length is used in computing the voltage drop, the drop determined will be greater than what would actually occur. The load center length of a feeder is that point in the feeder where, if the load were concentrated at that point, the voltage drop would be the same as the voltage drop to the farthest load in the actual feeder. To determine the load center length of a feeder with multiple taps, as shown in *Figure 1*, multiply each tap load by its actual physical routing distance from the supply end of the feeder. Add these products for all loads fed from the feeder and divide this sum by the sum of the individual loads. The resulting distance is the load center length (L) for the total load (I) of the feeder.

Example 1:

The length of a 240V, three-wire panel feeder is 145'. The total panel load consists of 180A of noncontinuous duty loads. No. 3/0 THHN copper conductors would normally be selected to supply this load. If they are used, will the voltage drop for this panel feeder exceed 2%?

Solution:

Use the first voltage drop formula to determine the voltage drop for the feeder. Look up the resistance for No. 3/0 AWG copper in **NEC Chapter 9, Table 8**. The DC resistance is 0.0766Ω/1,000 ft. Using the voltage drop formula:

$$VD = \frac{2 \times L \times R \times I}{1,000}$$

$$VD = \frac{2 \times 145' \times 0.0766Ω \times 180A}{1,000}$$

$$VD = 4V$$

Divide by applied voltage to find percent voltage drop:

4V ÷ 240V = 0.0167

0.0167 × 100 = 1.67%

1.67% < 2%

Since this percentage does not exceed 2%, the answer to the question is no, the voltage drop for this feeder using No. 3/0 AWG conductors will not exceed the recommended value. Note that in this example, the second formula could have been used, if desired. Either formula will yield similar results.

Example 2:

What size THHN copper conductors would be required for a continuous feeder load of 180A, 208V, single-phase for fixed utilization equipment? The total length of the feeder is 318'.

Solution:

Since this is a continuous load, the load is multiplied by 125% to determine the feeder ampacity.

$$180A \times 1.25 = 225A$$

This feeder will require a No. 4/0 THHN copper conductor to supply the continuous load. The connected load, however, is 180A, and this is the amount of current that will be used in the voltage drop calculations.

Use the second voltage drop formula as an example to determine the voltage drop for this feeder. Look up the area in CM for a No. 4/0 THHN copper conductor in **NEC Chapter 9, Table 8**. Using the voltage drop formula:

$$VD = \frac{2 \times L \times K \times I}{CM}$$

$$VD = \frac{2 \times 318' \times 12.9 \times 180A}{211,600}$$

$$VD = 6.98V$$

Divide by applied voltage to find percent voltage drop:

6.98V ÷ 208V = 0.0336

0.0336 × 100 = 3.36%

3.36% > 3%

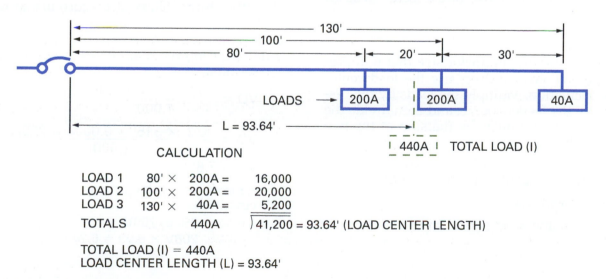

Figure 1 Load center length calculation.

This percentage exceeds 3%; therefore, No. 4/0 AWG copper would not meet the *NEC®* recommendations for voltage drop. Using the area in CM for 250 kcmil (250,000 CM) conductors, the formula becomes:

$$VD = \frac{2 \times L \times K \times I}{CM}$$

$$VD = \frac{2 \times 318' \times 12.9 \times 180A}{250,000}$$

$$VD = 5.91V$$

Divide by applied voltage to find percent voltage drop:

$$5.91V \div 208V = 0.0284$$
$$0.0284 \times 100 = 2.84\%$$

$$2.84\% < 3\%$$

No. 250 kcmil conductors would meet the *NEC®* voltage drop recommendations for this feeder.

Example 3:

What size THHN copper conductors would be required for a tapped 240V, single-phase, three-wire feeder consisting of continuous fixed 240V utilization equipment loads of 180A at 30' and, at the end, a noncontinuous fixed load of 100A at 450'?

Solution:

First, calculate the total load of the feeder by finding 125% of the continuous load and 100% of the noncontinuous load, and then adding the results. This is calculated as follows:

$$Tap\ 1 = 180A \times 1.25 = 225A$$
$$Tap\ 2 = 100A \times 1 = 100A$$

$$Total\ load = Tap\ 1 + Tap\ 2$$
$$Total\ load = 225A + 100A = 325A$$

The total calculated load of the feeder is 325A. Per **NEC Table 310.16**, the minimum conductor size required for this feeder would be 400 kcmil copper.

To determine the conductor size that will prevent excessive voltage drop, use the load center length calculation. Multiply the loads by the distance from the feeder source, and then divide the sum of the two products by the sum of the two loads as follows:

$$Tap\ 1 = 30' \times 180A = 5,400$$
$$Tap\ 2 = 450' \times 100A = 45,000$$

Add the products:

$$5,400 + 45,000 = 50,400$$

Divide by the sum of the loads:

$$\frac{50,400}{180A + 100A} = \frac{50,400}{280A} = 180'\ for\ load\ center$$

Using the second single-phase voltage drop formula, substitute values using the load center length, 280A, and 400 kcmil (400,000 circular mils, or CM) to check the voltage drop:

$$\frac{2 \times 180' \times 12.9 \times 280A}{400,000\ CM} = 3.25V$$

The maximum recommended voltage drop is 240V x 3% or 7.2V; therefore, 400 kcmil conductors are more than adequate. Note that if the full length of the feeder (450') had been used in this example instead of the load center length, the voltage drop would be 8.1V and 500 kcmil conductors would have to be selected, which is excessive for this application.

Voltage drop for balanced three-phase circuits (with negligible reactance and a power factor of 1.0) can be calculated using the two voltage drop formulas by substituting the square root of 3 (1.732) for the value of 2 in the first two formulas, which results in the following formulas:

Formula 3:

$$VD = \frac{\sqrt{3} \times L \times R \times I}{1,000}$$

Formula 4:

$$VD = \frac{\sqrt{3} \times L \times K \times I}{CM}$$

Example 4:

What is the voltage drop of a 208V, three-phase (3Ø) feeder with a noncontinuous load of 230A, a distance from the circuit breaker to the load of 315', and using No. 4/0 copper conductors?

Solution:

Use the third voltage drop formula to determine voltage drop for this three-phase feeder. Look up the resistance for No. 4/0 copper conductor in **NEC Chapter 9, Table 8**. Using the voltage drop formula:

$$VD = \frac{\sqrt{3} \times L \times R \times I}{1,000}$$

$$VD = \frac{\sqrt{3} \times 315' \times 0.0608 \times 230A}{1,000}$$

$$VD = 7.63V$$

The maximum voltage drop permitted for a 208V circuit is:

$$3\%\ of\ 208V =$$
$$208V \times 0.03 = 6.24V$$

In this problem, 7.63V is more than the maximum allowable voltage drop, so this circuit will not meet the minimum *NEC®* recommendation for voltage drop, and the conductor size should be increased to compensate for the voltage drop.

Example 5:

What size THHN copper conductors would be required for a panelboard used to supply continuous nonlinear loads of 100A for a 208V, three-phase, four-wire feeder circuit with a length of 150' from the source to the panelboard?

Solution:

Since the feeder is supplying nonlinear loads, the neutral wire becomes a current-carrying conductor per *NEC Section 310.15(E)(3)*, and therefore, there are four current-carrying conductors. This will require the ampacity of the wire to be adjusted by 80% per *NEC Table 310.15(C)(1)*:

Actual current = rated current × derating %

Since the actual current is known, solve the equation for the rated current:

Actual current ÷ derating % = rated current

Substituting values yields:

$$100A \div 0.8 = 125A$$

With the computed value of 125A, the minimum size for the THHN copper conductor per the 90°C column of *NEC Table 310.16* is No. 2. However, because this is a feeder conductor, it must also meet the requirements of *NEC Section 215.2(A)(1)*. This requires a minimum ampacity of 125% for continuous loads before the application of any adjustment factors. Multiply the actual load of 100A by 125% to obtain 125A for the required conductor ampacity. Since the actual load is 100A, you are limited to ampacity values of 60°C per *NEC Section 110.14(C)*. Refer to *NEC Table 310.16* and select No. 1/0 copper based on the ampacity requirements of 125A.

Apply the fourth formula, substituting values (using the actual connected load in amps) to check that the voltage drop does not exceed the recommended 3% of 208V, or 6.24V.

$$VD = \frac{\sqrt{3} \times L \times K \times I}{CM}$$

$$VD = \frac{\sqrt{3} \times 150' \times 12.9 \times 100A}{105,600 \text{ CM}} = 3.17V$$

The No. 1/0 AWG conductors will be adequate for this application.

1.1.3 Temperature Limitations for Terminating Conductors

NEC Section 110.14(C) includes temperature limitations for terminating conductors. The ampacity rating of conductors must be selected and coordinated so that the ampacity rating of the conductor does not exceed the temperature rating of any component in the feeder circuit. If the feeder size is less than or equal to 100A, the 60°C rating for the feeder conductor must be used unless all feeder components are rated and listed for use at a higher temperature. If the feeder size is larger than 100A, the 75°C rating for the feeder conductor must be used unless all feeder components are rated and listed for use at a higher temperature. For example, if the termination rating on a 125A fuse or circuit breaker is 75°C, type THHN conductors may be used for the feeder; however, the 75°C rating for the conductor size must be used. Conductors with higher temperature ratings may be used for ampacity adjustment, correction, or both. As you work through this module, keep in mind that, although type THHN conductors are often used in example problems, the limitations of *NEC Section 110.14(C)* still apply.

Example:

A 200A feeder is required to serve a new panelboard. The terminals in the panelboard are rated at 75°C, and the terminals in the fusible switch protecting the feeder are rated at 60°C. What is the minimum size THHN copper conductor required to supply the panelboard?

Solution:

Per *NEC Table 310.16*, No. 2/0 THHN is rated at 195A and *NEC Section 240.4(B)* would permit connection to the 200A fuse. However, because the fuse terminals are rated at 60°C, the rating must be taken from the 60°C column. The minimum allowable size taken from the 60°C column for the 200A load is No. 4/0. The answer is No. 4/0 THHN copper conductor.

1.2.0 Calculating Feeder Ampacity

Feeder ampacity for single-phase circuits (with negligible reactance and a power factor of 1.0) is calculated by dividing VA by the circuit voltage:

$$I = \frac{VA}{V}$$

For example, the ampacity of a single-phase 240V load rated at 21,600VA is determined by dividing 21,600VA by 240V.

$$21,600VA \div 240V = 90A$$

Feeder ampacity for balanced three-phase circuits (wye or delta with negligible reactance and a power factor of 1.0) is calculated by dividing VA by the product of the line-to-line circuit voltage and the square root of 3 (1.732), as follows:

$$I = \frac{VA}{V \times \sqrt{3}} \quad or \quad I = \frac{VA}{V \times 1.732}$$

For example, the ampacity of a 208V, three-phase load rated at 15,200VA is determined by dividing 15,200VA by (208V x 1.732).

$$I = \frac{VA}{V \times \sqrt{3}}$$

$$I = \frac{15,200kVA}{208V \times \sqrt{3}} = 42.19A$$

The ampacity of a 144kVA three-phase load at 480V is 173.21A.

$$I = \frac{VA}{V \times \sqrt{3}}$$

$$I = \frac{144,000VA}{480V \times \sqrt{3}} = 173.21A$$

Example 1:
What is the ampacity of a single-phase resistive load with a nameplate rating of 42.5kW, 240V?
Solution:
Multiply 42.5kW by 1,000 to determine VA (since watts = volt-amperes), and then divide the result by 240V.

Convert nameplate power rating to VA:
   42.5kW × 1,000 = 42,500VA

Divide by voltage to find current:
   I = VA ÷ V
   I = 42,500VA ÷ 240V = 177.08A

Example 2:
What is the ampacity of a three-phase panel feeder with a calculated load of 135kVA, 208V?
Solution:
Multiply 135kW by 1,000 to determine VA (since watts = volt-amperes). Then, since this is a three-phase circuit, divide the result by the product of the voltage and the square root of three (1.732):

Convert nameplate power rating to VA:
   135kVA × 1,000 = 135,000VA

Find current:

$$I = \frac{VA}{V \times \sqrt{3}} = \frac{135,000VA}{208V \times 1.732} = 374.72A$$

Transformers

Transformer secondary conductors are considered taps and as such are limited in length to no more than 10' (3 m) or 25' (7.5 m), depending on the application, when located inside buildings. See *NEC Section 240.21(C)*.

Example 3:
What size THWN copper conductors are required to supply the primary of a 75kVA, three-phase transformer? The primary line-to-line voltage is 480V.
Solution:
Multiply 75kVA by 1,000 to determine total VA and divide the result by (480V x 1.732) to determine the transformer primary current. Using *NEC Table 310.16*, determine the conductor size:

Convert nameplate power rating to VA:
   75kVA × 1,000 = 75,000VA

Find current:

$$I = \frac{VA}{V \times \sqrt{3}} = \frac{75,000VA}{480V \times 1.732} = 90.21A$$

Per *NEC Table 310.16*, No. 3 THWN copper (rated at 100A) is required if both the transformer and the panel are marked for 75°C terminations. If not, then No. 1 AWG copper would be required.
Example 4:
What is the secondary current for a 1,500kVA three-phase transformer with a voltage rating of 480V/277Y?
Solution:
Multiply 1,500kVA by 1,000 to determine total VA and divide the result by (480V x 1.732) to determine the transformer primary current.

Convert nameplate power rating to VA:
   1,500kVA × 1,000 = 1,500,000VA

Find current:

$$I = \frac{VA}{V \times \sqrt{3}} = \frac{1,500,000VA}{480V \times 1.732} = 1,804.22A$$

The secondary current for this transformer = 1,804.22A.

1.3.0 Tap Rules

When the overcurrent protection for a conductor is located at the load end instead of at the supply side of a conductor, the *NEC®* generally considers that to be a tap. This means that almost all secondary sides of transformers are classified as tap conductors, since there is no overcurrent device within the transformer itself. The most frequently used tap rules are the 10' (3 m) tap rule and the 25' (7.5 m) tap rule (*NEC Section 240.21*).

1.3.1 The 10' (3 m) Tap Rule

Generally, smaller conductors may be tapped from larger conductors, providing that the total length of the tap conductor does not exceed 10' (3 m). However, there are very specific *NEC®* requirements for each tap rule, including the 10' (3 m) tap rule [*NEC Section 240.21(B)(1)*]. It is important to note that, when applying the 10' (3 m) tap rule, the total length of the tap conductors (from their source of supply to their termination point in an overcurrent device) cannot exceed 10' (3 m). The *NEC®* requirements for the 10' (3 m) tap rule state that, in order for this rule to be applied, all of the following conditions must be met:

- The total length of the tap conductors must not exceed 10' (3 m).
- The tap conductors must be rated at no less than the calculated load(s) to be served or the device to which the tap conductors are connected.
- The tap conductors cannot extend beyond the switchboard, panel, or control devices that they supply.
- The tap conductors must be enclosed in a raceway.
- The rating of the line side overcurrent device supplying the tap conductors shall not exceed 1,000% of the rating of the tap conductors.

1.3.2 The 25' (7.5 m) Tap Rule

NEC Section 240.21(B)(2) lists specific conditions for the application of the 25' (7.5 m) tap rule. It is important to note that, when applying the 25' (7.5 m) tap rule, the total length of the tap conductors (from their source of supply to their termination point in an overcurrent device) cannot exceed 25' (7.5 m). The *NEC®* requirements for the 25' (7.5 m) tap rule state that, in order for this rule to be applied, all of the following conditions must be met:

- The total length of the tap conductors must not exceed 25' (7.5 m).

- The tap conductors must be rated at no less than one-third the capacity of the overcurrent device protecting the feeder.
- The tap conductors must terminate in a single circuit breaker or a single set of fuses that will limit the load to the conductor rating.
- The tap conductors must be enclosed in a raceway or otherwise suitably protected from physical damage.

1.3.3 Taps Supplying a Transformer – Primary and Secondary Not Over 25' (7.5 m)

NEC Section 240.21(B)(3) lists specific conditions for the application of the transformer 25' (7.5 m) tap rule. It is important to note that, when applying the 25' (7.5 m) tap rule, the total length of the tap conductors including one primary conductor and one secondary conductor (from their source of supply to their termination point in an overcurrent device) cannot exceed 25' (7.5 m). The requirements of this rule apply only when the primary conductors that supply a transformer are not protected by an overcurrent device at their rated ampacity. The *NEC®* requirements for this transformer 25' (7.5 m) tap rule state that, in order for this rule to be applied, all of the following conditions must be met:

- The primary conductors must have an ampacity that is at least one-third the capacity of the overcurrent device protecting the primary conductors.
- The secondary conductors must have an ampacity not less than the ratio of the primary to the secondary voltage multiplied by one-third the size of the overcurrent device protecting the feeder conductors.
- The total length of the primary and secondary conductors must not exceed 25' (7.5 m).
- Both primary and secondary conductors must be protected from physical damage by being enclosed in an approved raceway or other approved means.
- The secondary conductors must terminate in a single circuit breaker or set of fuses that limit the load to not more than conductor ampacity as permitted by *NEC Section 310.15*.

1.3.4 The 100' (30 m) Tap Rule

Use of the 100' (30 m) tap rule is limited to installations in a high-bay manufacturing building with wall heights over 35' (11 m). *NEC Section 240.21(B)(4)* lists specific conditions for the application of the 100' (30 m) tap rule. This tap rule may be used in buildings where only qualified persons will service and maintain the systems.

The length of this tap is limited to 25' (7.5 m) horizontally, and the total length of the tap cannot exceed 100' (30 m). In addition, the *NEC®* requires that the following conditions be met:

- The ampacity of the tap conductors must be at least one-third the ampacity of the feeder overcurrent protection device.
- The tap conductors must terminate in a single circuit breaker or set of fuses that will limit the load to the conductor ratings.
- The tap conductors must be protected from physical damage.
- The tap conductors must not be spliced.
- The tap conductors must be a minimum size of No. 6 copper or No. 4 aluminum.
- The tap conductors must not penetrate walls, floors, or ceilings.
- The tap to the feeder must be made at least 30' (9 m) above the floor.

1.3.5 Outside Taps of Unlimited Length

NEC Section 240.21(B)(5) states that where the conductors are located outside of a building, except at the point of load termination, there is no limit to the length of the tap conductors provided that all of the following conditions are met:

- The conductors are suitably protected from physical damage in an approved manner.
- The conductors terminate in a single circuit breaker or set of fuses that limit the load to the ampacity of the conductors. Additional loads may be supplied from the breaker or set of fuses.
- The overcurrent device for the conductors is an integral part of the disconnecting means or is located immediately adjacent to the disconnecting means.
- The disconnecting means is installed at a readily accessible location complying with one of the following:
  - Outside of a building or structure
  - Inside, nearest the point of entrance of the conductors
  - Where installed in accordance with *NEC Section 230.6*, nearest the point of entrance of the conductors

1.3.6 Transformer Secondary Conductors

NEC Section 240.21(C) provides the installation and length requirements for the secondary conductors of transformers. Although there are many similarities in these tap rules, there are a few differences. The 10' (3 m) tap rule for a transformer secondary requires that the ampacity of the secondary conductors be at least 10% of the rating of the overcurrent device on the primary side multiplied by the primary to secondary transformer voltage ratio. The outside taps of unlimited length for transformers are similar to the tap rules previously discussed. To install secondary conductors longer than 10' (3 m) inside buildings, the requirements in *NEC Section 240.21(C)(3)* or *NEC Section 240.21(C)(6)* must be met.

NEC Section 240.21(C)(3) will permit secondary conductors from a transformer in lengths up to 25' (7.5 m) in industrial installations only, when the following conditions are met:

- Conditions of maintenance and supervision ensure that only qualified persons service the system.
- The ampacity of the secondary conductors is not less than the secondary current rating of the transformer, and the sum of the ratings of the overcurrent devices does not exceed the ampacity of the secondary conductors.
- All overcurrent devices are grouped.
- The secondary conductors are protected from physical damage by approved means.

NEC Section 240.21(C)(6) will permit secondary conductors from a transformer in lengths up to 25' (7.5 m) when these conditions are met:

- The ratio of the primary to the secondary voltage is multiplied by one-third the size of the overcurrent device protecting the feeder conductors.
- The secondary conductors terminate in a single circuit breaker or set of fuses that limit the load to not more than the conductor ampacity that is permitted by *NEC Section 310.15*.
- The conductors are protected from physical damage by approved means.

1.3.7 Tap Calculations

Example 1:
Using the 10' (3 m) tap rule, what size THWN copper conductors are required for a 40hp, 460V, three-phase motor load tapped from a 400A feeder when all terminations are marked as suitable for 75°C connections?

Solution:
According to *NEC Table 430.250*, the FLA for that motor is 52A. All motor loads are calculated at 125%, so 52A × 1.25 = 65A.

Using *NEC Table 310.16*, determine the THWN copper conductor size needed to supply 65A. No. 6 THWN copper is rated for 65A. Assuming all conditions are met for application of the 10' (3 m) tap rule, No. 6 THWN copper conductors are required.

Example 2:

What size THWN copper conductors are required for a 25' (7.5 m) tap supplying a load of 100A from a feeder protected by a 600A overcurrent device?

Solution:

To apply the 25' (7.5 m) tap rule, the tap conductors must be rated at least one-third the capacity of the feeder overcurrent device.

$$600A \div 3 = 200A$$

The conductors must be rated for at least 200A, even though the supplied load is substantially lower. Using *NEC Table 310.16*, determine the THWN copper conductor size. No. 3/0 THWN copper is rated for 200A.

1.4.0 Demand Factors

The total load placed upon certain feeders may be reduced by applying demand factors. The *NEC®* permits the reduction of these loads by using the demand factors listed in the following *NEC®* tables for a dwelling unit and an other-than-residential (nondwelling) unit.

- *NEC Table 220.42*, General lighting load demand factors
- *NEC Table 220.44*, Demand factors for non-dwelling receptacle loads
- *NEC Table 220.54*, Demand factors for household electric clothes dryers
- *NEC Table 220.55*, Demand loads for household cooking appliances
- *NEC Table 220.56*, Demand factors for kitchen equipment – Other than dwelling unit(s)
- *NEC Section 220.82*, Optional calculation – Dwelling unit
- *NEC Section 220.83*, Optional calculation for additional loads in existing dwelling unit
- *NEC Table 220.84*, Optional calculations – Demand factors for multi-family dwelling units
- *NEC Table 220.86*, Optional method – Demand factors for feeders and service-entrance conductors for schools
- *NEC Section 220.87*, Optional method for determining existing loads
- *NEC Table 220.88*, Optional method – Demand factors for service-entrance and feeder conductors for new restaurants
- *NEC Table 220.102*, Method for computing farm loads
- *NEC Table 220.103*, Method for computing total farm load

Calculations utilizing these demand factors will be applied throughout this module.

Example:

The total general-purpose receptacle load in an office building is 26,800VA. Calculate the demand load.

Solution:

Per *NEC Table 220.44*, the first 10kVA is calculated at 100% and the remaining load is calculated at 50%. Subtract 10,000VA from the receptacle load in this example, and multiply the result by 50%. Add the result to the 10,000VA to determine the demand receptacle load.

$$26,800VA - 10,000VA = 16,800VA$$
$$16,800VA \times 0.5 = 8,400VA$$
$$8,400VA + 10,000VA = 18,400VA$$

The total general-purpose receptacle load is calculated to be 18,400VA.

1.4.1 Demand Factors for Neutral Conductors

The neutral conductor of electrical systems generally carries only the current imbalance of the phase conductors. For example, in a single-phase feeder circuit with one phase conductor carrying 50A and the other carrying 40A, the neutral conductor would carry 10A. Since the neutral in many cases will not be required to carry as much current as the phase conductors, the *NEC®* allows us to apply a demand factor. Refer to *NEC Section 220.61* for specific information. Also, *NEC Section 310.15(E)(1)* allows the exclusion of a neutral conductor carrying only unbalanced current when determining any adjustment factor for three or more conductors in a conduit.

> **NOTE**
>
> In certain circumstances such as nonlinear loads, data processing equipment, and other similar equipment, you cannot apply a demand factor on the neutral conductors because these types of equipment produce harmonic currents that increase the heating effect in the neutral conductor. Also see *NEC Section 220.61(C)(2)*.

1.4.2 Service Conductor Considerations

Sizing of overhead service conductors, underground service conductors, and primary feeder conductors as determined in the examples in the remainder of this module are based only on the calculated demand load for the premises served. Except in a few instances, derating factors for temperature, number of conductors, or voltage drop due to conductor length have not been considered. In actual situations, these factors would have to be included when sizing the conductors.

Depending on the service point of the utility serving a particular location and, in some cases, the type of premises involved, sizing of a service drop or service lateral may also be required, and the same adjustment factors will apply to their conductors as well.

1.4.3 Lighting Loads

NEC Section 220.12 provides specific information for determining lighting loads for non-dwelling occupancies. NEC Table 220.12, General Lighting Loads by Non-Dwelling Occupancy, provides unit load per ft²/m² according to the type of occupancy.

NEC Section 220.14(J) provides information for dwelling occupancies. A dwelling unit has a unit lighting load of 3VA per square foot (33VA per square meter).

To determine the lighting load, the area of the occupancy is calculated using outside dimensions. For a dwelling, this area does not include an open porch, a garage, or any unused or unfinished space not adaptable for future use.

Figure 2 is a block diagram that illustrates the basic steps in performing load calculations. This chart may be useful as a guide when calculating the loads of various types of electrical installations.

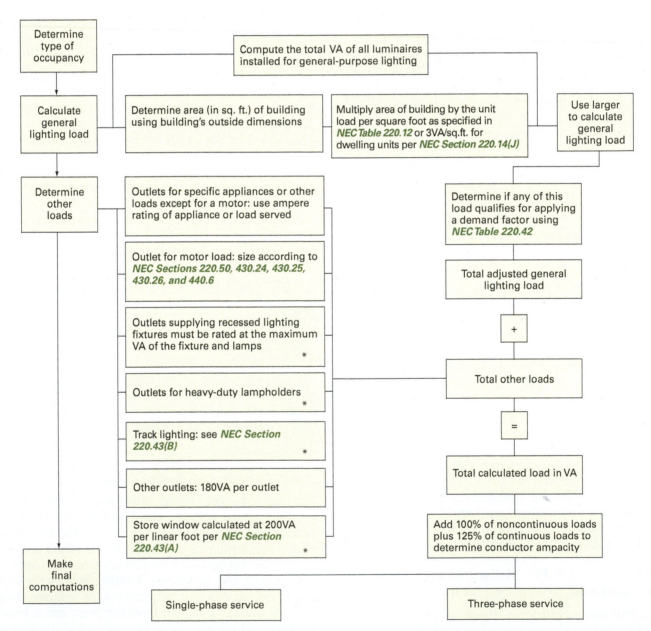

* Lighting not used for general illumination. Typical applications are task or display accent lighting.

Figure 2 Basic steps in load calculation.

1.0.0 Section Review

1. Per *NEC Table 310.15(C)(1)*, the ampacity of 15 current-carrying conductors must be must be adjusted by
   _____.

   a. 80%
   b. 70%
   c. 50%
   d. 40%

2. What is the ampacity of a single-phase resistive load with a nameplate rating of 38.4kW, 240V?

   a. 160A
   b. 240A
   c. 384A
   d. 580A

3. A high-bay manufacturing building with a wall height of 50' has a maximum horizontal tap length of
   _____.

   a. 10' (3 m)
   b. 25' (7.5 m)
   c. 50' (15 m)
   d. 100' (30 m)

4. The total general-purpose receptacle load in an office building is 16,400VA. The demand load is _____.

   a. 4,100VA
   b. 8,200VA
   c. 13,200VA
   d. 16,400VA

2.0.0 SERVICE CALCULATIONS

Objective

Make service calculations for residential installations.

a. Calculate the minimum service size for simple electrical installations.
b. Make service calculations for single-family dwellings.
c. Make service calculations for multi-family dwellings.

Trade Term

Service equipment: The necessary equipment, usually consisting of a circuit breaker or switch and fuses, and their accessories, connected to the load end of service conductors to a building or other structure, or an otherwise designated area, and intended to constitute the main control and means of cutoff of the supply.

The following steps describe the method of determining the service load for a given type of occupancy. Note that these steps are very general, and may change slightly depending on the situation; specific examples will be provided in later sections of this module.

Step 1 Compute the floor area of the building using the outside dimensions of the building. This area is used to determine the general lighting load for the particular occupancy. *NEC Table 220.12* (for non-dwelling units) and *NEC Section 220.14(J)* (for dwelling units) allow different general lighting loads based upon the use of the area. For example, an office in a warehouse building would require a multiplication factor of 1.3VA/ft² (14VA/m²), while another calculation of the general lighting load for a warehouse floor area requires a multiplication factor of 1.2VA/ft² (13VA/m²). If the use of the area is in question, the authority having jurisdiction (*NEC Section 90.4*) will make the determination of the use of the building as it relates to the *NEC®*.

Step 2 Multiply the floor area determined in Step 1 by the unit load (VA), and calculate the general lighting load per *NEC Table 220.12* (for non-dwelling units) and *NEC Section 220.14(J)* (for dwelling units).

Compare this value to the total installed lighting used for general illumination, and use the larger of the two loads.

Step 3 Calculate the lighting loads used for other than general illumination, such as for task lighting, exterior lighting, or accent lighting. Refer to the applicable sections of the *NEC®*:
- *NEC Section 220.14(D)*, *Luminaires*
- *NEC Section 220.14(E)*, *Heavy-Duty Lampholders*
- *NEC Section 220.14(F)*, *Sign and Outline Lighting*
- *NEC Sections 220.14(G) and 220.43(A)*, *Show Windows*
- *NEC Section 220.43(B)*, *Track Lighting*

Step 4 Calculate the receptacle loads. Refer to the applicable sections of the *NEC®*:
- *NEC Section 220.14(A)*, *Specific Appliances or Loads*
- *NEC Section 220.14(H)*, *Fixed Multi-Outlet Assemblies*
- *NEC Section 220.14(I)*, *Receptacle Outlets*

Step 5 Apply applicable demand factors. For specific applications, refer to the following sections of the *NEC®*:
- *NEC Section 220.44*, *Receptacle Loads, Other than Dwelling Units*
- *NEC Article 610*, *Cranes and Hoists*
- *NEC Article 620*, *Elevators, Dumbwaiters, Escalators, Moving Walks, Platform Lifts, and Stairway Chairlifts*
- *NEC Article 630*, *Electric Welders*
- *NEC Article 660*, *X-Ray Equipment*
- *NEC Article 695*, *Fire Pumps* (and other articles as needed for specific equipment or locations)

Step 6 Calculate all the motor loads in accordance with *NEC Section 220.50*.
- Use *NEC Section 430.17* to determine the largest motor in the calculation.
- Add together 125% of the FLA of the largest motor plus 100% of the FLA for all other motors.

Step 7 Calculate both the heating and air conditioning loads, and use only the largest of the two in the total load calculation (*NEC Section 220.60*).

Step 8 Add all of the loads. The resultant load in volt-amps is divided by the system voltage for single-phase service or by the system voltage times the square root of three for three-phase services.

Step 9 Add 100% of the noncontinuous loads and 125% of the continuous loads to determine conductor ampacity.

2.1.0 Calculations for Simple Electrical Installations

The total calculated load of an occupancy is determined by adding all the individual loads after any demand factors have been applied. Generally, the total calculated load for a particular occupancy determines the minimum size service it should have (*NEC Section 220.40*). The following sections provide examples of how to make these calculations in a variety of situations.

2.1.1 Rural Pump House Sample Calculation

A floor plan for a small rural pump house is shown in *Figure 3*. This pump house may be used on a farm to supply water for livestock at a large distance from buildings containing electricity. Therefore, a separate service must be supplied for this pump house.

The total loads for this facility consist of the following:

- Shallow well pump with a 1/3hp, 115V single-phase motor
- One wall switch-controlled lighting fixture containing one 60W lamp

The load and service size are calculated using the following steps:

Step 1 *NEC Table 220.12* does not list a small pump house. Therefore, use *NEC Section 220.14(A)* for the calculation.

Step 2 *NEC Section 430.6(A)(1)* requires that where the current rating of a motor is used to determine the ampacity of conductors, the values given in *NEC Tables 430.247 through 430.250*, including notes, shall be used instead of the actual current rating marked on the motor nameplate. *NEC Table 430.248* shows that the listed ampacity for a 1/3hp single-phase 115VAC motor is 7.2A.

Also, according to *NEC Section 430.22*, a single motor shall have an ampacity of no less than 125% of the motor full-load current rating. Therefore, the motor load is calculated as follows:

Calculate 125% of the ampacity:
$$7.2A \times 1.25 = 9.0A$$

Multiply ampacity by source voltage to find motor load:
$$9.0A \times 120V = 1,080VA$$

This, added to the 60W lamp load (60VA), gives a total connected load of 1,140VA.

Step 3 Determine the size of the service-entrance conductors by dividing the total connected load by the voltage:
$$1,140VA \div 120V = 9.5A$$

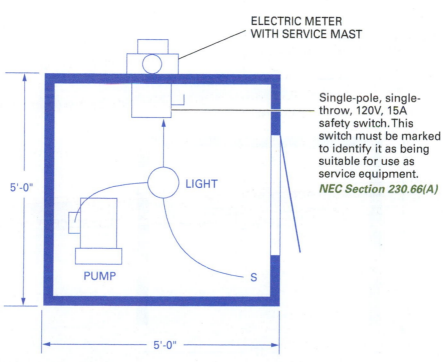

ELECTRIC METER
WITH SERVICE MAST

Single-pole, single-throw, 120V, 15A safety switch. This switch must be marked to identify it as being suitable for use as service equipment.
NEC Section 230.66(A)

LIGHT

5'-0"

PUMP

S

5'-0"

Figure 3 Floor plan of a small pump house.

Step 4 Check *NEC Section 230.42(A)* for the minimum size of the service-entrance conductors. Check *NEC Section 230.79* for the minimum rating of the service disconnecting means. The service-entrance conductors must have an ampacity that is not less than the load served and not less than the minimum required disconnecting means. The disconnect in this case would be 15A.

Since the total connected load for the pump house is less than 10A and the single branch circuit feeding the pump and light fixture need only be No. 14 AWG copper (15A), a No. 14 AWG copper or No. 12 aluminum conductor will qualify for the service-entrance conductors.

NEC Section 230.23 gives the minimum size of service drop conductors, and *NEC Section 230.31* gives the minimum size of service lateral conductors. In addition to being able to serve the load, service drops and service laterals must have a minimum mechanical strength.

2.1.2 Roadside Vegetable Stand

Another practical application of *NEC Sections 230.42 and 230.79* is provided in *Figure 4*, which shows a floor plan of a typical roadside vegetable stand. Since *NEC Table 220.12* does not list this

> **NOTE**
>
> Many inspection jurisdictions and utility companies now require a fault current study prior to issuance of a permit. The fault current study may result in the use of a conductor size that is larger than the minimum *NEC®* requirement.

facility, use *NEC Section 220.14* for the calculation. The load and service size are calculated using the following steps:

Step 1 Determine the lighting load. Two fluorescent fixtures are used to illuminate the 9 x 12 prime area. Since each fixture contains two 40W fluorescent lamps, and the ballast is rated at 0.83A (which translates to approximately 100VA), the total connected load for each fixture is 100VA, or a total of 200VA for both fixtures.

Step 2 Determine the remaining loads. The only other electrical outlets in the stand consist of two receptacles: one furnishes power to a refrigerator with a nameplate full-load rating of 12.2A (total VA = 120V × 12.2A = 1,464VA); the other furnishes power for an electric cash register rated at 300VA and an electronic calculator rated at 200VA (total VA for other loads is 300VA + 200VA = 500VA).

Step 3 Determine the total connected load:

$$
\begin{array}{rr}
\text{Fluorescent fixtures} = & 200\text{VA} \\
\text{Receptacle for refrigerator} = & 1{,}464\text{VA} \\
+\ \text{Receptacle for other loads} = & 500\text{VA} \\
\hline
\text{Total calculated load} = & 2{,}164\text{VA}
\end{array}
$$

Step 4 Identify and total the continuous loads; in this example, the continuous loads are the fluorescent fixtures and the refrigerator:

$$200\text{VA} + 1{,}464\text{VA} = 1{,}664\text{VA}$$

Step 5 Determine the size and rating of the service-entrance conductors. Add 100% of the noncontinuous loads and 125% of the continuous loads, and then divide

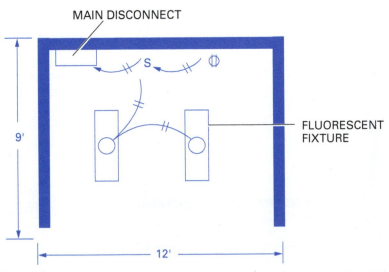

Figure 4 Roadside vegetable stand.

the result by the circuit voltage to find the current rating:

$$500VA + (1,664VA \times 1.25) = 2,580VA$$
$$2,580VA \div 240V = 10.75A$$

Step 6 Check *NEC Section 230.42* for the minimum size of the service-entrance conductors. Check *NEC Section 230.79* for the minimum rating of the service disconnecting means. No. 10 AWG copper or No. 8 aluminum is the minimum size allowed by the *NEC®* for the service-entrance conductors on this project. The disconnecting means in this instance would be 30A.

Refer to *NEC Section 230.23* for the minimum size of service drop conductors or *NEC Section 230.31* for the minimum size of service lateral conductors when they are not under the jurisdiction of the serving utility.

2.2.0 Calculations for Single-Family Dwellings

A floor plan of a single-family dwelling is shown in *Figure 5*. This building is constructed on a concrete slab with no basement or crawlspace. There is an unfinished attic above the living area and an open carport just outside the kitchen entrance. Appliances include a 12kW electric range (12,000W = 12,000VA) and a 4.5kW water heater (4,500W = 4,500VA). There is also a washer and a 5.5kW dryer (5,500W = 5,500VA) in the utility room. Gas heaters are installed in each room with no electrical requirements. What size service entrance should be provided for this residence if no other information is specified?

Figure 6 shows a typical form used for service load calculations. A blank form is included in the Appendix for future use. The steps necessary to perform these calculations are as follows:

Step 1 Compute the area of the occupancy per *NEC Section 220.11*. The general lighting load (including general receptacle load) is determined by finding the area of the occupancy using the outside dimensions of the structure. Remember that this does not include open porches, garages, or unused or unfinished spaces that are not adaptable for future use. For example, an unfinished basement would be included in this area measurement because it might later be converted into living space. The area is used to calculate the general lighting load in Step 2. Using the floor plan and an architect's scale, measure the longest width of the building using the outside dimensions. In this case, it is 33 ft. The longest length of the building is 48 ft. The product of these two measurements is:

$$33 \text{ ft} \times 48 \text{ ft} = 1,584 \text{ ft}^2$$

The area of the carport within this total area is:

$$12 \text{ ft} \times 19.5 \text{ ft} = 234 \text{ ft}^2$$

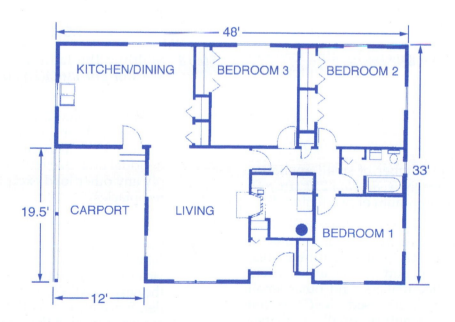

Figure 5 Single-family dwelling.

General Lighting Load						Phase		Neutral
Square footage of the dwelling	[1]	1,350	@ 3VA =	[2]	4,050			
Kitchen small appliance circuits	[3]	2	@ 1500 =	[4]	3,000			
Laundry branch circuit	[5]	1	@ 1500 =	[6]	1,500			
Subtotal of general lighting loads per *NEC Section 220.42*				[7]	8,550			
Subtract 1st 3000VA per *NEC Table 220.42*				[8]	3,000	@ 100% =	[9]	3,000
Remaining VA times 35% per *NEC Table 220.42*				[10]	5,550	@ 35% =	[11]	1,943
Total demand for general lighting loads =						[12] 4,943	[13]	4,943

Fixed Appliance Loads (nameplate or *NEC®* FLA of motors) per *NEC Section 220.53*					
Hot water tank, 4.5kVA, 240V	[14]	4,500			
	[15]				
	[16]				
	[17]				
	[18]				
	[19]				
Subtotal of fixed appliances [20] 4,500					
If 3 or less fixed appliances take @ 100% =	[21]	4,500	[22]	0	
If 4 or more fixed appliances take @ 75% =	[23]		[24]		

Other Loads per *NEC Section 220.14*					
Electric range per *NEC Table 220.55* [neutral @ 70 % per *NEC Section 220.61(B)(1)*]	[25]	8,000	[26]	5,600	
Electric dryer per *NEC Table 220.54* [neutral @ 70% per *NEC Section 220.61(B)(1)*]	[27]	5,500	[28]	3,850	
Electric heat per *NEC Section 220.51*					
Air conditioning per *NEC Section 220.14(A)*	Omit smaller load per *NEC Section 220.60*	[29]		[30]	
Largest Motor = 0	@ 25% per *NEC Section 430.24* =	[31]	0	[32]	0
Total VA Demand =		[33]	22,943	[34]	14,393
VA/240V = Amps =		[35]	96	[36]	60
Service OCD and Minimum Size Grounding Electrode Conductor (*NEC Table 250.66*) =		[37]	100	[38]	8 AWG
AWG per *NEC Section 310.12* and *Table 310.16*		[39]	3 AWG	[40]	6 AWG

Figure 6 Completed calculation form.

The net living area is then calculated by subtracting the carport area from the total area:

$$1,584 \text{ ft}^2 - 234 \text{ ft}^2 = 1,350 \text{ ft}^2$$

Step 2 Determine the general lighting load using *NEC Section 220.14(J)*. Multiply the resulting area found in Step 1 by the unit load for dwelling units. In this example, it is 3VA/ft². The lighting load is now:

$$1,350 \text{ ft}^2 \times 3\text{VA/ft}^2 = 4,050\text{VA}$$

Step 3 Multiply the number of small appliance branch circuits [*NEC Section 210.11(C)(1)*] by 1,500VA (minimum of two required).

The *NEC®* requires at least two 120V, 20A small appliance branch circuits to be installed in each kitchen, dining area, breakfast nook, and similar areas where toasters, coffee makers, and other small appliances will be used. *NEC Section 220.52* also requires small appliance branch circuits to be rated at 1,500VA

for each kitchen area served. Since the sample residence has only one kitchen area, the load for these circuits would be:

$$2 \times 1,500\text{VA} = 3,000\text{VA}$$

Step 4 Multiply the laundry receptacle [*NEC Section 210.11(C)(2)*] circuit by 1,500VA per *NEC Section 220.52*.

An additional 20A branch circuit must be provided for the exclusive use of each laundry area. This circuit must be an individual branch circuit and must not supply any other load except for the laundry receptacles.

$$1 \times 1,500\text{VA} = 1,500\text{VA}$$

Step 5 Calculate the lighting load demand. In *NEC Table 220.42*, note that the demand factor is 100% for the first 3,000VA, 35% for the next 117,000VA, and 25% for the remainder. Since all residential electrical outlets are never used at the same time, the *NEC®* allows the use of demand factors.

First 3,000VA at 100%:
    3,000VA × 1 = 3,000VA

Remainder at 35%:
    Remainder = 8,550 − 3,000 = 5,550VA
    5,550VA × 0.35 = 1,942.5VA

Add results to find net general lighting and small appliance load:
    3,000VA + 1,942.5VA = 4,942.5VA

Step 6 Calculate the loads for other appliances (*NEC Section 220.53*), such as a water heater, disposal, dishwasher, furnace blower motor, etc.

Water heater calculated at 100% =
4,500VA

Step 7 Calculate the load for the range and other cooking appliances (*NEC Section 220.55 and Table 220.55*).

Electric range (using demand factor) =
8,000VA

Step 8 Calculate the load for the clothes dryer (*NEC Section 220.54 and Table 220.54*). Note that this load must be either 5,000VA or the nameplate rating of the dryer, whichever is larger. If a gas dryer is used, it will be part of the 1,500VA demand in Step 4.

Clothes dryer (nameplate rating) =
5,500VA

> **NOTE**
>
> Four or more other appliances may have a demand factor of 75% applied.

Step 9 Calculate the larger of the heating or air conditioning loads (*NEC Section 220.60*). (Negligible in this example.)

> **NOTE**
>
> If a gas forced air furnace is used, the blower motor will have been included in Step 6. If some form of electric heat is used (baseboard, heat pump, or furnace), this load will be compared to the air conditioning load and the larger load then added to the calculations (*NEC Section 220.60*).

Step 10 Calculate 25% of the largest motor load (*NEC Section 430.24*). (None in this example.)

Step 11 Total all the loads.

General lighting and appliance load: 4,942.5VA
Electric range (using demand factor): 8,000.0VA
Clothes dryer: 5,500.0VA
Water heater: 4,500.0VA
Total load = 22,942.5VA

Step 12 Convert the volt-amps to amperes. The total load amperage is derived by dividing the total volt-amps of the load by the circuit voltage, as follows (in this example, the system voltage is 240V):

22,942.5VA ÷ 240V = 95.6A

Step 13 Determine the required service size (*NEC Section 230.42*).

Typical electric service for residential use is 120/240V, three-wire, single-phase. Services are sized in amperes after the total load in volt-amperes is determined.

For this ampere load, the minimum 100A service would be required [*NEC Section 230.79(C)*].

2.2.1 Sizing the Neutral Conductor

The neutral conductor in a three-wire, single-phase service carries only the unbalanced load between the two hot legs. Since there are several 240V loads in the example residence, these 240V loads will be balanced and therefore reduce the load on the service neutral conductor. Consequently, in most cases, the service neutral does not have to be as large as the ungrounded (hot) conductors.

In this example, the water heater does not have to be included in the neutral conductor calculation since it is strictly 240V with no 120V loads. The clothes dryer and electric range, however, have 120V lights that will unbalance the current between phases. *NEC Section 220.61(B)* allows a demand factor of 70% for these two appliances. Using this information, the neutral conductor may be sized accordingly:

General lighting and appliance load =   4,942.5VA
Electric range (using demand factor):
        8,000VA × 0.70 =   5,600VA
Clothes dryer (using demand factor):
        5,500VA × 0.70 =   3,850VA
        Total = 14,392.5VA

To find the total line-to-line amperes, divide the total volt-amperes by the voltage between lines.

14,392.5VA ÷ 240V = 59.97A *or* 60A

The service-entrance conductors have now been calculated and must be rated at 100A with a neutral conductor rated for at least 60A.

Applying *NEC Section 310.12* for a single-phase residential service in conjunction with *NEC Table 310.16* shows that a No. 3 AWG copper or No. 1 AWG aluminum or copper-clad aluminum is sufficient for a 100A service.

When sizing the grounded conductor for services, all of the provisions stated in *NEC Sections 215.2(A)(2), 220.61, and 230.42* must be met, along with other applicable sections.

2.2.2 Single-Family Dwelling Standard Calculation

To calculate the service size for the example dwelling, first determine the square footage. This dwelling consists of 1,624 ft² of living area served by 120/240 single-phase. The following loads must be accounted for:

- 12kW range
- 5.5kW dryer
- 1,250VA dishwasher
- ¾hp, 120V disposal
- 1hp, 240V pump
- ⅓hp, 120V blower for gas furnace

Step 1 Calculate the general lighting and small appliance loads.

> *General lighting load:*
> 1,624 sq ft × 3VA = 4,872VA
> *Small appliance load:*
> 1,500VA × 2 = 3,000VA
> *Laundry load* = 1,500VA
> Total = 9,372VA

Step 2 Apply demand factors for general lighting and small appliances (*NEC Table 220.42*).

> *First 3,000VA at 100%:*
> 3,000VA × 1 = 3,000VA
>
> *Remainder at 35%:*
> Remainder = 9,372 − 3,000 = 6,372VA
> 6,372VA × 0.35 = 2,230VA
>
> *Add results to find net general lighting and small appliance load:*
> 3,000VA + 2,230VA = 5,230VA

Step 3 Determine the range load per *NEC Table 220.55, Column C* = 8,000VA.

Step 4 Determine the dryer load per *NEC Table 220.54* = 5,500VA.

Step 5 Calculate the fixed appliance loads.

> *Dishwasher* = 1,250VA
> *Disposal (NEC Table 430.248):*
> 13.8A × 120V = 1,656VA
> *Pump (NEC Table 430.248):*
> 8A × 240V = 1,920VA
> *Blower (NEC Table 430.248):*
> 7.2A × 120V = 864VA
> Total fixed appliance loads = 5,690VA

Step 6 Since four or more fixed appliances exist, apply the demand factor for fixed appliance loads (*NEC Section 220.53*).

> *Total fixed appliance loads:*
> 5,690VA × 0.75 = 4,268VA

Step 7 Combine the calculated loads.

> *Total load:*
> 5,230VA
> 8,000VA
> 5,500VA
> + 4,268VA
> 22,998VA

Step 8 Use *NEC Section 430.17* to determine the largest motor:

> 25% of largest motor load =
> 1,656VA × 0.25 = 414VA

Step 9 Combine the computed loads.

> *Total load:*
> 22,998VA + 414VA = 23,412VA

The minimum size service for this dwelling is determined by dividing the total calculated VA by the line-to-line service voltage.

> 23,412VA ÷ 240V = 97.55A

NEC Section 230.79(C) requires a minimum service size of 100A for a single-family dwelling.

2.2.3 Standard Calculation for Single-Family Dwelling Using Electric Heat

This section provides an example load calculation for a single-family residence using electric heat. In this example, the dwelling consists of 1,775 ft² of living area. The home has the following loads:

- 8.75kW range
- 5.5kW dryer
- 1,000VA dishwasher
- ¾hp, 120V disposal
- ½hp, 120V attic fan
- ⅓hp, 120V blower for electric furnace
- 20kW electric furnace
- Central air conditioning unit with a nameplate rating of 25A at 240V

Step 1 Calculate the general lighting and small appliance loads.

General lighting:
1,775 ft² × 3VA = 5,325VA

Small appliance load:
1,500VA × 2 = 3,000VA

Laundry load = 1,500VA

Total = 9,825VA

Step 2 Apply demand factors for general lighting and small appliances (*NEC Table 220.42*).

First 3,000VA at 100%:
3,000VA × 1 = 3,000VA

Remainder at 35%:
Remainder = 9,825 − 3,000 = 6,825VA
6,825VA × 0.35 = 2,389VA

Add results to find net general lighting and small appliance load:
3,000VA + 2,389VA = 5,389VA

Step 3 Determine the range load (*NEC Table 220.55, Column B*).

8,750VA × 0.8 = 7,000VA

Step 4 Determine the dryer load (*NEC Table 220.54*).

5,500VA × 1 = 5,500VA

Step 5 Calculate the fixed appliance loads.

Dishwasher = 1,000VA

Disposal (NEC Table 430.248):
13.8A × 120V = 1,656VA

Furnace blower (NEC Table 430.248):
7.2A × 120V =   864VA

Attic fan (NEC Table 430.248):
9.8A × 120V = 1,176VA

Total fixed appliance loads = 4,696VA

Step 6 Since four or more fixed appliances exist, apply the demand factor for fixed appliance loads (*NEC Section 220.53*).

Total fixed appliance loads =
4,696VA × 0.75 = 3,522VA

Step 7 Calculate the larger of the heating and cooling loads (use *NEC Section 220.60* for noncoincidental loads to omit the smaller load).

Heating load = 20,000VA
Cooling load = 25A × 240V = 6,000VA

Larger of the two loads: 20,000VA
(since 20,000VA > 6,000VA)

Step 8 Use *NEC Section 430.17* to determine the largest motor.

25% of largest motor load =
1,656VA × 0.25 = 414VA

Step 9 Combine the computed loads.

Total load:
5,389VA
7,000VA
5,500VA
3,522VA
20,000VA
+   414VA
41,825VA

The minimum size service for this dwelling is determined by dividing the total calculated VA by the line-to-line service voltage. The line-to-line voltage is 240V, single phase.

41,825VA ÷ 240V = 174.27A

The next larger standard size overcurrent device is 175A. Although a 175A device would satisfy the *NEC®* requirements, it is more likely that a 200A service would be installed.

2.2.4 Optional Calculation for a Single-Family Dwelling

The *NEC®* provides an alternative method of computing the load for a dwelling unit in **NEC Section 220.82**. Using this method will almost always result in a calculation that is lower than the standard method when electric heat is used.

For example, suppose you have a single-family dwelling with 1,624 ft² of living area and the following loads:

- 12kW range
- 5.5kW dryer
- 1,250VA dishwasher
- ½hp, 120V disposal at 1,176VA
- ¾hp, 120V attic fan at 1,656VA
- 14.4kW electric furnace
- ⅓hp, 120V (864VA) blower motor for electric furnace
- Air conditioning unit with a nameplate rating of 6,000VA

Step 1 Calculate the general lighting, small appliance, and fixed appliance loads. Use full nameplate values for each.

General lighting:
1,624 ft² × 3VA = 4,872VA

Small kitchen appliance load:
1,500VA × 2 = 3,000VA
Laundry load = 1,500VA
Range = 12,000VA
Dryer = 5,500VA
Dishwasher = 1,250VA
Disposal = 1,176VA
Attic fan = 1,656VA
Blower = 864VA
Subtotal load = 31,818VA

Step 2 Apply the demand factor in *NEC Section 220.82(B)*.

First 10kVA at 100%:
10,000VA × 1 = 10,000VA

Remainder at 40%:
Remainder = 31,818 − 10,000 = 21,818VA
21,818VA × 0.4 = 8,727VA

Subtotal for general loads:
10,000VA + 8,727VA = 18,727VA

Step 3 Compute the demands permitted for electric heating and air conditioning per *NEC Section 220.82(C)*. Use the largest of the computed demand loads.

Air conditioner at 100%:
6,000VA × 1 = 6,000VA

Central electric heating at 65%:
14,400VA × 0.65 = 9,360VA

9,360VA > 6,000VA

Step 4 Add together the demand for general loads and the largest of the heating or air conditioning load.

Subtotal general loads: 18,727VA
Largest of heating and AC loads: 9,360VA
Total load = 28,087VA

The minimum size service required for this dwelling is determined by dividing the total calculated VA by the line-to-line service voltage. The line-to-line voltage is 240V, single-phase.

28,087VA ÷ 240V = 117A

The minimum service required for this dwelling would be 125A.

2.3.0 Calculations for Multi-Family Dwellings

The service load for multi-family dwellings is not simply the sum of the individual dwelling unit loads because of demand factors that may be applied when either the standard calculation or the optional calculation is used to compute the service load.

When the standard calculation is used to compute the service load, the total lighting, small appliance, and laundry loads as well as the total load from all electric ranges and clothes dryers are subject to the application of demand factors. Additional demand factors may be applied to the portion of the neutral load contributed by electric ranges and the portion of the total neutral load greater than 200A.

When the optional calculation is used, the total connected load is subject to the application of a demand factor that varies according to the number of individual units in the dwelling.

A summary of the calculation methods for designing wiring systems in multi-family dwellings and the applicable *NEC®* references are shown in *Figure 7.* The selection of a calculation method for computing the service load is not affected by the method used to design the feeders to the individual dwelling units.

The rules for computing the service load are also used for computing a main feeder load when the wiring system consists of a service that supplies main feeders that, in turn, supply a number of sub-feeders to individual dwelling units.

Standard Calculation: Feeder sized per *NEC Article 220, Part III*
Optional Calculation: Per *NEC Section 220.84*, if each dwelling unit has:
 1. Single feeder
 2. Electric cooking equipment
 3. Electric space heating or air conditioning, or both

MAIN SERVICE

APARTMENT PANELS

SERVICE EQUIPMENT

APARTMENT PANELS

APARTMENT FEEDERS

Standard Calculation: Feeder sized per *NEC Article 220, Part III*
Optional Calculation: Use *NEC Section 220.84* if each dwelling unit has a single feeder.

Figure 7 *NEC®* requirements for multi-family dwellings.

2.3.1 Multi-Family Dwelling Sample Calculation

A 30-unit apartment building consists of 18 one-bedroom units of 650 ft² each, 6 two-bedroom units of 775 ft² each, and 6 three-bedroom units of 950 ft² each. The kitchen in each unit contains the following loads:

- 7.5kW electric range
- 1,250VA dishwasher
- ⅓hp, 120V garbage disposal at 864VA
- Air conditioning unit rated at 3,600VA

The entire building is heated by a boiler that uses three electric pump motors: a 2hp squirrel-cage motor and two ½hp squirrel-cage induction motors, all 240V single-phase, two-wire. There is a community laundry room in the building with four washers with a nameplate rating of 12.5A at 120V (1,500VA) and four clothes dryers, with the latter rated at 4,500VA each. The house lighting consists of 40 incandescent fixtures rated at 60W each. There are three general-purpose receptacle outlets for servicing of the boilers and laundry room. The building is furnished with a 120/240V, single-phase, three-wire service. What size service is required?

Step 1 Calculate the general lighting and small appliance loads in each apartment.

General lighting:
650 ft² × 3VA = 1,950VA
1,950VA × 18 units =   35,100VA

General lighting:
775 ft² × 3VA = 2,325VA
2,325VA × 6 units =   13,950VA

General lighting:
950 ft² × 3VA = 2,850VA
2,850VA × 6 units =   17,100VA

Small appliance circuits:
1,500VA × 2 × 30 units =   90,000VA
Total = 156,150VA

Step 2 Apply the demand factors using *NEC Table 220.42*.

First 3,000VA (out of 120,000) at 100%:
3,000VA × 1 = 3,000VA

Next 117,000VA (out of 120,000) at 35%:
117,000VA × 0.35 = 40,950VA

Remaining 36,150VA at 25%:
31,150VA × 0.25 = 9,038VA

Net general lighting & small appliance load:
3,000VA
40,950VA
+ 9,038VA
52,988VA

Step 3 Calculate the appliance loads using *NEC Section 220.53*.

Dishwashers:
30 × 1,250VA × 0.75 = 28,125VA
Disposals:
30 × 864VA × 0.75 = 19,440VA

Step 4 Calculate the range load using *NEC Table 220.55*.

Range load (*NEC Table 220.55, Column C*) =
45,000VA

> **NOTE**
>
> The range load could have been calculated using *NEC Table 220.55*, Column B, Note 3. However, the value obtained (54kVA) is higher, so Column C was used.

Step 5 Calculate the air conditioning load [*NEC Sections 220.60 and 220.14(A)*].

Air conditioners =
30 × 3,600VA × 100% = 108,000VA

Step 6 Calculate the house load.

2hp boiler (NEC Table 430.248):
$$12A \times 240V = 2{,}880VA$$

Two ½hp boiler motors:
$$4.9A \times 240V \times 2 = 2{,}352VA$$

Laundry:
$$1{,}500VA \times 4 = 6{,}000VA$$

Dryers (NEC Table 220.54):
$$5{,}000 \ (NEC\ Section\ 220.54) \times 4 \times 1 = 20{,}000VA$$

House lighting:
$$60W \times 40 = 2{,}400VA$$

General-purpose receptacles:
$$180VA \times 3 = \underline{540V}$$

Total house load = 34,172VA

Step 7 Determine the largest motor load and multiply the FLA or VA by 25% (*NEC Section 430.24*).

Boiler motor =
$$2{,}880VA \times 25\% = 720VA$$

Step 8 Determine the total calculated load.

Net general lighting &	
small appliance load =	52,988VA
Dishwasher load =	28,125VA
Disposal load =	19,440VA
Range load =	45,000VA
Air conditioning load =	108,000VA
House load =	34,172VA
25% of largest motor load =	720VA
Total load =	288,445VA

Add 25% of continuous loads:
$$2{,}400VA \text{ (house lighting)} \times 0.25 = \underline{600VA}$$

Total adjusted load = 289,045VA

Step 9 Determine the total load in amperes in order to select service conductors, equipment, and overcurrent protective devices.

$$289{,}045VA \div 240V = 1{,}204A$$

Step 10 Select the service-entrance conductors, **service equipment**, and overcurrent protection.

> **NOTE**
>
> Since the exact calculated load is not a standard size, the rating of the service for this apartment building would be a minimum of 1,600A; this is the next largest standard size and is permitted by the *NEC*®. See *NEC Sections 240.4 and 240.6*.

2.3.2 Multi-Family Dwelling Calculation Using Optional Method

NEC Section 220.84 permits an alternative method of computing the load for multi-family dwellings. The conditions that must be met to use this calculation are that each individual dwelling must be supplied by only one feeder, have electric cooking (or an assumed cooking load of 8kW), and have electric heating, air conditioning, or both. The demand factors of *NEC Table 220.84* are not permitted to be applied to the house loads.

For example, suppose you have a 30-unit apartment building consisting of 18 one-bedroom units of 650 ft[2] each; six two-bedroom units of 775 ft[2] each; and six three-bedroom units of 950 ft[2] each. The kitchen in each unit contains one 7.5kW electric range, one 1,250VA dishwasher, and one ⅓hp, 120V garbage disposal at 864VA. Each unit has an air conditioning unit rated at 3,600VA.

The entire building is heated by a boiler that uses three electric pump motors: a 2hp squirrel-cage motor and two ½hp squirrel-cage induction motors, all 240V single-phase, two-wire. There is a community laundry room in the building with four washers and four clothes dryers, with the latter rated at 4,500VA each. The house lighting consists of 40 incandescent fixtures rated at 60W each. There are three general-purpose receptacle outlets for servicing of the boilers and laundry room. The building is to be furnished with a 120/240V, single-phase, three-wire service.

Step 1 Calculate the lighting and small appliance loads [*NEC Sections 220.84(C)(1) and (2)*].

General lighting:
650 ft² × 3VA = 1,950VA
1,950VA × 18 units = 35,100VA

General lighting:
775 ft² × 3VA = 2,325VA
2,325VA × 6 units = 13,950VA

General lighting:
950 ft² × 3VA = 2,850VA
2,850VA × 6 units = 17,100VA

Small appliance circuits:
1,500VA × 2 × 30 units = 90,000VA
Total = 156,150VA

Step 2 Calculate the appliance and motor loads [*NEC Sections 220.84(C)(3), (4), and (5)*].

Dishwashers:
1,250VA × 30 units = 37,500VA

Disposals:
864VA × 30 units = 25,920VA

Ranges:
7,500VA × 30 units = 225,000VA

Air conditioners:
3,600VA × 30 units = 108,000VA
Total appliance & motor loads = 396,420VA

Step 3 Calculate the apartment unit loads.

156,150VA + 396,420VA = 552,570VA

Step 4 Apply the demand factor (*NEC Table 220.84*).

33% of 552,570VA =
552,570VA × 0.33 = 182,348VA

Step 5 Calculate the total house load [*NEC Section 220.84(B)*].

2hp boiler (NEC Table 430.248):
12A × 240V = 2,880VA
Two ½hp boiler motors:
4.9A × 240V × 2 = 2,352VA
Laundry:
1,500VA × 4 = 6,000VA
Dryers (NEC Table 220.54):
5,000 (*NEC Section 220.54*) × 4 × 1 = 20,000VA
House lighting:
60W × 40 = 2,400VA
General-purpose receptacles:
180VA × 3 = 540V
Total house load = 34,172VA

Step 6 Determine the total load.

182,348VA
34,172VA
25% of continuous load = 600VA
Total load = 217,120VA

Using the optional method, the current would be:

217,120VA ÷ 240V = 905A

The next standard size of service equipment greater than 905A is 1,000A, so this apartment building would require a 1,000A service. Note that this is less than the 1,600A required when using the standard method of calculation for the same building.

2.0.0 Section Review

1. A backyard pool pump house is not listed in *NEC Table 220.12*. The *NEC®* section that applies to load calculations for this facility is _____.

   a. *NEC Section 215.2(B)*
   b. *NEC Section 220.5(A)*
   c. *NEC Section 220.14(A)*
   d. *NEC Section 220.16(A)*

2. Which of the following is considered an occupancy area and must be included in residential load calculations?

   a. Unfinished basement
   b. Garage
   c. Open porch
   d. Carport

3. An apartment building has a total general lighting and small appliance load of 170,000VA. Using the standard calculation, the net load after the application of demand factors is _____.

   a. 127,500VA
   b. 85,000VA
   c. 59,500VA
   d. 56,450VA

3.0.0 COMMERCIAL OCCUPANCY CALCULATIONS

Objective

Make service calculations for commercial installations.

  a. Size commercial and industrial lighting loads.
  b. Calculate loads for retail stores.
  c. Calculate loads for office buildings.
  d. Calculate loads for restaurants.
  e. Calculate loads for hotels and motels.
  f. Perform optional calculations for schools.
  g. Size shore power circuits for marinas and boatyards.
  h. Make farm load calculations.
  i. Size motor circuits.

Trade Term

Separately derived system: An electrical source, other than a service, having no direct connection(s) to circuit conductors of any other electrical source other than those established by grounding and bonding connections.

Calculating load requirements for commercial occupancies is based on specific *NEC*® requirements that relate to the loads present. The basic approach is to separate the loads into the following:

- Lighting
- Receptacles
- Motors
- Appliances
- Other special loads

In general, all loads for commercial occupancies should be considered continuous unless specific information is available to the contrary. Smaller commercial establishments normally use single-phase, three-wire services; larger projects almost always use a three-phase, four-wire service. Many installations have secondary feeders supplying panelboards, which, in turn, supply branch circuits operating at different voltages. This requires separate calculations for the feeder and branch circuit loads for each voltage. The rating of the main service is based on the total load with the load values transformed according to the various circuit voltages, if necessary.

Demand factors also apply to some commercial establishments. For example, the lighting loads in hospitals, hotels, motels, and warehouses are subject to the application of demand factors. In restaurants and similar establishments, the load of electric cooking equipment is subject to a demand factor if there are more than three cooking units. Optional calculation methods to determine feeder or service loads for schools and similar occupancies are also provided in the *NEC*®.

Special occupancies, such as mobile homes and recreational vehicles, require the feeder or service load to be calculated in accordance with specific *NEC*® requirements. The service for mobile home parks and recreational vehicle parks is also designed based on specific *NEC*® requirements that apply only to those locations. The feeder or service load for receptacles supplying shore power for boats in marinas and boatyards is also specified in the *NEC*®.

When transformers are not involved, a relatively simple calculation involving only one voltage results. If step-down transformers are used, the transformer itself must be protected by an overcurrent device that may also protect the circuit conductors in most cases.

Switches and panelboards used for the distribution of electricity within a commercial building are also subject to *NEC*® rules. These requirements could affect the number of feeders required when a large number of lighting or appliance circuits are needed. See *NEC Article 408*.

3.1.0 Commercial and Industrial Lighting Loads

The type of occupancy determines the methods employed to calculate loads. Industrial and commercial buildings are two of these categories addressed in the *NEC*®. Buildings that fall into these categories are occupied by businesses that manufacture, process, market, or distribute goods. Loads are calculated based on the type of occupancy and the electrical requirements of the installed equipment.

Standard calculations used to compute the load and determine the service equipment requirement are arranged differently from those for dwelling units. There are specific types of loads that must be considered when making load calculations of commercial and industrial occupancies. Demand factors may be required for certain types of loads.

The following are *NEC*® references for feeder and service loads:

- *NEC Section 215.2*, *Minimum Rating and Size*

- *NEC Section 220.42, General Lighting*
- *NEC Section 220.43, Show-Window and Track Lighting*
- *NEC Section 220.44, Receptacle Loads – Other Than Dwelling Units*
- *NEC Section 220.50, Motors*
- *NEC Section 220.51, Fixed Electric Space Heating*
- *NEC Section 220.56, Kitchen Equipment – Other Than Dwelling Units*
- *NEC Section 220.60, Noncoincident Loads*
- *NEC Section 220.61, Feeder or Service Neutral Load*

Each load must be designated as either continuous or noncontinuous. Most commercial loads are generally considered to be continuous. One exception would be a storage warehouse that is occupied on an infrequent basis. Demand factors may be applied to various noncontinuous loads according to the respective section of the *NEC®*. For example, *NEC Section 220.44* permits both general-purpose receptacles at 180VA per outlet and fixed multi-outlet assemblies to be adjusted for the demand factors given in *NEC Tables 220.42 and 220.44*.

3.1.1 Lighting Loads

Lighting loads are divided into seven different categories to take into account the following:

- *NEC Table 220.12, General Lighting Loads by Non-Dwelling Occupancy*
- *NEC Section 220.14(D), Luminaires*
- *NEC Section 220.14(E), Heavy-Duty Lampholders*
- *NEC Section 220.14(F), Sign and Outline Lighting*
- *NEC Section 220.14(G), Show Windows*
- *NEC Section 220.18(B), Inductive and LED Lighting Loads*
- *NEC Section 220.43(B), Track Lighting*

These lighting loads must be designated as either continuous or noncontinuous. Some of these lighting loads may have demand factors applied. The lighting loads should be calculated before any other loads are determined.

3.1.2 General Lighting Loads

The general lighting load is the main lighting used for general illumination within the building and is in addition to any lighting installed for accent, display, task, show windows, or signs. It is computed based on the type of occupancy and unit loads (in VA) from *NEC Table 220.12*. For types of occupancies not listed in *NEC Table 220.12*, the general lighting load should be calculated according to the *NEC®* sections listed previously.

The minimum load for general lighting must be compared to the actual installed lighting used within a building, and the larger of the two computations must be used. In most cases, due to energy codes, the installed lighting will generally be lower than that required by *NEC Table 220.12*.

Any of these loads that are continuous duty are multiplied by 125% to determine the overcurrent protection device and conductor size. All others are calculated at 100%. Noncontinuous loads may have a demand factor applied in accordance with *NEC Section 220.42* and *NEC Table 220.42*.

Example:
Suppose a warehouse building has 57,600 ft² of floor area.

Step 1 From *NEC Table 220.12*:

$$57,600 \text{ ft}^2 \times 1.2 \text{VA/ft}^2 = 69,120 \text{VA}$$

Step 2 Apply demand factors per *NEC Table 220.42*.

First 12,500VA at 100%:

$$12,500 \text{VA} \times 1 = 12,500 \text{VA}$$

Remaining 56,620VA at 50%:

$$56,620 \text{VA} \times 0.5 = \underline{28,310 \text{VA}}$$

$$\text{Net lighting load} = 40,810 \text{VA}$$

3.1.3 Show Window Loads

NEC Section 220.43(A) requires that the lighting load for show window feeders be calculated at 200VA per linear foot of show window (660VA per linear meter).

Example:
Suppose a show window is 55' in length. What is the feeder load?

Solution:

$$\text{Load} = 55' \times 200 \text{VA per linear ft.} = 11,000 \text{VA}$$

> **NOTE**
>
> Any loads that are continuous duty are multiplied by 125% to determine the overcurrent protection device and conductor size. All others are calculated at 100%.

3.1.4 Track Lighting Loads

Track lighting loads are calculated by allowing 150VA for every two feet (600 mm) of track or fraction thereof. Refer to *NEC Section 220.43(B)*.

Example 1:
The lighting load for 140' of track lighting is determined as follows:

$$140' \div 2 = 70'$$
$$70' \times 150 \text{VA} = 10,500 \text{VA}$$

Example 2:

The lighting load of 21' of track lighting is determined as follows:

$$21' \div 2 = 10.5' \text{ (round fractions up)} = 11'$$
$$11' \times 150VA = 1,650VA$$

> **NOTE**
>
> Any loads that are continuous duty are multiplied by 125% to determine the overcurrent protection device and conductor size. All others are calculated at 100%.

3.1.5 Outside Lighting Loads

Outside lighting loads are calculated by multiplying the rating in VA by the number of fixtures. Refer to *NEC Section 220.14(D)*.

Example:

Determine the load for 30 outside lighting fixtures, each rated at 250VA, including the ballast. These lights are rated at 240V.

Solution:
$$250VA \times 30 = 7,500VA$$

> **NOTE**
>
> Any loads that are continuous duty are calculated at 125% (multiplied by 1.25) to determine the overcurrent protection device and conductor size. All others are calculated at 100%.

3.1.6 Sign and Outline Lighting Loads

NEC Section 600.5(A) requires that occupancies with a ground floor entry accessible to pedestrians have a 20A outlet installed for each tenant space in an accessible location. This must be a 20A branch circuit that supplies no other load.

NEC Section 220.14(F) states that the load for the branch circuit installed for the supply of exterior signs or outline lighting must be computed at a minimum of 1,200VA. However, if the actual rating of the sign lighting is greater than 1,200VA, the actual load of the sign shall be used.

> **NOTE**
>
> Any loads that are continuous duty are calculated at 125% (multiplied by 1.25) to determine the overcurrent protection device and conductor size. All others are calculated at 100%.

3.2.0 Loads for Retail Stores

Figure 8 shows a small store building with a show window in front. Note that the storage area has four general-purpose duplex receptacles, while the retail area has 14 wall-mounted duplex receptacles and two floor-mounted receptacles for a total of 16 in this area. These, combined with the storage area receptacles, bring the total to 20 general-purpose duplex receptacles that do not supply a continuous load. What are the conductor sizes for the service entrance if a 120/240V single-phase service will be used?

Step 1   Determine the total area of the building by multiplying length by width.

$$50' \times 80' = 4,000 \text{ ft}^2$$

Step 2   Calculate the lighting load using *NEC Table 220.12*; according to this table, the load is 1.9VA per square foot.

$$4,000 \text{ ft}^2 \times 1.9VA = 7,600VA$$

Step 3   Determine the total volt-amperes for the 20 general-purpose duplex receptacles.

$$20 \times 180VA = 3,600VA$$

Step 4   Calculate the load for the 30' show window on the basis of 200VA per linear foot.

$$30' \times 200VA = 6,000VA$$

Step 5   Allow one 20A outlet for sign or outline lighting if the store is on the ground floor [*NEC Section 600.5(A)*].

If the actual load of the sign is not known, a minimum load of 1,200VA is used in the calculation per *NEC Section 220.14(F)*.

Step 6   Calculate the total load in volt-amperes.

Noncontinuous load:
Receptacle load = 3,600VA

Continuous loads:

Lighting load =	7,600VA
Show window =	6,000VA
Sign =	1,200VA
Total continuous load =	14,800VA

Calculate at 125%:
$$14,800VA \times 1.25 = 18,500VA$$

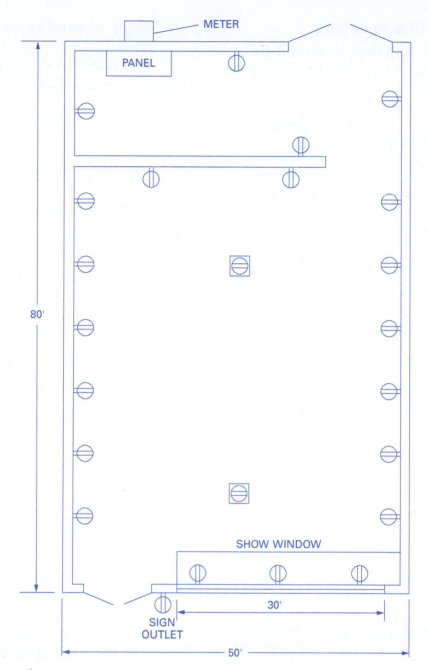

Figure 8 Floor plan of a retail store.

Total load:

Total VA =
continuous loads + noncontinuous load
Total VA =
3,600VA + 18,500VA = 22,100VA

Step 7 Calculate the service size in amperes.

22,100VA ÷ 240V = 92A

Consequently, the service-entrance conductors must be rated for no less than 100A. The conductors would be No. 1 AWG copper or No. 1/0 AWG aluminum using the 60° column of *NEC Table 310.16* per *NEC Section 110.14(C)(1)(a)* for circuits under 100A.

3.3.0 Loads for Office Buildings

A 20,000 ft² office building is served by a 480Y/277V, three-phase service. The building contains the following loads:

- 10,000VA, 208V, three-phase sign
- 100 duplex receptacles supplying continuous loads rated at 180VA each
- 30' long show window
- 12kVA, 208Y/120V, three-phase electric range
- 10kVA, 208Y/120V, three-phase electric oven
- 20kVA, 480V, three-phase water heater
- Seventy-five 150W, 120V incandescent outdoor lighting fixtures
- Two hundred 125VA input, 277V LED lighting fixtures
- 7.5hp, 480V, three-phase motor for fan coil unit
- 40kVA, 480V, three-phase electric heating unit
- 60A, 480V, three-phase air conditioning unit

> **NOTE**
>
> The ratings of the service equipment, transformers, and feeders are to be determined, along with the required size of the service grounding conductor. Circuit breakers are used to protect each circuit, and THWN copper conductors are used throughout the electrical system.

A one-line diagram of the electrical system is shown in *Figure 9*. Note that the incoming three-phase, four-wire, 480Y/277V main service terminates into a main distribution panel containing six overcurrent protective devices. Because there are only six circuit breakers in this enclosure, no main circuit breaker or disconnect is required per *NEC Section 230.71(B)*. Five of these circuit breakers protect feeders and branch circuits to 480/277V equipment, while the sixth circuit breaker protects the feeder to a 480/208Y/120V transformer. The secondary side of this transformer feeds a 208/120V lighting panel with all 120V loads balanced. Start at the loads connected to the 208Y/120V panel and perform the required calculations.

Step 1 Calculate the load for the 100 receptacles.

$$100 \times 180VA = 18,000VA$$

Apply demand factor from *NEC Table 220.44*:

First 10,000VA at 100%:

$$10,000VA \times 1 = 10,000VA$$

Remaining 8,000VA at 50%:

$$8,000VA \times 0.5 = \underline{4,000VA}$$

$$Total = 14,000VA$$

Step 2 Calculate the load for the show window using 200VA per linear foot.

$$200VA \times 30' = 6,000VA$$

Step 3 Calculate the load for the incandescent outside lighting.

$$75 \times 150VA = 11,250VA$$

Step 4 Calculate the load for the 10kVA sign.

$$10kVA \times 1,000 = 10,000VA$$

Step 5 Calculate the load for the 12kVA range (*NEC Section 220.56*).

$$12kVA \times 1,000 = 12,000VA$$

Step 6 Calculate the load for the 10kVA oven (*NEC Section 220.56*).

$$10kVA \times 1,000 = 10,000VA$$

Step 7 Determine the sum of the loads on the 208/120V lighting panel.

Noncontinuous loads:

$$Range = 12,000VA$$
$$Oven = \underline{10,000VA}$$
$$Total\ noncontinuous\ load = 22,000VA$$

Continuous loads:

$$Receptacles = 14,000VA$$
$$Show\ window = 6,000VA$$
$$Outside\ lighting = 11,250VA$$
$$Sign = \underline{10,000VA}$$
$$Total\ continuous\ load = 41,250VA$$

Total load:

$$Total\ feeder\ load =$$
$$continuous\ loads + noncontinuous\ loads$$

$$Total\ feeder\ load =$$
$$22,000VA + 51,563VA = 73,563VA$$

Step 8 Determine the feeder rating for the subpanel.

$$\frac{73,563VA}{208V \times 1.732} = 204.20A$$

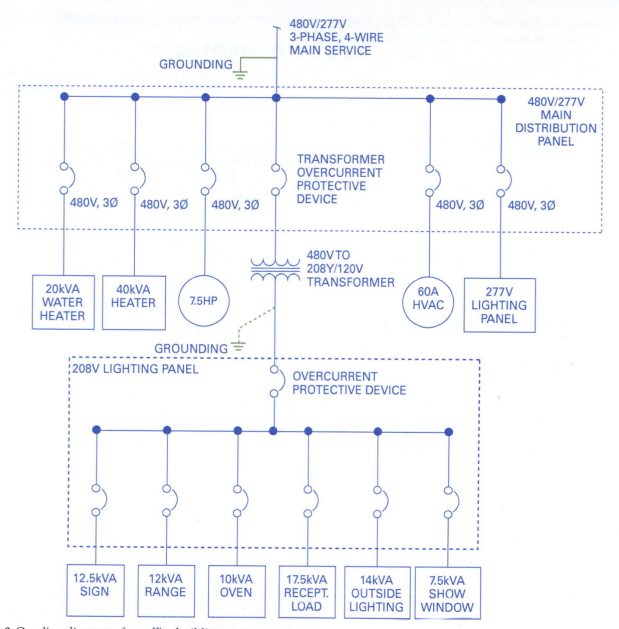

Figure 9 One-line diagram of an office building.

Step 9 Refer to ***NEC Table 310.16*** and find that 4/0 THWN conductor (rated at 230A) is the closest conductor size that will handle the load. A 225A breaker will protect this feeder.

    The 208Y/120V feeder is a **separately derived system** from the 480V transformer and is grounded by means of a grounding electrode conductor (***NEC Table 250.66***) that must be at least a No. 2 copper conductor based on the No. 4/0 copper feeder conductors.

Step 10 The transformer is sized to accommodate the computed load of 73,563VA. Select the overcurrent protective device for the transformer.

$$I = \frac{VA}{V \times \sqrt{3}}$$

$$I = \frac{73,563VA}{480V \times \sqrt{3}} = 88.5A$$

    Per ***NEC Table 450.3(B)*** and because a secondary protective device is used, the maximum setting of the transformer primary overcurrent protective device can be up to 250% or 2.5 × 88.5 = 221A; that is, it may not exceed this rating. The secondary protective device maximum setting is 125% of the 208V load:

$$I = \frac{VA}{V \times \sqrt{3}}$$

$$I = \frac{73,563VA}{208V \times \sqrt{3}} = 204A$$

Calculate at 125%:
    204A × 1.25 = 255A

The largest permissible standard overcurrent protective device is 250A. If this is used, the feeder conductors would need to be sized for this rating as well, instead of the size determined in Step 9.

3.3.1 Calculations for the Primary Feeder

For the feeder calculations, start with the lighting load. *NEC Table 220.12* requires a minimum general lighting load for office buildings to be based on $1.3VA/ft^2$ ($14VA/m^2$). Since the building has already been sized at 20,000 ft², the minimum VA for lighting may be determined as follows:

    20,000 ft² × 1.3VA = 26,000VA

The actual connected LED lighting fixture load, however, is as follows:

    200 × 125VA = 25,000VA

Since this connected load is less than the *NEC®* requirement, it is neglected in the calculation, and the 1.3VA/ft² (14VA/sq m) value is used. Therefore, the total load on the 277V lighting panel may be determined as follows:

$$I = \frac{VA}{V \times \sqrt{3}}$$

$$I = \frac{26,000VA}{480V \times \sqrt{3}} = 31.29A$$

The overcurrent protective device protecting the feeder panel will be sized in accordance with *NEC Section 215.2(A)(1)*.

    1.25 × 31.29A = 39.13A

It would be rated for 40A, the closest standard fuse or circuit breaker size [*NEC Section 240.6(A)*]. However, the smallest main lugs only panelboard is 125A. A 100A feeder was chosen by the engineer. No. 3 AWG copper or No. 1 AWG aluminum will be used for this feeder.

The feeder load for the water heater in this example is calculated as follows:

    20kVA × 1,000 = 20,000VA

NEC Section 220.51 permits the load on feeders and services for electric space heating to be computed at 100%. Therefore, the load on the feeder for the heating is computed at 40kVA as follows:

    40kVA × 1,000 = 40,000VA

NEC Section 220.50 requires that motor loads be computed in accordance with *NEC Sections 430.24, 430.25, and 430.26*. Therefore, the FLA of the fan coil motor (11A) taken from *NEC Table 430.250* must be multiplied by 1.25 since this is the largest (only) motor.

    11A × (480V × $\sqrt{3}$) × 1.25 = 11,431.5VA

NEC Section 220.50 requires that hermetic refrigerant motor compressors be sized in accordance with *NEC Section 440.6*. Therefore, the load for the air conditioning unit is calculated as follows:

    60A × (480V × $\sqrt{3}$) = 49,883VA

3.3.2 Main Service Calculations

When performing the calculations for the main service, assume that all three-phase loads are balanced and may be computed in terms of volt-amperes or in terms of amperes. Calculation of the loads in amperes simplifies the selection of the main overcurrent protective device and service conductors. A summary of this calculation is shown in *Table 1*.

Table 1 Main Service Calculation Example

Type of Load	Computed Load	Neutral
208/120V system	88.5A	0
277V lighting panel	39.13A	39.13A
Water heater	24A	0
Electric heater (neglected)	0	0
7.5hp motor	11A	0
Air conditioner	60A	0
25% of largest motor	2.75A	0
Service load	225.38A	39.13A

3.4.0 Loads for Restaurants

The combined load of three or more cooking appliances and other equipment for a commercial kitchen may be reduced in accordance with *NEC®* demand factors. This provision would apply to restaurants, bakeries, and similar locations. For example, a small restaurant is supplied by a 120/208V, four-wire, three-phase service. The restaurant has the following loads:

- 1,000 ft² area lighted by 120V lamps
- Ten duplex receptacles
- 20A, 208V, three-phase motor compressor
- 5hp, 208V, three-phase roof ventilation fan, running continuously, protected by an inverse time circuit breaker
- More than six units of kitchen equipment with a total connected load of 80kVA (all units are 208V, three-phase equipment)
- Two 20A sign circuits

The main service uses type THHN copper conductors and is calculated as shown in *Figure 10*. Lighting and receptacle loads contribute 30.6A to either phase A or C and 30.6A to the neutral. The 80kVA kitchen equipment load is subject to the application of a 65% demand factor (per *NEC Table 220.56*), which reduces it to a demand load of 52kVA. This load requires a minimum ampacity of 144.3A per phase at 208V. The load of the three-phase motors and 25% of the largest motor load bring the service load total to 241.6A for phase A or C and 186A for phase B.

If the phase conductors are three 250 kcmil THHN copper conductors, the grounding electrode conductor and the neutral conductor must each be at least a No. 2 copper conductor.

The fuses are selected in accordance with the *NEC®* rules for motor feeder protection. The ungrounded conductors, therefore, are protected at 300A each.

NEC Section 220.88 allows an optional method for the service calculation of a new (or completely rewired) restaurant in lieu of the standard method.

According to *NEC Table 220.88*, add all electrical loads, including heating and cooling loads, to compute the total connected load. Then select a single demand factor from the column that applies to the total computed load, and multiply the load by that factor to obtain the total demand load used for sizing the primary feeder or service-entrance conductors.

3.5.0 Loads for Hotels and Motels

The portion of the feeder or service load contributed by general lighting in hotels and motels without provisions for cooking by tenants is subject to the application of demand factors. In addition, the receptacle load in the guest rooms is included in the general lighting load at 1.7VA/ft² (18VA/m²). However, the demand factor (load reduction) does not apply to any area where the entire lighting is likely to be used at one time, such as the dining room or a ballroom (see *NEC Table 220.42*).

Example:
Determine the 120/240V feeder load contributed by general lighting in a 100-unit motel. Each guest room is 240 ft² in area.

Solution:
The general lighting load is:

$$1.7VA \times 240 \text{ ft}^2 \times 100 \text{ units} = 40,800VA$$

The demand lighting load (per *NEC Table 220.42*) is:

First 20,000 at 60% = 12,000VA
40,800 − 20,000 = 20,800 at 50% = 10,400VA
Total = 22,400VA

This load would be added to any other loads on the feeder or service to compute the total capacity required.

3.6.0 Optional Calculations for Schools

NEC Section 220.86 provides an optional method for determining the feeder or service load of a school equipped with electric space heating, air conditioning, or both. This optional method applies to the building load, not to feeders within the building.

The optional method for schools basically involves determining the total connected load in volt-amperes, converting the load to volt-amperes per square foot/square meter, and applying the demand factors from *NEC Table 220.86*. If both air conditioning and electric space heating loads are present, only the larger of the loads is to be included in the calculation.

A school building has 200,000 ft² of floor area and a 480Y/277V service. The electrical loads are as follows:

- Interior lighting at 3VA/ft² = 600,000VA
- 300kVA power load
- 100kVA water heating load

Service Loads	Line A, C	Neutral	Line B

Note: Unbalanced loads are calculated entirely on either phase A or C for a worst-case neutral (and conductor) load (*NEC Section 220.61*).

	Line A, C	Neutral	Line B
A. 208/120V loads	15.6	15.6	0

Lighting = $\dfrac{1.25 \times 1.5\text{VA per sq. ft.} \times 1{,}000 \text{ sq. ft.}}{120\text{V}}$ = 15.6A

	Line A, C	Neutral	Line B
Receptacles = $\dfrac{180\text{VA} \times 10}{120\text{V}}$ = 15.0A	15.0	15.0	0
B. Three-phase loads Kitchen equipment (6 or more units) =	144.3	0	144.3

$$\frac{80{,}000\text{W} \times .65}{\sqrt{3} \times 208\text{V}} = 144.3\text{A}$$

	Line A, C	Neutral	Line B
C. 20A three-phase motor compressor Dual-element fuse rating = 1.75 × 20A = 35A Use 35A (*NEC Table 430.52*)	20	0	20
D. Three-phase 5hp fan (16.7A) (*NEC Table 430.250*) Breaker rating = 2.5 × 16.7A = 41.75A Use 45A inverse-trip breaker (*NEC Table 430.52*)	16.7	0	16.7
E. 25% of largest motor load = .25 × 20A = 5A	5	0	5
F. Sign circuit = $\dfrac{1{,}200\text{VA}}{120\text{V}}$ = 10A each × 2 = 20A × 125% = 25A	25	25	0
	241.6A	**55.6A**	**186A**

Service Load
NEC Table 310.16
NEC Table 250.66
NEC Section 310.15(E)(1)

NEC Section 230.90

NEC Section 240.6(A)

1. Conductors: Use No. 250 kcmil THHN copper at 75°C for ungrounded conductors; use No. 2 THHN copper conductor for neutral (neutral based on size of grounding electrode conductor).

2. Overcurrent protective device:
   Phases A and C = 45A (largest motor device) + 15.6A + 15.0A + 144.3A + 20A + 25A = 264.9A
   Use standard size 300A fuses.

3. Grounding electrode conductor required to be No. 2 copper.

Figure 10 Restaurant service specifications.

- 100kVA cooking load
- 100kVA miscellaneous loads
- 200kVA air conditioning load
- 300kVA heating load

The service load in volt-amperes is to be determined by the optional calculation method for schools.

The combined connected load is 1,500,000VA. (This excludes the 200kVA air conditioning load since the heating load is larger.) Based on the 200,000 ft² of floor area, the load per square foot is:

$$\frac{1,500,000VA}{200,000 \text{ ft}^2} = 7.5VA/ft^2$$

The demand factor for the portion of the load up to and including 3VA per square foot is 100%. The remaining 4.5VA per square foot in the example is added at a 75% demand factor for a total load of 1,275,000VA.

$$(200,000 \times 3VA/ft^2)$$
$$+ (200,000 \times 4.5kVA/ft^2 \times 0.75)$$
$$= 1,275,000VA$$

To size the service-entrance conductors, *NEC Section 230.42* must be applied to this demand.

$$1,275,000VA \times 1.25 = 1,593,750VA$$

Assuming a relatively short, rigid conduit overhead service where voltage drop will be negligible, calculate the size of THHW service-entrance conductors as follows for the 480Y/277V, four-wire, three-phase load:

$$\text{Load Current} = \frac{1,593,750VA}{480V \times \sqrt{3}} = 1,917A$$

NEC Section 240.4(C) requires that the conductors have an ampacity that is not less than the rating of the overcurrent device when the device is rated more than 800A. The nearest standard rating of a nonadjustable trip overcurrent device needed for a 1,917A load is 2,000A. The conductors for this service must have an ampacity of at least 2,000A.

Because no standard size copper conductor can carry this load, the service will be accomplished by a number of parallel, four-wire runs in separate conduit between the service equipment and the service drop. The desired result is to establish the minimum number of runs with the minimum sized conductors to accommodate the load. To accomplish this, divide the load by a number of different runs and select a wire size from *NEC Table 310.16* that is very close to or exceeds the resulting current load.

$$\text{Option 1} = \frac{2,000A}{4 \text{ runs}} = 500A, 900 \text{ kcmil conductor}$$

$$\text{Option 2} = \frac{2,000A}{5 \text{ runs}} = 400A, 600 \text{ kcmil conductor}$$

$$\text{Option 3} = \frac{2,000A}{6 \text{ runs}} = 333A, 400 \text{ kcmil conductor}$$

$$\text{Option 4} = \frac{2,000A}{7 \text{ runs}} = 286A, 350 \text{ kcmil conductor}$$

Evaluate the options:

- *Option 1* – Discarded because the size of the conductors makes them difficult to install and also because it provides more capacity than needed.
- *Option 2* – Discarded because of the handling problem due to conductor size.
- *Option 3* – Selected because it meets the load capacity desired with the minimum number of runs.
- *Option 4* – Although this option meets the load requirements, it is discarded because Option 3 meets the requirement with fewer runs.

The overhead service entrance is therefore minimally sized as six paralleled four-wire runs of 400 kcmil XHHW-2 copper conductor. [Type XHHW-2 was selected because it can be used in either wet or dry locations. See *NEC Table 310.4(A)*.] At this point, if the length of the run had been known, the voltage drop could be checked using the appropriate formula to see if the drop exceeded 2%. If so, the next larger standard wire could be selected.

Note that Option 4 with 350 kcmil conductors could be installed to allow for future needs. Also note that an alternate solution is to bring the higher line voltage to a small high-voltage substation inside or adjacent to the building, where it would be stepped down to an intermediate voltage and/or then stepped down to 480V and busbarred directly to switchboard equipment. This is a fairly common practice, especially in medium-to-large industrial establishments.

3.7.0 Shore Power Circuits for Marinas and Boatyards

The wiring systems for marinas and boatyards are designed using the same *NEC®* rules as for other commercial occupancies, except for the application of several special rules dealing primarily with the design of circuits supplying power to boats (*NEC Article 555*).

The smallest sized locking and grounding-type receptacle that may be used to provide shore power for boats is 30A. Each single receptacle that supplies power to boats must be supplied by an individual branch circuit with a rating corresponding to the rating of the receptacle [*NEC Sections 555.33(A)(3) and 555.33(A)(4)*].

The feeder or service ampacity required to supply the receptacles depends on the number of receptacles and their rating, but demand factors may be applied that will reduce the load of five or more receptacles (*NEC Table 555.6*).

For example, a feeder supplying ten 30A shore power receptacles in a marina requires a minimum ampacity of:

$$10 \times 30A \times 120V = 36,000VA$$
$$36,000VA \times 0.8 = 28,800VA$$
$$28,800VA \div 240V = 120A$$

This is the minimum required by the *NEC®* unless individual watthour meters are provided for each shore power outlet.

3.8.0 Farm Load Calculations

NEC Article 220, Part V provides a separate method for computing farm loads (other than the dwelling). Tables of demand factors are provided for use in computing the feeder loads of individual buildings as well as the service load of the entire farm. See *NEC Sections 220.102 and 220.103*. See *NEC Article 547* for specific code requirements for agricultural buildings.

The demand factors may be applied to the 120/240V feeders for any building or load (other than the dwelling) that is supplied by two or more branch circuits. All loads that operate without diversity—that is, the entire load is on at one time—must be included in the calculation at 100% of the connected load. All other loads may be included at reduced demands. The load to be included at 100% demand, however, cannot be less than 125% of the largest motor and no less than the first 60A of the total load. In other words, if the nondiverse and largest motor load is less than 60A, a portion of the other loads will have to be included at 100% in order to reach the 60A minimum.

After the loads from individual buildings are computed, it may be possible to reduce the total farm load further by applying additional demand factors.

For example, a farm has a dwelling and two other buildings supplied by the same 120/240V service. The electrical loads are as follows:

- *Dwelling* = 100A load as computed by the calculation method for dwellings.

- *Building No. 1* = 5kVA continuous lighting load operated by a single switch; 10hp, 240V motor; and 21kVA of other loads.
- *Building No. 2* = 2kVA continuous load operated by a single switch and 15kVA of other loads.

Determine the individual building loads and the total farm load, as illustrated in *Figure 11*. The nondiverse load for building No. 1 consists of the 5kVA lighting load and the 10hp motor for a total of 83.3A. This value is included in the calculation at the 100% demand factor. Since the requirement for adding at least the first 60A of load at the 100% demand factor has been satisfied, the next 60A of the 87.5A from all other loads are added at a 50% demand factor and the remainder of 27.5A (87.5 – 60) is added at a 25% demand factor.

In the case of building No. 2, the nondiverse load is only 8.3A; therefore, 51.7A of other loads must be added at the 100% demand factor in order to meet the 60A minimum.

Using the method given for computing the total farm load, the service load is:

Largest load at 100% (Building No. 1) =	120.2A
Second largest at 75% (Building No. 2) =	49.1A
Total for both buildings =	169.3A
Dwelling =	100.0A
Total load in amperes =	269.3A

The total service load of 269A requires the ungrounded service-entrance conductors to be at least 300 kcmil THHW copper conductors. The neutral load of the dwelling is assumed to be 100A, which brings the total farm neutral load to 207A.

3.9.0 Motor Circuits

Two or more motors may be connected to the same feeder circuit instead of routing an individual circuit to each motor. This saves money and is just as efficient if properly designed. *NEC Section 430.24* covers this type of installation. For two or more motors connected to the same feeder conductors, the rule is to compute the sum of all motors plus 25% of the largest motor. Note the exceptions in *NEC Section 430.24* regarding motors with duty cycle ratings.

Care must be taken when designing circuits such as this to prevent several motors connected to the same circuit from starting at the same time. If not designed correctly, this would result in a severe voltage dip that could affect the operation of other equipment or prevent one or more of the motors from starting.

Building No. 1 Feeder Load	240V Load	Neutral
Lighting (5kVA nondiverse load) = 5,000VA/240V	20.8	20.8
10hp motor = 1.25 × 50A	62.5	-0-
Total motor and nondiverse load	83.3	20.8
Other loads = 21,000VA/240V	87.5	87.5
Application of demand factors		
Motor and nondiverse loads @ 100%	83.3	20.8
Next 60A of other loads @ 50%	30.0	30.0
Remainder of other loads (87.5 – 60) @ 25%	6.9	6.9
Feeder Load	**120.2A**	**57.7A**
Building No. 2 Feeder Load		
Lighting (2kVA nondiverse load) = 2,000VA/240V	8.3	8.3
Other loads = 15,000VA/240V	62.5	62.5
Application of demand factors		
Nondiverse load @ 100%	8.3	8.3
Remainder of first 60 (60 – 8.3) @ 100%	51.7	51.7
Remainder of other loads (62.5 – 51.7) @ 50%	5.4	5.4
Total Farm Load	**65.4A**	**65.4A**
Application of demand factors		
Largest load (Building No. 1) @ 100%	120.2	57.7
Next largest load (Building No. 2) @ 75%	49.1	49.1
Farm load (less dwelling)	169.3	106.8
Farm dwelling load	100.0	100.0
Total Farm Load	**269.3A**	**206.8A**

Figure 11 Summary of farm calculations.

Example:

The following three-phase, 208V motors are supplied by the same feeder circuit: 20hp, 10hp, and 5hp. Determine the total load and minimum size conductors to supply these motors.

Step 1 Determine the FLA of each motor. Refer to *NEC Table 430.250*.

20hp = 59.4A
10hp = 30.8A
5hp = 16.7A

Step 2 Take 125% of the largest motor FLA.

59.4A × 1.25 = 74.25

Step 3 Add all loads.

74.25A + 30.8A + 16.7A = 121.75A

Step 4 Find the minimum size THWN copper conductor from *NEC Table 310.16*. A No. 1 AWG, type THWN copper conductor is rated at 130A and will be suitable for this feeder circuit.

3.0.0 Section Review

1. The feeder load for a show window that is 35' in length is _____.

   a. 3,500VA
   b. 7,000VA
   c. 8,400VA
   d. 10,500VA

2. The general lighting load of a hospital per *NEC Table 220.12* is _____.

   a. 1.2VA/ft²
   b. 1.6VA/ft²
   c. 2.2VA/ft²
   d. 2.6VA/ft²

3. A 15,000 ft² office building has a minimum lighting load of _____.

   a. 10,200VA
   b. 19,500VA
   c. 25,000VA
   d. 32,500VA

4. Per *NEC Table 220.56*, the demand factor for eight pieces of commercial kitchen equipment is _____.

   a. 90%
   b. 80%
   c. 70%
   d. 65%

5. The demand lighting load for a 10,000 ft² motel with no dining room or ballroom is _____.

   a. 10,200VA
   b. 20,700VA
   c. 26,400VA
   d. 30,200VA

6. A school has a 150kVA air conditioning load and a 200kVA heating load. The value to be used when sizing the combined connected load is _____.

   a. the smaller of the two loads (150kVA)
   b. the larger of the two loads (200kVA)
   c. 125% of the larger load (200kVA) or 250kVA
   d. the sum of both loads (150kVA + 200kVA = 350kVA)

7. A feeder supplying six 30A shore power receptacles requires a minimum ampacity of _____.

   a. 43,200VA
   b. 34,560VA
   c. 21,600VA
   d. 19,440VA

8. Which of the following is *true* when calculating farm loads?

   a. Only the motor loads are counted.
   b. Only the nonmotor loads are counted.
   c. At least the first 60A must be counted at 100%.
   d. All loads are counted at 75% demand.

9. A 10hp (30.8A) motor and a 15hp (46.2A) motor are connected to a three-phase, 208V feeder circuit. The total load is _____.

   a. 77A
   b. 84.7A
   c. 88.55A
   d. 96.25A

1. A feeder circuit supplying a lighting and appliance panelboard is sized by adding _____% of the noncontinuous load and _____% of the continuous load.

   a. 80; 100
   b. 100; 100
   c. 100; 110
   d. 100; 125

2. What is the ampacity for each of eight No. 2 THWN copper current-carrying conductors installed in a single conduit?

   a. 57.5A
   b. 80.5A
   c. 91A
   d. 92A

3. What is the rating for a No. 250 kcmil THHN copper conductor in an ambient temperature of 101°F?

   a. 232.05A
   b. 263.9A
   c. 278.4A
   d. 290A

4. What is the rated ampacity per phase for six No. 3/0 THHN copper current-carrying conductors with two conductors per phase (parallel) and all conductors installed in the same conduit?

   a. 280A
   b. 315A
   c. 320A
   d. 360A

5. The voltage drop percentage for a panel feeder with a noncontinuous load of 180A, using 3/0 THWN copper conductors at 480V, three-phase, and a length of 248' is _____.

   a. 1.23%
   b. 1.28%
   c. 1.42%
   d. 2.03%

6. What is the required size of THHN copper conductors for a feeder serving two 208Y/120V, four-wire, three-phase, 75°C branch circuit panelboards, one with a continuous load of 150A and the other with a noncontinuous load of 180A?

   a. 250 kcmil
   b. 400 kcmil
   c. 500 kcmil
   d. 600 kcmil

7. What is the percent voltage drop for a 208V, three-phase feeder serving a load of 200A at a distance of 290' using 4/0 THWN copper conductors?

   a. 1.7%
   b. 2.94%
   c. 3.39%
   d. 4.83%

8. For a THHN feeder rated at 225A, which column from *NEC Table 310.16* must be used unless all components in the feeder circuit are rated for a higher temperature?

   a. 60°F
   b. 75°C
   c. 90°C
   d. 140°C

9. What size THWN copper conductors are required to supply a 120/208V, three-phase panelboard with 20.5kVA of noncontinuous load and 25kVA of continuous load?

   a. 250 kcmil
   b. No. 1
   c. 1/0
   d. 4/0

10. What is the load, in amps, for a single-phase, 240V feeder supplying a load calculated at 23,800VA?

    a. 57.25A
    b. 66.06A
    c. 99.17A
    d. 114.42A

11. What is the load, in amps, for a three-phase, 208V feeder supplying a load calculated at 96.75kVA?

    a. 116.37A
    b. 268.56A
    c. 403.13A
    d. 465.14A

12. What is the load, in amps, for a three-phase, 480V feeder supplying a load calculated at 112.5kVA?

    a. 135.32A
    b. 312.27A
    c. 468.75A
    d. 540.87A

13. Using the 10' (3 m) tap rule, what size copper THWN conductor is required for a tap from a 400A, 480V, three-phase feeder to serve a 150A load?

    a. No. 1
    b. No. 2
    c. 1/0
    d. 2/0

14. Using the 25' (7.5 m) tap rule, what is the minimum size copper THWN conductor required for a tap from a 400A, 480V, three-phase feeder?

    a. No. 1
    b. No. 2
    c. 1/0
    d. 2/0

15. What is the minimum general lighting load, in VA per square foot, required by the *NEC®* for a single-family dwelling?

    a. 1
    b. 2
    c. 3
    d. 3½

16. The lighting load for dwellings and other *NEC®*-listed occupancies is generally determined by _____.

    a. volt-amperes per square foot
    b. total connected load for the entire building
    c. volt-amperes per linear foot
    d. anticipated connected load for the building

17. Using *NEC Table 220.12*, what is the minimum general lighting load, in VA per square foot, required by the *NEC®* for an office building?

    a. 1.2VA/ft²
    b. 1.3VA/ft²
    c. 1.5VA/ft²
    d. 1.6VA/ft²

18. When determining the building area for load calculations, building measurements are taken using the _____.

    a. outside dimensions of the building
    b. inside wall dimensions of the building
    c. width and length of each individual room
    d. property line perimeters

19. When both heating and cooling systems are used in a building, which of the following statements is *true*?

    a. The total load for both systems must be combined and then multiplied by a factor of 125% (1.25).
    b. The larger of the two loads is used in the calculation.
    c. The smaller of the two loads is used in the calculation.
    d. The total load for both systems must be multiplied by a factor of 150% (1.50).

20. When calculating the service size for a single-family dwelling using the standard calculation, what part of the total general lighting and small appliance loads must be calculated at 100%?

    a. First 3,000VA
    b. First 10,000VA
    c. First 15,000VA
    d. First 30,000VA

1. Services are frequently sized larger than the required minimum size in order to account for _____.

2. List four situations in which feeder conductor size is adjusted.

   _____

   _____

   _____

   _____

3. A continuous load must be multiplied by _____ to determine feeder ampacity.

4. What five conditions must be met in order for the 10' (3 m) tap rule to be applied?

   _____

   _____

   _____

   _____

   _____

5. What four conditions must be met in order for the 25' (7.5 m) tap rule to be applied?

   _____

   _____

   _____

   _____

6. Name two of the applications in which you would not be able to apply a demand factor for a neutral conductor.

   _____

7. Where in the *NEC*® would you find VA values for determining lighting loads for various occupancies?

   _____

8. True or False? When calculating the load for the main service, you must include both the heating and air conditioning loads.

   _____

9. *NEC Section 230.79(C)* requires a minimum service of _____ for a single-family dwelling.

10. *NEC Section 220.43(A)* requires that the lighting load for show window feeders be calculated at _____ per linear foot of show window.

11. When calculating the lighting load for an office building, you must use 1.3VA/ft² (14VA/m²) or the actual connected load, whichever is the _____.

12. *NEC Section 220.51* permits the load on feeders and services for electric space heating to be computed at _____.

13. Per *NEC Table 220.56*, a restaurant kitchen with more than six pieces of equipment has a demand factor of _____.

14. The general lighting load in hotel guest rooms is _____.

15. For two or more motors connected to the same feeder conductors, the rule is to compute the sum of all motors plus _____.

LOAD CALCULATION FORM

General Lighting Load				Phase	Neutral
Square footage of the dwelling	[1]	@ 3VA =	[2]		
Kitchen small appliance circuits	[3]	@ 1500 =	[4]		
Laundry branch circuit	[5]	@ 1500 =	[6]		
Subtotal of general lighting loads per *NEC Section 220.42*		[7]			
Subtract 1st 3000VA per *NEC Table 220.42*		[8]	@ 100% =	[9]	
Remaining VA times 35% per *NEC Table 220.42*		[10]	@ 35% =	[11]	
Total demand for general lighting loads =				[12]	[13]
Fixed Appliance Loads (nameplate or *NEC*® FLA of motors) per *NEC Section 220.53*					
Hot water tank, 4.5kVA, 240V		[14]			
		[15]			
		[16]			
		[17]			
		[18]			
		[19]			
Subtotal of fixed appliances		[20]			
If 3 or less fixed appliances take @ 100% =				[21]	[22]
If 4 or more fixed appliances take @ 75% =				[23]	[24]
Other Loads per *NEC Section 220.14*					
Electric range per *NEC Table 220.55* [neutral @ 70 % per *NEC Section 220.61(B)(1)*]				[25]	[26]
Electric dryer per *NEC Table 220.54* [neutral @ 70% per *NEC Section 220.61(B)(1)*]				[27]	[28]
Electric heat per *NEC Section 220.51*					
Air conditioning per *NEC Section 220.14(A)*		*Omit smaller load per NEC Section 220.60*		[29]	[30]
Largest Motor =		@ 25% per *NEC Section 430.24* =		[31]	[32]
Total VA Demand =				[33]	[34]
VA/240V = Amps =				[35]	[36]
Service OCD and Minimum Size Grounding Electrode Conductor *(NEC Table 250.66)* =				[37]	[38]
AWG per *NEC Section 310.12* and *Table 310.16*				[39]	[40]

Trade Terms Introduced in This Module

Building: A structure that stands alone or that is separated from adjoining structures by firewalls.

Dwelling unit: A single unit providing complete and independent living facilities for one or more persons, including permanent provisions for living, sleeping, cooking, and sanitation. A one-family dwelling consists solely of one dwelling unit. A two-family dwelling consists solely of two dwelling units. A multi-family dwelling is a building containing three or more dwelling units.

Feeder: All circuit conductors between the service equipment, the source of a separately derived system, or other supply source and the final branch circuit overcurrent device.

Garage: A building or portion of a building in which one or more self-propelled vehicles can be kept for use, sale, storage, rental, repair, exhibition, or demonstration purposes.

Overhead service conductors: The overhead conductors between the service point and the first point of connection to the service-entrance conductors at the building or other structure.

Separately derived system: An electrical source, other than a service, having no direct connection(s) to circuit conductors of any other electrical source other than those established by grounding and bonding connections.

Service: The conductors and equipment for delivering electric energy from the serving utility to the wiring system of the premises served.

Service conductors: The conductors from the service point to the service disconnecting means.

Service drop: The overhead conductors between the utility electric supply system and the service point.

Service equipment: The necessary equipment, usually consisting of a circuit breaker or switch and fuses, and their accessories, connected to the load end of service conductors to a building or other structure, or an otherwise designated area, and intended to constitute the main control and means of cutoff of the supply.

Service lateral: The underground conductors between the utility electric supply system and the service point.

Service point: The point of connection between the facilities of the serving utility and the premises wiring.

Switchboard: A large single panel, frame, or assembly of panels on which are mounted on the face, back, or both, switches, overcurrent and other protective devices, buses, and usually instruments. Switchboards are generally accessible from the rear as well as from the front and are not intended to be installed in cabinets.

Tap conductors: As defined in *NEC Article 240*, a tap conductor is a conductor, other than a service conductor, that has overcurrent protection ahead of its point of supply that exceeds the value permitted for similar conductors that are protected as described in *NEC Section 240.4*.

Underground service conductors: The underground conductors between the service point and the first point of connection to the service-entrance conductors in a terminal bus.

Additional Resources

This module presents thorough resources for task training. The following resource material is suggested for further study.

National Electrical Code® Handbook, Latest Edition. Quincy, MA: National Fire Protection Association.

Figure Credits

Section Review Answer Key

SECTION 1.0.0

Answer	Section Reference	Objective
1. c	1.1.1	1a
2. a	1.2.0	1b
3. b	1.3.4	1c
4. c	1.4.0	1d

SECTION 2.0.0

Answer	Section Reference	Objective
1. c	2.1.1	2a
2. a	2.2.0	2b
3. d	2.3.1	2c

SECTION 3.0.0

Answer	Section Reference	Objective
1. b	3.1.3	3a
2. b	3.2.0	3b
3. b	3.3.1	3c
4. d	3.4.0	3d
5. a	3.5.0	3e
6. b	3.6.0	3f
7. d	3.7.0	3g
8. c	3.8.0	3h
9. c	3.9.0	3i

Section Review Calculations

Section 1.0.0

Question 2

Since this is a single-phase circuit, divide the nameplate power rating by the circuit voltage to find the feeder ampacity:

Convert nameplate power rating to VA:
$$38.4\text{kW} \times 1{,}000 = 38{,}400\text{VA}$$

Divide by voltage to find current:
$$I = \text{VA} \div V$$
$$I = 38{,}400\text{VA} \div 240\text{V} = 160\text{A}$$

The ampacity is **160A**.

Question 4

Per *NEC Table 220.44*:

First 10kVA at 100%: 16,400VA − 10,000VA = 6,400VA
6,400VA × 50% = 3,200VA
3,200VA + 10,000VA = 13,200VA

The demand load is **13,200VA**.

Section 2.0.0

Question 3

Apply the demand factors using *NEC Table 220.42*:

First 3,000VA (out of 120,000VA) at 100% = 3,000VA
Next 117,000VA (out of 120,000VA) at 35% = 40,950VA
Remaining 50,000VA at 25% = 12,500VA
Net general lighting and small appliance load = 56,450VA

The net load is **56,450VA**.

Section 3.0.0

Question 1

Multiply the length by 200VA:

Load = 35' × 200VA per linear ft = 7,000VA

The feeder load is **7,000VA**.

Question 3

Since the minimum lighting load for an office building is 1.3VA per square foot, multiply the square footage by 1.3VA/ft²:

Minimum lighting load = $1.3\text{VA/ft}^2 \times 15{,}000\text{VA} = 19{,}500\text{VA}$

The feeder load is **19,500VA**.

Question 5

Multiply the square footage by 1.7VA/ft², and find 50% of the result:

Demand lighting load = $1.7\text{VA/ft}^2 \times 10{,}000\text{VA} = 17{,}000\text{VA} \times 60\% = 10{,}200\text{VA}$

The demand lighting load is **10,200VA**.

Question 7

Calculate the feeder load, and apply the demand factor per *NEC Table 555.6*:

Feeder load = $6 \times 30\text{A} \times 120\text{V} = 21{,}600\text{VA} \times 0.9$ (per *NEC Table 555.6*) = 19,440VA

The minimum ampacity is **19,440VA**.

Question 9

Calculate the total load as follows:

Total load = (largest motor FLA × 1.25) + FLA of all other motor(s)
Total load = $(46.2\text{A} \times 1.25) + 30.8\text{A} = 88.55\text{A}$

The total load is **88.55A**.

This page is intentionally left blank.

NCCER CURRICULA — USER UPDATE

NCCER makes every effort to keep its textbooks up-to-date and free of technical errors. We appreciate your help in this process. If you find an error, a typographical mistake, or an inaccuracy in NCCER's curricula, please fill out this form (or a photocopy), or complete the online form at **www.nccer.org/olf**. Be sure to include the exact module ID number, page number, a detailed description, and your recommended correction. Your input will be brought to the attention of the Authoring Team. Thank you for your assistance.

Instructors – If you have an idea for improving this textbook, or have found that additional materials were necessary to teach this module effectively, please let us know so that we may present your suggestions to the Authoring Team.

NCCER Product Development and Revision

13614 Progress Blvd., Alachua, FL 32615

Email: curriculum@nccer.org
Online: www.nccer.org/olf

❏ Trainee Guide   ❏ Lesson Plans   ❏ Exam   ❏ PowerPoints   Other _____

Craft / Level: _____ Copyright Date: _____

Module ID Number / Title: _____

Section Number(s): _____

Description: _____

_____

_____

_____

Recommended Correction: _____

_____

_____

_____

Your Name: _____

Address: _____

_____

Email: _____ Phone: _____

This page is intentionally left blank.

Health Care Facilities

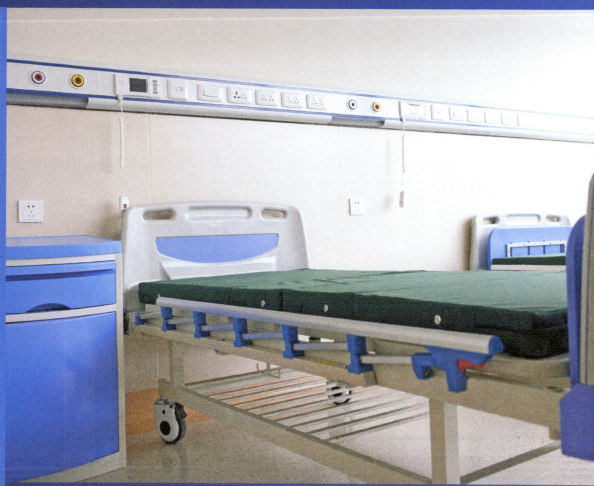

OVERVIEW

In health care facilities, reliable electrical systems and adequate backup power are vital for the safety of patients. For this reason, these systems are strictly regulated in the *NEC*®, as well as in state and local codes. This module covers the installation, alarm system, and backup system requirements of electrical systems in health care facilities, including the requirements for life safety and critical circuits.

Module 26402-20

26402-20 V10.0

Objectives

When you have completed this module, you will be able to do the following:

1. List the types of health care facilities and their power requirements.
   a. Identify types of essential electrical systems used in health care facilities.
   b. Identify types of distribution systems used in health care facilities.
2. Describe the categories and branch portions of the distribution circuits.
   a. Describe the operation and applications of hospital-grade receptacles.
   b. Identify the receptacle requirements for general care (Category 2) spaces.
   c. Identify the receptacle requirements for critical care (Category 1) spaces.
   d. Identify the grounding requirements for receptacles and fixed electrical equipment.
3. List the required wiring methods in health care facilities.
   a. Identify the wiring requirements for inhalation anesthetizing locations.
   b. Identify the wiring requirements for low-voltage equipment and instruments.
   c. Identify the wiring requirements for X-ray installations.
   d. Identify the requirements for communication, signaling, data, and fire alarm systems installed in patient care areas.
   e. Identify the requirements for isolated power systems.

Performance Tasks

This is a knowledge-based module. There are no Performance Tasks.

Trade Terms

Equipotential ground plane
Inpatient

Outpatient
Throwover

Industry Recognized Credentials

If you are training through an NCCER-accredited sponsor, you may be eligible for credentials from NCCER's Registry. The ID number for this module is 26402-20. Note that this module may have been used in other NCCER curricula and may apply to other level completions. Contact NCCER's Registry at 888.622.3720 or go to **www.nccer.org** for more information.

> **NOTE**
>
> NFPA 70®, *National Electrical Code*® and *NEC*® are registered trademarks of the National Fire Protection Association, Quincy, MA.

Contents

1.0.0   Types of Health Care Facilities 1
   1.1.0   Essential Electrical Systems 2
     1.1.1   Type 1 EES 4
     1.1.2   Type 2 EES 4
     1.1.3   Other Health Care Facilities 6
   1.2.0   Electrical Distribution Systems 6
     1.2.1   Double-Ended System 7
     1.2.2   Alternate Power Source 7
     1.2.3   Ground Fault Protection 7
     1.2.4   Additional Grounding and Bonding Requirements.......... 9

2.0.0   Devices Used in Health Care Facilities 10
   2.1.0   Hospital-Grade Receptacles 10
     2.1.1   Isolated Ground Receptacles 10
     2.1.2   Tamper-Resistant Receptacles 10
   2.2.0   General Care (Category 2) Spaces... 10
   2.3.0   Critical Care (Category 1) Spaces 12
   2.4.0   Grounding Receptacles and Fixed Equipment 13

3.0.0   Wiring in Health Care Facilities 17
   3.1.0   Inhalation Anesthetizing Locations... 17
     3.1.1   Wiring and Equipment within
          Hazardous Anesthetizing Locations................................. 17
     3.1.2   Wiring and Equipment Installed
          Above Hazardous Anesthetizing Locations...................... 17
     3.1.3   Wiring in Unclassified Anesthetizing Locations 18
     3.1.4   Grounding in Anesthetizing Locations.............................. 18
     3.1.5   Grounded Power Systems in Anesthetizing Locations.... 19
   3.2.0   Low-Voltage Equipment and Instruments 19
   3.3.0   X-Ray Installations.. 20
     3.3.1   Connection to Supply Circuit... 20
     3.3.2   Disconnecting Means... 20
     3.3.3   Rating of Supply Conductors and Overcurrent Protection 20
     3.3.4   Control Circuit Conductors 21
     3.3.5   Transformers and Capacitors 21
     3.3.6   High Tension X-Ray Cable 21
     3.3.7   Guarding and Grounding... 21
   3.4.0   Communication, Signaling, Data, and Fire Alarm Systems.......... 21
   3.5.0   Isolated Power Systems 22
     3.5.1   Installation 23
     3.5.2   Line Isolation Monitors 23

Figures

Figure 1   Nonessential and essential loads 2

Figure 2   EES branches.. 3

Figure 3   Double-ended substation 7

Figure 4   Typical diesel engine generator 7

Figure 5   Schematic diagram of an alternate power
source with transfer switching for an EES 8

Figure 6   Example of circuit color-coding for a headwall 11

Figure 7   Hospital-grade receptacles 11

Figure 8   Patient bed location receptacles 12

Figure 9   Typical receptacle configuration in a critical care space.............. 14

Figure 10 Common signaling devices for nurse call systems..................... 22

Figure 11 Isolated power panel... 23

This page is intentionally left blank.

SECTION ONE

1.0.0 TYPES OF HEALTH CARE FACILITIES

Objective

List the types of health care facilities and their power requirements.
  a. Identify types of essential electrical systems used in health care facilities.
  b. Identify types of distribution systems used in health care facilities.

Trade Terms

Inpatient: A patient admitted into the hospital for an overnight stay.

Outpatient: A patient who receives services at a hospital but is not admitted for an overnight stay.

Throwover: A transfer switch used for supplying temporary power.

Like homes, health care facilities run on electrical power. However, unlike homes, a loss of power at a hospital can result in lost lives. Imagine the impact of an unexpected power failure during a delicate surgery. The loss of electrical service to a health care facility can be devastating.

Severe weather conditions, such as hurricanes, floods, tornados, and earthquakes, can interrupt electrical service. Fires, explosions, and electrical failures in the facility due to shorts or overloads can also prevent the successful transfer of power. A well-planned electrical system is required in order to prevent or limit the internal disruption of power in these facilities. Loss of power can be corrected in seconds depending on the system employed. Systems should be designed to cope with the longest probable power outage. Safeguards are needed to ensure the proper operation and maintenance of electrical circuitry and mechanical components in vital areas.

> **NOTE**
>
> This is based on NFPA 70®, *National Electrical Code®(NEC®)*, and NFPA 99, *Standard for Health Care Facilities*. However, state and local codes vary and may use different standards than those listed here.

In addition to NFPA 70®, *National Electrical Code® (NEC®)*, and NFPA 99, The Joint Commission (formerly The Joint Commission on Accreditation of Health Care Organizations, or JCAHO) addresses emergency electrical power systems in their standards *EC.7.20* and *EC.7.40*, and addresses emergency procedures for utility system disruptions in standard *EC.7.10*. The standards for health care facilities are much more rigidly enforced than are those for other facilities. This is especially true for fireproofing wall or floor penetrations because patients cannot be evacuated quickly in case of fire. The Joint Commission inspects each health care facility at the time of occupancy, and then performs an extensive on-site inspection at least once every three years. Always check the most recent national, state, and local codes as well as Joint Commission requirements before beginning any installation.

The installation of electrical systems in health care facilities is governed by **NEC Article 517** and NFPA 99. (*OSHA Standard 29 CFR 1910.307* also applies, but this standard supports the *NEC®*.) *Health care facilities* are defined as places where medical, dental, or nursing care is provided. This includes hospitals, nursing homes, limited care facilities, ambulatory facilities, psychiatric hospitals, dental and medical clinics, and stand-alone day surgery centers. These types of facilities may be permanent construction, such as multilevel hospitals, or movable, such as mobile medical clinics.

To understand the electrical code requirements for health care facilities, it is important to be clear about the definitions for these facilities. The *NEC®* definitions for various facilities are as follows:

- *Ambulatory health care occupancy* – An occupancy used to provide services or treatment simultaneously to four or more patients that provides, on an **outpatient** basis, one or more of the following:
  - Treatment for patients that renders a patient incapable of taking action for self-preservation under emergency conditions without the assistance of others.
  - Anesthesia that renders a patient incapable of taking action for self-preservation under emergency conditions without the assistance of others.
  - Emergency or urgent care for patients who, due to the nature of their injury or illness, are incapable of taking action for self-preservation under emergency conditions without the assistance of others.

- *Hospital* – A building or portion thereof used on a 24-hour basis for the medical, psychiatric, obstetrical, or surgical care of four or more patients.
- *Limited care facility* – A building or portion thereof used on a 24-hour basis for the housing of four or more persons who are incapable of self-preservation because of age; physical limitation due to accident or illness; or limitations such as intellectual disability, mental illness, or chemical dependency.
- *Nursing home* – A building or portion of a building used on a 24-hour basis for the housing and nursing care of four or more persons who, because of mental or physical incapacity, might be unable to provide for their own needs and safety without the assistance of another person.

> **NOTE**
>
> To qualify as any of the facilities defined here, they must be used simultaneously for four or more patients.

> **NOTE**
>
> In addition to the *NEC®* and NFPA 99, some parts of a health care facility may fall under *IEEE Standard 241, Recommended Practice for Electrical Systems in Commercial Buildings.*

Within a health care facility, there are various categories for patient care spaces. A *patient care space* is defined as any portion of a health care facility where patients undergo examination or treatment. Normally, business offices, corridors, lounges, day rooms, dining rooms, and similar areas are not classified as patient care spaces. Categories of patient care spaces include the following:

- *Category 1: Critical care spaces* – Space in which failure of equipment or a system is likely to cause major injury or death of patients, staff, or visitors.
- *Category 2: General care space* – Space in which failure of equipment or a system is likely to cause minor injury to patients, staff, or visitors.
- *Category 3: Basic care space* – Space in which failure of equipment or a system is not likely to cause injury to the patients, staff, or visitors but can cause patient discomfort.
- *Category 4: Support space* – Space in which failure of equipment or a system is not likely to have a physical impact on patient care.

1.1.0 Essential Electrical Systems

In a health care facility, power is divided into two categories: the *nonessential loads* and the *essential electrical system (EES)*. The relationship between these categories is illustrated in *Figure 1*. Nonessential loads are those that are not necessary for the safety of building occupants or for safe and effective patient care; examples of nonessential loads include decorative lighting and televisions. The EES consists of systems or devices that are needed for the safety of all occupants, such as fire alarms, exit lighting, and hallway lighting, and for patient care, such as respirators, cardiac bypass equipment, and suction devices. When the normal electrical source is lost, the facility's alternate source, shown in *Figure 1*, must be able to supply the needs of the EES.

The EES is divided into the equipment branch, the life safety branch, and the critical branch. The equipment branch consists of those items that are needed to operate the facility but are not directly related to patient care, such as the facility's telephone system. The life safety branch and critical branch consists of those items that are essential to patient care. The life safety branch supplies power to those items needed to ensure the safety of building occupants as well as patients. It includes exit signs, stairway lighting, alarm system, and communication systems. The critical branch supplies power to areas that are directly related to patient care. These include pharmacies, emergency rooms, intensive care units, and so on (*Figure 2*).

The requirements for the EES are based on the care provided at the facility. The levels of EES are Type 1 and Type 2. Type 1 is the most stringent.

Typically, normal power for a system is supplied by electric companies, while an on-site power source (battery system or generators) is used as an alternate power source. Naturally, this depends on the specifics of the health care facility. However, when normal power is from on-site power generators, the alternate power source can be either the utility or other generators. In hospitals, when the normal power for a system is

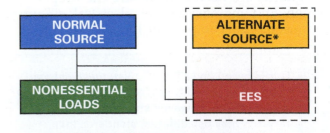

*ALTERNATE SOURCE IS PART OF THE EES.

Figure 1   Nonessential and essential loads.

Hurricane Katrina

After Hurricane Katrina devastated New Orleans, patients and staff in hospitals and other health care facilities suffered through days without electrical power before they were rescued. As a result, The Joint Commission has revised a number of their standards related to electrical power.

Always check the applicable standards for your type of facility before beginning work on a project.

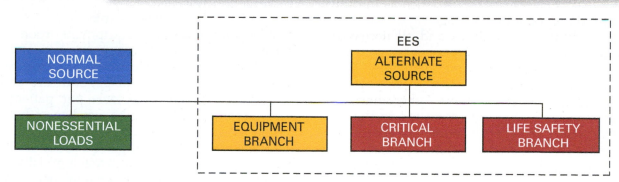

Figure 2 EES branches.

supplied by the utility, a generator is required as the alternate power source. Some facilities that do not admit patients on life support equipment may be able to use battery systems as their alternate power source.

In hospitals and similar facilities, the equipment branch consists of electrical equipment and circuits for three-phase distribution including manual transfer or delayed automatic devices to feed essential equipment loads as described in NFPA 99 and *NEC Section 517.35*. The life safety branch and critical branch includes distribution equipment and circuitry, and the automatic transfer devices needed to transfer from normal power to emergency power. To improve system reliability, each circuit must be installed separately from all other electrical circuits.

The *NEC*® and NFPA 99 both require that the emergency system automatically restore power within 10 seconds of power interruption or loss. These standards also define the electrical loads to be served by the life safety branch and critical branch. In addition, *NEC Section 517.34(A)(10)* allows the installation of additional task illumination, receptacles, and selected power circuits needed for the effective operation of the hospital on the critical branch of the emergency system. This permits some flexibility in customizing the system to fit the needs of the hospital. Essential electrical systems for hospitals are made up of three separate branches to supply lighting and

power service that has been designated as essential for life safety and effective hospital operation during the time the normal electrical service is interrupted. See *NEC Section 517.31(A)*. These three branches are life safety, critical, and equipment. These three systems are categorized as follows:

- The life safety and critical branches are limited to circuits essential to life safety and critical patient care.
- The equipment branch must supply major electric equipment necessary for patient care and basic hospital operation.

The number of transfer switches used to transfer power from the normal source to the alternate source is selected based on reliability, design, and load considerations. Each branch of the Type 1 EES must be served by one or more transfer switches, as shown in *NEC Informational Note Figure 517.31(a)*. One transfer switch may serve one or more branches or systems in a facility with a maximum demand on the Type 1 EES of 150kVA, as shown in *NEC Informational Note Figure 517.31(b)*.

Naturally, this is assuming that the transfer switch has sufficient capacity to serve the combined additional loads and that the alternate source of power is large enough to support the simultaneous transfer of all three branches when normal power is lost.

1.1.1 Type 1 EES

Type 1 EES is required in any facility with a critical care unit or that admits patients requiring electrical life support equipment. Type 1 must automatically restore operation within 10 seconds.

The life safety branch of a Type 1 EES supplies power for the following equipment, lighting, and receptacles per *NEC Section 517.33*:

- Illumination means of egress, such as lighting required for corridors, passageways, stairways, and landings at exit doors and all necessary ways of approach to exits
- Exit signs and exit directional signs
- Fire alarm systems
- Alarm and alerting systems (other than fire alarm systems) shall be connected to the life safety branch or critical branch
- Hospital communications systems, where used for issuing instructions during emergency conditions
- Task illumination battery charger for battery-powered lighting unit(s) and selected receptacles at the generator set and essential transfer switch locations
- Elevator cab lighting, control, communications, and signal systems
- Automatically operated doors used for building egress

The critical branch of a Type 1 EES may consist of one or more branches that are designed to serve a limited number of receptacles and locations to reduce the load demand and the chances of a fault condition. Receptacles in general patient care corridors are allowed on the critical branch. However, they must be identified as part of the critical branch (either labeled or color-coded), and indicate panelboard and circuit number in compliance with NFPA 99 and *NEC Section 517.18(A)*.

Per *NEC Section 517.34(A)*, the essential electrical system's critical branch supplies power for task illumination, selected receptacles, fixed equipment, and special power circuits feeding the following locations and functions related to patient care:

- Critical care (Category 1) spaces that use anesthetizing gases: task illumination, selected receptacles, and fixed equipment

- Patient care spaces: task illumination and selected receptacles in infant nurseries, medication preparation areas, pharmacy dispensing areas, selected acute nursing areas, psychiatric bed areas (omit receptacles), ward treatment rooms, and nurses' stations unless adequately lighted by corridor luminaires
- Additional specialized patient care task illumination and receptacles, where needed
- Nurse call systems
- Blood, bone, and tissue banks
- Telecommunications equipment rooms and closets
- Task illumination, selected receptacles, and selected power circuits for general care beds (at least one duplex receptacle per patient bedroom), angiographic labs, cardiac catheterization labs, coronary care units, hemodialysis rooms or areas, emergency room treatment areas (selected), human physiology labs, intensive care, and postoperative recovery rooms (selected)
- Additional task illumination, receptacles, and selected power circuits needed for effective hospital operation

> **CAUTION**
> Equipment to transfer patient corridor lighting from general illumination circuits to night illumination circuits is allowed as long as only one of two circuits can be selected, and both circuits cannot be off at the same time.

> **NOTE**
> Single-phase fractional horsepower exhaust fan motors that are interlocked with three-phase motors on the equipment may be connected to the critical branch per *NEC Section 517.34(A) (10)*.

1.1.2 Type 2 EES

Many nursing homes and limited care facilities require Type 2 EES. These facilities do not have critical care and intensive care units, but they do provide patient care that is equivalent to hospital inpatient care. Therefore, the branches of the emergency system in Type 2 EES are similar to Type 1.

Code Blue

Sometimes communication systems are used to call a code blue alarm. When a code blue is called, trained staff responds with a crash cart, which contains special equipment and medication that is used to revive a patient whose heart has stopped beating. Some hospitals use a wall-mounted alarm device such as the one shown here. It is included in the life safety branch of a Type 1 EES.

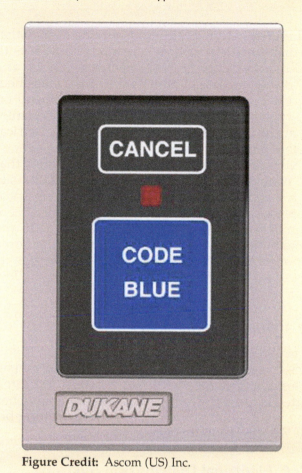

Figure Credit: Ascom (US) Inc.

NFPA 99 requires an alternate source of power for any health care facilities where patients are treated in primarily the same manner as in hospitals, even though the facility may not be designated a hospital. Nursing homes and limited care facilities that are adjoining a hospital may use the essential electrical systems supplied by the hospital, but both facilities must have their own transfer devices.

Essential electrical systems for limited care facilities and nursing homes must have two separate branches that are able to supply the lighting and power required for the protection of life safety and the effective operation of the facility during the interruption of normal electrical service. These two branches are the life safety branch and the equipment branch. The number of transfer switches used is based on reliability, design, and load considerations. Every branch of the Type 2 EES must be served by one or more transfer switches as shown in *NEC Informational Note Figure 517.42(a)*. A single transfer switch will serve one or more branches or systems in a facility with a maximum demand on the Type 2 EES of 150kVA, as shown in *NEC Informational Note Figure 517.42(b)*.

The life safety and equipment branches must be completely independent of all other equipment and wiring. These circuits cannot enter the same cabinets, boxes, or raceways with other wiring except in transfer switches or in exit or emergency lighting fixtures (or their attached junction boxes), as long as the fixtures are supplied from two sources. See *NEC Section 517.42(D)*.

The life safety branch must be installed and connected to the alternate power source so that designated functions are automatically restored to operation within 10 seconds after the normal source interruption. The life safety branch supplies power for the following equipment, lighting, and receptacles:

- Communication systems, where used for dissemination of instructions during emergencies
- Task illumination battery chargers for emergency battery-powered lighting units and selected receptacles at the generator set location
- Illumination of exits, such as lighting required for corridors, passageways, stairways, landings at exit doors, and all approaches to exits
- Exit direction and exit signs
- Alarm and alerting systems, including fire alarms and alarms required for systems used for the piping of nonflammable medical gases
- Elevator cab communication, control, lighting, and signal systems
- Lighting in recreation and dining areas must be sufficient to illuminate exits.

The equipment branch must be installed and connected to the alternate power source so that the equipment listed in *NEC Section 517.44(A)* is automatically restored to operation at appropriate time-lag intervals following the restoration of the life safety branch to operation. Its arrangement must also provide for the additional connection of equipment listed in *NEC Section 517.44(B)* by either delayed automatic or manual operation.

The following equipment may be connected to the equipment branch and arranged for delayed automatic connection to the alternate power source:

> **CAUTION**
>
> Equipment to transfer patient corridor lighting from general illumination circuits to night illumination circuits is allowed as long as only one of two circuits can be selected, and both circuits cannot be off at the same time.

- Patient care spaces: task illumination and selected receptacles in medication preparation areas, pharmacy dispensing areas, and nurses' stations unless adequately lighted by corridor luminaires
- Sump pumps and other equipment required to operate for the safety of major apparatus and associated control systems and alarms
- Smoke control and stair pressurization systems
- Kitchen hood supply and/or exhaust systems, if required to operate during a fire in or under the hood
- Supply, return, and exhaust ventilating systems for airborne infectious isolation rooms

Per *NEC Section 517.44(B)*, the following equipment may be connected to the critical equipment branch and arranged for either delayed automatic or manual connection to the alternate power source:

- Equipment to provide heating for patient rooms, except where:
  - The outside design temperature is higher than +20°F (−6.7°C).
  - The outside design temperature is lower than +20°F (−6.7°C), but selected areas are heated for the comfort of all confined patients.
  - The facility is served by a dual source of normal power.
- In instances in which disruption of power would result in elevators stopping between floors, throwover (transfer) switches must be provided to allow temporary operation of any elevator for the release of passengers. For elevator cab lighting, control, and signal system requirements, see *NEC Section 517.43(F)*.

- Additional illumination, receptacles, and equipment may be connected only to the critical equipment branch.
- Where one switch serves multiple systems as permitted in *NEC Section 517.43*, transfer for all loads shall be nondelayed and automatic.

> **CAUTION**
>
> No function other than those listed above may be connected to the life safety branch.

Limited care facilities and nursing homes that furnish inpatient critical care must comply with the requirements of *NEC Sections 517.29 and 517.30*. Regardless of how the facility is classified, the category of electrical system will depend on the type of patient care. Where care is evidently inpatient hospital care, a hospital-type electrical system must be installed.

Nursing homes and limited care facilities that are contiguous with or located on the same site as a hospital may have their essential electrical systems supplied by that of the hospital. Where a limited care facility or nursing home is in the same building with a hospital, the nursing home is not required to have its own EES if it gets its power from the hospital. This rule does not allow sharing of transfer devices.

1.1.3 Other Health Care Facilities

In facilities that do not require a Type 1 or Type 2 EES, emergency power may be provided by batteries in accordance with *NEC Section 700.12(C)*. Batteries are required to maintain the emergency load for $1\frac{1}{2}$ hours without the voltage falling below $87\frac{1}{2}$ percent of normal.

1.2.0 Electrical Distribution Systems

Providing a health care facility with reliable electrical power begins with careful system design followed by proper installation. To ensure continuous electrical service from the utility provider, hospitals and certain other health care facilities are placed on a spot network. Spot networks are considered the most reliable and flexible arrangement. Ideally, these feeders are installed in parallel so that a fault in one feeder will not affect or produce faults in another. Each feeder must be sized to carry the full load of the system. This type of installation helps to ensure an adequate amount of power during high use periods.

1.2.1 Double-Ended System

A double-ended substation is typically used when transformation over 750kVA is needed. The substation shown in *Figure 3* uses a normally open (NO) tie circuit breaker interlocked with the main circuit breakers so that not all three can be closed simultaneously. When a single mains feeder has loss, the tie circuit breaker may be closed, manually or automatically, placing additional load on the remaining feeder. The substation feeders and equipment can be medium-voltage equipment to reduce current rating requirements, or low-voltage equipment. In either case, each of the utility feeders and circuit breakers used must be rated to carry the full load of the facility in the event that one of the mains sources is lost. Double-ended substations have additional benefits, such as lower fault current and the ability to differentiate between motor loads that require specific voltage regulation.

If the double-ended substation design uses an NO, electrically operated tie circuit breaker that automatically closes upon loss of either of the incoming mains feeders, then control and protective relaying must be added to prevent the bus tie circuit breaker from closing when a mains circuit breaker has tripped due to overload or short circuit conditions.

Utility companies may allow double-ended substations to operate with a closed tie and mains circuit breakers. Double-ended substations with mains and normally closed (NC) ties have distinct advantages such as better voltage regulation, transfer capabilities when power loss occurs, immediate switching when a power source is lost, and greater reliability of system operation. However, the disadvantages are greater fault current, greater cost, and a more complex design. The decision to use a closed tie is made by the designer of the system.

1.2.2 Alternate Power Source

For a Type 1 or 2 EES, one or more diesel engine generators are normally used (*Figure 4*). In cases in which multiple generators are used, one is a designated backup generator. Generators may be housed internally or externally to the health facility. If housed in a nonheated environment where temperatures fall below about 30°F (−1.1°C), the starting batteries must be heated and cold-weather diesel engine starting equipment must be used in order to meet the 10-second startup requirement for restoration of power. *Figure 5* is a schematic diagram showing the arrangement of

an alternate power source with transfer switching for an EES. As noted in the figure, delayed switching of the EES equipment system may be used if the alternate power source cannot tolerate the total EES load at startup.

1.2.3 Ground Fault Protection

Phase overcurrent device settings are primarily established by the load requirements. They are set to be insensitive to full-load and inrush current and to provide selectivity between downstream and upstream devices. However, phase overcurrent devices cannot distinguish between normal

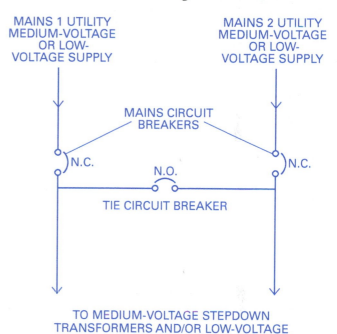

Figure 3 Double-ended substation.

Figure 4 Typical diesel engine generator.

load current and a low-magnitude ground fault short circuit current of the same magnitude. For this reason, ground fault detection is used in addition to the phase overcurrent devices to provide protection.

NEC Sections 215.10 and 230.95 describe requirements for ground fault protection at the feeder. In health care facilities, additional ground fault protection may be required downstream. *NEC Section 517.17(B)* requires two levels of ground fault protection in health care facilities to improve the selectivity between the feeder and building devices. The wiring configuration and estimated ground currents must be carefully analyzed before selecting appropriate ground fault protective devices.

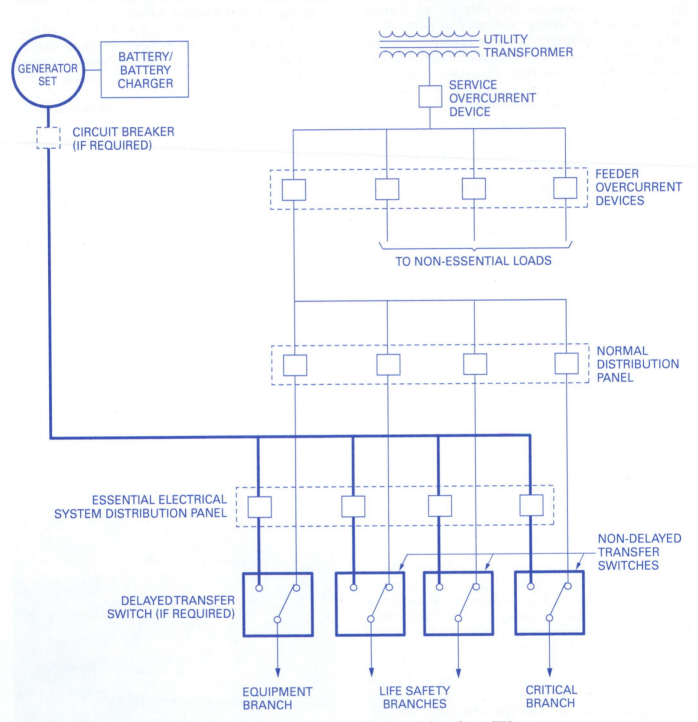

Figure 5 Schematic diagram of an alternate power source with transfer switching for an EES.

1.2.4 Additional Grounding and Bonding Requirements

Besides the usual grounding and bonding required by *NEC Article 250*, health care facilities require additional precautions for patient care spaces. As specified in *NEC Section 517.14*, where branch circuit panelboards for normal and essential power serve the same patient care vicinity, they must be connected together with an insulated copper conductor no smaller than No. 10 AWG. In addition, where two or more panelboards that serve the same patient care vicinity are served from two different transfer switches, their equipment grounding terminals must be connected with an insulated conductor no smaller than No. 10 AWG.

NEC Section 517.19(E) specifies that where a grounded electrical distribution system is used, and feeder metal raceway or Type MC or MI cable that qualifies as an equipment grounding conductor per *NEC Section 250.118* is installed, the grounding of a panelboard or switchboard must be ensured by one of the following:

- A grounding bushing and continuous copper bonding jumper sized per *NEC Section 250.122*, with the jumper connected to the junction enclosure or the ground bus of the panel
- Connection of the feeder raceways or Type MC or MI cable to threaded hubs or bosses on the terminating enclosures
- Other approved devices such as bonding-type locknuts or bushings (standard locknuts shall not be used for bonding)

1.0.0 Section Review

1. The essential electrical system used in a limited care facility is a Type _____.

    a. 0 EES
    b. 1 EES
    c. 2 EES
    d. 3 EES

2. Ground fault detection is used in addition to the phase overcurrent devices to provide protection against _____.

    a. inrush current
    b. full-load current
    c. low-magnitude short circuit current
    d. high-magnitude short circuit current

2.0.0 DEVICES USED IN HEALTH CARE FACILITIES

Objective

Describe the categories and branch portions of the distribution circuits.

a. Describe the operation and applications of hospital-grade receptacles.
b. Identify the receptacle requirements for general care (Category 2) spaces.
c. Identify the receptacle requirements for critical care (Category 1) spaces.
d. Identify the grounding requirements for receptacles and fixed electrical equipment.

Trade Term

Equipotential ground plane: A mass of conducting material that, when bonded together, provides a low impedance to current flow over a wide range of frequencies.

Generally, hospitals have two sources of available power: normal and emergency. Devices on emergency power must be easy to identify. This reduces the time wasted in locating receptacles during a power outage. These devices may be identified with a distinctive color such as red, or by labeling. (When color-coding is used, it must be the same throughout the facility.) Color-coding is easier and less expensive than labeling, and it may prevent later confusion. For example, if painters remove the labeled cover plates, there is no longer a distinction between devices. In addition, since receptacles in critical care spaces must have panelboard and circuit number labels, the device cover plates tend to become cluttered. *Figure 6* shows an example of color-coding used to identify various circuits in a patient headwall. Lighted emergency power receptacles are useful in patient areas that do not have emergency lighting because they make the receptacle easier to locate in the near-darkness of a power outage.

Devices in a hospital should be mounted so they are easy to use. Because a large number of hospital patients spend time in wheelchairs, particular attention should be given to these

requirements. All receptacles should be mounted approximately 24" (600 mm) above the floor for the convenience of patients and staff. Always refer to the job plans and specifications.

2.1.0 Hospital-Grade Receptacles

Hospital-grade receptacles (*Figure 7*) are listed as suitable for use in health care facilities. Per *NEC Article 517*, health care facilities include (but are not limited to) hospitals, nursing homes, limited care facilities, clinics, medical and dental offices, and ambulatory care centers, whether permanent or movable. Hospital-grade receptacles are often marked with a green dot on the outer face of the receptacle.

Hospital-grade receptacles meet nationally recognized testing laboratory criteria for superior electrical characteristics and mechanical strength. This is needed in a medical environment to ensure that equipment plugs fit tightly into the receptacles and the receptacles are durable.

2.1.1 Isolated Ground Receptacles

Isolated ground receptacles are used where separation of the device ground and the building ground is desired. They are normally used with digital electronic equipment, including computer cash registers, computer peripherals, and digital processing equipment. They prevent transient voltages on the ground system from causing malfunctions in digital circuits. Isolated ground receptacles shall not be installed in a patient care vicinity per *NEC Section 517.16(A)*.

2.1.2 Tamper-Resistant Receptacles

Tamper-resistant receptacles prevent foreign objects from coming into contact with energized components inside the receptacle by only operating when both the hot and neutral blades of a plug are simultaneously inserted. These receptacles must be used throughout psychiatric and designated general care pediatric locations. When choosing tamper-resistant receptacles, be careful not to use the type that makes and breaks contacts where life support apparatus may be required.

2.2.0 General Care (Category 2) Spaces

It is difficult to prevent a conductive path from a grounded object to a patient's body in a health care facility. The path may be created accidentally or through instruments directly connected to the patient that act as a source of electric current. This hazard worsens as more equipment is used near a patient, such as in surgical suites and intensive care units. Methods of reducing the likelihood of electric shock to a patient include the following:

- Increasing the resistance of the conductive circuit that includes the patient
- Insulation of exposed surfaces that may become energized

- Lowering the potential difference between exposed conductive surfaces in the patient's vicinity
- Using a combination of the above methods

NEC Section 517.18(A) requires that every patient bed location be supplied with at least two branch circuits, one from the normal system and the other from the critical system. All branch circuits from the normal system must originate in the same panelboard. The branch circuit serving

Figure 6 Example of circuit color-coding for a headwall.

Figure 7 Hospital-grade receptacles.

patient bed locations shall not be part of a multi-wire branch circuit. The *NEC*® lists three exceptions to this requirement:

- Branch circuits serving only special-purpose outlets or receptacles, such as portable X-ray outlets, do not have to be served from the same distribution panel or panels.
- Patient beds located in clinics; medical and dental offices; outpatient facilities; psychiatric, substance abuse, and rehabilitation hospitals; nursing homes; and limited care facilities meeting the requirements of *NEC Section 517.10(B) (2)* are exempt from this requirement.
- A general care (Category 2) patient bed location served from two separate transfer switches on the emergency system is not required to have circuits from the normal system.

In addition to two branch circuits, each patient bed location must have a minimum of eight receptacles per *NEC Section 517.18(B)(1)*. Receptacles may be of the single, duplex, or quadruplex types, or any combination of the three. Some receptacles are supplied as part of a prefabricated patient headwall unit, as shown in *Figure 8*. All receptacles must be hospital-grade and grounded by means of an insulated copper equipment grounding conductor sized in accordance with *NEC Table 250.122*.

NEC Section 517.18(C) requires that all receptacles meant to supply patient care spaces of designated general care (Category 2) pediatric locations or spaces (bathrooms, playrooms, activity rooms, and patient care spaces) must be tamper-resistant or use a tamper-resistant cover.

(A) RECEPTACLES ABOVE BED

(B) RECEPTACLES BEHIND BED

Figure 8   Patient bed location receptacles.

critical branch so that patients will not be without electrical power when there is a fault. At least one branch circuit from the critical branch must supply an outlet only at that bed location. All branch circuits from the normal system must be from a single panelboard. Critical branch receptacles must be identified as such and labeled with the supply panelboard and circuit number. The branch circuit serving patient bed locations shall not be part of a multiwire branch circuit.

NEC Section 517.19(B)(1) requires that each patient bed location in a critical care space be provided with a minimum of 14 receptacles, at least one of which must be connected to the normal system branch circuit required in *NEC Section 517.19(A)* or a critical branch circuit supplied by a different transfer switch than the other receptacles at the same location. Per *NEC Section 517.19(C)*, operating rooms shall be provided with 36 receptacles with at least 12 connected

> **NOTE**
>
> This does not apply to psychiatric, substance abuse, and rehabilitation hospitals meeting the requirements of *NEC Section 517.10(B) (2)*. In addition, psychiatric security rooms are not required to have receptacle outlets. *NEC Section 517.19(D)* requires that an equipment bonding jumper no smaller than No. 10 AWG be used to connect the grounding terminal of all grounding-type receptacles of a patient headwall to the patient grounding point if provided.

2.3.0 Critical Care (Category 1) Spaces

NEC Section 517.19(A) requires that every patient bed location in critical care spaces be supplied by at least two branch circuits, one or more from the normal system and one or more from the

Patient Headwall Column

This patient headwall column supplies both an oxygen line and electrical power. This combination requires the use of a protective bollard to guard it against damage. When devices are combined in an assembly such as this one, a nationally recognized testing laboratory may be required to come in and label the entire assembly. Check with the authority having jurisdiction in your area.

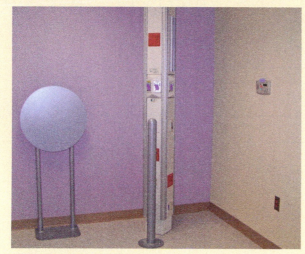

Figure Credit: Mike Powers

- Use only hospital-grade permissible power connectors.
- Use no less than two duplex outlets on each side of the patient's bed. The outlets must be protected by overcurrent protective devices in accordance with the *NEC*®.
- Furnish power outlets within a few feet of the junction boxes and equipment support locations.
- Allocate a grounding system in accordance with NFPA 99 and the *NEC*®.
- Plan and design the system to keep power cables to a minimum length. This contributes both to physical and electrical safety.
- Ensure adequate ventilation in all areas where patient monitoring equipment is to be installed.

> **NOTE**
>
> Branch circuits serving only special-purpose receptacles or equipment in critical care spaces may be served by other panelboards; also, when the location is served by two individual transfer switches on the emergency system, it does not require a branch on the normal system. See *NEC Section 517.19(A), Exceptions*.

2.4.0 Grounding Receptacles and Fixed Equipment

NEC Section 517.13 pertains to equipment grounding conductors for receptacles and fixed electrical equipment in patient care spaces. These areas require redundant grounding. All branch circuits serving patient care spaces must have an effective ground-fault current path by installation in a grounded metal raceway or a cable having a metallic armor or sheath assembly. The metal raceway, metallic armor, or sheath assembly must itself qualify as an equipment grounding conductor per *NEC Section 250.118*. Type MI and MC cable must have a health care facility rating and be clearly marked as such and terminated using listed fittings.

to the normal or critical branch. They may be of the single, duplex, or quadruplex types, or any combination. All receptacles must be listed as hospital-grade and connected to the reference grounding point by means of an insulated copper equipment grounding conductor.

A typical receptacle configuration in a critical care space is shown in the diagram in *Figure 9*. The normal circuits must be supplied from the same panel (L-1). The critical branch circuits may be supplied from different panels (EML-1 and EML-2). However, the critical branch circuit to patient bed location A cannot supply critical branch receptacles for patient bed location B.

The required power level for each bedside is typically 90VA for single-outlet receptacles and 180VA for duplex receptacles. The following recommendations are for bedside stations:

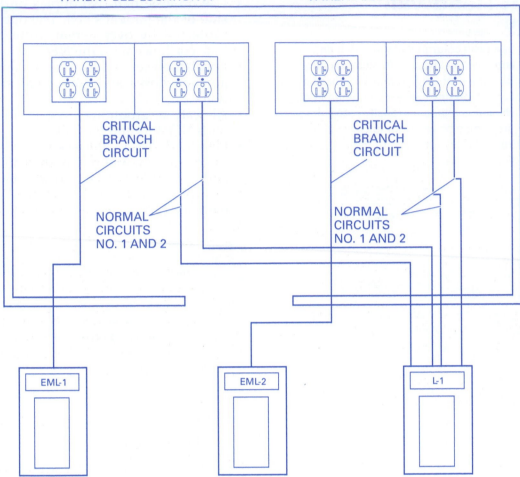

Figure 9 Typical receptacle configuration in a critical care space.

Per **NEC Section 517.13(B)**, the grounding terminals of receptacles, the metal boxes and enclosures containing receptacles, and all non-current-carrying conductive surfaces of fixed electric equipment likely to become energized that are subject to personal contact and operating at over 100V must be connected to an insulated copper equipment grounding conductor. Metal faceplates may be connected to the equipment grounding conductor by means of metal mounting screws securing the faceplate to a grounded outlet box or grounded wiring device. This permits metal faceplates to be grounded without having to run a separate equipment grounding conductor to the faceplate. The grounding conductor must be sized in accordance with **NEC Table 250.122** and installed in a metal raceway with the branch circuit conductors supplying these receptacles or fixed electric equipment.

The insulated copper conductor is installed either with the branch circuit conductors or as part of listed cable with a metallic armor or sheath assembly. The reason for redundant grounding using an insulated ground wire within a grounded raceway or grounded sheathed cable is so that an effective ground-fault current path exists if an insulated grounding conductor or device fails. **NEC Section 517.13(B)(1)** includes the following exceptions:

- For other than isolated ground receptacles, an insulated equipment bonding jumper that directly connects to the equipment grounding conductor is permitted to connect the box and receptacle(s) to the equipment grounding conductor.

- Luminaires more than $7\frac{1}{2}'$ (2.3 m) above the floor and switches located outside of the patient care vicinity shall be permitted to be connected to an equipment grounding return path complying with *NEC Section 517.13(A) or (B)*. (These are both unlikely to contact the patient or any equipment connected to the patient.)

Metal-sheathed cable assemblies are not typically authorized for essential electrical circuits because *NEC Section 517.31(C)(3)* requires such wiring to be protected by installation in a non-flexible metal raceway or encased in concrete. Type MC cable is allowed where it is placed in listed prefabricated medical headwalls, fished into existing walls or ceilings where it is not likely to be damaged or casually accessed, located in listed office furniture, or where the equipment requires a flexible connection.

Equipotential grounding is one method of grounding used for sensitive equipment. An *equipotential ground plane* is a mass of conducting material that, when bonded together, provides a low impedance to current flow over a wide range of frequencies.

Equipotential ground planes shield adjacent sensitive circuits or equipment, contain electromagnetic (EM) noise fields between their source (cable, for example) and the plane, increase filtering effectiveness of contained EM fields, and provide a low-impedance return path for radio frequency (RF) noise current.

Equipotential plane structures include metallic screens or sheet metal under floor tile, the supporting grids of raised access flooring (such as in computer rooms), conductive grids rooted in or attached to a concrete floor, and ceiling grids above sensitive equipment.

An equipotential reference plane can be used within a section of a single sensitive equipment enclosure, among various pieces of interconnected equipment, or over an entire facility. In all cases, the equipotential plane is bonded to both the local building ground and the grounding electrode conductor.

Within sensitive equipment cabinets, all signal return leads, backplanes, and other significant components must be connected via short (less than 5% to 10% of the wavelength of the highest frequency) conductors to the equipment chassis that forms the equipotential plane. All similar equipment-level equipotential planes should be coupled to the grounding electrode conductor as well as the room-level equipotential plane by way of multiple short conductors. The room-level equipotential plane must in turn be linked to one or more building-level equipotential planes via multiple short conductors. This process is continued until the total sensitive electronic equipment system is interconnected to one large continuous equipotential plane. It is best to interlink conductors with single thin-wire cross sections to minimize their impedance at higher frequencies.

> **NOTE**
>
> Special grounding requirements are required in wet procedure locations. See *NEC Section 517.20*. These locations are designated by the facility governing body. Patient bathrooms are not considered wet procedure locations. They may be *NEC*®-defined wet locations.

Hospital-Grade AC Cable

This AC cable has an internal insulated ground wire that is acceptable for use in patient care spaces. It is identified by a green metal sheath.

Figure Credit: Ed Cockrell

Electrical Risks to Patients

One of the most dangerous situations is when a patient has an external direct conductive path to the heart, such as a cardiac catheter. In this situation, the patient may be electrocuted at current levels so low that additional protection in the design of equipment, catheter insulation, and control of operational practices is necessary. Patients are at increased risk of electric shock when their body resistance is compromised either accidentally or by needed medical procedures. Situations such as catheter insertion or incontinence may make a patient much more susceptible to the effects of an electric current.

2.0.0 Section Review

1. Hospital-grade receptacles are often marked with a(n) _____.

   a. purple dot
   b. orange dot
   c. green dot
   d. blue dot

2. Which of the following is *true* with regard to electrical safety in patient care spaces?

   a. The possibility of electric shock can be reduced by reducing the resistance of the conductive circuit that includes the patient.
   b. All branch circuits from the normal system must originate in different panelboards.
   c. Every circuit supplying a patient bed location must be supplied from the emergency system.
   d. This hazard worsens as more equipment is used near a patient, such as in surgical suites and intensive care units.

3. Every patient bed location in a critical care space must be supplied with at least _____.

   a. 8 receptacles
   b. 10 receptacles
   c. 12 receptacles
   d. 14 receptacles

4. Sensitive equipment in health care facilities can be shielded using _____.

   a. equipotential grounding
   b. metallic paint
   c. nonmetallic spacers
   d. white noise

3.0.0 WIRING IN HEALTH CARE FACILITIES

Objective

List the required wiring methods in health care facilities.

a. Identify the wiring requirements for inhalation anesthetizing locations.

b. Identify the wiring requirements for low-voltage equipment and instruments.

c. Identify the wiring requirements for X-ray installations.

d. Identify the requirements for communication, signaling, data, and fire alarm systems installed in patient care areas.

e. Identify the requirements for isolated power systems.

Health care facilities can include hazardous (classified) locations and unclassified locations. Each of these locations has specific wiring requirements.

3.1.0 Inhalation Anesthetizing Locations

Anesthetizing locations are divided into hazardous (classified) locations, where flammable or nonflammable anesthetics may be interchangeably used, and unclassified locations where only nonflammable anesthetics are used. Although flammable anesthetics are virtually obsolete in the United States, *NEC Section 517.60* and NFPA 99 still cover them because they are in use in other countries that use these codes. In areas where flammable anesthetics are used, the entire location is considered a Class I, Division 1 location up to a level 5' (1.5 m) above the floor. The area remaining up to the structural ceiling is considered to be above a hazardous (classified) location.

3.1.1 Wiring and Equipment within Hazardous Anesthetizing Locations

Per *NEC Section 517.61(A)*, wiring and equipment *within* hazardous (classified) anesthetizing locations must comply with the following:

- Except as permitted by *NEC Section 517.160*, each power circuit within, or partially within, a flammable anesthetizing location as referred to in *NEC Section 517.60* must be isolated from any distribution system by the use of an isolated power system.

- Isolated power system equipment must be listed for the purpose, and the system must be designed and installed in accordance with *NEC Section 517.160*.

- In hazardous locations referred to in *NEC Section 517.60*, all fixed wiring and equipment and all portable equipment, including lamps and other utilization equipment operating at more than 10V between conductors, must comply with *NEC Sections 501.1 through 501.25*, *NEC Sections 501.100 through 501.150*, and *NEC Sections 501.30(A) and (B)* for Class I, Division 1 locations. All such equipment must be specifically approved for the hazardous atmospheres involved.

- Where a box, fitting, or enclosure is partially, but not entirely, within a hazardous (classified) location, the entire box, fitting, or enclosure is considered within the hazardous (classified) location.

- Attachment plugs and receptacles in hazardous locations must be listed for use in Class I, Group C hazardous locations and have provision for the connection of a grounding conductor.

- Flexible cords used in hazardous locations for connection to portable utilization equipment, including lamps operating at more than 8V between conductors, must be approved for extra-hard usage in accordance with *NEC Table 400.4* and include an additional grounding conductor. Storage devices must be provided for flexible cords. These storage devices must not subject the cord to a bending radius of less than 3" (75 mm).

3.1.2 Wiring and Equipment Installed Above Hazardous Anesthetizing Locations

Per *NEC Section 517.61(B)*, wiring and equipment installed *above* hazardous (classified) anesthetizing locations must comply with the following:

- Wiring above a hazardous location referred to in *NEC Section 517.60* must be installed in rigid metal conduit, electrical metallic tubing, intermediate metal conduit, Type MI cable, or Type MC cable that employs a continuous gas/vapor-tight metal sheath.

- Equipment that may produce sparks, arcs, or particles of hot metal, such as lamps and lampholders for fixed lighting, switches, generators, motors, cutouts, or other equipment having sliding contacts, must be of the totally enclosed type or constructed in such a manner as to prevent the escape of sparks or hot metal particles. (This does not apply to wall-mounted receptacles installed above the hazardous location.)
- Surgical and other luminaires must conform to *NEC Section 501.130(B)*. See also *NEC Section 517.61(B)(3), Exceptions*, which state that the surface temperature limitations set forth for fixed lighting in *NEC Section 501.130(B)(1)* do not apply, and integral or pendant switches that are located above and cannot be lowered into the hazardous (classified) location(s) are not required to be explosion proof.
- Only listed seals may be used per *NEC Section 501.15. NEC Section 501.15(A)(4)* applies to both the horizontal and vertical boundaries of the defined hazardous (classified) locations.
- Attachment plugs and receptacles positioned over hazardous (classified) anesthetizing locations must be listed as hospital grade for services of prescribed voltage, frequency, rating, and number of conductors with provision for the connection of the grounding conductor. This requirement applies to attachment plugs and receptacles of the two-pole, three-wire grounding type for single-phase, 120V, nominal AC service.
- Plugs and receptacles rated at 250V for connection of 50A and 60A, AC medical equipment for use above hazardous (classified) locations must be arranged so that the 60A receptacle will accept either the 50A or the 60A plug. The 50A receptacles must be designed so as not to accept a 60A attachment plug. The plugs must be of the two-pole, three-wire design with a third contact connecting to the insulated (green or green with yellow stripe) equipment grounding conductor of the electrical system.

3.1.3 Wiring in Unclassified Anesthetizing Locations

Per *NEC Section 517.61(C)*, wiring in unclassified anesthetizing locations must comply with the following:

- Wiring serving unclassified locations, as defined in *NEC Section 517.60*, must be installed in a metal raceway system or cable assembly.

The metal raceway system, cable armor, or sheath assembly must qualify as an equipment grounding conductor in accordance with *NEC Section 250.118*. Type MC and MI cable must have an outer metal armor or sheath that is identified as an acceptable equipment grounding conductor. Per *NEC Section 517.61(C)(1), Exception*, pendant receptacle installations employing at least Type SJO or equivalent hard or extra-hard usage, flexible cords suspended not less than 6' (1.8 m) from the floor are not required to be installed in a metal raceway or cable assembly.

- Receptacles and attachment plugs installed and used in unclassified locations must be listed as hospital grade for services of prescribed voltage, frequency, rating, and number of conductors with provision for connection of the grounding conductor. This requirement applies to the two-pole, three-wire grounding type for single-phase 120V, 208V, or 240V nominal AC service.
- Plugs and receptacles rated at 250V for connection of 50A and 60A, AC medical equipment for use in unclassified locations must be arranged so that the 60A receptacle will accept either the 50A or the 60A plug. The 50A receptacles must be designed so as not to accept a 60A attachment plug. The plugs must be of the two-pole, three-wire design with a third contact connecting to the insulated (green or green with a yellow stripe) equipment grounding conductor of the electrical system.

3.1.4 Grounding in Anesthetizing Locations

In any anesthetizing location, all metal raceways and metal-sheathed cable and all normally non-current-carrying conductive portions of fixed electric equipment must be connected to an equipment grounding conductor per *NEC Section 517.62*. Grounding in Class I locations must comply with *NEC Section 501.30*.

Grounding specifications and requirements for anesthetizing areas apply only to metal-sheathed cable, metal raceways, and electric equipment. It is not a requirement to ground tables, carts, or any other nonelectrical items. However, in anesthetizing locations where flammables are present, portable carts and tables usually have an acceptable resistance to ground through conductive tires and wheels and conductive flooring to avoid the buildup of static electrical charges. See NFPA 99 for information on static grounding.

3.1.5 Grounded Power Systems in Anesthetizing Locations

As a secondary backup, at least one battery-powered emergency lighting unit must be installed in every anesthetizing location per *NEC Section 517.63(A)*. This light may be wired ahead of the switch and connected to a critical lighting circuit. Without this backup, failure of the emergency circuit feeder that supplies the operating room with power would likely place the room in complete darkness, which could have fatal results.

Per *NEC Section 517.63(B)*, branch circuits supplying only listed, stationary, diagnostic, and therapeutic equipment, permanently installed above the hazardous (classified) area or in unclassified locations may be fed from a normal grounded service (single- or three-phase system), provided that:

- Wiring for isolated and grounded circuits does not occupy the same cable or raceway.
- All conductive surfaces of the equipment are connected to an equipment grounding conductor.
- Equipment (except enclosed X-ray tubes and the leads to the tubes) is located at least 8' (2.5 m) above the floor or outside the anesthetizing location, and switches for the grounded branch circuit are located outside the hazardous (classified) location. (This does not apply to unclassified locations.)

Per *NEC Section 517.63(C)*, fixed lighting branch circuits feeding only fixed lighting may be supplied by a normal grounded service, provided that:

- The fixtures are located at least 8' (2.5 m) above the floor and switches are wall-mounted and located above hazardous (classified) locations. (This does not apply to unclassified locations.)
- All conductive surfaces of fixtures are connected to an equipment grounding conductor.
- Wiring for circuits supplying power to fixtures does not occupy the same raceway or cable as circuits supplying isolated power.

Wall-mounted remote control stations may be installed in any anesthetizing location for the operation of remote control switching operating at 24V or less per *NEC Section 517.63(D)*.

An isolated power system and its grounded supply circuit may be located in an anesthetizing location, providing it is installed above a hazardous (classified) location or in an unclassified location and is listed for the purpose. See *NEC Section 517.63(E)*.

Except as permitted above, each power circuit within, or partially within, a flammable anesthetizing location as referred to in *NEC Section 517.60* must be isolated from any distribution system supplying other-than-anesthetizing locations.

3.2.0 Low-Voltage Equipment and Instruments

Low-voltage equipment that is typically in contact with patients or has exposed current-carrying elements must be moisture-resistant and operate on 10V or less or be approved as intrinsically safe or double-insulated equipment. Refer to *NEC Section 517.64*.

Electrical power to low-voltage equipment must be supplied from an individual portable isolation transformer (autotransformers cannot be used). This equipment must be connected to an isolated power circuit receptacle by means of an appropriate cord and attachment plug, a standard low-voltage isolation transformer installed in an unclassified location, individual dry-cell batteries, or common batteries made up of storage cells located in an unclassified location.

Isolating-type transformers for supplying low-voltage circuits must have an approved means of insulating the secondary circuit from the primary circuit. In addition, the case and core must be connected to an equipment grounding conductor.

Impedance and resistance devices are allowed to control low-voltage equipment. However, they cannot be used to limit the maximum available voltage supplied to the equipment.

Battery-supplied appliances must not be capable of being charged during operation unless the charging circuitry uses an integral isolating-type transformer. In addition, receptacles and attachment plugs placed on low-voltage circuits must be of a type that does not permit interchangeable connection with circuits of higher voltage.

3.3.0 X-Ray Installations

The installation of X-ray equipment requires specialized training in order for the equipment both to operate as intended and to protect patients and personnel from unintended exposure to radiation. Any equipment for new X-ray installations and all used or reconditioned X-ray equipment transported to and reinstalled in a new location must be of an approved type. *NEC Article 517, Part V* covers X-ray equipment.

3.3.1 Connection to Supply Circuit

X-ray equipment, whether fixed or stationary, must be connected to the power supply in such a way that the wiring methods meet the requirements of the *NEC®*. Equipment supplied by a branch circuit rated at no more than 30A may be supplied through a suitable attachment plug and hard-service cable or cord.

Individual branch circuits are not required for portable, mobile, and transportable medical X-ray equipment requiring a capacity of no more than 60A per *NEC Section 517.71(B)*. Circuits and equipment functioning on a supply circuit over 1,000V must comply with the requirements of *NEC Article 490*.

3.3.2 Disconnecting Means

Per *NEC Section 517.72(A)*, a disconnecting means rated for at least 50% of the input required for the momentary rating or 100% of the input required for the long-term rating of the X-ray equipment (whichever is greater) must be provided in the supply circuit.

The disconnecting means must be in a location readily accessible from the X-ray control. A grounding-type attachment plug and receptacle of proper rating may serve as a disconnecting means for equipment supplied by a 120V branch circuit of 30A or less.

3.3.3 Rating of Supply Conductors and Overcurrent Protection

For diagnostic equipment, the amperage of supply branch circuit conductors and the current rating of overcurrent protective devices must be no less than 50% of the momentary rating or 100% of the long-term rating, whichever is greater. See *NEC Section 517.73(A)(1)*.

The amperage rating of supply feeders and the current rating of overcurrent protective devices feeding two or more branch circuits supplying X-ray units must be no less than 50% of the momentary demand rating of the largest unit, plus 25% of the momentary demand rating of the next largest unit, plus 10% of the momentary demand rating of each additional unit. Where concurrent bi-plane tests are undertaken with the X-ray units, the supply conductors and overcurrent protective devices must be 100% of the momentary demand rating of each X-ray unit.

Box Shielding

If an outlet or switch box must be installed in an X-ray area, lead shielding must be molded around the box. This ensures the integrity of the overall X-ray shield in the area. Some electricians peel the lead backing from scrap pieces of lead-shielded gypsum wallboard panels and save it for shielding the boxes in the area.

The minimum conductor sizes for branch and feeder circuits are governed by voltage regulation requirements. In specific installations, manufacturers typically specify the minimum distribution transformer and conductor sizes required, the disconnect rating, and overcurrent protection requirements.

For therapeutic equipment, the amperage rating of conductors and the rating of overcurrent protective devices must not be less than 100% of the current rating of medical X-ray therapy equipment. The amperage rating of the branch circuit conductors and the ratings of disconnecting means and overcurrent protection for X-ray equipment are usually designated by the manufacturer for the specific installation.

3.3.4 Control Circuit Conductors

The number of control circuit conductors installed in a raceway must be determined in accordance with *NEC Section 300.17*. No. 18 AWG or No. 16 AWG fixture wires as specified in *NEC Section 725.49* and flexible cords are permitted for the control and operating circuits of X-ray and auxiliary equipment where protected by overcurrent devices of no more than 20A. See *NEC Section 517.74(B)*.

3.3.5 Transformers and Capacitors

Transformers and capacitors that are a part of X-ray equipment are not required to comply with *NEC Articles 450 and 460*. However, capacitors must be mounted within enclosures of insulating material or grounded metal. See *NEC Section 517.76*.

3.3.6 High Tension X-Ray Cable

Cable with a grounded shield connecting X-ray tubes and image intensifiers may be installed in cable trays or troughs along with X-ray equipment control and power supply conductors without the necessity of barriers to separate the wiring. See *NEC Section 517.77*.

3.3.7 Guarding and Grounding

All high-voltage parts, including X-ray tubes, must be mounted within grounded enclosures per *NEC Section 517.78(A)*. The link from high-voltage equipment to the X-ray tubes and other high-voltage devices must be made using high-voltage shielded cable. Air, oil, gas, or other suitable insulating material must be used to insulate the high voltage from the grounded enclosure.

All cable providing low voltages and connecting to oil-filled units that is not completely sealed (such as transformers, condensers, oil coolers, and high-voltage switches) must have insulation of the oil-resistant type. See *NEC Section 517.78(B)*.

3.4.0 Communication, Signaling, Data, and Fire Alarm Systems

Communication, signaling, data, and fire alarm systems, as well as systems operating on less than 120V in patient care spaces, must conform to the same insulation and isolation requirements as electrical distribution systems. Those systems not located in patient care spaces must adhere to requirements in *NEC Articles 640, 725, 760, and 800,* as applicable. Nurse call systems (*Figure 10*) may use nonelectrified patient control (emergency pull-cords, for example) as an alternative isolation method.

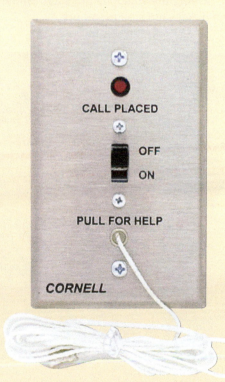

Pull-Cord Station

The emergency pull-cord station shown here is a type of nonelectrified nurse call device.

Figure Credit: Cornell Communications, Inc.

3.5.0 Isolated Power Systems

Isolated power systems are used to prevent power interruption due to a single line-to-ground fault and to isolate power from any electrical ground in certain critical spaces. *Figure 11* shows an isolated power panel that may be installed in a corridor outside an operating room or other locations where power interruption cannot be tolerated. These panels are also available with various types of receptacles and up to seven ground jacks on the panel face for use in intensive care and coronary care units, cardiac catheterization labs, neonatal units, emergency rooms, critical care spaces, and recovery rooms.

Visibility

Many nurse call master stations have video touch screens that allow staff to quickly access information. This screen shows pending calls from patients and indicates that the responsible staff member for each call has been notified. The lower left portion of the screen shows that the staff members will be reminded if the call signal has not been turned off within a set time. At the bottom right of the screen, the staff locator system shows the location of selected staff members.

Figure Credit: Ascom (US) Inc.

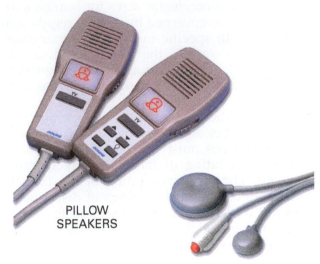

PILLOW
SPEAKERS

CALL BUTTONS

CALL STATIONS

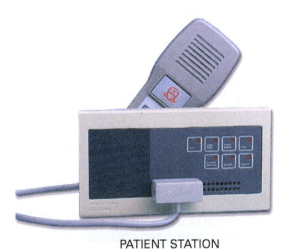

PATIENT STATION

Figure 10 Common signaling devices for nurse call systems.

Figure 11   Isolated power panel.

3.5.1 Installation

Per *NEC Section 517.160(A)*, each isolated power circuit must be controlled by a switch or a circuit breaker having a disconnecting pole in each isolated circuit conductor to simultaneously disconnect all power. Isolation is achieved by using one or more isolation transformers, motor generator sets, or electrically isolated batteries. Conductors of isolated power circuits shall not be installed in cables, raceways, or other enclosures containing conductors of another system.

Circuits feeding the primaries of isolation transformers must not exceed 600V between conductors and must be furnished with the proper overcurrent protection. The secondary voltage of isolation transformers must not exceed 600V between conductors of each circuit. All circuits fed from these secondaries must be ungrounded and have an approved overcurrent device of the proper rating in each conductor. If an electrostatic shield is present, it must be connected to the reference grounding point.

An isolated branch circuit supplying an anesthetizing location must not supply any other location except as noted in *NEC Sections 517.160(A) (4)(a) and (b)*.

Per *NEC Section 517.160(A)(5)*, the isolated circuit conductors must be identified as follows:

- *Isolated Conductor No. 1* – Orange with at least one distinctive colored stripe other than white, green, or gray along the entire length of the conductor

- *Isolated Conductor No. 2* – Brown with at least one distinctive colored stripe other than white, green, or gray along the entire length of the conductor
- *Isolated Conductor No. 3 (for three-phase conductors)* – Yellow with at least one distinctive colored stripe other than white, green, or gray along the entire length of the conductor

No wire-pulling compounds that increase the dielectric constant of a conductor may be used on the secondary conductors of the isolated power supply. Special lubricants are available for this application.

To meet impedance requirements, it may be desirable to limit the size of the isolation transformer to 10kVA and to use conductor insulation with low leakage. To reduce leakage from line to ground, steps should be taken to minimize the length of branch circuit conductors and use conductor insulation that meets the requirements of *NEC Section 517.160(A)(6), Informational Note No. 2*.

3.5.2 Line Isolation Monitors

Every isolated power system must be provided with an approved, continually functioning line isolation monitor that indicates possible leakage or fault current from either isolated conductor to ground, in addition to the usual control and protective devices. See *NEC Section 517.160(B)(1)*.

Grounding systems are the primary protection devices for patients. The ungrounded secondary of an isolation transformer reduces the maximum current in the grounding system in the event of a single fault between either of the isolated power conductors and ground. The line isolation monitor provides a warning when a single fault occurs, or when excessively low impedance to ground develops, which might expose the patient to an unsafe condition should an additional fault occur. Excessive current in the grounding conductors will not result from an initial fault. A hazard exists if a second fault occurs before the first fault is cleared.

Monitors must include a green signal lamp, visible to individuals in the anesthetizing area, which remains illuminated when the system is sufficiently isolated from ground. An adjacent red signal lamp must illuminate with an audible warning siren when the total hazard current (consisting of possible capacitive and resistive leakage current) from either isolated conductor to ground reaches a threshold value of 5mA under normal line voltage conditions. The line isolation monitor should not alarm for any hazard fault current of 3.7mA or less or a total hazard current of less than 5mA. The audible siren may sound in another area if desired.

A line isolation monitor must have enough internal impedance so that when it is properly connected to the isolated system, the maximum internal current that will flow through the line isolation monitor when any point of the isolated system is grounded is 1mA.

An ammeter connected in series to verify the total hazard current of the system (the fault hazard current and monitor hazard current) must be mounted in a visible place on the line isolation monitor with the Alarm On zone at approximately the center of the scale. Locate the ammeter where it is visible to persons in the anesthetizing location.

Like all major electrical installations, health care facilities require extensive testing and performance verification. Typical testing forms are included in the *Appendix*.

3.0.0 Section Review

1. Any location or room in which flammable anesthetics or volatile flammable disinfecting agents are stored is considered a Class I, Division 1 location _____.

   a. up to a level 3' (900 mm) above the floor
   b. up to a level 5' (1.5 m) above the floor
   c. up to a level 7½' (2.3 m) above the floor
   d. from floor to ceiling

2. *NEC Section 517.64* applies to _____.

   a. low-voltage equipment and instruments
   b. control circuit conductors
   c. isolated power systems
   d. essential electrical systems

3. *NEC Article 490* applies to X-ray circuits and equipment over _____.

   a. 60V
   b. 120V
   c. 480V
   d. 1,000V

4. A type of nonelectrified patient control device is a(n) _____.

   a. bed speaker
   b. call button
   c. emergency pull-cord
   d. pillow speaker

5. The secondary voltage of isolation transformers in health care facilities must NOT exceed _____.

   a. 120V between conductors of each circuit
   b. 240V between conductors of each circuit
   c. 480V between conductors of each circuit
   d. 600V between conductors of each circuit

1. Electrical work in health care facilities is governed by _____.

   a. *NEC Article 511*
   b. *NEC Article 517*
   c. *NEC Article 550*
   d. *NEC Article 820*

2. Which of the following provide around-the-clock services to patients?

   a. Hospitals, nursing homes, ambulatory health care facilities, and limited care facilities
   b. Nursing homes, ambulatory health care facilities, and limited care facilities
   c. Ambulatory health care facilities, hospitals, and limited care facilities
   d. Limited care facilities, hospitals, and nursing homes

3. The life safety branch of the essential electrical system includes _____.

   a. intensive care units
   b. patient care spaces
   c. pharmacies
   d. exit signs

4. The maximum time allowed for transfer switches to return to operation with an emergency power system is _____.

   a. 5 seconds
   b. 10 seconds
   c. 2 minutes
   d. 10 minutes

5. The maximum demand of a facility with only one transfer switch in an essential electrical system is _____.

   a. 10kVA
   b. 50kVA
   c. 100kVA
   d. 150kVA

6. Which type of facility must have a Type 2 EES?

   a. Doctor's office
   b. Nursing home
   c. Dialysis center
   d. Surgical center

7. Throwover switches are often used to allow temporary operation of _____.

   a. elevators
   b. suction devices
   c. heating systems
   d. pharmacies

8. In a Type 2 EES, which of the following must have alternate power replaced within 10 seconds of a power failure?

   a. Storage areas
   b. Medication preparation areas
   c. Exit signs
   d. Heating equipment

9. Where branch circuit panelboards for normal and essential power serve the same patient care vicinity, they must be connected together with an insulated copper conductor no smaller than No. _____.

   a. 10 AWG
   b. 12 AWG
   c. 14 AWG
   d. 16 AWG

10. Outlets that are on emergency power are often identified by _____.

    a. color
    b. stripes
    c. flags
    d. arrows

11. Hospital-grade tamper-resistant receptacles _____.

    a. provide power only when the hot and neutral blades of a plug are simultaneously inserted
    b. can be used when an isolated ground is needed
    c. are used in wet areas only
    d. are obsolete

12. Hospital-grade tamper-resistant receptacles must be used in _____.

    a. psychiatric locations
    b. critical care units
    c. wet areas
    d. laboratories

13. Each patient bed in the general patient care space must have a minimum of _____.

    a. two receptacles
    b. four receptacles
    c. six receptacles
    d. eight receptacles

14. In an area where flammables are employed, a Class I, Division 1 location extends upward to a level of _____.

    a. 2' (600 mm) above the floor
    b. 5' (1.5 m) above the floor
    c. 10' (3 m) above the floor
    d. 15' (4.5 m) above the floor

15. In order to use a suitable attachment plug and hard-service cable, X-ray equipment supplied by a branch circuit must be rated at no more than _____.

    a. 15A
    b. 30A
    c. 45A
    d. 60A

1. Areas in which failure of equipment or a system is likely to cause major injury or death to patients or caregivers are known as _____.

2. A facility for the treatment of chemical dependency is most likely to be defined as a(n) _____.

3. Patients on life support must be in an area served by a Type _____.

4. In a Type 2 EES, the life safety branch must have power restored within _____ second(s) after normal source interruption.

5. Name some typical applications of isolated ground receptacles.

    _____

6. *NEC Section 517.19(D)* requires that an equipment bonding jumper no smaller than _____ AWG be used to connect the grounding terminal of all grounding-type receptacles of a patient headwall to the patient grounding point.

7. True or False? In critical care spaces, each bed location requires at least one dedicated branch circuit from the critical branch, in addition to a branch from the normal system.

    _____

8. All branch circuits serving patient care spaces must have an effective ground-fault current path by installation in a(n) _____.

9. True or False? Nurse call systems may use nonelectrified patient control as an alternative isolation method.

    _____

10. Isolated power systems are used to prevent power interruption due to a single line-to-ground fault and to _____.

Trade Terms Introduced in This Module

Equipotential ground plane: A mass of conducting material that, when bonded together, provides a low impedance to current flow over a wide range of frequencies.

Inpatient: A patient admitted into the hospital for an overnight stay.

Outpatient: A patient who receives services at a hospital but is not admitted for an overnight stay.

Throwover: A transfer switch used for supplying temporary power.

TYPICAL TESTING AND PERFORMANCE VERIFICATION FORMS

BED TOWER EXPANSION
Part 3 – Bed Tower Expansion & Renovation
SECTION 16020
TESTS AND PERFORMANCE VERIFICATION

PART 1 - GENERAL

1.1   RELATED DOCUMENTS

General: Drawings and general provisions of the Contract, including General and Supplementary Conditions and Division 1 Specification sections, apply to work specified in this section.

1.2   DESCRIPTION

A.   Time: Perform verification work as required to show that the System is operating correctly in accordance with contract documents and manufacturer's literature. All verification shall be done after 3-day full operational period.

B.   Submission: Submit check out memos and completed testing results of all systems, cable, equipment, devices, etc., for acceptance prior to being energized or utilized.

1.3   QUALITY ASSURANCE

A.   Compliance: Testing shall comply to the following standards;

1.   NEMA
2.   ASTM
3.   NETA
4.   ANSI C2
5.   ICEA
6.   NFPA

1.4   QUALIFICATIONS OF TESTING FIRM

A.   Qualification: The testing firm shall be an independent testing organization which can function as an unbiased testing authority, professionally independent of the manufacturers, supplier, and installers of equipment or systems evaluated by the testing firm.

B.   Experience: The testing firm shall be regularly engaged in the testing of electrical equipment devices, installations, and systems.

C.   Accreditation: The testing firm shall meet OSHA criteria for accreditation of testing laboratories, Title 29, Part 1907, or be a Full Member company of the International Electrical Testing Association.

D.   Certification: The lead, on-site, technical person shall be currently certified by the International Electrical Testing Association (IETA) or National Institute for Certification in Engineering Technologies (NICET) in electrical power distribution system testing.

16020 - 1

E. Personnel: The testing firm shall utilize engineers and technicians who are regularly employed by the firm for testing services.

F. Proof of Qualifications: The testing firm shall submit proof of the above qualifications when requested.

G. Companies: Must use NETA-certified pre-qualified testing firms.

PART 2 - TEST

2.1 EQUIPMENT

A. Instruments: Supply all instruments required to read and record data. Calibration date shall be submitted on test reports. All instruments shall be certified per NETA standards.

B. Adjustments: Adjust system to operate at the required performance levels and within all tolerances as required by NETA Standards.

2.2 APPLICATIONS

A. Switchboards, Panelboards and Mechanical Equipment Feeders: After feeders are in place, but before being connected to devices and equipment, test for shorts, opens, and for intentional and unintentional grounds.

B. Ratings 250V or Less: Cables 250V or less in size #1/0 and larger shall be meggered using an industry approved "megger" with 500V internal generating voltage. Readings shall be recorded and submitted to the Engineer for acceptance prior to energizing same. Submit (5) copies of tabulated megger test values for all cables.

C. Cables 600 Volts or Less: Cables 600 volts or less in size #1/0 and larger shall be meggered using an industry approved "megger" with 1000V internal generating voltage. Readings shall be recorded and submitted to the Engineer for acceptance prior to energizing same. If values are less than recommended NETA values notify Engineer. Submit 5 copies of tabulated megger test values for all cables.

D. Ratings 600 Volts or Less: Cables 600 volts or less in size #1/0 AWG and larger shall be meggered using an industry approved "megger" with 500V internal generating voltage. Readings shall be recorded and submitted to the Engineer for acceptance prior to energizing same. Submit 5 copies of tabulated megger test values for all cables.

E. Ratings Above 600V: Cables above 600 volts in all sizes shall first be meggered, using an industry approved "megger" having 2500V internal generating voltage. When proper readings are obtained, the cables shall be "hi-potted" using (5) potentials and periods as recommended by NETA, cable manufacturer for the type and voltage class of cables installed. Do not exceed (80%) of factory test voltage. Readings ("megger" and "hi-pot") shall be recorded and submitted to the Engineer, for acceptance prior to energizing it. Submit 5 copies of tabulated megger test values for all cables.

16020 - 2

F. Main circuit breakers and feeder circuit breakers 200 amps and greater shall be tested using primary injection testing as per NETA Specifications. Reports to include manufacturer's time current curve number and trip time. Submit five (5) copies to the Engineer at substantial completion.

G. Transformers (75) KVA and larger. Perform Insulation resistance test and turns ratio test. Submit five (5) copies to Engineer at substantial completion.

2.3   MOTORS

A. Procedure: Test run each motor, (25 HP) and larger. Tabulate and submit 5 copies of the Test Information at substantial completion for final inspection. Refer to form at the end of this Section.

B. Provisions: With the system energized, line-to-line voltage and line current measurements shall be made at the motors under full load conditions. The condition shall be corrected when measured values deviate plus or minus 10% from the nameplate ratings.

C. Insulation: Test the insulation resistance's of all motor windings to ground with an appropriate test instrument as recommended by the motor manufacturer, before applying line voltage to the motors. If these values are less than the manufacturer's recommended values, notify the contractor providing the motor for correction before initial start up.

D. Power Factor: Check power factor of all motors (5 HP) and larger while driving its intended load, and at all operating speeds.

2.4   GROUNDS

A. Electrode Ground: The resistance of electrodes (main service, generators, transformer, etc.) shall not exceed 10 ohms and shall be measured before equipment is placed in operation. Testing shall be performed on all grounding electrode installations. Testing of main ground shall be (3) point method in accordance with IEEE No. 81 Section 9.04 Standard.9. (2) point method for distribution equipment. Testing to be completed before service energized. Submit all ground test readings to the Engineer in tabulated format at substantial completion.

B. Electrode Ground: The resistance of electrodes (main service, generators, transformer, etc.) shall not exceed 10 ohms and shall be measured by The Contractor before equipment is placed in operation. Testing shall be performed on all grounding electrode installations. Testing shall be 3 point method in accordance with IEEE Standard 81. Submit all ground test readings to the Engineer in tabulated format at substantial completion.

C. Electrode Ground: The resistance of electrodes (main service, generators, transformer, etc.) shall not exceed 5 ohms and shall be measured by The Contractor before

16020 - 3

equipment is placed in operation. Testing shall be performed on all grounding electrode installations. Testing shall be 2-point method in accordance with IEEE Standard 81. Submit all ground test readings to the Engineer in tabulated format at substantial completion.

2.5   EQUIPOTENTIAL GROUND

A.   Equipotential Ground: Test all metal conductive surfaces likely to become energized within patient care areas. Test all large conductive surfaces likely to become energized within a volume defined as 6 foot from the patient bed horizontally or 7 foot 6 inches vertically.

    1.   Large metal surfaces not likely to be energized, which do not require testing:
        a.   Window frames
        b.   Door frames
        c.   Floor drains
        d.   Moveable metal cabinets

    2.   Test Method:
        a.   Use impedance and voltage measurements
        b.   Utilize established ground bus or ground bar in panel serving area.
        c.   Measure voltage from reference point to conductive surfaces and all receptacle ground contacts.
        d.   Measure impedance between reference point and receptacle ground contacts.
        e.   Check for proper polarity.
        f.   Identify the reference ground for each room on the ground test report. Provide a blue dot label with a permanent adhesive backing located on the bottom center of the reference ground outlet cover.

    3.   Maximum Acceptable Values:
        a.   Voltage: 20 mV plus or minus 20 percent
        b.   Impedance: 0.1 ohm plus or minus 20 percent
        c.   Quiet ground impedance: 0.2 ohm plus or minus 20 percent

    4.   Equipment:
        a.   Millivolt meter with 1 Kohm impedance and proper frequency response, in accordance with NFPA 99.
        b.   Polarity tester

    5.   Ground Test Report. Complete ground test report included at the end of this specification section, and make available copies of such to engineer and inspecting authority at final inspection.

2.6   PRIMARY CABLE

A.   Utilize D.C. proof test: Test new cable from primary circuit utility interface to utilization equipment, after complete installation.

16020 - 4

1. Defer test until all terminations have been completed. Disconnect permanently connected switches and other devices, except for connection to power grid.

2. Apply a D.C. voltage for 5 minutes from the appropriate test equipment to the conductor of the cable to be tested, and the low potential or grounded terminal connected to the cable shield.

3. Record leakage currents after 15, 20, 30, 45, and 60 seconds and at one minute intervals. Apply test voltage in accordance with IPCEA Standards.

B. Repair or Replacement: Repair or replace as directed any cables or terminals not meeting or exceeding minimum standards. Retest replaced cable or repaired splices.

2.7 DRY-TYPE TRANSFORMERS

A. Required Factory Tests: Required factory tests shall be as follows;

1. Ratio
2. Polarity
3. Losses
   a. No load
   b. Full load
4. Resistance Measurements
5. Impedance
6. Temperature
7. Impulse Strength
8. Sound Level
9. Exciting Current
10. Low-frequency Dielectric Strength
11. ANSI Point and Curve

B. Submission: Submit test results with shop drawings.

2.8 EMERGENCY SYSTEM

A. General: Submit emergency system tests in accordance with NFPA 110. Refer to emergency section of the specification for additional information.

PART 3 - EXECUTION

3.1 SUBMITTALS

A. Equipotential Ground Test Report: Complete report form at the end of this specification.

B. Cable Test Report: Submit Cable Test Report in Triplicate.

C. Transformer Test Report: Indicate comparative data of ANSI and NEMA Standards. Indicate all characteristic values as specified herein. Certified copies of tests on electrically duplicate units are acceptable.

16020 - 5

D. Check out Memos: Complete all information on forms at the end of this specification, project information, and certificate of completed demonstration memo. Submit data for examination and acceptance prior to final inspection request.

E. Tabulated Data: Submit data on 8-1/2 x 11 inch sheets with names of the personnel who performed the test.

F. Final: Submit accepted memos before a request for final inspection.

3.2 QUANTITIES

A. Quantity: Submit 5 copies of the check out memo on each major item of equipment. Insert accepted memos in each brochure with the performance verification information and submittal data.

END OF SECTION

16020 - 6

FACILITY NAME:_____ PROJECT NAME:_____ AHCA LOG NO.

DATE: _____TESTED BY:

MAXIMUM TEST INTERVALS: NAME:
GENERAL CARE - 12 MOS.
CRITICAL CARE - 6 MOS. COMPANY:
WET LOCATIONS - 12 MOS.

GROUND TEST REPORT

TYPE METER USED AND EXTERNAL NETWORK IF USED:

NOTE: MAXIMUM READINGS PERMITTED: 20 MV NEW CONSTRUCTION
0.1 OHM NEW CONSTRUCTION

Room No.	AREA TYPE Description (C) = CRITICAL CARE (G) = GENERAL CARE	VOLTAGE MEASUREMENT			IMPEDANCE MEASUREMENT		REMARKS - IF VOLTAGE READINGS MORE THAN 20MV IN EXISTING CONST. NOTE TESTS & INVESTIGATION REQUIRED.
		NO. OF RECEPTS.	NO. OF OTHER	MAX. READING IN MILLIVOLTS	NO. OF RECEPTS.	MAX READING IN OHMS	

16020 - 7

PROJECT NAME: _____

MOTOR TEST INFORMATION

Name of Checker: _____

Date Checked: _____

(a)   Name and identifying mark of motor   _____

(b)   Manufacturer   _____

(c)   Model Number   _____

(d)   Serial Number   _____

(e)   RPM   _____

(f)   Frame   _____

(g)   Code Letter   _____

(h)   Horsepower   _____

(i)   Nameplate Voltage and Phase   _____

(j)   Nameplate Amps   _____

(k)   Actual Voltage   _____

(l)   Actual Amps   _____

(m)   Starter Manufacturer   _____

(n)   Starter Size   _____

(o)   Heater Size, Catalog No. and Amp Rating   _____

(p)   Manufacturer of dual-element fuse   _____

(q)   Amp rating of fuse   _____

(r)   Power Factor at _____ Speed   _____
      (For variable speed motors provide
      recording chart over operating range)

16020 - 8

TABULATED DATA

VOLTAGE AND AMPERAGE READINGS

SWITCHGEAR OR PANELBOARD

FULL LOAD AMPERAGE READINGS:

DATE

TIME

PHASE

      A

      B

      C

      N

FULL LOAD VOLTAGE READINGS:

DATE

TIME

PHASE   A TO N _____ A TO B

             B TO N _____ A TO C

             C TO N _____ B TO C

NO LOAD VOLTAGE READINGS

DATE

TIME

PHASE   A TO N _____ A TO B

             B TO N _____ A TO C

             C TO N _____ B TO C

_____ ENGINEER'S REPRESENTATIVE

_____ CONTRACTOR'S REPRESENTATIVE

16020 - 9

Additional Resources

This module presents thorough resources for task training. The following reference material is recommended for further study.

National Electrical Code® Handbook, Latest Edition. Quincy, MA: National Fire Protection Association.

Standard for Health Care Facilities (NFPA 99), Latest Edition. Quincy, MA: National Fire Protection Association.

Figure Credits

Section Review Answer Key

Section 1.0.0

Answer	Section Reference	Objective
1. c	1.1.2	1a
2. c	1.2.3	1b

Section 2.0.0

Answer	Section Reference	Objective
1. c	2.1.0	2a
2. d	2.2.0	2b
3. d	2.3.0	2c
4. a	2.4.0	2d

Section 3.0.0

Answer	Section Reference	Objective
1. d	3.1.0	3a
2. a	3.2.0	3b
3. d	3.3.1	3c
4. c	3.4.0	3d
5. d	3.5.1	3e

NCCER CURRICULA — USER UPDATE

NCCER makes every effort to keep its textbooks up-to-date and free of technical errors. We appreciate your help in this process. If you find an error, a typographical mistake, or an inaccuracy in NCCER's curricula, please fill out this form (or a photocopy), or complete the online form at **www.nccer.org/olf**. Be sure to include the exact module ID number, page number, a detailed description, and your recommended correction. Your input will be brought to the attention of the Authoring Team. Thank you for your assistance.

Instructors – If you have an idea for improving this textbook, or have found that additional materials were necessary to teach this module effectively, please let us know so that we may present your suggestions to the Authoring Team.

NCCER Product Development and Revision

13614 Progress Blvd., Alachua, FL 32615

Email: curriculum@nccer.org
Online: www.nccer.org/olf

❑ Trainee Guide ❑ Lesson Plans ❑ Exam ❑ PowerPoints Other _____

Craft / Level: _____ Copyright Date: _____

Module ID Number / Title: _____

Section Number(s): _____

Description: _____

_____

_____

_____

_____

Recommended Correction: _____

_____

_____

_____

_____

Your Name: _____

Address: _____

_____

Email: _____ Phone: _____

This page is intentionally left blank.

Standby and Emergency Systems

OVERVIEW

In places of assembly, power systems provide important functions such as lighting, evacuation routes and others during emergency situations. This module explains the *NEC®* installation requirements for electric generators and storage batteries used during such emergency situations.

Module 26403-20

Objectives

When you have completed this module, you will be able to do the following:

1. Differentiate between emergency and standby systems and identify their primary components.
   a. Identify emergency and standby power system components.
   b. Explain the principles of transfer switch operation and their configuration/sizing considerations.
2. Describe battery and UPS system types and explain their maintenance requirements.
   a. Describe the different types of batteries used.
   b. Explain the maintenance requirements of batteries and their charging considerations.
   c. Identify single- and double-conversion UPS systems.
3. Describe the *NEC*® requirements for emergency/standby power and lighting systems.
   a. Describe legally required standby systems.
   b. Describe the alternate power requirements for health care facilities.
   c. Describe the alternate power requirements for places of assembly.
   d. Describe emergency lighting requirements and devices for public buildings.

Performance Tasks

This is a knowledge-based module. There are no Performance Tasks.

Trade Terms

Armature
Electrolyte

Exciter
Synchronous generator

Industry Recognized Credentials

If you are training through an NCCER-accredited sponsor, you may be eligible for credentials from NCCER's Registry. The ID number for this module is 26403-20. Note that this module may have been used in other NCCER curricula and may apply to other level completions. Contact NCCER's Registry at 888.622.3720 or go to **www.nccer.org** for more information.

Contents

1.0.0   Standby and Emergency Systems 1

    1.1.0   System Components 2

      1.1.1   Engine-Driven Generator Sets 2

      1.1.2   Generator Selection Considerations and Sizing 4

      1.1.3   Generator Set Maintenance and Service 5

    1.2.0   Transfer Switches 6

      1.2.1   Transfer Switch Operation 6

      1.2.2   Transfer Switch Configurations and Sizing 7

      1.2.3   Automatic Sequential Paralleling 8

2.0.0   Storage Batteries and UPS Systems 10

    2.1.0   Storage Batteries 10

      2.1.1   Lead-Acid Batteries 10

      2.1.2   Nickel Cadmium Batteries 10

    2.2.0   Battery Maintenance 10

      2.2.1   Battery Maintenance Overview 11

      2.2.2   Battery Maintenance Guidelines 13

      2.2.3   Battery and Battery Charger Operation 15

    2.3.0   Uninterruptible Power Supply Systems 15

      2.3.1   Double-Conversion UPS Systems 16

      2.3.2   Single-Conversion UPS Systems 16

3.0.0   *NEC*® Requirements for Emergency Systems 20

    3.1.0   Legally Required Standby Systems 20

    3.2.0   Health Care Facilities 21

    3.3.0   Places of Assembly 24

    3.4.0   Emergency Lighting 24

      3.4.1   Battery-Powered Emergency Lighting 24

      3.4.2   Emergency Lighting Units 25

Figures

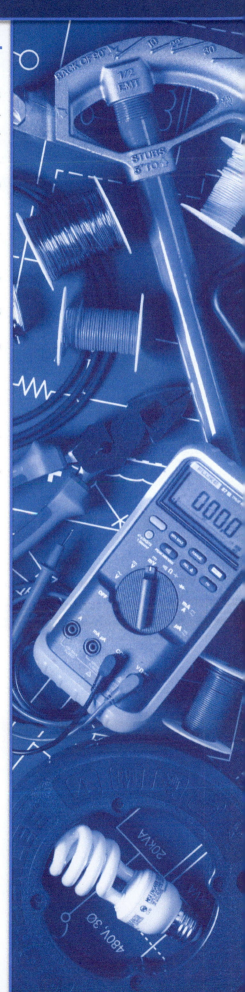

Figure 1   One-line diagram of an electrical distribution system
            incorporating emergency power system components 3

Figure 2   Cross-sectional view of a four-pole AC generator 4

Figure 3   Self-exciter generator schematic 4

Figure 4   Simplified emergency/standby power
            distribution with automatic transfer switch 7

Figure 5   Typical automatic sequential paralleling emergency system........ 9

Figure 6   *NEC*® requirements governing storage batteries
            for legally required standby systems 11

Figure 7   Chemical reaction in a lead-acid battery 11

Figure 8   Chemical reaction in a nickel cadmium battery 11

Figure 9   Stationary battery in float operation... 15

Figure 10   Simplified block diagram of
            a double-conversion UPS system 16

Figure 11   Simplified block diagram of
            a single-conversion UPS system 19

Figure 12   Two-way bypass isolation system 21

Figure 13   Summary of *NEC*® installation requirements
            governing legally required standby systems................................ 22

Figure 14   Essential electrical systems must be served
            by one or more transfer switches 23

Figure 15   Summary of *NEC*® requirements for
            emergency circuit identification.. 24

Figure 16   Typical emergency lighting units 25

This page is intentionally left blank.

1.0.0 STANDBY AND EMERGENCY SYSTEMS

Objectives

Differentiate between emergency and standby systems and identify their primary components.

a. Identify emergency and standby power system components.

b. Explain the principles of transfer switch operation and their configuration/sizing considerations.

Trade Terms

Armature: The assembly of windings and metal core laminations in which the output voltage is induced. It is the rotating part in a revolving field generator.

Exciter: A device that supplies direct current (DC) to the field coils of a synchronous generator, producing the magnetic flux required for inducing output voltage in the stator coils.

Synchronous generator: An AC generator having a DC exciter. Synchronous generators are used as stand-alone generators for emergency and standby power systems and can also be paralleled with other synchronous generators and the utility system.

Emergency systems are defined by *NEC Article 700* as those systems legally required and classified as such by municipal, state, federal, and other codes, or by the government agency having jurisdiction. These systems are intended to automatically supply illumination and/or power to designated areas and equipment in the event of failure of the normal supply or in the event of an accident to elements of a system intended to supply, distribute, and control power and illumination for safety of human life.

Emergency systems are typically installed in places of assembly where artificial illumination is required for safe exiting and for panic control in buildings subject to occupancy by larger numbers of people, such as hotels, theaters, hospitals, sports arenas, health care facilities, and similar institutions. They may also be used to provide power for such functions as:

- Ventilation where essential to maintain life

- Fire detection and alarm system operation
- Elevator operation
- Fire pump operation
- Public safety communication system operation
- Industrial processes where current interruption would produce serious life safety or health hazards
- Similar functions

In the event of failure of the normal power supply, emergency lighting, emergency power, or both must be available within the time required for the application, but not to exceed 10 seconds.

Legally required standby systems are defined by *NEC Article 701* as those systems required and classified as such by municipal, state, federal, and other codes, or by the government agency having jurisdiction. These systems are intended to automatically supply power to selected loads (other than those classified as emergency systems) in the event of failure of the normal source.

Legally required standby systems are typically installed to serve loads such as heating and refrigeration systems, communication systems, ventilation and smoke removal systems, sewage disposal, lighting systems, and industrial processes that, when stopped during any interruption of the normal electrical supply, could create hazards or hamper rescue or fire-fighting operations.

The requirements for legally required standby systems are basically the same as for emergency systems; however, there are some differences. Upon loss of normal power, legally required systems are required to be able to supply standby power in 60 seconds or less, instead of 10 seconds or less as required for emergency systems.

Optional standby systems are defined by *NEC Article 702* as those systems intended to protect public or private facilities or property where life safety does not depend on the performance of the system. Optional standby systems are intended to supply site-generated power to selected loads either automatically or manually. Optional standby systems are typically installed to provide an alternate source of electric power for such facilities as industrial and commercial buildings, farms, and residences. These systems serve loads such as heating and refrigeration systems, data processing and communication systems, and industrial processes that, when stopped during any power outage, could cause discomfort, serious interruption of the process, or damage to the product or process.

1.1.0 System Components

Figure 1 shows a one-line diagram of a typical electrical distribution system that incorporates a basic emergency power system (shown as heavy lines). The components that form an emergency or standby power system are basically the same. They include the following:

- An engine-driven AC generator set
- Automatic transfer switches

1.1.1 Engine-Driven Generator Sets

Many designs and sizes of engine-driven synchronous generator sets are used as power sources for emergency and standby systems.

An AC generator (alternator) converts rotating mechanical energy, typically supplied by an air- or water-cooled diesel or gas engine, into electrical energy. The AC generator consists of a rotor (*Figure 2*) also called the armature, and stator, with the shaft of the rotor being driven by the engine. When the engine is running, it turns the rotor, which carries the generator field windings. These field windings are energized (or excited) by a DC source called the exciter, which is connected to the positive (+) and negative (–) ends of

the field windings. The generator is constructed such that the lines of force of the magnetic field cut perpendicularly across the stator windings when the engine turns the rotor, causing a voltage to be induced into the stator windings. This voltage reverses each time the polarity changes. For the four-pole generator shown in *Figure 2*, this would occur twice during each revolution. Typically, a generator has four times as many winding slots as shown and is wound to obtain a sinusoidally alternating single- or three-phase output.

The voltage induced in each stator winding depends on the strength of the magnetic field and the velocity with which the lines of force cut across the windings. Therefore, in order to vary the output voltage of a generator of a given

Typical Engine-Generator Set

A large water-cooled engine-generator set is shown here. This unit is diesel-powered and is designed for long periods of operation. Critical applications may require one or more additional units as backup(s). These are commonly referred to as *N + 1*, *N + 2*, or *N + X* designs. Codes normally require that generators installed in habitable buildings be located inside a separate fire-rated room protected by a fire suppression system.

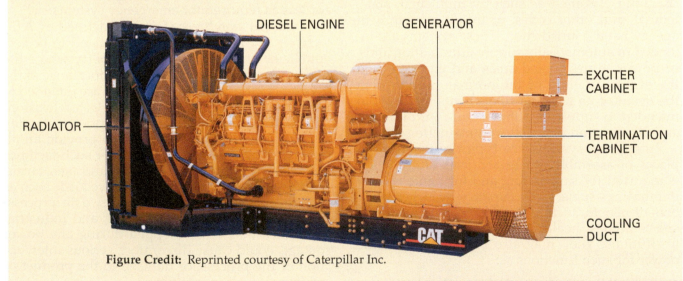

Figure Credit: Reprinted courtesy of Caterpillar Inc.

diameter and operating speed, it is necessary to vary the strength of the magnetic field. This is controlled by the voltage regulator, which controls the output of the exciter.

The excitation system for a typical generator is controlled via a voltage regulator (*Figure 3*). The voltage regulator senses the generator output, compares it to a reference value, and then supplies a regulated DC output to the exciter field windings. The exciter produces an AC output in the exciter rotor, which is on the rotating engine-driven generator shaft. This exciter output is rectified by diodes, also on the generator shaft, to supply DC for the main rotor (generator field). The voltage regulator increases or decreases the exciter voltage as it senses changes in the output generator voltage due to changes in load, thus increasing or decreasing the generator field strength. The resultant generator output is directly proportional to the field strength.

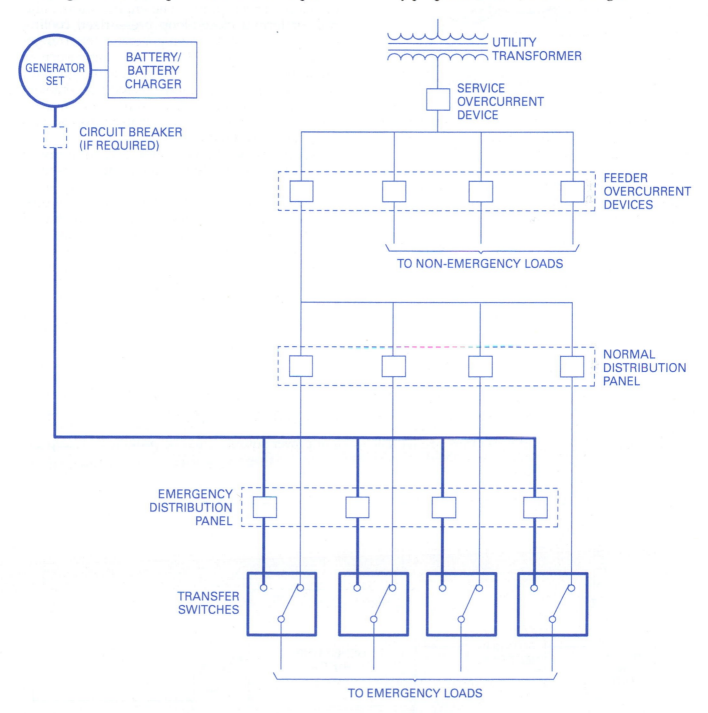

Figure 1 One-line diagram of an electrical distribution system incorporating emergency power system components.

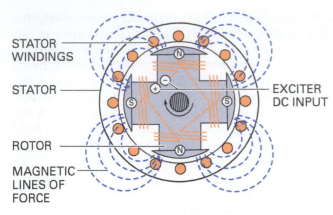

Figure 2 Cross-sectional view of a four-pole AC generator.

1.1.2 Generator Selection Considerations and Sizing

A wide variety of generator sets are available. A generator set selected for a specific emergency or standby power application must have the same voltage output and frequency as the normal building power supply system. Generator engines are generally four-cycle units having one to six cylinders, depending on size and capacity. In larger units, ignition current is generally supplied by starting batteries. The type of fuel used by the engine that drives the generator is an important consideration in the selection of a generator set. Fuels that may be used include LP gas, natural gas, gasoline, and diesel fuel. Factors that affect the selection of the fuel to be used include the availability of the fuel and local regulations governing the storage of the fuel. Problems of fuel storage can be a deciding factor in the selection and types of generators used. If natural gas is used, its Btu heating content must be greater than 1,000 Btus per cubic foot (37 megajoules per cubic meter). Diesel-powered generators are widely used because they require less maintenance and have a longer life.

Smaller generator sets are available with either air or liquid cooling systems. Larger units typically use liquid-cooled engines. Liquid-cooled engines used with generator sets operate in basically the same way as a liquid-cooled automobile engine. They are cooled by pumping a mixture of water and antifreeze through passages in the engine cylinder block and head(s) by means of an engine-driven pump. The engine, pump, and radiator form a closed-loop pressurized cooling system. The most common generator set arrangement has a mounted radiator and engine-driven fan to cool the coolant and ventilate the generator room. Alternative methods include a remotely located radiator, or a local or remotely located liquid-to-liquid heat exchanger used instead of a forced air-cooled radiator.

Depending on the generator size, compressed air and battery starting systems can be used to start generator engines. Compressed air starting systems can be used on larger generator sets. Battery starting systems are generally preferred for most applications involving generator sets smaller than 2,000kW because they are considered to be as reliable as compressed air systems, but are much less expensive. Battery starter systems used with generator sets are usually 12V or 24V. The batteries must have enough capacity to provide the cranking motor current indicated on the generator set specification sheet. The batteries are normally either the lead-acid or nickel cadmium type. They must be designated for this use or may have to be approved by the authority having jurisdiction. For emergency power systems, a float-type battery charger, powered by the normal power source, must be provided to keep the

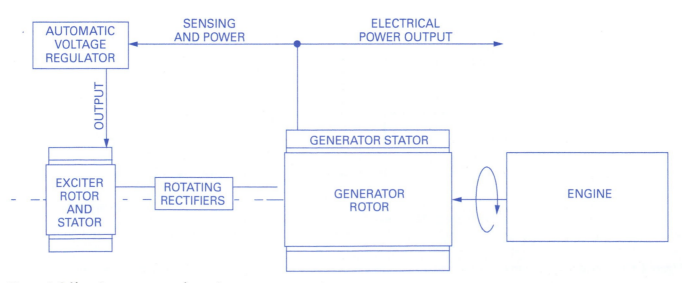

Figure 3 Self-exciter generator schematic.

batteries fully charged during standby without damage. As the battery approaches full charge, the charging current automatically tapers off to zero amperes or to a steady-state load on the battery. A high-output engine-driven alternator and automatic voltage regulator (part of the generator set) are provided to recharge the batteries during generator set operation. Batteries and battery charging systems are described in more detail later in this module.

The sizing of a generator set for a particular application is typically done using computer software developed specifically for that purpose or procedures outlined in the product literature produced by most generator set manufacturers. As a minimum, the generator set must be sized to supply the maximum starting (surge) demands and steady-state running loads of the connected equipment. Once the starting and running loads have been determined, it is common to add a margin factor of up to 25% for future expansion or to select a generator set of the next largest standard rating. A large connected load that does not run during usual power outages, such as a fire pump, can serve as part of the margin factor. For fuel efficiency, the running load should stay within approximately 50% to 80% of the generator set kW rating.

In applications where the voltage and frequency dip performance specifications are stringent, it is often necessary to oversize a generator set. Applications with the following types of loads and similar devices may also require that the generator be oversized:

- Static uninterruptible power supplies (UPSs)
- Variable frequency drives
- Battery charging rectifiers
- Medical diagnostic imaging equipment
- Fire pump(s)

1.1.3 Generator Set Maintenance and Service

The failure of an emergency or standby generator to start and run could lead to loss of life, personal injury, or property damage. For this reason, a comprehensive maintenance and service program

Cogeneration

If an application requires an around-the-clock electric demand and a substantial heat requirement, combined heat and power generation (cogeneration) using turbine- or engine-generators may offer substantial energy savings. In other cases, some electric utilities/suppliers offer peak-sharing rate arrangements by using distributed generation facilities. Under distributed generation (electric cogeneration), intermittently operated generators, located at a customer site, are automatically started and paralleled with the utility supplier on their demand. Either natural gas or diesel engine-powered generators may be furnished by the utility supplier and/or the customer, but they are usually operated and maintained by the customer. In many cases, idle standby-power systems can be converted to revenue-producing assets. With peak sharing, the utility/supplier pays monthly credits or grants special interruptible pricing that effectively caps the customer's energy price and provides profit from the sale of excess power. Payback on the generating equipment is often three years or less.

Figure Credit: Reprinted courtesy of Caterpillar Inc.

carried out on a scheduled basis is absolutely necessary. This maintenance and service should be performed when scheduled and as recommended by the generator set manufacturer.

1.2.0 Transfer Switches

Transfer switches are a primary part of a building's emergency or standby power system. They are made in both automatic and nonautomatic/remote types.

1.2.1 Transfer Switch Operation

An automatic transfer switch automatically transfers electrical loads from a normal source (utility power) to an emergency source (generator set) when the normal source fails or its voltage is substantially reduced. The transfer switch automatically retransfers the load back to the normal power source when it is restored.

> **WARNING!**
>
> Always follow lockout procedures even if the circuit is de-energized.

The typical automatic transfer switch consists of two major component areas, an electrically operated double-throw transfer switch and related control panel circuits. *Figure 4* shows a basic application of an automatic transfer switch.

As shown, voltage inputs from the normal power source and the generator set (when running) are applied to the voltage monitor and voltage/frequency monitor circuits, respectively, of the control panel. The transfer switch operates to perform the following functions:

- It initiates starting the emergency generator engine via contacts in the transfer switch control panel when the utility voltage drops to an unacceptable level.
- It maintains connection of the load circuits to the normal power source during the starting period to provide use of any existing service on the normal source.
- It measures the output voltage and frequency of the voltage being generated by the generator set by using a voltage/frequency-sensitive monitor. Note that combined frequency and voltage monitoring protects against transferring loads to a generator set with an unacceptable output.
- It automatically switches the load circuits to the generator set when both its voltage and frequency are about normal.
- It provides a visual indication and auxiliary contact for remote indication when the generator set is supplying the loads.
- It automatically switches back to the utility power source when it is deemed to be stable and normal.

Time delays are provided in the transfer switch control circuits to program the operation of the switch. One such delay, typically adjustable

Micro-Processor-Controlled Automatic Transfer Switches

Microprocessor-controlled automatic transfer switches are commonly available for voltages up to 600VAC and thousands of amperes of load-switching capability. A typical unit can be programmed to start an engine-generator when the voltage drops to a predetermined level (usually 80% of nominal). When the emergency source reaches 90% of rated voltage and 95% of rated frequency, the unit switches and locks the load to the emergency source. When the outside line voltage is restored to a predetermined point as sensed by the unit (usually 90%), the unit switches and locks the load to the outside source. The microprocessor control allows programming the unit to switch when the outside line and emergency source are in phase and at or near zero phase difference. Optional programming allows momentary disconnection of a large motor load during switching, along with staged motor restarts, if necessary, after switching. If necessary, the unit also allows a delayed transition on transfer to ensure decay of rotating motor or transformer fields for UPS system sequencing. The unit can be programmed to activate load-shedding functions prior to switching.

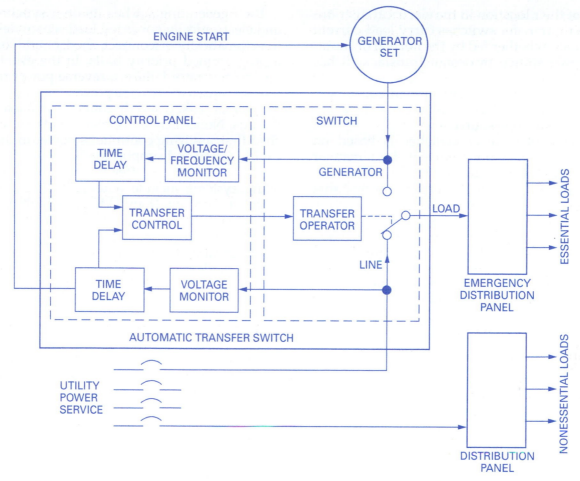

Figure 4 Simplified emergency/standby power distribution with automatic transfer switch.

between 0 and 15 seconds in duration, is provided to avoid unnecessary starting and transfer of the generator set power source for power interruptions of short duration. If the duration of the interruption exceeds the time delay, the transfer switch initiates generator set engine startup.

Once the load has been transferred to the generator set, another timer delays switching back (retransfer) to the normal source for an adjustable delay period (typically up to 30 minutes) until the source has had time to stabilize. This timer also serves the function of allowing the generator set to operate under load for a preselected minimum time, thus assuring continued good performance of the engine and its starting system. An intermediate bypass of this transfer time delay is automatic should the generator set stop inadvertently during this delay.

A third time delay, usually adjustable from 0 to 10 minutes, is provided to allow for generator set engine cooldown. During cooldown, the generator set runs unloaded after the load has been switched back to the utility power. Note that running an unloaded engine for more than 10 minutes is not recommended because it can cause deterioration in engine performance.

The operation of a nonautomatic/remote transfer switch is initiated either by an operator at the transfer switch or by an external signal from a remote control. The nonautomatic/remote transfer switch does not include an automatic controller, but can be slaved to an automatic transfer switch with field wiring interconnections.

1.2.2 Transfer Switch Configurations and Sizing

Common transfer switch configurations are three-pole with a solid neutral for single- and three-phase three-wire or four-wire circuits, or four-pole with a switched neutral for use with separately derived three-phase four-wire systems.

Because of their location in the electrical distribution system, transfer switches carry load current continuously, whether fed by the normal or the alternate power source. By design, transfer switches must be highly reliable because failure to operate as intended could leave critical load equipment without power, even though power may be available from at least one source.

The sizing of transfer switches is based on the calculated continuous current, in a manner similar to sizing the circuit conductors. Use transfer equipment continuous current ratings that equal the calculated continuous current for the load equipment or the next higher rating. If future load expansion is a consideration, a transfer switch rating based on the maximum rating of the circuit overcurrent protective device may be selected. The available fault current and upstream overcurrent device may also affect sizing of the transfer equipment.

1.2.3 Automatic Sequential Paralleling

Automatic sequential paralleling of two or more electric generating sets performing as a single power source offers important advantages to the user, such as increased power capability, greater versatility in installation and operation, and, most importantly, the greater reliability of a multiple-unit system. A typical automatic sequential paralleling emergency system for hospitals is illustrated in *Figure 5*. This system consists of engine-generator sets, paralleling switchboards, a totalizing board, and automatic transfer switches that monitor the total system and control the operation of each individual generating set.

If utility power is interrupted, an automatic transfer switch initiates starting of all three generator sets (A, B, and C) simultaneously. When the first of the three generating sets reaches operating voltage, a special emergency bus directs the power to the top-priority requirement areas, such as operating rooms and life safety and support systems.

With automatic sequential paralleling, each set generates power as soon as it starts, and any one can be the lead unit or the first on the bus. Speed in assuming the major or critical load is a major advantage of this system. With all sets in the system answering a request for power, there is a much higher probability of at least one of them assuming the emergency load sooner than a single-set system.

Each generating set has its own synchronizer and electronically governed load sensor designed to automatically disconnect, shed, or add loads on a programmed priority basis. In the event of an engine-generator failure, a reverse power relay in that set senses the reverse power condition, opens its circuit breaker, and disconnects that load from the bus. Noncritical equipment loads are usually shed first, assuring continuous power to the critical life safety and support loads.

Other advantages offered by sequential paralleling systems include automatic operation and system expansion. If an addition is built or power needs increase, the system can be expanded. A component can be serviced with only a limited loss of power rather than total power loss, as is the case with a single-unit source. Furthermore, a sequentially paralleled system usually operates at higher overall efficiencies than individual generators.

> **WARNING!**
>
> The failure of an emergency or standby system may result in loss of life or other hazards. For this reason, *NEC Sections 700.3(C) and 701.3(C)* require maintenance in accordance with manufacturer's instructions and industry standards.

Automatic Paralleling and Transfer Switchgear

The automatic paralleling and transfer switchgear shown here was designed for two large engine-generator sets. It serves as a standby system for the large data processing center of an insurance company.

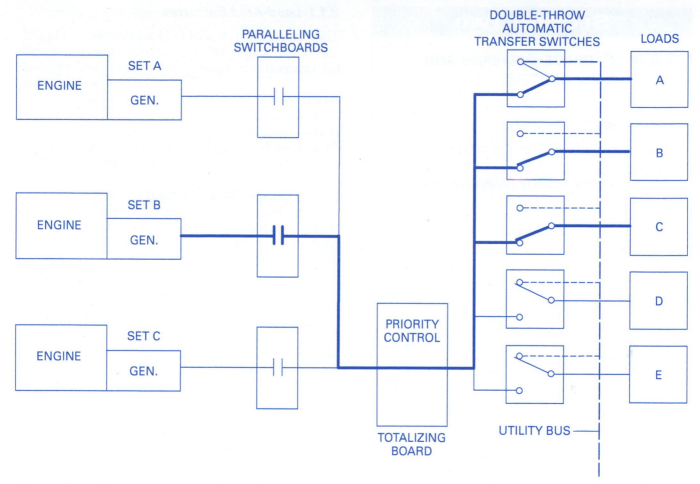

Figure 5 Typical automatic sequential paralleling emergency system.

1.0.0 Section Review

1. If natural gas is used to supply a generator, its Btu content must be greater than _____.

    a. 750 Btus/cubic foot
    b. 1,000 Btus/cubic foot
    c. 1,100 Btus/cubic foot
    d. 1,200 Btus/cubic foot

2. The third and final time delay used in an automatic transfer switch is used to provide for _____.

    a. source power stabilization
    b. temporary power failures
    c. engine cooldown
    d. brownouts

2.0.0 STORAGE BATTERIES AND UPS SYSTEMS

Objective

Describe battery and UPS system types and explain their maintenance requirements.

a. Describe the different types of batteries used.
b. Explain the maintenance requirements of batteries and their charging considerations.
c. Identify single- and double-conversion UPS systems.

Trade Term

Electrolyte: A substance in which the conduction of electricity is accompanied by chemical action (for example, the paste that forms the conducting medium between the electrodes of a dry cell or storage cell).

Storage batteries are widely used in starting systems for generator sets, emergency lighting, and many other applications. The demand for reliable battery backup is a priority in any standby and/or emergency system. No matter what type of equipment the battery backs up, the cost of failure is unacceptable. It is therefore a top priority to apply, operate, and maintain batteries correctly.

2.1.0 Storage Batteries

NEC Section 701.12(C) specifies that a storage battery of suitable rating and capacity is required to supply and maintain at least $87\frac{1}{2}$% of the system voltage for the total load of circuits supplying legally required standby power for a period of at least $1\frac{1}{2}$ hours. Batteries, whether acid or alkali types, must be designed and constructed to meet the service requirements of emergency service and must be compatible with the charger for that particular installation. Some *NEC®* regulations governing storage batteries used in legally required standby systems are shown in *Figure 6*.

2.1.1 Lead-Acid Batteries

There are many types of lead-acid batteries. Regardless of the design, with all lead batteries the chemical reaction that takes place to store and produce electricity is the same (*Figure 7*). The lead-acid battery is a secondary or storage type of battery. In secondary batteries, electricity or power must be put into the battery to start the chemical reaction. When the battery is being charged, power is stored chemically in the electrodes and electrolyte of the battery. When the battery is called upon to provide power to its load, the resulting chemical reaction is reversed, discharging the battery, and the flow of electrons in the external conductors or posts of the battery is from the negative electrode to the positive electrode. This constitutes the actual direction of the flow of electrons, as opposed to the conventional direction of current flow from positive to negative.

2.1.2 Nickel Cadmium Batteries

Nickel cadmium batteries have been widely used for years. The nickel cadmium battery is the most rugged of the battery technologies, mainly because of its durability in extreme-temperature environments. Its reliability has been demonstrated both in arctic and tropical regions. However, the initial cost of the nickel cadmium battery may be difficult to justify for many applications. It is in the extremes of cold or heat, and where maintenance practices are not all that they might be, that the nickel cadmium battery is well worth the investment.

The performance of the nickel cadmium battery in extreme temperatures results from the chemical composition of its electrolyte. The electrolyte is an alkali solution of potassium hydroxide, which remains unaltered during the charge/discharge reaction. Like the lead-acid battery, the nickel cadmium battery is also a secondary or storage-type battery. The chemical reaction that takes place to store and produce electricity in a nickel cadmium battery is shown in *Figure 8*. Nickel cadmium batteries fall into three main categories: pocket plate, fiber plate, and sintered plate.

2.2.0 Battery Maintenance

Proper maintenance of storage batteries is extremely important in order to have a reliable emergency/standby system. Battery maintenance includes the cells as well as charging systems and ventilation systems.

Battery voltage is computed on the basis of 2.0V per cell for lead-acid types and 1.2V per cell for alkali types.

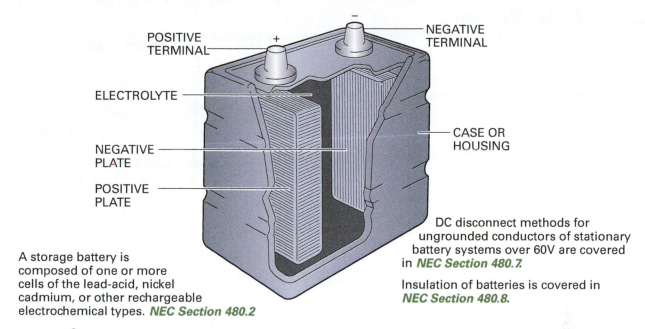

A storage battery is composed of one or more cells of the lead-acid, nickel cadmium, or other rechargeable electrochemical types. **NEC Section 480.2**

DC disconnect methods for ungrounded conductors of stationary battery systems over 60V are covered in **NEC Section 480.7.**

Insulation of batteries is covered in **NEC Section 480.8.**

Figure 6 NEC® requirements governing storage batteries for legally required standby systems.

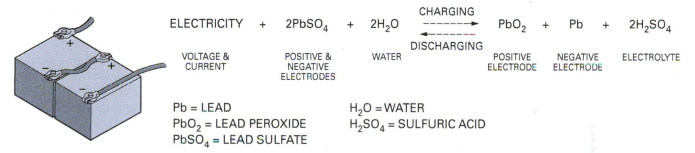

Figure 7 Chemical reaction in a lead-acid battery.

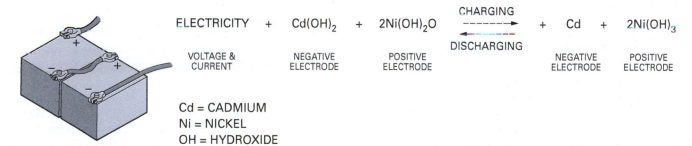

Figure 8 Chemical reaction in a nickel cadmium battery.

2.2.1 Battery Maintenance Overview

It is very important to note the differences in the charging or float voltage for each type of battery. Setting and maintaining proper float voltage is critical to battery life. Overcharging a lead-acid battery will destroy the positive plate, evident by a characteristic dark sediment at the bottom of the cell jar. Undercharging a lead-acid battery will destroy the negative plate, producing a light gray sediment and a crystal-like sparkle on the plates. The nominal open circuit voltage of a lead-acid cell is 2.0V, and of a nickel cadmium cell is 1.3V,

Lead-Acid Battery Construction

The internal construction of a typical standard lead-acid battery is shown here. Note the anti-flame propagation vent caps and the built-in lifting brackets.

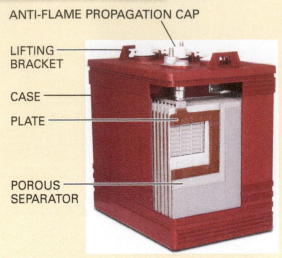

ANTI-FLAME PROPAGATION CAP

LIFTING BRACKET

CASE

PLATE

POROUS SEPARATOR

Figure Credit: Trojan Battery Company

but the charging voltage is higher. An equalizing charge is even higher than a charging float voltage. An equalizing charge is applied to a battery when any individual cell voltage falls below a maintenance-specified minimum voltage. The equalizing overcharge tends to bring all cells back to float voltage when complete.

When a lead-acid battery is being discharged, the sulfuric acid of the battery is consumed and water is produced. During maintenance, this is observed and noted to determine the battery's state of charge. This is done by using a hydrometer to read the specific gravity of the sulfuric acid. The specific gravity of water is 1.0, whereas the specific gravity of sulfuric acid is near 1.15 when a battery is discharged and 1.28 when it is near full charge. Note that measuring the specific gravity on nickel cadmium batteries is not required because the electrolyte (potassium hydroxide) does not change during charge or discharge.

Adverse temperature environments affect the life and performance of lead-acid batteries. High temperatures will reduce the life of the lead-acid battery significantly. A general rule of thumb is that for every 15°F (8.3°C) above 77°F (25°C), the life of the battery will be reduced by 50%. At low temperatures, the performance of discharge quickly falls off, and at extremely low temperatures, the battery can freeze once it goes into discharge.

The nickel cadmium battery provides excellent durability in temperature extremes. Nickel cadmium batteries can operate in temperatures as low as −40°F (−40°C) without damage or the threat of freezing. They have some degradation in life expectancy at high temperatures, but not to the extent of lead-acid batteries. The general rule

of thumb for nickel cadmium batteries is that every 15°F above 77°F (8.3°C above 25°C) results in a 20% reduction in expected life.

Corrosion of the intercell connections in a bank of battery cells causes high resistance and even open circuits when the battery is being discharged under load. These problems have the potential to cause major system failures and battery fires. An intercell resistance test during maintenance using a digital low-resistance ohmmeter is used to detect these high-resistance connections. It is common practice to use a reading of greater than 20% deviation from the average reading to initiate corrective action. Retorque or replace the weak connections upon detection to prevent conduction path failures in the battery bank.

A battery bank can be effectively maintained with routine checks and inspections. Visually inspecting the bank's physical condition for damage or corrosion as well as electrolyte level, along with keeping a monitored record of specific gravity, voltage, and temperature readings is a necessary part of any successful battery maintenance program. These readings will provide a measure of the state of charge of the battery. The state of charge, however, is not a complete indication of battery life or capacity. Battery capacity is the single most important criterion for replacing a battery.

The method of determining battery capacity is by performing a load test to measure the amount of power a fully charged battery is capable of delivering over a specific time period. Batteries are rated in amp-hours. A load test simply measures the percent of time that the battery actually carries the load versus the time the battery is supposed to carry the load based on its amp-hour rating. The test is carried out down to a minimum operating voltage, typically 1.75V per cell. Any battery that exhibits less than 80% time capacity is considered failed and should be replaced. Various types of load boxes and monitors are used to perform load tests.

> **WARNING!**
>
> Batteries are hazardous and must be disposed of properly.

2.2.2 Battery Maintenance Guidelines

Much of the maintenance performed on batteries is done with the batteries in service. Because of this, the methods used should follow the protective procedures outlined in the appropriate standards and the battery manufacturer's instructions. In addition, these methods should preclude circuit interruption or arcing in the vicinity of the battery. One of the reasons for this is that hydrogen is liberated from all vented cell batteries while they are on float or equalize charge, as well as when they are nearing completion of a recharge. Valve-regulated lead-acid cells will also liberate hydrogen whenever their vents open. Therefore, adequate ventilation by natural or mechanical means must be available to minimize hydrogen accumulation. In addition, the battery area should be a mandated nonsmoking area.

> **NOTE**
>
> There are specific manufacturer procedures to be followed for receiving, handling, and installing battery systems and chargers. This includes receiving, inspection, storage, filling of lead-acid batteries, interconnection, and initial charge. Manufacturer services are often used to oversee these steps.

All work performed on a battery shall be done only with the proper and safe tools and with the appropriate protective equipment, including:

- Chemical-resistant goggles and face shield
- Acid-resistant or alkali-resistant gloves
- Chemical-resistant protective aprons and over-shoes
- Portable or stationary water facilities for rinsing eyes and skin in case of contact with electrolyte
- Acid- or alkali-neutralizing solution
- Insulated tools

Sealed Valve-Regulated Lead-Acid (VRLA) Batteries

High-amperage modular battery units are available that incorporate absorbed glass mat (AGM) battery technology with a 20-year design life when used in float service. The units can be connected in parallel and in series to obtain the desired voltage and amperage. The units use sealed valve-regulated lead-acid (VRLA) batteries that usually require no special venting under normal operating conditions. However, explosive gases may be vented during use if the batteries are overcharged, damaged, or subjected to a high-temperature environment.

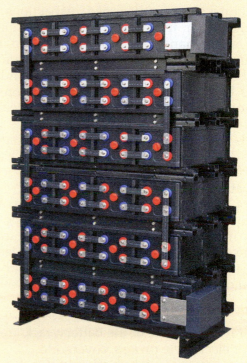

Figure Credit: Storage Battery Systems, LLC.

- Class C fire extinguisher or other type as recommended by the battery manufacturer
- Respirator, if required

> **WARNING!**
>
> Some battery manufacturers do not recommend the use of CO_2 (Class C) fire extinguishers due to the possibility of thermal shock and possible cracking of the battery containers. Always refer to the manufacturer's instructions for safety and maintenance procedures.

Standard UPS Batteries

A UPS battery bank with vented standard lead-acid batteries is shown here. Codes normally require that these types of batteries and battery banks be located in a dedicated fire-rated room vented to the outdoor air and protected by a fire suppression system. Some codes may require a special explosion-resistant room with reinforced walls, ceilings, floors, and steel doors. Continuous outdoor air exchange for the room is monitored, and all battery charging must be shut down automatically if the ventilation fails. UPS units using sealed batteries do not normally require this type of facility.

Figure Credit: Tim Ely

Guidelines for maintenance inspections and tests follow. For more detailed information on battery maintenance scheduling, refer to *IEEE Standard 446*. Anyone performing maintenance should be familiar with the appropriate standards and the battery manufacturer's instructions for the battery to be maintained. This is particularly true for valve-regulated lead-acid cells since some testing is recommended that is designed to determine if dry-out of the cell is occurring. Dry-out can be caused by a number of factors, such as high float voltage and high temperature.

As applicable to the types of batteries involved, the following visual inspections should be performed in accordance with the battery manufacturer's recommendations and/or the facility schedule. A typical maintenance interval for each task is shown here.

Monthly checks:

- General appearance of the battery/rack area
- Dirt or electrolyte on jars, covers, etc.
- Charger voltage and current output
- Electrolyte level (vented lead-acid and vented nickel cadmium cells)
- Jar or cover for cracks or leaks
- Jar or post seals (valve-regulated lead-acid cells)
- Excessive jar or cover distortion (valve-regulated lead-acid cells)
- Evidence of corrosion

Annual checks:

- Flame arrester clogged (vented lead-acid and vented nickel cadmium cells)
- Detailed rack inspection
- Insulating covers on racks

- Seismic rack part and spacers
- Detailed cell inspection (vented lead-acid cells)
- Plates for cracks, sulfate, and hydration (vented lead-acid cells)
- Abnormal sediment accumulation (vented lead-acid cells)
- Jar and post seals (other than valve-regulated lead-acid cells)
- Excessive jar or cover distortion (other than valve-regulated lead-acid cells)
- Excessive gassing (vented lead-acid cells)
- Signs of vibration

As applicable to the types of batteries involved, the following measurements should be performed in accordance with the battery manufacturer's recommendations and/or the user facility schedule. A typical maintenance interval for each task is shown.

Monthly checks:

- Battery float voltage
- Pilot cell voltage (vented lead-acid cells)
- Pilot cell electrolyte temperature (vented lead-acid cells)
- Pilot cell electrolyte specific gravity (vented lead-acid cells)
- Ventilation equipment adequacy

Quarterly checks:

- Individual cell voltage
- Individual cell electrolyte specific gravity (vented lead-acid cells)
- Pilot cell electrolyte temperature (vented nickel cadmium, vented lead-acid, and valve-regulated lead-acid cells)

- Intercell connection resistance (valve-regulated lead-acid cells)
- Cell impedance, conductance, and resistance (valve-regulated lead-acid cells)
- Ambient temperature (vented lead-acid and valve-regulated lead-acid cells)

Annual checks:

- AC ripple current and voltage (valve-regulated lead-acid cells)
- Intercell connection torque (vented nickel cadmium cells)
- Intercell connection resistance (vented lead-acid cells)
- Battery rack torque
- Ground connections

2.2.3 Battery and Battery Charger Operation

In addition to providing voltage to start generator sets as previously described, standby batteries are widely used in many other emergency power applications, particularly to power emergency lighting systems. Standby batteries used in these applications are normally operated in the float mode of operation where the battery, battery charger, and load are connected in parallel (*Figure 9*). The charging equipment should be of a size that provides all the power normally required by the loads plus enough additional power to keep the battery at full charge. Relatively large intermittent loads will draw power from the battery. This power is restored to the battery by the charger when the intermittent load ceases. If AC input power to the system is lost, the battery instantly carries the full load. If the battery and charger are properly matched to the load and to each other, there is no discernible voltage dip when the system goes to full battery operation.

When AC charging power is restored, a constant potential charger will deliver more current than would be necessary if the battery were fully charged. Some constant potential chargers will automatically increase their output voltage to the

equalizer setting after power to the charger is restored. Similarly, a constant current charger may automatically increase its output to the high rate. The battery charger must be sized such that it can supply the load and restore the battery to full charge within an acceptable time. The increased current delivered by a constant potential charger during this restoration period will decrease as the battery approaches full charge.

The battery charger is an important part of the emergency power system, and consideration must be given to redundant chargers on critical systems. The charger should be operated only with the battery connected to the DC bus. Without this connection, excessive ripple on the system could occur, which could affect the connected equipment, causing improper operation or failure of the equipment. If a system must be able to operate without the battery connected, then a device called a battery eliminator should be used. Also note that a battery charger's output must be derated for altitude when it is installed at locations above 3,300' and for temperature when the ambient temperature exceeds 122°F (50°C). The battery charger manufacturer can provide these derating factors.

2.3.0 Uninterruptible Power Supply Systems

A static uninterruptible power supply (UPS) system is an electronically controlled, solid-state power control system designed as an alternate source to provide regulated AC power to critical loads in the event of a partial or total failure of the normal source of power, typically the power supplied by the electric utility. Today, the term *UPS* commonly refers to a system consisting of a dedicated stationary-type battery, rectifier/battery charger, static inverter, and accessories such as a transfer switch. Note that in the past, the term *UPS* was commonly used to describe a type of static inverter often used in large DC systems.

There is a wide diversity in UPS configurations that provide for voltage regulation, line conditioning, lightning protection, redundancy, electromagnetic interference (EMI) management, extended run time, load transfer to other units, and other such features required to protect the critical loads against failure of the normal AC source and against other power system disturbances. UPS systems are available in a number of designs and sizes, ranging from less than 100W to several megawatts. They may provide single-phase or

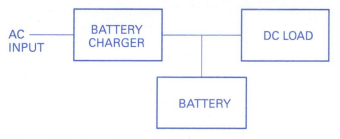

Figure 9 Stationary battery in float operation.

three-phase power at frequencies of 50Hz, 60Hz, or 400Hz. In addition, static UPS systems may provide power of which the output purity may vary from a near-perfect sine wave with less than 1% total harmonic distortion to a power wave that is essentially a square wave. There are two basic types of UPS systems used in a variety of configurations: double-conversion UPS systems and single-conversion UPS systems.

2.3.1 Double-Conversion UPS Systems

In double-conversion UPS systems, the incoming AC power is first rectified and converted to DC (*Figure 10*). The DC is then supplied as input power to a DC-to-AC converter (inverter). The inverter output is AC, which is used to power the critical loads. This type of system is the static electrical equivalent to the motor-generator set. A battery is connected in parallel with the DC input to the inverter and provides continuous power to it any time the incoming line is outside of its specification or fails. Switching to the battery is automatic, with no break in either the input to the inverter or the output from it.

The double-conversion system has several advantages:

- Excellent frequency stability
- High degree of isolation from variations in incoming line voltage and frequency
- Zero transfer time possible
- Quiet operation
- Can provide a sinusoidal output waveform with low distortion

In lower-power UPS applications (0.1kVA to 20kVA), the double-conversion UPS can have some of the following disadvantages:

- Lower efficiency
- Large DC power supply required (typically 1.5 times the full-load rating of the UPS)
- Poor noise isolation from line to load
- Possible shortened service life due to greater heat dissipation

2.3.2 Single-Conversion UPS Systems

In the single-conversion system (*Figure 11*), the incoming line to the UPS is not rectified to produce DC power to provide input to the inverter. The normal AC power is supplied directly to the critical loads through a series inductor or transformer. The normal AC power also supplies a small battery charger used to maintain the UPS batteries in a fully charged condition. Thus, the battery is only used when the inverter requires the battery's output to supplement or replace the normal power source.

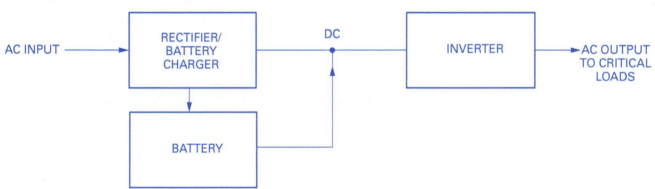

Figure 10 Simplified block diagram of a double-conversion UPS system.

Typical UPS Unit

The UPS unit shown here is a double-conversion 800kVA unit with a 12-pulse rectifier and an insulated gate bipolar transistor, pulsewidth modulated (IGBT-PWM) inverter. It provides computer-grade power with excellent generator compatibility. High efficiency is maintained under light loading by the control system that optimizes the switching frequency for the load level and profile.

Figure Credit: Schneider Electric

Think About It
Sizing a UPS System

What other factors besides the direct load should be taken into account when sizing a UPS system?

Sealed VRLA Battery UPS Systems

The UPS system shown here has operating characteristics similar to the 800kVA unit. The difference is that this particular unit is equipped with sealed VRLA batteries in a battery cabinet attached to the UPS cabinet. Sealed battery UPS systems can be located in most areas without special ventilation requirements.

SEALED VRLA BATTERIES

Figure Credit: Tim Ely

UPS

BATTERY CABINET

Stored-Energy Flywheel UPS Systems

Battery-free, double-conversion UPS systems use a type of motor-generator and flywheel technology to provide 15 seconds of power at 100% rated load during power outages. These types of systems are sometimes referred to as ride-through UPS systems. In most cases, 15 seconds is sufficient to start and stabilize emergency engine-generators, including backup generators, and bring them online for the load, as well as to re-establish the UPS capability. Within the 15-second time frame, there is enough time to wait a few seconds to avoid unnecessary engine-generator starts resulting from momentary outages. The startup time for an engine-generator is normally five to six seconds, with another five or six seconds required to stabilize the output. If the primary engine-generators start, any backup units will be shut down after a predetermined delay. The engine-generator starting voltage and current are normally furnished directly from the UPS so that engine starting batteries can be eliminated.

This type of UPS system is highly reliable and economical because it is free of all batteries that deteriorate over time and because it is a simple rotating electrical device that requires little maintenance. The motor-generator also provides power conditioning by isolating the facility power from the utility power, thus eliminating power line transients and noise.

The unit shown here is a 600kVA UPS with two paralleled motor-generator cabinets. The integrated motor-generator/flywheel devices in the cabinets are very compact, vertically mounted devices that occupy the lower third of each cabinet. The unit furnishes regulated output power with harmonics cancellation. It is microprocessor-controlled and can be scheduled for automatic, non-interfering periodic operational tests and engine-generator operational tests.

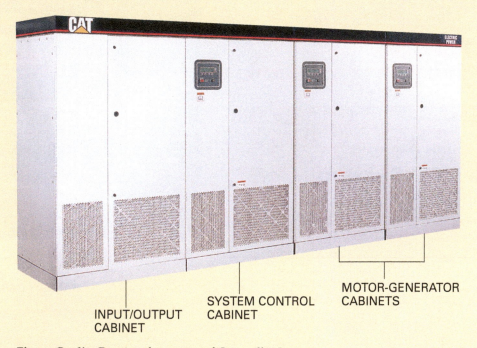

MOTOR-GENERATOR CABINETS

SYSTEM CONTROL CABINET

INPUT/OUTPUT CABINET

Figure Credit: Reprinted courtesy of Caterpillar Inc.

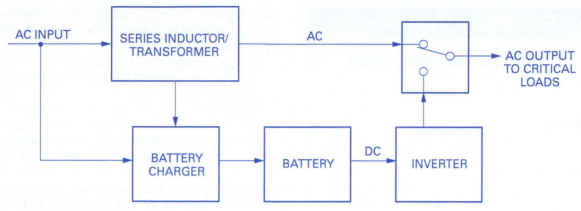

Figure 11 Simplified block diagram of a single-conversion UPS system.

2.0.0 Section Review

1. A storage battery is required to supply and maintain at least 87½% of the system voltage for the total load of circuits supplying legally required standby power for a period of at least _____.

   a. 30 minutes
   b. 45 minutes
   c. 1 hour
   d. 1½ hours

2. A battery charger's output must be derated for temperature when the ambient temperature exceeds _____.

   a. 86°F
   b. 100°F
   c. 122°F
   d. 140°F

3. One disadvantage of a low-power, double-conversion UPS system is _____.

   a. increased distortion
   b. increased transfer time
   c. poor frequency stability
   d. lower efficiency

3.0.0 *NEC*® REQUIREMENTS FOR EMERGENCY SYSTEMS

Objective

Describe the *NEC*® requirements for emergency/standby power and lighting systems.

   a. Describe legally required standby systems.
   b. Describe the alternate power requirements for health care facilities.
   c. Describe the alternate power requirements for places of assembly.
   d. Describe emergency lighting requirements and devices for public buildings.

Several areas of the *NEC*® deal with emergency standby systems, along with the means of generating or producing electricity for such systems. In general, *NEC Article 700* deals with emergency systems. However, there are other *NEC*® requirements that are directly related to the sources of power used for emergency systems. Branch circuits for emergency power are covered to some extent in *NEC Chapters 2 and 3*, transfer switches are dealt with in *NEC Section 700.5*, and storage batteries used for emergency electrical systems are covered in *NEC Section 700.12(C)* and *NEC Article 480*. Consequently, several areas of the *NEC*® will have to be referred to when designing, installing, or maintaining these systems.

3.1.0 Legally Required Standby Systems

NEC Article 701 covers installation requirements of legally required standby electrical systems. The article's intent is to provide a system that is electrically safe and covers circuits and equipment intended to supply, distribute, and control electricity to required facilities for illumination or power (or both) when the normal electrical supply is interrupted.

This article covers only those systems that are permanently installed in their entirety, including the power source.

In general, legally required standby systems are typically installed to serve loads such as heating and refrigeration systems, communication systems, ventilation and smoke removal systems, sewage disposal, lighting systems, and industrial processes that, when stopped during any interruption of the normal electrical supply, could create hazards or hamper rescue or fire-fighting operations.

A legally required standby system must have adequate capacity and rating for the supply of all equipment intended to be operated at one time. Furthermore, it must be suitable for the maximum available fault current at its terminals.

The alternate power source is permitted to supply legally required standby and optional standby system loads when automatic selective load pickup and load shedding are provided as needed to ensure adequate power to the legally required standby circuits.

Transfer equipment, including automatic transfer switches, must be identified for standby use and approved by the authority having jurisdiction. Transfer equipment must also be designed and installed to prevent the inadvertent interconnection of normal and alternate sources of supply in any operation of the transfer equipment. Consequently, transfer equipment, including transfer switches, must operate in such a manner that all ungrounded conductors of one source of supply are completely disconnected before any ungrounded conductors of the second source are connected.

A means may be provided to bypass and isolate the transfer switch equipment. However, where bypass isolation switches are used, inadvertent parallel operation must be avoided (*Figure 12*).

Where possible, audible and visual signal devices are usually provided to indicate the following:

- Derangement of the standby source
- A load on the standby source
- A malfunctioning battery charger

A sign must be placed at the service entrance indicating the type and location of on-site legally required standby power sources. Furthermore, where the grounded conductor that is connected to the emergency source is connected to a grounding electrode conductor at a location remote from the emergency source, there must be a sign at the grounding location to identify all emergency and normal sources connected at that location.

Current must be supplied to legally required standby systems so that in the event of a normal power failure, legally required standby power will be available within the time required for the application, but not to exceed 60 seconds. As described earlier in this module, *NEC Section 701.12(C)* specifies that a storage battery of suitable rating and capacity is required to supply and maintain at least $87\frac{1}{2}$% of system voltage for the total load of the circuits supplying legally

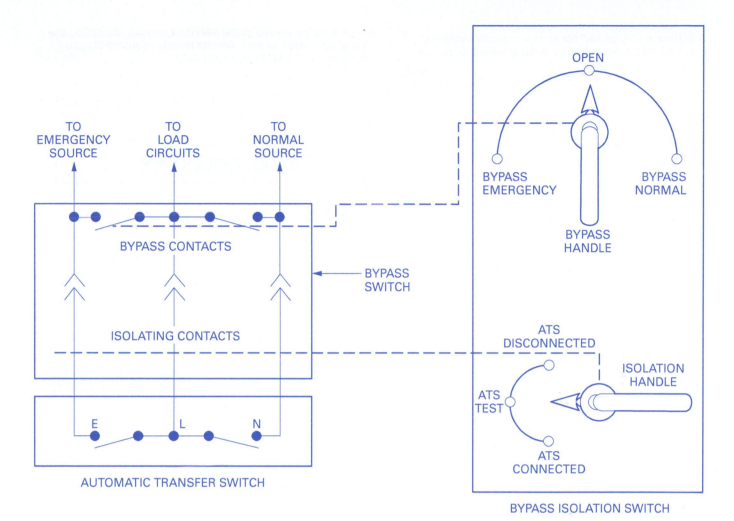

Figure 12 Two-way bypass isolation system.

required standby power for a period of at least 1½ hours. Generator sets suitable for use as a power supply for legally required standby systems were covered earlier in this module. However, a few additional notes are in order. In general, the prime mover (usually an engine) must be acceptable to the authority having jurisdiction and sized in accordance with *NEC Section 701.4*—that is, it must have adequate capacity and rating for the supply of all equipment intended to be operated at one time.

A means must be provided for automatic starting when the normal service fails. Furthermore, the system must be controlled so that the standby power is automatically transferred to all required electrical circuits. A short-time feature permitting a 15-minute setting must be provided to avoid retransfer in case of short-time re-establishment of the normal service.

Refer to *Figure 13* for a summary of other *NEC®* requirements governing legally required standby systems. Refer to the listed *NEC®* sections for greater detail. Also be aware that when other *NEC®* sections apply, they should be adhered to where applicable.

3.2.0 Health Care Facilities

NEC Article 517 requires an alternate power source consisting of one or more generator sets or battery systems (where permitted) to be installed in health care facilities in order to serve the essential electrical system. Essential electrical systems for hospitals must comprise three separate branches capable of supplying a limited amount of lighting and power service that is considered essential for life safety and effective hospital operation during the time the normal electrical ser-

Think About It

Emergency Systems

Why can't other loads also be connected to an emergency system?

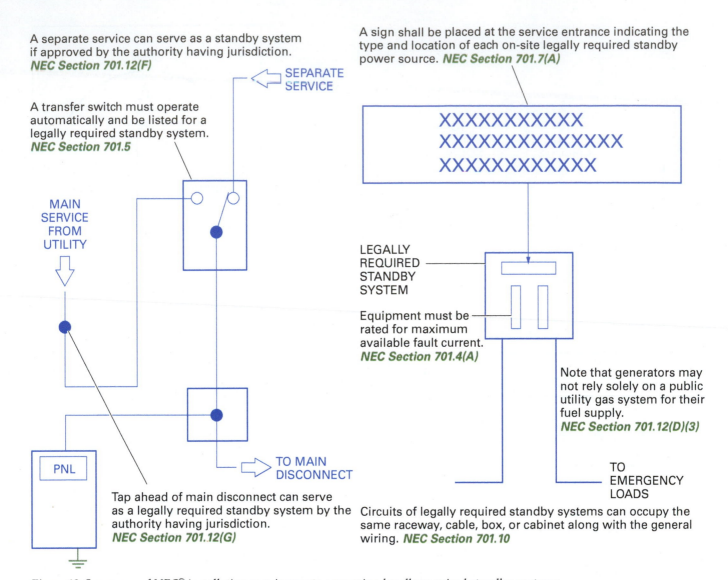

A separate service can serve as a standby system if approved by the authority having jurisdiction. *NEC Section 701.12(F)*

A transfer switch must operate automatically and be listed for a legally required standby system. *NEC Section 701.5*

SEPARATE SERVICE

A sign shall be placed at the service entrance indicating the type and location of each on-site legally required standby power source. *NEC Section 701.7(A)*

XXXXXXXXXX
XXXXXXXXXXXXX
XXXXXXXXXXX

MAIN SERVICE FROM UTILITY

LEGALLY REQUIRED STANDBY SYSTEM

Equipment must be rated for maximum available fault current. *NEC Section 701.4(A)*

Note that generators may not rely solely on a public utility gas system for their fuel supply. *NEC Section 701.12(D)(3)*

PNL

TO MAIN DISCONNECT

Tap ahead of main disconnect can serve as a legally required standby system by the authority having jurisdiction. *NEC Section 701.12(G)*

TO EMERGENCY LOADS

Circuits of legally required standby systems can occupy the same raceway, cable, box, or cabinet along with the general wiring. *NEC Section 701.10*

Figure 13 Summary of *NEC*® installation requirements governing legally required standby systems.

vice is interrupted. These three branches are the life safety branch, the critical branch, and the equipment branch. The equipment branch must supply the major electrical equipment necessary for patient care and basic hospital operation.

Each patient bed location must be supplied by at least two branch circuits, one or more from the critical branch and one or more circuits from the normal electrical system. The receptacles supplied from the critical branch must be identified and must also indicate the panelboard and circuit number supplying them. All receptacles used in patient bed locations must be listed as hospital grade.

The number of transfer switches to be used must be based upon reliability, design, and load considerations. Each branch of the essential electrical system must be served by one or more transfer switches, as shown in *Figure 14*. One transfer switch may serve one or more branches in a facility with a maximum demand on the essential electrical system of 150kVA.

The life safety and critical branches of the essential electrical system must be kept entirely independent of all other wiring and equipment and must not enter the same raceways, boxes, or cabinets with each other or other wiring, except as follows:

- In transfer switches
- In exit or emergency lighting fixtures supplied from two sources
- In a common junction box attached to exit or emergency lighting fixtures supplied from two sources

Testing of the system is required upon installation and periodically thereafter. A written record must be kept of all testing and maintenance.

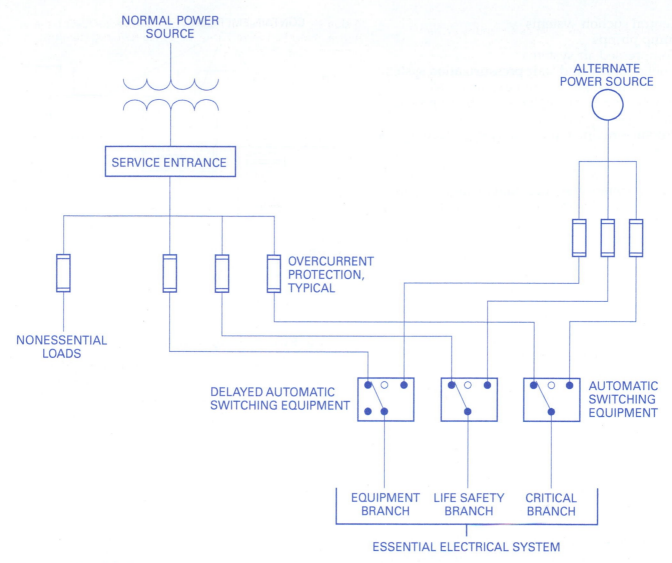

Figure 14 Essential electrical systems must be served by one or more transfer switches.

The wiring of the equipment branch is permitted to occupy the same raceways, boxes, or cabinets of other circuits that are not part of the essential electrical system.

The essential electrical system must have adequate capacity to meet the demand for the operation of all functions and equipment to be served by each branch.

The life safety branch must supply power for the following:

- Lighting required for corridors, passageways, stairways, and landings at all exit doors
- Exit signs
- Alarm and alerting systems
- Communication systems
- Generator set accessories
- Elevators
- Automatic doors

The critical branch must supply power for task illumination, fixed equipment, selected receptacles, and special power circuits serving the following areas and functions related to patient care:

- Anesthetizing locations
- Isolated power systems in special environments
- Patient care spaces
- Nurse call system
- Blood, bone, and tissue banks
- Telephone equipment room and closets
- Task illumination, receptacles, and selected power circuits needed for effective hospital operation

Per *NEC Section 517.35(A) and (B)*, the equipment branch must supply power for the following with a delayed connection:

- Central suction systems
- Sump pumps
- Compressed air systems
- Smoke control and stair pressurization systems
- Kitchen hood supply and exhaust systems
- Supply, return, and exhaust ventilating systems
- Heating equipment
- Pressure maintenance for fire protection systems

NEC Section 700.10(A) also requires labels, signs, or some other permanent marking to be used on all enclosures containing emergency circuits to readily identify them as components of an emergency system (*Figure 15*).

All boxes and enclosures for emergency circuits must be painted red, marked with red labels saying Emergency Circuits, or marked in some other manner clearly identifiable to electricians and maintenance personnel.

3.3.0 Places of Assembly

NEC Article 518 covers assembly occupancies—that is, all buildings or portions of buildings or structures designed or intended for assembly of 100 or more persons. Such facilities include assembly halls, restaurants, dance halls, courtrooms, auditoriums, churches, and similar locations. The installation and control of emergency electrical systems in these facilities must comply with *NEC Article 700, Emergency Systems*. All equipment must be approved for use on emergency systems. The authority having jurisdiction must conduct or witness a test on the complete system upon installation and periodically afterward. A written record must be kept of such tests and maintenance.

All boxes and enclosures (including transfer switches, generators, and power panels) for emergency circuits must be permanently marked so they can be readily identified as a component of an emergency circuit or system. These boxes and enclosures for emergency circuits must be painted red, marked with red labels saying Emergency Circuits, or marked in some other manner clearly identifiable to electricians and maintenance personnel.

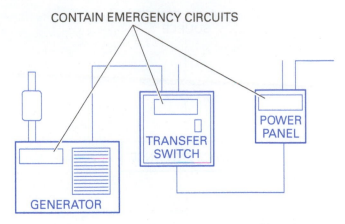

CONTAIN EMERGENCY CIRCUITS

GENERATOR · TRANSFER SWITCH · POWER PANEL

Labels, signs, or some other permanent marking must be used on all enclosures containing emergency circuits to readily identify them as components of an emergency system. *NEC Section 700.10(A)*

Figure 15 Summary of *NEC®* requirements for emergency circuit identification.

3.4.0 Emergency Lighting

The loads on emergency circuits for light and power will vary with the building type. However, in almost all areas throughout the United States, all public buildings must have exit lights and emergency white lights (to enable occupants to see while exiting the building) connected to an emergency electric system. In general, the following items must be fed from an emergency circuit in public buildings:

- All exit lights
- Sufficient emergency white lights
- Fire alarm systems
- Life safety systems
- Sprinkler system water pumps

3.4.1 Battery-Powered Emergency Lighting

Emergency lighting systems are used to provide the required lighting to a facility for egress or other purposes such as security when the normal electrical supply is interrupted. The purpose for this lighting is to illuminate evacuation routes and exits so that the chance of personal injury or loss of life can be minimized in the event of a fire or other emergency that can cause the illumination in an area to be lost or reduced to a level where evacuation may be difficult or impossible.

Batteries are used in self-contained emergency lighting units and central emergency lighting systems; these may be DC batteries only or UPSs.

Specific requirements for emergency lighting are defined in *NEC Sections 700.15 through 700.26*. The required backup time for batteries for emergency lighting is $1\frac{1}{2}$ hours (90 minutes), although some systems may be designed to have a longer backup time. Also, if emergency lighting depends upon the changing of sources (for instance, from AC to DC), the change must occur automatically with no appreciable interruption (less than 10 seconds). In areas that are normally lighted solely by high-intensity discharge (HID) systems, any battery-powered emergency lighting must continue to function once power is restored, until the HID fixtures provide illumination. The system design must also consider a single failure of any individual lighting element, such as a lamp failure, so that any space requiring emergency lighting cannot be left in total darkness.

The emergency lighting fixtures may be incandescent, fluorescent, or LED. They may vary in style from the industrial type suitable for use in classified areas to decorator types used in offices, convention centers, etc. Emergency lighting fixtures also include exit signs. The lamps illuminate stairs, stairway landings, aisles, corridors, ramps, passageways, exit doors, and any other areas requiring emergency lighting.

3.4.2 Emergency Lighting Units

Emergency lighting units, also called unit equipment, are the most popular type of emergency lighting equipment in use today (*Figure 16*). *NEC Sections 701.12(I) and (J)* define the requirements for unit equipment. The unit itself consists of a rechargeable battery, a battery charger, and one or more lamps, provisions for remote lamps, or both. The unit has the necessary controls to maintain the battery fully charged while the normal source of power is available and to automatically energize the lamps upon the loss of the normal power sup-

ply. Most units have the capability for automatic testing and self-diagnostics. The controls will de-energize the lamps immediately or after a time delay (for HID lamps) upon restoration of the normal power supply. If the unit is used with a fluorescent fixture, it will also have a high-frequency inverter and controls intended for that type of fixture.

These units can be supplied with any of the lead-acid or pocket-plate nickel cadmium batteries described earlier in this module. In addition, the units may be supplied with chargers that have an automatic temperature compensation of float/charge voltage. This feature is a must when a valve-regulated lead-acid battery is supplied with the unit. The selection of the battery type depends on the installation itself. For example, pocket-plate nickel cadmium batteries may be used in high-temperature areas, such as near an industrial boiler. On the other hand, valve-regulated lead-acid or sealed sintered-plate batteries may be used in an office area. All of these batteries require periodic maintenance and testing.

The batteries used in emergency lighting units are normally either 6VDC or 12VDC. For the lead-acid cells, the electrolyte specific gravity is normally 1.225 or greater, and the cell capacity is often expressed in ampere-hours at the 10-hour or 20-hour discharge rate of 1.75V per cell at 77°F (25°C).

Figure 16 Typical emergency lighting units.

3.0.0 Section Review

1. A legally required standby system is likely to include loads such as _____.

    a. lighting systems
    b. commercial kitchen equipment
    c. fire alarms
    d. data processing equipment

2. The life safety branch of a hospital essential electrical system supplies power to _____.

    a. patient care areas
    b. nurse call systems
    c. corridor lighting
    d. anesthetizing locations

3. Wiring considerations for a church are covered in _____.

    a. *NEC Article 430*
    b. *NEC Article 480*
    c. *NEC Article 500*
    d. *NEC Article 518*

4. The required backup time for emergency lighting is _____.

    a. 15 minutes
    b. 30 minutes
    c. 60 minutes
    d. 90 minutes

1. In the event of a normal power supply failure, standby power for an emergency system must be available within _____.

   a. 10 seconds of the failure
   b. 30 seconds of the failure
   c. 60 seconds of the failure
   d. 5 minutes of the failure

2. Legally required standby systems are covered in _____.

   a. *NEC Article 700*
   b. *NEC Article 701*
   c. *NEC Article 702*
   d. *NEC Article 705*

3. The field windings of an AC generator are located on the _____.

   a. rotor of the generator
   b. stator of the generator
   c. transfer switch of the generator
   d. exciter of the generator

4. A stator is _____.

   a. the rotating element of a motor or generator
   b. the stationary part of a motor or generator
   c. a device that supplies direct current
   d. not a part of a motor or generator

5. All of the following are parts of an AC generator, *except* a(n) _____.

   a. stator
   b. rotor
   c. transfer switch
   d. exciter

6. When sizing a generator set once the starting and running loads have been determined, it is common to add a margin factor for future expansion of up to _____.

   a. 10%
   b. 15%
   c. 25%
   d. 33%

7. To prevent engine deterioration, avoid running an unloaded engine for longer than _____.

   a. 1 minute
   b. 10 minutes
   c. 30 minutes
   d. 60 minutes

8. The specific gravity of sulfuric acid when a battery is discharged is near _____.

   a. 1.10
   b. 1.15
   c. 1.20
   d. 1.28

9. How often should the intercell connections of a vented lead-acid battery bank be resistance tested with a digital low-resistance ohmmeter?

   a. Monthly
   b. Quarterly
   c. Annually
   d. Bi-annually

10. Each of the following is an advantage of a double-conversion UPS system, *except* _____.

    a. zero transfer time is possible
    b. quiet operation
    c. excellent efficiency in all applications
    d. excellent frequency stability

11. In a single-conversion UPS system, the incoming AC utility power is first applied to the _____.

    a. battery charger
    b. inverter
    c. critical load
    d. battery charger and transformer

12. Which of the following covers the installation of emergency systems?

    a. *NEC Article 518*
    b. *NEC Article 700*
    c. *NEC Article 703*
    d. *NEC Section 250.34*

13. The maximum time allowed for a legally required standby generator system to be started and switched online in the event of a power failure is _____.

    a. 20 seconds
    b. 40 seconds
    c. 60 seconds
    d. 80 seconds

14. What is the minimum number of essential electrical system branches that must be provided in hospitals?

    a. One
    b. Two
    c. Three
    d. Four

15. All of the following may contain both the life safety branch and critical branch wiring of the essential electrical system used in hospitals, *except* _____.

    a. raceways
    b. transfer switches
    c. exit lighting fixtures supplied from two sources
    d. emergency lighting fixtures supplied from two sources

Supplemental Exercises

1. In the event of a normal power supply failure, standby power for a legally required standby system must be available within _____ of the failure.

2. In an AC generator, the voltage induced into each stator winding depends on the strength of the _____ and the _____ with which the lines of force cut across the windings.

3. The excitation system for a typical generator is controlled via a(n) _____.

4. Which of the following time delays are NOT incorporated in an automatic transfer switch control circuit?
   a. Time delay to avoid unnecessary starting and transfer of generator power
   b. Time delay for switching (retransfer) back to the normal power source
   c. Time delay for generator engine cooldown
   d. Time delay for engine cooling fan operation

5. The _____ battery is the most rugged of battery technologies, mainly because of its durability in extreme temperature environments.

6. The nominal open circuit voltage is _____ for a lead-acid cell and _____ for a nickel cadmium cell.

7. An equalizing charge is _____ than a charging float voltage.

8. Any battery that exhibits less than _____ percent time capacity is considered to have failed a load test.

9. For lead-acid batteries, a general rule of thumb is that for every 15°F above 77°F, the life of the battery will be reduced by _____ percent.

10. When AC charging power is restored, a constant potential charger will deliver _____ current than is necessary if the battery were fully charged.

11. Uninterruptible power supplies may provide _____ and _____ power at 60Hz.

12. Where in the *NEC®* would you find information on emergency systems?

    _____.

13. In order to meet the requirements of *NEC Section 701.12(C)*, storage batteries are required to supply and maintain at least _____ percent of the system voltage for the total load of the circuits for a period of at least _____ hour(s).

14. What is the minimum number of people required under *NEC Article 518* for a space to be considered a place of assembly?
    a. 50
    b. 75
    c. 100
    d. 150

15. Which of the following is an approved method of marking boxes and enclosures for emergency circuits?
    a. Painted yellow and white
    b. Painted green with yellow stripes
    c. Painted red
    d. Painted blue and white

Trade Terms Introduced in This Module

Armature: The assembly of windings and metal core laminations in which the output voltage is induced. It is the rotating part in a revolving field generator.

Electrolyte: A substance in which the conduction of electricity is accompanied by chemical action (for example, the paste that forms the conducting medium between the electrodes of a dry cell or storage cell).

Exciter: A device that supplies direct current (DC) to the field coils of a synchronous generator, producing the magnetic flux required for inducing output voltage in the stator coils.

Synchronous generator: An AC generator having a DC exciter. Synchronous generators are used as stand-alone generators for emergency and standby power systems and can also be paralleled with other synchronous generators and the utility system.

Additional Resources

This module presents thorough resources for task training. The following resource material is suggested for further study.

Liquid-Cooled Generator Sets Application Manual, Latest Edition. Minneapolis, MN: Cummins Onan.
National Electrical Code® Handbook, Latest Edition. Quincy, MA: National Fire Protection Association.
OT III Transfer Switches Application Manual, Latest Edition. Minneapolis, MN: Cummins Onan.

Figure Credits

© iStockphoto.com/thexfilephoto
Hubbell Incorporated, Figure 16

Section Review Answer Key

SECTION 1.0.0

Answer	Section Reference	Objective
1. b	1.1.2	1a
2. c	1.2.1	1b

SECTION 2.0.0

Answer	Section Reference	Objective
1. d	2.1.0	2a
2. c	2.2.3	2b
3. d	2.3.1	2c

SECTION 3.0.0

Answer	Section Reference	Objective
1. a	3.1.0	3a
2. c	3.2.0	3b
3. d	3.3.0	3c
4. d	3.4.1	3d

NCCER CURRICULA — USER UPDATE

NCCER makes every effort to keep its textbooks up-to-date and free of technical errors. We appreciate your help in this process. If you find an error, a typographical mistake, or an inaccuracy in NCCER's curricula, please fill out this form (or a photocopy), or complete the online form at **www.nccer.org/olf**. Be sure to include the exact module ID number, page number, a detailed description, and your recommended correction. Your input will be brought to the attention of the Authoring Team. Thank you for your assistance.

Instructors – If you have an idea for improving this textbook, or have found that additional materials were necessary to teach this module effectively, please let us know so that we may present your suggestions to the Authoring Team.

NCCER Product Development and Revision
13614 Progress Blvd., Alachua, FL 32615

Email: curriculum@nccer.org
Online: www.nccer.org/olf

❏ Trainee Guide   ❏ Lesson Plans   ❏ Exam   ❏ PowerPoints   Other _____

Craft / Level: _____ Copyright Date: _____

Module ID Number / Title: _____

Section Number(s): _____

Description: _____

_____

_____

_____

_____

Recommended Correction: _____

_____

_____

_____

_____

Your Name: _____

Address: _____

_____

Email: _____ Phone: _____

This page is intentionally left blank.

Basic Electronic Theory

OVERVIEW

Electronics is a science that deals with the behavior and effect of electron flow in specific substances (which are used to create electronic devices), such as semiconductors. Electronics can be distinguished from electricity in terms of the voltages and currents used. This module explains the function and operation of basic electronic devices, including semiconductors, diodes, rectifiers, and transistors.

Module 26404-20

Objectives

When you have completed this module, you will be able to do the following:

1. Describe electronic fundamentals.
   a. Explain basic electronic theory.
   b. Explain semiconductor fundamentals.
2. Identify and describe semiconductor devices.
   a. Describe the operation and uses of diodes.
   b. Describe the operation and uses of transistors.
   c. Describe the operation and uses of semiconductor switching devices.

Performance Tasks

Under the supervision of the instructor, you should be able to do the following:

1. Test a transistor to determine whether it is an NPN or PNP.
2. Identify the cathode on three different styles of SCRs, using the shape or markings for identification.

Trade Terms

Avalanche breakover
Diac
Field-effect transistor (FET)
Forward bias

Junction field-effect transistor (JFET)
N-type material
P-type material
Rectifier

Reverse bias
Semiconductors
Silicon-controlled rectifier (SCR)
Solid-state device
Triac

Industry Recognized Credentials

If you are training through an NCCER-accredited sponsor, you may be eligible for credentials from NCCER's Registry. The ID number for this module is 26404-20. Note that this module may have been used in other NCCER curricula and may apply to other level completions. Contact NCCER's Registry at 888.622.3720 or go to **www.nccer.org** for more information.

> **NOTE**
>
> NFPA 70®, *National Electrical Code*® and *NEC*® are registered trademarks of the National Fire Protection Association, Quincy, MA.

Contents

1.0.0 Electronic Fundamentals...1

  1.1.0 Basic Electronic Theory 1

  1.2.0 Semiconductor Fundamentals 2

    1.2.1 Conductors 2

    1.2.2 Insulators 3

    1.2.3 Semiconductors 3

2.0.0 Semiconductor Devices 5

  2.1.0 Diodes 5

    2.1.1 Rectifiers 6

    2.1.2 Diode Identification..6

    2.1.3 Light-Emitting Diodes 8

  2.2.0 Transistors 10

    2.2.1 NPN Transistors..11

    2.2.2 PNP Transistors 12

    2.2.3 Identifying Transistor Leads 12

    2.2.4 Field-Effect Transistors..13

    2.2.5 Junction Field-Effect Transistors....................................13

  2.3.0 Semiconductor Switching Devices..13

    2.3.1 Silicon-Controlled Rectifiers 13

    2.3.2 Diacs 14

    2.3.3 Triacs 15

Figures and Tables

Figure 1   Structure of an atom 1

Figure 2   Electron in orbit around the nucleus 1

Figure 3   Atom of copper (a conductor) 3

Figure 4   Atom of an insulator 3

Figure 5   Atom of silicon (a semiconductor).....................................3

Figure 6   Material structure of a diode 5

Figure 7   Forward and reverse bias 6

Figure 8   Half-wave rectifier 7

Figure 9   Transformer center-tap full-wave rectifier 7

Figure 10 Bridge rectifier 7

Figure 11 Three-phase rectifier 8

Figure 12 Methods used to identify diodes 9

Figure 13 Testing a diode with an ohmmeter 9

Figure 14 Process of electroluminescence in an LED........................11

Figure 15 Schematic symbols for an LED and photo diode................11

Figure 16 Material arrangement in an NPN transistor 11

Figure 17 NPN transistor characteristics and schematic symbol 12

Figure 18 PNP transistor characteristics and schematic symbol.........12

Figure 19 Various transistors 12

Figure 20 Lead identification of transistors 13

Figure 21 Junction field-effect transistor symbols 13

Figure 22 SCR characteristics and symbol 14

Figure 23 Basic diac construction and symbols 15

Figure 24 Basic triac construction and symbol 15

Table 1 Diode Characteristics 9

This page is intentionally left blank.

1.0.0 ELECTRONIC FUNDAMENTALS

Objective

Describe electronic fundamentals.
  a. Explain basic electronic theory.
  b. Explain semiconductor fundamentals.

Trade Terms

N-type material: A material created by doping a region of a crystal with atoms from an element that has more electrons in its outer shell than the crystal.

P-type material: A material created when a crystal is doped with atoms from an element that has fewer electrons in its outer shell than the natural crystal. This combination creates empty spaces in the crystalline structure. The missing electrons in the crystal structure are called *holes* and are represented as positive charges.

Semiconductors: Materials that are neither good insulators nor good conductors. Such materials contain four valence electrons and are used in the production of solid-state devices.

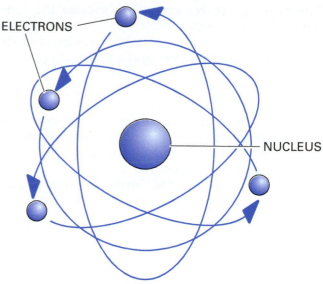

Figure 1 Structure of an atom.

E lectronics is a science that deals with the behavior and effect of electron flow in specific substances such as **semiconductors**, which will be discussed shortly. Electronics can be distinguished from electricity in terms of the voltages and currents used. Electrical circuit AC voltages are measured in hundreds of volts, and currents are measured in tens of amps. In electronic circuits, low-level DC voltages such as 5V, 10V, 15V, and 24V are common. Current is measured in milliamps (mA) and even microamps (μA).

1.1.0 Basic Electronic Theory

The atom consists of a central nucleus composed of protons and neutrons, surrounded by orbiting electrons, as shown in *Figure 1*. The nucleus is relatively large when compared with the orbiting electrons, just as the sun is larger than its orbiting planets.

In an atom, the orbiting electrons are held in place by the electric force between the negative electrons and positive protons, together with the centrifugal force that results from the motion of

the electrons rotating around the nucleus. It is similar to how the Earth's gravity keeps its satellite (the moon) from drifting off into space. The law of charges states that opposite charges attract and like charges repel. The positively charged protons in the nucleus, therefore, attract the negatively charged electrons. If this force of attraction were the only one in effect, the electrons would be pulled closer and closer to the nucleus and eventually be absorbed into the nucleus. However, this force of attraction is balanced by the centrifugal force that results from the motion of the electrons around the nucleus (*Figure 2*). The law of centrifugal force states that a spinning object will pull away from its center point.

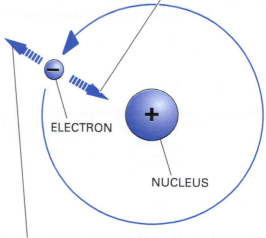

The positive charge of the proton in the nucleus attracts the negatively charged electron.

The force of attraction is balanced by the resulting centrifugal force, holding the electron in orbit.

Figure 2 Electron in orbit around the nucleus.

The faster an object spins, the greater the centrifugal force becomes.

Because the protons and orbital electrons of an atom are equal in number and equal and opposite in charge, they neutralize each other electrically. Each atom is normally electrically neutral—that is, it exhibits neither a positive nor a negative charge. However, under certain conditions, an atom can become unbalanced by losing or gaining electrons. If an atom loses a negatively charged electron, the atom will exhibit a positive charge and is then referred to as a *positive ion*. Similarly, an atom that gains an additional negatively charged electron becomes negatively charged itself and is then called a *negative ion*. In either case, an unbalanced condition is created in the atom, causing the formerly neutral atom to become charged. When one atom is charged and there is an unlike charge in another nearby atom, electrons can flow between the two. This flow of electrons is an electrical current.

1.2.0 Semiconductor Fundamentals

Semiconductors are the basis for what is known as *solid-state electronics*. Solid-state electronics is, in turn, the basis for all modern microminiature electronics, such as the tiny integrated circuits and microprocessor chips used in computers. The ability to control the amount of conductivity in semiconductors makes them ideal for use in integrated circuits. In order to understand how semiconductors work, it is first necessary to review the principles of conductors and insulators.

1.2.1 Conductors

Conductors are so-called because they readily carry electrical current. Good electrical conductors are usually also good heat conductors. In each atom, there is a specific number of electrons that can be contained in each orbit or shell. The outer shell of an atom is the valence shell, and the electrons contained in the valence shell are known as *valence electrons*. If the valence shell is not full, these electrons can be easily knocked out of their orbits and become free electrons.

Conductors are materials that have only one or two valence electrons in their atoms, as shown in *Figure 3*. An atom that has only one valence electron makes the best conductor because the electron is loosely held in orbit and is easily released to create current flow.

Gold and silver are excellent conductors, but they are too expensive to use on a large scale. However, in special applications requiring high conductivity, contacts may be plated with gold or silver. You would be most likely to find such conductors in precision devices where small currents are common and a high degree of accuracy is essential.

Copper is the most widely used conductor because it has excellent conductivity, while being much less expensive than precious metals such as gold and silver. Copper is used as the conductor in most types of wire and provides the current path on printed circuit boards. Aluminum is also used as a conductor, but it is not as good as copper.

Electron Orbits or Shells

Diagrams of atoms (like the one in *Figure 1*) usually depict fixed orbits for electrons. However, this is a classical and convenient method used for notation only and is also inaccurate. Modern quantum mechanics indicates that electrons occupy specific areas of space around an atom, depending on their energy level, rather than fixed orbits. The specific areas are called *orbitals*, and the electrons in a particular orbital space may be passing around the nucleus in any number of random circular or elliptical paths within that space. Each electron at a particular energy level is defined in a wave function called an *atomic orbital*. The wave function is obtained by solving a quantum mechanics equation known as the *Schrödinger equation*. As shown here, when the areas of all the theoretical orbitals for all atoms are combined, they appear as a cloud around the nucleus of an atom with the cloud being denser toward the center.

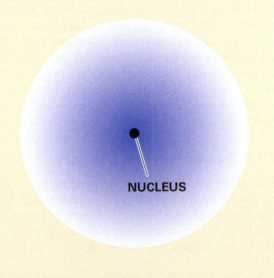

NUCLEUS

The sizes and proportions of atomic particles, as well as the distances between them, are also portrayed inaccurately in most diagrams (like *Figure 1*) for easier notation and visibility. However, just like a roadmap that doesn't depict the actual surface of roadways, these crude schematics suffice in showing general information.

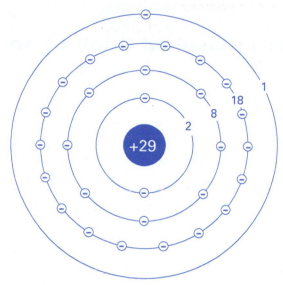

Figure 3 Atom of copper (a conductor).

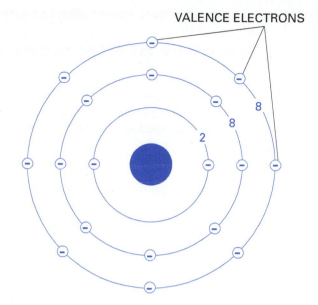

Figure 4 Atom of an insulator.

1.2.2 Insulators

As you already know, insulators are materials that resist (and sometimes totally prevent) the passage of electrical current. Rubber, glass, and some plastics are common insulators. The atoms of insulating materials are characterized by having more than four valence electrons in their atomic structures. *Figure 4* shows the structure of an insulator atom. Note that it has eight valence electrons; this is the maximum number of electrons for the outer shell of an atom. Therefore, this atom has no free electrons and will not easily pass electric current.

1.2.3 Semiconductors

Semiconductors are materials that are neither good conductors nor good insulators. Semiconductors, such as germanium and silicon, have less free electrons in their outer shell than an insulator, but more than a conductor. Silicon (*Figure 5*) is a commonly-used semiconductor because of its ability to withstand heat.

Semiconductors are valuable in electronic circuits because their conductivity can be readily controlled. Since they are electrically neutral, semiconductors can be made to have positive or negative characteristics by adding certain impurities. This process is known as *doping*.

For example, when a substance with five valence electrons (such as arsenic or antimony) is

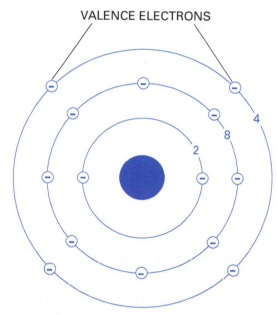

Figure 5 Atom of silicon (a semiconductor).

added to the semiconductor, the semiconductor material will no longer be electrically neutral. Instead, it will take on a negative charge. This is known as an **N-type material**.

When substances with three valence electrons (such as indium or gallium) are added to the semiconductor material, the material takes on a positive charge. This is known as a **P-type material**.

The Number of Electrons in Each Classical Shell of an Atom

The first (innermost) shell of an atom has a maximum of two electrons, and the second shell has a maximum of eight electrons. Unless it is the outermost shell, the third shell can have up to 18 electrons and the fourth and fifth shells can have up to 32 electrons. The maximum number of electrons in the outermost shell cannot exceed eight.

1.0.0 Section Review

1. In an atom, the orbiting electrons are held in place by _____.

   a. the Earth's gravity
   b. an electric force
   c. the law of charges
   d. an electrical current

2. Which of the following is NOT a good conductor?

   a. Gold
   b. Silver
   c. Copper
   d. Glass

2.0.0 SEMICONDUCTOR DEVICES

Objective

Identify and describe semiconductor devices.

a. Describe the operation and uses of diodes.
b. Describe the operation and uses of transistors.
c. Describe the operation and uses of semiconductor switching devices.

Performance Tasks

1. Test a transistor to determine whether it is an NPN or PNP.
2. Identify the cathode on three different styles of SCRs, using the shape or markings for identification.

Trade Terms

Avalanche breakover: A form of electrical current multiplication that can occur in insulating and semiconductor materials. It could be considered a "jail-break" of electron flow, leading to large currents in materials that otherwise would allow little or no current flow.

Diac: A three-layer diode designed for use as a trigger in AC power control circuits, such as those using triacs.

Field-effect transistor (FET): A transistor that controls the flow of current through it with an electric field.

Forward bias: Forward bias exists when voltage is applied to a solid-state device in such a way as to allow the device to conduct easily.

Junction field-effect transistor (JFET): A field-effect transistor formed by combining layers of semiconductor material.

Rectifier: A device or circuit commonly used to change AC voltage into DC voltage, or as a solid-state switch.

Reverse bias: A condition that exists when voltage is applied to a device in such a way that it causes the device to act as an insulator.

Silicon-controlled rectifier (SCR): A device that is used mainly to convert AC voltage into DC voltage. To do so, however, the gate of the SCR must be triggered before the device will conduct current.

Solid-state device: An electronic component made from semiconductor material. Such devices have all but replaced the vacuum tube in electronic circuits.

Triac: A bidirectional triode thyristor that functions as an electrically controlled switch for AC loads.

Diodes and transistors are among the most common types of semiconductor components. Semiconductor diodes are formed by joining two layers of semiconductor material to form a PN junction, while transistors are formed using three layers.

2.1.0 Diodes

A diode is made by joining a piece of P-type material with a piece of N-type material, as shown in *Figure 6*. The contacting surface is called the *PN junction*. Other semiconductor devices are used to perform switching functions.

Diodes allow current to flow in one direction but not in the other. This unidirectional current capability is the distinguishing feature of the diode. The activity occurring at the PN junction of the materials is responsible for the unidirectional characteristic of the diode.

Current does not normally flow across a PN junction. However, if a voltage of the correct polarity is applied, current will flow. This is known

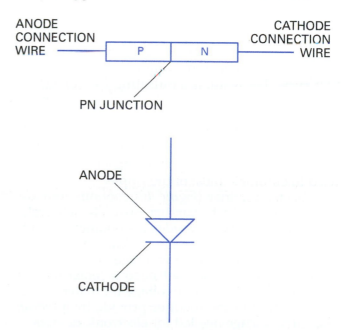

Figure 6 Material structure of a diode.

as **forward bias** and is shown in *Figure 7*. Note that the positive side of the voltage source is connected to the P material (the anode) and the negative side to the N material (the cathode). Also shown in *Figure 7* is a PN junction with **reverse bias**. Note that the polarity of the applied voltage is opposite that of the forward-biased junction. No current will flow in the reverse-biased arrangement unless the reverse-bias voltage exceeds the breakover voltage of the junction. This approach is used to create solid-state switching devices.

Modern systems rely heavily on electronic controls, which use low-level DC voltages. Some systems also use special controls powered by DC motors when very precise control is required. The electricity furnished by the power company is AC; it must be converted to DC to be suitable for most electronic circuits. The process of converting AC to DC is known as *rectification* (or *rectifying*) and is accomplished using **rectifier** diodes.

As shown in *Figure 7*, a diode conducts current only when the voltage at its anode is positive with respect to the voltage at its cathode (forward bias). When the voltage at the anode is negative with respect to the cathode (reverse bias), current will not flow unless the voltage is so high that it overwhelms the diode. Most circuits using diodes are designed so that the diode will not conduct current unless the anode is positive with respect to the cathode.

2.1.1 Rectifiers

Diodes are commonly used to form rectifiers, which convert AC to DC. In a half-wave rectifier (*Figure 8*), the single diode conducts current only when the AC applied to its anode is on its positive half-cycle. The result is a pulsating DC voltage.

A full-wave rectifier with a special center-tapped transformer is shown in *Figure 9*. In this circuit, one of the diodes conducts on each half-cycle of the AC input. This produces a smoother pulsating DC voltage. Filter capacitors can be used to eliminate most of the ripple.

A bridge rectifier (*Figure 10*) contains four diodes, two of which conduct on each half-cycle. The bridge rectifier provides a smoother DC output and is the type most commonly used in electronic circuits. An advantage of the bridge rectifier is that it does not need a center-tapped transformer. A filter and voltage regulator added to the output of the rectifier provide the precise, stable DC voltage needed for electronic devices.

In a three-phase power system, the three-phase rectifier shown in *Figure 11* is used. It contains six diodes to produce high-efficiency DC power.

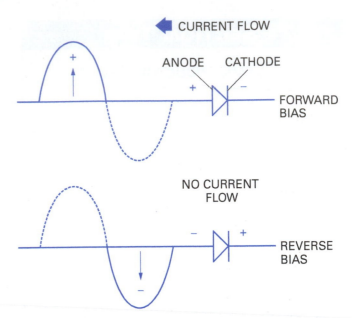

Figure 7 Forward and reverse bias.

2.1.2 Diode Identification

Most manufacturers' **solid-state device** catalogs list hundreds of semiconductor diodes, along with the specifications of each. *Table 1* shows the specifications for two sample diodes. Note that the diodes are designated 1N34A and 1N58A. Manufacturers have agreed upon certain standard designations for diodes having the same characteristics—the same as for wire sizes and types of insulation.

In *Table 1*, the peak inverse voltage (PIV) is the reverse bias at which **avalanche breakover** occurs. The ambient temperature rating is the range of temperatures over which the diode will operate and still maintain its basic characteristics. Forward current values are given for both the average current (that current at which the diode is usually operated) and the peak current (that current which, if exceeded, will damage the diode). The only difference between these two diodes is in the peak inverse voltage. Therefore, the 1N34A could be substituted for the 1N58A in applications involving signals of less than 60V peak-to-peak.

In general, there are two basic types of diodes: the silicon diode and the germanium diode. In most cases, silicon diodes have higher PIV and current ratings and wider temperature ranges than germanium diodes. This is why silicon diodes are the most commonly encountered type in motor and HVAC solid-state controls and power supplies.

PIV ratings for silicon can be in the neighborhood of 1,000V, whereas the maximum value for germanium is closer to 400V. Silicon can be used for applications in which the temperature may rise to about 200°C (392°F), whereas germanium

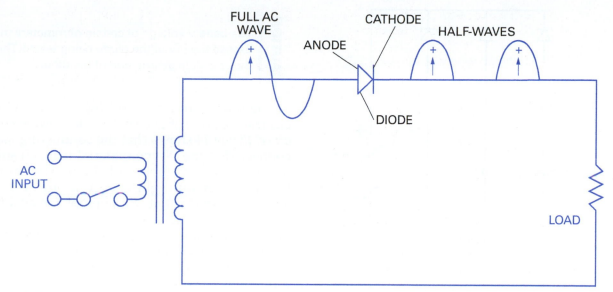

Figure 8 Half-wave rectifier.

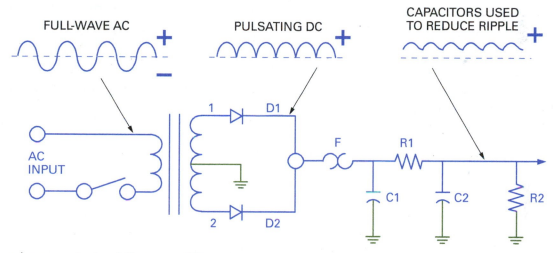

Figure 9 Transformer center-tap full-wave rectifier.

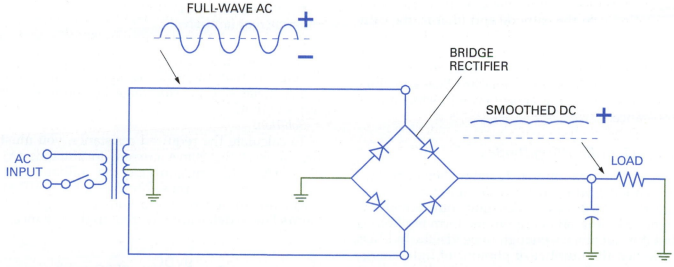

Figure 10 Bridge rectifier.

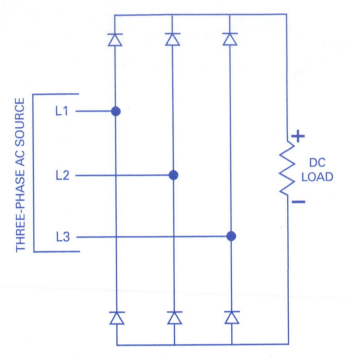

Figure 11 Three-phase rectifier.

has a much lower maximum rating of 100°C (212°F). The disadvantage of silicon, however, is that higher forward bias voltage is required to reach operational conditions.

Diodes have many functions in solid-state devices. For example, silicon diodes are most commonly used for constructing rectifiers, and germanium diodes are used mainly in signal and control circuits.

There are several ways in which diodes are marked to indicate the cathode and anode (*Figure 12*). Note that in some cases diodes are marked with the schematic symbol, or there may be a band at one end to indicate the cathode. Other types of diodes use the shape of the diode housing to indicate the cathode end (that is, the cathode end is either beveled or enlarged to ensure proper identification). When in doubt, the polarity of a diode may be checked with an ohmmeter, as shown in *Figure 13*. A forward bias will show a low resistance; a reverse bias will show a high resistance.

2.1.3 Light-Emitting Diodes

Diodes that give off visible light when energized are called *light-emitting diodes* (*LEDs*). All energized forward-bias diodes emit some amount of light by giving off energy in the form of photons, but it is not always enough to be visible. In LEDs, however, the number of photons of light energy emitted is sufficient to create a visible light source.

The process of giving off light by applying an electrical source of energy is called *electroluminescence*. *Figure 14* shows that the conducting surface connected to the P-type material is much smaller to permit the emergence of the maximum number of photons of light energy in an LED. Note also in *Figure 15* that the symbol for an LED is similar to that of a conventional diode except that an arrow is pointing away from the diode.

Another solid-state device activated by light is known as a *photo diode*. The schematic symbol for the photo diode is exactly like that of a standard LED except that the arrow is reversed, as shown in *Figure 15*. A photo diode must have light in order to operate. It acts similarly to a conventional switch—that is, light turns on the circuit, and the absence of light opens the circuit.

When used in a circuit, an LED is generally operated at about 20mA or less. So, to limit the amount of current that flows through the LED, a current-limiting resistor must be connected in series with the LED. The amount of resistance needed can be calculated using the voltage of the power supply, along with the following form of Ohm's law:

$$R = \frac{E}{I}$$

Where:

R = resistance (ohms)
E = voltage or emf (volts)
I = current (amperes)

Example:
An LED is to be connected to a 9VDC circuit. To ensure that the LED is operated at about 20mA, what size resistor must be connected in series with it?

Solution:
To calculate the required resistance, you must first convert the 20mA current to amperes (A). Since 1A = 1,000mA, divide the milliamps by 1,000 to convert to amps (20mA ÷ 1,000 = 0.02A). Then, use the current and source voltage with Ohm's law to determine the required resistance:

$$R = \frac{9VDC}{0.02A} = 450\Omega$$

Table 1 Diode Characteristics

Type	Peak Inverse Voltage (PIV)	Ambient Temperature Range (°C)	Forward Peak (mA)	Current Average (mA)	Capacitance (µF)
1N34A	60	−50 to +75	150	50	150
1N58A	100	−50 to +75	150	50	150

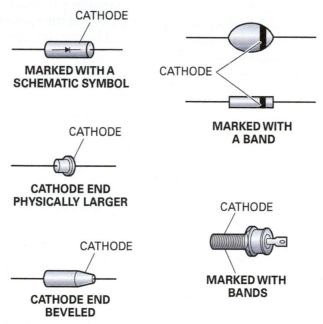

Figure 12 Methods used to identify diodes.

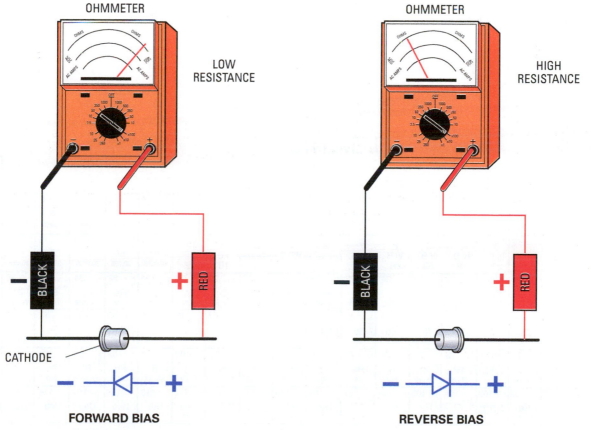

Figure 13 Testing a diode with an ohmmeter.

Therefore, a 450Ω resistor should be used to limit the current flow through the LED. If there is no available resistor of the calculated size, use the closest available size without going under. (The higher the resistance, the more current is limited. Thus, connecting a resistor slightly larger than the calculated value will ensure that the current does not exceed 20mA.)

LEDs are used as pilot lights on electronic equipment and as numerical displays. Many programmable HVAC controls use LEDs to indicate when a process is in operation. LEDs are also used in the opto-isolation circuit of solid-state relays for both motor controls and HVAC control systems.

2.2.0 Transistors

A transistor is made by joining three layers of semiconductor material. Regardless of the type of solid-state device being used, it is made by the joining of P-type and N-type materials to form

The Periodic Table

Some versions of the periodic table of elements, such as the portion of one table shown here, contain a vertical list of numbers beside each element. The list represents the numbers of electrons in each of the classical shells of the atom for the element. The total of the electrons for an element equals the atomic number above the element symbol. The period number for each row corresponds to the number of classical shells that exist for the elements in the row. The group numbers across the top with an A or B are an older method of notation for each column of elements.

The group number A elements are the representative elements that have a corresponding number of electrons in the outer shell with a corresponding valence even though an inner shell may be incomplete (except helium [He]). These representative elements become more stable from left to right as the number of electrons in the outer shell increases.

The group number B elements are the transition elements. These elements sometimes have incomplete inner shells as well as incomplete outer shells. For these elements, the common valence is generally represented by the group number.

Be aware that in quantum mechanics theory, the classical main shells shown in the table are made up of various energy-level sub-shells and the electrons that are allocated to the main shell are assigned to the various sub-shells.

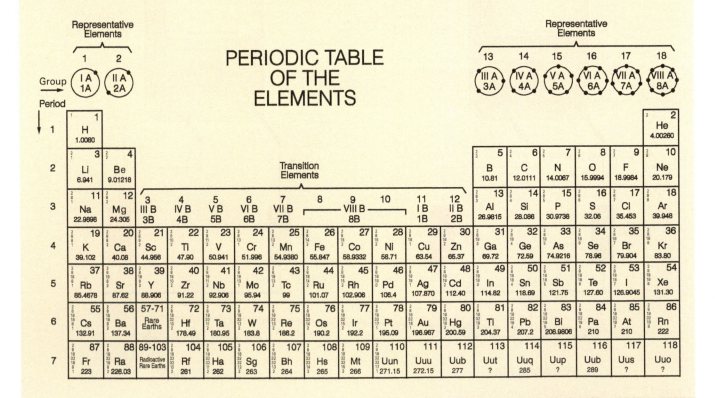

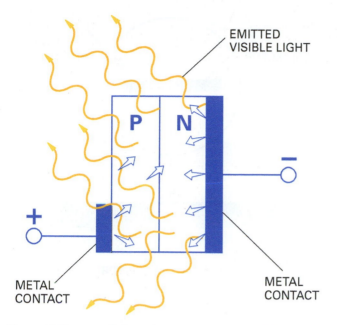

Figure 14 Process of electroluminescence in an LED.

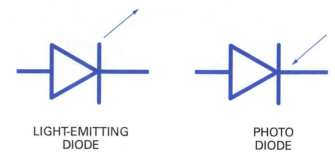

LIGHT-EMITTING DIODE   PHOTO DIODE

Figure 15 Schematic symbols for an LED and photo diode.

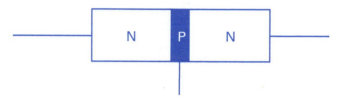

Figure 16 Material arrangement in an NPN transistor.

either an NPN or PNP transistor. Each type has two PN junctions. A diagram of an NPN transistor is shown in *Figure 16*.

Control voltages are applied to the center layer, which is known as the *base*. These voltages control when and how much the transistor conducts. You will not encounter many transistors as discrete components. In today's self-contained electronic integrated circuit devices and microprocessors, there may be thousands of transistors. They are so tiny that they can only be seen with a high-powered microscope.

Years ago, when computers were first invented, they required a roomful of electronic equipment. With the evolution of microminiature circuits, it became possible to perform the same work with a single tiny microprocessor chip containing thousands of microscopic electronic devices such as amplifiers and switches.

2.2.1 NPN Transistors

By sandwiching a very thin piece of P-type germanium between two slices of N-type germanium, an NPN transistor is formed. A transistor made in this way is called a *junction transistor*. The symbol for this type of transistor showing the three elements (emitter, base, and collector) is illustrated in *Figure 17*.

The current flow is in the opposite direction of the arrow. Because the arrow points away from the base, current flows from the emitter to the base or from the N-type material to the P-type material.

Silicon Power Rectifier Diodes

Silicon power rectifier diodes are used in high-current DC applications including uninterruptible power supplies (UPSs), welders, and battery chargers, such as those used for electric forklifts. The stud-mounted silicon power rectifier diode shown here is rated for 300A and a PIV of 600V. The cathode is the threaded section, and it is screwed into or secured with a nut to an insulated heat sink that is at the DC potential. The flexible anode lead is connected to the AC power source. These diodes are also made with the anode as the threaded portion and the cathode as the flexible lead.

Figure Credit: International Rectifier Corp.

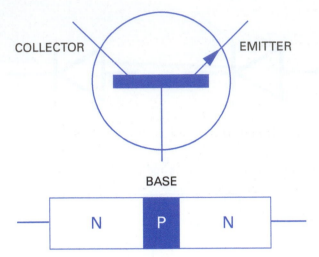

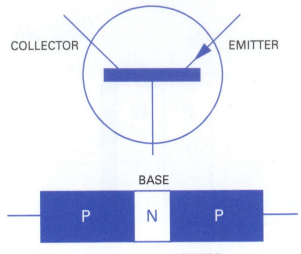

Figure 17 NPN transistor characteristics and schematic symbol.

JUNCTION TRANSISTOR

2.2.2 PNP Transistors

A PNP transistor is formed by placing N-type germanium between two slices of P-type germanium. The schematic symbol for the PNP transistor is almost identical to that of the NPN transistor. The only difference is the direction of the emitter arrow. In the NPN transistor, it points away from the base and in the PNP transistor, it points toward the base.

Electron flow in a PNP transistor is from the N-type germanium to the P-type germanium, and since the arrow in *Figure 18* points toward the base, the electron flow is in the opposite direction of the arrow—from base to emitter. This is the reverse of the NPN transistor discussed previously.

Note also the point-contact transistor shown in *Figure 18*. Junction and point-contact transistors are almost identical in operation. The main difference is in the method of assembly.

2.2.3 Identifying Transistor Leads

Transistors are manufactured in a variety of configurations. Those with studs and heat sinks, as shown in *Figure 19 (A)* and *Figure 19 (B)*, are high-power devices. Those with a small can (top hat) or plastic body, as shown in *Figure 19 (C)*, are low- to medium-power devices.

Whenever possible, transistor casings will have some marking to indicate which leads are connected to the emitter, collector, or base of the transistor. A few of the methods commonly used are indicated in *Figure 20*.

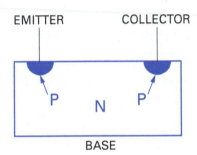

POINT-CONTACT TRANSISTOR

Figure 18 PNP transistor characteristics and schematic symbol.

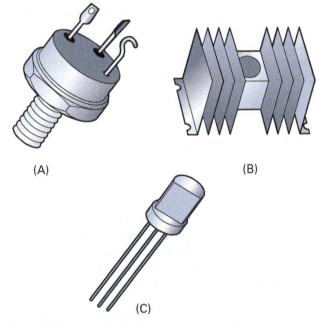

Figure 19 Various transistors.

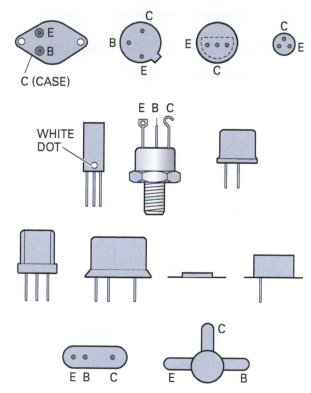

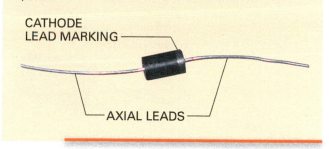

Figure 20 Lead identification of transistors.

2.2.4 Field-Effect Transistors

A **field-effect transistor (FET)** controls the flow of current with an electric field. There are two basic types of FETs: the **junction field-effect transistor (JFET)**, and the metal-oxide semiconductor field-effect transistor (MOSFET). A MOSFET is also referred to as an *insulated gate field-effect transistor (IGFET)*.

2.2.5 Junction Field-Effect Transistors

There are two types of JFETs: an N-channel and a P-channel. *Figure 21* shows the schematic symbols for these two types and also denotes the terms for each JFET lead.

The difference between these two devices is in the polarity of the voltage to which the transistor is connected. In an N-channel device, the drain is connected to the more positive voltage and the source and gate are connected to a more negative voltage. A P-channel device has the drain connected to a more negative voltage, and the source and gate are connected to a more positive voltage.

2.3.0 Semiconductor Switching Devices

The **silicon-controlled rectifier (SCR)**, along with the **diac**, and the **triac**, belongs to a class of semiconductors known as *thyristors*. One common

Axial-Lead Silicon Diodes

One of the most common diodes used is the tubular, axial-lead silicon diode. It is used in many low-voltage rectifiers for electronic circuits. The diode shown here is rated at 1A with a PIV of 1,400V.

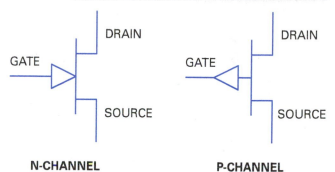

Figure 21 Junction field-effect transistor symbols.

characteristic of all thyristors is that they act as an open circuit until a triggering current is applied to their gate. When that happens, the thyristor acts as a low-resistance current path from anode to cathode. It will continue to conduct, even if the gate signal is removed, until either the current is reduced below a certain level or the thyristor is turned off. These devices are used for a variety of purposes, including:

- Emergency lighting circuits
- Lamp dimmers
- Motor speed controls
- Ignition systems

2.3.1 Silicon-Controlled Rectifiers

The SCR is similar to a diode except that it has three terminals. Like a common diode, current will flow through the SCR in only one direction. However, in addition to needing the correct voltage polarity at the anode and cathode, the SCR also requires a gate voltage of the same polarity as the voltage applied to the anode. When fired, the

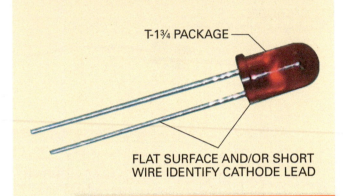
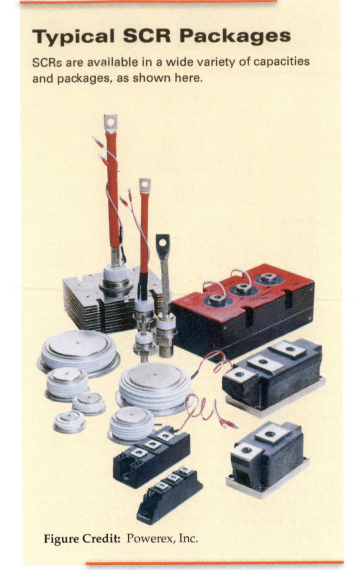
SCR will remain on until the cathode-to-anode current falls below a value known as the *holding current*. When the SCR is off, another positive gate voltage must be applied before it will start conducting again.

The SCR is made from four adjoining layers of semiconductor material in a PNPN arrangement, as shown in *Figure 22*. The SCR symbol is also shown. Note that it is the same as the diode symbol, except for the addition of a gating lead.

There is also a light-activated version of the SCR known as an *LASCR*. Its symbol is the same as that of the regular SCR, with the addition of two diagonal arrows representing light, similar to the photo diode.

The ability of an SCR to turn on at different points in the conducting cycle can be used to vary the amount of power delivered to a load. This type of variable control is called *phase control*. With such control, the speed of an electric motor, the brilliance of a lamp, or the output of an electric resistance heating unit can be controlled.

2.3.2 Diacs

A diac can be thought of as an AC switch. Because the diac is bidirectional, current will flow through it on either half of the AC waveform. One major use of the diac is as a control for a triac. Although the diac is not gated, it will not conduct until the applied voltage exceeds its breakover voltage.

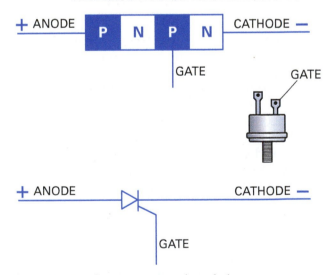

Figure 22 SCR characteristics and symbol.

The basic construction and symbols for a diac are shown in *Figure 23*. Note that the diac has two symbols, either of which may be found on schematic diagrams.

2.3.3 Triacs

The triac can be viewed as two SCRs turned in opposite directions with a common gate (*Figure 24*). It could also be viewed as a diac with a gate terminal added. One important distinction, however, is that the voltage applied across the triac does not have to exceed a breakover voltage in order for conduction to begin.

Like SCRs, triacs are used in phase control applications to control the average power applied to loads. Examples include light dimmers and photocell light switches.

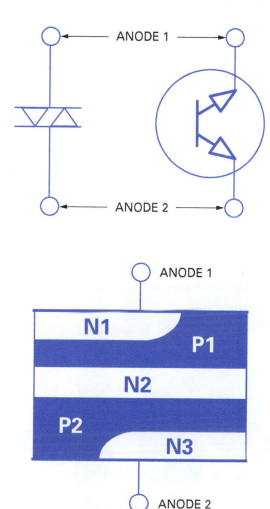

Figure 23 Basic diac construction and symbols.

Typical Low-Voltage Bridge Rectifiers

The low-voltage bridge rectifiers shown here are rated for 4A and 25A with a PIV of 50V. The metallic heat sink mating surface of the 25A rectifier must be coated with a heat sink compound and then bolted to a grounded heat sink. The heat sink compound helps rapidly transmit heat from the rectifier to the heat sink. In these rectifiers, the cases and heat sink are insulated from the AC and DC voltages.

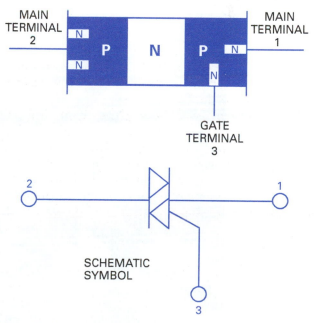

Figure 24 Basic triac construction and symbol.

Common Transistors

Transistors are available in a variety of shapes and sizes, as shown here.

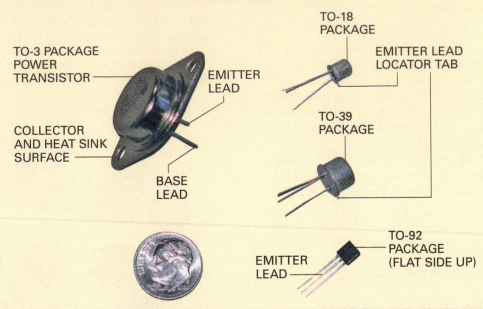

TO-3 PACKAGE POWER TRANSISTOR

EMITTER LEAD

COLLECTOR AND HEAT SINK SURFACE

BASE LEAD

TO-18 PACKAGE

EMITTER LEAD LOCATOR TAB

TO-39 PACKAGE

EMITTER LEAD

TO-92 PACKAGE (FLAT SIDE UP)

Photovoltaic Systems

GOING GREEN

This photovoltaic array is part of a large power system designed to provide 15 megawatts of power. Constructed at Nellis Air Force Base in Nevada, this solar farm covers 140 acres of land and consists of about 70,000 solar panels. Because the panels track the sun throughout the day, they produce about 30 percent more power than fixed systems.

Figure Credit: U.S. Air Force photo by Airman 1st Class Nadine Y. Barclay

Transistor Electron Flow and Current Flow

In a transistor, electrons always flow against the direction of the arrow and current flows in the direction of the arrow.

Typical Triac

A typical triac and its associated heat sink are shown here. This particular version has no visual method of determining lead identification other than the emitter lead locator dot. The manufacturer's literature supplied with the device must be used to identify the lead numbers.

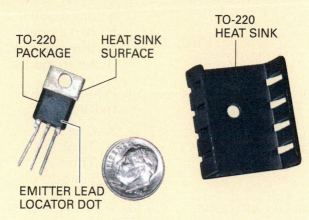

2.0.0 Section Review

1. How many diodes are used in a bridge rectifier?

    a. 2
    b. 3
    c. 4
    d. 5

2. The elements of a transistor are _____.

    a. emitter, base, and collector
    b. emitter, base, and LED
    c. collector, emitter, and LED
    d. collector, base, and rectifier

3. SCRs, diacs, and triacs belong to a class of devices known as _____.

    a. thermistors
    b. transistors
    c. rectifiers
    d. thyristors

1. An electron has a _____.
   a. positive charge
   b. negative charge
   c. neutral charge
   d. variable charge

2. The conductor commonly used to print circuits on a printed circuit board is _____.
   a. aluminum
   b. silver
   c. ceramic
   d. copper

3. An atom with five or more valence electrons is _____.
   a. an insulator
   b. easily combined with water
   c. a semiconductor
   d. a conductor

4. Which of the following can be added as an impurity to a semiconductor material to create an N-type semiconductor?
   a. Arsenic
   b. Germanium
   c. Gallium
   d. Indium

5. A PN junction solid-state device is more commonly known as a _____.
   a. transistor
   b. diode
   c. half-wave rectifier
   d. full-wave rectifier

6. What bias has been applied to a PN junction if the cathode is positive with respect to the anode?
   a. Negative bias
   b. Positive bias
   c. Forward bias
   d. Reverse bias

7. A common purpose of a rectifier (PN junction) is to _____.
   a. convert DC into AC
   b. step down voltage
   c. convert AC into DC
   d. step up voltage

8. A half-wave rectifier is constructed of _____.
   a. one diode
   b. two diodes
   c. three diodes
   d. seven diodes

9. What is the total number of PN junctions in a transistor?
   a. One
   b. Two
   c. Three
   d. Four

10. What are the two basic types of transistors?
    a. AND and OR
    b. ASCII and SCICA
    c. Digital and analog
    d. NPN and PNP

11. What are the two materials normally used to construct transistors?
    a. L and O materials
    b. P and N materials
    c. A and D materials
    d. Silver and aluminum

12. The term *thyristor* includes which of the following groups of components?
    a. Transistors, FETs, and SCRs
    b. Diacs, triacs, and SCRs
    c. Diodes, diacs, and SCRs
    d. Diacs, triacs, and op-amps

13. How many layers of semiconductor material are used to construct an SCR?

    a. One
    b. Two
    c. Three
    d. Four

14. Which of the following is a characteristic of a diac?

    a. It will conduct on either half of the AC waveform.
    b. The voltage applied to the anode and the gate junction must be of the same polarity.
    c. It will continue conducting after the gating voltage is removed.
    d. It has no breakover voltage rating.

15. A triac is equivalent to _____.

    a. three diacs
    b. two SCRs
    c. six SCRs
    d. three diodes

1. The outer shell of an atom is called a(n) _____.

2. Conductor atoms normally have _____ valence electrons.

3. Insulator atoms normally have _____ valence electrons.

4. Two semiconductor materials used in solid-state devices are _____ and _____.

5. A(n) _____ material is a semiconductor that has taken on a negative charge with the addition of other substances.

6. A(n) _____ is made by joining a piece of P-type material with a piece of N-type material.

7. A diode allows flow in _____ direction(s).

8. Diodes are commonly used to convert AC to DC in a process known as _____.

9. Which bias has been applied to a PN junction if the cathode is negative with respect to the anode?

    _____

10. The beveled or enlarged end of a diode marks the _____ end.

11. What are the two basic types of transistors?

    _____

12. The three elements of an NPN transistor are the _____, _____, and _____.

13. Which semiconductor material comprises the base of a PNP transistor?

    _____

14. Thyristors are a class of _____.

15. An SCR is made of _____ adjoining layers of semiconductor material.

Trade Terms Introduced in This Module

Avalanche breakover: A form of electrical current multiplication that can occur in insulating and semiconductor materials. It could be considered a "jail-break" of electron flow, leading to large currents in materials that otherwise would allow little or no current flow.

Diac: A three-layer diode designed for use as a trigger in AC power control circuits, such as those using triacs.

Field-effect transistor (FET): A transistor that controls the flow of current through it with an electric field.

Forward bias: Forward bias exists when voltage is applied to a solid-state device in such a way as to allow the device to conduct easily.

Junction field-effect transistor (JFET): A field-effect transistor formed by combining layers of semiconductor material.

N-type material: A material created by doping a region of a crystal with atoms from an element that has more electrons in its outer shell than the crystal.

P-type material: A material created when a crystal is doped with atoms from an element that has fewer electrons in its outer shell than the natural crystal. This combination creates empty spaces in the crystalline structure. The missing electrons in the crystal structure are called *holes* and are represented as positive charges.

Rectifier: A device or circuit commonly used to change AC voltage into DC voltage, or as a solid-state switch.

Reverse bias: A condition that exists when voltage is applied to a device in such a way that it causes the device to act as an insulator.

Semiconductors: Materials that are neither good insulators nor good conductors. Such materials contain four valence electrons and are used in the production of solid-state devices.

Silicon-controlled rectifier (SCR): A device that is used mainly to convert AC voltage into DC voltage. To do so, however, the gate of the SCR must be triggered before the device will conduct current.

Solid-state device: An electronic component constructed from semiconductor material. Such devices have all but replaced the vacuum tube in electronic circuits.

Triac: A bidirectional triode thyristor that functions as an electrically controlled switch for AC loads.

Additional Resources

This module presents thorough resources for task training. The following resource material is suggested for further study.

Electronics, Allan R. Hambley. Second Edition. 2000. New York, NY: Pearson Education, Inc.
National Electrical Code® Handbook, Latest Edition. Quincy, MA: National Fire Protection Association.

Figure Credits

© iStockphoto.com/Krasyuk, Module Opener

Section Review Answer Key

Section 1.0.0

Answer	Section Reference	Objective
1. b	1.1.0	1a
2. d	1.2.2	1b

Section 2.0.0

Answer	Section Reference	Objective
1. c	2.1.1	2a
2. a	2.2.1	2b
3. d	2.3.0	2c

This page is intentionally left blank.

NCCER CURRICULA — USER UPDATE

NCCER makes every effort to keep its textbooks up-to-date and free of technical errors. We appreciate your help in this process. If you find an error, a typographical mistake, or an inaccuracy in NCCER's curricula, please fill out this form (or a photocopy), or complete the online form at **www.nccer.org/olf**. Be sure to include the exact module ID number, page number, a detailed description, and your recommended correction. Your input will be brought to the attention of the Authoring Team. Thank you for your assistance.

Instructors – If you have an idea for improving this textbook, or have found that additional materials were necessary to teach this module effectively, please let us know so that we may present your suggestions to the Authoring Team.

NCCER Product Development and Revision

13614 Progress Blvd., Alachua, FL 32615

Email: curriculum@nccer.org
Online: www.nccer.org/olf

❏ Trainee Guide   ❏ Lesson Plans   ❏ Exam   ❏ PowerPoints   Other _____

Craft / Level: _____ Copyright Date: _____

Module ID Number / Title: _____

Section Number(s): _____

Description: _____

_____

_____

_____

_____

Recommended Correction: _____

_____

_____

_____

_____

Your Name: _____

Address: _____

_____

Email: _____ Phone: _____

This page is intentionally left blank.

Fire Alarm Systems

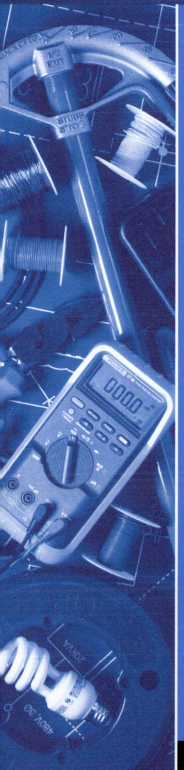

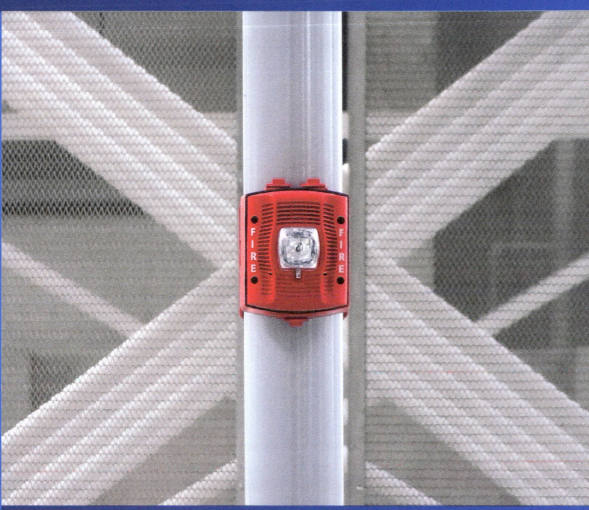

OVERVIEW

Fire alarms provide an essential service that protects both human life and property from the effects of fire. Fire alarms can be complex systems made up of many different technologies. Numerous codes govern fire alarms to ensure that they operate in useful and predictable ways. This module explores the technologies, codes, and wiring approaches used to assemble a fire alarm system. Installation and troubleshooting techniques are also examined.

Module 26405-20

Objectives

When you have completed this module, you will be able to do the following:

1. Describe the various codes and standards that relate to fire alarm systems.
   a. Explain how codes and standards are implemented and list organizations responsible for their creation and maintenance.
   b. List the various NFPA codes and standards that apply to fire alarm systems.
2. Describe the basic types of fire alarm systems and their primary components.
   a. Describe the basic types of fire alarm communication systems.
   b. Describe the primary components of fire alarm systems.
3. Describe fire alarm control panels and their primary features.
   a. Describe fire alarm control panels and their power source requirements.
   b. Explain how users interface with the control panel.
   c. Define and describe initiating circuits and panel outputs.
4. Identify and describe approaches to fire alarm notification and communication/monitoring.
   a. Describe visual and audible notification devices and systems.
   b. Describe important considerations in the use of fire alarm notification signals.
   c. Describe communication and monitoring options for fire alarm systems.
5. Describe fire alarm system installation guidelines and requirements.
   a. Describe the general wiring requirements.
   b. Describe the general installation requirements for wiring and various components.
   c. Describe the installation guidelines for totally protected premises.
   d. Describe the installation guidelines for fire alarm-related systems and devices.
   e. Describe how to troubleshoot fire alarm systems.

Performance Task

Under the supervision of your instructor, you should be able to do the following:

1. Connect selected fire alarm system(s).

Trade Terms

Addressable device	Authority having	Circuit	Digital Alarm
Air sampling detector	jurisdiction (AHJ)	Class A circuit	Communicator System
Alarm	Automatic fire alarm	Class B circuit	(DACS)
Alarm signal	system	Coded signal	Digital Alarm
Alarm verification	CABO	Control unit	Communicator
Americans with	Ceiling	Digital Alarm	Transmitter (DACT)
Disabilities Act (ADA)	Ceiling height	Communicator	End-of-line (EOL) device
Approved	Certification	Receiver (DACR)	Fault
Audible signal	Chimes		Fire

Flame detector
General alarm
Ground fault
Heat detector
Horn
Indicating device
Initiating device
Initiating device circuit
   (IDC)
Labeled
Light scattering
Listed
Maintenance
Multiplexing
National Fire Alarm Code®

National Fire Protection
   Association (NFPA)
Noise
Non-coded signal
Notification device
   (appliance)
Obscuration
Path (pathway)
Photoelectric smoke
   detector
Positive alarm sequence
Power supply
Projected beam smoke
   detector
Protected premises

Public Switched
   Telephone Network
Rate compensation
   detector
Rate-of-rise detector
Remote supervising
   station fire alarm
   system
Reset
Signal
Signaling line circuits
   (SLCs)
Smoke detector
Spacing
Spot-type detector

Stratification
Supervisory signal
System unit
Transmitter
Trouble signal
Visible notification
   appliance
Wavelength
Wide Area Telephone
   Service (WATS)
Zone

Industry Recognized Credentials

If you are training through an NCCER-accredited sponsor, you may be eligible for credentials from NCCER's Registry. The ID number for this module is 26405-20. Note that this module may have been used in other NCCER curricula and may apply to other level completions. Contact NCCER's Registry at 888.622.3720 or go to **www.nccer.org** for more information.

NOTE

NFPA 70®, *National Electrical Code*® and *NEC*® are registered trademarks of the National Fire Protection Association, Quincy, MA.

Contents

1.0.0 Codes and Standards 1

  1.1.0 Standards Organizations 1

  1.2.0 The National Fire Protection Association 2

    1.2.1 NFPA Codes 2

    1.2.2 NFPA Standards 3

2.0.0 Fire Alarm Systems Overview 5

  2.1.0 Fire Alarm Communication Systems 6

    2.1.1 Conventional Hardwired Systems 6

    2.1.2 Multiplex Systems 6

    2.1.3 Addressable Intelligent Systems 7

  2.2.0 Fire Alarm System Equipment 9

    2.2.1 Fire Alarm Initiating Devices 9

    2.2.2 Conventional versus Addressable Commercial Detectors 10

    2.2.3 Automatic Detectors 10

    2.2.4 Heat Detectors 11

    2.2.5 Smoke Detectors 12

    2.2.6 Other Types of Detectors 16

    2.2.7 Manual (Pull Station) Fire Detection Devices 18

    2.2.8 Auto-Mechanical Fire Detection Equipment 20

3.0.0 Control Panels 23

  3.1.0 Control Panel Overview 24

    3.1.1 FACP Primary and Secondary Power 24

    3.1.2 FACP Listings 24

  3.2.0 User Control Points 25

    3.2.1 Keypads 25

    3.2.2 Touch Screens 25

    3.2.3 Telephone/Computer Control 25

  3.3.0 FACP Initiating Circuits and Outputs 25

    3.3.1 Initiating Circuit Zones 26

    3.3.2 Alarm Verification 26

    3.3.3 FACP Labeling 26

    3.3.4 Types of FACP Alarm Outputs 27

4.0.0 Notification, Communication, and Monitoring 28

  4.1.0 Notification Appliances 29

    4.1.1 Visual Notification Devices 29

    4.1.2 Audible Notification Devices 29

    4.1.3 Voice Evacuation Systems 30

  4.2.0 Signal Considerations 30

  4.3.0 Communication and Monitoring 33

    4.3.1 Monitoring Options 33

    4.3.2 Digital Communicators 33

    4.3.3 Cellular Backup 35

Contents (continued)

5.0.0   Installation Guidelines...37

  5.1.0   General Wiring Requirements...37

  5.2.0   Installation Requirements..38

    5.2.1   Access to Equipment..38

    5.2.2   Fire Alarm Circuit Identification................................38

    5.2.3   Power-Limited Circuits in Raceways 38

    5.2.4   Mounting of Detectors 38

    5.2.5   Outdoor Wiring 38

    5.2.6   Fire Seals..39

    5.2.7   Wiring in Air Handling Spaces..................................39

    5.2.8   Wiring in Hazardous Locations.................................40

    5.2.9   Remote Control Signaling Circuits...........................40

    5.2.10 Cables Running Floor to Floor..................................40

    5.2.11 Cables Running in Raceways 40

    5.2.12 Cable Spacing 40

    5.2.13 Elevator Shafts...40

    5.2.14 Terminal Wiring Methods 40

    5.2.15 Conventional Initiation Device Circuits.....................40

    5.2.16 Notification Appliance Circuits 41

    5.2.17 Primary Power Requirements....................................43

    5.2.18 Secondary Power Requirements...............................43

  5.3.0   Total Premises Fire Alarm System Installation 43

    5.3.1   Manual Fire Alarm Box (Pull Station) Installation.....43

    5.3.2   Flame Detector Installation 44

    5.3.3   Smoke Chamber, Smoke Spread, and Stratification 45

    5.3.4   General Precautions for Detector Installation 48

    5.3.5   Spot Detector Installations on Flat, Smooth Ceilings.................49

    5.3.6   Photoelectric Beam Smoke Detector Installations on Flat, Smooth Ceilings...50

    5.3.7   Spot Detector Installations on Irregular Ceilings 51

    5.3.8   Notification Appliance Installation 54

    5.3.9   Fire Alarm Control Panel Installation Guidelines 56

  5.4.0   Fire Alarm-Related Systems..56

    5.4.1   Ancillary Control Relay...56

    5.4.2   Duct Smoke Detectors..57

    5.4.3   Elevator Recall 59

    5.4.4   Special Door Locking Arrangements.........................60

    5.4.5   Suppression Systems..60

    5.4.6   Supervision of Suppression Systems.......................61

  5.5.0   Troubleshooting 62

    5.5.1   Alarm System Troubleshooting Guidelines 62

    5.5.2   Addressable System Troubleshooting Guidelines 65

Figures and Tables

Figure 1   Typical conventional hardwired system 7
Figure 2   Typical multiplex system 7
Figure 3   Typical addressable or analog addressable system 8
Figure 4   Detection versus stages of fires 9
Figure 5   Typical commercial automatic sensors 10
Figure 6   Fusible link detector 11
Figure 7   Quick metal detector 12
Figure 8   Bimetallic detector 12
Figure 9   Rate-of-rise with fusible link detector 13
Figure 10   Rate-of-rise with bimetallic detector 13
Figure 11   Typical photoelectric smoke detector 14
Figure 12   Typical ionization smoke detector 14
Figure 13   Ionization detector 14
Figure 14   Ionization action 15
Figure 15   Light-scattering detector operation 15
Figure 16   Light obscuration principle 15
Figure 17   Typical duct detector installations 16
Figure 18   Cloud chamber smoke detector 17
Figure 19   Rate compensation detector 18
Figure 20   Restorable semiconductor line-type heat detector 18
Figure 21   Non-restorable fusible line-type heat detector 18
Figure 22   UV flame detector (top view) 19
Figure 23   IR flame detector (top view) 19
Figure 24   Typical pull stations 20
Figure 25   Key-operated pull station 20
Figure 26   Wet sprinkler system 21
Figure 27   Dry sprinkler system 22
Figure 28   Typical intelligent, addressable control panel 23
Figure 29   Fire alarm control panel inputs and outputs 23
Figure 30   Alphanumeric keypad and display 25
Figure 31   Typical ceiling-mounted and wall-mounted strobe devices 29
Figure 32   Typical bell with a strobe light 29
Figure 33   Typical horn with a strobe light 30
Figure 34   Voice evacuation system 30
Figure 35   Typical voice evacuation messages 31
Figure 36   Typical speakers 31
Figure 37   RJ31-X connection device 34
Figure 38   Line seizure 34
Figure 39   Cellular backup system 35
Figure 40   Building control circuits 40
Figure 41   Correct wiring for devices with EOL terminations 41
Figure 42   Typical Class B, Style B initiation circuit 41
Figure 43   Typical Class B, Style C initiation circuit 42
Figure 44   Typical Class A, Style D initiation circuit 42
Figure 45   Typical Class A, Style E initiation circuit 42

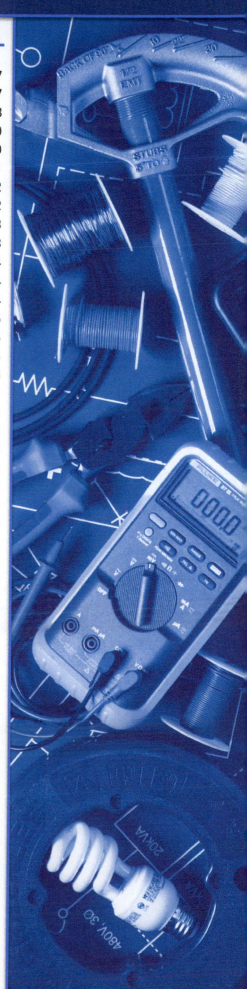

Figures and Tables (continued)

Figure 46 Typical notification appliance circuits.................................. 43
Figure 47 Pull station location and mounting height............................. 44
Figure 48 Maximum horizontal pull station distance from an exit 44
Figure 49 UV detector response 45
Figure 50 IR detector response 45
Figure 51 Reduced spacing required for a barrier................................. 46
Figure 52 Detailed grid definition 47
Figure 53 Smoke spread across a beamed or joist ceiling..................... 47
Figure 54 Smoke stratification 47
Figure 55 Smoke stratification countermeasure 48
Figure 56 Stack effect in a high-rise building....................................... 48
Figure 57 Locating the first column.. 50
Figure 58 Locating the first row 50
Figure 59 First and second detector locations...................................... 50
Figure 60 Maximum straight-line single-beam smoke detector
          coverage (ceiling view).. 51
Figure 61 Coverage for a two-mirror beam detector installation............. 51
Figure 62 Determining degree of slope.. 52
Figure 63 Protection of an FACP 56
Figure 64 Music system control in normal state 57
Figure 65 Non-ducted multiple AHU system 58
Figure 66 Example in which all AHUs require detectors on the
          supply side... 58
Figure 67 Typical remote duct indicator.. 59
Figure 68 Phase 1 recall... 59
Figure 69 Phase 2 recall... 59
Figure 70 Elevator lobby detector.. 61
Figure 71 System troubleshooting 63
Figure 72 Alarm output troubleshooting 64
Figure 73 Auxiliary power troubleshooting 64

Table 1   Heat Detector Temperature Ratings 13
Table 2   Typical Average Ambient Sound Levels 32
Table 3   Typical Sound Loss at 1,000Hz.. 33
Table 4   Secondary Power Duration Requirements.............................. 44
Table 5   Visual Notification Devices Required for Corridors
          Not Exceeding 20' (6 m).. 54
Table 6   Room Spacing for Wall-Mounted Visual
          Notification Appliances.. 55
Table 7   Room Spacing for Ceiling-Mounted Visual
          Notification Appliances.. 55
Table 8   Conversions.. 58

1.0.0 CODES AND STANDARDS

Objective

Describe the various codes and standards that relate to fire alarm systems.

a. Explain how codes and standards are implemented and list organizations responsible for their creation and maintenance.

b. List the various NFPA codes and standards that apply to fire alarm systems.

Trade Terms

Alarm: In fire systems, a warning of fire danger.

Automatic fire alarm system: A system in which all or some of the circuits are actuated by automatic devices, such as fire detectors, smoke detectors, heat detectors, and flame detectors.

Circuit: The conductors or radio channel as well as the associated equipment used to perform a definite function in connection with an alarm system.

Fire: A chemical reaction between oxygen and a combustible material where rapid oxidation may cause the release of heat, light, flame, and smoke.

Initiating device: A manually or automatically operated device, the normal intended operation of which results in a fire alarm or supervisory signal indication from the control unit. Examples of alarm signal initiating devices are thermostats, manual boxes (stations), smoke detectors, and water flow devices. Examples of supervisory signal initiating devices are water level indicators, sprinkler system valve-position switches, pressure supervisory switches, and water temperature switches.

Maintenance: Repair service, including periodic inspections and tests, required to keep the protective signaling system and its component parts in an operative condition at all times. This is used in conjunction with replacement of the system and its components when for any reason they become undependable or inoperative.

National Fire Alarm Code®: This is the update of the NFPA standards book that contains the former NFPA 71, NFPA 72®, and NFPA 74 standards, as well as the NFPA 1221 standard. The _NFAC®_ was adopted and became effective May 1993.

> **National Fire Protection Association (NFPA):** The NFPA administers the development and publishing of codes, standards, and other materials concerning all phases of fire safety.

Fire **alarm** systems can make the difference between life and death. In a **fire** emergency, a quick and accurate response by a fire alarm system can reduce the possibility of deaths and injuries dramatically. However, practical field experience is necessary to master this trade. Remember to focus on doing the job properly at all times. Failure to do so could have tragic consequences.

Always seek out and abide by any National Fire Protection Association (NFPA), federal, state, and local codes that apply to each and every fire alarm system installation.

Different types of buildings and applications will have different fire safety goals. Therefore, each building will have a different fire alarm system to meet those specific goals. An **automatic fire alarm system** can be configured to provide fire protection for life safety, property protection, or mission protection.

Life safety fire protection is concerned with protecting and preserving human life. In a fire alarm system, life safety can be defined as providing warning of a fire situation. This warning occurs early enough to allow notification of building occupants, ensuring sufficient time for their safe evacuation.

Fire alarm systems are generally taken for granted. People go about their everyday business with little or no thought of the safety devices that monitor for abnormal conditions. After all, no one expects a fire to happen. When it does, every component of the life safety fire alarm system must be working properly in order to minimize the threat to life.

1.1.0 Standards Organizations

Fire alarm system equipment and its installation are regulated and controlled by various national, state, and local codes. Industry standards have also been developed as a means of establishing a common level of competency. These codes and standards are set by a variety of different associations, agencies, and laboratories that consist of fire alarm professionals across the country. Depending on the application and installation, you will need to follow the standards established by one or more of these organizations.

In the United States, local and state jurisdictions will select a model code and supporting documents that detail applicable standards. If necessary, they will amend the codes and standards as they see fit. Some amendments may be substantial changes to the code, and in certain situations, the jurisdictions will author their own sets of codes. Once a jurisdiction adopts a set of codes and standards, it becomes an enforceable legal document within that jurisdiction.

The following list details some of the many national organizations responsible for setting the industry standards:

- *Underwriters Laboratories (UL)* – Establishes standards for fire equipment and systems.
- *The National Fire Protection Association (NFPA)* – Establishes standards for fire systems. The association publishes the *National Fire Alarm Code*® (NFPA 72®), the *Life Safety Code*® (NFPA 101®), and the *Uniform Fire Code*™ (NFPA 1). These codes are the primary reference documents used by fire alarm system professionals.
- *Factory Mutual (FM)* – Establishes standards for fire systems.
- *National Electrical Manufacturers' Association (NEMA)* – Establishes standards for equipment.
- *The Federal Bank Protection Act* – Establishes fire alarm equipment and system standards for banks.
- *The Defense Intelligence Agency (DIAM-50-3)* – Establishes standards for military and intelligence installations.
- *The National Institute for Certification in Engineering Technologies (NICET)* – Establishes standardized testing of fire alarm designers and installers.

1.2.0 The National Fire Protection Association

The National Fire Protection Association (NFPA) is responsible for setting the national standards and codes for the fire alarm industry. Consisting of fire alarm representatives from all areas of business and fire protection services, the NFPA responds to the ever-changing needs of society by using a democratic process to form consensus standards that are acceptable to all members. The NFPA reviews equipment and system performance criteria and input from experienced industry professionals to set an acceptable level of protection for both life and property.

Installing and Servicing Fire Alarm Systems

In certain jurisdictions, the person who installs and/or services fire alarm systems must be a licensed fire alarm specialist.

Property and Mission Protection

The design goal of most fire alarm systems is life safety. However, some systems are designed with a secondary purpose of protecting either property or the activities (mission) within a building. The goal of both property and mission protection is the early detection of a fire so that firefighting efforts can begin while the fire is still small and manageable. Fire alarm systems with a secondary goal of property protection are commonly used in museums, libraries, storage facilities, and historic buildings in order to minimize damage to the buildings or their contents. Systems with a secondary goal of mission protection are commonly used where it is essential to avoid business interruptions, such as in hospitals, financial businesses, security control rooms, and telecommunication centers.

1.2.1 NFPA Codes

The following four widely adopted national codes are specified by the NFPA:

- *National Electrical Code*® (NFPA 70®) – The *National Electrical Code*® (*NEC*®) covers all of the necessary requirements for all electrical work performed in a building. The Fire Alarm Systems portion of the code (*NEC Article 760*) details the specific requirements for wiring and equipment installation for fire protection signaling systems. Specifications include installation methods, connection types, circuit identification, and wire types (including gauges and insulation). The *NEC*® places restrictions on the number and types of circuit combinations that can be installed in the same enclosure.
- *National Fire Alarm Code*® (NFPA 72®) – The recommended requirements for installation of fire alarm systems and equipment in residential and commercial facilities are covered in this code. These installation requirements include

NFPA Codes

The three NFPA code books required for installing alarm systems are shown here. Nearly every requirement of NFPA 101®, *Life Safety Code*®, has resulted from the analysis of past fires in which human lives have been lost. The three code books shown were current editions when this module was published. Code books are revised and changed periodically, typically every three or four years. For this reason, you should always make sure that you are using the current edition of any code book.

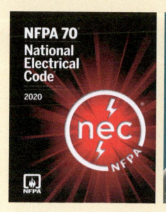 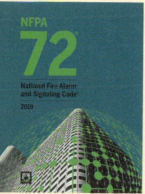 

the **initiating device** (sensor) and notification appliance (visual or audible) of a system. Inspection, testing, and **maintenance** requirements for fire alarm systems and equipment are also covered.

- *Life Safety Code*® (NFPA 101®) – This document is focused on the preservation and protection of human life, as opposed to property. Life safety requirements are detailed for both new construction and existing structures. Specifically, necessary protection for unique building features and construction are detailed. In addition, chapters are organized to explain when, where, and for what applications fire alarm systems are required, the necessary means of initiation and occupant notification, and the means by which to notify the fire department. This code also details any equipment exceptions to these requirements.
- *Uniform Fire Code*™ (NFPA 1) – This code was established to help fire authorities continually develop safeguards against fire hazards.

A chapter of this code is dedicated to fire protection systems. Information and requirements for testing, operation, installation, and periodic preventive maintenance of fire alarm systems are included in this portion of the code.

1.2.2 NFPA Standards

The NFPA also publishes specific standards that are used by fire alarm system professionals. These include:

- NFPA 75, *Standard for Protection of Electronic Computer/Data Processing Equipment*
- NFPA 80, *Standard for Fire Doors and Other Opening Protectives*
- NFPA 90A, *Standard for the Installation of Air Conditioning and Ventilating Systems*
- NFPA 90B, *Standard for the Installation of Warm Air Heating and Air Conditioning Systems*
- NFPA 92A, *Standard for Smoke-Control Systems Utilizing Barriers and Pressure Differences*

1.0.0 Section Review

1. Which organization establishes standards for equipment?

   a. NFPA
   b. DIAM-50-3
   c. NEMA
   d. NICET

2. NFPA 101® is also known as the _____.

   a. *NEC*®
   b. *Life Safety Code*®
   c. *Uniform Fire Code*®
   d. *NFAC*®

2.0.0 FIRE ALARM SYSTEMS OVERVIEW

Objective

Describe the basic types of fire alarm systems and their primary components.

a. Describe the basic types of fire alarm communication systems.
b. Describe the primary components of fire alarm systems.

Trade Terms

Addressable device: A fire alarm system component with discrete identification that can have its status individually identified or that is used to individually control other functions.

Air sampling detector: A detector consisting of piping or tubing distribution from the detector unit to the area or areas to be protected. An air pump draws air from the protected area back to the detector through the air sampling ports and piping or tubing. At the detector, the air is analyzed for fire products.

Alarm signal: A signal indicating an emergency requiring immediate action, such as an alarm for fire from a manual station, water flow device, or automatic fire alarm system.

Ceiling: The upper surface of a space, regardless of height. Areas with a suspended ceiling would have two ceilings: one visible from the floor, and one above the suspended ceiling.

Class A circuit: Class A refers to an arrangement of supervised initiating devices, signaling line circuits, or indicating appliance circuits (IAC) that prevents a single open or ground on the installation wiring of these circuits from causing loss of the system's intended function. It is also commonly known as a *four-wire circuit*.

Class B circuit: Class B refers to an arrangement of initiating devices, signaling lines, or indicating appliance circuits that does not prevent a single open or ground on the installation wiring of these circuits from causing loss of the system's intended function. It is commonly known as a *two-wire circuit*.

End-of-line (EOL) device: A device used to terminate a supervised circuit. An EOL is normally a resistor or a diode placed at the end of a two-wire circuit to maintain supervision.

Flame detector: A device that detects the infrared, ultraviolet, or visible radiation produced by a fire. Some devices are also capable of detecting the flicker rate (frequency) of the flame.

Heat detector: A device that detects abnormally high temperature or rate-of-temperature rise.

Horn: An audible signal appliance in which energy produces a sound by imparting motion to a flexible component that vibrates at some nominal frequency.

Light scattering: The action of light being reflected or refracted off particles of combustion, for detection in a modern-day photoelectric smoke detector. This is called the *Tyndall effect*.

Listed: Equipment or materials included in a list published by an organization acceptable to the authority having jurisdiction that is concerned with product evaluation and whose listing states either that the equipment or materials meets appropriate standards or has been tested and found suitable for use in a specified manner.

Multiplexing: A signaling method that uses wire path, cable carrier, radio, fiber optics, or a combination of these techniques, and characterized by the simultaneous or sequential (or both simultaneous and sequential) transmission and reception of multiple signals in a communication channel including means of positively identifying each signal.

Notification device (appliance): Any audible or visible signal employed to indicate a fire, supervisory, or trouble condition. Examples of audible signal appliances are bells, horns, sirens, electronic horns, buzzers, and chimes. A visible indicator consists of an incandescent lamp, strobe lamp, mechanical target or flag, meter deflection, or the equivalent. Also called an *indicating device*.

Obscuration: A reduction in the atmospheric transparency caused by smoke, usually expressed in percent per foot.

Path (pathway): Any conductor, optic fiber, radio carrier, or other means for transmitting fire alarm system information between two or more locations.

Photoelectric smoke detector: A detector employing the photoelectric principle of operation using either the obscuration effect or the light-scattering effect for detecting smoke in its chamber.

Power supply: A source of electrical operating power, including the circuits and terminations connecting it to the dependent system components.

Projected beam smoke detector: A type of photoelectric light-obscuration smoke detector in which the beam spans the protected area.

Rate compensation detector: A device that responds when the temperature of the air surrounding the device reaches a predetermined level, regardless of the rate of temperature rise.

Rate-of-rise detector: A device that responds when the temperature rises at a rate exceeding a predetermined value.

Reset: A control function that attempts to return a system or device to its normal, non-alarm state.

Signal: A status indication communicated by electrical or other means.

Signaling line circuits (SLCs): A circuit or path between any combination of circuit interfaces, control units, or transmitters over which multiple system input signals or output signals (or both input signals and output signals) are carried.

Smoke detector: A device that detects visible or invisible particles of combustion.

Spacing: A horizontally measured dimension related to the allowable coverage of fire detectors.

Supervisory signal: A signal indicating the need for action in connection with the supervision of guard tours, the fire suppression systems or equipment, or the maintenance features of related systems.

Transmitter: A system component that provides an interface between the transmission channel and signaling line circuits, initiating device circuits, or control units.

Trouble signal: A signal initiated by the fire alarm system or device that indicates a fault in a monitored circuit or component.

Wavelength: The distance between peaks of a sinusoidal wave. All radiant energy can be described as a wave having a wavelength. Wavelength serves as the unit of measure for distinguishing between different parts of the spectrum. Wavelengths are measured in microns, nanometers, or angstroms.

Zone: A defined area within the protected premises. A zone can define an area from which a signal can be received, an area to which a signal can be sent, or an area in which a form of control can be executed.

Fire alarm systems are primarily designed to detect and warn of abnormal conditions, alert the appropriate authorities, and operate the necessary facility safety devices to minimize fire danger. Through a variety of manual and automatic system devices, a fire alarm system links the sensing of a fire condition with people inside and outside the building. The system communicates to fire professionals that action needs to be taken.

2.1.0 Fire Alarm Communication Systems

Although specific codes and standards must always be followed when designing and installing a fire alarm system, many different types of systems that employ different types of technology can be used. Most of these types of fire alarm systems are categorized by the means of communication between the detectors and the fire alarm control panel (FACP). Three major types of fire alarm systems are:

- Conventional hardwired
- Multiplex
- Addressable intelligent

2.1.1 Conventional Hardwired Systems

The simplest of all fire alarm systems is a hardwired system using a conventional initiation device, such as a heat detector, smoke detector, or pull station, and a notification device (appliance), such as a bell, horn, or light. Large buildings or areas that are being protected are usually divided into more than one zone to identify the specific area where a fire is detected. A conventional hardwired system is limited to zone detection only, with no means of identifying the specific detector that initiated the alarm. A typical hardwired system might look like *Figure 1* with either two- or four-wire initiating or notification device circuits. In two-wire circuits, power for the devices is superimposed on the alarm circuits. In four-wire circuits, the operating power is supplied to the devices separately from the signal or alarm circuits. In either two- or four-wire systems, the end-of-line (EOL) device is used by the FACP to monitor circuit integrity.

2.1.2 Multiplex Systems

Multiplex systems are similar to hardwired systems in that they rely on zones for fire detection. The difference, however, is that multiplexing allows multiple signals from several sources to be sent and received over a single communication line. Each signal can be uniquely identified. This results in reduced control equipment, less wiring infrastructure, and a distributed power supply. *Figure 2* shows a simplified example of a multiplex system.

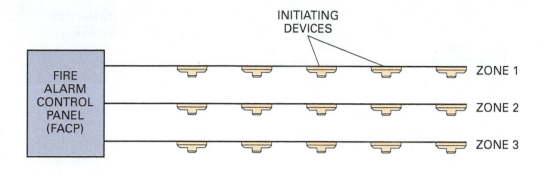

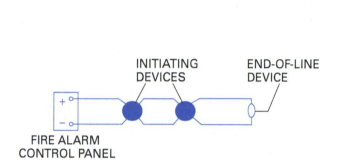

TWO-WIRE INITIATING CIRCUITS

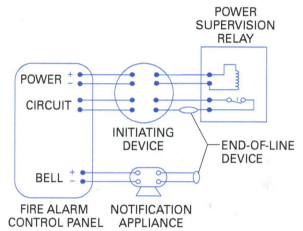

FOUR-WIRE INITIATING CIRCUITS

Figure 1 Typical conventional hardwired system.

2.1.3 Addressable Intelligent Systems

Two different versions of addressable intelligent systems are available. They are addressable and analog addressable systems. Analog addressable systems are more sophisticated than addressable systems.

An addressable system uses advanced technology and detection equipment for discrete identification of an **alarm signal** at the detector level. An addressable system can pinpoint an alarm location to the precise physical location of the initiating detector. The basic idea of an addressable system is to provide identification or control of individual initiation, control, or notification devices on a common circuit. Each component on **signaling line circuits (SLCs)** has an identification number or address. The addresses are usually assigned using switches or other similar

means. *Figure 3* is a simplified representation of an addressable or analog addressable fire alarm system.

The fire alarm control panel (FACP) constantly polls each device using a signaling line circuit (SLC). The response from the device being polled verifies that the wiring **path (pathway)** is intact (wiring supervision) and that the **addressable device** is in place and operational. Most addressable systems use at least three states to describe the status of the device: normal, trouble, and alarm. Smoke detection devices make the decision internally regarding their alarm state just like conventional smoke detectors. Output devices, like relays, are also checked for their presence and in some cases for their output status. Notification output modules also supervise the wiring to the horns, strobes, and other devices,

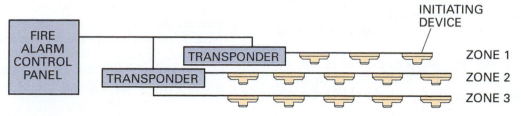

Figure 2 Typical multiplex system.

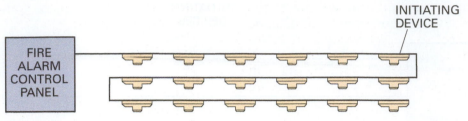

Figure 3 Typical addressable or analog addressable system.

as well as the availability of the power needed to run the devices in case of an alarm. When the FACP polls each device, it also compares the information from the device to the system program. For example, if the program indicates device 12 should be a contact transmitter but the device reports that it is a relay, a problem exists that must be corrected. Addressable fire alarm systems have been made with two, three, and four conductors. Generally, systems with more conductors can handle more addressable devices. Some systems may also contain multiple SLCs. These are comparable to multiple zones in a conventional hardwired system.

Analog systems take the addressable system capabilities much further and change the way the information is processed. When a device is polled, it returns much more information than a device in a standard addressable system. For example, instead of a smoke detector transmitting that it is in alarm status, the device actually transmits the level of smoke or contamination present to the fire alarm control panel. The control panel then compares the information to the levels detected in previous polls. A slow change in levels (over days, weeks, or months) indicates that a device is dirty or malfunctioning. A rapid change, however, indicates a fire condition. Most systems have the capability to compensate for the dirt buildup in the detectors. The system will adjust the detector sensitivity to the desired range. Once the dirt buildup exceeds the compensation range, the system reports a trouble condition. The system can also administer self-checks on the detectors to test their ability to respond to smoke. If the airflow around a device is too great to allow proper detection, some systems will generate a trouble report.

The information in some systems is transmitted and received in a totally digital format. Others transmit the polling information digitally but receive the responses in an analog current-level format.

The panel, not the device, performs the actual determination of the alarm state. In many systems, the light-emitting diode (LED) on the detector is turned on by the panel and not by the detector. This ability to make decisions at the panel also allows the detector sensitivity to be adjusted at the panel. For instance, an increase in the ambient temperature can cause a smoke detector to become more sensitive, and the alarm level sensitivity at the panel can be adjusted to compensate. Sensitivities can even be adjusted based on the time of day or day of the week. Other detection devices can also be programmed to adjust their own sensitivity.

The ability of an analog addressable system to process more information than the three elementary alarm states found in simpler systems allows the analog addressable system to provide pre-alarm signals and other information. In many devices, five or more different signals can be received.

Most analog addressable systems operate on a two-conductor circuit. Most systems limit the number of devices to about one hundred. Because of the high data rates on these signaling line circuits, capacitance also limits the conductor lengths. Always follow the manufacturer's installation instructions to ensure proper operation of the system.

In addressable or analog addressable systems, there are two basic wiring or circuit types (classes). Class B is the most common and has six different styles. Class A, with four styles, provides additional reliability but is not normally required and is generally more expensive to install. Various codes will address what circuit types are required. The performance requirements for each of the SLC styles is detailed in NFPA 72®. These basic wiring/circuit types are described as follows:

- **Class B circuit** – A Class B signaling line circuit for an addressable system essentially requires that two conductors reach each device on the circuit by any means as long as the wire type and physical installation rules are followed. It does not require wiring to pass in and out of each device in a series arrangement. With a circuit capable of 100 devices, it is permissible to go in 100 different directions from the panel. Supervision occurs because the fire alarm control panel polls and receives information from each device. The route taken is not important. A break in the wiring will result in the loss of communication with one or more devices.

- **Class A circuit** – A Class A signaling line circuit for an addressable system requires that the conductors loop into and out of each device. At the last device, the signaling line circuit is returned to the control panel by a different route. The control panel normally communicates with the devices via the outbound circuit but has the ability to communicate to the back side of a break through the return circuit. The panel detects the fact that a complete loop no longer exists and shifts the panel into Class A mode. All devices remain in operation. Some systems even have the ability to identify which conductor has broken and how many devices are on each side of the break.

- *Hybrid circuit* – A hybrid system may consist of a Class A main trunk with Class B spur circuits in each area. A good example would be a Class A circuit leaving the control panel, entering a junction box on each floor of a multistory building, and then returning to the control panel by a different route. The signaling line circuits on the floors are wired as Class B from the junction box. This provides good system reliability and keeps costs down.

2.2.0 Fire Alarm System Equipment

The equipment used in fire alarm systems is generally held to higher standards than typical elec-

trical equipment. The main components of a fire alarm system include:

- Alarm initiating devices
- Control panels
- Primary (main) and secondary (standby) power supplies
- Notification appliances
- Communications and monitoring

2.2.1 Fire Alarm Initiating Devices

Fire alarm systems use initiating devices to report a fire and provide supervisory or trouble reports. Some of the initiating devices used to trigger a fire alarm are designed to sense the signs of fire automatically (automatic sensors). Some report level-of-fire conditions only. Others rely on people to see the signs of fire and then activate a manual device. Automatic sensors (detectors) are available that sense smoke, heat, and flame. Manual initiating devices are usually some form of pull station.

Figure 4 is a graphic representation of the application of various detectors for each stage of a fire. However, detection may not occur at any specific point in time within each stage, nor will the indicated sensors always provide detection in the stages represented. For example, alcohol fires can produce flame followed by heat while producing few or no visible particles.

Besides triggering an alarm, some fire alarm systems use the input from one or more detectors or pull stations to trigger fire suppression equipment such as a carbon dioxide (CO_2), dry chemical, or water deluge system.

It is mandatory that devices used for fire detection be **listed** for the purpose for which they will be used.

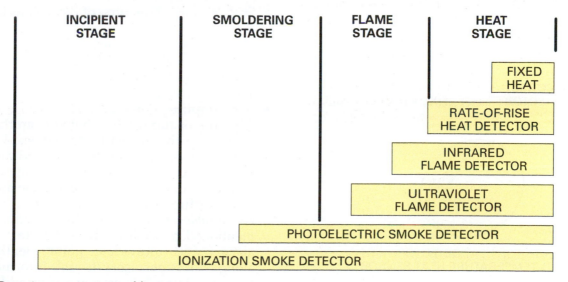

Figure 4 Detection versus stages of fires.

Common UL listings for fire detection devices are as follows:

- *UL 38* – Manually actuated signaling box
- *UL 217* – Single- or multiple-station smoke detector
- *UL 268* – System-type smoke detector
- *UL 268A* – Duct smoke detector
- *UL 521* – Heat detector

2.2.2 Conventional versus Addressable Commercial Detectors

Figure 5 shows some typical commercial detectors. Commercial detectors differ from stand-alone residential detectors in that a number of commercial detectors are usually wired in parallel and connected to a fire alarm control panel. In a typical conventional commercial fire alarm installation, a number of separate zones with multiple automatic sensors are used to partition the installation into fire zones.

IONIZATION OR PHOTOELECTRIC SMOKE DETECTOR

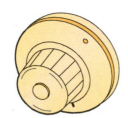

PHOTOELECTRIC SMOKE DETECTOR WITH FIXED-TEMPERATURE HEAT SENSOR

RATE-OF-RISE HEAT DETECTOR WITH FIXED-TEMPERATURE HEAT SENSOR

Figure 5 Typical commercial automatic sensors.

As described previously, newer commercial fire systems use addressable automatic detectors, pull stations, and notification appliances to supply a coded identification signal. The systems also provide an individual supervisory or trouble signal to the FACP, as well as any fire alarm signal, when periodically polled by the FACP. This provides the fire control system with specific device location information to pinpoint the fire, along with detector, pull station, or notification appliance status information. In many new commercial fire systems, analog addressable detectors (non-automatic) are used. Instead of sending a fire alarm signal when polled, these devices communicate information about the fire condition (level of smoke, temperature, etc.) in addition to their identification, supervisory, or trouble signals. In the case of these detectors, the control panel's internal programming analyzes the fire condition data from one or more sensors, using recent and historical information from the sensors, to determine if a fire alarm should be issued.

2.2.3 Automatic Detectors

Automatic detectors can be divided into the following types:

- *Line detector* – Detection is continuous along the entire length of the detector in this type of detection device. Typical examples may include certain older pneumatic rate-of-rise tubing detectors, the projected beam smoke detector, and heat-sensitive cable.
- *Spot detector* – This type of device has a detecting element that is concentrated at a particular location. Typical examples include bimetallic detectors, fusible alloy detectors, some types of rate-of-rise detector, certain smoke detectors, and thermoelectric detectors.
- Air sampling detector – This device consists of piping or tubing distributed from the detector unit to the area(s) to be protected. An air pump draws air from the protected area back to the detector through air sampling ports and piping or tubing. At the detector, the air is analyzed for fire products.
- *Addressable or analog addressable detector* – As mentioned previously, these detectors provide alarms and individual point identification, along with supervisory/trouble information, to the control panel. In certain analog addressable

Addressable and Analog Addressable Smoke Detectors

This is a typical addressable smoke detector. Addressable detectors send the fire alarm control panel (FACP) alarm status, detector location, and supervisory/trouble information. An analog addressable smoke detector performs the same functions; however, it has the additional capability of being able to report to the FACP information about the level of smoke that it is detecting.

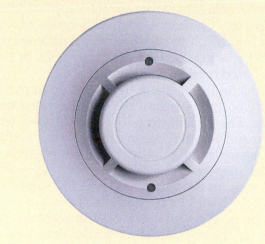

Figure Credit: System Sensor

detectors, adjustable sensitivity of the alarm signal can be provided. In other analog addressable detectors, an alarm signal is not generated. Instead, only a level of detection signal is fed to the control panel. Alarm sensitivity can be adjusted, and the level of detection can be analyzed from the panel (based on historical data) to reduce the likelihood of false alarms in construction areas or areas of high humidity on a temporary or permanent basis. Because all detectors are continually polled, a T-tap splice is permitted with some signaling line circuit styles. T-taps are not permitted with all styles.

2.2.4 Heat Detectors

Two major types of heat detectors are in general use. One is a rate-of-rise detector that senses a 15°F/8°C-per-minute increase in room/area temperature. The other is one of several versions of fixed-temperature detectors that activate if the room/area exceeds the rating of the sensor. Generally, rate-of-rise sensors are combined with fixed-temperature sensors in a combination heat

detector. Heat detectors are generally used in areas where property protection is the only concern or where smoke detectors would be inappropriate.

Fixed-temperature heat detectors activate when the temperature exceeds a preset level. Detectors are made to activate at different levels. The most commonly used temperature settings are 135°F, 190°F, and 200°F (57°C, 88°C, and 93°C respectively). The three types of fixed-temperature heat detectors are:

- *Fusible link* – The fusible link detector (*Figure 6*) consists of a plastic base containing a switch mechanism, wiring terminals, and a three-disc heat collector. Two sections of the heat collector are soldered together with an alloy that will cause the lower disc to drop away when the rated temperature is reached. This moves a plunger that shorts across the wiring contacts, causing a constant alarm signal. After the detector is activated, a new heat collector must be installed to **reset** the detector to an operating condition.

- *Quick metal* – The operation of the quick metal detector (*Figure 7*) is very simple. When the surrounding air reaches the prescribed temperature (usually 135°F/57°C or 190°F/88°C), the quick metal begins to soften and give way. This allows spring pressure within the device to push the top portion of the thermal element out of the way, causing the alarm contacts to close. After the detector has been activated, either the detector or the heat collector must be replaced to reset the detector to an operating condition.

- *Bimetallic* – In a bimetallic detector (*Figure 8*), two metals with different rates of thermal expansion are bonded together. Heat causes the two metals to expand at different rates, which causes the bonded strip to bend. This action closes a normally open circuit, which signals

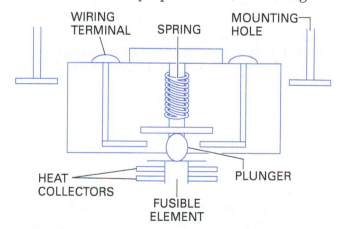

Figure 6 Fusible link detector.

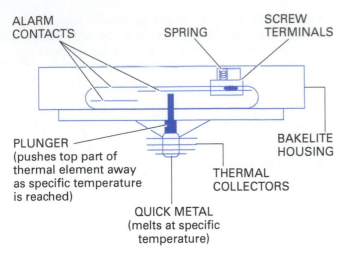

Figure 7 Quick metal detector.

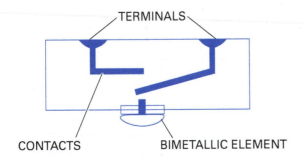

Figure 8 Bimetallic detector.

an alarm. Detectors with bimetallic elements automatically self-restore when the temperature returns to normal.

Combination heat detectors contain two types of heat detectors in a single housing. A fixed-temperature detector reacts to a preset temperature, and a rate-of-rise detector reacts to a rapid change in temperature even if the temperature reached does not exceed the preset level.

- *Rate-of-rise with fusible link detector* – (*Figure 9*) Rate-of-rise operation occurs when air in the chamber (1) expands more rapidly than it can escape from the vent (2). The increasing pressure moves the diaphragm (3) and causes the alarm contacts (4 and 5) to close, which results in an alarm signal. This portion of the detector automatically resets when the temperature stabilizes. A fixed-temperature trip occurs when the heat causes the fusible alloy (6) to melt, which releases the spring (7). The spring depresses the diaphragm, which closes the alarm contacts (4 and 5). If the fusible alloy is melted, the center section of the detector must be replaced.
- *Rate-of-rise with bimetallic detector* – (*Figure 10*) Rate-of-rise operation occurs when the air in

the air chamber (1) expands more rapidly than it can escape from the valve (2). The increasing pressure moves the diaphragm (3) and causes the alarm contact (4) to close. A fixed-temperature trip occurs when the bimetallic element (5) is heated, which causes it to bend and force the spring-loaded contact (6) to mate with the fixed contact (7).

As with all automatic detectors, heat sensors are also rated for a listed spacing as well as a temperature rating (*Table 1*). The listed spacing is typically 50' × 50' (15m × 15m). The applicable rules and formulas for proper spacing (from NFPA 72® and other applicable standards) are then applied to the listed spacing. Caution is advised because it is difficult or impossible to distinguish the difference in the listed spacing for two different types of heat sensors based on their appearance.

The maximum ceiling temperature must be 20°F (11°C) or more below the detector rated temperature. The difference between the rated temperature and the maximum ambient temperature for the space should be as small as possible to minimize response time.

Heat sensors are not considered life safety devices. They should be used in areas that are unoccupied or in areas that are environmentally unsuitable for the application of smoke detectors. To every extent possible, heat sensors should be limited to use as property protection devices.

2.2.5 Smoke Detectors

There are two basic types of smoke detectors: photoelectric detectors (*Figure 11*), which sense the presence of large smoke particles from smoldering fires, and ionization detectors (*Figure 12*), which sense the smaller smoke particles from fast-burning fires. For residential use, both types are available combined in one detector housing for maximum coverage. For commercial use, either type is also available combined with a heat detector. As mentioned previously, the newer commercial fire systems use smoke detectors that are analog addressable units, which do not signal an alarm. Instead, they return a signal that represents the level of detection to the FACP for further analysis.

An ionization detector (*Figure 13*) uses the change in the electrical conductivity of air to detect smoke. An alarm is indicated when the amount of smoke in the detector rises above a certain level. The detector has a very small amount of radioactive material in the sensing chambers. As shown in *Figure 14* (A), the radioactive material ionizes the air in the measuring and reference chambers,

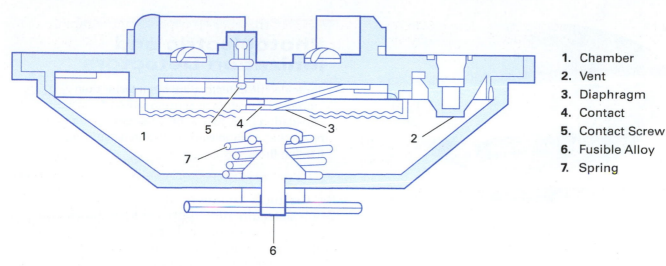

1. Chamber
2. Vent
3. Diaphragm
4. Contact
5. Contact Screw
6. Fusible Alloy
7. Spring

Figure 9 Rate-of-rise with fusible link detector.

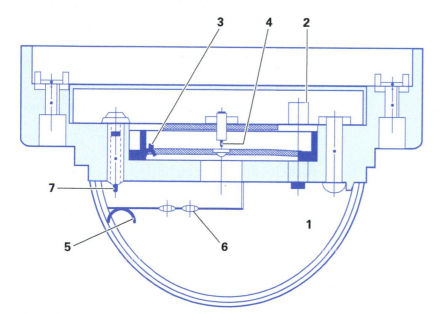

1. Air Chamber
2. Breather Valve
3. Diaphragm Assembly
4. Rate-of-Rise Contact
5. Bimetal Fixed-Temperature Element
6. Contact Spring
7. Fixed-Temperature Contact

Figure 10 Rate-of-rise with bimetallic detector.

Table 1 Heat Detector Temperature Ratings

Temperature Classification	Temperature Rating Range	Maximum Ceiling Temperature	Color Code
Low	100°F–134°F	80°F	No color
Ordinary	135°F–174°F	100°F	No color
Intermediate	175°F–249°F	150°F	White
High	250°F–324°F	225°F	Blue
Extra high	325°F–399°F	300°F	Red
Very extra high	400°F–499°F	375°F	Green
Ultra high	500°F–575°F	475°F	Orange

allowing the air to conduct current through the space between two charged electrodes. When smoke particles enter the measuring chamber, they prevent the air from conducting as much current [*Figure 14* (B)]. The detector activates an alarm when the conductivity decreases to a set level. The detector compares the current drop in the main chamber against the drop in the

Figure 11 Typical photoelectric smoke detector.

Figure 12 Typical ionization smoke detector.

reference chamber. This allows it to avoid alarming when the current drops due to surges, radio frequency interference (RFI), or other factors.

There are two basic types of **photoelectric smoke detector**: light-scattering detectors and beam detectors. Light-scattering detectors use the reflective properties of smoke to detect the smoke. The principle of **light scattering** is used for the most common single-housing detectors. Beam detectors rely upon smoke to block enough light to cause an alarm. Early smoke detectors operated on the light **obscuration** principle. Today, photoelectric smoke detectors are used primarily to sense smoke in large open areas with high ceilings. In some cases, mirrors are used to direct the beam in a desired path; however, the mirrors reduce the overall range of the detector. Most commercial photoelectric smoke detectors require an external power source.

Light-scattering detectors are usually spot detectors that contain a light source and a photosensitive device arranged so that light rays do not normally fall on the device (*Figure 15*). When smoke particles enter the light path, the light hits the particles and scatters, hitting the photosensor, which signals the alarm.

Projected beam smoke detectors that operate on the light obscuration principle (*Figure 16*) consist of a light source, a light beam focusing device, and a photosensitive device. Smoke obscures or blocks part of the light beam, which reduces the amount of light that reaches the photosensor, signaling an alarm. The prime use of the light obscuration principle is with projected beam-type smoke detectors that are employed in the protection of large, open high-bay warehouses,

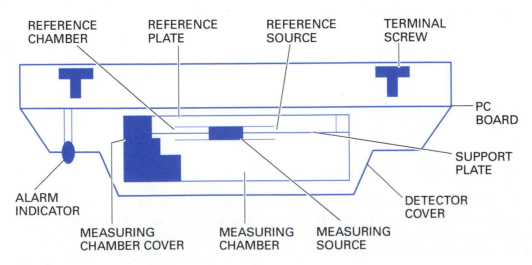

Figure 13 Ionization detector.

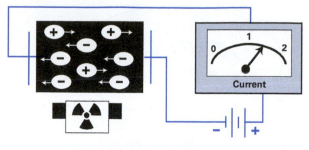

(A) CLEAN AIR

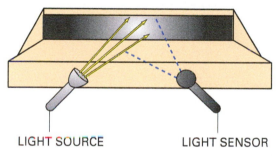

(B) SMOKE PRESENT

Figure 14 Ionization action.

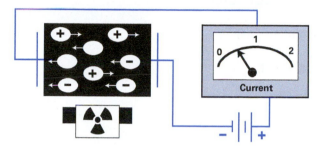

LIGHT SOURCE   LIGHT SENSOR

PULSED LIGHT

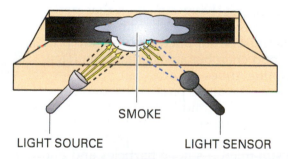

SMOKE

LIGHT SOURCE   LIGHT SENSOR

SCATTERED LIGHT

Figure 15 Light-scattering detector operation.

high-ceiling churches, and other large areas. A version of a beam detector was used in older style duct detectors.

After a number of fires where smoke spread through building duct systems and contributed to numerous deaths, building codes began to require the installation of duct detectors. Two

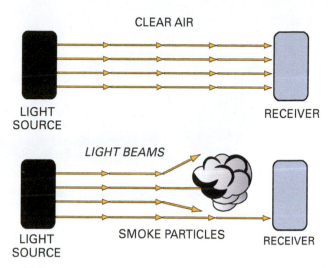

Figure 16 Light obscuration principle.

Projected Beam Smoke Detectors

The transmitter and receiver for a projected beam smoke detector are shown here. These units are designed for use in atriums, ballrooms, churches, warehouses, museums, factories, and other large, high-ceiling areas where conventional smoke detectors cannot be easily installed.

Figure Credit: Honeywell Security and Fire

versions of duct detectors are shown in *Figure 17*. The projected beam detector is an older style of duct detector that may be encountered in existing installations.

Duct detectors enable a system to control the spread of smoke within a building by turning off the HVAC system, operating exhaust fans, closing doors, or pressurizing smoke compartments in the event of a fire. This prevents smoke, fumes, and fire by-products from circulating through the

---

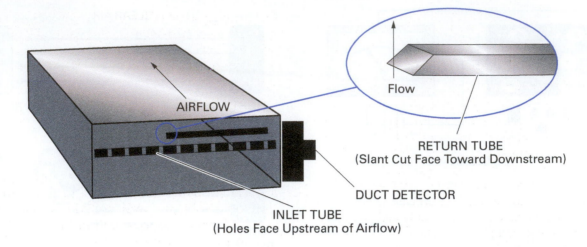

RETURN TUBE
(Slant Cut Face Toward Downstream)

DUCT DETECTOR

INLET TUBE
(Holes Face Upstream of Airflow)

SAMPLING TUBE STYLE

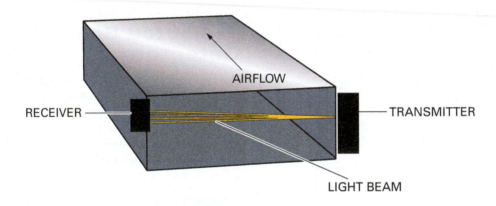

PROJECTED LIGHT BEAM STYLE

Figure 17 Typical duct detector installations.

ductwork. Because the air in the duct is either at rest or moving at high speed, the detector must be able to sense smoke in either situation. A typical duct detector has a listed airflow range within which it will function properly. It is not required to function when the duct fans are stopped.

At least one duct detector is available with a small blower for use in very low airflow applications. Duct detectors must not be used as a substitute for open area protection because smoke may not be drawn from open areas of the building when the HVAC system is shut down. Duct detectors can be mounted inside the duct with an access panel or outside the duct with sampling tubes protruding into the duct.

As mentioned previously, the primary function of a duct detector is to turn off the HVAC system to prevent smoke from being circulated. However, NFPA 90A and NFPA 90B require that duct detectors be tied into a general fire alarm system if the building contains one. If no separate fire alarm system exists, then remote audio/

visual indicators, triggered by the duct detectors, must be provided in normally occupied areas of the building. Duct detectors that perform functions other than the shutdown of HVAC equipment must be supplied with backup power.

Cloud chamber smoke detectors (*Figure 18*) use sampling tubes to draw air from several areas (zones). The air is passed through several chambers where humidity is added and pressure is reduced with a vacuum pump. Reducing the pressure causes water droplets to form around the sub-micron smoke particles and allows them to become visible. A light beam is passed through the droplets and measured by a photoelectric detector. As the number of droplets increases, the light reaching the detector is reduced, which initiates an alarm signal.

2.2.6 Other Types of Detectors

This section describes special-application detectors. These include such devices as the rate compensation detector, semiconductor line-type

Duct Smoke Detectors

This is a typical duct smoke detector with the cover removed. It is designed for HVAC applications. The manufacturer of this unit makes duct detectors based on either photoelectric technology or ionization technology. Up to ten of these units can be networked together. When one detector goes into alarm, all of the other detectors on the network activate in order to control the connected fans, blowers, and dampers. However, only the duct detector that initiated the alarm shows an alarm indication to identify it as the source of the alarm.

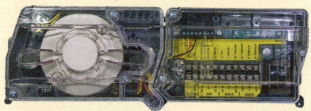

Figure Credit: System Sensor

heat detector, fusible line-type heat detector, ultraviolet **flame detector**, and infrared flame detector.

A rate compensation detector (*Figure 19*) is a device that responds when the temperature of the surrounding air reaches a predetermined level, regardless of the rate of temperature rise. The detector also responds if the temperature rises quickly over a short period. This type of detector has a tubular casing of metal that expands lengthwise as it is heated. A contact closes as it reaches a certain stage of elongation. These detectors are self-restoring. Rate compensation detectors are more complex than either fixed or rate-of-rise detectors. They combine the principles of both to compensate for thermal lag. When the air temperature is rising rapidly, the unit is designed to respond almost exactly at the point when the air temperature reaches the unit's rated temperature. It does not lag while it absorbs the heat and rises to that temperature. Because of the precision associated with their operation, rate compensation detectors are well suited for use in areas where thermal lag must be minimized.

A restorable semiconductor line-type heat detector (*Figure 20*) uses a semiconductor material and a stainless steel capillary tube. The capillary tube contains a coaxial center conductor separated from the tube wall by a temperature-sensitive semiconductor thermistor material. Under normal conditions, a small current (below the alarm threshold) flows. As the temperature rises, the resistance of the semiconductor thermistor decreases. This allows more current to flow and initiates the alarm. When the temperature falls, the current flow decreases below the alarm threshold level, which stops the alarm. The wire is connected to special controls or modules that establish the alarm threshold level and sense the current flow. Some of these control devices can pinpoint the location in the length of the wire where the temperature change occurs. Line-type heat detectors are commonly used in cable trays, conveyors, electrical switchgear, warehouse rack storage, mines, pipelines, hangars, and other similar applications.

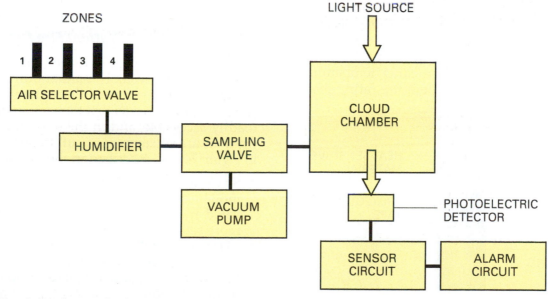

Figure 18 Cloud chamber smoke detector.

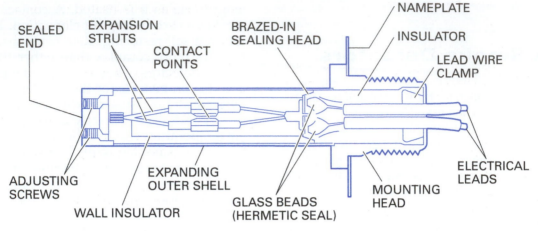

Figure 19 Rate compensation detector.

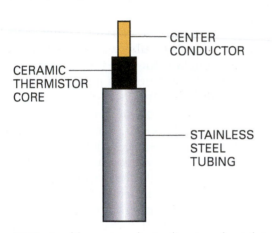

Figure 20 Restorable semiconductor line-type heat detector.

A non-restorable fusible line-type heat detector (*Figure 21*) uses a pair of steel wires in a normally open circuit. The conductors are held apart by heat-sensitive insulation. The wires, under tension, are enclosed in a braided sheath to make a single cable. When the temperature limit is reached, the insulation melts, the two wires contact, and an alarm is initiated. The melted and fused section of the cable must be replaced following an alarm to restore the system. The wire is available with different melting temperatures for the insulation. The temperature rating should

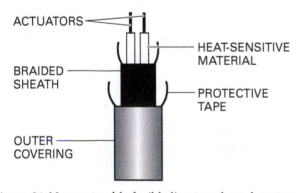

Figure 21 Non-restorable fusible line-type heat detector.

be approximately 20°F (11°C) above the ambient temperature. Special controls or modules are available that can be connected to the wires to pinpoint the fire location where the wires are shorted together.

An ultraviolet (UV) flame detector (*Figure 22*) uses a solid-state sensing element of silicon carbide, aluminum nitrate, or a gas-filled tube. The UV radiation of a flame causes gas in the element or tube to ionize and become conductive. When sufficient current flow is detected, an alarm is initiated.

An infrared (IR) flame detector (*Figure 23*) consists of a filter and lens system that screens out any unwanted radiant-energy **wavelength** and focuses the incoming energy on light-sensitive components. These flame detectors can respond to the total IR content of the flame alone or to a combination of IR with flame flicker of a specific frequency. They are used indoors and have filtering systems or solar sensing circuits to minimize unwanted alarms from sunlight.

2.2.7 Manual (Pull Station) Fire Detection Devices

When required by code, manual pull stations are required to be distributed throughout a commercial monitored area so that they are unobstructed, readily accessible, and in the normal path of exit from the area. Examples of pull stations are shown in *Figure 24* and described in the following list:

- *Single-action pull stations* – A single action or motion operates these devices. They are activated by pulling a handle that closes one or more sets of contacts and generates the alarm.
- *Glass-break pull stations* – In these devices, a glass rod, plate, or special element must be broken to activate the alarm. This is accomplished using a handle or hammer that is an integral

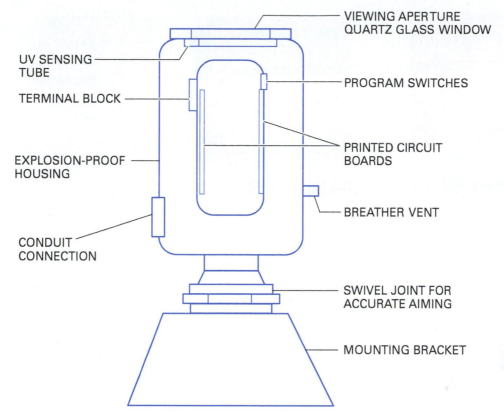

Figure 22 UV flame detector (top view).

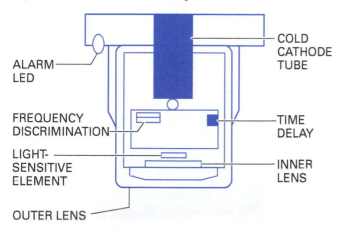

Figure 23 IR flame detector (top view).

part of the station. When the alarm is activated, one or more sets of contacts are closed and an alarm is actuated. Usually, the plate, rod, or element must be replaced to return the unit to service, although some stations will operate without the rod or plate.

- *Double-action pull stations* – Double-action pull stations require the user to lift a cover or open a door before operating the pull station. Two discretely independent actions are required to operate the station and activate the alarm. Using a stopper-type cover that allows the alarm to be tripped after a cover is lifted may turn a single-action pull station

---

Glass-Break Pull Station

When installing a glass-break pull station, consideration must be given to where the glass will go when broken and the danger it may pose to children and the disabled.

---

into a double-action pull station. According to Underwriters Laboratories, the stopper-type device is listed as an accessory to manual stations and is permissible for use as a double-action pull station device. With certain types of double-action stations, there have been instances when people have confused the sound of a tamper alarm that sounds when the cover is lifted with the sound of fire alarm activation. The sounding of this tamper alarm sometimes causes them to fail to activate the pull station.

- *Key-operated pull stations* – Applications for key-operated pull stations (*Figure 25*) are restricted. Key-operated stations are permitted in certain occupancies where facility staff members may be in the immediate area and where use by other occupants of the area is not desirable. Typical situations would include certain detention and correctional facilities and some health care facilities, particularly those that provide mental health treatment.

SINGLE-ACTION
PULL STATION

GLASS-BREAK COVER
FOR PULL STATION

DOUBLE-ACTION
PULL STATION

Figure 24 Typical pull stations.

Figure 25 Key-operated pull station.

2.2.8 Auto-Mechanical Fire Detection Equipment

This section describes various types of auto-mechanical fire detection equipment. It covers wet and dry sprinkler systems and water flow alarms.

A wet sprinkler system (*Figure 26*) consists of a permanently piped water system under pressure, using heat-actuated sprinklers. When a fire occurs, the sprinkler heads exposed to high heat open and discharge water individually in an attempt to control or extinguish the fire. They are designed to automatically detect and control a fire and protect a structure. Once a sprinkler head is activated, some type of water flow sensor signals a fire alarm. When activated, a sprinkler system may cause water damage. Wet systems should not be used in spaces subject to freezing.

A dry sprinkler system (*Figure 27*) consists of heat-operated sprinklers that are attached to a piping system containing air under pressure. Normally, air pressure in the pipes holds a water valve closed, which keeps water out of the piping system. When heat activates a sprinkler head, the

Manual Fire Alarm Station Reset

This manual fire alarm is operated by pulling on the pull cover. This engages a latching mechanism that prevents the pull cover from being returned to the closed position. The only way the station cover can be reset to the closed position is by using the appropriate reset key.

Figure Credit: Honeywell Security and Fire

open sprinkler head causes the air pressure to be released. This allows the water valve to open, and water flows through the pipes and out to the activated sprinkler head. Once a sprinkler head is activated and water starts to flow, a water flow sensor signals a fire alarm. Because the pipes are dry until a fire occurs, these systems may be used in spaces subject to freezing.

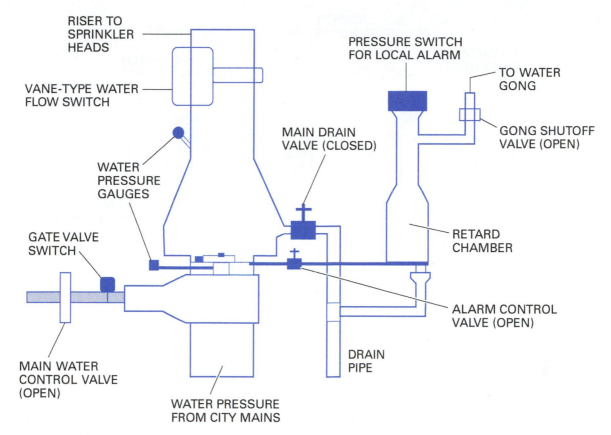

RISER TO SPRINKLER HEADS

VANE-TYPE WATER FLOW SWITCH

WATER PRESSURE GAUGES

GATE VALVE SWITCH

MAIN WATER CONTROL VALVE (OPEN)

WATER PRESSURE FROM CITY MAINS

PRESSURE SWITCH FOR LOCAL ALARM

TO WATER GONG

GONG SHUTOFF VALVE (OPEN)

MAIN DRAIN VALVE (CLOSED)

RETARD CHAMBER

ALARM CONTROL VALVE (OPEN)

DRAIN PIPE

Figure 26 Wet sprinkler system.

> **WARNING!**
>
> If the air pressure drops and the system fills with water, the system must be drained. Consult a qualified specialist to restore the system to normal operation.

When a building sprinkler head is activated by the heat generated by a fire, the sprinkler head allows water to flow. As shown in *Figure 26* and *Figure 27*, pressure-type or vane-type water flow switches are installed in the sprinkler system along with local alarm devices. The water flow switches detect the movement of water in the system. Activation of these switches by the movement of water causes an initiation signal to be sent to an FACP that signals a fire alarm.

In addition to devices that signal sprinkler system activation, other devices may be used to monitor the status of the system using an FACP. For instance, the position of a control valve may be monitored so that a **supervisory signal** is sent whenever the control valve is turned to shut off the water to the sprinkler system. If this valve is turned off, no water can flow through the sprinkler system, which means the system is inactive. In some systems, water pressure from the municipal water supply may not be strong enough to push enough water to all parts of a building.

Water Flow Detectors

This water flow detector is typical of those used with wet-pipe sprinkler systems. Water flow through the associated pipe deflects the detector's vane. This activates the internal switch contacts, initiating an alarm or auxiliary indication. Water flow in the pipe can be caused by the opening of one or more sprinkler heads because of a fire, the opening of a test valve, or a leaking or ruptured pipe.

Figure Credit: System Sensor

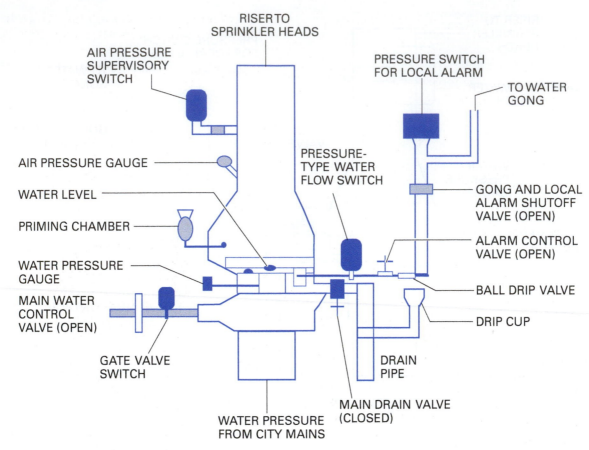

AIR PRESSURE SUPERVISORY SWITCH

RISER TO SPRINKLER HEADS

PRESSURE SWITCH FOR LOCAL ALARM

TO WATER GONG

AIR PRESSURE GAUGE

WATER LEVEL

PRIMING CHAMBER

PRESSURE-TYPE WATER FLOW SWITCH

GONG AND LOCAL ALARM SHUTOFF VALVE (OPEN)

ALARM CONTROL VALVE (OPEN)

WATER PRESSURE GAUGE

MAIN WATER CONTROL VALVE (OPEN)

BALL DRIP VALVE

DRIP CUP

GATE VALVE SWITCH

DRAIN PIPE

WATER PRESSURE FROM CITY MAINS

MAIN DRAIN VALVE (CLOSED)

Figure 27 Dry sprinkler system.

In these cases, a fire pump is usually required. When the fire pump runs, a supervisory signal is sent indicating that the fire pump is activated. If the pump runs to maintain system pressure and does not shut down within a reasonable time, a site visit may be required. When water is scarce or unavailable, an on-site water tank may be required. Supervisory signals may be generated when the temperature of the water drops to a level low enough to freeze or if the water level or pressure drops below a safe level. In a dry system, a supervisory signal is generated if the air pressure drops below a usable level. If fire pump power is monitored, lack of power or power phase reversal will cause a trouble signal.

2.0.0 Section Review

1. Which fire alarm system is the simplest of all systems?

   a. Multiplex
   b. Conventional hardwired
   c. Class B
   d. Addressable intelligent

2. Which kind of detector can sense along its entire length?

   a. An AHJ detector
   b. An air sampling detector
   c. A spot detector
   d. A line detector

3.0.0 CONTROL PANELS

Objective

Describe fire alarm control panels and their primary features.

a. Describe fire alarm control panels and their power source requirements.
b. Explain how users interface with the control panel.
c. Define and describe initiating circuits and panel outputs.

Trade Terms

Alarm verification: A feature of a fire control panel that allows for a delay in the activation of alarms upon receiving an initiating signal from one of its circuits. Alarm verification must not be longer than three minutes, but can be adjustable from 0 to 3 minutes to allow supervising personnel to check the alarm. Alarm verification is commonly used in hotels, motels, hospitals, and institutions with large numbers of smoke detectors.

Approved: Acceptable to the authority having jurisdiction.

Authority having jurisdiction (AHJ): The authority having jurisdiction is the organization, office, or individual responsible for approving equipment, installations, or procedures in a particular locality.

Labeled: In the context of fire alarm control panels, tags that identify various zones and sensors by descriptive names. Panels may include dual identifiers, one of which is meaningful to the occupants, while the other is helpful to firefighting personnel.

Positive alarm sequence: An automatic sequence that results in an alarm signal, even when manually delayed for investigation, unless the system is reset.

System unit: The active subassemblies at the central station used for signal receiving, processing, display, or recording of status change signals. The failure of one of these subassemblies causes the loss of a number of alarm signals by that unit.

Today, many different companies are manufacturing hundreds of different control panels, and more are developed each year.

Figure 28 shows a typical intelligent, addressable control panel. The control panel shown has a voice command center added for use in high-rise buildings. It provides for automatic evacuation messages, firefighter paging, and two-way communication to a central station through a telephone network. Although many control panels may have unique features, all control panels perform some specific basic functions. As shown in *Figure 29*, they detect problems through the sensor devices connected to them, and they sound alerts or report these problems to a central location.

Figure 28 Typical intelligent, addressable control panel.

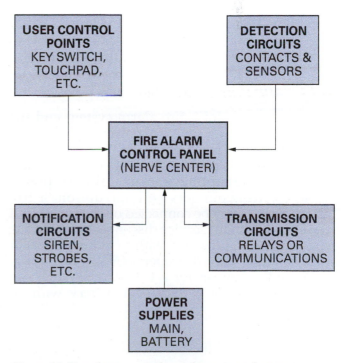

Figure 29 Fire alarm control panel inputs and outputs.

3.1.0 Control Panel Overview

A control panel can allow the user to reprogram the system and, in some cases, to activate or deactivate the system or zones. The user may also change the sensitivity of detectors in certain zones of the system. On panels that can be used for both intrusion and fire alarms, controls are provided to allow the user to enter or leave the monitored area without setting off the system. In addition, controls allow some portions of the system (fire, panic, holdup, etc.) to remain armed 24 hours a day, 365 days a year. Control panels also provide the alarm user, responding authority, or inspector with a way to silence bells or control other system features.

Control panels are usually equipped with LED, alphanumeric, or graphic displays to communicate information such as active zones or detectors to an alarm user or monitor. These types of displays are usually included on the FACP, but can also be located remotely if required.

The control panel organizes all the components into a working and functional system. The control panel is often referred to as the "brain" of the system. The control panel coordinates the actions that the system takes in response to messages it gets from initiating devices or, in some cases, notification devices that are connected to it. Depending on the inputs from the devices, it may activate the notification devices and may also transmit data to a remote location via transmission circuits. In most cases, the control panel also conditions the power it receives from the building power system or from a backup battery so that it can be used by the fire alarm system.

3.1.1 FACP Primary and Secondary Power

Primary power for a fire alarm system and the FACP is normally a source of power provided by a utility company. However, primary power for systems and panels can also be supplied from emergency uninterruptible backup primary power systems. To avoid service interruption, a fire alarm system may be connected on the line side of the electrical main service disconnect switch. The circuit must be protected with a circuit breaker or fuse no larger than 20 amperes (A).

Secondary power for a fire alarm system can be provided by a battery or a battery with an approved backup generator. Secondary power must be immediately supplied to the fire alarm system in the event of a primary power failure. As defined by NFPA 72®, standby time for operation of the system on secondary power must be no less than 24 hours for central, local, proprietary, voice communications, and household systems and no less than 60 hours for auxiliary and remote systems. The battery standby time may be reduced if a properly configured battery-backup generator is available. During an alarm condition, the secondary power must operate the system under load for a minimum of 4, 5, or 15 minutes after expiration of the standby period, depending on the type of system. Any batteries used for secondary power must be able to be recharged within 48 hours.

3.1.2 FACP Listings

It is mandatory that fire alarm control panels be listed for the purpose for which they will be used. Common UL listings for FACPs are *UL 864* for the fire alarm control panel and *UL 985* for the household fire warning system unit.

In combination fire/burglary panels, fire circuits must be on or active at all times, even if the burglary control is disarmed or turned off. Some codes prohibit using combination intrusion and fire alarm control panels in commercial applications.

A fire alarm control panel listed for household use by the UL may never be installed in any commercial application unless it also has the appropriate commercial listing, has been granted equivalency by the authority having jurisdiction (AHJ) in writing, or is specifically permitted by a document that supersedes the reference standards. The NFPA defines a household as a single- or two-family residential unit, and this term applies to systems wholly within the confines of the unit. Except for monitoring, no initiation from, or notification to, locations outside the residence are permitted. Although an apartment building or condominium is considered commercial, a household system may be installed within an individual living unit. However, any devices that are outside the confines of the individual living unit, such as manual pull stations, must be connected to an FACP listed as commercial.

3.2.0 User Control Points

User controls allow the alarm user to turn all or portions of the system on or off. User controls also allow the alarm user to monitor the system status, including the sensors or zones that are active, trouble reports, and other system parameters. They also allow the user to reset indicators of system events such as alarms or indications of trouble with phone lines, equipment, or circuits. In today's systems, many devices can be used to control the system, including keypads, key switches, touch screens, telephones (including wireless phones), and computers.

3.2.1 Keypads

Two general types of keypads are in use today: alphanumeric (*Figure 30*) and LED. Either type can be mounted at the control or at a separate location. Both allow the entry of a numerical code to program various system functions. An alphanumeric keypad combines a keypad that is similar to a pushbutton telephone dial with an alphanumeric display that is capable of showing letters and numbers. LED keypads use a similar keypad, but display information by lighting small LEDs.

Figure 30 Alphanumeric keypad and display.

3.2.2 Touch Screens

Touch screens allow the user to access multiple customized displays in graphic or menu formats. This enables rapid and easy interaction with the system.

3.2.3 Telephone/Computer Control

Since most fire alarm systems are connected to telephones for notification purposes, and telephones are often in locations where user control devices have traditionally been located, some control panel manufacturers have incorporated ways for the telephone to function as a user interface. When telephones or cell phones are used as an interface for the system, feedback on system status and events is given audibly over the phone. Because the system is connected to the telephone network, a system with this feature can be monitored and controlled from anywhere a landline or cell phone call can be made. Most new systems can also be connected to a computer either locally or via a telephone modem or a high-speed data line to allow control of events and receipt of information.

3.3.0 FACP Initiating Circuits and Outputs

An initiating circuit monitors the various types of initiating devices. A typical initiating circuit

Intelligent Systems

An intelligent system uses analog devices to communicate with an intelligent, addressable control panel typical of the one shown in *Figure 28*. The control panel individually monitors the value or status reported by the analog sensors and makes normal, alarm, or trouble decisions. The specific control panel in *Figure 28* is a software-controlled unit capable of monitoring 1,980 individually identifiable and controllable detection/control points. It is field programmable from the front panel keyboard.

can monitor for three states: normal, trouble, and alarm. Initiating circuits can be used for monitoring fire devices, non-fire devices (supervisory), or devices for watch patrol or security. The signals from the initiating devices can be separately indicated at the premises and remotely monitored at a central station.

3.3.1 Initiating Circuit Zones

The term *zone* as it relates to modern fire alarm systems can have several definitions. Building codes restrict the location and size of a zone to enable emergency response personnel to quickly locate the source of an alarm. The traditional use of a zone in a conventional system was that each initiating circuit was a zone. Today, a single device or multiple initiating circuits or devices may occupy a zone. Generally, a zone may not cover more than one floor, exceed a certain number of square feet, or exceed a certain number of feet in length or width. In addition, some codes require different types of initiating devices to be on different initiating circuits within a physical area or zone. With modern addressable systems, each device is like a zone because each is displayed at the panel. It is common in addressable systems to group devices of several types into a zone. This is usually accomplished by programming the panel rather than by hardwiring the devices. Ultimately, the purpose of a zone is to provide the system monitor with information as to the location of a system alarm or problem. Local codes provide the specific guidelines.

3.3.2 Alarm Verification

To reduce false alarms, most FACPs allow for a delay in the activation of notification devices upon receiving an alarm signal from an initiating device. In some conventional systems using a positive alarm sequence feature, alarm delay is usually adjustable for a period up to three minutes, but must not exceed three minutes. This allows supervising or monitoring personnel to investigate the alarm. If a second detector activates during the investigation period, a fire alarm is immediately sounded. In other conventional systems, verification can be of the reset-and-resample type: The panel resets the detector and waits 10 to 20 seconds for the sensor to retransmit the alarm. If there is an actual fire, the sensor should detect it both times, and an alarm is then activated. In addressable systems, alarm verification can be of the wait-and-check type: the panel notes the initiation signal and then waits for about 10 to 20 seconds to see if the initiation signal remains constant before activating an alarm.

Another verification method for large areas with multiple sensors is to wait until two or more sensors are activated before an alarm is initiated. In this method, known as *cross-zoning*, a single detector activation may set a one minute or less pre-alarm condition warning signal at a manned monitoring location. If the pre-alarm signal is not cancelled within the allowed time, or if a second detector activates, a fire alarm is sounded. The pre-alarm signals are not required by the NFPA. In cross-zoning, two detectors must occupy each space regardless of the size of the space, and each detector may cover only half of the normal detection area. Cross-zoning along with other methods of alarm verification cannot be used on the same devices. Some building codes require alarm verification or some equivalent on all smoke detection devices.

Verification is commonly used in hotels, motels, hospitals, and other institutions with large numbers of detectors. It may also be used in monitored fire alarm systems for households to reduce false alarms.

3.3.3 FACP Labeling

Effective system design and installation requires that zones (or sensors) be labeled in a way that makes sense to all who will use or respond to the system. In some cases, this may require two sets of labels: one for the alarm user and another for the police and fire authorities. A central station operator who will be talking to both the police and the alarm user should be aware of both sets of labels. Labeling for the alarm user is done using names familiar to the user (Johnny's room, kitchen, master bedroom, and so on). Labeling for the police and fire authorities is done from the perspective of looking at the building from the outside (first floor east, basement rear, second floor west, and so on).

3.3.4 Types of FACP Alarm Outputs

Most panels provide one or more of the following types of outputs:

- *Relay or dry contacts* – Relay contacts or other dry contacts are electrically isolated from the circuit controlling them, which provides some protection from external spikes and surges. Additional protection is sometimes needed. Always check the contact ratings to determine how much voltage and current the contacts can handle. Check with the manufacturer to determine if this provides adequate protection.
- *Built-in siren drivers* – Built-in siren drivers use a transistor amplifier circuit to drive a siren speaker. Impedance must be maintained within the manufacturer's specifications. Voltage drop caused by long wire runs of small conductors can greatly reduce the output level of the siren.
- *Voltage outputs* – Voltage outputs are not isolated from the control, and some protection may be required. Spikes and surges, such as those generated by solenoid bells or the back EMF generated when a relay is de-energized, can be a problem.
- *Open collector outputs* – Open collector outputs are outputs directly from a transistor and have limited current output. They are often used to drive a very low-current device or relay. Filtering may be required. Great caution should be used not to overload these outputs, as the circuit board normally has to be returned to the manufacturer if an overload of even a very short duration occurs.

3.0.0 Section Review

1. What part of a fire alarm system coordinates the actions that the system must take in response to its other components?

   a. The AHU
   b. The AHJ
   c. The FACP
   d. The NFPA

2. Many fire alarm systems provide graphical control through a(n) _____.

   a. voice interface
   b. touch screen
   c. local uplink
   d. alphanumeric keypad

3. An alarm delay can be up to _____.

   a. one minute
   b. two minutes
   c. three minutes
   d. five minutes

4.0.0 NOTIFICATION, COMMUNICATION, AND MONITORING

Objective

Identify and describe approaches to fire alarm notification and communication/monitoring.

a. Describe visual and audible notification devices and systems.
b. Describe important considerations in the use of fire alarm notification signals.
c. Describe communication and monitoring options for fire alarm systems.

Trade Terms

Americans with Disabilities Act (ADA): An act of Congress intended to ensure civil rights for physically challenged people.

Audible signal: An audible signal is the sound made by one or more audible indicating appliances, such as bells, chimes, horns, or speakers, in response to the operation of an initiating device.

CABO: Council of American Building Officials.

Chimes: A single-stroke or vibrating audible signal appliance that has a xylophone-type striking bar.

Coded signal: A signal pulsed in a prescribed code for each round of transmission.

Digital Alarm Communicator Receiver (DACR): A system component that will accept and display signals from digital alarm communicator transmitters (DACTs) sent over public switched telephone networks.

Digital Alarm Communicator System (DACS): A system in which signals are transmitted from a digital alarm communicator transmitter (DACT) located at the protected premises through the public switched telephone network to a digital alarm communicator receiver (DACR).

Digital Alarm Communicator Transmitter (DACT): A system component at the protected premises to which initiating devices or groups of devices are connected. The DACT will seize the connected telephone line, dial a preselected number to connect to a DACR, and transmit signals indicating a status change of the initiating device.

General alarm: A term usually applied to the simultaneous operation of all the audible alarm signals on a system, to indicate the need for evacuation of a building.

Indicating device: Any audible or visible signal employed to indicate a fire, supervisory, or trouble condition. Examples of audible signal appliances are bells, horns, sirens, electronic horns, buzzers, and chimes. A visible indicator consists of an incandescent lamp, strobe lamp, mechanical target or flag, meter deflection, or the equivalent. Also called a *notification device (appliance)*.

Noise: A term used in electronics to cover all types of unwanted electrical signals. Noise signals originate from numerous sources, such as fluorescent lamps, walkie-talkies, amateur and CB radios, machines being switched on and off, and power surges. Today, equipment must tolerate increasing amounts of electrical interference, and the quality of equipment depends on how much noise it can ignore and withstand.

Non-coded signal: A signal from any indicating appliance that is continuously energized.

Protected premises: The physical location protected by a fire alarm system.

Public Switched Telephone Network: An assembly of communications facilities and central office equipment, operated jointly by authorized common carriers, that provides the general public with the ability to establish communications channels via discrete dialing codes.

Remote supervising station fire alarm system: A system installed in accordance with the applicable code to transmit alarm, supervisory, and trouble signals from one or more protected premises to a remote location where appropriate action is taken.

Wide Area Telephone Service (WATS): Telephone company service that provides reduced costs for certain telephone call arrangements. In-WATS or 800-number service calls can be placed from anywhere in the continental United States to the called party at no cost to the calling party. Out-WATS is a service whereby, for a flat-rate charge, dependent on the total duration of all such calls, a subscriber can make an unlimited number of calls within a prescribed area from a particular telephone terminal without the registration of individual call charges.

Fire alarms are all about communication. Obviously, their first priority is to notify people in the protected facility. The system performs this task through devices designed to get people's attention. Using various kinds of signals, these devices communicate the nature of the emergency. But fire alarms must also communicate with outside facilities, such as external monitoring services. They accomplish this task through other technology, much of it digital.

4.1.0 Notification Appliances

Notification to building occupants of the existence of a fire is the most important life safety function of a fire alarm system. There are two primary types of notification: audible and visible. Concerns resulting from the **Americans with Disabilities Act (ADA)** have prompted the introduction of olfactory (sense of smell) and tactile (sense of touch) types of notification as well.

4.1.1 Visual Notification Devices

Strobes are high-intensity lights that flash when activated. They can be separate devices or mounted on or near the audible device. Strobe lights are more effective in residential, industrial, and office areas where they don't compete with other bright objects. Because ADA requirements dictate clear or white xenon (or equivalent) strobe lights, most interior visual notification devices are furnished with clear strobe lights (*Figure 31*). Strobe devices can be wall-mounted or ceiling-mounted and are usually combined with an audible notification device. When more than two strobes are visible from any location, the strobes must be synchronized to avoid random flashing, which can be disorienting and may actually cause seizures in certain individuals. Some strobes are brighter than others to accommodate different applications. These units may be rated in candelas.

4.1.2 Audible Notification Devices

Audible alarm devices are noise-making devices, such as sirens, bells, or horns, that are used as part of a local alarm system to indicate an alarm condition. In some cases, low-current **chimes** or buzzers are also used. An **audible signal** used for fire alarms in a facility that contains other sound-producing devices must produce a unique sound pattern so that a fire alarm can be recognized. If more than one audible signal is used in a facility, they must be synchronized to maintain any

sound pattern. NFPA 72® specifies that a standard signal known as *Temporal-Three* is required by most fire alarm systems, including household systems. This signal has three $\frac{1}{2}$-second tones, a pause, and then a repeat of the pattern until the alarm is manually reset. Only a fire alarm may use this signal.

Some bells (*Figure 32*) use an electrically-vibrated clapper to repeatedly strike a gong. These types of solenoid-operated bells can also produce electrical **noise**, which will interfere with controls unless proper filters are used. Solenoid-operated bells draw relatively high current, typically 750 to 1,500 milliamps (mA). Another type of bell is a motor-driven bell. In this type of bell, a motor drives the clapper and produces louder sounds than a solenoid-operated bell. Electrical interference is eliminated and less current is required to operate the bell.

Self-contained sirens are combinations of speakers and sound equipment. They produce siren sounds and can be used for voice announcements. Self-contained siren packaging saves installation time. If more than one siren is required, they must be synchronized.

Figure 32 Typical bell with a strobe light.

CEILING-MOUNTED   WALL-MOUNTED

Figure 31 Typical ceiling-mounted and wall-mounted strobe devices.

Figure 34 Voice evacuation system.

A horn (*Figure 33*) usually consists of a continuously vibrating membrane or a piezoelectric element. In a supervised notification circuit, all devices are polarized, allowing current to flow in one direction only. A buzzer uses less power and consists of a continuously vibrating membrane like a horn. Chimes are electronic devices, like self-contained sirens, and have a very low current draw.

4.1.3 Voice Evacuation Systems

With a voice evacuation system (*Figure 34*), building occupants can be given instructions in the event of an emergency. A voice evacuation system can be used in conjunction with an FACP or as a stand-alone unit with a built-in power supply and battery charger. Voice announcements (*Figure 35*) can be made to inform occupants what the problem is or how to evacuate. In many cases, the voice announcements are prerecorded and selected as required by the system. The announcements can

also be made from a microphone located at the FACP or at a remote panel. *Figure 36* shows several types of speakers used for voice evacuation systems. Temporal-Three signaling is not used with zoned voice evacuation systems.

4.2.0 Signal Considerations

Closed doors may drop the audible decibel (dBA) sound levels of alarms below those required to wake children or hearing-impaired adults. Air conditioners, room humidifiers, and other equipment may cause noise levels to increase. Also, sirens, horns, and bells may fail or be disabled by the fire. Multiple fire sounders provide redundancy and, if properly placed, will provide ample decibel levels to wake all sleeping occupants.

The following are descriptions of the types of code-authorized signals supplied to various notification devices from a typical fire control panel:

- *General signal* – General signals operate throughout the entire building. Evacuation signals require a distinctive signal. The code requires that a Temporal-Three signal pattern be used that is in accordance with *ANSI Standard S3.41* (and *ISO 8201*), NFPA 101®, and NFPA 72®. Temporal-Three signals consist of three short 1/2-second tones with 1/2-second pauses between the tones. This is followed by a 1-second silent period, and then the process is repeated.
- *Attendant signal* – Attendant signaling is used where assisted evacuation is required due to such factors as age, disability, and restraint. Such systems are commonly used in conjunction with coded chimes throughout the area or a coded voice message to advise staff personnel of the location of the alarm source. A coded message might be announced such as, "Mr. Green, please report to the third floor

Figure 33 Typical horn with a strobe light.

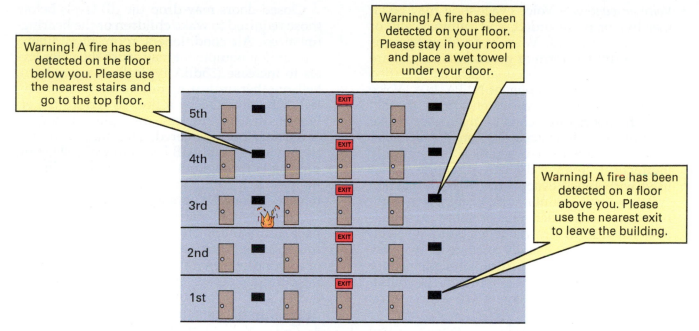

Figure 35 Typical voice evacuation messages.

(A)   (B)   (C)

Figure 36 Typical speakers.

nurses' station." No **general alarm** needs to be sounded, and no indicating appliances need to be activated throughout the area. The attendant signal feature requires the approval of the AHJ.

- *Pre-signal* – A pre-signal is an attendant signal with the addition of human action to activate a general signal. It is also used when the control delays the general alarm by more than one minute. The pre-signal feature requires the approval of the AHJ. A signal to remote locations must activate upon an initial alarm signal.

- *Positive alarm sequence* – When an alarm is initiated, the general alarm is not activated for 15 seconds. If a staff person acknowledges the alarm within the 15-second window, the general alarm is delayed for three minutes so that the staff can investigate. Failure to manually acknowledge the initial attendant signal will automatically cause a general alarm. Failure to abort the acknowledged signal within 180 seconds will also automatically cause a general

alarm. If any second selected automatic detector is activated during the delay window, the system will immediately cause a general alarm. An activation of a manual station will automatically cause a general alarm. This is an excellent technique for the prevention of unwanted general alarms; however, trained personnel are an integral part of these systems. The positive alarm sequence feature requires the approval of the AHJ.

Digital Voice Messages

Digital voice messages used with various voice evacuation systems are stored in nonvolatile read-only memory integrated circuit chips. These stored messages are used in applications where clear and defined messages of evacuation are required. One major fire alarm manufacturer of voice evacuation systems has more than 100 pre-stored evacuation messages available.

- *Voice evacuation* – Voice evacuation may be either live or prerecorded, and automatically or manually initiated. Voice evacuation systems are a permitted form of general alarm and are required in certain occupancies, as specified in the applicable chapter of NFPA 101®. Voice evacuation systems are often zoned so that only the floors threatened by the fire or smoke are immediately evacuated. Zoned evacuation is used where total evacuation is physically impractical. Temporal-Three signaling is not used with zoned evacuation. Voice evacuation systems must maintain their ability to communicate even when one or more zones are disabled due to fire damage. Buildings with voice evacuation usually include a two-way communication system (fireman's phone) for emergency communications. High-rise buildings are required to have a voice evacuation system.

- *Alarm sound levels* – Where a general alarm is required throughout the premises, audible signals must be clearly heard above maximum ambient noise under normal occupancy conditions. Public area audible alarms should be 75dBA at 10' (3 m) and a maximum of 120dBA at minimum hearing distance. Public area audible alarms must be at least 15dBA above ambient sound level, or a maximum sound level that lasts over 60 seconds measured at 5' (1.5 m) above finished floor (AFF) level. Audible alarms to alert persons responsible for implementing emergency plans (guards, monitors, supervisory personnel, or others) must be between 45dBA and 120dBA at minimum hearing distance. If an average sound level greater than 105dBA exists, the use of a visible signal is required. Typical average ambient sound levels from NFPA 72® are given in *Table 2*.

> **NOTE**
>
> The data in *Table 2* is merely a guide. Actual conditions may vary.

Closed doors may drop the dB levels below those required to wake children or the hearing-impaired. Air conditioners, room humidifiers, and similar equipment may cause the noise levels to increase (55dBA typical). A siren, horn, bell, or other **indicating device** (appliance) may fail or be disabled by a fire. To reduce the effect of disabled devices, multiple fire sounders should be considered. This may also help in providing ample dB levels to wake sleeping occupants throughout the structure (70dBA at the pillow is commonly accepted as sufficient to wake a sleeping person).

All residential alarm sounding devices must have a minimum dB rating of 85dBA at 10' (3 m). An exception to this rule is when more than one sounding device exists in the same room. A sounding device in the room must still be 85dBA at 10' (3 m), but all additional sounding devices in that room may have a rating as low as 75dBA at 10' (3 m). Sound intensity doubles with each 3dB gain and is reduced by one-half with each 3dB loss. Doubling the distance to the sound source will cause a 6dB loss. Some other loss considerations are given in *Table 3*.

- *Coded and non-coded signals* – A **coded signal** is a signal that is pulsed in a prescribed code for each round of transmission. For example, four pulses would indicate an alarm on the fourth floor. A minimum of three rounds and a minimum of three impulses are required for an alarm signal. Temporal-Three is not a coded signal and is only intended to be a distinct, general fire alarm signal. A **non-coded signal** is a signal that is energized continuously by the control. It may pulse, but the pulsing will not be designed to indicate any code or message.

Coded signals are usually used in manually operated devices on which the act of pulling a lever causes the transmission of not less than three rounds of coded alarm signals. These devices are similar to the non-coded type, except that instead of a manually operated switch, a mechanism to rotate a code wheel is utilized.

Table 2 Typical Average Ambient Sound Levels

Area	Sound Level (dBA)	Area	Sound Level (dBA)
Mechanical rooms	85	Educational occupancies	45
Industrial occupancies	80	Underground structures	40
Busy urban thoroughfares	70	Windowless structures	40
Urban thoroughfares	55	Mercantile occupancies	40
Institutional occupancies	50	Places of assembly	40
Vehicles and vessels	50	Residential occupancies	35
Business occupancies	45	Storage	30

Table 3 Typical Sound Loss at 1,000Hz

Area	Loss (dBA)
Stud wall	41
Open doorway	4
Typical interior door	11
Typical fire-rated door	20
Typical gasketed door	24

Rotation of the code wheel, in turn, causes an electrical circuit to be alternately opened and closed, or closed and opened, thus sounding a coded alarm that identifies the location of the box. The code wheel is cut for the individual code to be transmitted by the device and can operate by clockwork or an electric motor.

Clockwork transmitters can be prewound or can be wound by the pulling of the alarm lever. Usually, the box is designed to repeat its code four times before automatically coming to rest. Prewound transmitters must sound a trouble signal when they require rewinding. Solid-state electronic coding devices are also used in conjunction with the fire alarm control panel to produce coded sounding of the system's audible signaling appliances.

- *Visual appliance signals* – Notification signals for occupants to evacuate must be by audible and visible signals in accordance with NFPA 72® and CABO/ANSI A117.1. However, exceptions to this rule are possible. Under the existing NFPA 101® building chapter, only audible signals are required in premises where:

  - No hearing-impaired occupant is ever present under normal operation

  - In hotels and apartments where special rooms are made available to the hearing-impaired

  - Where the AHJ approves alternatives to visual signals (ADA codes may or may not allow these exceptions)

4.3.0 Communication and Monitoring

Communication is a means of sending information to personnel who are too far away to directly see or hear a fire alarm system's notification devices. It is the transmission and reception of information from one location, point, person, or piece of equipment to another. Understanding the information that a fire alarm system communicates makes it easier to determine what is happening at the alarm site. Knowing how that information gets from the alarm site to the monitoring site is helpful if a problem occurs somewhere in between.

4.3.1 Monitoring Options

There are several options for monitoring the signals of an alarm system. They include the following:

- *Central station* – A location, normally run by private individuals or companies, where operators monitor receiving equipment for incoming fire alarm system signals. The central station may be a part of the same company that sold and installed the fire alarm system. It is also common for the installing company to contract with another company to do the monitoring on its behalf.

- *Proprietary* – A facility similar to a central station except that the notification devices are located in a constantly staffed room maintained by the property owner for internal safety operations. The personnel may respond to alarms, alert local fire departments when alarms are activated, or both.

- *Certified central station* – Monitoring facilities that are constructed and operated according to a standard and are inspected by a listing agency to verify compliance. Several organizations, including the UL, publish criteria and list those central stations that conform to those criteria.

4.3.2 Digital Communicators

Digital communicators use standard telephone lines or wireless telephone service to send and receive data. Costs are low using this method because existing voice lines may be used, eliminating the need to purchase additional communication lines. Standard voice-grade telephone lines are also easier to repair than special fire alarm communication lines.

Digital communicators are connected to a standard, voice-grade telephone line through a special connecting device called the *RJ31-X* (*Figure 37*). The RJ31-X is a modular telephone jack into which a cord from the digital communicator is plugged. The RJ31-X separates the telephone company's equipment from the fire alarm system equipment and is approved by the Federal Communications Commission (FCC).

Although using standard telephone lines has several advantages, problems may arise if the customer and the fire alarm system both need the phone at the same time. In the event of an alarm, a technique known as *line seizure* gives the fire alarm system priority. The digital communicator is connected to the phones and can control or seize the line whenever it needs to send a signal. If the customer is using the telephone when the alarm system needs to send a signal, the digital communicator will disconnect the customer until the alarm signal has been sent. Once the signal is sent, the customer's phones are reconnected.

The typical sequence that occurs when the digital communicator for an alarm system is activated is shown in *Figure 38*. When an alarm is to be sent, the digital communicator energizes the seizure relay. The activated relay disconnects the house phones and connects the communicator to the Telco line. After the communicator detects a signal called the *kiss-off tone*, it de-energizes the seizure relay. This disconnects the communicator and reconnects the house phones.

In combination fire and security systems, the fire alarm supersedes the security alert. The RJ31-X will send the fire alarm signal before sending the security alert signal.

The RJ31-X is a modular connection and, like a standard telephone cord, it can unplug easily. If the cord remains unplugged from the jack, the digital communicator will be disconnected from the telephone line until the cord is reconnected. This creates a problem because the digital communicator cannot reach the digital receiver without the telephone line. This reduces the effectiveness of the alarm system, even though local notification devices are activated.

The NFPA uses the following terms to refer to digital communications:

- **Digital Alarm Communicator Receiver (DACR)** – This is a system component that will accept and display signals from the **Digital Alarm Communicator Transmitter (DACT)** sent over the **Public Switched Telephone Network**.
- **Digital Alarm Communicator System (DACS)** – This is a system in which signals are transmitted from a DACT (located in the secured area) through the public switched telephone network to a DACR.

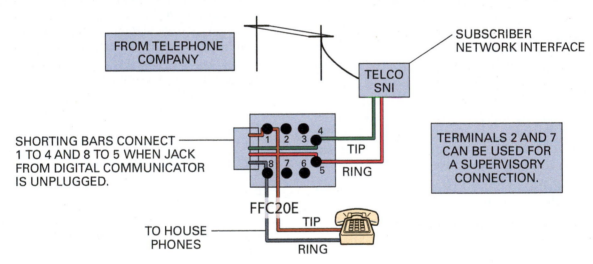

Figure 37 RJ31-X connection device.

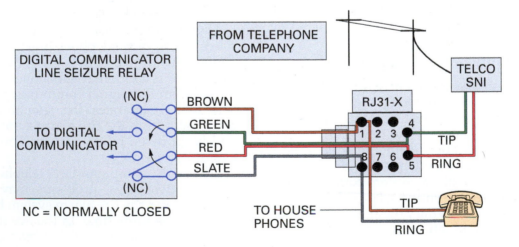

Figure 38 Line seizure.

- *Digital Alarm Communicator Transmitter (DACT)* – This is a device that sends signals over the public switched telephone network to a DACR.

The following conditions have been established by NFPA 72® with regard to digital communicators:

- They can be used as a **remote supervising station fire alarm system** when acceptable to the AHJ.
- Only loop start and not ground start lines can be used.
- The communicator must have line seizure capability.
- A failure-to-communicate signal must be shown if ten attempts are made without getting through.
- They must connect to two separate phone lines at the **protected premises**. Exception: The secondary line may be a radio system (this does not apply to household systems).
- Failure of either phone line must be annunciated at the premises within four minutes of the failure.
- If long distance telephone service, including **Wide Area Telephone Service (WATS)** is used, the second telephone number shall be provided by a different long distance provider, where available.

- Each communicator shall initiate a test call to the central station at least once every 24 hours (this does not apply to household systems).

4.3.3 Cellular Backup

Some fire alarm systems that rely on phone lines for communications have a cellular backup system that utilizes wireless phone technology to restore communications in the event of a disruption in normal telephone line service (*Figure 39*).

Telephone Line Problems

The RJ31-X can be helpful in determining the nature of a problem with the telephone line. One method of determining if the fire alarm system has caused a problem with the telephone lines is to unplug the cord to the RJ31-X. This disconnects the fire alarm system and restores the connections of the telephone line. If the customer's telephones function properly when the cord is unplugged, the fire alarm system may be causing the problem. If the problem with the telephone line persists with the cord to the RJ31-X unplugged, the source of the problem is not the alarm system.

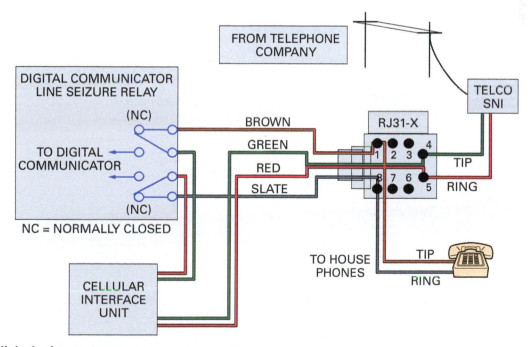

Figure 39 Cellular backup system.

4.0.0 Section Review

1. The ADA specifies that fire alarm strobe lights must be clear or _____.

   a. red
   b. white
   c. blue
   d. yellow

2. At minimum hearing distance, an audible alarm should be no louder than _____.

   a. 10dBA
   b. 50dBA
   c. 75dBA
   d. 120dBA

3. Digital communicators connect to telephone lines with a(n) _____.

   a. RJ31-X
   b. RJ11
   c. RJ45
   d. TIA-568B

5.0.0 INSTALLATION GUIDELINES

Objective

Describe fire alarm system installation guidelines and requirements.

  a. Describe the general wiring requirements.
  b. Describe the general installation requirements for wiring and various components.
  c. Describe the installation guidelines for totally protected premises.
  d. Describe the installation guidelines for fire alarm-related systems and devices.
  e. Describe how to troubleshoot fire alarm systems.

Performance Task

  1. Connect selected fire alarm system(s).

Trade Terms

Ceiling height: The height from the continuous floor of a room to the continuous ceiling of a room or space.

Certification: A systematic program using randomly selected follow-up inspections of the certified system installed under the program, which allows the listing organization to verify that a fire alarm system complies with all the requirements of the NFPA 72® code. A system installed under such a program is identified by the issuance of a certificate and is designated as a certificated system.

Control unit: A device with the control circuits necessary to furnish power to a fire alarm system, receive signals from alarm initiating devices (and transmit them to audible alarm indicating appliances and accessory equipment), and electrically supervise the system installation wiring and primary (main) power. The control unit can be contained in one or more cabinets in adjacent or remote locations.

Fault: An open, ground, or short condition on any line(s) extending from a control unit, which could prevent normal operation.

Ground fault: A condition in which the resistance between a conductor and ground reaches an unacceptably low level.

Initiating device circuit (IDC): A circuit to which automatic or manual signal-initiating devices such as fire alarm manual boxes (pull stations), heat and smoke detectors, and water flow alarm devices are connected.

Spot-type detector: A device in which the detecting element is concentrated at a particular location. Typical examples are bimetallic detectors, fusible alloy detectors, certain pneumatic rate-of-rise detectors, certain smoke detectors, and thermoelectric detectors.

Stratification: The phenomenon in which the upward movement of smoke and gases ceases due to a loss of buoyancy.

Visible notification appliance: A notification appliance that alerts by the sense of sight.

This section contains general installation information applicable to all types of fire alarm systems. For specific information, always refer to the manufacturer's instructions, the building drawings, and all applicable local and national codes.

5.1.0 General Wiring Requirements

NEC Article 760 specifies the wiring methods and special cables required for fire protective signaling systems. The following special cable types are used in protective signaling systems:

- Power-limited fire alarm (FPL) cable
- Power-limited fire alarm riser (FPLR) cable
- Power-limited fire alarm plenum (FPLP) cable
- Nonpower-limited fire alarm (NPLF) circuit cable
- Nonpower-limited fire alarm riser (NPLFR) circuit riser cable
- Nonpower-limited fire alarm plenum (NPLFP) circuit cable

In addition to *NEC Article 760*, the following *NEC®* articles cover other items of concern for fire alarm system installations:

- *NEC Sections 110.11 and 300.6(A), (B), and (C), Corrosive, Damp, or Wet Locations*
- *NEC Section 300.21, Spread of Fire or Products of Combustion*
- *NEC Section 300.22, Ducts, Plenums, and Other Air Handling Spaces*
- *NEC Articles 500 through 516 and 517, Part IV, Locations Classified as Hazardous*
- *NEC Article 695, Fire Pumps*

- *NEC Article 725, Remote-Control and Signaling Circuits (Building Control Circuits)*
- *NEC Article 770, Fiber Optics*
- *NEC Article 800, General Communications Systems*
- *NEC Article 801, Communications Circuits*
- *NEC Article 810, Radio and Television Equipment*

In addition to the *NEC*®, some AHJs may specify requirements that modify or add to the *NEC*®. It is essential that a person or firm engaged in fire alarm work be thoroughly familiar with the *NEC*® requirements, as well as any local requirements for fire alarm systems.

Fire alarm circuits must be installed in a neat and workmanlike manner. Cables must be supported by the building structure in such a manner that the cables will not be damaged by normal building use. One way to determine accepted industry practice is to refer to nationally recognized standards such as *Commercial Building Telecommunications Wiring Standard, ANSI/EIA/TIA 568; Commercial Building Standard for Telecommunications Pathways and Spaces, ANSI/EIA/TIA 569*; and *Residential and Light Commercial Telecommunications Wiring Standard, ANSI/EIA/TIA 570*.

5.2.0 Installation Requirements

Fire alarm devices will function as the manufacturer intended only when they are properly installed. Mounting, wiring, spacing, and circuit connections must all be correct or the system may not alarm reliably or at all. Fire alarm codes as well as the manufacturer's instructions provide the specific details that govern installation. The following sections summarize many of the key ideas behind a good installation.

5.2.1 Access to Equipment

Access to equipment must not be blocked by an accumulation of wires and cables that prevents removal of panels, including suspended ceiling panels.

5.2.2 Fire Alarm Circuit Identification

Fire alarm circuits must be identified at the control and at all junctions as fire alarm circuits. Junction boxes must be clearly marked as fire junction boxes to prevent confusion with commercial light and power. AHJs differ on what constitutes clear marking. Check the requirements before starting work. The following are examples of some AHJ-acceptable markings:

- Red painted cover

- The words Fire Alarm on the cover
- Red painted box
- The word Fire on the cover
- Red stripe on the cover

5.2.3 Power-Limited Circuits in Raceways

Power-limited fire circuits must not be run in the same cable, raceway, or conduit as high-voltage circuits. Examples of high-voltage circuits are electric light, power, and nonpower-limited fire (NPLF) circuits. When NPLF cables must be run in the same junction box, they must be run in accordance with the *NEC*®, including maintaining a ¼" (6 mm) spacing from Class 1, power, and lighting circuits.

5.2.4 Mounting of Detectors

Observe the following precautions when mounting detectors:

- Circuit conductors are not supports and must not be used to support the detector.
- Plastic masonry anchors should not be used for mounting detectors to gypsum drywall, plaster, or drop ceilings.
- Toggle or winged expansion anchors should be the minimum used for gypsum drywall.
- The best choice for mounting is an electrical box. Most equipment is designed to be fastened to a standard electrical box.

All fire alarm devices should be mounted to the appropriate electrical box as specified by the manufacturer. If not mounted to an electrical box, fire alarm devices must be mounted by other means as specified by the manufacturer. The precaution against using plastic masonry anchors does not mean that a plastic anchor designed for use in drywall cannot be used in drywall. Masonry plastic anchors do not open as far on the end as drywall plastic anchors. Always read and follow the manufacturer's instructions; doing so is necessary to meet the requirements of the UL listing.

5.2.5 Outdoor Wiring

Fire alarm circuits extending beyond one building are governed by the *NEC*®. The *NEC*® sets the standards on the size of cable and methods of fastening required for cabling. Some manufacturers prohibit any aerial wiring. The *NEC*® also specifies clearance requirements for cable from the ground. Overhead spans of open conductors and open multi-conductor cables of not over 600V, nominal, must be at least 10' (3 m) above finished grade, sidewalks, or from any platform or projec-

Firestopping Materials

Shown here are just a few examples of the wide variety of firestopping materials on the market. Firestopping and fireproofing are not the same thing. Firestopping is intended to prevent the spread of fire and smoke from room to room through openings in walls and floors. Fireproofing is a thermal barrier that causes a fire to burn more slowly and retards the spread of fire.

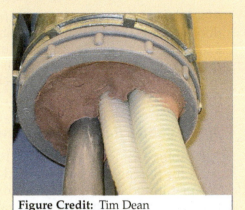

Figure Credit: Tim Dean

(A) FIRESTOPPING PUTTY

Figure Credit: Hilti North America

(B) FIRESTOPPING SLEEVES

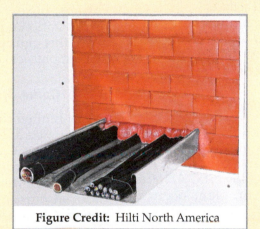

Figure Credit: Hilti North America

(C) FIRESTOPPING BRICKS AND FOAM

tion from which they might be reached, where the supply conductors are limited to 150V to ground and accessible to pedestrians only. Additional requirements apply for areas with vehicle traffic. In addition, the *NEC®* states that fire alarm wiring can be attached to the building, but must be protected against physical damage as afforded by baseboards, door frames, ledges, etc. per *NEC Section 760.130(B)(1)*.

5.2.6 Fire Seals

Electrical equipment and cables must not be installed in a way that might help the spread of fire. The integrity of all fire-rated walls, floors, partitions, and ceilings must be maintained. An approved sealant or sealing device must be used to fill all penetrations. Any wall that extends from the floor to the roof or from floor-to-floor should be considered a firewall. In addition, raceways and cables that go from one room to another through a fire barrier must be sealed.

5.2.7 Wiring in Air Handling Spaces

Wiring in air handling spaces requires the use of approved wiring methods, including:

- Special plenum-rated cable
- Flexible metal tubing (Greenfield)
- Electrical metallic tubing (EMT)
- Intermediate metallic conduit (IMC)
- Rigid metallic conduit (hard wall or Schedule 80 conduit)

Standard cable tie straps are not permissible in plenums and other air handling spaces. Ties must be plenum rated. Bare solid copper wire used in short sections as tie wraps may be permitted by most AHJs. Fire alarm equipment is permitted to be installed in ducts and plenums only to sense the air. All splices and equipment must be contained in approved fire-resistant and low-smoke-producing boxes.

5.2.8 Wiring in Hazardous Locations

The *NEC*® includes requirements for wiring in hazardous locations. Some areas that are considered hazardous are listed below:

- *NEC Article 511*, *Commercial Garages, Repair and Storage*
- *NEC Article 513*, *Aircraft Hangars*
- *NEC Article 514*, *Motor Fuel Dispensing Facilities*
- *NEC Article 515*, *Bulk Storage Plants*
- *NEC Article 516*, *Spray Application, Dipping, Coating, and Printing Processes*
- *NEC Article 517*, *Health Care Facilities*
- *NEC Article 518*, *Assembly Occupancies*
- *NEC Article 520*, *Theaters and Similar Locations*
- *NEC Article 545*, *Manufactured Buildings*
- *NEC Article 547*, *Agricultural Buildings*

5.2.9 Remote Control Signaling Circuits

Building control circuits (*Figure 40*) are normally governed by *NEC Article 725*. However, circuit wiring that is both powered and controlled by the fire alarm system is governed by *NEC Article 760*. A common residential problem occurs when using a system-type fire alarm that is a combination burglar and fire alarm. Many believe that the keypad is only a burglar alarm device. If the keypad is also used to control the fire alarm, it is a fire alarm device, and the cable used to connect it to the control panel must comply with

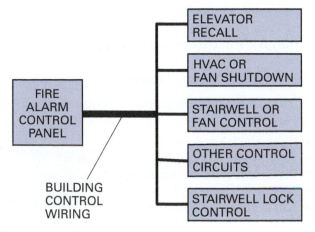

Figure 40 Building control circuits.

NEC Article 760. Also, motion detectors that are controlled and powered by the combination fire alarm and burglar alarm power must be wired in accordance with *NEC Article 760* (fire-rated cable).

5.2.10 Cables Running Floor to Floor

Riser cable is required when wiring runs from floor to floor. The cable must be labeled as passing a test to prevent fire from spreading from floor to floor. An example of riser cable is FPLR. This requirement does not apply to one- and two-family residential dwellings.

5.2.11 Cables Running in Raceways

All cables in a raceway must have insulation rated for the highest voltage used in the raceway. Power-limited wiring may be installed in raceways or conduit, exposed on the surface of a ceiling or wall, or fished in concealed spaces. Cable splices or terminations must be made in listed fittings, boxes, enclosures, fire alarm devices, or utilization equipment. All wiring must enter boxes through approved fittings and be protected against physical damage.

5.2.12 Cable Spacing

Power-limited fire alarm circuit conductors must be separated at least 2" (50 mm) from any electric light, power, Class 1, or nonpower-limited fire alarm circuit conductors. This is to prevent damage to the power-limited fire alarm circuits from induced currents caused by the electric light, power, Class 1, or nonpower-limited fire alarm circuits.

5.2.13 Elevator Shafts

Wiring in elevator shafts must directly relate to the elevator and be installed in rigid metallic conduit, rigid nonmetallic conduit, EMT, IMC, or up to 6' (1.8 m) of flexible conduit.

5.2.14 Terminal Wiring Methods

The wiring for circuits using EOL terminations must be done so that removing the device causes a trouble signal (*Figure 41*).

5.2.15 Conventional Initiation Device Circuits

There are three styles of Class B and two styles of Class A conventional initiation device circuits listed in NFPA 72®. Various local codes address what circuit types are required. NFPA 72® de-

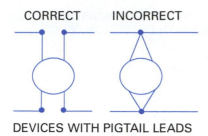

CORRECT   INCORRECT

DEVICES WITH PIGTAIL LEADS

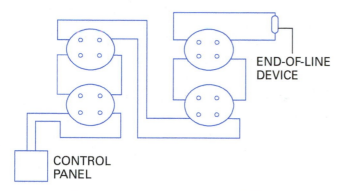

END-OF-LINE
DEVICE

CONTROL
PANEL

Figure 41 Correct wiring for devices with EOL
terminations.

scribes how they are to operate. A brief explanation of some of the circuits follows:

- *Class B, Style A* – Fire alarm control panels (FACPs) using this style are no longer made in the United States, but a few systems remain in operation. A single (wire) open and a **ground fault** are the only types of trouble that can be indicated, and the system will not receive an alarm in a ground fault condition. An alarm is initiated with a wire-to-wire short.
- *Class B, Style B* – (*Figure 42*) The FACP is required to receive an alarm from any device up to a break with a single open. An alarm is initiated with a wire-to-wire short. A trouble signal is generated for a circuit ground or open using an end-of-line (EOL) device that is usually a resistor. An alarm can also be received with a single ground fault on the system.
- *Class B, Style C* – (*Figure 43*) This style, while used in the United States, is more common

in Europe. An open circuit, ground, or wire-to-wire short will cause a trouble indication. Devices or detectors in this type of circuit require a device (normally a resistor) in series with the contacts in order for the panel to detect an alarm condition. The panel will receive an alarm signal with a single ground fault on the system. The current-limiting resistor is normally lower in resistance than the end-of-line resistor.

- *Class A, Style D* – (*Figure 44*) In this type of circuit, an open or ground will cause a trouble signal. Shorting across the initiation loop will cause an alarm. Activation of any initiation device will result in an alarm, even when a single break or open exists anywhere in the circuit, because of the back loop circuit. The loop is returned to a special condition circuit, so there is no end of line and, therefore, no EOL device.
- *Class A, Style E* – (*Figure 45*) This style is an enhanced version of Style D. An open circuit, ground, or wire-to-wire short is a trouble condition. All devices require another device (normally a resistor) in series with the contacts to generate an alarm. Activation of any of the initiating devices will result in an alarm even if the initiation circuit has a single break or the system has a single ground.

The style of circuit can affect the following:

- The maximum quantity of each type of device permitted on each circuit
- The maximum quantity of circuits allowed for a fire alarm control panel/communicator
- The maximum quantity of buildings allowed for a signaling line circuit (SLC)
- The maximum quantity of signaling circuits and buildings allowed for a monitoring station

5.2.16 Notification Appliance Circuits

The following classes and styles of circuits are used for notification circuits (*Figure 46*):

- *Class B, Style W* – Class B, Style W is a two-wire circuit with an end-of-line device. Devices will

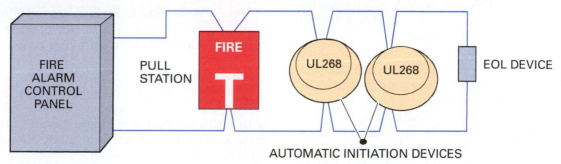

FIRE
ALARM
CONTROL
PANEL

PULL
STATION

FIRE
T

UL268   UL268

EOL DEVICE

AUTOMATIC INITIATION DEVICES

Figure 42 Typical Class B, Style B initiation circuit.

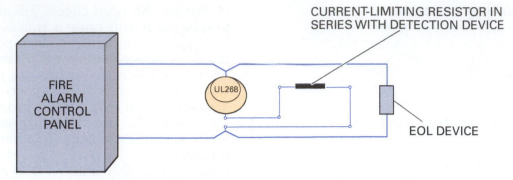

Figure 43 Typical Class B, Style C initiation circuit.

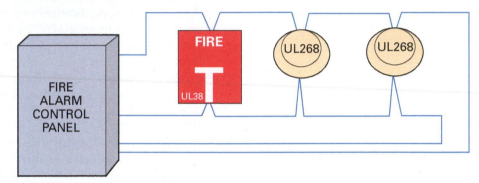

Figure 44 Typical Class A, Style D initiation circuit.

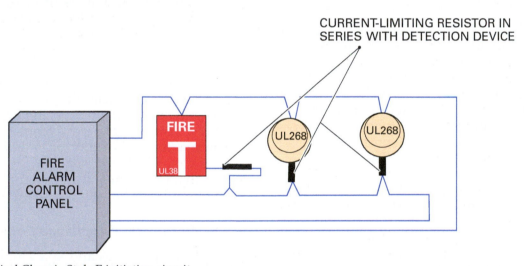

Figure 45 Typical Class A, Style E initiation circuit.

operate up to the location of a **fault**. A ground may disable the circuit.

- *Class B, Style X* – Class B, Style X is a four-wire circuit. It has alarm capability with a single open, but not during a ground fault.
- *Class B, Style Y* – Class B, Style Y is a two-wire circuit with an end-of-line device. Devices on this style of circuit will operate up to the location of a fault. Ground faults are indicated differently than other circuit troubles.
- *Class A, Style Z* – Class A, Style Z is a four-wire circuit. All devices should operate with a single ground or open on the circuit. Ground

faults are indicated differently than other circuit troubles.

Style X is similar to Style Z, except that during a ground fault, Style X will not operate and the panel will not be able to tell what type of trouble exists. Style W is similar to Style Y, except that during a ground fault, Style W will not operate and the panel will not be able to tell what type of trouble exists. Only Style Z operates all devices with a single open or a single ground fault. Only Style Z is a Class A notification appliance circuit (NAC).

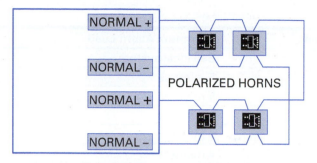

CLASS B, STYLE X OR CLASS A, STYLE Z

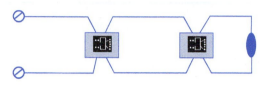

NOTIFICATION APPLIANCES

CLASS B, STYLES W AND Y

Figure 46 Typical notification appliance circuits.

All styles of circuits will indicate a trouble alarm at the premises with a single open and/or a single ground fault. Styles Y and Z have alarm capability with a single ground fault. Styles X and Z have alarm capability with a single open. In Class B circuits, the panel monitors whether or not the wire is intact by using an EOL device (wire supervision). Electrically, the EOL device must be at the end of the indicating circuit; however, Style X is an exception. Examples of EOL devices are resistors to limit current, diodes for polarity, and capacitors for filtering.

5.2.17 Primary Power Requirements

If more power is needed than can be supplied by one 20A circuit, additional circuits may be used, but none may exceed 20A. To help prevent system damage or false alarms caused by electrical surges or spikes, surge protection devices should be installed in the primary power circuits unless the fire alarm equipment has self-contained surge protection. In addition, some jurisdictions may require breaker locks and/or other means of identifying circuit breakers for FACPs.

5.2.18 Secondary Power Requirements

An approved generator supply or backup batteries must be used to supply the secondary power of a fire alarm system. The secondary power system must, upon loss of primary power, immediately keep the fire alarm functioning for at least as long as indicated in *Table 4*.

Network Command Center

This PC-based network command center is used to display event information from local or wide area network devices in a text or graphic format. When a device initiates an alarm, the appropriate graphic floor plan is displayed along with operator instructions. Other capabilities of this command center include event history tracking and fire panel programming/control.

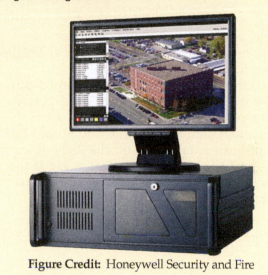

Figure Credit: Honeywell Security and Fire

5.3.0 Total Premises Fire Alarm System Installation

This section covers the requirements for the proper installation, testing, and certification of fire alarm systems and related systems for totally protected premises.

5.3.1 Manual Fire Alarm Box (Pull Station) Installation

The following guidelines apply to the installation of a manual fire alarm box (pull station):

- Manual pull stations must be *UL 38*-listed or the equivalent. Multi-purpose keypads cannot be used as fire alarm manual pull stations unless *UL 38*-listed for that purpose.
- Manual pull stations must be installed in the natural path of escape, near each required exit from an area, in occupancies that require manual initiation. Ideally, they should be located near the doorknob edge of the exit door. In any case, they must be no more than 5' (1.5 m) from the exit (*Figure 47*). In most cases, manual pull stations are installed with the actuators at the

Table 4 Secondary Power Duration Requirements

NFPA Standard	Maximum Normal Load	Maximum Alarm Load
Central station	24 hours	See local system
Local system	24 hours	5 minutes
Auxiliary systems	60 hours	5 minutes
Remote stations	60 hours	5 minutes
Proprietary systems	24 hours	5 minutes
Household system	24 hours	4 minutes
Emergency voice alarm communications systems	24 hours	15 minutes maximum load 2 hours emergency operation

heights of 42" to 54" (1.1 m to 1.4 m) to conform to local ADA requirements. Most new pull stations are supplied with Grade II Braille on them for the visually impaired.

- The force required to operate manual pull stations must be no more than 5 foot-pounds (6.8 newton-meters).
- A manual pull station must be within 200' (61 m) of horizontal travel on the same floor from any part of the building (*Figure 48*). If the distance is exceeded, additional pull stations must be installed on the floor.

5.3.2 Flame Detector Installation

When installing UV or IR flame detectors, the manufacturer's instructions and NFPA 72® should be consulted. The following should be observed when installing these detectors:

- *UV flame detectors*
  - Response is based on the distance from the fire, angle of view, and fire size.
  - While some units have a 180° field of view, sensitivity drops substantially with angles of more than 45° to 50°. Normally, the field of view is limited to less than 90° (*Figure 49*).
  - If used outdoors, the unit must be listed for outdoor use.
  - A UV detector is considered solar blind, but in order to prevent false alarms, it should never be aimed near or directly at any path that the sun can take.
  - The unit must never be aimed into areas where electric arc welding or cutting may be performed.

- *IR flame detectors*
  - Response varies depending on the angle of view. At 45°, the sensitivity drops to 60% of the 0° sensitivity (*Figure 50*).
  - IR detectors cannot be used to detect alcohol, liquefied natural gas, hydrogen, or magnesium fires.
  - IR detectors work best in low light level installations. High light levels desensitize the units. Discriminating units can tolerate up to ten footcandles (108 lux) of ambient light. Non-discriminating units can tolerate up to two footcandles (22 lux) of ambient light.
  - The units must never be used outdoors.

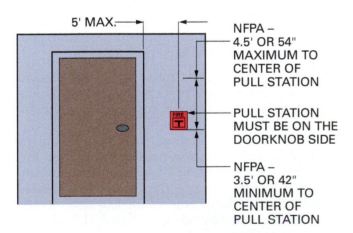

Figure 47 Pull station location and mounting height.

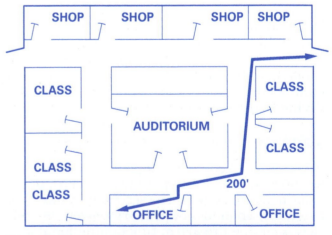

Figure 48 Maximum horizontal pull station distance from an exit.

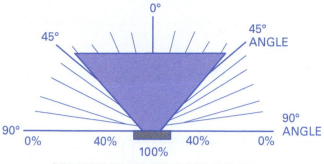

PERCENT OF RELATIVE SENSITIVITY

Figure 49 UV detector response.

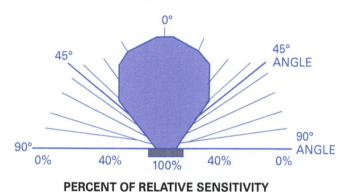

PERCENT OF RELATIVE SENSITIVITY

Figure 50 IR detector response.

5.3.3 Smoke Chamber, Smoke Spread, and Stratification

This section defines the term *smoke chamber*, and describes the smoke spread and stratification phenomena.

Before automatic smoke detectors can be installed, the area of coverage known as the *smoke chamber* must be defined. The smoke chamber is the continuous, smoke-resistant, perimeter boundary of a room, space, or area to be protected by one or more automatic smoke detectors between the upper surface of the floor and the lower surface of the ceiling. For the purposes of determining the area to be protected by smoke detectors, the smoke chamber is not the same as a smoke-tight compartment. It should be noted that some rooms that may have a raised floor and a false ceiling actually have three smoke chambers: the chamber beneath a raised floor, the chamber between the raised floor and the visible ceiling above, and the chamber between the room's visible ceiling and the floor above (or the lower portion of the roof). Few cases require detection in all these areas. However, some computer equipment rooms with a great deal of electrical power and communications cable under the raised floor may require detection in the chamber under the floor in addition to the chamber above the raised floor.

Wireless Smoke Detection System

The components that form a wireless smoke detection system are shown here. This type of system can be used in situations where the building design or installation costs make hard wiring of the detectors impractical or too expensive. The heart of this system is the translation unit, called a *gateway*. It can communicate with up to four remote receiver units. The receivers can monitor radio frequency signals from up to 80 wireless smoke detectors. Each receiver unit transmits the status of the related wireless detectors via communication wiring to the translator. The translator then communicates this status to an intelligent FACP via a signaling line circuit loop. (Note that some jurisdictions do not permit the use of these systems.)

Figure Credit: Honeywell Security and Fire

The simplest example of a smoke chamber would be a room with the door closed. If no intervening closed door exists between this and adjoining (communicating) rooms, the line denoting the barrier becomes less clear. The determining factor would be the depth of the wall section above the open archway or doorway, as follows:

- An archway or doorway that extends more than 18" (450 mm) down from the ceiling greatly delays smoke travel and is considered a boundary, just like a regular wall from floor to ceiling. A doorway that extends more than 4" (100 mm), but less than 18" (450 mm), must be considered a smoke barrier, and spacing of detectors must be reduced to ⅔ of the detectors' listed spacing distance on either side of the opening (*Figure 51*).
- Open grids above doors or walls that allow free flow of air and smoke are not considered barriers. To be considered an open grid, the

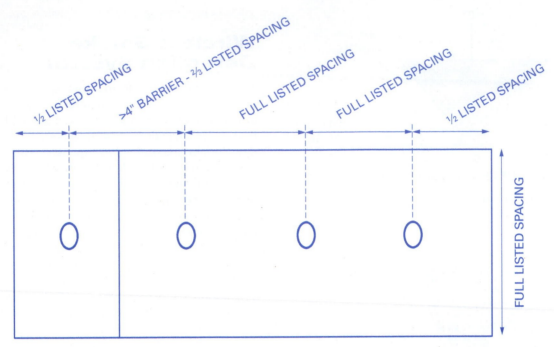

Figure 51 Reduced spacing required for a barrier.

opening must meet all of the requirements defined and demonstrated in *Figure 52*.

- Smoke doors, which are kept open by hold-and-release devices and meet all of the standards applicable thereto, may also be considered boundaries of a smoke chamber.

- If the space between the top of a low wall and the ceiling is less than 18" (450 mm), the wall is treated as if it extends to the ceiling and is a barrier. Smoke will still be able to travel to the other side of the wall, but it may be substantially delayed. This will delay notification. In this case, detector spacing must be reduced to ²⁄₃ of the listed spacing on either side of the wall. If the space between the top of a low wall and the ceiling is 18" (450 mm) or more, the wall is not considered to substantially affect the smoke travel.

In a fire, smoke and heat rise in a plume toward the ceiling because that air is lighter than the surrounding denser, cooler air. In an area with a relatively low, flat, smooth ceiling, the smoke and heat quickly spread across the entire ceiling, triggering smoke or heat detectors. When the ceiling is irregular, smoke and heated air will tend to collect, perhaps stratifying near a peak or collecting in the bays of a beamed or joist ceiling.

In the case of a beamed or joist ceiling, the smoke and heated air will fill the nearest bays and begin to overflow to adjacent bays (*Figure 53*). The process continues until smoke and heat reach either a smoke or heat detector in sufficient quantity to cause activation. Because each bay must fill before it overflows, a substantial amount of time

may pass before a detector placed at its maximum listed spacing activates. To reduce this time, the spacing for the detectors is reduced.

When smoke must rise a long distance, it tends to cool off and become denser. As its density becomes equal to that of the air around it, it stratifies, or stops rising (*Figure 54*). As a fire grows, heat is added, and the smoke will eventually rise to the ceiling. However, a great deal of time may be lost, the fire will be much larger, and a large quantity of toxic gases will be present at the floor level. Due to this delay, alternating detectors are lowered at least 3' (0.9 m) (*Figure 55*). The science of stratification is very complex and requires a fire protection engineer's evaluation to determine if stratification is a factor and, if so, to determine how much to lower the detectors to compensate for this condition.

Some conditions that cause stratification include:

- Uninsulated roofs that are heated by the sun, creating a heated air thermal block
- Roofs that are cooled by low outside temperatures, cooling the gases before they reach a detector
- HVAC systems that produce a hot layer of ceiling air
- Ambient air that is at the same temperature as the fire gases and smoke

There are no clear, set rules regarding which environments will and will not be susceptible to stratification. The factors and variables involved are beyond the scope of this course. For further

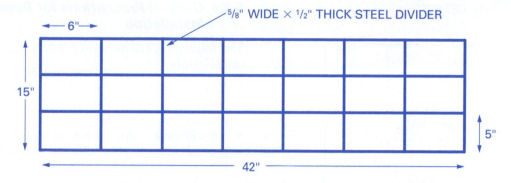

⁵⁄₈" WIDE × ¹⁄₂" THICK STEEL DIVIDER

6"

15"

5"

42"

GRID OPENINGS MUST BE AT LEAST ¼" IN THE LEAST DIMENSION.
OPENINGS ARE 5⅜" (6 MINUS ⅝") × 4⅜" (5 MINUS ⅝").
REQUIREMENT MET.

THE THICKNESS OF THE MATERIAL DOES NOT EXCEED THE LEAST DIMENSION.
OPENINGS ARE 5⅜" (6 MINUS ⅝") × 4⅜" (5 MINUS ⅝"). THICKNESS IS ½".
REQUIREMENT MET.

THE OPENINGS CONSTITUTE AT LEAST 70% OF THE AREA OF THE PERFORATED MATERIAL.
OPENINGS ARE 5⅜" (6 MINUS ⅝") × 4⅜" (5 MINUS ⅝").
5⅜" × 4⅜" = 23½ sq. in.

21 OPENINGS × 23½ sq. in. = TOTAL OPENING = 493½ sq. in.
15" × 42" = 630 sq. in.   630 sq. in. × 0.70 = 441 sq. in.

REQUIREMENT MET.

Figure 52 Detailed grid definition.

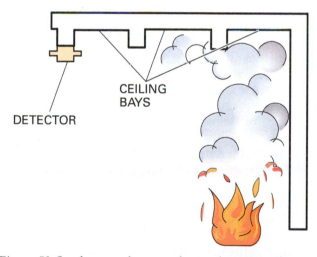

Figure 53 Smoke spread across a beamed or joist ceiling.

DETECTOR

CEILING BAYS

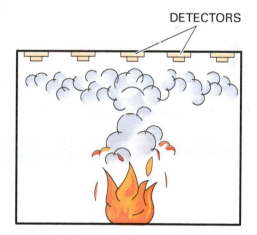

Figure 54 Smoke stratification.

DETECTORS

reading on the subject, contact the NFPA reference department for a bibliography. Note that only alternating detectors are suspended for stratification. Some fires can result in stratification caused by superheated air and gases reaching the ceiling prior to the smoke, causing a barrier that holds the smoke below the detectors. Smoke detectors with integral heat detectors should be effective in such cases, without the need to suspend alternating detectors below the ceiling.

Stratification can also occur in high-rise buildings because of the stack effect principle. This occurs when smoke rises until it reaches a neutral pressure level, where it begins to stratify (*Figure 56*). Stack effect can move smoke from the source to distant parts of the building. Penetrations left unsealed can be a major contributor in such smoke migration. Stack effect factors include:

- Building height
- Air tightness
- Air leakage between floors

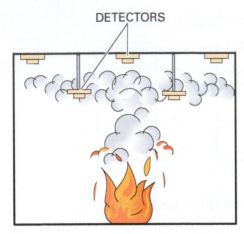

DETECTORS

Figure 55 Smoke stratification countermeasure.

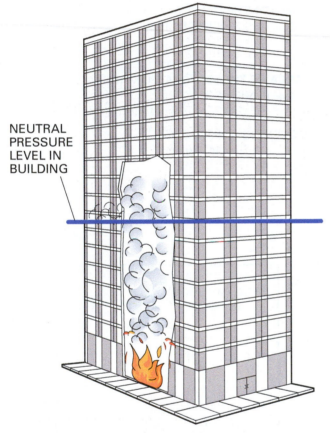

NEUTRAL PRESSURE LEVEL IN BUILDING

Figure 56 Stack effect in a high-rise building.

- Interior/exterior temperature differential
- Vertical openings
- Wind force and direction

Some factors that influence stack effect are variable, such as the weather conditions and which interior and exterior doors might be open. For this reason, the neutral pressure level may not be the same in a given building at different times. One item is very clear concerning stratification: penetrations made in smoke barriers, particularly those in vertical openings such as shafts, must be sealed.

5.3.4 *General Precautions for Detector Installation*

The following are general precautions for detector installation:

- *Recessed mounting* – Detectors must not be recess-mounted unless specially listed for recessed mounting.
- *Air diffusers* – Air movement can have a number of undesirable effects on detectors. The introduction of air from outside the fire area can dilute the smoke, which delays activation. The air movement may create a barrier, which delays or prevents smoke from reaching the detector chamber. Smoke detectors are listed for a specific range of air velocity. This means that they are expected to function with air (and smoke) moving through the detection chamber at any speed within the listed range. Detectors may not function properly if air movement is above or below the listed velocity. Air movement can be measured with an instrument called a *velocimeter*. Unless the airflow exceeds the listed velocity, spot-type smoke detectors must be a minimum of 3' (0.9 m) from air diffusers.
- *Problem locations and sources of false alarms* – Detectors, especially smoke detectors, can register false alarms as a result of the following:

  - *Electrical interference*: Keep detector locations away from fluorescent lights and radio transmitters, including cellular phones. Electrical noise or radio frequencies radiated from these devices can cause false alarms.
  - *Heating equipment*: High temperatures, dust accumulation, improper exhaust, and incomplete combustion from heating equipment are problems for detectors.
  - *Engine exhaust*: Exhaust from engine-powered forklifts, vehicles, and generators is a potential problem for ionization-type smoke detectors.
  - *Solvent and chemical fumes*: Cleaning solvents and adhesives are a problem for ionization detectors.
  - *Other gases and fumes*: Fumes from machining operations, paint spraying, industrial or battery gases, curing ovens/dryers, cooking equipment, sawing/drilling/grinding operations, and welding/cutting operations cause problems for smoke detectors.
  - *Extreme temperatures*: Avoid very hot or cold environments for smoke detector locations. Temperatures below 32°F (0°C) can cause false alarms, and temperatures above 120°F

Installation

When installing fire alarm systems, install the last initiating/notification device in the circuit first, followed by the next-to-last device, and so on while working toward the FACP. After installing each circuit device, immediately check the wiring for continuity, grounds, and/or shorts. This recommended method helps find wiring problems so that they can be corrected as the work progresses, thereby eliminating possible time-consuming troubleshooting after the system installation is completed.

(49°C) can prevent proper smoke detector operation. Extreme temperatures affect beam, ionization, and photoelectric detectors.

- *Dampness or humidity*: Smoke detectors must be located in areas where the humidity is less than 93%. In ionization detectors, dampness and high humidity can cause tiny water droplets to condense inside the sensing chamber, making it overly sensitive and causing false alarms. In photoelectric detectors, humidity can cause light refraction and loss of current flow, either of which can lead to false alarms. Common sources of moisture to avoid include slop sinks, steam tables, showers, water spray operations, humidifiers, and live steam sources.
- *Lightning*: Nearby lightning can cause electrical damage to a fire alarm system. It may also cause electrical noise or spikes to be induced in the alarm system wiring or detectors, resulting in false alarms. Surge arrestors installed in the system's primary power supply to protect the system, in conjunction with alarm verification, can reduce the chance of system damage and false alarms.
- *Dusty or dirty environments*: Dust and dirt can accumulate on a smoke detector's sensing chamber, making it overly sensitive. Avoid areas where fumigants, fog, dust, or mist-producing materials are consistently used.
- *Outdoor locations*: Dust, air currents, and humidity typically affect outdoor structures, including sheds, barns, stables, and other open structures. This makes outdoor structures unsuitable for smoke detectors.
- *Insect-infested areas*: Insects in a smoke detector can cause a false alarm. Good bug screens on a detector can prevent most adult insects from entering the detector. However, newly hatched insects may still be able to enter.

Troubleshooting

Troubleshooting older systems requires that you trace and check each detector in order to isolate the source of an alarm. The fire alarm panels of newer systems, such as the one shown here, have fault isolation messages and LCD display panels to help pinpoint which detector has alarmed or has a problem.

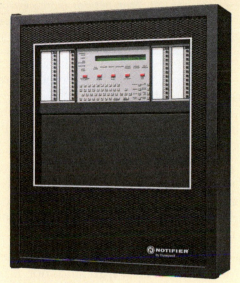

Figure Credit: Honeywell Security and Fire

An insecticide strip next to the detector may help solve the problem, but it may also cause false alarms because of fumes. Check with the manufacturer for the use of an approved strip. Ionization detectors are less prone to false alarms from insects.

- *Construction*: Smoke detectors must not be installed until after construction cleanup unless required by the AHJ for protection during construction. Detectors that are installed prior to construction cleanup must be cleaned or replaced. Prior to 1993, the code permitted covering smoke detectors to protect them from dirt, dust, or paint mist. However, the covers did not work very well and resulted in clogged, oversensitive, and damaged detectors.

5.3.5 Spot Detector Installations on Flat, Smooth Ceilings

Flat, smooth ceilings are defined as ceilings that have a slope equal to or less than 1.5" per foot (or with a pitch of less than 1.5 degrees) and do not have open joists or beams.

The following applies to conventional spot detector installation on flat, smooth ceilings:

- When a smoke detector manufacturer's specifications do not specify a particular spacing, a 30' (9 m) spacing guide may be applied.

> **NOTE**
>
> This section assumes detectors listed for 30' (9 m) spacing mounted on a smooth and flat ceiling of less than 10' (3 m) in height, where 15' (4.6 m) is one-half the listed spacing. Spacing is reduced as indicated in later sections when the ceiling height exceeds 10' (3 m), or the ceiling is not smooth and flat as defined by the code.

- The distance between heat detectors must not exceed their listed spacing.
- There must be detectors located within one-half the listed spacing measured at right angles from all sidewalls.

In the following example, the detector locations for a simple room that is 30' × 60' (9 m × 18 m) long will be determined. The locations are determined by the intersection of columns and rows marked on a sketch of the ceiling:

- The first column is located by a line that is parallel to the end wall and not more than one-half the listed spacing from the end wall (*Figure 57*).
- The first row is located by a line parallel to the sidewall that is not more than one-half the listed spacing from that sidewall (*Figure 58*).
- The first detector is located at the intersection of the row and column lines.
- The second column is located by a line that is parallel to the opposite end wall and not more than one-half the listed spacing from the end wall.
- The second detector is located at the intersection of the row and second column lines, provided that the distance between the first and

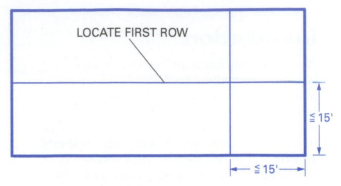

Figure 58 Locating the first row.

second detectors does not exceed the listed spacing (*Figure 59*). The only time that the full listed spacing of any detector is used is when measuring from one detector to the next.

5.3.6 Photoelectric Beam Smoke Detector Installations on Flat, Smooth Ceilings

Two configurations of beam detectors are used for open area installations. Beam smoke detectors can be installed as straight-line devices or as angled-beam devices that employ mirrors. When used as angled-beam devices, the beam length is reduced for the number of mirrors used, as specified in NFPA 72®. When installing either type of beam smoke detector, always follow the manufacturer's instructions.

Photoelectric beam smoke detectors have two listings: one for width of coverage and one for the minimum/maximum length of coverage (beam length), as shown in *Figure 60*. The beam length is listed by the manufacturer and, in most cases, so is the width coverage spacing (S). When a manufacturer does not list the width coverage, the 60' (18 m) guideline specified in NFPA 72® must be used as the width coverage. The distance of the beam from the ceiling should normally be between 4" and 12" (100 mm and 300 mm). However, NFPA allows a greater distance to compensate for stratification.

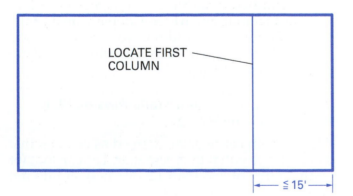

Figure 57 Locating the first column.

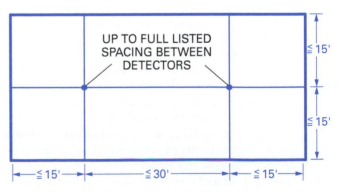

Figure 59 First and second detector locations.

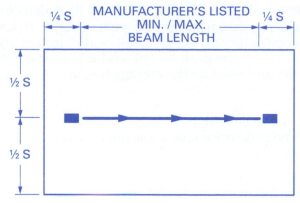

Figure 60 Maximum straight-line single-beam smoke detector coverage (ceiling view).

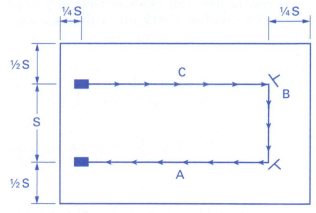

Figure 61 Coverage for a two-mirror beam detector installation.

Always make sure the beam does not cross any expansion joint or other point of slippage that could eventually cause misalignment of the beam. In large areas, parallel beam detectors may be installed, separated by no more than their listed spacing (S).

The use of mirrors to direct the light beam from the transmitter in a path other than a single straight line is permitted if the manufacturer's directions are followed and the mirror used is listed for the model of the beam detector being installed. The use of a mirror will require a reduction in the listed beam length of the detector. When using a mirror, the beam length is the distance between the transmitter and receiver (measured from the transmitter to the mirror, plus the distance from the mirror to the receiver). This distance is reduced to 66% of the manufacturer's listed beam length for a single mirror, and to 44% of the manufacturer's listed beam length for two mirrors. Maximum coverage for a two-mirror beam detector is shown in *Figure 61*. Width spacing (S) is defined as previously stated for single-beam and double-beam installations. The planner should also allow for additional installation and service time when mirrors are used. As with beam detectors without mirrors, make sure that the light beam does not cross a building expansion joint.

5.3.7 Spot Detector Installations on Irregular Ceilings

Any ceiling that is not flat and smooth is considered irregular. For heat detectors, ceilings of any description that are over 10' (3 m) above floor level are considered irregular. Heat detectors located on any ceiling over 10' (3 m) above finished floor (AFF) level must have their spacing reduced below their listed spacing in accordance with NFPA 72®. This is because hot air is diluted by the surrounding cooler air as it rises, which reduces its

Beam-Type Detectors

Objects other than smoke can interfere with beam-type detectors. For example, balloons released by children in malls have been known to activate the fire protection system.

temperature. Smoke detector spacing is not adjusted for high ceilings.

Irregular ceilings include sloped, solid joist, and beam ceilings:

- *Sloped ceilings* – A sloped ceiling is defined as a ceiling having a slope of more than 1.5" per foot or a pitch equal to or above 1.5 degrees. Any smooth ceiling with a slope equal to or less than 1.5" per foot or with a pitch below 1.5 degrees is considered a flat ceiling. Sloped ceilings are usually shed or peaked types.

    - *Shed*: A shed ceiling is defined as having the high point at one side with the slope extending toward the opposite side.
    - *Peaked*: A peaked ceiling is defined as sloping in two directions from its highest point. Peaked ceilings include domed or curved ceilings.

- *Solid joist or beam ceilings* – Solid joist or beam ceilings are defined as being spaced less than 3' (0.9 m) center-to-center with a solid member extending down from the ceiling for a specified distance. For heat detectors, the solid member must extend more than 4" (100 mm) down from the ceiling. For smoke detectors, the solid member must extend down more than 8" (200 mm).

Shed ceilings having a rise greater than 1.5"/1' run (pitch of 1.5 degrees and above) must have the first row of detectors (heat or smoke) located on the ceiling within 3' (0.9 m) of the high side of the ceiling (measured horizontally from the sidewall).

Heat detectors must have their spacing reduced in areas with high ceilings (over 10'/3 m). Smoke detector spacing is not adjusted due to the ceiling height.

For a roof slope of less than 30°, determined as shown in *Figure 62*, all heat detector spacing must be reduced based on the height at the peak. For a roof slope of greater than 30°, the average ceiling height must be used for all heat detectors other than those located in the peak. The average ceiling height can be determined by adding the high sidewall and low sidewall heights together and dividing by two. In the case of slopes greater than 30°, the spacing to the second row is measured from the first row and not the sidewall for shed ceilings. See NFPA 72® for heat detector spacing reduction based on peak or average ceiling heights.

Once you have determined that the ceiling you are working with is a shed ceiling, use the following guidelines for determining detector placement:

- Place the first row of detectors within 3' (0.9 m) of the high sidewall (measured horizontally).
- Use the listed spacing (adjusted for the height of the heat detectors on ceilings over 10'/3 m) for each additional row of detectors (measured from the detector location, not the sidewall).
- For heat detectors on ceilings over 10' (3 m) high, adjust the spacing per NFPA 72®.

DETERMINING IF A SLOPE IS GREATER THAN 30°

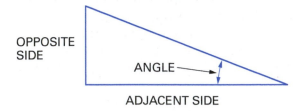

OPPOSITE SIDE

ANGLE

ADJACENT SIDE

If the opposite side divided by the adjacent side is > 0.5774, the angle is > 30°.

The tangent of the smallest angle of a right triangle equals the opposite side divided by the adjacent side.

The sum of all the angles of a triangle equals 180°.

The tangent of a 30° angle is 0.5774.

Figure 62 Determining degree of slope.

- If the slope is less than 30°, all heat detector spacing is based on the peak height. If the slope is more than 30°, peak heat detectors are based on the peak height, and all other detectors are based on the average height.
- Additional columns of detectors that run at right angles to the slope of the ceiling do not need to be within 3' (0.9 m) of the end walls.
- Smoke detector spacing is not adjusted for ceiling height.

To calculate the average ceiling height, use one of the following methods:

Method 1:

Step 1 Subtract the height of the low ceiling from the height of the high ceiling.

Step 2 Divide the above result by 2.

Step 3 Subtract the result from the height of the high ceiling.

Method 2:

Step 1 Add the height of the low ceiling to the height of the high ceiling.

Step 2 Divide the result by 2.

A ceiling must be sloped as defined by the code in order to be considered a peaked ceiling. It must slope in more than one direction from its highest point. Because domed and curved ceilings do not have clean 90° lines to clearly show that the ceiling slopes in more than one direction, they should be viewed from an imaginary vertical center line. From this perspective, it can be seen if the ceiling slopes in more than one direction.

- The first row of detectors (heat or smoke) on a peaked ceiling should be located within 3' (0.9 m) of the peak, measured horizontally. The detectors may be located alternately on either side of the peak, if desired.
- Regardless of where in the 3' (0.9 m) space on either side of the peak the detector is located, the measurement for the next row of heat detectors is taken from the peak (measured horizontally, not from the heat detector located near the peak). This is different from the way detector location is determined for shed ceilings.
- Heat detectors must have their spacing reduced in areas with ceilings over 10' (3 m). Smoke detector spacing is not adjusted for ceiling height.
- As with sloped ceilings, use of the peak height for just the row of heat detectors at the peak, or for all the ceiling heat detectors, depends on whether or not the ceiling is sloped more than 30°. Use the peak height to adjust heat

detector spacing of all the room's heat detectors when the slope of the ceiling is less than 30°. When the slope is greater than 30°, use the peak height for only the heat detectors located within 3' (0.9 m) of the peak. Use the average height for all other heat detectors in the room.

Smoke detectors should be located on a peaked ceiling such that:

• A row of detectors is within 3' (0.9 m) of the peak on either side or alternated from side to side.
• The next row of detectors is within the listed spacing of the peak measured horizontally. Do not measure from the detectors within 3' (0.9 m) of the peak. It does not matter where within 3' (0.9 m) of the peak the first row is; the next row on each side of the peak is installed within the listed spacing of the detector from the peak.
• Additional rows of detectors are installed using the full listed spacing of the detector.
• The sidewall must be within one-half the listed spacing of the last row of detectors.
• Columns of detectors are installed in the depth dimension of the above diagrams using the full listed spacing (the same as on a smooth, flat ceiling).
• There is no need to reduce smoke detector spacing for ceilings over 10' (3 m).

Solid joists are defined for heat detectors as being solid members that are spaced less than 3' (0.9 m) center-to-center and that extend down from the ceiling more than 4" (100 mm). Solid joists are defined for smoke detectors as being solid members that are spaced less than 3' (0.9 m) center-to-center and that extend down from the ceiling more than 8" (200 mm).

Heat detector spacing at right angles to the solid joist is reduced by 50%. In the direction running parallel to the joists, standard spacing principles are applied. If the ceiling height exceeds 10' (3 m), spacing is adjusted for the high ceiling in addition to the solid joist as defined in NFPA 72®.

Smoke detector spacing at right angles to the solid joists is reduced by one-half the listed spacing for joists 12" (300 mm) or less in depth and ceilings 12' (3.7 m) or less in height. In the direction running parallel to the joists, standard spacing principles are applied. If the ceiling height is over 12' (3.7 m) or the depth of the joist exceeds 12" (300 mm), a spot-type detector must be located on the ceiling in every pocket. Additional reductions for sloped joist ceilings also apply as defined in NFPA 72®.

Beamed ceilings consist of solid structural or solid nonstructural members projecting down from the ceiling surface more than 4" (100 mm) and spaced more than 3' (0.9 m) apart center-to-center as defined by NFPA 72®.

• *Heat detector installation:*
  – If the beams project more than 4" (100 mm) below the ceiling, the spacing of spot-type heat detectors at right angles to the direction of beam travel must not be more than two-thirds the smooth ceiling spacing.
  – If the beams project more than 12" (300 mm) below the ceiling and are more than 8' (2.4 m) on center, each bay formed by the beams must be treated as a separate area.
  – Reductions of heat detector spacing in accordance with NFPA 72® are also required for ceilings over 10' (0.9 m) in height.

• *Smoke detector installation:*
  – For smoke detectors, if beams are 4" to 12" (100 mm to 300 mm) in depth with an AFF level of 12' (3.7 m) or less, the spacing of spot-type detectors in the direction perpendicular to the beams must be reduced 50%, and the detectors may be mounted on the bottoms of the beams.
  – If beams are greater than 12" (300 mm) in depth with an AFF level of more than 12' (3.7 m), each beam bay must contain its own detectors. There are additional rules for sloped beam ceilings as defined in NFPA 72®.
  – The spacing of projected light beam detectors that run perpendicular to the ceiling beams need not be reduced. However, if the projected light beams are run parallel to the ceiling beams, the spacing must be reduced per NFPA 72®.

Ceiling surfaces referred to in conjunction with the locations of initiating devices are as follows:

• *Beam construction* – Ceilings having solid structural or solid nonstructural members projecting down from the ceiling surface more than 4" (100 mm) and spaced more than 3' (0.9 m) center to center.
• *Girders* – Girders support beams or joists and run at right angles to the beams or joists. When the tops of girders are within 4" (100 mm) of the ceiling, they are a factor in determining the number of detectors and are to be considered as beams. When the top of the girder is more than 4" (100 mm) from the ceiling, it is not a factor in detector location.

5.3.8 Notification Appliance Installation

There are several types of notification appliances:

- Audible devices such as bells, horns, chimes, speakers, sirens, and mini-sounders
- Visual devices such as strobe lights
- Tactile (sense of touch) devices such as bed shakers and special sprinkling devices
- Olfactory (sense of smell) devices, which are also accepted by the code

The appropriate devices for the occupancy must be determined. All devices must be listed for the purposes for which they are used. For example, *UL 1480* speakers are used for fire protective signaling systems, and *UL 1638* visual signaling appliances are used for private mode emergency and general utility signaling. While NFPA 72® recognizes tactile and olfactory devices, it does not specify installation requirements. If an occupant requires one of these types of notification, use equipment listed for the purpose and follow the manufacturer's instructions. As always, consult the local AHJ.

The following guidelines apply to the installation of notification devices:

- Ensure that notification devices are wired using the applicable circuit style (Class A or B).
- In Class B circuits, the panel supervises the wire using an end-of-line device. Electrically, the EOL device must be at the end of the indicating circuit. Examples of EOL devices are resistors to limit current, diodes for polarity, and capacitors for filtering.
- It is extremely important that any polarized devices be installed correctly. The polarization of the leads or terminals of these devices are marked on the device or are noted in the manufacturer's installation data. If the leads of a polarized notification device are reversed, the panel will not detect the problem, and the device will not activate. Moreover, the device will act as an EOL device, preventing the panel from detecting breaks between the device and the actual EOL device. Because it is very easy to accidentally wire a device backwards, testing every device is extremely important. The general alarm should be activated and every notification device checked for proper operation.
- Sidewall-mounted audible notification devices must be mounted at least 90" (2.3 m) AFF level or at least 6" (150 mm) below the finished ceiling. Ceiling-mounting and recessed appliances are permitted.
- A **visible notification appliance** must be mounted at the minimum height of 80" to 96"

(2 m to 2.4 m) AFF (NFPA 72®) or, for ADA requirements, either 80" (2 m) AFF or 6" (150 mm) below the ceiling. In any case, the device must be within 16' (4.9 m) of the pillow in a sleeping area. Combination audible/visible appliances must follow the requirements for visible appliances. Non-coded visible appliances should be installed in all areas where required by NFPA 101® or by the local AHJ. Consult ADA codes for the required illumination levels in sleeping areas.

- *Visual notification appliance spacing in corridors* – Table 5 provides spacing requirements for corridors less than 20' (6 m) wide. For corridors and rooms greater than 20' (6 m), refer to *Table 6* and *Table 7*. In corridor applications, visible appliances must be rated at not less than 15 candelas (cd). Per NFPA 72®, visual appliances must be located no more than 15' (4.6 m) from the end of the corridor with a separation of no more than 100' (30 m) between appliances. Where there is an interruption of the concentrated viewing path, such as a fire door or elevation change, the area is to be considered as a separate corridor.
- *Visual notification appliance spacing in other applications* – The light source color for visual appliances must be clear or nominal white and must not exceed 1,000cd (NFPA 72®). In addition, special considerations apply when more than one visual appliance is installed in a room or corridor. NFPA 72® specifies that the separation between appliances must not exceed 100' (30 m). Visible notification appliances must be installed in accordance with *Table 6* and *Table 7*, using one of the following:

  - A single visible notification appliance
  - Two visible notification appliances located on opposite walls
  - More than two appliances for rooms 80' × 80' (24 m × 24 m) or larger (must be spaced a minimum of 55' (16.8 m) from each other)

Table 5 Visual Notification Devices Required for Corridors Not Exceeding 20' (6 m)

Corridor Length (in ft)	Minimum Number of 15cd Appliances Required
0–30	1
31–130	2
131–230	3
231–330	4
331–430	5
431–530	6

Table 6 Room Spacing for Wall-Mounted Visual Notification Appliances

Maximum Room Size (in ft)	Minimum Required Light Output in Candelas (cd)		
	One Light per Room	Two Lights per Room*	Four Lights per Room**
20 × 20	15	Not allowable	Not allowable
30 × 30	30	15	Not allowable
40 × 40	60	30	Not allowable
50 × 50	95	60	Not allowable
60 × 60	135	95	Not allowable
70 × 70	185	95	Not allowable
80 × 80	240	135	60
90 × 90	305	185	95
100 × 100	375	240	95
110 × 110	455	240	135
120 × 120	540	305	135
130 × 130	635	375	185

* Locate on opposite walls

** One light per wall

Table 7 Room Spacing for Ceiling-Mounted Visual Notification Appliances

Maximum Room Size (in ft)	Maximum Ceiling Height (in ft)*	Minimum Required Light Output for One Light (cd)**
20 × 20	10	15
30 × 30	10	30
40 × 40	10	60
50 × 50	10	95
20 × 20	20	30
30 × 30	20	45
40 × 40	20	80
50 × 50	20	95
20 × 20	30	55
30 × 30	30	75
40 × 40	30	115
50 × 50	30	150

* Where ceiling heights exceed 30', visible signaling appliances must be suspended at or below 30' or wall mounted in accordance with NFPA 72®.

** This table is based on locating the visible signaling appliance at the center of the room. Where it is not located at the center of the room, the effective intensity (cd) must be determined by doubling the distance from the appliance to the farthest wall to obtain the maximum room size.

5.3.9 Fire Alarm Control Panel Installation Guidelines

The guidelines for installing a fire alarm control panel are as follows:

- When not located in an area that is continuously occupied, all fire alarm control equipment must be protected by a smoke detector, as shown in *Figure 63*. If the smoke detector is not designed to work properly in that environment, a heat detector must be used. It is not necessary to protect the entire space or room.

- Detector spacing must be adjusted if the ceiling over the control equipment is irregular in one or more respects. This is considered protection against a specific hazard under NFPA 72® and does not require the entire chamber (room) containing the control equipment to be protected under the 0.7 rule.

- A means of silencing audible notification appliances from an FACP must be protected against unauthorized use. Most FACPs are located inside locked metal cabinets as shipped from the manufacturer. If the silencing switch is key-actuated or is locked within the cabinet, this provision should be considered satisfied. If the silencing means is within a room that is restricted to authorized use only, no additional measures should be required. The NFPA codes do not clearly define nor specify unauthorized or authorized use.

- FACP connections to the primary light and power circuits must be on a dedicated branch circuit with overcurrent protection rated at 20A

UL 1971 Listing

A *UL 1971* listing for a visual notification appliance indicates that it meets the ADA hearing-impaired requirements.

or less. Any connections to the primary power circuit on the premises connected after the distribution panel (circuit breaker box) must be directly related to the fire alarm system. No other use is permitted. This requirement does not necessitate a direct tap into the power circuit ahead of the distribution panel, although connecting ahead of the main disconnect is acceptable with listed service equipment.

- The power disconnection means (circuit breaker) must be clearly identified as a fire alarm circuit control.

5.4.0 Fire Alarm-Related Systems

This section discusses various fire alarm-related systems, as well as the installation guidelines for each system.

5.4.1 Ancillary Control Relay

Ancillary functions, commonly called *auxiliary functions*, include such controls as elevator capture (recall), elevator shaft pressurization, HVAC system shutdown, stairwell pressurization, smoke management systems, emergency lighting, door unlocking, door hold-open device control, and building music system shutoff. For example, sound systems are commonly powered down by the fire alarm system so that the evacuation signal may be heard. In the normal state (*Figure 64*), an energized relay completes the power circuit to the device being controlled.

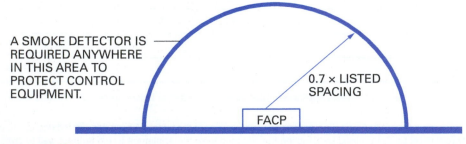

A smoke detector protecting control equipment must meet NFPA 72® spacing and placement standards, but the entire space (room) containing the FACP need not be protected.

A SMOKE DETECTOR IS REQUIRED ANYWHERE IN THIS AREA TO PROTECT CONTROL EQUIPMENT.

0.7 × LISTED SPACING

FACP

Figure 63 Protection of an FACP.

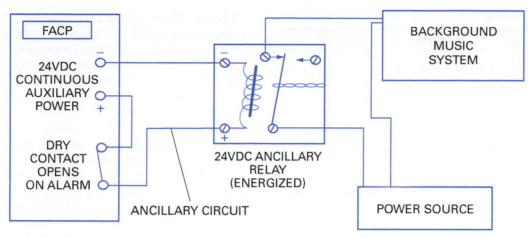

Figure 64 Music system control in normal state.

> **NOTE**
> The circuit from the relay to the background music system is a remote control signaling circuit (see *NEC Article 725*). If this circuit is both powered by and controlled by the fire alarm, the circuit is a fire alarm circuit (see *NEC Article 760*).

5.4.2 Duct Smoke Detectors

Duct smoke detectors are not simply conventional detectors applied to HVAC systems. Conventional smoke detectors are listed for open area protection (OAP) under *UL 268*. Duct smoke detectors are normally listed for a slightly higher velocity of air movement and are tested under *UL 268A*. The primary function of a duct smoke detector is to turn off the HVAC system. This prevents the system from spreading smoke rapidly throughout the building and stops the system from providing a forced supply of oxygen to the fire. Duct smoke detectors are not intended primarily for early warning and notification. Relevant fire alarm-related provisions pertaining to HVAC systems can be found in NFPA 90A and B.

NFPA 90A covers the following types of buildings and systems:

- HVAC systems serving spaces over 25,000 cubic feet in volume (708 cubic meters)

> **NOTE**
> No duct smoke detector requirements are found under *NFPA Standard 90B*.

- Buildings of Type III, IV, or V construction that are over three stories in height (NFPA 220)
- Buildings that are not covered by other applicable standards

- Buildings that serve occupants or processes not covered by other applicable standards

NFPA 90B covers the following types of systems:

- HVAC systems that service one- or two-family dwellings
- HVAC systems that service spaces not exceeding 25,000 cubic feet in volume (708 cubic meters)

When determining the location of duct detectors prior to installation, use the following guidelines:

- In HVAC units over 2,000 cubic feet per minute (cfm), or 57 cubic meters per minute, duct detectors must be installed on the supply side.
- Duct detectors must be located downstream of any air filters and upstream of any branch connection in the air supply.
- Duct detectors should be located upstream of any in-duct heating element.
- Duct detectors must be installed at each story prior to the connection to a common return and prior to any recirculation of fresh air inlet in the air return of systems over 15,000 cfm (425 cubic meters per minute) serving more than one story.
- Return air system smoke detectors are not required when the entire space served by the HVAC system is protected by a system of automatic smoke detectors and when the HVAC is shut down upon activation of any of the smoke detectors.

The approximations given in *Table 8* are useful when the protected-premises personnel do not know the cfm rating of an air-handling unit, but do know either the tonnage or British thermal unit (Btu) rating. Additionally, the cfm rating may

Table 8 Conversions

Capacity Rating	CFM
1 ton	400
12,000 Btus	400
5 tons	2,000
60,000 Btus	2,000
37.5 tons	15,000
450,000 Btus	15,000

not always appear on air handling unit (AHU) nameplates or in building specifications.

When more than one air handling unit (AHU) is used to supply air to a common space, and the return air is drawn from this common space, the total capacity of all units must be used in determining the size of the HVAC system (*Figure 65*). This formal interpretation makes clear the fact that interconnected air handling units should be viewed as a system, as opposed to treating each AHU individually. When multiple AHUs serve a common space, the physical location of each AHU, relative to others interconnected to the same space, is irrelevant to the application of the formal interpretation.

The common space being served by multiple air handling units need not be contiguous (connected) for the formal interpretation to apply. In *Figure 66*, AHUs #3 and #4 serve common space and must be added together for consideration (1,100 + 1,100 = 2,200). Because 2,200 is greater than 2,000, duct smoke detectors are required on both. The same is true for AHUs #1 and #2.

Duct smoke detectors may not be used as substitutes where open area detectors are required. This is because the HVAC unit may not be running when a fire occurs. Even if the fan is always on, the HVAC is not a listed fire alarm device.

Duct smoke detectors must automatically stop their respective fans upon detecting smoke. It is also acceptable for the fire alarm control panel to stop the fan(s) upon activation of the duct detector. However, fans that are part of an engineered smoke control or management system are an exception and are not always shut down in all cases when smoke is detected.

When a fire alarm system is installed in a building, all duct smoke detectors in that building must be connected to the fire alarm system as either initiating devices or as supervisory devices. The code does not require the installation of a building fire alarm system. It does require

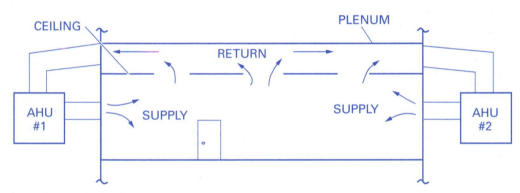

Figure 65 Non-ducted multiple AHU system.

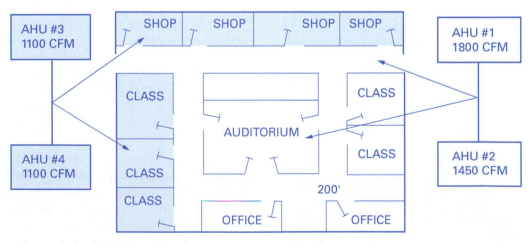

Figure 66 Example in which all AHUs require detectors on the supply side.

that the duct smoke detectors be connected to the building fire alarm, if one exists. Duct smoke detectors, properly listed and installed, will accomplish their intended function when connected as initiating or supervisory devices.

When the building is not equipped with a fire alarm system, visual and audible alarm and trouble signal indicators (*Figure 67*) must be installed in a normally occupied area. Smoke detectors whose sole function is stopping fans do not require standby power.

AHUs often come with factory-installed duct detectors. If they are installed in a building with a fire alarm system, they must be connected to the fire alarm system, and the detector may require standby power. Consult with the AHJ concerning how many duct detectors can be placed in a zone. If it cannot be determined which detector activated after the power failure, the alarm indicator on the detector may be considered an additional function by the local AHJ.

Duct detectors for fresh air or return air ducts should be located six to ten duct widths from any openings, deflectors, sharp bends, or branch connections. This is necessary to obtain a representative air sample and to reduce the effects of stratification and dead air space.

5.4.3 Elevator Recall

Elevator recall is intended to route an elevator car to a non-fire floor, open its doors, and then put the car out of normal service until reset (Phase 1 recall). It is usually assumed that the recall floor will be the level of primary exit discharge or a grade level exit. In the event that the primary level of exit discharge is the floor where the fire has been detected, the system should route the car

to a predetermined alternate floor, open its doors, and go out of normal service (Phase 2 recall). The operation of the elevators must be in accordance with *ANSI A17.1*.

Phase 1 recall (*Figure 68*) is the method of recalling the elevator to the floor that provides the highest probability of safe evacuation (as determined by the AHJ) in the event of a fire emergency.

Phase 2 recall (*Figure 69*) is the method of recalling the elevator to the floor that provides the next highest probability of safe evacuation (as determined by the AHJ) in the event of a fire emergency. Phase 2 recall is typically activated by the smoke detector in the lobby where the elevator would normally report during a Phase 1 recall.

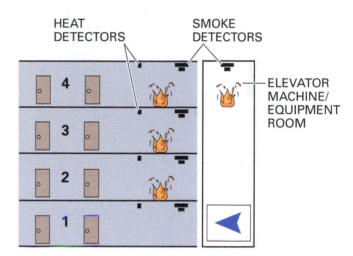

PHASE 1 RECALL:
Activation of any of these detectors sends all cars to the designated level.

Figure 68 Phase 1 recall.

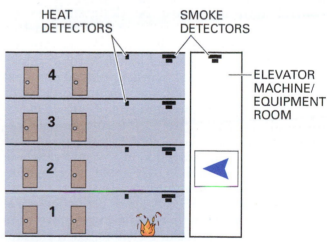

PHASE 2 RECALL:
Activation of the designated level detector sends all cars to the alternate floor.

Figure 69 Phase 2 recall.

Figure 67 Typical remote duct indicator.

Hoistway and elevator machine room smoke detectors must report an alarm condition to a building fire alarm control panel as well as an indicator light in the elevator car itself. However, notification devices may not be required to be activated if the control panel is in a constantly attended location. In facilities without a building fire alarm system, these smoke detectors must be connected to a dedicated fire alarm system control unit that must be designated as an Elevator Recall Control and Supervisory Panel.

Elevator recall must be initiated only by elevator lobby, hoistway, and machine room smoke detectors. Activation of manual pull stations, heat detectors, duct detectors, and any smoke detector not mentioned previously must not initiate elevator recall. In many systems, it would be inappropriate for the fire alarm control panel to initiate elevator recall. An exception is any system where the control function is selectable or programmable and the configuration limits the recall function to the specified detector activation only.

Caution should be used when using two-wire smoke detectors to recall elevators. Each elevator lobby, elevator hoistway, and elevator machine room smoke detector must be capable of initiating elevator recall when all other devices on the same initiating device circuit (IDC) have been manually or automatically placed in the alarm condition.

Unless the area encompassing the elevator lobby is a part of a chamber being protected by smoke detectors, such as the continuation of a corridor, the smoke detectors serving the elevator lobby (*Figure 70*) may be applied to protect against a specific hazard per NFPA 72®. The elevator lobby, hoistway, or associated machine room detectors may also be used to activate emergency control functions as permitted by NFPA 101®.

Where ambient conditions prohibit installation of automatic smoke detection, other appropriate automatic fire detection must be permitted. For example, a heat detector may protect an elevator door that opens to a high-dust or low-temperature area. Always refer to local codes.

5.4.4 Special Door Locking Arrangements

No lock, padlock, hasp, bar, chain, or other device intended to prevent free egress may be used at any door on which panic hardware or fire exit hardware is required (NFPA 101®). Note that these requirements apply only to exiting from the building. There is no restriction on locking doors against unrestricted entry from the exterior.

Upon activation of the fire alarm, stair enclosure doors must unlock to permit egress. They must also unlock to permit reentry into the building floors. This provision applies to buildings four stories and taller. The purpose of this requirement is to provide for an escape from fire and smoke entering a lower level of the stair enclosure and blocking safe egress from the stair enclosure. See NFPA 101® for additional details and exceptions.

5.4.5 Suppression Systems

There are dry and wet chemical extinguishing systems, and this section discusses them both.

Dry chemical extinguishing systems must be connected to the building fire alarm system if a fire alarm system exists in the structure. The extinguishing system must be connected as an initiating device. The standard (NFPA 17, *Dry Chemical Extinguishing Systems*) does not require the installation of a building fire alarm system if one was not required elsewhere. Also see NFPA 12 (*Carbon Dioxide Extinguishing Systems*), NFPA 12A (*Halon 1301 Fire Extinguishing Systems*), and NFPA 2001 (*Clean Agent Fire Extinguishing Systems*).

Wet chemical extinguishing systems must be connected to the building fire alarm system if a fire alarm system exists in the structure. As with

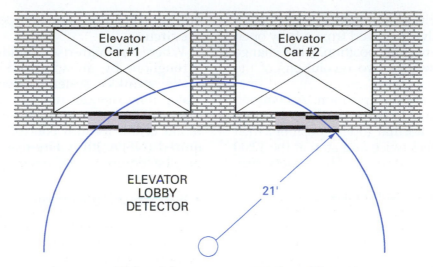

Assumes a 30' listed detector on a smooth and flat ceiling.
Target protection should be the extreme edges of the elevator door opening.

Figure 70 Elevator lobby detector.

dry chemical extinguishing systems, the standard (NFPA 17A, *Wet Chemical Extinguishing Systems*) does not require the installation of a building fire alarm system. Also, see NFPA 16 (*Foam-Water Sprinkler and Foam-Water Spray Systems*).

5.4.6 Supervision of Suppression Systems

Each of the occupancy chapters of NFPA 101® will specify the extinguishing requirements for that occupancy. The code will specify either an approved automatic sprinkler or an approved supervised automatic sprinkler system if a sprinkler system is required.

Where a supervised automatic sprinkler is required, various sections of NFPA 101® are applicable. These code sections begin with the phrase, "Where required by another section of this code." Use of the word *supervised* in the extinguishing requirement section of each occupancy chapter is the method used to implement these two provisions. Sprinkler systems do not automatically require electronic supervision. Supervised automatic sprinkler systems require a distinct supervisory signal. This signal indicates a condition that would impair the proper operation of the sprinkler system, and the signal is received at a location constantly attended by qualified personnel or at an approved remote monitoring facility. Water flow alarms from a supervised automatic sprinkler system must be transmitted to an approved monitoring station. When the supervised automatic sprinkler supervisory signal terminates on the protected premises in areas that are constantly attended by trained personnel, the supervisory signal is not required to be transmitted

to a monitoring facility. In such cases, only alarm and trouble signals need to be transmitted.

The following are some of the sprinkler elements that are required to be supervised where applicable:

- Water supply control valves
- Fire pump power (including phase monitoring)
- Fire pump running
- Water tank levels
- Water tank temperatures
- Tank pressures
- Air pressure of dry-pipe systems

A sprinkler flow alarm must be initiated within 90 seconds of the flow of water equal to or greater than the flow from the sprinkler head, or from the smallest orifice (opening) size in the system. In actual field verification activities, the 90 seconds is measured from the time water begins to flow into the inspector's test drain, and *not* from the time the inspector's test valve is opened. The smallest orifice is the size of the opening at the smallest sprinkler head.

Sprinkler systems that initiate a fire alarm system by a water flow switch must include at least one manual station that is located where required by the AHJ. Manual stations required elsewhere in the codes or standards can be considered to meet this requirement. Some occupancy chapters of NFPA 101® will allow the sprinkler flow switch to substitute for manual stations at all the required exits. If such an option is utilized, this provision requires that at least one manual station be installed where acceptable to the AHJ.

Each shutoff valve, also called an *outside screw and yoke (OS&Y) control valve*, must be supervised by a tamper switch. A distinctive signal must sound when the valve is moved from a fully open position to the off-normal position. This change must be detected within two revolutions of the hand wheel, or when the stem has moved one-fifth from its normal open position. A common verification practice is to mark the 12:00 position on the valve in its normal (open) position, then rotate the hand wheel twice and stop at the 12:00 position on the second pass. The supervisory signal must be initiated by the time the wheel reaches the end of the second pass.

Water flow and supervisory devices, in addition to their circuits, must be installed such that no unauthorized person may tamper with them, open them, remove them, or disconnect them without initiating a signal. Publicly accessible junction boxes must have tamper-resistant screws or tamper-alarm switches. Most water flow switch and valve tamper switch housings come from the manufacturer with tamper-resistant screws (hex or allen head), and they may be configured to signal when the housing cover is removed. If the device, circuit, or junction box requiring protection is in an area that is not accessible to unauthorized personnel, no additional protective measures should be required. Simply sealing, locking, or removing the handle from a valve is not sufficient to meet the supervision requirement.

Water flow devices that are alarm-initiating devices cannot be connected on the same initiating circuit as valve supervisory devices. This is commonly done in violation of the code by connecting the valve tamper switch in series with the initiating circuit's EOL device, resulting in a trouble signal when activated. This method of wiring does not provide for a distinctive visual or audible signal. This statement is true for most conventional systems. It should be noted that at least one known addressable system has a listed module capable of distinctly separating the two types of signals. These devices must be wired on the same circuit to meet the standard.

> **NOTE**
> Addressable devices are not connected to initiating circuits. They are connected to signaling line circuits.

A supervisory signal must be visually and audibly distinctive from both alarm signals and trouble signals. It must be possible to tell the difference between a fire alarm signal, a valve being off-normal (closed), and an open (broken wire) in the circuit.

Where a high-rise building is protected throughout by an approved, supervised automatic sprinkler system, valve supervision and water flow devices must be provided on each floor (NFPA 101®). In such buildings, a fire command center (central control system) is also required (NFPA 101®). Fire-resistive cable systems may be required (*NEC Article 728*).

5.5.0 Troubleshooting

The troubleshooting approach to any fire alarm system is basically the same. Regardless of the situation or equipment, some basic steps can be followed to isolate problems:

Step 1 Know the equipment. For easy reference, keep specification sheets and instructions for commonly used and serviced equipment. Become familiar with the features of the equipment.

Step 2 Determine the symptoms. Try to make the system perform or fail to perform as it did when the problem was discovered.

Step 3 List possible causes. Write down everything that could possibly have caused the problem.

Step 4 Check the system systematically. Plan activities so that problem areas are not overlooked, in order to eliminate wasted time.

Step 5 Correct the problem. Once the problem has been located, repair it. If it is a component that cannot be repaired easily, replace the component.

Step 6 Test the system. After the initial problem has been corrected, thoroughly check all the functions and features of the system to make sure other problems are not present that were masked by the initial problem.

5.5.1 Alarm System Troubleshooting Guidelines

A system troubleshooting chart is shown in *Figure 71*. In *Figure 72* an alarm output troubleshooting chart is shown, and in *Figure 73* an auxiliary power troubleshooting chart is shown.

The following guidelines provide information for resolving potential problems for specific conditions:

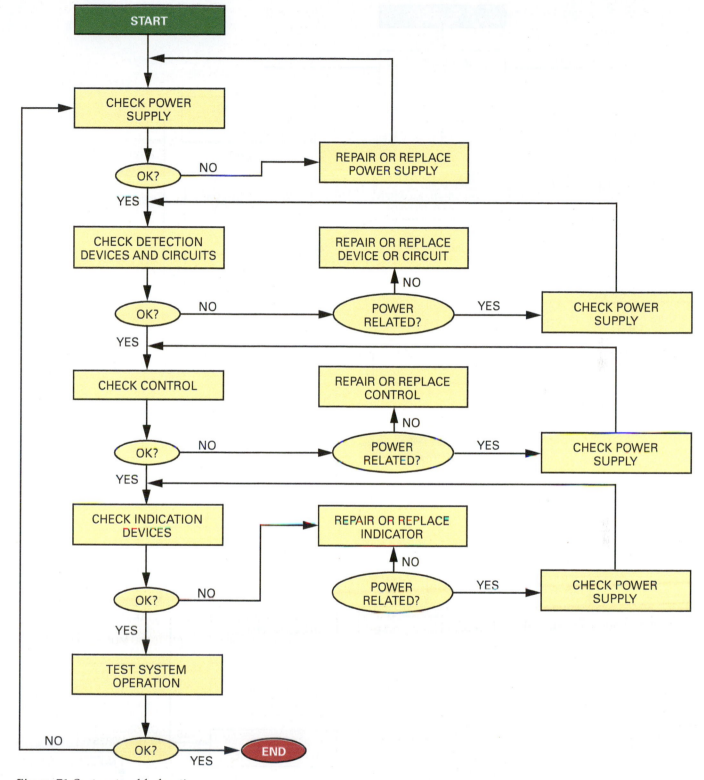

Figure 71 System troubleshooting.

- *Sensors* – Always check sensor power, connections, environment, and settings. Lack of detection can be caused by a loose connection, obstacles in the area of the sensor, or a faulty unit. If unwanted alarms occur, recheck the installation for changing environmental factors. If none are present, cover or seal the sensor to confirm that the alarm originated with the sensor. If alarms still occur, wiring problems, power problems or electromagnetic interference (EMI) could be the cause. If the alarm stops when the sensor is covered or sealed, the environment monitored by the sensor is the source of the problem. Replacing the unit

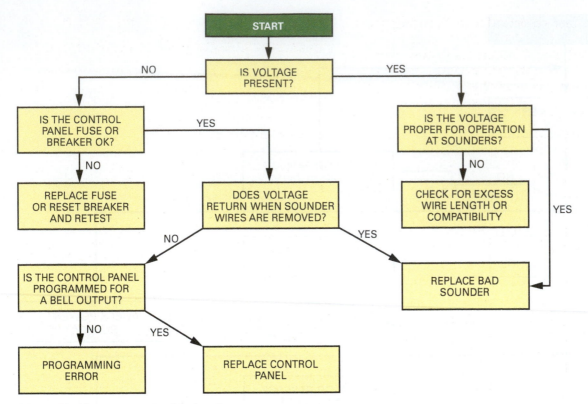

Figure 72 Alarm output troubleshooting.

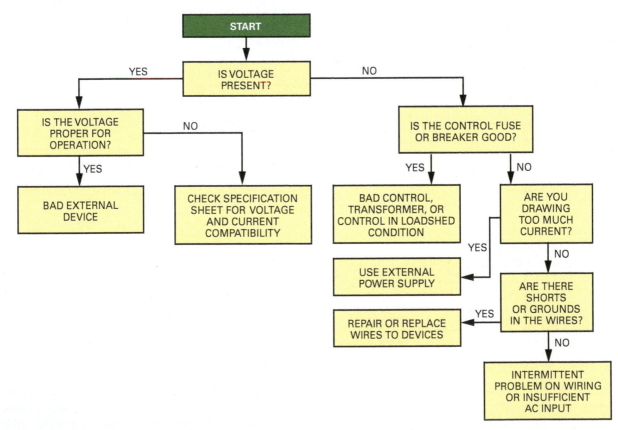

Figure 73 Auxiliary power troubleshooting.

should be the last resort after performing the above checks.

- *Open circuit problems* – Opens will cause trouble signals and account for the largest percentage of faults. Major causes of open circuits are loose connections, wire breaks (ripped or cut), staple cuts, bad splices, cold-solder joints, wire fatigue due to flexing, and defective wire.
- *Short circuit or ground fault problems* – Shorts or ground faults on a circuit will cause alarms or trouble signals. Events occurring past a short will not be seen by the system. Some of the common causes of shorts or ground faults are staple cuts, sharp edge cuts, improper splices, moisture, and cold-flow of wiring insulation.

5.5.2 Addressable System Troubleshooting Guidelines

There is a wide variety of designs, equipment, and configurations of addressable fire alarm systems. For this reason, some manufacturers of addressable systems provide troubleshooting data in their service literature, but other manufacturers do not. Most of the general guidelines for troubleshooting discussed earlier also apply to troubleshooting addressable systems. In the event general troubleshooting methods fail to find the problem and no troubleshooting information is provided by the manufacturer in the service literature for the equipment, the best thing to do is to contact the manufacturer and ask for technical assistance. This will prevent a needless loss of time.

Supervisory Switches

This supervisory switch is typical of those used with sprinkler systems to provide a tamper indication of valve movement. The switch is mounted on a valve with the actuator arm normally resting against the movable target indicator assembly. If the normal position of the valve is altered, the valve stem moves, forcing the actuator arm to operate the switch. The switch activates between the first and second revolutions of the control valve wheel.

Figure Credit: System Sensor

Fire Pumps

A fire pump system is necessary when the available water supply is not adequate, in pressure or volume, to supply the fire suppression water needs of a sprinkler or standpipe system. The fire pump system comprises several components, including the pump, a driver (typically electric or diesel), and a controller. A typical fire pump system is shown here.

Figure Credit: John Viola

5.0.0 Section Review

1. Which electrical code specifies wiring methods and cables for fire alarm systems?

    a. *NEC Article 760*
    b. *NEC Article 770*
    c. *NEC Article 800*
    d. *NEC Article 810*

2. Which of the following is an unacceptable means of mounting a detector on gypsum drywall?

    a. Winged expansion anchors
    b. Toggle expansion anchors
    c. An electrical box
    d. Plastic masonry anchors

3. Which of the following fires will an IR detector NOT be able to sense?

    a. Wood
    b. Alcohol
    c. Gasoline
    d. Paper

4. Duct smoke detectors are governed by _____.

    a. *UL 268A*
    b. *UL 1971*
    c. NFPA 75
    d. NFPA 80

5. What is the first step in troubleshooting fire alarm equipment?

    a. Perform systematic tests.
    b. List the symptoms.
    c. Know the equipment.
    d. Make the system fail.

1. The specific requirements for wiring and wiring equipment installation for fire protective signaling systems are covered in _____.
   a. NFPA 1
   b. NFPA 70®
   c. NFPA 72®
   d. NFPA 101®

2. A code that was established to help fire authorities continually develop safeguards against fire hazards is _____.
   a. NFPA 1
   b. NFPA 70®
   c. NFPA 72®
   d. NFPA 101®

3. A defined area within the boundaries of a fire alarm system is known as a(n) _____.
   a. section
   b. signal destination
   c. alarm unit area
   d. zone

4. A fire alarm system with zones that allow multiple signals from several sources to be sent and received over a single communication line is known as a(n) _____.
   a. addressable system
   b. conventional hardwired system
   c. zoned system
   d. multiplex system

5. An SLC that consists of a Class A main trunk with a Class B spur circuit is known as an SLC _____.
   a. dual circuit
   b. combined circuit
   c. twin circuit
   d. hybrid circuit

6. An automatic detector that draws air from the protected area back to the detector is called a(n) _____.
   a. line detector
   b. spot detector
   c. air sampling detector
   d. addressable detector

7. A heat detector that is NOT a fixed-temperature device is a _____.
   a. fusible link detector
   b. rate-of-rise detector
   c. quick metal detector
   d. bimetallic detector

8. Older type duct detectors used the _____.
   a. ionization principle
   b. light scattering principle
   c. light obscuration principle
   d. smoke rate compensation principle

9. The type of pull station that is restricted for use in special applications is the _____.
   a. single-action type
   b. glass-break type
   c. double-action type
   d. key-operated type

10. Primary power circuits supplying fire alarm systems must be protected by circuit breakers not larger than _____.
    a. 15A
    b. 20A
    c. 25A
    d. 30A

11. Fire warning system units for residential use must be listed in accordance with *UL Standard* _____.
    a. *780*
    b. *864*
    c. *985*
    d. *995*

12. Addressable systems commonly use a verification method called _____.
    a. positive alarm sequence
    b. wait and check
    c. reset and resample
    d. cross-zoning

13. The NFPA 72®–specified Temporal-Three sound pattern for a fire alarm signal is _____.

    a. three short ($\frac{1}{2}$ second), three long (1 second), three short, a pause, then a repeat
    b. three long (1 second), a pause, then a repeat
    c. three short ($\frac{1}{2}$ second), a pause, then a repeat
    d. three short ($\frac{1}{2}$ second), three long (1 second), three short, three long, a pause, then a repeat

14. Public area audible notification devices must have a minimum dBA rating of _____.

    a. 65 at 10' (3 m)
    b. 75 at 10' (3 m)
    c. 85 at 10' (3 m)
    d. 95 at 10' (3 m)

15. The use of an attendant signal requires the approval of the _____.

    a. NFPA
    b. manufacturer
    c. UL
    d. AHJ

16. Commercial fire alarm cables that run from floor to floor must be rated at least as _____.

    a. general-purpose cable
    b. water-resistant cable
    c. riser cable
    d. plenum cable

17. Power-limited fire alarm circuit conductors must be separated from nonpower-limited fire alarm circuit conductors by at least _____.

    a. 2" (50 mm)
    b. 6" (150 mm)
    c. 12" (300 mm)
    d. 24" (600 mm)

18. For smoke chamber definition purposes, an archway that extends down 18" (450 mm) or more from the ceiling is considered the same as a(n) _____.

    a. barrier
    b. boundary
    c. open grid
    d. beam pocket

19. The primary function of a commercial duct smoke detector is to _____.

    a. sound an alarm for duct smoke
    b. shut down an associated HVAC system
    c. detect external open-area fires
    d. disable a music system for voice evacuation purposes

20. In a supervised automatic sprinkler system, a visual or audible supervisory signal must be _____.

    a. different from an alarm signal
    b. the same as an alarm signal
    c. different from a trouble signal
    d. different from both the alarm and trouble signals

1. Provide the names of the NFPA codes listed below.
   NFPA 1: _____
   NFPA 70®: _____
   NFPA 72®: _____
   NFPA 101®: _____

2. True or False? Smoke detection devices in an analog addressable fire alarm system make the decision internally regarding their alarm state.

   _____

3. _____ is generally used as the actuating medium in a rate-of-rise detector.

4. The two basic types of smoke detectors are the _____ and the _____ types.

5. The melted and fused cable section in a(n) _____ detector must be replaced once it has been activated.

6. True or False? Primary power for a fire alarm system may be connected on the line side of the electrical main service disconnect switch.

   _____

7. A digital alarm communicator _____ accepts and displays signals from a digital alarm communicator _____ sent over a public switched telephone network.

8. True or False? Cables installed in plenums and other air handling spaces may be secured using standard cable tie wraps.

   _____

9. A manual fire alarm box (pull box) installed at an exit door should be installed near the _____.

10. True or False? Under normal conditions, smoke detectors should be installed after construction cleanup is completed.

    _____

Trade Terms Introduced in This Module

Addressable device: A fire alarm system component with discrete identification that can have its status individually identified or that is used to individually control other functions.

Air sampling detector: A detector consisting of piping or tubing distribution from the detector unit to the area or areas to be protected. An air pump draws air from the protected area back to the detector through the air sampling ports and piping or tubing. At the detector, the air is analyzed for fire products.

Alarm: In fire systems, a warning of fire danger.

Alarm signal: A signal indicating an emergency requiring immediate action, such as an alarm for fire from a manual station, water flow device, or automatic fire alarm system.

Alarm verification: A feature of a fire control panel that allows for a delay in the activation of alarms upon receiving an initiating signal from one of its circuits. Alarm verification must not be longer than three minutes, but can be adjustable from 0 to 3 minutes to allow supervising personnel to check the alarm. Alarm verification is commonly used in hotels, motels, hospitals, and institutions with large numbers of smoke detectors.

Americans with Disabilities Act (ADA): An act of Congress intended to ensure civil rights for physically challenged people.

Approved: Acceptable to the authority having jurisdiction.

Audible signal: An audible signal is the sound made by one or more audible indicating appliances, such as bells, chimes, horns, or speakers, in response to the operation of an initiating device.

Authority having jurisdiction (AHJ): The authority having jurisdiction is the organization, office, or individual responsible for approving equipment, installations, or procedures in a particular locality.

Automatic fire alarm system: A system in which all or some of the circuits are actuated by automatic devices, such as fire detectors, smoke detectors, heat detectors, and flame detectors.

CABO: Council of American Building Officials.

Ceiling: The upper surface of a space, regardless of height. Areas with a suspended ceiling would have two ceilings: one visible from the floor, and one above the suspended ceiling.

Ceiling height: The height from the continuous floor of a room to the continuous ceiling of a room or space.

Certification: A systematic program using randomly selected follow-up inspections of the certified system installed under the program, which allows the listing organization to verify that a fire alarm system complies with all the requirements of the NFPA 72® code. A system installed under such a program is identified by the issuance of a certificate and is designated as a certificated system.

Chimes: A single-stroke or vibrating audible signal appliance that has a xylophone-type striking bar.

Circuit: The conductors or radio channel as well as the associated equipment used to perform a definite function in connection with an alarm system.

Class A circuit: Class A refers to an arrangement of supervised initiating devices, signaling line circuits, or indicating appliance circuits (IAC) that prevents a single open or ground on the installation wiring of these circuits from causing loss of the system's intended function. It is also commonly known as a *four-wire circuit*.

Class B circuit: Class B refers to an arrangement of initiating devices, signaling lines, or indicating appliance circuits that does not prevent a single open or ground on the installation wiring of these circuits from causing loss of the system's intended function. It is commonly known as a *two-wire circuit*.

Coded signal: A signal pulsed in a prescribed code for each round of transmission.

Control unit: A device with the control circuits necessary to furnish power to a fire alarm system, receive signals from alarm initiating devices (and transmit them to audible alarm indicating appliances and accessory equipment), and electrically supervise the system installation wiring and primary (main) power. The control unit can be contained in one or more cabinets in adjacent or remote locations.

Digital Alarm Communicator Receiver (DACR): A system component that will accept and display signals from digital alarm communicator transmitters (DACTs) sent over public switched telephone networks.

Digital Alarm Communicator System (DACS): A system in which signals are transmitted from a digital alarm communicator transmitter (DACT) located at the protected premises through the public switched telephone network to a digital alarm communicator receiver (DACR).

Digital Alarm Communicator Transmitter (DACT): A system component at the protected premises to which initiating devices or groups of devices are connected. The DACT will seize the connected telephone line, dial a preselected number to connect to a DACR, and transmit signals indicating a status change of the initiating device.

End-of-line (EOL) device: A device used to terminate a supervised circuit. An EOL is normally a resistor or a diode placed at the end of a two-wire circuit to maintain supervision.

Fault: An open, ground, or short condition on any line(s) extending from a control unit, which could prevent normal operation.

Fire: A chemical reaction between oxygen and a combustible material where rapid oxidation may cause the release of heat, light, flame, and smoke.

Flame detector: A device that detects the infrared, ultraviolet, or visible radiation produced by a fire. Some devices are also capable of detecting the flicker rate (frequency) of the flame.

General alarm: A term usually applied to the simultaneous operation of all the audible alarm signals on a system, to indicate the need for evacuation of a building.

Ground fault: A condition in which the resistance between a conductor and ground reaches an unacceptably low level.

Heat detector: A device that detects abnormally high temperature or rate-of-temperature rise.

Horn: An audible signal appliance in which energy produces a sound by imparting motion to a flexible component that vibrates at some nominal frequency.

Indicating device: Any audible or visible signal employed to indicate a fire, supervisory, or trouble condition. Examples of audible signal appliances are bells, horns, sirens, electronic horns, buzzers, and chimes. A visible indicator consists of an incandescent lamp, strobe lamp, mechanical target or flag, meter deflection, or the equivalent. Also called a *notification device (appliance)*.

Initiating device: A manually or automatically operated device, the normal intended operation of which results in a fire alarm or supervisory signal indication from the control unit. Examples of alarm signal initiating devices are thermostats, manual boxes (stations), smoke detectors, and water flow devices. Examples of supervisory signal initiating devices are water level indicators, sprinkler system valveposition switches, pressure supervisory switches, and water temperature switches.

Initiating device circuit (IDC): A circuit to which automatic or manual signal-initiating devices such as fire alarm manual boxes (pull stations), heat and smoke detectors, and water flow alarm devices are connected.

Labeled: In the context of fire alarm control panels, tags that identify various zones and sensors by descriptive names. Panels may include dual identifiers, one of which is meaningful to the occupants, while the other is helpful to firefighting personnel.

Light scattering: The action of light being reflected or refracted off particles of combustion, for detection in a modern-day photoelectric smoke detector. This is called the *Tyndall effect*.

Listed: Equipment or materials included in a list published by an organization acceptable to the authority having jurisdiction that is concerned with product evaluation and whose listing states either that the equipment or materials meets appropriate standards or has been tested and found suitable for use in a specified manner.

Maintenance: Repair service, including periodic inspections and tests, required to keep the protective signaling system and its component parts in an operative condition at all times. This is used in conjunction with replacement of the system and its components when for any reason they become undependable or inoperative.

Multiplexing: A signaling method that uses wire path, cable carrier, radio, fiber optics, or a combination of these techniques, and characterized by the simultaneous or sequential (or both simultaneous and sequential) transmission and reception of multiple signals in a communication channel including means of positively identifying each signal.

National Fire Alarm Code®: This is the update of the NFPA standards book that contains the former NFPA 71, NFPA 72®, and NFPA 74 standards, as well as the NFPA 1221 standard. The *NFAC®* was adopted and became effective May 1993.

National Fire Protection Association (NFPA): The NFPA administers the development and publishing of codes, standards, and other materials concerning all phases of fire safety.

Noise: A term used in electronics to cover all types of unwanted electrical signals. Noise signals originate from numerous sources, such as fluorescent lamps, walkie-talkies, amateur and CB radios, machines being switched on and off, and power surges. Today, equipment must tolerate increasing amounts of electrical interference, and the quality of equipment depends on how much noise it can ignore and withstand.

Non-coded signal: A signal from any indicating appliance that is continuously energized.

Notification device (appliance): Any audible or visible signal employed to indicate a fire, supervisory, or trouble condition. Examples of audible signal appliances are bells, horns, sirens, electronic horns, buzzers, and chimes. A visible indicator consists of an incandescent lamp, strobe lamp, mechanical target or flag, meter deflection, or the equivalent. Also called an *indicating device.*

Obscuration: A reduction in the atmospheric transparency caused by smoke, usually expressed in percent per foot.

Path (pathway): Any conductor, optic fiber, radio carrier, or other means for transmitting fire alarm system information between two or more locations.

Photoelectric smoke detector: A detector employing the photoelectric principle of operation using either the obscuration effect or the light-scattering effect for detecting smoke in its chamber.

Positive alarm sequence: An automatic sequence that results in an alarm signal, even when manually delayed for investigation, unless the system is reset.

Power supply: A source of electrical operating power, including the circuits and terminations connecting it to the dependent system components.

Projected beam smoke detector: A type of photoelectric light-obscuration smoke detector in which the beam spans the protected area.

Protected premises: The physical location protected by a fire alarm system.

Public Switched Telephone Network: An assembly of communications facilities and central office equipment, operated jointly by authorized common carriers, that provides the general public with the ability to establish communications channels via discrete dialing codes.

Rate compensation detector: A device that responds when the temperature of the air surrounding the device reaches a predetermined level, regardless of the rate of temperature rise.

Rate-of-rise detector: A device that responds when the temperature rises at a rate exceeding a predetermined value.

Remote supervising station fire alarm system: A system installed in accordance with the applicable code to transmit alarm, supervisory, and trouble signals from one or more protected premises to a remote location where appropriate action is taken.

Reset: A control function that attempts to return a system or device to its normal, non-alarm state.

Signal: A status indication communicated by electrical or other means.

Signaling line circuits (SLCs): A circuit or path between any combination of circuit interfaces, control units, or transmitters over which multiple system input signals or output signals (or both input signals and output signals) are carried.

Smoke detector: A device that detects visible or invisible particles of combustion.

Spacing: A horizontally measured dimension related to the allowable coverage of fire detectors.

Spot-type detector: A device in which the detecting element is concentrated at a particular location. Typical examples are bimetallic detectors, fusible alloy detectors, certain pneumatic rate-of-rise detectors, certain smoke detectors, and thermoelectric detectors.

Stratification: The phenomenon in which the upward movement of smoke and gases ceases due to a loss of buoyancy.

Supervisory signal: A signal indicating the need for action in connection with the supervision of guard tours, the fire suppression systems or equipment, or the maintenance features of related systems.

System unit: The active subassemblies at the central station used for signal receiving, processing, display, or recording of status change signals. The failure of one of these subassemblies causes the loss of a number of alarm signals by that unit.

Transmitter: A system component that provides an interface between the transmission channel and signaling line circuits, initiating device circuits, or control units.

Trouble signal: A signal initiated by the fire alarm system or device that indicates a fault in a monitored circuit or component.

Visible notification appliance: A notification appliance that alerts by the sense of sight.

Wavelength: The distance between peaks of a sinusoidal wave. All radiant energy can be described as a wave having a wavelength. Wavelength serves as the unit of measure for distinguishing between different parts of the spectrum. Wavelengths are measured in microns, nanometers, or angstroms.

Wide Area Telephone Service (WATS): Telephone company service that provides reduced costs for certain telephone call arrangements. In-WATS or 800-number service calls can be placed from anywhere in the continental United States to the called party at no cost to the calling party. Out-WATS is a service whereby, for a flat-rate charge, dependent on the total duration of all such calls, a subscriber can make an unlimited number of calls within a prescribed area from a particular telephone terminal without the registration of individual call charges.

Zone: A defined area within the protected premises. A zone can define an area from which a signal can be received, an area to which a signal can be sent, or an area in which a form of control can be executed.

Additional Resources

This module is intended as a thorough resource for task training. The following reference works are suggested for further study.

NFPA 72®, *National Fire Alarm and Signaling Code*, National Fire Protection Association. Available at **www.nfpa.org**.

NFPA 101®, *Life Safety Code*, National Fire Protection Association. Available at **www.nfpa.org**.

Figure Credits

© iStockphoto.com/SandroBassi, Module Opener

System Sensor, Figures 11, 12, 31, 36A, 67

Brooks Equipment Company, LLC, Figure 24

Honeywell Security and Fire, Figures 28, 30, 34

Wheelock, Figures 32, 36B–C

Section Review Answer Key

Section 1.0.0

Answer	Section Reference	Objective
1. c	1.1.0	1a
2. b	1.2.1	1b

Section 2.0.0

Answer	Section Reference	Objective
1. b	2.1.1	2a
2. d	2.2.3	2b

Section 3.0.0

Answer	Section Reference	Objective
1. c	3.1.0	3a
2. b	3.2.2	3b
3. c	3.3.2	3c

Section 4.0.0

Answer	Section Reference	Objective
1. b	4.1.1	4a
2. d	4.2.0	4b
3. a	4.3.2	4c

Section 5.0.0

Answer	Section Reference	Objective
1. a	5.1.0	5a
2. d	5.2.4	5b
3. b	5.3.2	5c
4. a	5.4.2	5d
5. c	5.5.0	5e

This page is intentionally left blank.

NCCER CURRICULA — USER UPDATE

NCCER makes every effort to keep its textbooks up-to-date and free of technical errors. We appreciate your help in this process. If you find an error, a typographical mistake, or an inaccuracy in NCCER's curricula, please fill out this form (or a photocopy), or complete the online form at **www.nccer.org/olf**. Be sure to include the exact module ID number, page number, a detailed description, and your recommended correction. Your input will be brought to the attention of the Authoring Team. Thank you for your assistance.

Instructors – If you have an idea for improving this textbook, or have found that additional materials were necessary to teach this module effectively, please let us know so that we may present your suggestions to the Authoring Team.

NCCER Product Development and Revision

13614 Progress Blvd., Alachua, FL 32615

Email: curriculum@nccer.org
Online: www.nccer.org/olf

❏ Trainee Guide   ❏ Lesson Plans   ❏ Exam   ❏ PowerPoints   Other _____

Craft / Level: _____   Copyright Date: _____

Module ID Number / Title: _____

Section Number(s): _____

Description: _____

_____

_____

_____

Recommended Correction: _____

_____

_____

_____

_____

Your Name: _____

Address: _____

_____

Email: _____   Phone: _____

This page is intentionally left blank.

Specialty Transformers

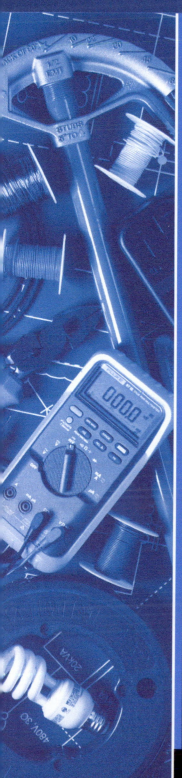

OVERVIEW

Transformers are used to increase or decrease voltage coming from a power source. They can be constructed in a variety of configurations for different applications. This module covers various types of transformers, and provides information on selecting, sizing, and installing them.

Module 26406-20

Trainees with successful module completions may be eligible for credentialing through the NCCER Registry. To learn more, go to **www.nccer.org** or contact us at 1.888.622.3720. Our website, **www.nccer.org**, has information on the latest product releases and training.

Your feedback is welcome. You may email your comments to **curriculum@nccer.org**, send general comments and inquiries to **info@nccer.org**, or fill in the User Update form at the back of this module.

This information is general in nature and intended for training purposes only. Actual performance of activities described in this manual requires compliance with all applicable operating, service, maintenance, and safety procedures under the direction of qualified personnel. References in this manual to patented or proprietary devices do not constitute a recommendation of their use.

26406-20 V10.0

SPECIALTY TRANSFORMERS

Objetives

When you have completed this module, you will be able to do the following:

1. Identify and describe various types of transformers.
   a. Identify common transformer types.
   b. Explain how three-phase transformers can be wired internally.
   c. Identify types of specialty transformers.
2. Identify instrument transformers.
   a. Identify and describe the use of current transformers.
   b. Identify and describe the use of potential transformers.
3. Define harmonics and explain how harmonic issues are identified and resolved.
   a. Describe the common sources of harmonics in office buildings and industrial plants.
   b. Explain how to survey a system to identify the source of harmonics.
   c. Explain how to resolve harmonics.

Performance Tasks

Under the supervision of the instructor, you should be able to do the following:

1. Identify various specialty transformers.
2. Connect a buck-and-boost transformer to a single-phase circuit so that it will first be in the boost mode and then in the buck mode. Record the voltage increase and decrease for each configuration.
3. Using a clamp-on ammeter, demonstrate the principles of a current transformer; identify the primary winding, and then calculate and measure the effects of increasing the number of turns (loops) in the primary winding.

Trade Terms

Ampere turns	Eddy currents	Isolation transformer
Autotransformer	Harmonic	Reactance
Bank	Hysteresis	
Core loss	Impedance	

Industry Recognized Credentials

If you are training through an NCCER-accredited sponsor, you may be eligible for credentials from NCCER's Registry. The ID number for this module is 26406-20. Note that this module may have been used in other NCCER curricula and may apply to other level completions. Contact NCCER's Registry at 888.622.3720 or go to **www.nccer.org** for more information.

> **NOTE**
>
> NFPA 70®, *National Electrical Code*® and *NEC*® are registered trademarks of the National Fire Protection Association, Quincy, MA.

Contents

1.0.0 Identifying Transformers 1

  1.1.0 Common Transformers 1

  1.2.0 Connections in Three-Phase Transformers 2

  1.3.0 Specialty Transformers 5

    1.3.1 Transformers with Multiple Secondaries 5

    1.3.2 Three-Winding Transformers 7

    1.3.3 Autotransformers 7

    1.3.4 Constant-Current Transformers 9

    1.3.5 Control Transformers 10

    1.3.6 Series Transformers 10

    1.3.7 Step-Voltage Regulators 11

    1.3.8 Buck-and-Boost Transformers 11

2.0.0 Identifying Instrument Transformers 15

  2.1.0 Current Transformers 15

  2.2.0 Potential Transformers 16

3.0.0 Identifying and Resolving Harmonics 19

  3.1.0 Culprits in Offices and Plants 21

    3.1.1 Neutral Conductors 21

    3.1.2 Circuit Breakers 21

    3.1.3 Busbars and Connecting Lugs 21

    3.1.4 Electrical Panels 22

    3.1.5 Telecommunications 22

    3.1.6 Transformers 22

    3.1.7 Generators 22

  3.2.0 Surveying for Harmonics 22

    3.2.1 Meters 23

    3.2.2 Crest Factor 24

  3.3.0 Resolving Harmonics 24

    3.3.1 Harmonics in Overloaded Neutrals 24

    3.3.2 Derating Transformers 24

Figures and Tables ———

Figure 1   Typical three-phase transformer 1
Figure 2   Typical liquid-filled, three-phase power transformer 2
Figure 3   Two types of current transformers 2
Figure 4   Voltage (potential) transformer 3
Figure 5   Common power transformer connections 4
Figure 6   Single-phase transformer with two secondaries 7
Figure 7   Wiring diagram of a typical autotransformer 7
Figure 8   *NEC*® grounding requirements for isolation
           and autotransformers 9
Figure 9   Regulation curves 11
Figure 10   Isolation transformer 12
Figure 11   Boost transformer connection 13
Figure 12   Typical line diagram of a transformer circuit 13
Figure 13   Polarity marks on transformers 15
Figure 14   Connection of instrument transformers 16
Figure 15   Voltage waveforms 20
Figure 16   Three-phase, delta-wye transformer configuration 22

Table 1   Harmonic Rates and Effects 21
Table 2   Current Measurements 26
Table 3   Neutral Loads 26
Table 4   Phase Currents and Neutral-to-Ground Voltages 27

This page is intentionally left blank.

SECTION ONE

1.0.0 IDENTIFYING TRANSFORMERS

Objective

Identify and describe various types of transformers.

a. Identify common transformer types.
b. Explain how three-phase transformers can be wired internally.
c. Identify types of specialty transformers.

Performance Tasks

1. Identify various specialty transformers.
2. Connect a buck-and-boost transformer to a single-phase circuit so that it will first be in the boost mode and then in the buck mode. Record the voltage increase and decrease for each configuration.

Trade Terms

Autotransformer: Any transformer in which the primary and secondary connections are made to a single winding. The application of an autotransformer is a good choice when a 480Y/277V or 208Y/120V, three-phase, four-wire distribution system is used.

Bank: An installed grouping of a number of units of the same type of electrical equipment, such as a bank of transformers, a bank of capacitors, or a meter bank.

Isolation transformer: A transformer that has no electrical metallic connection between the primary and secondary windings.

Reactance: The imaginary part of impedance; also, the opposition to alternating current due to capacitance (X_C) and/or inductance (X_L).

When the AC voltage needed for an application is lower or higher than the voltage available from the source, a transformer is used. The essential parts of a transformer are the primary winding (which is connected to the source) and the secondary winding (which is connected to the load), both wound on an iron core. The two windings are not physically connected.

The alternating voltage in the primary winding induces an alternating voltage in the secondary winding. The ratio of the primary and secondary voltages is equal to the ratio of the number of turns in the primary and secondary windings.

Transformers may step up the voltage applied to the primary winding and have a higher voltage at the secondary terminals, or they may step down the voltage applied to the primary winding and have a lower voltage at the secondary terminals.

A transformer can be constructed as a single-phase or three-phase apparatus. A three-phase transformer, such as the one shown in *Figure 1*, has three primary and three secondary windings, which may be connected in either a delta (Δ) or wye (Y) configuration. Combinations such as Δ-Δ, Δ-Y, Y-Δ, and Y-Y are possible. The first letter indicates the connection of the primary winding, and the second letter indicates that of the secondary winding.

> **NOTE**
>
> Transformers are applied in AC systems only and would not work in DC systems because the induction of voltage depends on the rate of change in the current.

A bank of three single-phase transformers can serve the same purpose as one three-phase transformer. The connections between the three primary windings and the three secondary windings are again delta or wye, and they are available in all combinations.

1.1.0 Common Transformers

The principle of operation is the same for all transformers, but the forms, connections, and auxiliary devices differ widely.

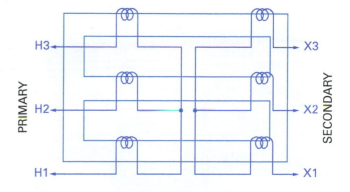

Figure 1 Typical three-phase transformer.

Large transformers used in transmission systems are called *power transformers*. They may step up the voltage produced in the generator to make it suitable for commercial transmission, or they may step down the voltage in a transmission line to make it practical for distribution. Power transformers are usually installed outdoors in generating stations and substations. The transformer shown in *Figure 2* is a liquid-filled power transformer with radiators that disperse the heat from the transformer.

An autotransformer is a transformer with only one winding, which serves as both a primary and a secondary. Autotransformers are economical, space saving, and especially practical if the difference between the primary and secondary voltages is relatively small.

Three-winding transformers have two secondary windings so that they can deliver two different secondary voltages.

In many applications of electrical measuring instruments, the voltage or current in the circuit to be measured is too high for the instruments. In such situations, instrument transformers are used to ensure safe operation of the instruments. An instrument transformer is either a current transformer, as shown in *Figure 3*, or a potential (voltage) transformer, as shown in *Figure 4*. For example, if the current in a line is approximately 100A and the ammeter is rated at 5A, a bar-type current transformer with a current ratio of 100:5 can be connected in series with the line. The secondary winding is then connected to the ammeter. The current in the secondary is proportional to the current in the line. With the toroid current transformer, an insulated line passes through the center, and proportional current is induced in the toroid winding for application to the meter.

BAR-TYPE CURRENT TRANSFORMER

TOROID (DONUT-TYPE) CURRENT TRANSFORMERS

Figure 3 Two types of current transformers.

Similarly, the potential transformer reduces the voltage in a high-voltage line to the 120V for which the voltmeter is normally rated. The potential transformer is always connected across the line to be measured. Sometimes it is important to indicate the polarity of the current in both the instrument and the transformer. The polarity is the instantaneous direction of current at a specific moment. Wires with the same polarity usually contain a small black block or cross in electrical diagrams, as shown in *Figure 4*.

1.2.0 Connections in Three-Phase Transformers

Various combinations of primary and secondary three-phase voltages are possible with the proper combination of delta and wye internal connections of three-phase transformers or with banks of single-phase transformers.

Some common power transformer connections are shown in *Figure 5*. In all examples, it is assumed that the primary line voltages between lines A, B, and C are all 1,000V and that the transformation ratio is 10:1. The primary connections are shown by the upper three windings, and the

Figure 2 Typical liquid-filled, three-phase power transformer.

BASIC SYMBOL

POTENTIAL
TRANSFORMER
WITH POLARITY
MARKS

600V CLASS VOLTAGE
TRANSFORMER

Figure 4 Voltage (potential) transformer.

Δ (delta) or Y (wye) next to the winding indicates how the winding is connected.

In the delta connections, as in *Figure 5 (A)* and *Figure 5 (D)*, the phase voltages of the primaries are the same as the line voltage or 1,000V. In the Y connections of the primaries, as in *Figure 5 (B)* and *Figure 5 (C)*, the phase voltages are 0.577 times the line voltages or 577V. The wye point (N) is indicated as a common point if the windings are wye connected.

The secondaries are shown in the lower row of windings. Each secondary phase winding has only $1/10$ of the turns used in the corresponding primary phase winding and, therefore, supplies a phase voltage that is $1/10$ of the primary phase voltage. In *Figure 5 (A)* and *Figure 5 (D)*, the secondary phase voltages are $1,000 \div 10 = 100V$, and in *Figure 5 (B)* and *Figure 5 (C)*, they are $577 \div 10 = 57.7V$. The secondaries are delta connected in *Figure 5 (A)* and *Figure 5 (C)*, and their line voltages or the voltages between the secondary line wires (a, b, and c) are the same as the secondary phase windings. When the secondaries are wye connected, as in *Figure 5 (B)* and *Figure 5 (D)*, they have a common wye point (N) and the secondary line voltages are 1.732 times the phase voltage. In *Figure 5 (B)*, the voltages between lines a, b, and c are $1.732 \times 57.7 = 100V$, and in *Figure 5 (D)*, they are $1.732 \times 100 = 173V$. In addition, a fourth secondary wire (n) is brought out from the wye point (N) in *Figure 5 (D)*, and the secondary voltage between any of the lines (a, b, or c and n) is equal to the secondary phase voltage or 100V.

As you become more experienced in reading electrical diagrams, you should be able to immediately recognize the differences between delta and wye connections in any three-phase system. If the three windings have a common point, it is a wye connection, and the line voltages are 1.732 times higher than the individual phase voltages. If the three windings build a closed path, it is a delta connection, and the line voltages are the same as the phase voltages.

A connection using two single-phase transformers for a three-phase system is shown in *Figure 5 (E)*. This is an open delta connection that provides a three-phase secondary with only two transformers.

Transformer Management Systems

Some manufacturers provide monitoring and diagnostic systems for large liquid-filled power transformers used in substations or in large industrial applications. These systems are used to obtain warnings of impending faults and as a predictor for scheduling maintenance shutdowns. The systems use data from sensors incorporated in the transformers. The sensors monitor the following conditions:

- Dissolved gas-in-oil
- Top and bottom oil temperature
- Gas pressure
- Ambient temperature
- Oil level
- Winding temperature
- Load and meter values
- Bushing activity

This data is transmitted via a network to a computer and used with IEEE or IEC analytical modeling software offline to access the condition of the equipment and diagnose impending failures. This includes moisture-in-insulation modeling, predictive insulation aging rate, cumulative aging, partial discharge activity, overflux (overvoltage), fault event recording, harmonic monitoring, cooling system efficiency, and other trending information.

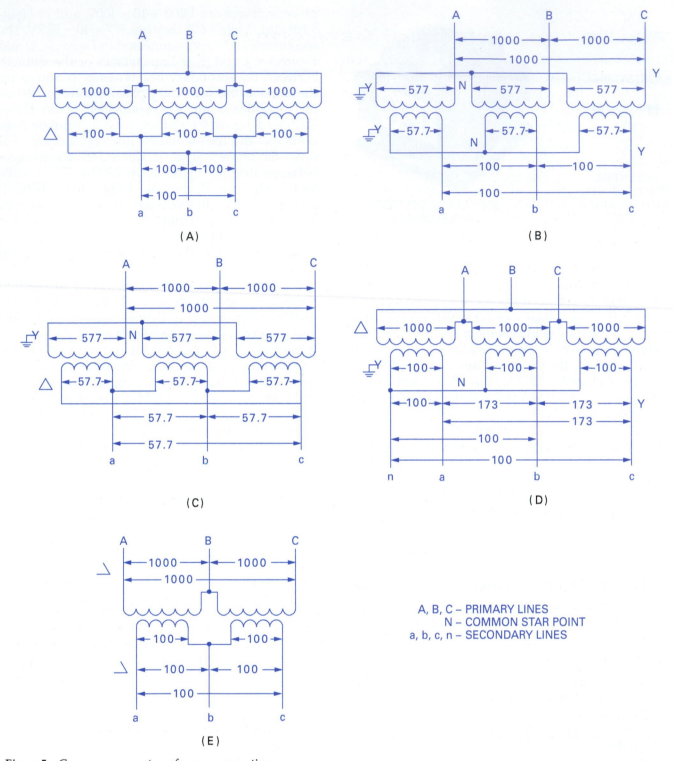

Figure 5 Common power transformer connections.

A, B, C – PRIMARY LINES
N – COMMON STAR POINT
a, b, c, n – SECONDARY LINES

Maintenance Testing of Transformers

In commercial or industrial applications involving a number of large liquid-filled or dry transformers that are not automatically monitored, manual periodic preventive maintenance testing and performance trending is normally instituted in a manner similar to that for large motors. These tests include liquid dielectric tests to measure the breakdown voltage of samples of insulating fluids from transformers and DC hi-pot tests of transformer winding insulation.

LIQUID DIELECTRIC TEST SET   COMPONENT TEST POT   DC HI-POT TEST SET

Figure Credit: Phenix Technologies, Inc.

1.3.0 Specialty Transformers

Specialty transformers make up a large class of transformers and autotransformers used for changing a line voltage to some particular value best adapted to the load device. The primary voltage is generally 600V or less. Examples of specialty transformers are sign-lighting transformers in which 120V is stepped down to 25V for low-voltage tungsten sign lamps; arc-lamp autotransformers in which 240V is stepped down to the voltage required for best operation of the arc; and transformers used to change 240V power to 120V for operating portable tools, fans, welders, and other devices. Also included in this specialty class are neon sign transformers that step 120V up between 2,000V and 15,000V for the operation of neon signs. Many special step-down transformers are used for small work, such as doorbells or other signaling systems, battery-charging rectifiers, and individual low-voltage lamps. Practically all specialty transformers are self-cooled and air-insulated. Sometimes the cases are filled with a special compound to prevent moisture absorption and to conduct heat to the enclosing structure.

Among the many transformers designed for a specific purpose are single-phase transformers with two secondaries, single-phase transformers with three windings, autotransformers, constant-current transformers, series transformers, rectifier transformers, network transformers, and step-voltage regulators.

1.3.1 Transformers with Multiple Secondaries

One common type of transformer is a single-phase transformer with two secondaries (*Figure 6*). The first secondary has terminals X_1 and X_2 and is connected internally in series with the other secondary, which has terminals X_3 and X_4. Terminals X_1 and X_4 are connected to outside leads, and terminals X_2 and X_3 are connected together, with the junction point connected to a third outside lead. Such transformers are commonly used

Cast-Coil Transformers

Some newer types of power transformers are made without using wire to wind the coils of the transformer. In the three-phase, step-down transformer shown here, strip foil technology is used to wind each of the primary coils. The primary coils are then placed in molds and encased with a mixture of epoxy resin and quartz powder under a high vacuum to remove any moisture. After heat curing, the coils are placed on the legs of the transformer core. Then, concentric secondary coils are wound using a sheet conductor. These secondary coils are encased in the same way as the primary coils. After curing, they are placed over the primary coils. Transformers made in this manner are capable of operating at temperatures ranging from –40°C to +180°C. They do not require liquid cooling or vaults and can be placed in NEMA Type 1 indoor or NEMA Type 3R outdoor enclosures. The epoxy-encased coils are highly resistant to caustic and humid environments.

CAST-COIL THREE-PHASE TRANSFORMER

STRIP-FOIL WINDING OF A PRIMARY COIL

SHEET-CONDUCTOR WINDING OF
A SECONDARY COIL

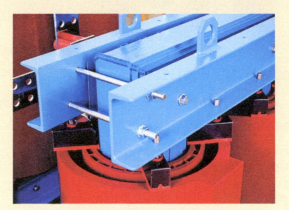

CONCENTRIC PRIMARY AND
SECONDARY CAST COILS

as distribution transformers where three-wire service is needed from a two-wire, single-phase supply. The rated secondary voltage (120V in *Figure 6*) is obtained between either outside lead and the middle lead or neutral. In addition, double voltage (240V in *Figure 6*) is available between the two outside leads. This higher voltage is usually needed for HVAC equipment, electric ranges, and dryers, while lamps and small appliances are operated on 120V.

1.3.2 Three-Winding Transformers

Another widely used type of transformer is a three-winding transformer. A third winding can be added to a transformer, and voltages are induced in this winding proportional to the number of turns, the same as in the other windings. As a matter of fact, there is theoretically no limit to the number of windings that may be provided in a transformer. Practically, however, there is a limit because of the greater complexity, and transformers are seldom provided with more than four windings.

Three-winding transformers are used when transmission voltage is stepped down to produce two secondary voltages. For example, a transformation may be desired from 230kV down to 69kV and 34.5kV to feed separate distribution networks.

Three-winding transformers are built in the same way as two-winding transformers, and they may be either the core or shell types. In the core construction, the additional winding is usually arranged so that it is concentric with the other two. In a shell transformer, the third winding is interleaved with the other two, which makes it more flexible than the core form.

1.3.3 Autotransformers

The usual transformer has two windings that are not physically connected. In an autotransformer, one of the windings is connected in series with the other, thereby forming the equivalent of a single winding, as shown in *Figure 7*. This illustration represents a step-up autotransformer, so called because the secondary voltage is higher than the voltage supplied to the primary.

The primary voltage (E_P) is applied to the primary or common winding. The secondary or series winding is connected in series with the primary at the junction terminal. This point may be obtained by a tap, which will divide a single winding into a primary and a secondary.

A voltage induced in the secondary winding adds to the voltage in the primary winding, and the secondary voltage (E_S) is higher than the applied voltage. The ratio of transformation depends on the turns ratio, as in a two-winding transformer.

The primary current (I_P) branches into current I_C through the common winding and the current I_S through the series winding, as indicated by the arrows in *Figure 7*. The values of currents I_C and I_S are inversely proportional to the ratio of turns in the two windings and the primary current ($I_P = I_S + I_C$). Because the currents I_S and I_C oppose each other, the secondary current is lower than the primary current.

Autotransformers may be used economically to connect individual loads requiring voltages other than those available in the distribution

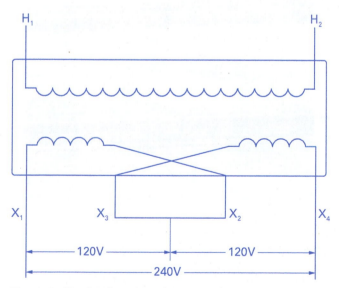

Figure 6 Single-phase transformer with two secondaries.

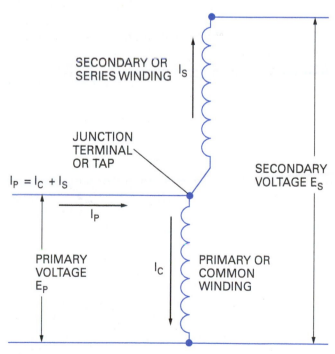

Figure 7 Wiring diagram of a typical autotransformer.

Multiple-Secondary Transformers

The two multiple-secondary transformers shown here are liquid-filled for cooling and insulation purposes and have internal, manually switched voltage taps. They are used for single-phase residential 120V/240V service.

(A) PAD-MOUNTED TRANSFORMER

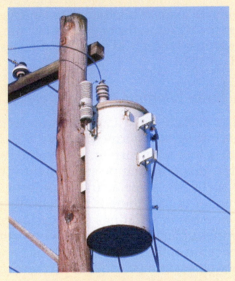

(B) POLE-MOUNTED TRANSFORMER

Figure Credit: Tim Dean

system. An example would be increasing the voltage on a branch circuit from a 120V/208V panel to 240V for more effective utilization of a motor or resistive load.

An autotransformer, however, cannot be used on a 240V or 480V, three-phase, three-wire delta system. A grounded neutral phase conductor must be available in accordance with *NEC Section 210.9*, which states that branch circuits shall not be supplied by autotransformers unless the system supplied has a grounded conductor that is electrically connected to a grounded conductor of the system supplying the autotransformer.

In general, the *NEC®* requires that separately derived alternating current systems be grounded. The secondary of an isolation transformer is a separately derived system. Therefore, it must be grounded in accordance with *NEC Section 250.30*. See *Figure 8*. In the case of an autotransformer, the grounded conductor of the supply is brought into the transformer to the common terminal and the ground is established to satisfy the *NEC®*.

Three-Winding Control Transformers

Small three-winding transformers may be used in control systems. The one shown here has separate 24V and 26V secondary windings and a 120V primary winding.

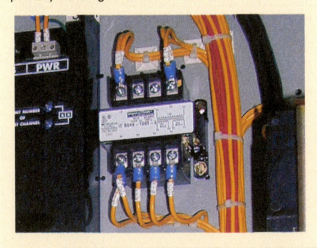

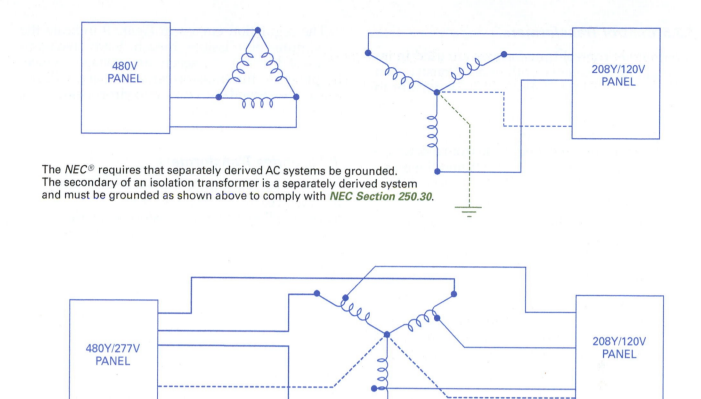

The *NEC®* requires that separately derived AC systems be grounded. The secondary of an isolation transformer is a separately derived system and must be grounded as shown above to comply with **NEC Section 250.30**.

In the case of autotransformers, the grounded conductor of the supply is connected to the transformer common terminal and the ground is established to satisfy **NEC Section 210.9**.

Figure 8 *NEC®* grounding requirements for isolation and autotransformers.

1.3.4 *Constant-Current Transformers*

Constant-current transformers are used to supply series airport lighting and street lighting circuits in which the current must remain constant while the number of lamps in series varies because of burnout or bypass switching. This type of transformer has a stationary primary coil connected to a source of alternating voltage and a movable secondary coil connected to the lamp circuit. To allow it to move freely, the secondary coil is suspended from a shaft and rod attached to a rocker arm on which hinges are attached. The tendency of the secondary coil to move downward due to gravity is opposed by both the counterweight and the magnetic repulsion between the coils caused by the current in them.

The constant secondary current is usually between 4A and 7.5A, depending on the current rating of the lamps. In order to maintain a constant secondary current, the voltage of the secondary must vary directly with the number of lamps in series. If the resistance of the secondary is reduced by decreasing the number of lamps in series because of burnout or bypass switching, the current in the transformer will momentarily increase. This will increase the force of repulsion between the coils, which will move apart. The increase in distance

between the coils increases the leakage reactance of the transformer and causes a greater voltage drop in the transformer, with a consequent decrease in the secondary voltage. The coil will continue to move until the current in the secondary winding reaches its original value, at which position the mechanical forces acting on the secondary coil are again balanced. Incandescent lamps used in series lighting circuits have a film cutout in the socket. When a lamp burns out, the full transformer secondary voltage appears between two spring contacts separated by a thin insulation film in the socket. The voltage punctures the film, causing the series circuit to be re-established.

Advantages of Autotransformers

Autotransformers are less costly than conventional two-winding transformers and have better voltage regulation and a better ratio of energy output to input (efficiency). They are used in motor starters so that a voltage lower than the line voltage may be applied during the starting period.

1.3.5 Control Transformers

A number of control transformers are used in industrial establishments, both in new construction and in existing installations. It is important to have a basic knowledge of how to size these transformers for any given application.

In general, the selection of a proper control circuit transformer must be made from a determination of the maximum inrush VA and the maximum continuous VA to which it is subjected. This data can be determined as follows:

Step 1   Determine the inrush and sealed VA of all coils to be used.

Step 2   Determine the maximum sealed VA load on the transformer.

Step 3   Determine the maximum inrush VA load on the transformer at 100% of the secondary voltage. Add this value to any sealed VA present at the time inrush occurs.

Step 4   Calculate the power factor of the VA load obtained in Step 3. The actual coil power factor should be used. If this value is unknown, an inrush power factor of 35% may be assumed.

Step 5   Select a transformer with a continuous VA rating equal to or greater than the value obtained in Step 2 and whose maximum inrush VA from Step 3 at the calculated load power factor falls on or below the corresponding curve in *Figure 9*.

The regulation curves in *Figure 9* indicate the maximum permissible inrush loads (volt-amperes at 100% of the secondary voltage), which, if applied to the transformer secondary, will not cause the secondary voltage to drop below 85% of the rated voltage when the primary voltage has been reduced to 90% of the rated voltage.

1.3.6 Series Transformers

A power or distribution transformer is normally used with each of its primary terminals connected across the line. When a transformer is used in series with the main line, the term *series transformer* is applied.

A common application of a series transformer is in an airport system that has runway lamps connected in series. A small transformer is used with one or more lamp(s). The primary is connected in series with the line, and the secondary is connected across the lamp(s). The secondary winding is automatically short-circuited if a lamp burns out. The short circuit is obtained by a film cutout. When the circuit opens and the voltage across the secondary rises, it punctures the film, thereby short-circuiting the transformer secondary.

Series transformers are used in load tap changing circuits and with step-voltage regulators to reduce the operating voltage to ground when it is too high for the tap changes or to reduce the current in the tap changer contacts when the current exceeds the tap changer rating. These series transformers usually serve the purpose of an auxiliary transformer mounted in the main transformer tank.

Control Power Transformers

The control power transformer shown here is rated for use in medium-voltage applications, including switchgear, with primary voltage ranges from 5kV to 34kV. Standard output voltages are 120V/240V for single-phase units and 208V/120V for three-phase units. Some versions can be mounted vertically to conserve space.

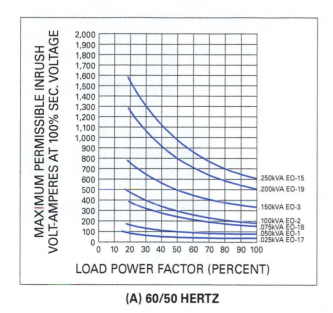

(A) 60/50 HERTZ

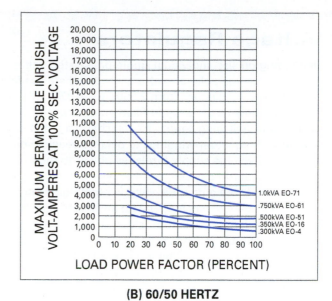

(B) 60/50 HERTZ

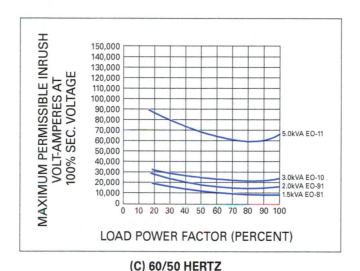

(C) 60/50 HERTZ

Figure 9 Regulation curves.

1.3.7 Step-Voltage Regulators

Regulators of the step-voltage type are transformers provided with load tap changers. They are used to raise or lower the voltage of a circuit in response to a voltage-regulating relay or another control device. Regulators are usually designed to provide secondary voltages ranging from 10% below the supply voltage to 10% above it, or a total change of 20% in 32 steps of ⅝% each.

1.3.8 Buck-and-Boost Transformers

A buck-and-boost transformer is used to control the voltage applied to low-voltage (12V–48V) equipment. Buck-and-boost transformers can be used to power uninterruptible power supply (UPS) systems for data processing equipment, control circuits, and lighting circuits.

Manufacturers of buck-and-boost transformers usually offer easy-to-use selector charts that allow you to quickly select a buck-and-boost transformer for practically any application. These charts may be obtained from electrical equipment suppliers or ordered directly from the manufacturer—often at no charge. Instructions accompanying these charts will enable anyone familiar with transformers and electrical circuits to use them. An overview of the principles involved in using buck-and-boost transformers is given here.

When reviewing the selector charts, it may surprise you to discover that these transformers can handle loads that are much greater than their nameplate ratings. For example, a typical 1kVA buck-and-boost transformer can easily handle an 11kVA load when the voltage boost is only 10%. In this case, an isolation transformer will be

Voltage Regulators

Depending on size, individual single-phase step-voltage regulators can be pole- or pad-mounted and are available in sizes up to 830kVA at voltages from 2,500V to 19,920V. The step-voltage regulator shown here is liquid-filled and has both remote and local digital switching control for tap changing. It uses a switching reactor with equalizer windings to balance reactor voltage. An internal voltage supply furnishes power to the switching motor and control devices. Another type of voltage-adjusting transformer is the automatic voltage regulator. It continuously senses and self-adjusts to maintain a selected output voltage level at ±1% for critical loads including computer, medical, communications, and industrial-process equipment. These are dry-type units with a variable-ratio autotransformer consisting of a rotor and stator. The rotor turns only 180° to add to or subtract from the supply voltage and is driven by a reversible motor. Because these units regulate by transformer action instead of impedance change, no waveform distortion occurs. The regulators are available in single- and three-phase versions in sizes up to 1,000kVA at voltages from 120V to 480V.

incorporated as part of an autotransformer to form a buck-and-boost transformer.

Assume that you have a 1kVA (1,000VA) isolation transformer that is designed to transform 208V to 20.8V (*Figure 10*). This results in a transformer winding ratio of 10:1. The primary current may be found using the following equation:

Primary current = 1,000VA ÷ 208V = 4.8A

Because the transformation ratio is 10:1, the secondary amperes will be 48A (4.8A x 10 = 48A), or the amperage may be determined using the following equation:

Secondary current = 1,000VA ÷ 20.8V = 48A

Figure 11 shows how the windings are connected in series to form an autotransformer.

Because you started with 208V at the source and now add 20.8V to it, the load is now 208V + 20.8V = 228.8V. To find the kVA rating of the system at the load, use the following equation:

$$kVA = (volts \div 1,000) \times amps$$
$$kVA = (228V \div 1,000) \times 48A$$
$$kVA = 0.228V \times 48A$$
$$kVA = 11kVA$$

The source conducts 10kVA, and 1kVA is transformed from the source. The total kVA rating is 10kVA + 1kVA = 11kVA.

The H_1H_2 winding is rated at 4.8A, and the X_1X_2 winding is rated at 48A. Therefore, the line current at 208V would be 4.8A + 48A = 52.8A. The input kVA rating is as follows:

$$kVA = (volts \div 1,000) \times amps$$
$$kVA = (208V \div 1,000) \times 52.8A$$
$$kVA = 0.208 \times 52.8A$$
$$kVA = 11kVA$$

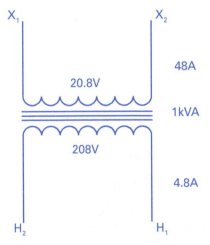

Figure 10 Isolation transformer.

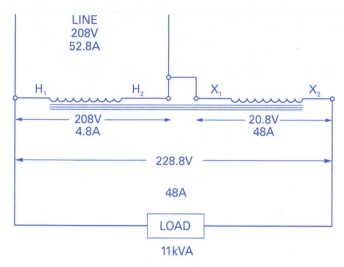

Figure 11 Boost transformer connection.

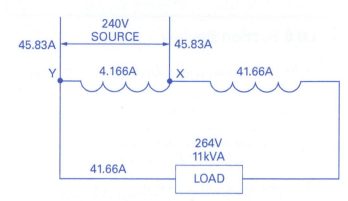

Figure 12 Typical line diagram of a transformer circuit.

The diagrams shown in *Figure 10* and *Figure 11* are usually simplified even more in a line diagram, as shown in *Figure 12*. Actually, all three wiring diagrams indicate the same thing, and following the connections on any of these drawings will produce the same results at the load. *Figure 12* shows the calculations for a 240V source.

It should now be evident how a 1kVA buck-and-boost transformer, when connected in the circuit as described previously, can actually carry 11kVA in its secondary winding.

Buck-and-Boost Transformers

A small buck-and-boost transformer is shown here. It is a single-phase compound-filled unit rated at 0.05kVA. However, depending on the percentage of voltage buck or boost required, it can be used in circuits with much higher loads.

Figure Credit: Electro-Mechanical Corp./Line Power/Federal Pacific

1.0.0 Section Review

1. A transformer with a single winding that serves as both the primary and the secondary is known as a(n) _____.

   a. control transformer
   b. autotransformer
   c. instrument transformer
   d. potential transformer

2. If the three windings in a three-phase transformer have a common point, it is a(n) _____.

   a. delta transformer
   b. delta-delta transformer
   c. wye transformer
   d. autotransformer

3. An arc-lamp is *most* likely to be supplied by a(n) _____.

   a. oil-insulated transformer
   b. autotransformer
   c. rectifier transformer
   d. power transformer

2.0.0 IDENTIFYING INSTRUMENT TRANSFORMERS

Objective

Identify instrument transformers.
 a. Identify and describe the use of current transformers.
 b. Identify and describe the use of potential transformers.

Performance Task

3. Using a clamp-on ammeter, demonstrate the principles of a current transformer; identify the primary winding, and then calculate and measure the effects of increasing the number of turns (loops) in the primary winding.

Trade Terms

Ampere turns: The product of amperes times the number of turns in a coil.

Impedance: The opposition to current flow in an AC circuit; impedance includes resistance (R), capacitive reactance (X_C), and inductive reactance (X_L). It is measured in ohms (Ω).

Instrument transformers are so named because they are usually connected to an electrical instrument, such as an ammeter, voltmeter, wattmeter, or relay. As mentioned earlier, there are two types of instrument transformers: current and potential (voltage).

2.1.0 Current Transformers

The primary winding of a current transformer is connected in series in a line connecting the power source and the load, and it carries the full-load current. The turns ratio is designed to produce a rated current of 5A (or some other specified value) in the secondary winding when the rated current flows in the primary winding. The current transformer provides a small current suitable for the current coil of standard instruments and proportional to the load current. The low-voltage, low-current secondary winding providing the current is grounded for safety and economy in the secondary wiring and instruments.

A potential transformer is connected from one power line to another. Its secondary winding provides a low voltage, usually up to 120V, that is proportional to the line voltage. This low voltage is suitable for the voltage coil of standard instruments. The low-voltage secondary winding is grounded for the safety of the secondary wiring and the instrument, regardless of the power line voltage.

When connecting instrument transformers to wattmeters, watt-hour meters, power factor meters, or other instruments, it is necessary to know the polarity of the leads. One primary lead and one secondary lead of the same polarity are clearly marked on all instrument transformers, usually by a white spot or white marker on the leads.

The direction of current in the two leads of the same polarity is such that, if it is toward the transformer in the marked primary lead, it is away from the transformer in the marked secondary lead. This is done to maintain the same phase sequence in the receiving device as is present on the power lines. In diagrams, the polarity mark is usually indicated in one of three ways, as shown in *Figure 13*.

A current transformer is always a single-phase transformer. If current transformers are used in a three-phase system, one current transformer is inserted into each phase line between the power supply and the instrument. A current transformer is considered a low kVA device. A connection of a current transformer into a single-phase, two-wire line is shown in *Figure 14*.

The primary winding of a current transformer must be connected in series with one of the main power lines; thus, the main line load current flows through the primary winding. The secondary

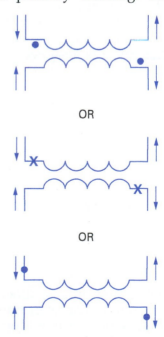

Figure 13   Polarity marks on transformers.

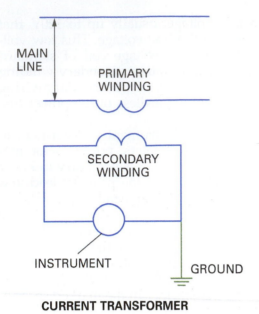

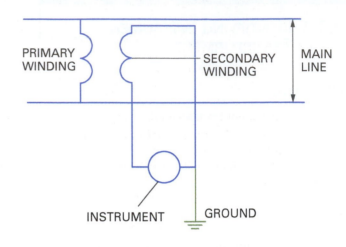

Figure 14 Connection of instrument transformers.

winding at the current transformer is connected to a current-responsive instrument. The secondary circuit is grounded for safety. Instruments that use a current transformer may include ammeters, wattmeters, watt-hour meters, power factor meters, some forms of relays, and trip coils of circuit breakers. One current transformer may feed through several devices in series.

Current transformers are built to standard IEEE burden ratings. *Burden* refers to the total load seen by the CT secondary. This load is the total **impedance** of conductors and instruments connected. For example, a CT with a B-1 designation is intended to work properly with no more than 1V impedance in the secondary circuit. This B-1 CT is rated to deliver 25VA at a secondary current of 5A. A CT with a B-8 designation is rated at 200VA at 5A. It is designed to work with an 8V impedance. When current flow is interrupted in the CT secondary, the secondary voltage can rise to dangerously high levels as the CT tries to deliver VA. Therefore, a shorting jumper should be placed across the secondary terminals when no device is connected.

> **CAUTION**
> CTs shipped as part of equipment usually come with a shorting jumper installed. Be sure to remove the shorting jumper after the connection to the relaying or metering circuit is verified.

Before disconnecting an instrument, the secondary of the current transformer must be short-circuited. If the secondary circuit is opened while the primary winding is carrying current, there will be no secondary **ampere turns** to balance the primary ampere turns. Therefore, the total primary current will become exciting current and magnetize the core to a high flux density, which will produce a high voltage across both the primary and secondary windings.

Because current transformers are designed for accuracy, the normal exciting current is only a small percentage of the full-load current. The voltage produced with the secondary open circuited is high enough to endanger the life of anyone coming in contact with the meters or leads. The high secondary voltage may also overstress the secondary insulation and cause a breakdown. Still other damage may be caused by operation with the secondary open circuited—the transformer core may become permanently magnetized, impairing the accuracy of the transformer. If this occurs, the core may be demagnetized by passing about 50% excess current through the primary, with the secondary connected to an adjustable high resistance that is gradually reduced to zero.

> **WARNING!**
> The secondary circuit of a current transformer should never be opened while the primary is carrying current.

2.2.0 Potential Transformers

Potential or voltage transformers are single-phase transformers. If used on three-phase circuits, sets of two or three potential transformers are applied.

The primary winding of a potential transformer is always connected across the main power lines.

A connection to a single-phase, two-wire circuit is shown in *Figure 14*. The primary of the potential transformer is connected across the main line, and the secondary is connected to an instrument. For safety, the secondary circuit is grounded.

The main circuit voltage exists across the primary winding. The secondary of the potential transformer is connected to one or more voltage-responsive devices, such as voltmeters, wattmeters, watt-hour meters, power factor meters, or some forms of relays or trip coils. The voltage across the secondary terminals of the potential transformer is always lower than the primary voltage and is rated at 120V. For example, if a potential transformer with a turns ratio of 12:1 is connected to a 1,440V main line, the voltage in the instrument connected across the secondary terminals will be 1,440V divided by 12 or 120V.

All potential transformers have a wound primary and a wound secondary. Mechanically, their construction is similar to that of wound current transformers. Their secondary thermal kVA rating seldom exceeds 1.5kVA.

Potential transformers are available in both oil-filled and dry types. In both types, the primary high-voltage leads are terminated at bushings and the housing contains the secondary low-voltage terminations. When supplied as part of a switchgear assembly, they occupy dedicated spaces.

In certain small services, both current and potential transformers are packaged into a metering unit that is a single enclosure. This reduces the assembly time in field installations.

Single-Phase Current Transformers

This is an assembly of three wound-type single-phase current transformers serving a three-phase feeder.

Figure Credit: Jim Mitchem

Potential Transformers

This potential transformer has a fused input and a primary rated for medium-voltage applications up to 34.5kV.

2.0.0 Section Review

1. Current transformers are built to standard _____.

   a. NEMA enclosure sizes
   b. NFPA ratings
   c. ASTM frame sizes
   d. IEEE burden ratings

2. The voltage across the secondary of a potential transformer is rated at _____.

   a. 24V
   b. 60V
   c. 120V
   d. 240V

3.0.0 IDENTIFYING AND RESOLVING HARMONICS

Objective

Define harmonics and explain how harmonic issues are identified and resolved.

a. Describe the common sources of harmonics in office buildings and industrial plants.
b. Explain how to survey a system to identify the source of harmonics.
c. Explain how to resolve harmonics.

Trade Terms

Core loss: The electric loss that occurs in the core of an armature or transformer due to conditions such as the presence of eddy currents or hysteresis.

Eddy currents: The circulating currents that are induced in conductive materials by varying magnetic fields; they are usually considered undesirable because they represent a loss of energy and produce excess heat.

Harmonic: An oscillation at a frequency that is an integral multiple of the fundamental frequency.

Hysteresis: The time lag exhibited by a body in reacting to changes in the forces affecting it; hysteresis is an internal friction.

Harmonics are the byproducts of modern electronics. They are especially prevalent wherever there are large numbers of personal computers (PCs), adjustable-speed drives, and other types of equipment that draw current in short pulses.

This equipment is designed to draw current only during a controlled portion of the incoming voltage waveform. Although this dramatically improves efficiency, it causes harmonics in the load current. This results in overheated transformers and neutrals, as well as tripped circuit breakers.

The problem is evident when you look at a waveform. A normal 60-cycle power line voltage appears on the oscilloscope as a near sine wave, as shown in *Figure 15 (A)*. When harmonics are present, the waveform is distorted, as shown in *Figure 15 (B)* and *Figure 15 (C)*. These waves are described as nonsinusoidal. The voltage and current waveforms are no longer simply related—hence the term *nonlinear*.

Finding the problem is relatively easy when you know what to look for and where to look. Harmonics are usually anything but subtle. This section gives you some basic pointers on how to find harmonics and some suggested ways to address the problem. However, in many cases, consultants must be called in to analyze the operation and design a plan for correcting the problem.

> **CAUTION**
>
> As part of a regular maintenance program, pay careful attention to overheating of the neutral conductors in distribution systems. Harmonics may cause deterioration of the insulation.

Harmonics are currents or voltages with frequencies that are integer multiples of the fundamental power frequency. For example, if the fundamental frequency is 60Hz, then the second harmonic is 120Hz, the third is 180Hz, and so on.

Harmonics are created by nonlinear loads that draw current in abrupt pulses rather than in a smooth sinusoidal manner. These pulses cause distorted current waveshapes, which in turn cause harmonic currents to flow back into other parts of the power system. This phenomenon is especially prevalent with equipment that contains diode/capacitor input or solid-state switched power supplies, such as personal computers, printers, and medical test equipment.

In a diode/capacitor, the incoming AC voltage is diode rectified and is then used to charge a large capacitor. After a few cycles, the capacitor is charged to the peak voltage of the sine wave (for example, 168V for a 120V line). The electronic equipment then draws current from this high DC voltage to power the rest of the circuit.

The equipment can draw the current down to a regulated lower limit. Typically, before reaching that limit, the capacitor is recharged to the peak in the next half cycle of the sine wave. This process is repeated over and over. The capacitor basically draws a pulse of current only during the peak of the wave. During the rest of the wave, when the voltage is below the capacitor residual, the capacitor draws no current.

Harmonics

If you were to listen to an ordinary 60-cycle power line, you would hear a monotone hum or buzz. When harmonics are present, you hear a different tune, rich with high notes.

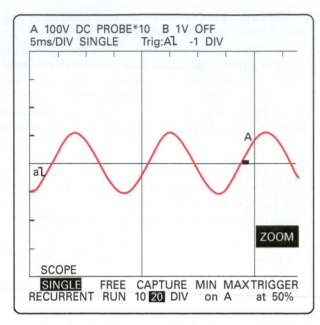

(A) NEAR SINE WAVE

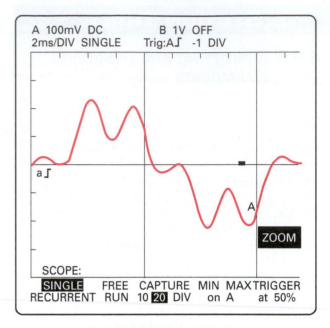

(B) DISTORTED WAVEFORM

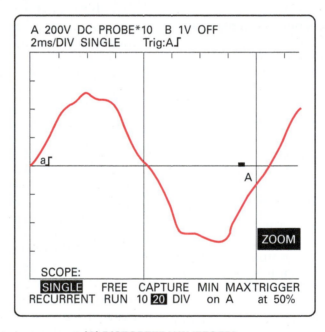

(C) DISTORTED WAVEFORM

Figure 15   Voltage waveforms.

The power line itself can be an indirect source of voltage harmonics. The harmonic current drawn by nonlinear loads acts in an Ohm's law relationship with the source impedance of the supplying transformer to produce the voltage harmonics. The source impedance includes both the supplying transformer and the branch circuit components. For example, a 10A harmonic current being drawn for a source impedance of 0.1Ω will generate a harmonic voltage of 1.0V. Any loads sharing a transformer or branch circuit with a heavy harmonic load can be affected by the voltage harmonics generated.

Many types of devices are very susceptible to voltage harmonics. The performance of the diode/capacitor power supply is critically dependent on the magnitude of the peak voltage. Voltage harmonics can cause flat-topping of the voltage waveform, lowering the peak voltage. In severe cases, the computer may reset due to insufficient peak voltage.

In an industrial environment, the induction motor and power factor correction capacitors can also be seriously affected by voltage harmonics.

Power correction capacitors can form a resonant circuit with the inductive parts of a power distribution system. If the resonant frequency is near that of the harmonic voltage, the resultant harmonic current can increase substantially, overloading the capacitors and blowing the capacitor fuses. Fortunately, the capacitor failure detunes the circuit and the resonance disappears.

Each harmonic has a name, frequency, and sequence. The sequence refers to phasor rotation with respect to the fundamental (F); that is, in an induction motor, a positive sequence harmonic would generate a magnetic field that rotates in the same direction as the fundamental. A negative sequence harmonic would rotate in the reverse direction. The first nine harmonics, along with their effects, are shown in Table 1.

3.1.0 Culprits in Offices and Plants

Harmonics have a significant effect in office buildings and industrial establishments. Symptoms of harmonics usually show up in the power distribution equipment that supports the nonlinear loads. There are two basic types of nonlinear loads: single phase and three phase. Single-phase loads are prevalent in offices; three-phase loads are widespread in industrial plants.

Each component of the power distribution system manifests the effects of harmonics a little differently, but they are all subject to damage and inefficient performance.

3.1.1 Neutral Conductors

In a three-phase, four-wire system, the neutral conductor can be severely affected by nonlinear loads connected to the 120V branch circuits. Under normal conditions for a balanced linear load, the fundamental 60Hz portions of the phase currents cancel in the neutral conductor.

In a four-wire system with single-phase, nonlinear loads, certain odd-numbered harmonics called *triplens*—odd multiples of the third harmonic: 3rd, 9th, 15th, and so on—do not cancel, but rather add together in the neutral conductor. In systems with many single-phase, nonlinear loads, the neutral current can actually exceed the phase current. The danger here is excessive overheating because there is no circuit breaker in the neutral conductor to limit the current as there are in the phase conductors.

Excessive current in the neutral conductor can also cause higher voltage drops between the neutral conductor and ground at the 120V outlet.

3.1.2 Circuit Breakers

Common thermal-magnetic circuit breakers use a bimetallic trip mechanism that responds to the heating effect of the circuit current. They are designed to respond to the true root-mean-square (rms) value of the current waveform and therefore will trip when they get too hot. This type of breaker has a better chance of protecting against harmonic current overloads.

A peak sensing electronic trip circuit breaker responds to the peak of the current waveform. As a result, it does not always respond properly to harmonic currents. Since the peak of the harmonic current is usually higher than normal, this type of circuit breaker may trip prematurely at a low current. If the peak is lower than normal, the breaker may fail to trip when it should.

3.1.3 Busbars and Connecting Lugs

Neutral busbars and connecting lugs are sized to carry the full value of the rated phase current. They can become overloaded when the neutral conductors are overloaded with the additional sum of the triplen harmonics.

Table 1 Harmonic Rates and Effects

Name	F	2nd	3rd	4th	5th	6th	7th	8th	9th
Frequency	60	120	180	240	300	360	420	480	540
Sequence	+	−	0	+	−	0	+	−	0

Sequence	Rotation	Effects (skin effect, eddy currents, etc.)
Positive	Forward	Heating of conductors and circuit breakers
Negative	Reverse	Heating as above, plus motor problems
Zero	None	Heating plus add-in neutral of three-phase, four-wire system

3.1.4 Electrical Panels

Harmonics in electrical panels can be quite noisy. Panels that are designed to carry 60Hz current can become mechanically resonant to the magnetic fields generated by high-frequency harmonic currents. When this happens, the panel vibrates and emits a buzzing sound at the harmonic frequencies.

3.1.5 Telecommunications

Telecommunications cable is commonly run right next to power cables. To minimize the inductive interference from phase current, telecommunications cables are run closer to the neutral wire. Triplens in the neutral conductor commonly cause inductive interference that can be heard on a phone line. This is often the first indication of a harmonics problem and gives you a head start in detecting the problem before it causes major damage.

3.1.6 Transformers

Commercial buildings commonly have a 120V/208V transformer in a delta-wye configuration, as shown in *Figure 16*. These transformers commonly feed receptacles in a commercial building. Single-phase, nonlinear loads connected to the receptacles produce triplen harmonics that algebraically add up in the neutral. When this neutral current reaches the transformer, it is reflected into the delta primary winding, where it circulates and causes overheating and transformer failures.

Another transformer problem results from core loss and copper loss. Transformers are normally rated for a 60Hz phase-current load only.

High-frequency harmonic currents cause increased core loss due to eddy currents and hysteresis, resulting in more heating than would occur at the same 60Hz current. These heating effects demand that transformers be derated for harmonic loads or replaced with specially designed transformers.

Nonlinear Loads

The diode/capacitor or solid-state switched power supplies found in office equipment typically consist of single-phase, nonlinear loads. In industrial plants, the most common causes of harmonic currents are three-phase, nonlinear loads. These include electronic motor drives, uninterruptible power supplies (UPSs), HID lighting, and welding machines.

3.1.7 Generators

Standby generators are subject to the same types of overheating problems as transformers. Because they provide emergency backup for harmonics-producing loads such as data processing equipment, they are often even more vulnerable. In addition to overheating, certain types of harmonics produce distortion at the zero crossing of the current waveform, which causes interference and instability in the generator control circuits.

3.2.0 Surveying for Harmonics

A quick survey will help to determine whether you have a harmonics problem and where it is located. The survey procedure should include the following:

Step 1 Take a walking tour of the facility and look at the types of equipment in use. If there are many personal computers and printers, adjustable-speed motors, solid-state heater controls, and certain types of fluorescent lighting, there is a good chance that harmonics are present.

Step 2 Locate the transformers feeding the non-linear loads and check for excess heating. Also, make sure that the cooling vents are unobstructed.

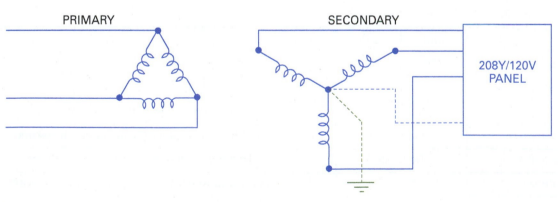

Figure 16   Three-phase, delta-wye transformer configuration.

K-Rated and Zig-Zag Transformers

Specially designed transformers that reduce or compensate for harmonic current include K-rated and zig-zag transformers. To prevent overheating, K-rated three-phase transformers for delta-to-wye connections are sized to handle 100% of the normal 60Hz load plus a specified nonlinear load defined by a K number. Manufacturers' specifications define the K numbers for various types of loads. If the various harmonics are known, a specific transformer can be custom built. The inherent phase shift of a K-rated transformer will cancel 5th and 7th harmonics but not the triplens. Because triplens are not cancelled, the neutral of the secondary is normally oversized at 200% of the maximum current rating of one of the phase connections. A zig-zag phase shift transformer cancels triplens and other harmonics at the load side of the transformer by using six windings in the secondary. An extra winding for each phase is connected in series with another phase to produce a phase shift. This type of transformer works well if the loads are balanced. A drawback is that zig-zag transformers use more material in the windings and are heavy. Other techniques such as active filters or ferroresonant transformers are also used for harmonic current suppression in certain applications.

Step 3 Use a true rms meter to check transformer currents.

- Verify that the voltage ratings for the test equipment are adequate for the transformer being tested.
- Measure and record the transformer secondary currents in each phase and in the neutral (if used).
- Calculate the kVA delivered to the load, and compare it to the nameplate rating. If harmonic currents are present, the transformer can overheat even if the kVA delivered is less than the nameplate rating.
- If the transformer secondary is a four-wire system, compare the measured neutral current to the value predicted from the imbalance in the phase currents. (The neutral current is the vector sum of the phase currents and is normally zero if the phase currents are balanced in both amplitude and phase.) If the neutral current is unexpectedly high, triple harmonics are likely and the transformer may need to be derated.
- Measure the frequency of the neutral current. 180Hz would be a typical reading for a neutral current consisting of mostly third harmonics.

Step 4 Survey the subpanels that feed harmonic loads. Measure the current in each branch neutral, and compare the measured value to the rated capacity for the wire size used. Check the neutral busbar and feeder connections for heating or discoloration.

Step 5 Neutral overloading in receptacle branch circuits can sometimes be detected by measuring the neutral-to-ground voltage at the receptacle. Measure the voltage when the loads are on. A reading of 2V or less is normal. Higher voltages can indicate trouble, depending on the length of the run, quality of the connections, and other factors.

Step 6 Measure the frequency. A result of 180Hz would suggest a strong presence of harmonics; 60Hz would suggest that the phases are out of balance. Pay special attention to undercarpet wiring and modular office panels with integrated wiring that use a neutral shared by three-phase conductors. Because the typical loads in these two areas are computers and office machines, they are often trouble spots for overloaded neutrals.

3.2.1 Meters

Having the proper equipment is crucial to diagnosing harmonics problems. The type of equipment used varies with the complexity of measurements required.

To determine whether you have a harmonics problem, you need to measure the true rms value and the instantaneous peak value of the waveshape. For this test, you need a true rms clamp-on multimeter or a handheld digital multimeter that makes true rms measurements and has a high-speed peak hold circuit.

The term *true rms* refers to the root-mean-square or equivalent heating value of a current or voltage waveshape. True distinguishes the measurement from those taken by average responding meters.

The vast majority of low-cost, portable clamp-on ammeters are average responding. These instruments give correct readings for pure sine waves only and typically read low when confronted with a distorted current waveform. The result is a reading that can be as much as 50% low.

True rms meters give correct readings for any waveshape within the instrument's crest factor and bandwidth specifications.

3.2.2 Crest Factor

The crest factor of a waveform is the ratio of the peak value to the rms value. For a sine wave, the crest factor is 1.414. A true rms meter has a crest factor specification. This specification relates to the level of peaking that can be measured without errors.

A quality true rms handheld digital multimeter has a crest factor of 3.0 at full scale. This is more than adequate for most power distribution measurements. At half scale, the crest factor is double. For example, a meter may have a crest factor specification of 3.0 when measuring 400VAC and a crest factor of 6.0 when measuring 200VAC.

The crest factor can be easily calculated using a true rms meter with a peak function or a crest function. A crest factor other than 1.414 indicates the presence of harmonics. In typical single-phase cases, the greater the difference from 1.414, the greater the harmonics. For voltage harmonics, the typical crest factor is below 1.414 (that is, a flat-top waveform). For single-phase current harmonics, the typical crest factor is well above 1.414. Three-phase current waveforms often exhibit the double-hump waveform; therefore, the crest factor comparison method should not be applied to three-phase load currents.

> **NOTE**
>
> Most true rms meters cannot be used for signals below 5% of scale because of the measurement noise problem. Use a lower range if it is available.

After you have determined that harmonics are present, you can make a more in-depth analysis of the situation using a harmonics analyzer.

3.3.0 Resolving Harmonics

The following are some suggested ways of addressing some typical harmonics problems. Before taking any measures, you should consult a power quality expert to analyze the problem and design a plan tailored to your specific situation.

3.3.1 Harmonics in Overloaded Neutrals

In a three-phase, four-wire wye system, the 60Hz portion of the neutral current can be minimized by balancing the loads in each phase. *NEC Section 220.61(C)* prohibits reduced sizing of neutral conductors serving nonlinear loads. The triplen harmonics neutral current can be reduced by adding harmonic filters at the load. If neither of these solutions is practical, you can pull in extra neutrals (ideally one neutral for each phase) or you can install an oversized neutral to be shared by the three conductors.

In new construction, undercarpet wiring and modular office partition wiring should be specified with individual neutrals and possibly an isolated ground separate from the safety ground.

3.3.2 Derating Transformers

One way to protect a transformer from harmonics is to limit the amount of load placed on it. This is called derating the transformer. The most rigorous derating method is described in *ANSI/IEEE Standard C57.110-2018*. This method is somewhat impractical because it requires extensive loss data from the transformer manufacturer, plus a complete harmonics spectrum of the load current.

The Computer and Business Equipment Manufacturers' Association (CBEMA) recommends a second method that involves several straightforward measurements using common test equipment. It appears to give reasonable results for 208V/120Y receptacle transformers that supply the low-frequency odd harmonics (3rd, 5th, and 7th) commonly generated by computers and office machines.

The test equipment must be capable of taking both the true rms phase current and the instantaneous peak phase current for each phase of the secondary.

Think About It

Four-Wire Systems

What condition might cause some conductors of a four-wire system to feel hot to the touch?

Power Quality Analyzers

Portable power quality analyzers like the one shown here can be used to determine harmonics, as well as to make power measurements (kW, VA, and VAR), power factor and displaced power factor measurements, voltage and current readouts and waveforms, inrush current and duration recording, transient measurements, and sag and swell recording. Extensive power quality measurements can be made continuously using the software provided with most energy management systems.

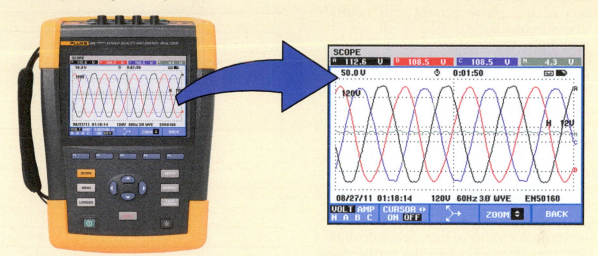

Figure Credit: Fluke Corporation, reproduced with permission

To determine the transformer harmonics derating factor (THDF), take the peak and true rms current measurements for the three-phase conductors. If the phases are not balanced, average the three measurements and plug that value into the following equation:

$$\text{THDF} = \frac{1.414 \times \text{true rms phase current}}{\text{instantaneous peak phase current}}$$

This equation generates a value between 0 and 1.0 (typically between 0.5 and 0.9). If the phase currents are purely sinusoidal (undistorted), the instantaneous current peaks at 1.414 times the true rms value and the derating factor is 1.0. If that is the case, no derating is required.

However, with harmonics present, the transformer rating is the product of the nameplate kVA rating times the THDF:

$$\text{Derated kVA} = \text{nameplate kVA} \times \text{THDF}$$

For example, a modern office building dedicated primarily to computer software development contains a large number of PCs and other electronic office equipment. These electronic loads are fed by a 120V/208V transformer configured with a delta primary and a wye secondary. The PCs are fairly well distributed throughout the building, except for one large room that contains several machines. The PCs in this room, used exclusively for testing, are served by several branch circuits.

The transformer and main switchgear are located in a ground floor electrical room. An inspection of this room immediately reveals two symptoms of high-harmonic currents:

- The transformer is generating a substantial amount of heat.
- The main panel emits an audible buzzing sound. The sound is not the chatter commonly associated with a faulty circuit breaker, but rather a deep, resonant buzz that indicates the mechanical parts of the panel itself are vibrating.

The ductwork installed directly over the transformer to carry off some of the excess heat keeps the room temperature within reasonable limits.

Current measurements (see *Table 2*) are taken on the neutral and on each phase of the transformer secondary using both a true rms multimeter and an average responding unit.

A 600A, clamp-on current transformer accessory is connected to each meter to allow the meters to make high-current readings.

The presence of harmonics is obvious by a comparison of the phase-current and neutral-current measurements. As *Table 2* shows, the neutral current is substantially higher than any of the phase currents, even though the phase currents are relatively well balanced. The average responding meter consistently shows readings that are approximately 20% low on all phases. Its neutral current readings are only 2% low.

The waveforms explain the discrepancy. The phase currents are badly distorted by large amounts of triplen harmonics, while the neutral current is not affected. The phase current readings listed in *Table 2* clearly demonstrate why true rms measurement capability is required to accurately determine the value of harmonic currents.

The next step is to calculate the transformer harmonic derating factor, or THDF, as explained previously.

The results indicate that, with the level of harmonics present, the transformer should be derated to 72.3% of its nameplate rating to prevent overheating. In this case, the transformer should be derated to 72.3% of its 225kVA rating, or 162.7kVA.

The actual load is calculated to be 151.3kVA. Although this figure is far less than the nameplate rating, the transformer is operating close to its derated capacity.

Next, a subpanel that supplies branch circuits for the 120V receptacles is also examined. The current in each neutral is measured and shown in *Table 3*.

When a marginal or overloaded conductor is identified, the associated phase currents and the neutral-to-ground voltage at the receptacle are also measured. When a check of neutral No. 6 reveals 15A in a conductor rated for 16A, the phase currents of the circuits (No. 25, No. 27, and No. 29) that share that neutral are also measured (*Table 4*).

Table 3 Neutral Loads

Neutral Conductor Number	Current (Amps)
01	5.0
02	11.3
03	5.0
04	13.1
05	12.4
06	15.0
07	1.8
08	11.7
09	4.5
10	11.8
11	9.6
12	11.5
13	11.3
14	6.7
15	7.0
16	2.3
17	2.6

Note that each of the phase currents of these three branch circuits is substantially less than 15A and the same phase conductors have significant neutral-to-ground voltage drops.

In the branch circuits that have high neutral currents, the relationship between the neutral and the phase currents is similar to that of the transformer secondary. The neutral current is higher than any of the associated phase currents. The danger here is that the neutral conductors could become overloaded and not offer the warning signs of tripped circuit breakers.

The recommendations are:

- Refrain from adding additional loads to the receptacle transformer unless steps are taken to reduce the level of harmonics.
- Pull extra neutrals into the branch circuits that are heavily loaded.
- Monitor the load currents on a regular basis using true rms test equipment.

Table 2 Current Measurements

Conductor Name	Multimeter (true rms)	Average Responding Multimeter	Instantaneous Peak Current
Phase 1	410A	328A	804A
Phase 2	445A	346A	892A
Phase 3	435A	355A	828A
Neutral	548A	537A	762A

Table 4 Phase Currents and Neutral-to-Ground Voltages

Circuit Number	Phase Current	Neutral-to-Ground Voltage Drop at Receptacle
25	7.8A	3.75V
27	9.7A	4.00V
29	13.5A	8.05V

3.0.0 Section Review

1. In a three-phase, four-wire system, overheating in the neutral can be caused by odd multiples of the _____.

   a. F harmonic
   b. second harmonic
   c. third harmonic
   d. seventh harmonic

2. When surveying an installation, the presence of harmonics may be indicated by a _____.

   a. frequency of 180Hz
   b. receptacle neutral-to-ground voltage of 2V or less
   c. frequency of 120Hz
   d. neutral current of 0V

3. When calculating the THDF, a purely sinusoidal waveform has a derating factor of _____.

   a. 0
   b. 1
   c. 3
   d. 9

1. What is the name given to the large transformers used in transmission systems to step voltage up and down?

   a. Control transformers
   b. Power transformers
   c. Instrument transformers
   d. Potential transformers

2. The purpose of the two secondary windings in a three-winding transformer is _____.

   a. so the transformer may be used on systems with different primary voltages
   b. to produce a nonmagnetic field
   c. to produce a high-impedance field
   d. to produce two secondary voltages

3. How many windings are normally found in an autotransformer?

   a. One
   b. Two
   c. Three
   d. Four

4. What is a transformer called when it is connected in series with the main line?

   a. Reactance transformer
   b. Series transformer
   c. High-voltage transformer
   d. Low-ampere transformer

5. Regulators are usually designed to provide a secondary voltage range of _____.

   a. +/- 10% of the supply voltage
   b. +/- 15% of the supply voltage
   c. +/- 25% of the supply voltage
   d. +/- 35% of the supply voltage

6. What type of transformer is normally able to handle a load that is much greater than its nameplate rating?

   a. Potential transformer
   b. Reactance transformer
   c. Current transformer
   d. Buck-and-boost transformer

7. Which of the following is always a single-phase transformer?

   a. Power transformer
   b. Reactance transformer
   c. Current transformer
   d. Autotransformer

8. What type of transformer is normally used with metering equipment when the current is too high for the metering equipment?

   a. Potential transformer
   b. Reactance transformer
   c. Current transformer
   d. Autotransformer

9. Which type of transformer is normally used with metering equipment when the line voltage is too high for the metering equipment?

   a. Potential transformer
   b. Reactance transformer
   c. Series transformer
   d. Autotransformer

10. The secondary thermal kVA rating of a conventional potential transformer seldom exceeds _____.

    a. 1kVA
    b. 1.5kVA
    c. 1,500kVA
    d. 15,000kVA

11. Potential transformers are available in _____.

    a. oil-filled and hydraulic types
    b. oil-filled and dry types
    c. dry and chemical types
    d. dry and coiled types

12. Each harmonic has a _____.

    a. name only
    b. name and frequency only
    c. name, frequency, and sequence
    d. name, frequency, and rotation

13. Neutral busbars and connecting lugs are sized to carry _____.

    a. 75% of the rated phase current
    b. 80% of the rated phase current
    c. 90% of the rated phase current
    d. 100% of the rated phase current

14. The problem of harmonics is often first detected by the presence of inductive interference in _____.

    a. lighting systems
    b. heating systems
    c. telecommunications systems
    d. emergency power systems

15. For a sine wave, the crest factor is _____.

    a. 1.000
    b. 1.414
    c. 2.212
    d. 2.414

Supplemental Exercises

1. The four different connections possible in a three-phase transformer are _____.

2. True or False? A typical transformer has two windings that are physically connected.

   _____

3. Power transformers are usually installed outdoors in _____.

4. True or False? Autotransformers are especially practical if the difference between the primary and secondary voltages is relatively small.

   _____

5. In a three-phase, delta-connected transformer, how does the line voltage compare to the phase voltage?

   _____

6. A constant-current transformer has a(n) _____ primary coil and a(n) _____ secondary coil.

7. In general, the selection of a proper control circuit transformer must be made from a determination of the _____ to which it is subjected.

8. True or False? Buck-and-boost transformers can handle loads that are much greater than their nameplate ratings.

   _____

9. The two types of instrument transformers are _____.

10. The polarity in an instrument transformer identifies _____.

11. The primary winding of a potential transformer is always connected across the _____.

12. Symptoms of harmonics in an office building or industrial establishment usually show up in the _____ equipment that supports the nonlinear loads.

13. In a three-phase, four-wire system, the _____ is *most likely* to be severely affected by nonlinear loads.

14. What trip device is designed to respond to the true rms value of the current waveform?

   _____

15. A buzzing sound and/or vibration of an electrical panel can be caused by _____.

Trade Terms Introduced in This Module

Ampere turns: The product of amperes times the number of turns in a coil.

Autotransformer: Any transformer in which the primary and secondary connections are made to a single winding. The application of an autotransformer is a good choice when a 480Y/277V or 208Y/120V, three-phase, four-wire distribution system is used.

Bank: An installed grouping of a number of units of the same type of electrical equipment, such as a bank of transformers, a bank of capacitors, or a meter bank.

Core loss: The electric loss that occurs in the core of an armature or transformer due to conditions such as the presence of eddy currents or hysteresis.

Eddy currents: The circulating currents that are induced in conductive materials by varying magnetic fields; they are usually considered undesirable because they represent a loss of energy and produce excess heat.

Harmonic: An oscillation at a frequency that is an integral multiple of the fundamental frequency.

Hysteresis: The time lag exhibited by a body in reacting to changes in the forces affecting it; hysteresis is an internal friction.

Impedance: The opposition to current flow in an AC circuit; impedance includes resistance (R), capacitive reactance (X_C), and inductive reactance (X_L). It is measured in ohms (Ω).

Isolation transformer: A transformer that has no electrical metallic connection between the primary and secondary windings.

Reactance: The imaginary part of impedance; also, the opposition to alternating current due to capacitance (X_C) and/or inductance (X_L).

Additional Resources

This module presents thorough resources for task training. The following resource material is suggested for further study.

C57.110-2018 – IEEE Recommended Practice for Establishing Liquid-Filled and Dry-Type Power and Distribution Transformer Capability When Supplying Nonsinusoidal Load Currents, 2018. Available online at **standards.ieee.org**

National Electrical Code® Handbook, Latest Edition. Quincy, MA: National Fire Protection Association.

Figure Credits

Fluke Corporation, reproduced with permission, Module Opener

Section Review Answer Key

Section 1.0.0

Answer	Section Reference	Objective
1. b	1.1.0	1a
2. c	1.2.0	1b
3. b	1.3.0	1c

Section 2.0.0

Answer	Section Reference	Objective
1. d	2.1.0	2a
2. c	2.2.0	2b

Section 3.0.0

Answer	Section Reference	Objective
1. c	3.1.1	3a
2. a	3.2.0	3b
3. b	3.3.2	3c

This page is intentionally left blank.

NCCER CURRICULA — USER UPDATE

NCCER makes every effort to keep its textbooks up-to-date and free of technical errors. We appreciate your help in this process. If you find an error, a typographical mistake, or an inaccuracy in NCCER's curricula, please fill out this form (or a photocopy), or complete the online form at **www.nccer.org/olf**. Be sure to include the exact module ID number, page number, a detailed description, and your recommended correction. Your input will be brought to the attention of the Authoring Team. Thank you for your assistance.

Instructors – If you have an idea for improving this textbook, or have found that additional materials were necessary to teach this module effectively, please let us know so that we may present your suggestions to the Authoring Team.

NCCER Product Development and Revision
13614 Progress Blvd., Alachua, FL 32615

Email: curriculum@nccer.org
Online: www.nccer.org/olf

❏ Trainee Guide   ❏ Lesson Plans   ❏ Exam   ❏ PowerPoints   Other _____

Craft / Level: _____   Copyright Date: _____

Module ID Number / Title: _____

Section Number(s): _____

Description: _____

_____

_____

_____

_____

Recommended Correction: _____

_____

_____

_____

_____

Your Name: _____

Address: _____

_____

Email: _____   Phone: _____

This page is intentionally left blank.

Advanced Controls

OVERVIEW

Control systems are what regulate and direct the behavior of devices within an electrical system. They vary in complexity and consist of a variety of components, which provide different types of control. This module discusses applications and operating principles of various control system components, such as solid-state relays, reduced-voltage starters, and adjustable-frequency drives. It also covers basic troubleshooting procedures.

Module 26407-20

Trainees with successful module completions may be eligible for credentialing through the NCCER Registry. To learn more, go to **www.nccer.org** or contact us at 1.888.622.3720. Our website, **www.nccer.org**, has information on the latest product releases and training.

Your feedback is welcome. You may email your comments to **curriculum@nccer.org**, send general comments and inquiries to **info@nccer.org**, or fill in the User Update form at the back of this module.

This information is general in nature and intended for training purposes only. Actual performance of activities described in this manual requires compliance with all applicable operating, service, maintenance, and safety procedures under the direction of qualified personnel. References in this manual to patented or proprietary devices do not constitute a recommendation of their use.

26407-20 V10.0

Objectives

When you have completed this module, you will be able to do the following:

1. Describe the various types of relays used in motor control circuits.
   a. Identify and describe solid-state relays and their uses.
   b. Identify timing relays and explain their uses.
2. Explain how reduced-voltage starting is accomplished.
   a. Describe the use and selection of conventional reduced-voltage motor starting methods.
   b. Describe solid-state reduced-voltage motor starting methods.
3. Specify an adjustable frequency drive (AFD) for a given application.
   a. Describe the types of AFDs, along with their operation and components.
   b. Compare the uses, benefits, and problems when selecting an AFD.
4. Compare motor braking methods.
   a. Describe DC injection braking.
   b. Identify dynamic and regenerative braking methods.
   c. Describe friction braking.
5. Describe how to maintain motor controls for peak operation.
   a. Apply the precautions associated with solid-state controls.
   b. Implement the preventive maintenance procedures associated with solid-state controls.
6. Describe how to troubleshoot motor controls.
   a. Explain basic troubleshooting methods.
   b. Describe the electrical troubleshooting methods used to check control circuits and devices.

Performance Task

Under the supervision of the instructor, you should be able to do the following:

1. Identify and connect various control devices.

Trade Terms

Base speed
Carrier frequency
Closed-circuit transition
Dashpot
Fan Affinity Laws
Heat sink
Insulated gate bipolar transistors (IGBTs)

Open-circuit transition
Slip
Solid-state overload relay (SSOLR)
Solid-state relay (SSR)
Thermal conductivity
Thyristors

Industry Recognized Credentials

If you are training through an NCCER-accredited sponsor, you may be eligible for credentials from NCCER's Registry. The ID number for this module is 26407-20. Note that this module may have been used in other NCCER curricula and may apply to other level completions. Contact NCCER's Registry at 888.622.3720 or go to **www.nccer.org** for more information.

> **NOTE**
>
> NFPA 70®, *National Electrical Code*® and *NEC*® are registered trademarks of the National Fire Protection Association, Quincy, MA.

Contents

1.0.0 Identifying Relays 1
  1.1.0 Solid-State Relays 1
    1.1.1 Solid-State Relay Operation 1
    1.1.2 Electromechanical Relays versus Solid-State Relays 2
    1.1.3 Two-Wire and Three-Wire SSR Control 4
    1.1.4 Connecting SSRs to Switch Multiple Outputs 4
    1.1.5 SSR Temperature Considerations 4
    1.1.6 Overvoltage and Overcurrent Protection 5
    1.1.7 SSOLRs 5
  1.2.0 Timing Relays 9
    1.2.1 Solid-State Timing Relays 10
    1.2.2 Timing Relay Applications 11
2.0.0 Reduced-Voltage Starting Methods 14
  2.1.0 Conventional Methods 14
    2.1.1 Autotransformer Reduced-Voltage Motor Starting Control 14
    2.1.2 Part-Winding Reduced-Voltage Motor Starting Control 16
    2.1.3 Wye-Delta Reduced-Voltage Motor Starting Control 16
  2.2.0 Solid-State Reduced-Voltage Motor Starting Control 17
3.0.0 Adjustable-Frequency Drives 24
  3.1.0 Types of Adjustable Frequency Drives 24
    3.1.1 AFD Forms and Types 25
    3.1.2 AFD Components 25
  3.2.0 AFD Uses and Benefits 27
    3.2.1 AFD Selection 27
    3.2.2 AFD Programming 29
    3.2.3 AFD Application Precautions 30
4.0.0 Motor Braking Methods 35
  4.1.0 DC Injection Braking 35
  4.2.0 Dynamic and Regenerative Braking 36
    4.2.1 Dynamic Braking 36
    4.2.2 Regenerative Braking 37
  4.3.0 Friction Braking 37
5.0.0 Working with Motor Controls 40
  5.1.0 Precautions for Working with Solid-State Controls 40
  5.2.0 Motor Control Maintenance 41
6.0.0 Motor Control Troubleshooting 44
  6.1.0 Basic Troubleshooting Practices 44
    6.1.1 Customer Interface 44
    6.1.2 Physical Examination 44
    6.1.3 Basic System Analysis 44
    6.1.4 Manufacturer Resources 45
    6.1.5 Troubleshooting Motor Control Circuits and Components 45
  6.2.0 Universal Troubleshooting Procedures 47

Contents (continued)

6.2.1 Input Voltage/Voltage Imbalance Measurements..........................49

6.2.2 Fuse Checks..50

6.2.3 Circuit Breaker Checks...50

6.2.4 Troubleshooting Control Circuits..53

6.2.5 Control Transformer Checks..53

6.2.6 Contactor/Relay Coil Resistance Checks......................................54

6.2.7 Relay/Contactor Contact Checks..55

6.2.8 Troubleshooting Circuit Boards..55

Figures and Tables

Figure 1 Solid-state relay with heat sink..1

Figure 2 Block diagram of an optically isolated solid-state relay....................2

Figure 3 Two-wire and three-wire SSR control..5

Figure 4 Multiple SSRs controlled in parallel or in series to switch a three-phase load..6

Figure 5 Solid-state overload relay..7

Figure 6 Typical programmable solid-state overload relay..............................8

Figure 7 Timed-contact symbols and timing..10

Figure 8 Typical solid-state plug-in timing relay..10

Figure 9 Plug-in solid-state relay connection diagrams and sequence charts..12

Figure 10 Time-delayed starting for three motors..13

Figure 11 Autotransformer reduced-voltage starting characteristics............15

Figure 12 Typical autotransformer reduced-voltage starting circuit..............15

Figure 13 Part-winding reduced-voltage starting characteristics...................16

Figure 14 Wye-delta reduced-voltage starting characteristics........................17

Figure 15 Typical wye-delta reduced-voltage starting circuit.........................18

Figure 16 Solid-state reduced-voltage starting characteristics......................19

Figure 17 Simplified solid-state reduced-voltage starting circuit...................19

Figure 18 Reduced-voltage starting controller..20

Figure 19 Starting characteristics of solid-state reduced-voltage controllers..21

Figure 20 Block diagram of a typical adjustable-frequency drive..................26

Figure 21 Variable-torque load...27

Figure 22 Constant-torque load...27

Figure 23 Constant-horsepower load...28

Figure 24 AFD operating frequency range..28

Figure 25A AFD application checklist (1 of 2)...31

Figure 25B AFD application checklist (2 of 2)...32

Figure 26 AFD cable...33

Figure 27 Circuit for DC braking of an AC motor..36

Figure 28 Simplified schematic of an AC drive using dynamic braking.......37

Figure 29 Friction brake solenoid connection...38

Figure 30 ESD protection bag and tote box..41

Figure 31 Sample portion of a fault troubleshooting table 46
Figure 32 Basic motor control circuit 47
Figure 33 Three-phase input voltage checks 51
Figure 34 Fuse checks 52
Figure 35 Circuit breaker checks 52
Figure 36 Isolating to a faulty control circuit component 54
Figure 37 Control transformer checks 54
Figure 38 Coil resistance checks 55
Figure 39 Checking the continuity of contactor/relay contacts 55

Table 1   Comparison Chart for EMR and SSR Technology 3
Table 2   EMR and SSR Performance 4
Table 3   Comparison of Reduced-Voltage Controllers 22
Table 4   Advantages and Disadvantages of Reduced-Voltage Controllers 22
Table 5   Key Features of Different AFD Types 26
Table 6   AFD Use with Different NEMA Motor Designs 29
Table 7A   Motor Control Troubleshooting Chart (1 of 2) 48
Table 7B   Motor Control Troubleshooting Chart (2 of 2) 49

This page is intentionally left blank.

1.0.0 IDENTIFYING RELAYS

Objective

Describe the various types of relays used in motor control circuits.

a. Identify and describe solid-state relays and their uses.
b. Identify timing relays and explain their uses.

Trade Terms

Dashpot: A device that uses a gas or fluid to absorb energy from or retard the movement of the moving parts of a relay, circuit breaker, or other electrical or mechanical device.

Heat sink: A metal mounting base that dissipates the heat of solid-state components mounted on it by radiating the heat into the surrounding atmosphere.

Solid-state overload relay (SSOLR): An electronic device that provides overload protection for electrical equipment.

Solid-state relay (SSR): A switching device that has no contacts and switches entirely by electronic means.

Thermal conductivity: The ability of a material or substance to transfer heat away from a given point of contact.

Thyristors: Bi-stable semiconductor devices that can be switched off and on. Thyristors turn on with a quick pulse of control current. They turn off only when the working current is interrupted elsewhere in the circuit. SCRs and triacs are forms of thyristors.

Relays are a type of switching device that starts and stops the flow of current to an electrical device or appliance. This essential functionality makes relays key components of all electrical control systems. Electromechanical relays and contactors are still in use, but they are gradually being replaced by solid-state devices.

1.1.0 Solid-State Relays

The **solid-state relay (SSR)** has many advantages over the electromechanical relay (EMR). It has longer life, higher reliability, higher speed switching, and high resistance to shock and vibration. The absence of mechanical contacts eliminates contact bounce, arcing when the contacts open, and hazards from explosives and flammable gases. For this reason, SSRs (*Figure 1*) have replaced EMRs in many applications.

1.1.1 Solid-State Relay Operation

Solid-state relays can be used in many of the same control applications as EMRs and in many new applications involving electronic control circuit devices. An SSR is activated by the control circuit voltage applied to its input terminals. Depending on the SSR design, this input control voltage can be AC or DC. AC voltages typically range between 90VAC and 280VAC; DC voltages range between 3VDC and 32VDC. DC voltage inputs are normally applied from digitally controlled motor control circuits. Application of the AC or DC input control voltage to the relay can be done by the closure of a mechanical switch or from the activation of a solid-state device, such as a transistor or integrated circuit.

Internally an SSR consists of an input control circuit and an output circuit (*Figure 2*). The function of the input control circuit can be compared to that of the coil in an EMR in that it senses the input control signal and provides coupling between the input and output circuits. In most SSRs the coupling of the switching command to

CONTROL TERMINALS

OUTPUT TERMINALS

HEAT SINK

Figure 1 Solid-state relay with heat sink.

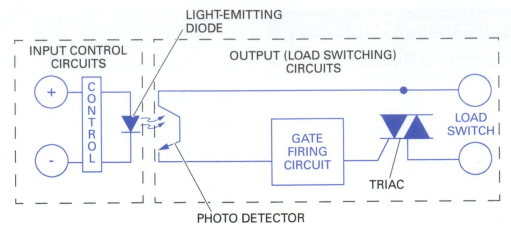

Figure 2 Block diagram of an optically isolated solid-state relay.

the output circuit is done optically using a light-emitting diode (LED) and photo detector. This type of SSR is called an *optically isolated SSR*. The optical isolation keeps voltage spikes or electrical noise generated on the load side of the SSR from being coupled into the input control circuit side of the SSR.

When the applied input control voltage level exceeds a predetermined threshold known as the *pickup voltage*, a light beam is generated by the LED and transmitted to the photo detector. This light is used to electronically activate the SSR output switch. The output switch will remain activated until the input control voltage level drops below the relay's minimum dropout voltage level, at which time the LED stops emitting light and the output switch is deactivated.

The output switching circuit electronically performs the same function as the mechanical contacts of an EMR. Depending on the design of the SSR and the type of load to be controlled, the output switching can be performed by power transistors, silicon-controlled rectifiers (SCRs), triacs, or thyristors. Power transistor outputs are typically used to control low-current loads while SCR-switched outputs are typically used to control high-current DC loads. Triac-switched outputs are typically used to control high-current AC loads, while thyristor-switched outputs are used to control either low- or high-current AC loads, depending on their circuit configuration. The switching circuit can operate in one of these four switching modes, depending on the SSR design:

- *Instant on* – This mode turns the load on immediately at any point on the AC sine wave when the control voltage is present. It turns the load off when the control voltage is removed and the current in the load crosses zero. Instant-on

relays are commonly used to control contactors, magnetic starters, and similar loads.

- *Zero switching* – This mode turns the load on when the control voltage is applied and the voltage at the load crosses zero. The relay turns the load off when the control voltage is removed and the current in the load crosses zero. Zero-switching relays are typically used to control heating elements, ovens, incandescent and tungsten lamps, and programmable controllers.
- *Peak switching* – This mode turns the load on when the control voltage is applied and the voltage at the load is at its peak amplitude. The relay turns the load off when the control voltage is removed and the current in the load crosses zero. Peak switching relays are typically used to control heavy inductive loads.
- *Analog switching* – In this type of relay, the amount of output voltage varies as a function of the input voltage. For any voltage within its input range (typically 0VDC to 5VDC), the output is a percentage of the available output voltage within its rated range. Analog switching relays are commonly used in closed-loop control applications.

1.1.2 Electromechanical Relays versus Solid-State Relays

Differences exist both in the terminology used to describe EMRs and SSRs and in their overall ability to perform certain functions. *Table 1* shows the equivalent or comparable terminology between an EMR and an SSR; *Table 2* indicates the ability of each device to perform certain functions. When making a choice between solid-state and electromechanical devices, compare the electrical and

Table 1 Comparison Chart for EMR and SSR Technology

Comparable or Equivalent Terminology for Electromechanical and Solid-State Relays	
Electromechanical Relays (EMRs)	**Solid-State Relays (SSRs)**
Coil voltage – The minimum voltage necessary to energize or operate the relay. Also referred to as the *pickup voltage*.	*Control voltage* – The minimum voltage required to gate or activate the control circuit of a solid-state relay. Generally, a maximum value is also specified.
Inrush VA – The amount of power necessary to energize or operate the relay coil.	*Control current* – The minimum current required to activate a solid-state control circuit. Generally, a maximum value is also specified.
Sealed VA – The power required to keep the relay energized or operating.	See *control current*.
Dropout voltage – The minimum voltage required to keep the relay energized.	See *control voltage*.
Pull-in time – The amount of time required to operate (open or close) the relay contacts after the coil voltage is applied.	*Turn-on time* – The elapsed time between the application of the control voltage and the application of the voltage to the load circuit.
Dropout time – The amount of time required for the relay contacts to return to their normal (unoperated) position after the coil voltage is removed.	*Turn-off time* – The elapsed time between the removal of the control voltage and the removal of the voltage from the load circuit.
Contact voltage rating – The maximum voltage rating that the contacts can switch safely.	*Load voltage* – The maximum output voltage handling capability of a solid-state relay.
Contact current rating – The maximum current rating that the contacts can switch safely.	*Load current* – The maximum output current handling capability of a solid-state relay.
Surge current – The maximum peak current that the contacts can withstand for short periods of time without damage.	*Surge current* – The maximum peak current that a solid-state relay can withstand for short periods of time without damage.
Contact voltage drop – The voltage drop across the relay contacts when the relay is operating (usually quite low).	*Switch-on voltage drop* – The voltage drop across a solid-state relay when the relay is operating (usually quite low).
Insulation resistance – The amount of resistance measured across relay contacts in the open position.	*Switch-off resistance* – The amount of resistance measured across a solid-state relay when turned off.
No equivalent	*Off state current leakage* – The amount of current leakage through a solid-state relay when turned off but still connected to the load voltage.
No equivalent	*Zero current turn-off* – Turn-off at essentially the zero crossing of the load current that flows through an SSR. A thyristor will turn off only when the current falls below the minimum holding current. If input control is removed when the current is at a higher value, turn-off will be delayed until the next zero current crossing.
No equivalent	*Zero voltage turn-on* – Initial turn-on occurs at a point near the zero crossing of the AC line voltage. If input control is applied when the line voltage is at a higher value, initial turn-on will be delayed until the next zero current crossing.

Table 2 EMR and SSR Performance

Advantages and Disadvantages of Electromechanical and Solid-State Relays		
General Characteristics	EMR	SSR
Arcless switching of the load	–	+
Electronic compatibility for interfacing	–	+
Effects of temperature	+	–
Shock and vibration resistant	–	+
Immunity to improper functioning due to transients	+	–
Radio frequency switching	+	–
Zero voltage turn-on	–	+
Acoustic noise	–	+
Selection of multi-pole, multi-throw switching capability	+	–
Contact bouncing	–	+
Ability to withstand surge currents	+	–
Response time	–	+
Voltage drop in load circuit	+	–
AC & DC switching with same contacts	+	–
Zero current turn-off	–	+
Leakage current	+	–
Minimum current turn-on	+	–
Life expectancy	–	+
Initial cost	+	–
Lifetime (total) cost	–	+

mechanical operating characteristics of each device with the application in which it is to be used.

Plus (+) indicates advantages; minus (–) indicates disadvantages.

Each device has its limitations on current and voltage, so the manufacturer's data sheets must be used to select the proper device for the intended load.

The electronic nature of the SSR and its input circuit lends itself to use in digitally controlled logic circuits. Another advantage of the SSR over the EMR is its ability to turn on and off very quickly because it has no moving mechanical parts.

1.1.3 Two-Wire and Three-Wire SSR Control

Two-wire control of an SSR is very similar to that of an EMR, as shown in *Figure 3 (A)*. For three-wire control there is no auxiliary contact in an SSR to seal in the relay until the STOP pushbutton is activated. However, addition of an SCR and a current-limiting resistor will achieve the same results, as shown in *Figure 3 (B)*. Pressing the START pushbutton turns power on to the gate input of the SCR, causing its state to flip to the ON position. It will stay on until pressing the STOP pushbutton returns it to the OFF position.

1.1.4 Connecting SSRs to Switch Multiple Outputs

SSRs are made with one to four switched outputs. If an application needs a higher current rating than a multi-pole output SSR can provide, higher-rated single-pole SSRs can have their input signal wired either in series or parallel. *Figure 4* shows both series and parallel connections for the control of a three-phase load such as a motor. When the inputs are wired in series, the voltage used must be higher than the sum of each input's minimum control voltage. Similarly, when the inputs are wired in parallel, the available input current must be higher than the sum of each SSR control current.

1.1.5 SSR Temperature Considerations

Every solid-state electronic device, including SSRs, will be damaged by excessive temperature. Manufacturers of SSRs specify the maximum permitted environmental temperature, typically 104°F (40°C). As a rule, for every 18°F (10°C) increase above this temperature the life span of the SSR is cut in half. This environmental heat comes from the surrounding area and from various

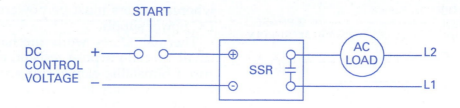

(A) TWO-WIRE CONTROL

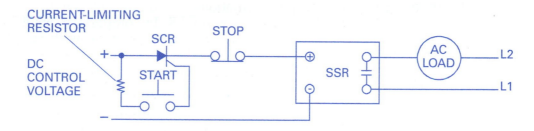

(B) THREE-WIRE CONTROL

Figure 3 Two-wire and three-wire SSR control.

components in the enclosure, such as the voltage drop and current flow through the SSR.

Because the heat created by the SSR is concentrated in a small area, manufacturers add a fan and/or a finned metal mounting base, called a **heat sink**, to carry the heat away. The heat sink is made from a metal with high **thermal conductivity** (or low thermal resistance), typically aluminum.

The manufacturer's installation instructions list the type and size of heat sink (if not provided), as well as installation instructions. Following these instructions is required by *NEC Section 110.3(B)*. Installation considerations for heat sinks include the following:

- Use the proper heat sink; if practical, it should have fins for greater heat dissipation.
- Make sure that the mounting surface between the heat sink and the SSR is flat and smooth.
- Ensure that the mounting area meets the manufacturer requirements for free space around the heat sink.
- Use thermal grease or a thermally conductive pad between the SSR and the heat sink to eliminate any air gaps and maximize heat dissipation.
- Make sure all mounting hardware is securely fastened.

1.1.6 Overvoltage and Overcurrent Protection

SSR components, such as power transistors, SCRs, and triacs, can be damaged by a shorted load. It is a good practice to install an overcurrent protection fuse to open the output circuit should the load current increase to a higher value than the nominal load current. This fuse should be an ultrafast type designed for use with semiconductor devices.

To protect power transistors, SCRs, triacs, and other devices from overvoltages resulting from transients induced on the load lines, install a peak voltage clamping device in the output circuit, such as a varistor, a Zener diode, or an energy-limiting resistor-capacitor device known as a *snubber*.

1.1.7 SSOLRs

A solid-state protective relay, commonly called a **solid-state overload relay (SSOLR)**, protects the motor against damage from overheating. Depending on their design, many SSOLRs also provide protection against phase loss, phase imbalance, phase reversal, and undervoltage. They do not protect against short circuits or ground faults. Many SSOLRs are designed to be retrofitted into equipment in which thermal overload protection (OLR) devices previously were used. Depending on the design, some SSOLRs are self-powered,

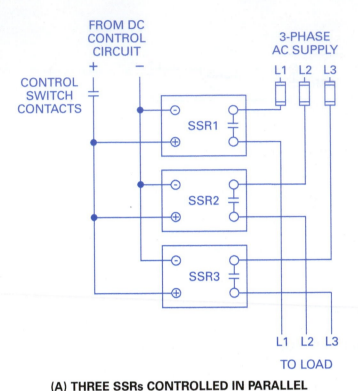

(A) THREE SSRs CONTROLLED IN PARALLEL

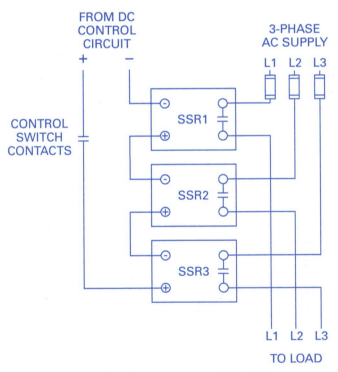

(B) THREE SSRs CONTROLLED IN SERIES

Figure 4 Multiple SSRs controlled in parallel or in series to switch a three-phase load.

whereas others must be powered by a separate DC power supply.

Thermal OLRs work mechanically, using a small special-purpose resistor, called a *heater*, to trip a bimetallic element or spring-loaded melting alloy contact to interrupt the motor control circuit. For each manufacturer, the heater selected is from among hundreds of sizes as listed within many dozens of tables, according to the specific starter, trip class, and motor service factor. With some brands there are up to four different orientations for mounting the heater within the thermal OLR. In comparison, SSOLRs come in a modest number of sizes, with simpler selection and installation. There are two types of SSOLRs: nonprogrammable and programmable.

Nonprogrammable SSOLRs

SSOLRs are made in many sizes and shapes for standalone mounting, starter mounting, and DIN-rail mounting. *Figure 5* shows an example of one manufacturer's self-powered, standalone nonprogrammable SSOLR and a related control diagram. Control power is derived from the three-phase inputs. Regardless of manufacturer and internal circuitry differences, all SSOLRs operate in much the same way. As shown, the SSOLR senses the current in each of the three motor leads via three built-in current transformer (CT) windings. In some models, the motor leads attach to a bar and terminals, already installed within the CT; while for others a hole through the CT is provided, through which 1, 2, or 3 loops of the motor leads can be run. The trip point for the SSOLR is set within the range shown on the adjustable trip current dial on the face of the unit, as directed in the manufacturer's instructions. This setting for the SSOLR is based on the motor full load amperage (FLA), service factor, and the number of CT loop turns being used. Settings on SSOLRs like the one shown are for motors with a service factor of 1.15 or more. For motors with a service factor of less than 1.15, use a setting of 0.9 times the motor FLA. If more than one loop of wire is run through the CT, an additional multiplier is the number of CT loops.

The SSOLR initiates a trip if the phase currents exceed 125% of the trip current dial setting. The time it takes to trip depends on the level of monitored current, the trip class (Class 10, Class 20, and so on), and the length of time since the last trip. A mechanically latched mechanism opens (unlatches) when a trip occurs. This action opens

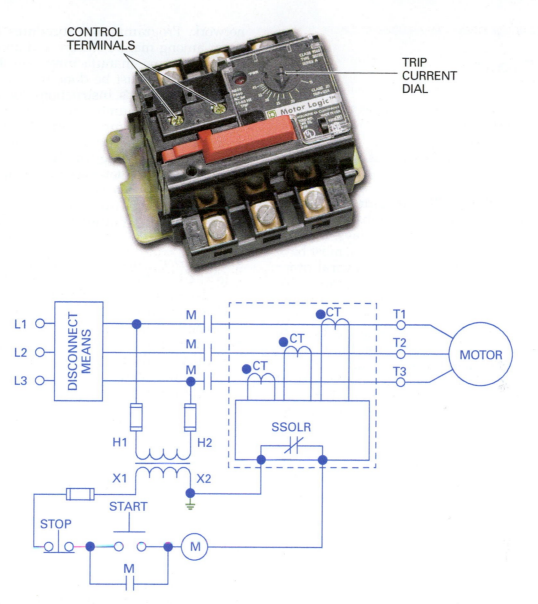

Figure 5 Solid-state overload relay.

the normally closed (NC) overload trip contact, interrupting current flow in the motor control circuit. The relay's phase loss/phase imbalance circuitry will also initiate a trip if a current imbalance of 25% or greater exists or if one of the phase currents is missing. To provide protection for lightly loaded motors, the phase loss/phase imbalance circuitry remains operational at currents below the trip current. Once tripped, the SSOLR is reset by pressing the RESET pushbutton on the face of the unit. Most manufacturers also have a remote reset module that can be installed. Typically, this would be used with cranes, hoists, and similar applications where the controls are mounted in a remote location, making them difficult to access.

CAUTION

If a lightly loaded motor loses one input phase while running, the current flow through the remaining motor leads is usually below the motor FLA, so a standard OL relay will not trip. However, the magnetic fields in the motor will be severely unbalanced and overheating of the motor will occur.

An SSOLR can be an instantaneous-trip device (instant removal with no time delay) or inverse-time, Class 10, 15, 20, 30, or other class device. Some have an adjustment to set the trip class. These models are usually shipped with trip Class 20 selected. Most SSOLRs are also equipped with LED indicators to show when power is applied and when a trip has occurred.

Trip Class

When the current flowing through the adjustable overload relay is 600% of the actual dial setting, the trip class is the number of seconds until it will trip.

Programmable SSOLRs

Programmable SSOLRs provide the same overload and phase loss/phase imbalance motor protection as nonprogrammable relays. Refer to *NEC Section 430.52(C)(5)*. In addition, most have an increased capability to detect several other types of motor-related faults. The programmable SSOLR is connected into the motor circuit in the same way as a nonprogrammable relay. However, most programmable SSOLRs require the use of an external power supply for their solid-state circuitry. Like nonprogrammable SSOLRs, programmable models sense the motor current via current transformer (CT) windings. For some models, these CT windings are internal to the relay. For others, they are mounted externally. External mounting typically occurs in motor circuit applications having high full-load currents (90A and above). Operation of the SSOLR is controlled by a microprocessor that can be programmed so the relay can be used in a wide variety of applications. *Figure 6* shows an example of one manufacturer's programmable SSOLR.

Programming of an SSOLR is done using controls on the front panel of the unit. However, many SSOLR manufacturers also have software programs available that allow an SSOLR to be programmed using a personal computer (PC) connected to the relay via a communications network. Programming procedures for SSOLRs differ among manufacturers and among models from the same manufacturer. For this reason, programming must be done in accordance with the manufacturer's instructions for the specific relay in use. The number and types of parameters that can be programmed differ from one SSOLR to another depending on their design. For the SSOLR shown, setting trip thresholds for the different circuit parameters is done using the MODE and DISPLAY selector switches on the front of the unit. The types of functions and trip thresholds that can be programmed with most programmable SSOLRs include the following:

- *Overcurrent trip class* – Selects the trip class (e.g., Class 10, 20, 30). The specific trip class is determined by the motor and application. Class 20 is used for most NEMA-rated general purpose motors.
- *Effective turns ratio of CTs* – Used by the SSOLR to calculate the true current based on the turns ratio of the internal or externally mounted CTs.
- *Low- and high-voltage trip points* – Typically set to ±10% of the motor nameplate rating.
- *Voltage imbalance trip point* – Sets the percentage of voltage imbalance allowed by the motor. Typically set for 5%.
- *Overcurrent trip point* – Typically set to between 110% and 120% of the motor FLA.
- *Undercurrent trip point* – Sets the level of acceptable undercurrent. Typically set for 80% of the motor FLA.
- *Undercurrent trip delay* – Determines how long the SSOLR will allow an undercurrent condition to exist before it trips.
- *Restart delay time after a trip* – Controls the elapsed time between motor restarts after a trip. Different settings allow for different restart times, depending on the type of fault that has occurred.
- *Number of restarts* – If enabled, this determines the number of automatic restarts that can be initiated. After the programmed number of restarts have been attempted and failed, a manual reset must be performed.
- *Ground fault current trip* – Determines when a Class II ground fault exists because degradation of the motor insulation is allowing current leakage. The threshold is typically set for 10% to 20% of the motor FLA. Class II is for motor protection only; it is not the same as ground fault protection for personnel.

Figure 6 Typical programmable solid-state overload relay.

When tripped by an overload, the programmable SSOLR is reset locally by pressing the RESET pushbutton on the face of the unit. If equipped with a remote reset module, it can be reset from a remote location.

1.2.0 Timing Relays

Timing relays open or close electric circuits in order to perform selected operations according to a timed program. Various methods have been used to create timing relays. Adding a bellows chamber and air metering valve to an EMR creates a pneumatic timing relay. Similarly, using silicone oil instead of air creates a dashpot timing relay. These are older relay types that may still be found mounted on or alongside other relays in older control cabinets. Using a motor and an adjustable dial pointer creates a motorized timing relay, which is rugged and often found in control enclosure doors because it is easy to set and usually displays the elapsed time. This section will focus on electronic (solid-state) timing relays because of their wide range of capabilities and applications, as well as their increasingly common usage.

A simple application involves the use of timing relays to control the sequential energization or de-energization of two or more motors. Most timing relays have adjustable time intervals, and often have more than one selectable timing function, allowing them to be used in more than one application. Before going further, it is first necessary to introduce terms commonly used when describing timing relays, as well as symbols used on schematic and ladder diagrams to identify timed contacts (*Figure 7*).

These terms include the following:

- *On-delay relay* – An on-delay relay provides the time delay after the relay is energized.
- *Off-delay relay* – An off-delay relay provides the time delay after the relay is de-energized.
- *Normally open, timed-closed (NOTC) contacts* – NOTC contacts (*Figure 7*) are on-delay, timed-closed contacts. The contacts are normally open. When the relay is energized, timing begins, and the contacts close after the specified delay time has elapsed. The contacts remain closed until the relay is de-energized, at which time the contacts are immediately opened.
- *Normally closed, timed-open (NCTO) contacts* – NCTO contacts are on-delay, timed-open contacts. The contacts are normally closed. When the relay is energized, timing begins, and the contacts are opened after the specified delay time has elapsed. The contacts close immediately when the relay is de-energized.
- *Normally open, timed-open (NOTO) contacts* – NOTO contacts are off-delay, timed-open contacts. When the relay is energized, the normally open contacts close immediately and stay closed. When the relay is de-energized, timing begins, and the contacts are opened after the specified delay time has elapsed.
- *Normally closed, timed-closed (NCTC) contacts* – NCTC contacts are off-delay, timed-closed contacts. When the relay is energized, the normally closed contacts open immediately and remain open. When the relay is de-energized, timing begins, and the contacts are closed after the specified delay time has elapsed.

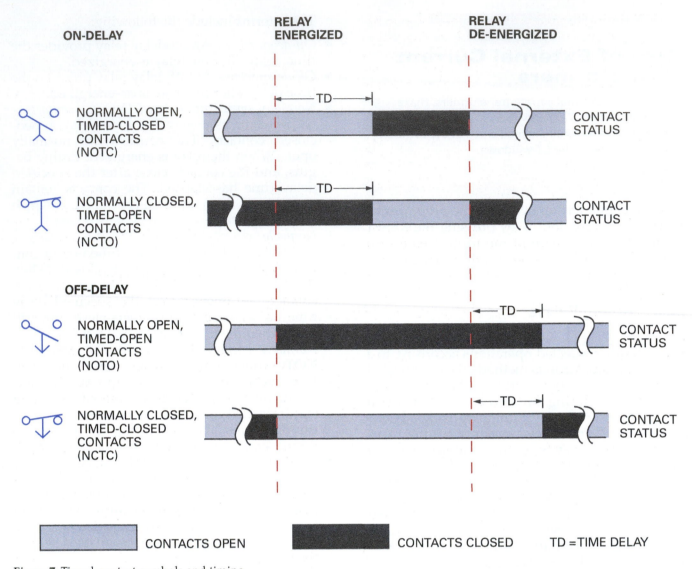

ON-DELAY

NORMALLY OPEN, TIMED-CLOSED CONTACTS (NOTC)

NORMALLY CLOSED, TIMED-OPEN CONTACTS (NCTO)

OFF-DELAY

NORMALLY OPEN, TIMED-OPEN CONTACTS (NOTO)

NORMALLY CLOSED, TIMED-CLOSED CONTACTS (NCTC)

RELAY ENERGIZED

RELAY DE-ENERGIZED

TD

CONTACT STATUS

CONTACTS OPEN   CONTACTS CLOSED   TD = TIME DELAY

Figure 7 Timed-contact symbols and timing.

1.2.1 *Solid-State Timing Relays*

The solid-state timing relay (SSTR) gets its name from the fact that its time delay and function is provided by solid-state electronic devices enclosed within the relay body. Early versions of SSTRs only provided a few timing functions and a limited number of timing ranges. With the trend to the increasing use of microprocessors, a single relay can be made with as many as 25 or more different selectable timing functions, 6 or more different timing ranges, and a timing accuracy of better than 0.1%. *Figure 8* is an example of a plug-in SSTR with 10 selectable timing functions, 6 timing ranges from 9.99 seconds to 999 hours, and 0.1% accuracy.

Figure 8 Typical solid-state plug-in timing relay.

SSTR timing functions are provided by two basic forms of control: supply voltage timing relays and contact-controlled timing relays. These relays provide four common timing functions: on delay, off delay, interval, and single shot. They operate as follows:

- *Supply-voltage timing relay* – A supply-voltage timing relay begins its operation with the application of voltage to its input power terminals or pins. The input power source and pilot devices that control this relay must be suitable for the voltage and current rating of the timing relay in use. The upper half of *Figure 9* shows delay-on-make (on delay) and interval timing modes for a typical supply voltage timing relay. For the delay-on-make mode, the output contacts switch state after the time delay and remain that way until the input power is turned off. For the interval mode, the output contacts switch state when the input power is turned on and switch back at the end of the interval; it will not repeat until the next time the input power is turned on.

- *Contact-controlled timing relay* – A contact-controlled timing relay normally has continuous power to its input power terminals and begins its operation with the closure of a control switch or pilot device (often called *S1*) between two terminals. The pilot devices that control this relay must be suitable for the low voltage of the control terminals and be wired only to them. The lower half of *Figure 9* shows delay-on-break (off delay) and single-shot timing modes typical for a contact-controlled timing relay. For the off-delay mode, the output contacts switch state when the control switch is closed and switch back only after the control switch has remained open for the set time delay. For the single-shot mode, the output contacts change state when the control switch is closed and switch back after the time delay has elapsed, regardless of the control switch status.

In addition to the 8-pin and 11-pin plug-in types, SSTRs are made in many other forms, sizes, and mounting methods, with some having a time display or a moving time dial. A wide variety of timing functions are also available, including repeat cycle, unequal repeat cycle, percentage, time totalizer, re-triggerable off-delay, or delayed interval.

Transistor- and Sensor-Controlled Timing Relays

Transistor-controlled and sensor-controlled timing relays are also available. A transistor-controlled timing relay is controlled by a transistor switch located in an external electronic device. A sensor-controlled timing relay is controlled by an external proximity or photoelectric sensor in which the relay itself supplies the power to operate the sensor.

1.2.2 Timing Relay Applications

There are hundreds of applications in which timing relays are used to provide timed control of equipment, operations, or processes. One common application involves the use of a timer to start multiple motors at preset intervals to avoid overloading a feeder circuit with motor inrush current. With large motors, a series of 20-second time delays between each motor startup is common. If the motors are on a conveyor system, the time delay could be as short as 0.5 seconds, with the downstream conveyor starting first. *Figure 10* shows the control wiring for starting three motors with time delays between each one. As shown, the overload relay contacts for all three motors are in series just before L_2, so if any one of them opens from an overload, all motor starters and timing relays will be de-energized.

When the START button is pressed, the circuit is completed from L_1 to L_2 through the STOP button, the START button, the motor starter coil M_1, and the closed overload contacts. The energizing of M_1 closes its NO auxiliary contacts to seal around the START button. Simultaneously, timing relay TR_1 receives its power and starts its on-delay timing while motor M_1 starts and gets up to speed. When TR_1 finishes its time period, the NO contacts close to energize motor starter M_2 and supply power to TR_2. While TR_2 is timing, motor M_2 is getting up to speed. When TR_2 finishes its time period, the NO contacts close to energize motor starter M_3.

When the STOP button is pressed, M_1 and TR_1 both turn off. The NO contacts on TR_1 open to de-energize M_2 and TR_2. The NO contacts on TR_2 open to de-energize M_3. This process of stopping the motors occurs almost instantaneously.

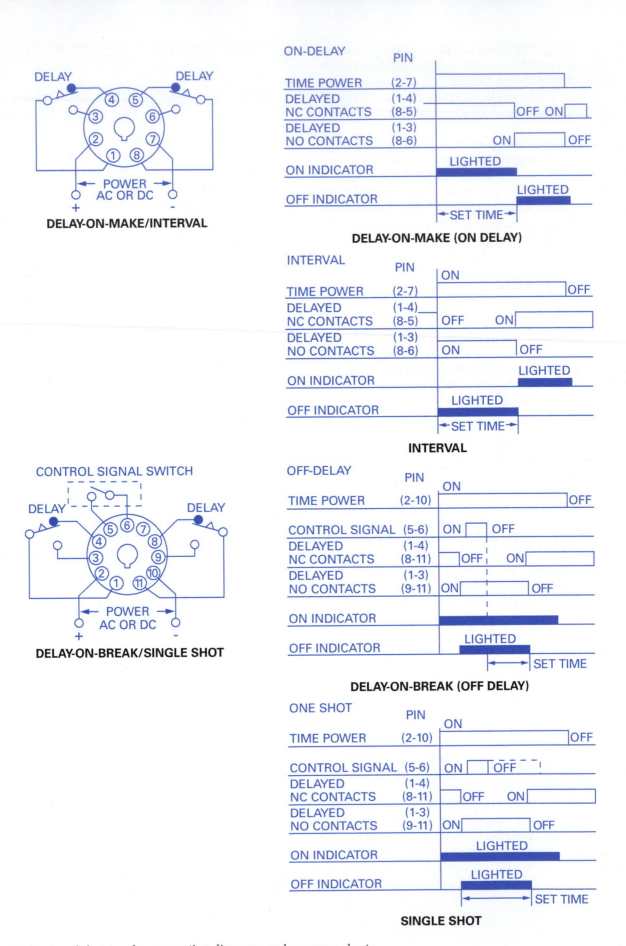

Figure 9 Plug-in solid-state relay connection diagrams and sequence charts.

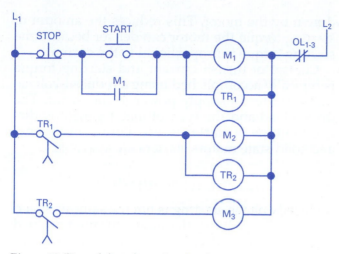

Figure 10 Time-delayed starting for three motors.

1.0.0 Section Review

1. In which of these switching modes does the amount of output voltage vary as a function of the input voltage?

   a. Instant-on switching
   b. Analog switching
   c. Peak switching
   d. Zero switching

2. The type of time-delay relay that provides the time delay after the relay is de-energized is a(n) _____.

   a. off-delay relay
   b. on-delay relay
   c. NOTO relay
   d. NCTC relay

2.0.0 REDUCED-VOLTAGE STARTING METHODS

Objective

Explain how reduced-voltage starting is accomplished.

a. Describe the use and selection of conventional reduced-voltage motor starting methods.
b. Describe solid-state reduced-voltage motor starting methods.

Trade Terms

Closed-circuit transition: A method of reduced-voltage starting in which the motor being controlled is never removed from the source of voltage while moving from one voltage level to another.

Open-circuit transition: A method of reduced-voltage starting in which the motor being controlled may be disconnected temporarily from the source of voltage while moving from one voltage level to another.

Reduced-voltage starting reduces the amount of current an induction motor draws when starting. There are several reasons for using reduced-voltage starting. One common reason is to reduce the large amount of current drawn from utility lines by across-the-line starting of large motors. This occurs because the inrush current drawn by a motor is typically five to six times its nameplate current rating; and for high-efficiency motors, this inrush current can be as high as ten or more times its nameplate rating. The voltage drop caused by this large current demand can stress the upstream service equipment or cause malfunctions in other equipment.

Another common reason for using reduced-voltage starting is to control motor torque at startup so that the torque is applied to a load gradually. This is needed for applications in which a high starting torque can damage gears, belts, and chain drives being driven by the motor or when damage to a product can occur from a sudden forceful start. Reducing the voltage applied to a motor at startup reduces the current

drawn by the motor. This reduces the amount of starting torque the motor can deliver because the starting torque is proportional to the current.

Control of inrush current and starting torque is typically accomplished using a reduced-voltage motor starter to supply power to the motor. The driven load and the type of motor generally dictate the type of starter to use. Both conventional and solid-state motors starters are available.

2.1.0 Conventional Methods

Reduced-voltage starting is not the same as speed control. By applying the motor torque to a load gradually, reduced-voltage starting acts as a buffer or shock absorber to the load during initial startup. A speed control operates to control the speed of a motor during its entire operation cycle. Three conventional methods of reduced-voltage starting are:

- Autotransformer
- Part-winding
- Wye-delta

2.1.1 Autotransformer Reduced-Voltage Motor Starting Control

Autotransformer reduced-voltage starters are widely used because of their efficiency and flexibility. See *NEC Section 430.109(D)*. All power taken from the line, except transformer loss, is transmitted to the motor to accelerate the load. Several different taps on the transformer allow for adjustment of the starting torque and inrush current to meet job requirements. Autotransformer reduced-voltage starters are typically used with hard-to-start loads such as reciprocating compressors, grinding mills, pumps, and similar devices. *Figure 11* shows a graph of the starting characteristics for an autotransformer reduced-voltage starting controller.

Depending on the manufacturer and model, there are several designs for autotransformer reduced-voltage starting. Some use three-coil autotransformers; others use two-coil autotransformers. Some perform the transition from reduced-voltage to full voltage using a **closed-circuit transition** method of motor connection. Others use an **open-circuit transition** method of motor connection. In a closed-circuit transition controller, the motor is never removed from the source of voltage while moving from one voltage level to another. In an open-circuit transition controller, the motor being controlled may be disconnected temporarily from the line while moving

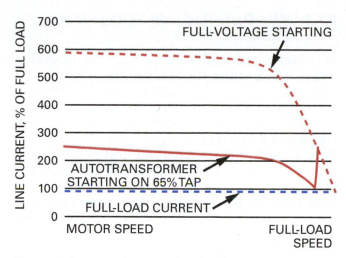

Figure 11 Autotransformer reduced-voltage starting characteristics.

from one voltage level to another. Controllers using the closed-circuit transition method are more expensive but are preferred because they cause the least amount of interference to the related electrical distribution system. This is because the motor is never disconnected from the line. Therefore, there is no interruption of line current with a resulting second inrush current during the transition period.

Figure 12 shows a schematic diagram for an induction motor being controlled by a closed-circuit transition type of autotransformer reducedvoltage controller. As shown, it includes a three-coil autotransformer and two three-pole contactors (1S and 2S) used to connect the autotransformer for reduced-voltage starting. It also includes a three-pole run contactor (R) used to bypass the autotransformer and connect the motor directly across the line for full-voltage running. The size of these three-pole contactors is determined by the maximum horsepower of the motor and its voltage and frequency requirements. The timing of the reduced-voltage interval is controlled by a timing relay (TR) that is adjustable from 1.5 to 15 seconds.

Pressing the START pushbutton energizes timing relay TR. This causes its two sets of normally open instantaneous contacts to close. One set of closed instantaneous contacts provides holding for the timer; the other set completes an electrical path, via the normally closed contacts of contactor R and the timed open contacts (TRTO) of the timer relay, to energize the coil of contactor 1S. When contactor 1S is energized, its power contacts close, connecting the ends of the autotransformers together. It also causes the contactor 1S auxiliary contacts to close, thereby energizing

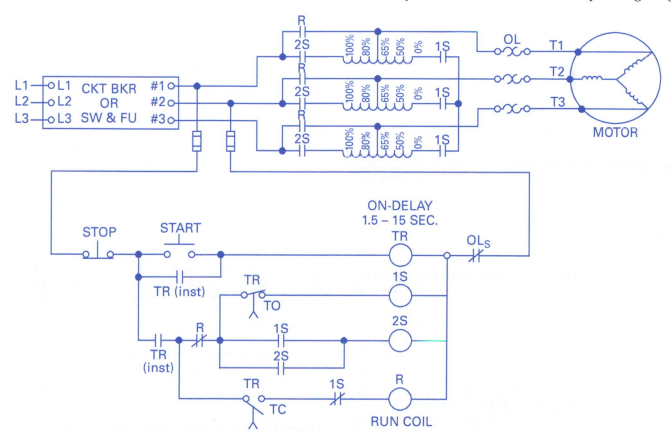

Figure 12 Typical autotransformer reduced-voltage starting circuit.

contactor 2S. The normally closed auxiliary contacts of 1S in the control circuit for contactor R open. This provides an electrical interlock to prevent contactors R and 1S from being energized at the same time. When coil 2S is energized, its power contacts close and connect the motor through the autotransformer taps to the power line, starting the motor at reduced inrush current and starting torque. The normally open auxiliary contacts of 2S close to provide a holding circuit for contactor 2S.

After a predetermined time (1.5 to 15 seconds), timer relay TR times out and its normally closed, timed open (NCTO) contacts open, causing contactor 1S to de-energize. This opens its power contacts and returns its normally closed auxiliary contacts to the closed position. Simultaneously, the normally open, timed closed (NOTC) contacts of timing relay TR close, causing contactor R to energize via the closed auxiliary contacts of contactor 1S. When contactor R energizes, its power contacts close, bypassing the autotransformer and connecting the motor to full line voltage. Its normally closed auxiliary contacts also open, causing contactor 2S to de-energize.

Note that during the transition from starting to full-line voltage, the motor was not disconnected from the circuit, indicating a closed-circuit transition method of operation. As long as the motor is running in the full-voltage condition, timing relay TR and contactor R remain energized. Only an overload or pressing the STOP pushbutton stops the motor and resets the circuit.

2.1.2 Part-Winding Reduced-Voltage Motor Starting Control

Part-winding reduced-voltage starting is an older method of reduced-voltage starting in which an induction motor is started by first applying power to part of the motor windings for starting, then after a short time delay, applying power to the remaining windings for normal running. Part-winding starters are used with part-winding motors. A part-winding motor is one that has two sets of windings that are connected in parallel during full-voltage operation. Partwinding starters have typically been used with low starting torque loads such as fans, blowers, and motor-generator sets. These starters can be used with nine-lead, dual-voltage motors on the lower voltage and with special part-winding motors designed for any voltage. Most motors will produce a starting torque equal to between $^1/_2$ and $^2/_3$ of NEMA standard values with half of the winding energized and draw about $^2/_3$ of normal line current inrush. *Figure 13* shows a graph of the typical starting characteristics for a part-winding, reduced-voltage starting mode of operation.

2.1.3 Wye-Delta Reduced-Voltage Motor Starting Control

Wye-delta reduced-voltage starting operates by first connecting the leads of the motor being controlled into a wye configuration for starting. Then, after a short time delay, the leads are reconfigured via switched contacts so that the motor is connected into a delta configuration for running. Both closed-circuit transition and open-circuit transition models are available. Wye-delta starters are used for controlling high-inertia loads with long acceleration times, such as with centrifugal compressors, centrifuges, and similar loads. When six- or twelve-lead, delta-connected motors are started wye-connected, approximately 58% of the line voltage is applied to each winding. The motor develops 33% of its full-voltage starting torque and draws 33% of its normal locked-rotor current from the line. When the motor has nearly accelerated, it is reconnected for normal delta operation. *Figure 14* shows a graph of the typical starting characteristics for a wye-delta, reduced-voltage starting mode of operation.

Figure 15 shows a schematic diagram for a wye-delta motor controlled by a closed-circuit transition, wye-delta type of reduced-voltage controller. As shown, it includes a three-pole contactor 1S that is used to short motor leads T4, T5, and T6 during starting to connect the motor in the wye configuration. A three-pole contactor 1M energizes motor leads T1, T2, and T3 for both wye and delta connections. A three-pole contactor 2M energizes motor leads T4, T5, and T6 during running to connect the motor in the delta configuration. It also includes a three-pole contactor 2S that connects resistors in series with the motor windings during the start-to-run transition period. Note that contactor 2S and the resistor bank are

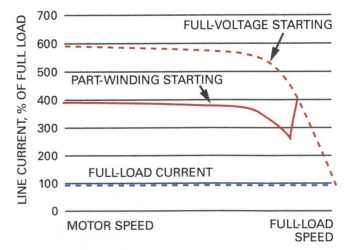

Figure 13 Part-winding reduced-voltage starting characteristics.

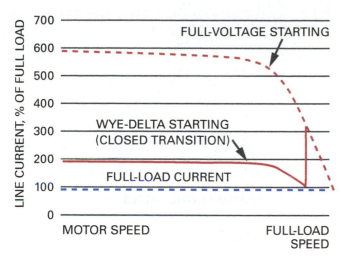

Figure 14 Wye-delta reduced-voltage starting
characteristics.

unique to the closed-circuit transition starter. An open-circuit transition starter does not have these components.

Pressing the START pushbutton energizes timing relay TR and contactor 1S, whose power contacts connect the motor in a wye configuration. Interlock 1S closes, energizing contactor 1M. The 1M power contacts energize the motor windings in a wye configuration. After a preset time, interval, timer contacts TRTC close, energizing contactor 2S and timing relay TRP. Interlock 2S opens, dropping out contactor 1S. The motor is now energized in series with the resistors. Interlock 1S closes, energizing contactor 2M, bypassing the resistors, and energizing the delta-connected motor at full voltage. Interlock 2M opens, de-energizing contactor 2S and timing relay TRP. Timing relay TRP opens the control circuit if the duty cycle for the transition resistors is exceeded. As long as the motor remains running in the full-voltage condition, contactors 1M and 2M and timing relay TR all remain energized. If an overload occurs or the STOP pushbutton is pressed, it de-energizes contactors 1M and 2M and the timer, removing the motor from the line and resetting the timer.

Wye-Delta Reduced-Voltage Controllers

When using a closed transition type of wye-delta reduced-voltage controller, make sure that there is adequate ventilation to remove the heat dissipated by the resistors.

2.2.0 Solid-State Reduced-Voltage Motor Starting Control

Solid-state, reduced-voltage starters perform the same function as electromechanical reduced-voltage starters, but they provide a smoother, stepless start and acceleration. For this reason, they are often referred to as *soft-start controllers*. Because they provide controlled acceleration, they are ideal for use with many loads, including conveyors, compressors, and pumps. Some models can also provide a soft stop when a sudden stop may cause system or product damage. For these reasons, as well as competitive costs, soft-start controllers are becoming the preferred replacement for other methods of reduced-voltage motor starting. *Figure 16* shows a graph of the typical starting characteristics for a solid-state (soft-start) reduced-voltage starting mode of operation.

There are many designs and models of solid-state, reduced-voltage controllers. *Figure 17* shows a simplified schematic diagram for a motor being controlled by a typical solid-state, reduced-voltage controller. A similar controller is shown in *Figure 18*. As shown, the controller consists of an electronic circuit board and six SCRs connected back-to-back. The typical circuit board contains a microprocessor and related input/output (I/O) circuitry that function both to control the operation of the SCRs and to protect the motor and starter circuits from damage. The current transformers sense the motor current and feed this information back to the control circuit for comparison and processing throughout the entire motor control process.

The use of six back-to-back SCRs provides control for the full cycle of the AC input voltage to the motor. Each single SCR controls one-half cycle of the AC input voltage on one motor lead; the six together provide full-wave control of the input voltage on all the motor leads. When first starting a motor, the gate on each SCR is turned on for a very short time just before the end of its half cycle (the SCR automatically turns off when the voltage supplying it crosses through zero at the end of its half cycle). By the end of the acceleration time, each SCR is turned on for its entire half cycle, so the motor sees full voltage (less SCR losses). This is shown for a single SCR in the top portion of *Figure 17*.

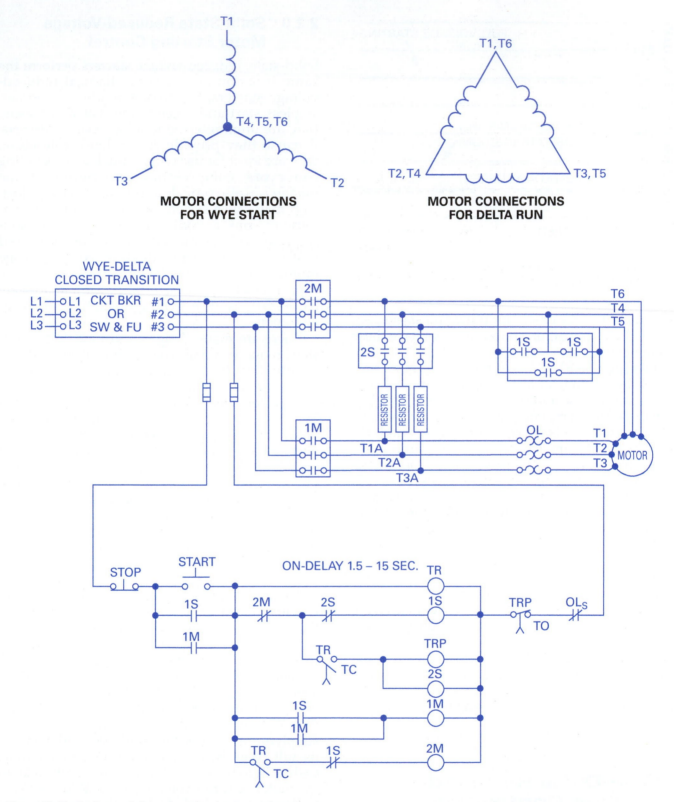

Figure 15 Typical wye-delta reduced-voltage starting circuit.

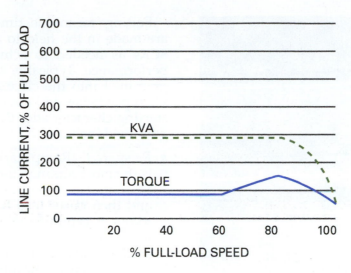

Figure 16 Solid-state reduced-voltage starting characteristics.

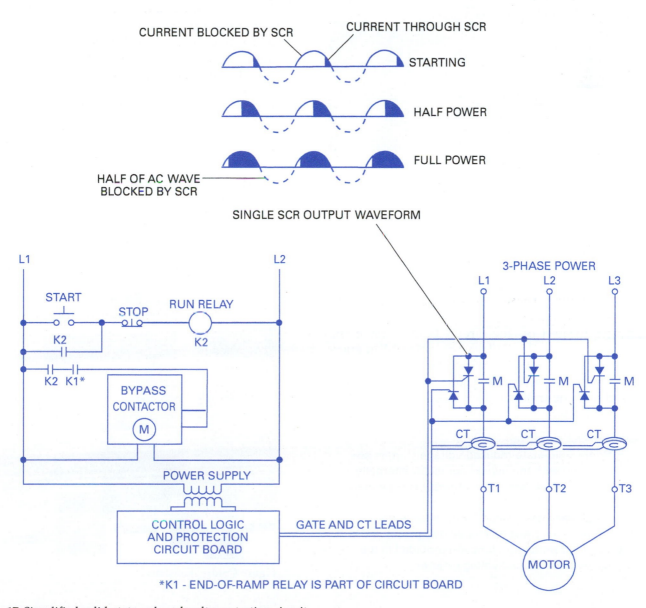

Figure 17 Simplified solid-state reduced-voltage starting circuit.

Figure 18 Reduced-voltage starting controller.

The circuit shown is also equipped with a bypass contactor (M) that is energized when the motor attains full voltage and speed at the end of the voltage ramp-up interval. This is common in some solid-state, reduced-voltage controller designs. The contacts of contactor M are connected in parallel with the SCRs. Contactor M is controlled by the closed contacts of the end-of-ramp relay K1 located on the circuit board inside the solid-state controller. When contactor M is energized at the end of the voltage ramp-up interval, its contacts close, causing the SCRs to be bypassed and the motor to be placed directly across the power line.

> **CAUTION**
>
> Some solid-state reduced-voltage starters are supplied with the bypass contactor internally connected. However, for others it is separately supplied and connected to the terminals specifically designated for it. These wiring connections must be followed to preserve the overload protection function provided in the solid-state reduced-voltage starter.

Settings for torque, time, and other parameters are made in the field to obtain switching of the SCRs as needed to achieve the desired motor performance. Typically, they are manually programmed into the circuit board microprocessor via dipswitches and circuit board controls. Some starting characteristics that normally can be programmed include:

- *Ramp start* – Ramp start, shown in *Figure 19 (A)*, is the most common form of soft start. It allows the initial torque value of the ramp to be set and then raises it to full voltage. The starting torque is typically adjusted from 1% to 85% of locked rotor torque but must be high enough so the motor starts to turn when started and not later; ramp time adjustment is from 0 to 180 seconds or more.

- *Kick start* – Kick start, shown in *Figure 19 (B)*, provides an initial boost of current to the motor to help break free the rotor and start the motor. The starting torque is typically adjusted from 1% to 85% of locked rotor torque; the duration of the kick start interval is from 0 to 2 seconds.

Solid-State Reduced-Voltage Starters

Solid-state reduced-voltage starters are available in a wide variety of sizes to suit various motor applications.

Figure Credit: Benshaw

- *Current limit* – Current limit, shown in *Figure 19* (C), is used when it is necessary to limit the maximum starting current because of long starting times or to protect the motor. The current limit is typically set in the range of 42% to 85% of the locked rotor current (250% to 500% of the motor full load current); the ramp time is typically from 0 to 180 seconds or more.
- *Soft stop* – Soft stop, shown in *Figure 19* (D), is used when an extended coast-to-stop period is needed. It is often used with high-friction loads when a sudden stop can cause system or product damage such as in hydraulic pumps. The stop ramp time is typically adjustable from 0 to 60 seconds.

A solid-state reduced-voltage controller typically also contains electronic circuits to protect the motor from overload and the starter from damage. Depending on the design, protection can be provided for shorted SCRs, over-temperature, current imbalance, undervoltage, phase loss, and/or phase reversal.

The following characteristics must be considered when selecting a reduced-voltage starter for use with a squirrel cage motor-driven load:

- Motor characteristics to satisfy the application starting requirements
- Source of power, including the effect of the motor starting current on the line voltage
- Load characteristics, including the effect of the motor starting torque on driven parts during acceleration
- Motor voltage
- Required startup torque
- Heater selection
- Enclosure type

Table 3 and *Table 4* provide a brief summary of the types of reduced-voltage controllers covered in this module and their characteristics.

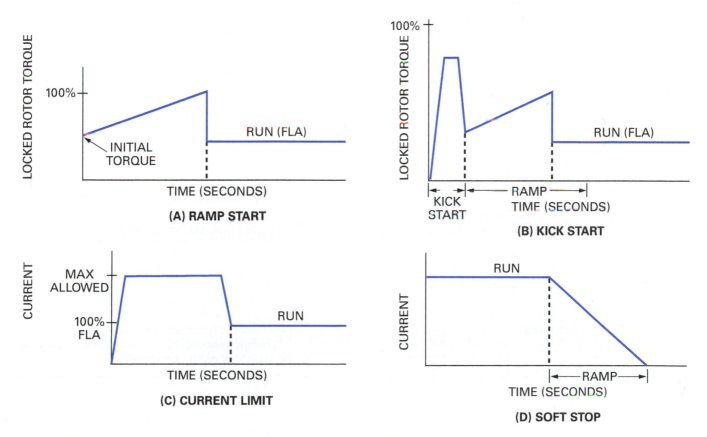

Figure 19 Starting characteristics of solid-state reduced-voltage controllers.

Table 3 Comparison of Reduced-Voltage Controllers

Type of Reduced-Voltage Controller	Starting Characteristics			Applications
	Voltage at Motor (%)	Line Current (%)	Torque (%)	
Autotransformer	80	64	64	Blowers, pumps, compressors, conveyors
	65	42	42	
	50	25	25	
Part-winding	100	65	48	Reciprocating compressors, pumps, blowers, fans
Wye-delta	100	33	33	Centrifugal compressors, centrifuges
Solid-state	Adjustable	Adjustable	Adjustable	Machine tools, hoists, packaging equipment, conveyor systems

Table 4 Advantages and Disadvantages of Reduced-Voltage Controllers

Type of Reduced-Voltage Controller	Advantages	Disadvantages
Autotransformer	• Provides maximum torque per ampere of line current • Starting characteristics easily adjusted • Different starting torques available through autotransformer taps • Suitable for relatively long starting periods • Motor current greater than line current during starting	• Most complex type due to precise sequencing of energization • Large physical size • Low power factor • Expensive in lower hp ratings
Part-winding	• Least expensive • Small physical size • Suitable for low or high voltage • Full acceleration in one step for most standard or special part-winding motors when 2/3 winding connection is used	• Unsuitable for high-inertia loads • Specific motor types required • Motor does not start when required load torque exceeds that developed by the motor when the first half of the motor is energized • Torque efficiency usually poor for high-speed loads
Wye-delta	• Suitable for high-inertia, long-accelerating loads • High torque efficiency • Ideal for stringent inrush current restrictions • Ideal for frequent starts	• Requires special motor • Low starting torque • Momentary inrush occurs during open transition when delta contactor is closed (open circuit transition modules)
Solid-state	• Voltage gradually applied during start for a soft-start condition • Adjustable acceleration time • Adjustable kick start, current limit, and soft stop	• More expensive than electromechanical types • Requires specialized installation and maintenance • Electrical transients can damage solid-state components • Requires good ventilation

2.0.0 Section Review

1. The type of reduced-voltage starter typically used with hard-to-start loads is _____.

    a. part-winding
    b. wye-delta
    c. thyristor
    d. autotransformer

2. The type of solid-state reduced-voltage starting control that provides an initial boost of current to the motor to help break the rotor free and start the motor is _____.

    a. ramp start
    b. current limit
    c. kick start
    d. soft stop

3.0.0 ADJUSTABLE-FREQUENCY DRIVES

Objective

Specify an adjustable frequency drive (AFD) for a given application.

a. Describe the types of AFDs, along with their operation and components.

b. Compare the uses, benefits, and problems when selecting an AFD.

Performance Task

1. Identify and connect various control devices.

Trade Terms

Base speed: The shaft speed at which the motor will develop rated horsepower at rated load and voltage, as stated on a motor nameplate.

Carrier frequency: In a pulse-width modulated (PWM) drive, the frequency at which the IGBT can be switched on or off to create an output waveform that mimics a sine wave of desired voltage and frequency. Carrier frequencies are seldom less than 500Hz nor more than 20,000Hz; but need to be at least 10 times the maximum drive output frequency.

Fan Affinity Laws: The mathematical relationship between fan speed, airflow rate, power used, and torque. Airflow is directly proportional to speed. Torque is proportional to the square of speed. Power used is proportional to the cube of speed. For example, decreasing fan speed by 20% decreases the power used by about 50%.

Insulated gate bipolar transistors (IGBTs): A type of transistor that has low losses and low gate drive requirements. This allows it to be operated at higher switching frequencies.

Slip: The speed difference between the synchronous speed (t) and the base speed. Induction motors develop their torque because of the slip. The amount of slip depends on motor design and motor load.

An adjustable-frequency drive (AFD) converts three-phase 50Hz or 60Hz input power to an adjustable frequency and voltage output to control the speed of an AC motor. AFDs are discussed in *NEC Article 430, Part X*. AFDs are also known as *variable frequency drives*, *variable speed drives*, *inverter drives*, or *adjustable speed drives*.

3.1.0 Types of Adjustable Frequency Drives

Squirrel cage induction motors are only available with base speeds determined by the frequency of the incoming power line, slip, and number of magnetic poles in the motor. In most cases, these limited choices of base speed do not directly match the needs of the equipment or machinery being driven by the motor. Various methods have been used to match the motor speed to the desired load speed. Mechanical methods change the way the motor is coupled to the load, such as variable pitch pulleys, eddy-current drives, sprockets and chains, constantly variable transmissions, hydraulic couplers, etc. Different motor designs that change the motor base speed have also been used, such as wound-rotor motors or Schrage motors. All of these mechanical and motor design methods introduce additional expense and require a lot of maintenance, although some are still in use. None of these methods provide optimum control automation, energy conservation, or ease of use.

Today, the most common method of matching motor speed to the needs of the equipment or machinery is to change the frequency coming to the motor using an AFD. AFDs are easier to use, work well with control automation, and can conserve energy.

The earliest AFDs used silicon-controlled rectifiers (SCRs) to create the output frequency to the motor in a six-step waveform. This output waveform worked with motors, but required a lot of filtering, with additional expense, weight, and space requirements. With the development of insulated gate bipolar transistors (IGBTs) in the early 1980s, a much wider range of applications in AFDs became possible. More recently, with the decreasing cost and greatly expanded capabilities of computer processors, the engineering equations for vector analysis and control of motor power became possible in real time, again increasing the range of possible applications. AFDs that use this are often called *vector drives*. If a

Adjustable-Frequency Drives

GOING GREEN

Adjustable-frequency drives are available for use with a wide range of motors and applications. These drives save energy by eliminating the losses associated with traditional speed control.

Figure Credit: Schneider Electric

feedback sensor on the motor gives the AFD accurate information on drive shaft rotation (via an encoder), known as a *feedback vector drive*, very precise motor control is possible. Today's AFDs can power motors as small as $\frac{1}{2}$hp up through medium voltage motors of 100,000hp or more.

3.1.1 AFD Forms and Types

The National Electrical Manufacturers Association (NEMA) describes three AFD drive classifications according to the relationship between their input and output.

- *Form FA* – AC input directly to AC output by synthesizing the output waveform from segments of the AC supply without an intermediate DC link. Types include cycloconverters and Matrix converters.
- *Form FB* – DC input to AC output. Uses include electric vehicles and some industrial machines.
- *Form FC* – AC input to DC intermediate bus to AC output. These are the most common and will be the focus of this module.

Although AFDs come in about six different types, three of them are likely to be seen on a regular basis. *Table 5* compares key features of each type. This module will focus on these AFD types and highlight the few places where they differ.

3.1.2 AFD Components

An AFD has four main sections:

- *AC input section* – The AC input section includes high-speed overcurrent protection, input voltage spike protection, and SCRs to power the DC bus.
- *DC bus* – The DC bus includes capacitors and inductors (LC filters) as necessary to smooth the DC voltage and provide robust power near the peak voltage of the incoming AC waveform. This is 1.4 times the nominal RMS voltage, so incoming power at 480V will supply a DC bus running close to 670V.
- *Logic section* – The logic section has specially designed computer chips, program memory, and input/output handling to control the inverter section. The logic section can receive commands and share drive conditions and performance with a remote computer or supervisory control and data acquisition (SCADA) system, or with a local operator interface.
- *Inverter section* – The inverter section includes the IGBTs that use the carrier frequency in a pulse-width modulation (PWM) method to chop the DC bus voltage by varying the widths of on/off pulses of positive or negative voltages to create three output waveforms for the three-phase motor. (For example, when the off portion of the output is wide compared to the on portion, the effective value of the output voltage applied to the motor will be low.)

See *Figure 20* for a block diagram illustrating the relationships between these sections and a graph of the output waveforms.

Depending on the specifications and application, an AFD installation can have any of the following optional components:

- An input filter section can be provided to keep the AFD from imposing voltage and current harmonics back onto the incoming feeders and utility power line. Filtering methods can include a line reactor, active components, or a drive isolation transformer. *IEEE Standard 519* limits total harmonic distortion at the service to 5%, but without filtering, the AFD can create harmonics of 50% or more on the incoming power line.

Table 5 Key Features of Different AFD Types

Control Methods	Volts/Hz	Sensorless Vector	Feedback Vector
Control criteria	Open loop	Open loop	Closed loop
Typical speed regulation	1%	0.5%	0.01%
Speed range at constant torque	10% – 100%	3% – 100%	0% – 100%
Multiple motor operation	Yes	No	No
Torque at zero speed	Impossible	Poor	Best
Replace existing DC drive	No	Maybe	Yes

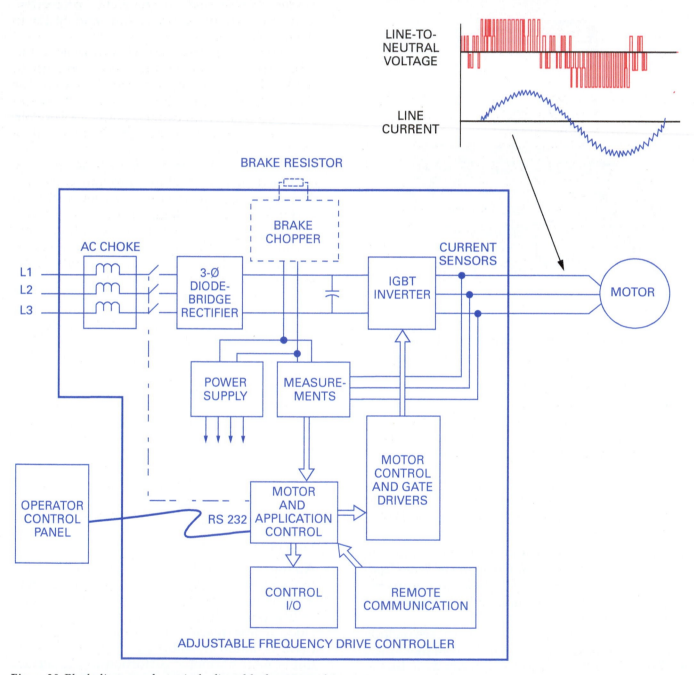

Figure 20 Block diagram of a typical adjustable-frequency drive.

- Power line regenerative components or a chopping IGBT and external power resistor can be used to prevent excessive DC bus voltage when the AFD is slowing or stopping the driven load.
- Output filter sections are available in three types: a load reactor, a delta voltage/delta time (dV/dT) filter, or a sine wave filter. These can allow a longer cable from the AFD to the motor and minimize high-frequency noise that can harm the motor or nearby electronics.

3.2.0 AFD Uses and Benefits

Powering a motor with an AFD provides the following benefits:

- *Controlled starting* – AFDs can limit starting current and reduce power line demand/disturbance on startup.
- *Controlled acceleration* – AFDs allow for soft-start, adjustable acceleration based on time or load, and reduced motor size for pure inertial load acceleration.
- *Adjustable operating speed* – Speed control allows a process to be optimized or changed, provides energy savings, and allows process control at reduced speeds set by a programmable controller.
- *Energy savings* – The **Fan Affinity Laws** show that replacement of a damper or throttling valve with a change in motor speed can produce direct energy savings of 25% to 50%. Motors are often oversized in order to allow for aging, wear, and similar changes. AFDs deliver only the power required and maintain a superior power factor on the incoming power line.
- *Adjustable torque limiting* – Current limiting can be used to provide fast, accurate torque control in order to protect the machinery, product, or process against damage.
- *Controlled stopping* – AFDs provide soft stop, timed stopping, and fast reversal with much less stress on an AC motor compared to plug reverse.
- *Operation above base speed* – Many motors designed to NEMA standards can be operated up to 1.5 times their base speed, with higher limits for 4- or 6-pole motors and lower limits for large motors.

3.2.1 AFD Selection

The first selection consideration is the load characteristics of the machine or equipment being powered. The three common types are variable torque, constant torque, and constant horsepower.

- *Variable-torque loads* – These are typically fans, blowers, pumps, and similar loads where a higher speed requires more power and more torque. These relationships follow the fan affinity laws and are graphed in *Figure 21*. A volt/Hz AFD works very well for variable-torque loads and the installation is cost effective.
- *Constant-torque loads* – These are speed-independent, friction-type loads, such as hoists, conveyors, printing presses, positive displacement pumps, and extruders. A machine with a large flywheel (such as a chipper) will act as a constant-torque load while it is being accelerated up to operating speed. With these loads, the torque remains constant over the speed range, but the horsepower increases as the speed increases. See the graph in *Figure 22*. A vector-type AFD is best for these types of loads.
- *Constant-horsepower loads* – With these loads as the speed increases, the torque decreases, so the horsepower stays the same. Typical machines include lathes, grinders, and many winders. Because the horsepower requirements are constant, an AFD will not provide energy savings, but can provide improved process control. See the graph in *Figure 23*.

Two less common types of loads are constant force (tension control) and wind/unwind. The use of AFDs with these loads provides many

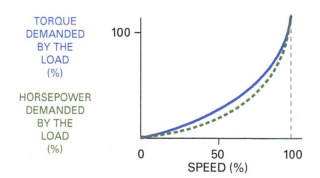

Figure 21 Variable-torque load.

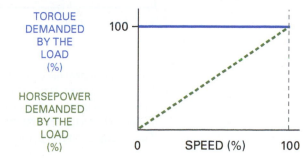

Figure 22 Constant-torque load.

benefits, but generally requires careful engineering support and evaluation.

Motors operate over a wide frequency range with minimal heat losses when the ratio of the input voltage to frequency matches the ratio of the nameplate voltage to frequency. This is a range with a constant volts-to-hertz ratio, as shown in *Figure 24*, and provides constant torque from the motor. The AFD cannot output a voltage above its rated output voltage, so when the motor is operated above its base speed, the motor is producing constant power. Motors without external cooling fans cannot be operated for very long at low speeds, and AFDs (other than vector types) cannot be used at less than about 10% of the incoming line frequency—this is the typical minimum programmed frequency for the AFD output. The red line on the graph in *Figure 24* shows these relationships.

The second selection consideration is variations in the load and environment; some of these may require the use of an AFD sized larger than the motor full load current.

- High-torque loads can require a larger AFD size during starting or running. AFD torque limits usually are 150% for constant-torque applications or 110% for variable-torque applications.
- An impact load can be a punch press with a flywheel or a conveyor that gets a heavy package dropped onto its surface. The AFD must have the capacity to overcome this sudden load change without an overload trip from the sudden change. Often a NEMA Design D motor is used here.
- Breakaway torque is what overcomes the higher friction that a non-moving load has compared to a moving one, such as a cement or slurry mixer. An AFD must be sized to provide the extra current needed to start the load.

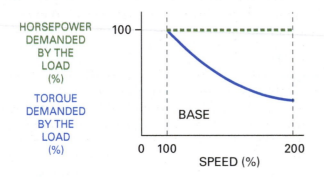

Figure 23 Constant-horsepower load.

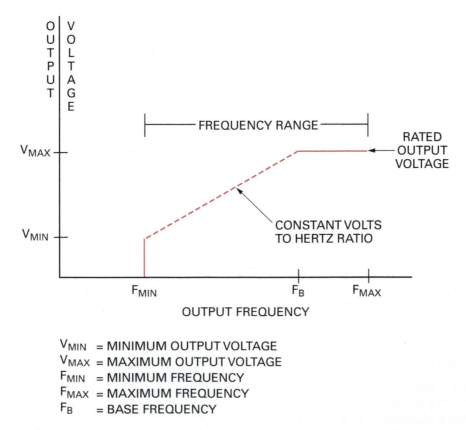

V_{MIN} = MINIMUM OUTPUT VOLTAGE
V_{MAX} = MAXIMUM OUTPUT VOLTAGE
F_{MIN} = MINIMUM FREQUENCY
F_{MAX} = MAXIMUM FREQUENCY
F_B = BASE FREQUENCY

Figure 24 AFD operating frequency range.

- Starting a spinning load can cause an AFD overload fault unless the AFD has been programmed to match the load speed when it starts. Fans and pumps may be turning backwards while off, and any load with high inertia may still be spinning while an AFD has suffered a momentary power failure.
- A limited number of AFDs can be powered from single-phase AC to run a three-phase motor. The reduced available single-phase input power requires choosing an AFD to that is larger than the motor rating.
- An AFD can be supplied with just a NEMA-1 terminal cover, but this does not protect it from weather, dust, contamination, or temperature extremes. Oversizing it may be needed if temperatures are likely to be above its rated environmental limit. If in an enclosure, a filtered fan, air conditioner, or other method for cooling may be needed.
- Matching the AFD to the motor design is important. As shown in *Table 6*, NEMA Design A, Design B, and Design E motors are best suited for use with an AFD, although an AFD can be used with the other motor designs.

3.2.2 AFD Programming

All AFDs are matched to their load by setting adjustable parameters to control the details of the AFD performance. Early models had 50 or more parameters, but today's much more capable AFDs

AFD Environment

Heat rejection is a major concern with AFDs. The heat produced in the AFD cabinet can be substantial and can cause failure of the AFD SCRs and IGBTs if their operating temperature limits are exceeded. For this reason, operating an AFD at higher altitudes often requires that the unit be derated, because the lower-density air reduces the amount of heat transfer.

can have up to 600 or more parameters. Follow the manufacturer's installation and/or operation manuals for programming. Common parameters that must be programmed include:

- *Motor nameplate data* – Voltage, frequency, base speed, current, service factor, and power factor.
- *Incoming power data* – Line voltage.
- *AFD input usage* – Input selection for enabling, starting, stopping, reversing, and fault resets.
- *Communications* – Selection of I/O method via wired terminals or communications link and use of an operator interface.
- *Setup* – Minimum and maximum frequency, skip frequencies (to avoid vibration), acceleration and deceleration time(s), short-term current limit, and speed reference source.

Table 6 AFD Use with Different NEMA Motor Designs

Motor Design Letter	Typical Driven Load Types	Slip	Locked rotor current	Efficiency	Use with AFD
Design A	Fans, blowers, MG sets, centrifugal pumps, and compressors	0.5 – 3	600% – 700%	High or medium	Good
Design B	Fans, blowers, MG sets, centrifugal pumps, and compressors	0.5 – 5	600% – 700%	High or medium	Very good
Design C	Conveyors, crushers, stirring, agitating, reciprocal pumps	1 – 5	600% – 700%	Medium	Poor, risk of overheating
Design D	High-peak loads with or without flywheels	5 – 8+	600% – 700%	Lower	Can be done, but more difficult
Design E	Fans, blowers, MG sets, centrifugal pumps, and compressors	0.5 – 3	800% – 1,000%	Highest	Good, but may need a larger drive size

NOTE: Current is % of full-load values. Slip is % below synchronous speed.

- *Output volt/Hz ratio* – For a constant-torque load, the linear volt/Hz ratio provides full output torque at high efficiency over a wide range of frequencies from the AFD. This matches induction motor design. However, for variable-torque loads, the squared volt/Hz ratio will save energy at lower AFD output frequencies by reducing the voltage to match the torque required by the load.
- *Operation above base frequency* – For AFD output frequencies above the base frequency (if permitted by the maximum frequency setting), the voltage output remains constant at the motor nameplate voltage, so motor power is constant, but motor torque is decreased.
- *IR compensation and voltage boost* – These help with starting a high-inertia or high-friction load. Moderate amounts of voltage boost are used for most loads, but the use of IR compensation instead of higher levels of voltage boost is best for more difficult loads.

In many cases, successful programming and startup of an AFD installation may require the skills and knowledge of trained professionals. See *Figure 25A* and *Figure 25B* for a checklist used by one manufacturer to gather the data needed to evaluate the requirements for an AFD system.

3.2.3 AFD Application Precautions

AFDs have a few special application considerations and precautions. These include the following:

- Any fan-cooled motor is at risk of overheating and failure when an AFD operates it at slower speeds (50% of base speed or less) for more than a few minutes. To prevent this, an externally powered fan must be used to maintain the needed cooling airflow. This is particularly true if the motor has a Class F temperature rise or a 1.0 service factor.
- *NEMA Standard MG31* specifies motor construction requirements for a motor that should withstand the voltage stresses the AFD imposes on its windings. The motor nameplate will be labeled INVERTER DUTY or have similar wording. Inverter duty motors are wound using magnet wire with a higher voltage insulation level. This is needed because the high-frequency PWM pulses from the AFD reflect at the first few turns of the motor winding, causing the voltage on these motor windings to be as high as 2.5 times the actual output voltage.

- Non-inverter duty motors operated above 240V are prone to premature failure due to the overvoltage imposed on their windings. The severity of this overvoltage can be reduced with the use of the lowest possible carrier frequency and special AFD power cables. This overvoltage can be reduced or even eliminated with the use of output filtering devices selected by a professional.

Adjustable-Frequency Drives

For retrofit AFD applications, many AFD manufacturers recommend that existing motors be replaced to ensure that the AFD unit is properly sized and is not affected by previous motor misuse or rewinding.

- AFD manufacturers typically recommend the use of special shielded cables between the AFD and the motor (see *Figure 26*). This minimizes the radiation of the high-frequency noise into nearby sensitive equipment or signal cables. Spacing, barriers, cable orientation, and termination restrictions must be followed, particularly in control enclosures.
- Motor grounding wires must be connected from the motor to the ground terminal on the AFD, with that terminal then being connected to the grounding terminals of the grounding electrode system. This reduces the magnitude of electronic noise that can be imposed by the AFD and its attached motor on the facility's grounding electrode system and nearby equipment.
- A motor powered by an AFD often has parasitic or stray voltages coupled onto the motor shaft. If these voltages are discharged through the bearings, they will cause electric discharge machining of the bearing races, typically in the form of parallel grooves (called *fluting*) and premature bearing failure. Certain types of AFD output filtering or the use of shaft grounding brushes will prevent this problem. Other methods sometimes available include the use of electrically insulated bearings or shaft couplings.

AFD APPLICATION CHECKLIST

Motor

New_____ Existing _____ Horsepower: _____ Base Speed: _____ Voltage: _____

FLA: _____ LRA: _____ NEMA Design: _____ Gearbox/Pulley Ratio: _____

Service Factor:_____

Load

Application: _____

Load Type: Constant Torque _____ Variable Torque _____ Constant Horsepower _____

Load inertia reflected to motor: _____

Required breakaway torque from motor: _____

Running load on motor: _____

Peak torques (above 100% running): _____

Shortest/longest required accel. time: _____ / _____ secs up to _____ Hz from zero speed

Shortest/longest required decel. time: _____ / _____ secs down to _____ Hz from max. speed

Operating speed range: _____ Hz to _____ Hz

Time for motor/load to coast to stop: _____ secs

AFD

Source of start/stop commands: _____

Source of speed adjustment: _____

Other operating requirements: _____

Will the motor ever be spinning when the AFD is started? _____

Is the load considered to be high inertia? _____

Is the load considered to be hard to start? _____

Distance from AFD to the motor: _____ feet

Type of AFD (V/Hz, Flux Vector, Closed Loop Vector): _____

Options desired: _____

_____

_____

Other special requirements/conditions: _____

Figure 25A AFD application checklist (1 of 2).

AFD APPLICATION CHECKLIST (Continued)

Power Supply

Supply Transformer ______ kVA and ______ % Z or short circuit current at drive input: ______ amps

(If the drive does not include a built-in line reactor and the available feeder short circuit current is more than 100 times the drive FLA rating, a 1% line reactor or a drive isolation transformer is required.)

Total horsepower of all drives connected to supply transformer or feeder _______ hp

Is a drive transformer or line reactor desired? ______________________________

Any harmonic requirements? _______ % Voltage THD: ________ % Current THD: ______ IEEE 519:______

Total non-drive load connected to the same feeder as drive(s): ________ amps

Service

Start-up Assistance: __________________ Customer Training: __________________

Preventive Maintenance: __________________ Spare Parts: __________________

Additional Issues

Will the AFD operate more than one motor? ______________________________

Will the power supply source ever be switched with the AFD running? __________________

Is starting or stopping time critical? ______________________________

Are there any peak torques or impact loads? ______________________________

Will user-supplied contactors be used on the input or output of the AFD? ________________

Does the user or utility system have PF capacitors that are being switched? ________________

Will the AFD be in a harsh environment or high altitude? __________________________

Does the utility system experience surges, spikes or other fluctuations? __________________

Figure 25B AFD application checklist (2 of 2).

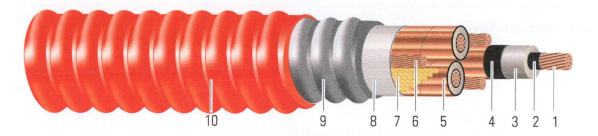

CONSTRUCTION:

1. **Conductor:** Class B compressed stranded bare copper per ASTM B3 and ASTM B8
2. **Conductor Shield:** Semi-conducting cross-linked copolymer
3. **Insulation:** 220 Mils No Lead Ethylene Propylene Rubber (NL-EPR) 133% Insulation Level,
4. **Insulation Shield:** Strippable semi-conducting cross-linked copolymer
5. **Copper Tape Shield:** Helically wrapped 5 mil copper tape with 25% overlap
6. **Grounding Conductor:** Three separate ground wires with a combined circular mil of 50% of the phase conductor. Class B compressed stranded bare copper per ASTM B3 and ASTM B8
7. **Filler:** Wax paper filler
8. **Binder:** Polypropylene tape
9. **Armor:** Continuous Corrugated Welded Aluminum Armor (Armor-X)
10. **Overall Jacket:** Polyvinyl Chloride (PVC)

Figure 26 AFD cable.

PWM AFD Outputs to a Motor over Long Cable Runs

When a PWM AFD drive output is connected to a motor over long cable runs, the difference in the motor and cable impedance can cause voltage spikes at the motor terminals. These spikes can greatly exceed 1,000V and have the potential to damage the motor insulation. For small motors, this typically occurs with cable lengths that exceed 33' (10 m) and, for larger motors, with cable lengths that exceed 100' (30 m). For situations involving the use of long cable runs, install a reflective wave trap or similar filter at the motor terminals to prevent overshoots.

3.0.0 Section Review

1. The combination of an LC filter and a bridge rectifier is used to _____.

   a. reduce noise coupling from nearby circuits
   b. develop a smooth DC voltage
   c. produce high-speed switching
   d. develop the AC waveform to drive the motor

2. Which of these actions should be taken if an AFD is used at lower speeds on an older motor with a Class F temperature rise or 1.0 service factor?

   a. Derate the motor load capacity by 18%.
   b. Reduce the input voltage to the motor.
   c. Use a bridge rectifier at the motor input.
   d. Install a separately powered cooling fan on the motor.

4.0.0 MOTOR BRAKING METHODS

Objective

Compare motor braking methods.
  a. Describe DC injection braking.
  b. Identify dynamic and regenerative braking methods.
  c. Describe friction braking.

Braking provides a means of quickly stopping an AC motor when it is necessary to stop it faster than is normally possible by removing the power and letting the motor coast to a stop. Braking can be accomplished in several ways. The method used depends on the application, available power, and circuit requirements. Three common methods of motor braking include the following:

- *DC injection braking of an AC motor* – Braking force is independent of motor speed. Requires a source of DC power. Braking energy is lost as heat within the motor.
- *Dynamic and regenerative braking* – Braking force decreases with decreasing motor speed. Requires connection to the motor controls. Dynamic braking energy is lost to resistors while regenerative braking energy is saved to the incoming power source.
- *Friction braking* – Braking force is independent of motor speed. Does not require any operating power sources. Braking energy is lost as heat within the brake components.

4.1.0 DC Injection Braking

DC injection braking utilizes a DC voltage applied to (injected into) the motor stator windings for a timed interval (see *Figure 27*). The application of direct current to the stator windings creates a stationary magnetic field. This field interacts with the magnetic fields in the rotor to create a braking torque on the rotor, which can bring the motor to a stop and keep it from rotating. The magnitude of the applied DC determines if the stopping action occurs very quickly (as for an emergency) or more gently (for safety). An advantage of DC injection braking is that motors can be stopped rapidly without wear on brake linings or drums. Since DC injection braking requires the availability and use of incoming power, it cannot be safely used for loads suspended from a crane or hoist—friction braking must be employed instead.

In the DC brake circuit of *Figure 27*, a typical motor starter circuit with its magnetic starter (M) and STOP/START pushbuttons is modified with the addition of an electrically and mechanically interlocked braking relay (BR), an off-delay timing relay (TR), and a DC power supply with a step-down transformer and full-wave bridge rectifier. These operate together as follows:

- Contactor M is de-energized by pressing the STOP pushbutton.
- Simultaneously, timing relay TR is de-energized, closing its normally open, timed open (NOTO) contacts.
- Braking relay BR is energized to apply the DC power from the rectifier to the motor.
- When the preset time interval in TR ends, the NOTO contacts open to de-energize braking relay BR.
- This removes the DC braking power from the motor and the START pushbutton can start the motor when pressed.

A transformer with tapped windings is often used to adjust the amount of braking voltage applied to the motor. Current-limiting resistors could be used for the same purpose. The higher the applied DC voltage, the greater the braking torque. This allows for a low or high amount of braking, depending on the application. The time interval set in timing relay TR must be longer than the actual time required for the motor to come to a complete stop—this must be done in the field.

The mechanical and electrical interlocking of starter M and braking relay BR is required because the AC and DC power supplies must never be connected to the motor simultaneously. Total interlocking should always be used on electrical braking circuits.

DC braking can be used with AC motors controlled by any of the reduced-voltage starters discussed earlier. DC braking can also be used with motors controlled by AFDs. In this case, the AFD already has all the components required—the only need is to set the handful of parameters to permit DC injection braking and set both the magnitude and duration of the DC injection.

Three limits exist in the use of DC braking of motors:

- The braking energy is lost as heat in the motor, which can cause overheating and damage to the motor. For applications with frequent use of dynamic braking, an external cooling fan may be required.

- If the braking is very rapid, the torque produced can damage the motor or the connected load, and the heat produced within the motor can damage or destroy the motor windings.
- For safety, the application may include a requirement for a mechanical brake to automatically activate upon loss of power.

4.2.0 Dynamic and Regenerative Braking

In both dynamic and regenerative braking, the power connections to the motor (either AC or DC) are reconfigured so the motor acts as a generator. With dynamic or friction braking, the energy stored in the motor and driven load is converted to external heat and is lost. A more energy-efficient approach is to use regenerative braking. Unlike DC injection braking (in which the braking force is constant regardless of motor speed), dynamic

and regenerative braking force is proportional to motor speed and therefore drops to zero when the motor is stopped. In applications requiring a definite stop to a zero speed, dynamic or regenerative braking can be used for most of the speed reduction (minimizing heating within the motor). DC braking is then used to ensure a positive stop.

4.2.1 Dynamic Braking

In addition to use of an AFD for DC injection braking, an AFD can also provide dynamic braking by use of a timed (ramped) stop for the motor and its driven load. When the output frequency from the drive to the motor is less than the speed of the motor, negative torque is developed within the motor. The motor then acts like a generator by converting mechanical power from the shaft into electrical power that is returned to the DC bus within the AFD. AFDs are made with an IGBT

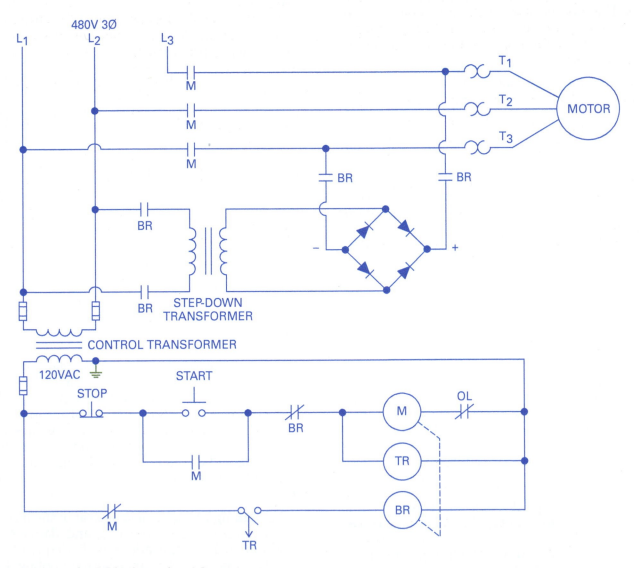

Figure 27 Circuit for DC braking of an AC motor.

switch connected to terminals which, in this application, are connected to an external braking resistor where the excess energy gets dissipated as heat.

Figure 28 shows a simplified diagram of an AFD wired to provide dynamic braking. The drive control logic will turn the IGBT switch on and off as necessary to keep the DC bus voltage from reaching an unsafe level. To function safely and properly the resistor must be large enough to handle the heat created within it by the braking action. It will be located outside the AFD enclosure and connected using high-temperature wire. The resistor has a high-temperature limit switch wired back to the drive control inputs. If the braking resistor fails, the DC bus voltage in the AFD will exceed its high limit and the AFD will trip offline, with the resulting loss of all motor control. Therefore, the resistor(s) used for dynamic braking must be capable of absorbing at least six times the stored rotational energy of the motor at its operating speed. Dynamic braking is commonly used with traction motors in rail transportation.

4.2.2 Regenerative Braking

Due to the growing demand for energy-efficient systems, AFD applications often include regenerative braking. Methods of regenerative braking include the following:

- *Regeneration to the power line* – AFDs that can do this are called *line regenerative*. They are made with active components in their input section instead of diodes. When the DC bus voltage reaches its upper safety limit, the AFDs logic returns the excess energy back to the incoming power line.

- *Sharing the DC bus* – Form FB drives (described earlier) are connected and get their power from a common DC bus with its own rectifier source. When one AFD is braking its load, the excess energy put onto the DC bus can be used by other AFDs on this bus or returned to the incoming power line.

- *Storage in batteries* – The DC bus is directly connected to batteries. When the motor is being operated as a generator, the excess energy from the motor is stored in the batteries. This method is used in electric and hybrid automobiles, resulting in considerable energy savings.

4.3.0 Friction Braking

Motor braking is also done using solenoid-operated friction brakes. Friction brakes are found on printing presses, small cranes, overhead doors, hoisting equipment, machine tool control, incline or decline conveyors, and other applications.

Many friction brakes used with motors are like those used with vehicles. They normally consist of two shoe- or pad-type friction surfaces that contact a wheel mounted on the motor shaft. Brake operation is controlled by a solenoid. The solenoid is energized when the motor is running, causing the brake shoes or pads to be retracted from the wheel. When the motor is turned off, the solenoid is de-energized, and the tension from a return spring in the brake mechanism brings the shoes in contact with the wheel, causing braking

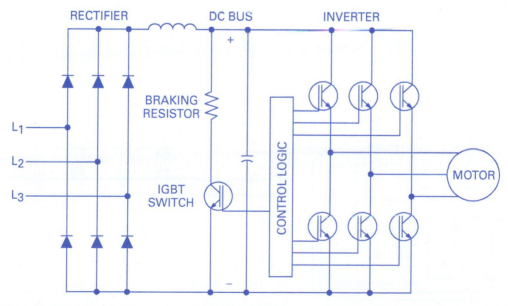

Figure 28 Simplified schematic of an AC drive using dynamic braking.

to occur as a result of the friction between the shoes and the wheel. The braking torque developed is directly proportional to the braking surface area and spring pressure. The spring pressure is adjustable on nearly all friction brakes.

In the diagram shown in *Figure 29*, the brake solenoid energizes and de-energizes when voltage is applied to and removed from the motor, respectively. To prevent improper brake activation, the brake solenoid is connected directly into the motor power circuit, not its control circuit. Because the brake solenoid in the motor is designed to release the brake when full motor voltage is applied, wiring to the solenoid must be run from a separate contactor when the motor is operated from any of the reduced-voltage starting methods or from an AFD.

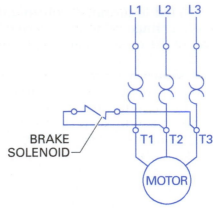

Figure 29 Friction brake solenoid connection.

Rural Applications of AFDs

In rural areas, three-phase power may not be available at an affordable cost. One contractor uses a special type of AFD known as an *active front end (AFE) drive* that acts as the DC supply source for three connected variable frequency drives (VFDs). An AFE saves energy by returning the regenerative energy back to the line.

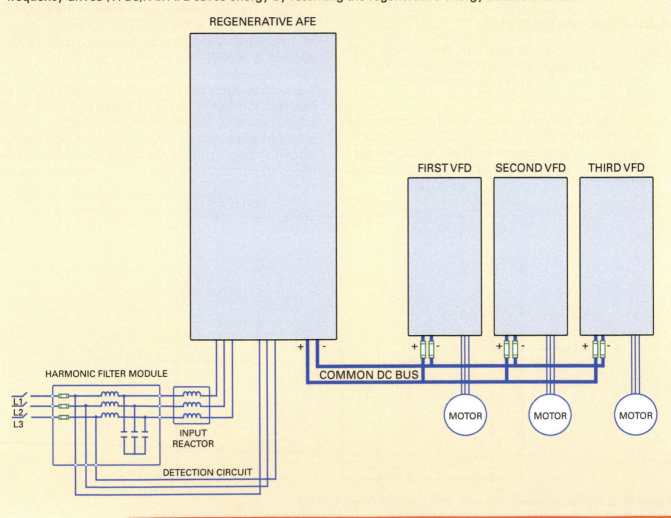

Friction brakes are positive-action, mechanical friction devices. Normal operation is such that when the power is removed from the motor, the brake is set. For this reason, it can be used as a holding brake and is described as *fail-safe*. The advantage of using a friction brake is lower cost. However, friction brakes require more maintenance than other braking methods. Like the brakes used in a vehicle, the shoes or pads must be replaced periodically.

In some applications with very frequent stops and starts, such as a conveyor system brake/spacer belt, a separately controlled clutch/brake assembly is installed on the motor output shaft. The motor will continue to run but the driven load will be disconnected and quickly stopped. This approach allows more frequent stops and starts than is permitted by the motor design.

4.0.0 Section Review

1. The amount of braking torque when using DC braking of an AC motor can be controlled by using a tapped transformer or by _____.

   a. installing an LC filter
   b. adding a current-limiting resistor
   c. adding a switching circuit
   d. replacing the brake pads

2. Braking energy is returned to the power line by _____.

   a. sharing the AFDs DC bus
   b. raising the DC bus voltage
   c. active components in the AFD input section
   d. connection to a battery

3. Friction brakes are controlled by _____.

   a. using a torque converter
   b. adding a current-limiting resistor
   c. a transformer
   d. a solenoid

5.0.0 WORKING WITH MOTOR CONTROLS

Objective

Describe how to maintain motor controls for peak operation.

a. Apply the precautions associated with solid-state controls.
b. Implement the preventive maintenance procedures associated with solid-state controls.

5.1.0 Precautions for Working with Solid-State Controls

> **WARNING!**
>
> Solid-state equipment may contain devices that have potentially hazardous leakage current in the off state. Power should be turned off and the device disconnected from the power source before working on the circuit or load.

> **CAUTION**
>
> Many solid-state devices can be damaged by electrostatic charges. These devices should be handled in the manner specified by the manufacturer. This is particularly true because higher density of the electronic components requires lower operating voltages, thus increasing the possibility of electrostatic damage (ESD).

Solid-state equipment and/or components can be damaged if improperly handled or installed. The following guidelines should be followed when handling, installing, or otherwise working with solid-state equipment or components:

- Microprocessors and other integrated circuit chips are very sensitive to, and can be damaged by, static electricity from sources such as lightning or people. When handling circuit boards, avoid touching the components, printed circuit, and connector pins. Always ground yourself before touching a board. Disposable grounding wrist straps are sometimes supplied with boards containing electrostatic-sensitive components. If a wrist strap is not supplied with the equipment, a wrist strap grounding system should be obtained and used. It allows you to handle static-sensitive components without fear of electrostatic discharge. Store unused boards inside their special metallized shielded storage bags or a conductive tote box (*Figure 30*).

- Solid-state devices can be damaged by the application of reverse polarity or incorrect phase sequence. Input power and control signals must be applied with the polarity and phase sequence specified by the manufacturer.

- The main source of heat in many solid-state systems is the energy dissipated in the output power devices. Ensure that the device is operated within the maximum and minimum ambient operating temperatures specified by the manufacturer. Follow the manufacturer's recommendations pertaining to the selection of enclosures and ventilation requirements. For existing equipment, make sure that ventilation passages are kept clean and open.

- Moisture, corrosive gases, dust, and other contaminants can all have adverse effects on a system that is not adequately protected against atmospheric contaminants. If these contaminants collect on printed circuit boards, bridging between the conductors may result in a malfunction of the circuit. This could lead to noisy and erratic control operation or a permanent malfunction. A thick coating of dust could also prevent adequate cooling of the board or heat sink, causing a malfunction.

- Excessive shock or vibration may cause damage to solid-state equipment. Follow all manufacturer instructions for mounting.

Electrical noise can also affect the operation of solid-state controls. Sources of electrical noise include machines with large, fast-changing voltages or currents when they are energized or de-energized, such as motor starters, welding equipment, adjustable speed drives, and other inductive devices. The following are basic steps to help minimize electrical noise:

- Maintain enough physical separation between electrical noise sources and sensitive equipment to ensure that the noise will not cause malfunctions or unintended actuation of the control.

- Maintain a physical separation between sensitive signal wires and electrical power and control conductors. This separation can be accomplished by using separate conduits, wiring trays, or methods recommended by the manufacturer.

- Use twisted-pair wiring in critical signal circuits and noise-producing circuits to minimize magnetic interference.
- Use shielded wire to reduce the magnitude of the noise coupled into the low-level signal circuit by electrostatic or magnetic coupling.
- Where signal and power wires or cables must cross, do so at right angles to minimize any magnetically induced voltages (noise) between the cables.
- Follow all requirements of the *National Electrical Code*® with respect to grounding. Additional grounding precautions may be required to minimize electrical noise. These precautions generally deal with ground loop currents arising from multiple ground paths. The manufacturer's grounding recommendations must also be followed.

5.2.0 Motor Control Maintenance

A well-planned and well-executed preventive maintenance program is necessary to achieve satisfactory operation of electrical and electronic motor control equipment. Preventive maintenance keeps the equipment running with little or no downtime. Ideally, routine preventive maintenance for a specific piece of equipment is based on the manufacturer's recommendations, the severity of use, and the surrounding environment.

Before any preventive maintenance work begins that will shut down or otherwise interrupt equipment operation, always notify and/or coordinate with the building manager, shift leader, foreman, or other responsible person.

> **WARNING!**
>
> Before any preventive maintenance tasks are performed, de-energize, lockout, and tagout all circuits and equipment in accordance with the prevailing lockout/tagout policy. If more than one incoming power source exists (such as when there is a separate control circuit, or when other forms of energy are involved) make sure to de-energize all such sources. Note that solid-state electronic equipment may contain devices that have potentially hazardous leakage current in the off state. Such equipment should be isolated from the power source by opening the related disconnect switch, rather than by simply turning off the device.

Preventive maintenance for motor control circuit components normally involves inspection and cleaning of the equipment. Control equipment should be kept clean and dry. Before opening the door or cover of a cabinet or enclosure, clean any foreign material, dirt, or debris from the outside top surfaces to avoid the risk of anything falling onto the equipment. After opening the cabinet or enclosure, inspect the equipment for the presence of any dust, dirt, moisture or evidence of moisture, or other contamination. If any is found, the cause must be determined and eliminated. This could indicate an incorrectly selected, deteriorated, or damaged enclosure; unsealed enclosure openings; internal condensation; condensate from an unsealed conduit; or improper operating procedures, such as operating the enclosure with the door open or the cover removed.

Ambient Temperature Surrounding Solid-State Devices

Do not overlook any sources of heat that might raise the ambient temperature at an enclosure containing solid-state devices. Consider heat sources such as power supplies, transformers, radiated heat, adjacent furnaces, and even sunlight.

Clean the parts of the control equipment by vacuuming or wiping with a dry cloth or soft brush. Use care not to damage delicate parts. Do not clean the equipment or components using compressed air because it may displace dirt, dust, or other debris into other parts or equipment, and the force may damage delicate components. Liquid cleaners, including spray cleaners, should not be used unless specified by the equipment manufacturer. This is because residues may cause damage or interfere with electrical or mechanical

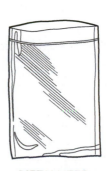 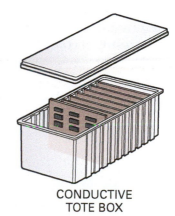

METALLIZED SHIELDING BAG   CONDUCTIVE TOTE BOX

Figure 30 ESD protection bag and tote box.

functions. If dust or dirt has accumulated on heat sinks and/or components that generate heat, it should be carefully removed because such accumulation can reduce the heat dissipation capability of these components and lead to premature failures. Ventilation passages should be open and cleaned of any dust and dirt.

Check the mechanical integrity of the equipment. Physical damage or broken parts can usually be found quickly by inspection. A general inspection should be made to look for loose, broken, missing, or badly worn parts. Components and wiring should be inspected for signs of overheating. The movement of mechanical parts, such as the armature and contacts of contactors, disconnect switches, circuit breaker operator mechanisms, and mechanical interlocks, should be checked for functional operation and freedom of motion. Any broken, deformed, or badly worn parts or assemblies should be replaced with manufacturer-recommended components. Any loose terminals or attaching hardware should be retightened securely to the torque specified by the equipment manufacturer.

The contacts of contactors should be checked for wear caused by arcing. During arcing, a small part of each contact melts, then vaporizes and is blown away. When the contacts are new, they are smooth and have a uniform silver color. As the device is used, the contacts become pitted, and the color may change to blue, brown, or black. These colors result from the normal formation of metal oxide on the contact surfaces and are not detrimental to contact life and performance. Contacts should be replaced under the following conditions:

- *Insufficient contact material* – When less than $1/64$" (0.4 mm) remains, replace the contacts.
- *Irregular surface wear* – This type of wear is normal. However, if a corner of the contact material is worn away and a contact may mate with the opposing contact support member, the contacts should be replaced. This condition can result in contact welding.

- *Pitting* – Under normal wear, contact pitting should be uniform. This condition occurs during arcing, as described above. The contacts should be replaced if the pitting becomes excessive and little contact material remains.
- *Curling of contact surfaces* – This condition results from severe service that produces high contact temperatures and causes separation of the contact material from the contact support member.

Preventive Maintenance

For programmable overloads, reduced-voltage starters, adjustable-frequency drives, and similar devices, refer to the user's manual/service manual supplied with the equipment for a description and schedule of specific preventive maintenance tasks that apply to the equipment.

Motor control mechanisms should not be lubricated unless recommended by the manufacturer. If lubrication is required, use only the recommended type and amount of lubricant. Remove any surplus lubricant to avoid the risk of establishing a tracking path across insulating surfaces, as well as the risk of excess lubricant migrating into areas that should not be lubricated.

5.0.0 Section Review

1. Shielded wire is used with solid-state devices in order to _____.

   a. prevent damage from static electricity
   b. reduce the amount of coupled noise
   c. maintain a physical separation between wires
   d. prevent damage to the wiring

2. Which of the following is a recommended method for cleaning solid-state devices?

   a. Compressed air
   b. Soap and water
   c. Dry cloth or soft brush
   d. Liquid spray cleaners

6.0.0 MOTOR CONTROL TROUBLESHOOTING

Objective

Describe how to troubleshoot motor controls.

a. Explain basic troubleshooting methods.
b. Describe the electrical troubleshooting methods used to check control circuits and devices.

Troubleshooting motor control systems requires a systematic approach. The process includes:

- Customer interface
- Physical examination of the system
- Basic system analysis with a clear understanding of the sequence of operation, interlocks, and limits
- Use of the equipment manufacturer's troubleshooting resources
- Troubleshooting the motor control circuit and components

6.1.0 Basic Troubleshooting Practices

The process of troubleshooting motor control circuit problems should always begin with obtaining all the information possible about the equipment problem or malfunction.

6.1.1 Customer Interface

Talking to and asking questions of a customer with first-hand knowledge of the problem is always recommended. This can provide valuable information on equipment operation that can aid in the troubleshooting process. Be sure to talk with and listen carefully to the machine or equipment operator(s) and not just to the person who requested the service call because the machine operator often knows the most about the problem. When interviewing the customer about a specific fault or problem, try to determine the following:

- The scope of the problem and when and how it began
- How often the problem occurs
- Whether the equipment has been worked on or tested recently
- Whether the equipment ever operated correctly

- Whether the problem occurred following a specific event, such as the addition of new electrical equipment, relocation of equipment, or new construction work

Before working on the motor control circuit equipment, always explain to the customer the course of action you intend to take and how it might affect the system, such as service interruptions or total system downtime. If the motor circuit with a problem is part of an emergency system (fire or backup), notify the police, fire department, or other off-site monitoring agencies or authorities that you will be working on the system.

6.1.2 Physical Examination

Problems can sometimes be identified by a visual inspection of the motor control circuit equipment, its wiring, and the adjacent area. For newly installed equipment, make sure it has been wired correctly. For nonoperating equipment, look for tripped circuit breakers, motor overloads, and signs of overheating. For operating equipment, check for odors of burning insulation, sounds of arcing, and signs that the unit is abnormally warm. For equipment with adjustable controls or mode selection switches, look for switches and/or controls that may be set to the wrong positions. For programmable equipment, check whether the selected operating and parameter inputs are correct.

6.1.3 Basic System Analysis

A proper diagnosis of a problem requires knowing what the motor control circuit and application equipment being driven by the motor should be doing when they are operating properly. If you are not familiar with how a system or unit should operate, study the manufacturer's service literature to familiarize yourself with the equipment modes and sequence of operation.

The second part of the diagnosis is to find out which symptoms are exhibited by the improperly operating system. This can be done by carefully listening to the customer's complaints and then analyzing the operation of the motor control circuit and application equipment. This usually means making electrical measurements at key points in the motor control circuit. Measured values can be compared with a set of previously recorded values or with values given in the manufacturer's service literature. When troubleshooting, avoid making assumptions about the cause of problems. Always secure as much information as possible before arriving at a diagnosis.

6.1.4 Manufacturer Resources

Today's electronic equipment can be very complex, and no one person is an expert on the wide variety of equipment being used to control motors. For this reason, a good troubleshooter should know how and where to locate information about the specific unit being serviced. This includes making phone calls to manufacturers' technical support centers, using manufacturers' website resources, and using manufacturers' user/service manuals. The user/service manuals supplied with solid-state programmable drive units and similar types of complex equipment contain invaluable operating and troubleshooting information, such as the following:

- Principles of operation
- Functional description of the circuitry
- Illustrations showing the location of each control or indicator and its function
- Illustrations and text describing menu-driven control panel displays
- Step-by-step procedures for the proper use and maintenance of the equipment
- Safety considerations and precautions that should be observed when operating the equipment
- Schematic diagrams, wiring diagrams, mechanical layouts, and parts lists for the specific unit

AFD units and similar microprocessor-controlled devices usually have built-in fault diagnostic circuits to aid in troubleshooting. When a fault occurs that interrupts the operation of the drive, these circuits give a visual indication that a fault trip has occurred. They also generate a fault code that can be selected for display on the unit control panel. When a fault occurs, look up the specific fault code displayed on the panel in a fault troubleshooting table in the user/service manual to help pinpoint the problem area (*Figure 31*).

6.1.5 Troubleshooting Motor Control Circuits and Components

A motor control circuit is a means of supplying power to and removing power from a motor. Most motor circuits consist of a combination of starting mechanisms, both automatic and manual. The simple motor control circuit shown in *Figure 32* is typical of most motor circuits. Troubleshooting electrical problems in motor control circuits may appear complex. However, it can be simplified if the electrical components are divided into smaller functional circuit areas based on the operations they perform. Motor control circuits can be divided into three functional circuit areas: power circuits, control circuits, and the load.

The power circuits, represented by the heavier lines shown in *Figure 32*, provide line power to the motor. Line voltages are usually 240VAC or 480VAC. As shown, the power circuits include the disconnecting means, the overcurrent protection devices (fuses or circuit breakers), the normally open contactor contacts, and the motor overload device. The load circuit is the AC motor.

The thinner lines represent the control circuit used in a magnetic-type starter. The control circuit is used to direct power to a magnetic contactor through the STOP/START station, thermal overload relay contacts, holding contacts, and the contactor coil. Note that the control circuit can contain several other types of pilot devices used to control the motor. The control circuit can be connected directly to the line, as shown in *Figure 32*, or it can be isolated from the power circuit by a step-down control transformer. Lower voltages, such as 24VAC, 120VAC, and 240VAC, are commonly used with a control transformer. The voltage rating for the coil on the motor starter indicates which control voltage is used.

Isolation to the faulty circuit area (power, control, or load) is based on an analysis of equipment operation and a process of elimination. Troubleshooting can be done in several ways. When troubleshooting a motor circuit when the motor is not working or has a problem, the contactor or motor starter is usually checked first. This is because it is the point where the incoming power, load, and control circuit are connected. Basic voltage readings should be taken at a contactor or motor starter to determine where the problem lies. The same basic procedure used to troubleshoot a motor starter works for contactors because a motor starter is a contactor with added overload protection.

The tightness of all terminals and busbar connections should be checked when troubleshooting control devices. Loose connections in the power circuit of contactors and motor starters cause overheating that can result in equipment malfunction or failure. Loose connections in the control circuit cause control malfunctions. Loose connections of grounding terminals can lead to electrical shock and cause electromagnetic interference.

Consider the following safety practices during troubleshooting and similar work on energized equipment:

- Use barriers or similar means to keep unqualified people safely away.
- Have an escape route.
- Always remain alert. Stop work if you are distracted by unrelated activities.
- Make sure that there is adequate lighting to work safely.
- Remove all conductive apparel (jewelry, metal headgear, etc.).

- Use insulated tools and equipment.
- Use protective shields if there is a possibility of exposure to energized parts.
- Have a definite plan and defined scope of work to perform.

Table 7A and *Table 7B* provide additional motor control troubleshooting information for motor control. Specific procedures for making electrical measurements and troubleshooting individual components in motor control circuits are provided later in this section. Be sure to consult the manufacturer's information for procedures specific to the application. Some general guidelines for isolating a fault in a motor control circuit are outlined in the following steps:

Step 1 Visually inspect the motor starter contactor and overload relay for signs of heat damage, arcing, or wear. Repair or replace any motor starter components that show visual signs of such damage.

FAULT CODES	FAULT	POSSIBLE CAUSES	CHECKING
F1	OVERCURRENT	FREQUENCY CONVERTER HAS MEASURED TOO HIGH A CURRENT IN THE MOTOR OUTPUT: – SUDDEN HEAVY LOAD INCREASE – SHORT CIRCUIT IN THE MOTOR CABLES – UNSUITABLE MOTOR	CHECK THE LOAD CHECK THE MOTOR SIZE CHECK THE CABLES
F2	OVERVOLTAGE	THE VOLTAGE OF THE INTERNAL DC-LINK OF THE FREQUENCY CONVERTER HAS EXCEEDED THE NOMINAL VOLTAGE BY 35%: – DECELERATION TIME IS TOO FAST – HIGH OVERVOLTAGE SPIKES AT UTILITY	ADJUST THE DECELERATION TIME
F3	GROUND FAULT	CURRENT MEASUREMENT HAS DETECTED THAT THE SUM OF THE MOTOR PHASE CURRENT IS NOT ZERO: – INSULATION FAILURE IN THE MOTOR OR THE CABLES	CHECK THE MOTOR CABLES
F4	INVERTER FAULT	FREQUENCY CONVERTER HAS DETECTED FAULTY OPERATION IN THE GATE DRIVERS OR IGBT BRIDGE: – INTERFERENCE FAULT – COMPONENT FAILURE	RESET THE FAULT AND RESTART AGAIN; IF THE FAULT OCCURS AGAIN, CONTACT YOUR DISTRIBUTOR
F5	CHARGING SWITCH	CHARGING SWITCH IS OPEN WHEN THE START COMMAND IS ACTIVE: – INTERFERENCE FAULT – COMPONENT FAILURE	RESET THE FAULT AND RESTART AGAIN; IF THE FAULT OCCURS AGAIN, CONTACT YOUR DISTRIBUTOR
F6	UNDERVOLTAGE	DC-BUS VOLTAGE HAS GONE BELOW 65% OF THE NOMINAL VOLTAGE: – MOST COMMON REASON IS FAILURE OF THE UTILITY SUPPLY	IN CASE OF TEMPORARY SUPPLY VOLTAGE BREAK, RESET THE FAULT AND START

Figure 31 Sample portion of a fault troubleshooting table.

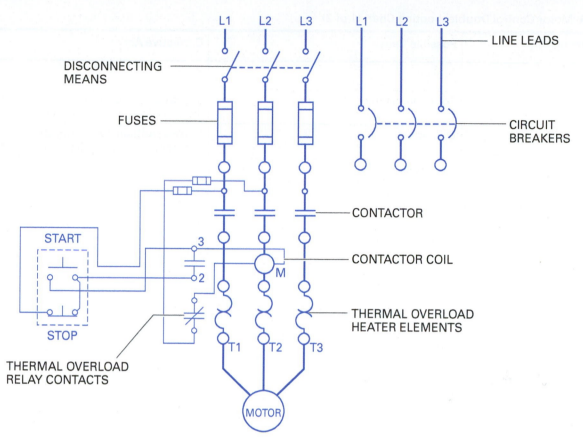

Figure 32 Basic motor control circuit.

Step 2 If there are no visual signs of damage, reset the overload relay and then press the START pushbutton to energize the motor starter contactor. If the related motor starts, observe the motor starter for several minutes. If the overload relay trips, troubleshoot the motor and related wiring downstream from the motor starter for causes of an overload.

Step 3 If the motor does not start after resetting the overload, check the voltage applied to the line side of the contactor. If the voltage reading is 0V or is not within 10% of the motor voltage rating, check the power wiring and components (fuses, circuit breaker, and disconnect switch) upstream of the motor starter.

Step 4 If voltage is applied to the line side of the contactor and it is at the correct level, press the START pushbutton to energize the contactor. If the contactor coil does not energize, troubleshoot the contactor coil to determine whether it is open. If the coil is not open, troubleshoot the contactor coil control circuit components and wiring for a problem.

Step 5 If the contactor coil energizes, check the voltage at the load side of the contactor. If the voltage reading is acceptable (within 10% of the motor voltage rating), the contacts are good. If there is no voltage reading or the reading is too low, turn the power off and replace the contactor contacts.

Step 6 If voltage is present at the load side of the contactor and it is at the correct level, check the output voltage from the overload relay. If the voltage reading is 0V, turn the power off and replace the overload relay. If the voltage reading is acceptable and the motor is not operating, troubleshoot the motor or power wiring downstream from the motor starter.

6.2.0 Universal Troubleshooting Procedures

There are several troubleshooting procedures that are common to most motor control circuits. These procedures often provide information that leads to detailed troubleshooting specific to the circuit. In many cases though, these common procedures reveal the specific source of the problem directly, leading to its resolution.

Malfunction	Possible Cause	Corrective Action
Constant chatter	Broken pole shader	Replace.
	Poor contact in control circuit	Improve contact or use holding circuit interlock (three-wire control).
	Low voltage	Correct voltage condition; check momentary voltage dip during starting.
Contactor welding or freezing	Abnormal inrush of current	Use larger contactor or check for grounds.
	Rapid jogging	Install larger device for jogging service.
	Insufficient contact pressure	Replace contact springs; check contact carrier for damage.
	Low voltage preventing magnet from sealing	Correct voltage condition; check momentary voltage dip during starting.
	Foreign matter preventing contacts from closing	Clean contacts with approved solvent.
	Short circuit	Remove fault and check to be sure fuse or circuit breaker size is correct.
Short contact life or tip overheating	Filing or dressing	Do not file silver-faced contacts; rough spots or discoloration will not harm contacts.
	Interrupt excessively high	Install larger device or check currents for grounds, shorts, or excessive motor currents; use silver-faced contacts.
	Excessive jogging	Install larger device rated for jogging.
	Weak contact pressure	Adjust or replace contact springs.
	Dirt or foreign matter on contact surfaces	Clean contacts with approved solvent.
	Short circuit	Remove fault and check for proper fuse or circuit breaker size.
	Loose connection	Clean and tighten.
	Sustained overload	Install larger device or check for excessive load current.
Coil overheating	Overvoltage or high ambient temperature	Check application and circuit.
	Incorrect coil	Check rating and if incorrect, replace with proper coil.
	Shorted turns caused by mechanical damage or corrosion	Replace coil.
	Undervoltage; failure of magnet to seal in	Correct system voltage.
	Dirt or rust on pole faces increasing air gap	Clean pole faces.
Overload relays tripping	Sustained overload	Check for grounds, shorts, or excessive currents.
	Loose connection on load wires	Clean and tighten.
	Incorrect heater	Replace relay with correct size heater unit.

Malfunction	Possible Cause	Corrective Action
Failure to trip causing motor burnout	Mechanical binding, dirt, corrosion, etc.	Clean or replace.
	Wrong heater or heaters omitted, and jumper wires used.	Check ratings; apply proper heater.
	Motor and relay at different temperatures.	Adjust relay rating accordingly.
	Wrong calibration or improper calibration adjustment.	Consult factory.
Magnetic and mechanical parts inoperative	Broken shading coil	Replace shading coil.
Noisy magnet humming	Magnet faces not mating	Replace magnet assembly; realign.
	Dirt or rust on magnet faces	Clean and realign.
	Low voltage	Check system voltage and voltage dips during starting.
Failure to pick up and seal	Low voltage	Check system voltage and voltage dips during starting.
	Coil open or shorted	Replace.
	Wrong coil	Check coil number.
	Mechanical obstruction	With power off, check for free movement of contact and armature assembly.
Failure to drop out	Gummy substance on pole faces	Clean with solvent.
	Voltage not removed	Check coil circuit.
	Worn or rusted parts causing binding	Replace parts.
	Residual magnetism due to lack of air gap in magnet path	Replace worn magnet parts.

6.2.1 Input Voltage/Voltage Imbalance Measurements

All motors and related motor control equipment are designed to operate within a specific range of system voltages, including a safety factor that is typically 10%. This safety factor is added to compensate for temporary supply voltage fluctuations that might occur. Continuous operation of a motor outside the intended range of voltages can damage the motor. Insufficient operating voltage can cause overload/overheating and possible failure of motors and other devices.

For a 230V rated motor this means the supply voltage should remain in the range of 207V to 253V (230V ±10%). If the motor is rated for 208V/230V, the supply voltage must be between 187V (10% below 208V) and 253V (10% above 230V).

Voltage imbalance is very important when working with three-phase equipment. A small imbalance in phase-to-phase voltage can result in a much greater current imbalance. With a current imbalance, more heat is generated in the motor windings. Both current and heat can cause nuisance overload trips and may cause motor failure. For this reason, the imbalance between any two legs of the voltage applied to a three-phase motor or system should not exceed 2%. If a voltage imbalance of more than 2% exists at the input to the equipment, correct the problem in the building or utility power distribution system before operating the equipment. *Figure 33* shows an example of how the amount of voltage imbalance is determined in a three-phase system.

A current imbalance may occur without a voltage imbalance. If the resistance of one leg is higher or lower, Ohm's law will predict a corresponding imbalance in currents. This could be because of a short between adjacent windings within the motor, or if an electrical terminal or contact becomes loose or corroded, resulting in a higher resistance in the leg. Current imbalance in a three-phase system is determined in the same way as voltage imbalance. Current imbalance in any one leg of a three-phase system should not exceed 10%.

Electrical Troubleshooting

Most electrical troubleshooting in motor control circuits can be done using a multimeter and an AC clamp-on ammeter. Some types of clamp-on instruments, like the one shown here, incorporate the functions of a clamp-on ammeter and multimeter into one device, allowing it to be used to measure current, voltage, and resistance.

Figure Credit:
Courtesy of Extech Instruments, a FLIR Company

6.2.2   Fuse Checks

Fuses or circuit breakers are normally the first components checked when a motor is totally inoperative. One way to test a fuse is by measuring continuity (*Figure 34*). To check the fuses, open the disconnect switch, then remove the fuses using an insulated fuse puller. Test the fuses for continuity using a volt-ohm-milliammeter (VOM) or digital multimeter (DMM) set to measure resistance. If the meter reads 0 ohms, indicating continuity, the fuse is usually good. If an open circuit (infinite resistance) exists across the fuse, it has failed. A blown fuse is usually caused by some abnormal overload condition, such as a short circuit within the equipment or an overloaded motor. These problems may not always occur when the motor starts, but they may occur after a significant amount of run time generates heat, or only when the load on the motor is increased. Replacing a blown fuse without locating and correcting the cause can result in damage to the motor or control equipment.

Fuses can also be tested with the circuit energized. Set the VOM/DMM to measure AC voltage on a range that is higher than the highest voltage expected. Place one VOM/DMM test lead on the line side of the L_1 fuse. Touch the other test lead to the load side of the L_2 fuse. If voltage is measured, the L_2 fuse is good; if not, the fuse is open. Repeat this procedure so that all fuses are measured with one test lead on the line side and the other test lead on the load side of a different fuse. This method tests one fuse at a time. If the measurement is performed with both test leads on the load side of the fuses, and the VOM/DMM shows no reading, you know that a fuse is blown but not which one.

> **CAUTION**
> When checking fuses, false readings may occur if there is another component, such as a control transformer, that may provide a false indication that the fuse is good due to the component's feedback.

6.2.3   Circuit Breaker Checks

Circuit breaker checks will also be necessary at times. To check a circuit breaker, set the circuit breaker to the OFF position. If required, and with proper arc flash protection, remove any panel that covers the circuit breaker to expose the body of the breaker and the wires connected to its terminals. Set up a VOM/DMM to measure AC voltage on a range that is higher than the highest voltage expected. Measure the voltage applied to the circuit breaker line terminals in the following manner (*Figure 35*):

- A to neutral or ground (single-pole breaker)
- A to B (two-pole breaker)
- A to B, B to C, and C to A (three-pole breaker)

After making these measurements, verify that the breaker is closed by first setting it to the OFF position, then setting it to the ON position. Measure the voltage at the circuit breaker load terminals:

- A1 to neutral or ground (single-pole breaker)
- A1 to B1 (two-pole breaker)
- A1 to B1, B1 to C1, and C1 to A1 (three-pole breaker)

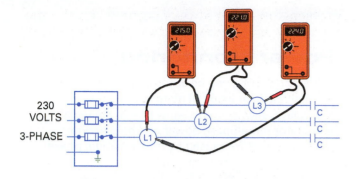

1)
PHASE	MEASURED READING
L1	215V
L2	221V
L3	224V

2) $\dfrac{\text{AVERAGE}}{\text{VOLTAGE}} = \dfrac{215 + 221 + 224}{3} = 220V$

3) INDIVIDUAL PHASE IMBALANCE FROM AVERAGE
   - L1 TO L2 = 220 − 215 = 5V
   - L2 TO L3 = 221 − 220 = 1V
   - L3 TO L1 = 224 − 220 = 4V

4) 5V = MAXIMUM IMBALANCE

5) $\%\ \text{IMBALANCE} = \dfrac{\text{MAXIMUM IMBALANCE}}{\text{AVERAGE VOLTAGE}} \times 100$

   $\%\ \text{IMBALANCE} = \dfrac{5V}{220V} \times 100 = 2.27\%$ (OUT OF BALANCE)

> MAXIMUM VOLTAGE IMBALANCE BETWEEN ANY TWO LEGS MUST NOT EXCEED 2%.

(A) CALCULATING VOLTAGE IMBALANCE

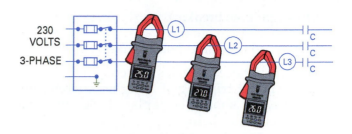

1)
PHASE	MEASURED READING
L1	25A
L2	27A
L3	26A

2) $\dfrac{\text{AVERAGE}}{\text{CURRENT}} = \dfrac{25 + 27 + 26}{3} = 26A$

3) INDIVIDUAL PHASE IMBALANCE FROM AVERAGE
   - L1 TO L2 = 26 − 25 = 1A
   - L2 TO L3 = 27 − 26 = 1A
   - L3 TO L1 = 26 − 26 = 0A

4) 1A = MAXIMUM IMBALANCE

5) $\%\ \text{IMBALANCE} = \dfrac{\text{MAXIMUM IMBALANCE}}{\text{AVERAGE CURRENT}} \times 100$

   $\%\ \text{IMBALANCE} = \dfrac{1A}{26A} \times 100 = 3.8\%$ (IN BALANCE)

> MAXIMUM CURRENT IMBALANCE BETWEEN ANY TWO LEGS MUST NOT EXCEED 10%.

(B) CALCULATING CURRENT IMBALANCE

Figure 33 Three-phase input voltage checks.

Troubleshooting

When troubleshooting, observe the following list of guidelines:

- Do practice safe testing procedures.
- Do change only one component at a time.
- Do have a complete understanding of the machine's function.
- Do make sure that test equipment is operational and calibrated before using it for troubleshooting.
- Do use the correct test equipment with the proper rating for the circuit being tested.
- Do not change wiring connections in circuits that have worked in the past.

The measured line and load voltages should be the same if the breaker contacts are closed. If the load side voltage is significantly lower than that measured at the line side of the circuit breaker, visually inspect the circuit breaker for loose wires and terminals or signs of overheating that might cause voltage drop. If no cause is found, the circuit breaker should be replaced. Remember however, that significant voltage drop is not always evident until a load is placed on the circuit breaker.

If the circuit breaker shows signs of overheating or trips when power is applied to the equipment, reset it, then check the current flow through the breaker using an AC clamp-on ammeter. Set up the AC clamp-on ammeter to measure AC current on a range that is higher than the highest current expected. Check the ampere rating marked on the breaker. It is usually stamped on the breaker lever or body. One wire at a time, measure the current flow in the wires connected to the circuit breaker output terminals:

- A1 (single-pole breaker)
- A1, B1 (two-pole breaker)
- A1, B1, and C1 (three-pole breaker)

If the circuit breaker trips at a current below its rating or is not tripping at a higher current, the circuit breaker should be replaced. Be sure that the breaker is not being tripped because of high ambient temperature or other causes at the load or in the feeder.

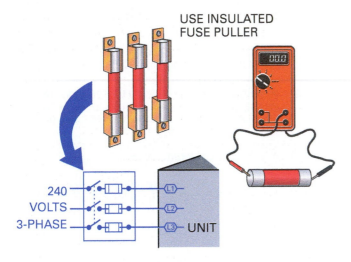

ZERO Ω READING = GOOD FUSE

MEASURABLE OR INFINITE RESISTANCE READING = BAD FUSE

CONTINUITY CHECK

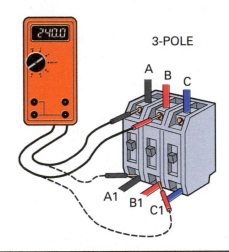

MEASURED INPUT AND OUTPUT VOLTAGES SHOULD BE THE SAME.

CIRCUIT BREAKER VOLTAGE CHECK

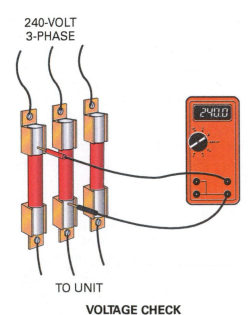

VOLTAGE CHECK

Figure 34 Fuse checks.

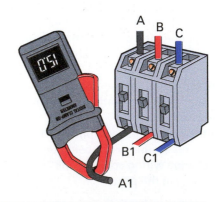

CIRCUIT BREAKER CURRENT CHECK

Figure 35 Circuit breaker checks.

Digital Logging Meters

Digital logging meters measure various values including voltage, current, resistance, frequency, and temperature. The logging meter shown here can be connected to a PC using special software for extended system monitoring and data storage. These devices provide valuable information when a problem occurs sporadically, making it difficult to diagnose.

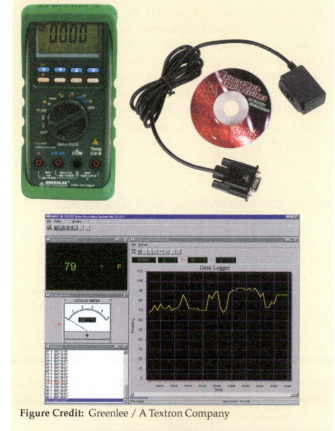

Figure Credit: Greenlee / A Textron Company

6.2.4 Troubleshooting Control Circuits

After the source of an electrical problem has been isolated to a control circuit and the control circuit voltage is known to be good, a series of voltage measurements can be made across the electrical devices in the control circuit to find the faulty device. As shown in *Figure 36*, the measurements can start from the line or control voltage side of the circuit and move toward the load device, typically a contactor coil or relay coil. Measurements are made until either no voltage is observed or until the voltage has been measured across all the devices in the circuit. Note that when there are many devices in the circuit being tested, the measurements can be made by starting at the mid-

point in the circuit (divide-in-half method) and then working toward either the source of control voltage or the load device, depending on whether the expected voltage was measured at the mid-point. As a result of making the voltage measurements, one of the two situations described below should exist.

- *Zero voltage* – At some point within the circuit, no voltage will be indicated on the VOM/DMM. This pinpoints an open set of switch or relay contacts between the present measurement point and the previous measurement point. Part (A) in *Figure 36* shows an example of this situation, in which the contacts of the low-pressure switch are open, preventing the contactor coil (C) from energizing. If the open is caused by a set of contactor or relay contacts, find out if the related contactor or relay coil is not being energized or is defective. Part (B) in *Figure 36* shows an example of this situation, in which the CR contacts in the control circuit are open, preventing the contactor (C) from energizing. These contacts close when the control relay coil (CR) is energized. Before assuming that the problem is caused by the open CR contacts, first troubleshoot the control circuit containing the CR relay coil to find out if the coil is energized or de-energized. If it is de-energized, further troubleshoot its control circuit to find out why. For example, if the contacts of the thermostat Cool switch are open, the CR relay coil will not be energized.
- *Voltage present at coil* – If voltage is measured at the contactor coil, and the contactor is not energized, the contactor coil is most likely defective. Turn off the power to the circuit, then disconnect the coil from the circuit and test it as described later in this section. Part (C) in *Figure 36* shows an example of a situation in which 24V is applied to the contactor coil (C), but it is not energized. In this case, the contactor coil is probably open.

6.2.5 Control Transformer Checks

Control transformers are checked by measuring the voltages across the secondary and primary windings (*Figure 37*). Typically, the secondary winding is measured first. The VOM/DMM should be set up to measure AC voltage on a range that is higher than the control voltage expected. If the voltage measured across the secondary winding is within 10% of the required voltage, the transformer is good. If no voltage is measured at the secondary winding, the voltage across the primary winding must be measured. Also check the secondary fuse (if provided) to see if it is blown.

If the voltage measured at the primary winding is within ±10% of the required voltage, the transformer is probably defective. This can be confirmed by performing a continuity check of the transformer primary and secondary windings with power off. If no voltage or low voltage is measured across the primary winding, the power supply voltage to the equipment should be checked. If the power supply voltage is okay, troubleshoot the circuit wiring between the power supply and control transformer primary winding.

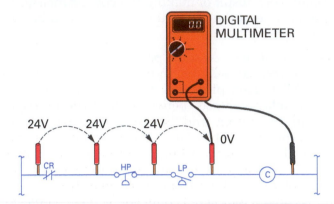

(A) OPEN LOW-PRESSURE SWITCH CONTACTS

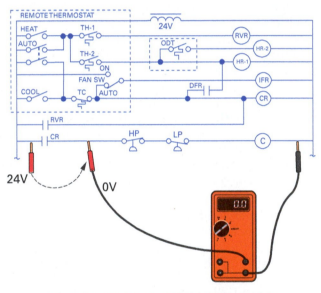

(B) OPEN CR RELAY CONTACTS – CHECK RELATED COIL CONTROL CIRCUIT

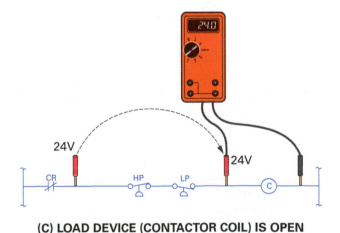

(C) LOAD DEVICE (CONTACTOR COIL) IS OPEN

Figure 36 Isolating to a faulty control circuit component.

6.2.6 Contactor/Relay Coil Resistance Checks

When the correct voltage is being applied to the coil of a contactor or relay and it is not energized, the device is probably faulty. After the coil of a contactor or relay has been identified as the probable cause of an electrical problem, it should be

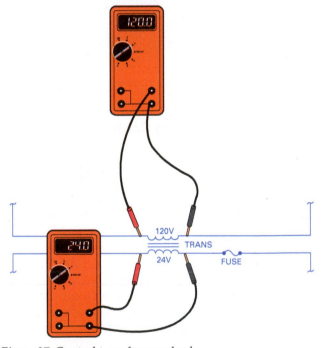

Figure 37 Control transformer checks.

tested. The best way to test a coil is by measuring the resistance across the terminals of the coil. Before measuring resistance, make sure to electrically isolate the coil from the remainder of the circuit by removing all power to the circuit and disconnecting at least one lead of the coil. This is important in order to get an accurate resistance reading. Otherwise, the meter may read the resistance of other components that are connected in parallel with the coil being measured. As shown in *Figure 38*, a reading of zero ohms indicates a shorted coil and a reading of infinite resistance indicates an open coil. In either case, replace the contactor or relay (or the coil, if replaceable). When the VOM/DMM indicates a measurable resistance, it usually indicates that the coil is good. If a low resistance is measured, connect one of the meter probes to ground or to the unit frame. Touch the other probe to each coil terminal. If a resistance is measured from either terminal to ground, replace the contactor or relay.

Measuring Coil Resistance

The actual resistance measured for a contactor or relay coil can vary widely depending on the device. Ideally, the exact resistance value for the device can be found in the manufacturer's service literature. Another way to judge whether the resistance reading is good is by comparing the resistance of the coil being tested with that of a similar coil that is known to be good.

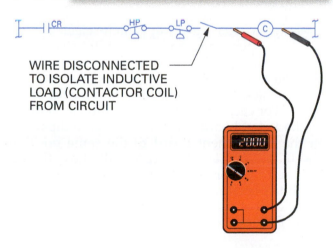

WIRE DISCONNECTED TO ISOLATE INDUCTIVE LOAD (CONTACTOR COIL) FROM CIRCUIT

MEASURABLE RESISTANCE = GOOD LOAD
ZERO RESISTANCE = SHORTED LOAD
INFINITE RESISTANCE = OPEN LOAD

Figure 38 Coil resistance checks.

6.2.7 Relay/Contactor Contact Checks

After a set of contactor or relay contacts has been identified as the probable cause of an electrical problem, the contacts can be tested to confirm their condition. With the power to the circuit turned off, the contacts can be tested by making a continuity measurement to determine whether the contacts are open or closed (*Figure 39*). If the contacts are open, the VOM/DMM indicates an infinite resistance reading. If the contacts are closed, the VOM/DMM indicates a short (0 ohms).

When testing relay contacts with the power off, remember that only the normally open or normally closed position of the contacts is being tested. It may have no bearing on the status of the contacts when the system is powered up.

When working with contactor/relay contacts, remember the following:

- Contacts that open when the contactor/relay is energized are called *normally closed (NC)* contacts.
- Contacts that close when the contactor/relay is energized are called *normally open (NO)* contacts.

6.2.8 Troubleshooting Circuit Boards

Solid-state motor control devices, such as adjustable-frequency drives, reduced-voltage starters, and similar equipment, contain printed circuit boards. When a board is installed and operating, it is unlikely to fail unless the failure is caused by some outside influence. Because of their complexity, many boards are difficult to troubleshoot and repair in the field. It often takes special automated test equipment at the manufacturer's plant to accomplish this task. For this reason, if a board fails, it is usually replaced with a new one. If the board is still under warranty, the manufacturer will generally replace the board at no

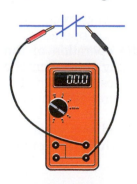

ZERO OHMS = CLOSED CONTACTS

INFINITE RESISTANCE = OPEN CONTACTS

Figure 39 Checking the continuity of contactor/relay contacts.

cost. Do not try troubleshooting components on a board unless instructed to do so. After a board has been worked on, the manufacturer has no way of knowing whether the board was damaged through a manufacturing error or because it has been altered. In these cases, many manufacturers will not replace the board without payment.

Checking for Proper Contact Closure

The contacts of an energized contactor can be checked for proper closure under loaded conditions by measuring for a voltage drop across the related line and load terminals of the contactor with a multimeter. The multimeter should indicate 0V across each set of power contacts if they are closing properly. If the meter indicates a voltage, this means a voltage drop is occurring across the contacts being measured because of a poor or high-resistance connection.

Most boards are expensive, so be certain that the board is defective before replacing it. A board should be replaced only when all the input signals to the board are known to be good, but the board fails to generate the proper output signal(s). Many boards contain a microprocessor that controls all sequences of operation for the drive control system. To determine the condition (functional or defective) of the board, it is necessary to check for the presence of the proper input signals. It is also necessary to verify that the board generates the proper output signal(s) in response to the input signal(s). Some boards have test points where measurements of key input and output signals can be made. The schematic or wiring diagram will identify the signal that should be available at each point.

To diagnose a board's condition correctly, you must understand the sequence of operation. If you are not familiar with the operation of the board used in the system, study the schematic and sequence of operation described in the manufacturer's service literature. When you understand what the system and/or board should be doing, you must find out what the board is actually doing, or which symptoms

are present in the unit. To aid in this analysis, many microprocessor-controlled boards have built-in diagnostic circuits that can run a check of motor functions and parameters and report their status. When troubleshooting, always use any available built-in diagnostic features and follow the manufacturer's related troubleshooting instructions.

Before replacing a board that appears to have failed, try to determine if an external cause might account for the problem. There are a variety of possibilities:

- Electrical noise (EMI/RFI) from communications equipment, switching power supplies, and other sources will cause microprocessors to fail or operate erratically. The source of interference should be removed, or the board should be shielded.
- As explained earlier, microprocessors and other integrated circuit chips can be damaged by static electricity from sources such as lightning or people. When handling circuit boards, avoid touching the components, printed circuits, and connector pins. Always ground yourself before touching a board. Store unused boards properly in electrostatically protected packaging.
- Voltage surges and excessive voltage can also cause board failures. Make sure that the applied voltage is within allowable limits and that the unit (and therefore the board) is properly grounded.
- Excessive heat can damage the board. The enclosure in which the board is installed should not be altered in any way that will restrict airflow to the board.

Many boards have dipswitch-selectable options or operating modes used to select site-specific parameters. When replacing a board with dipswitch-selectable options, make sure to set the dipswitch(es) to the correct position for the application. For each dipswitch, determine its position on the failed board, then set the same dipswitch on the replacement board to the same position. The manufacturer's literature will show the default settings for the dipswitches. Do not use a graphite pencil to set a dipswitch—it can deposit particles that may short the switch or cause unpredictable action.

6.0.0 Section Review

1. The difference between motor starters and contactors is _____.

    a. motor starters have ventilating fans to prevent overheating
    b. motor starters provide additional overload protection
    c. contactors have built-in rectifier circuits
    d. the contacts on motor starters have gold-plated contacts

2. The voltage imbalance between any two legs of a three-phase voltage source should NOT exceed _____.

    a. 2%
    b. 5%
    c. 7%
    d. 10%

1. Which of the following statements is *true* of solid-state relays?

   a. They have a shorter life than electromechanical relays.
   b. They have higher reliability than electromechanical relays.
   c. They have slower switching speeds than electromechanical relays.
   d. They can be operated at higher temperatures than electromechanical relays.

2. Which of the following is a characteristic of SSRs, but NOT of EMRs?

   a. Zero current turn-off
   b. Surge current
   c. Coil voltage
   d. Dropout time

3. When connecting a solid-state relay for three-wire control of a load, what is typically used to seal in or provide holding for the START pushbutton after it has been pressed?

   a. A triac
   b. An auxiliary solid-state relay contact
   c. An SCR
   d. An insulated gate bipolar transistor (IGBT)

4. The trip current setting for a solid-state protective relay (overcurrent relay) is based on the _____.

   a. number of loops on the current transformer
   b. motor FLA, service factor, and number of current loops
   c. motor FLA
   d. motor service factor

5. The programmable threshold that determines how long a solid-state overload relay will allow an undercurrent condition to exist before it trips is known as the _____.

   a. undercurrent trip delay
   b. undervoltage trip point
   c. ground fault current trip
   d. undercurrent trip point

6. A timing relay that provides the timed delay after the relay is energized is called a(n) _____.

   a. delay-on-break relay
   b. on-delay relay
   c. timed-closed relay
   d. off-delay relay

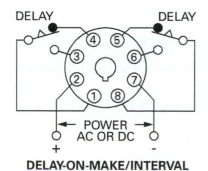

DELAY-ON-MAKE/INTERVAL

ON-DELAY	PIN		
TIME POWER	(2-7)		
DELAYED NC CONTACTS	(1-4) (8-5)		OFF ON
DELAYED NO CONTACTS	(1-3) (8-6)	ON	OFF
ON INDICATOR		LIGHTED	
OFF INDICATOR			LIGHTED
		←SET TIME→	

DELAY-ON-MAKE (ON DELAY)

INTERVAL	PIN		
TIME POWER	(2-7)	ON	OFF
DELAYED NC CONTACTS	(1-4) (8-5)	OFF ON	
DELAYED NO CONTACTS	(1-3) (8-6)	ON	OFF
ON INDICATOR			LIGHTED
OFF INDICATOR		LIGHTED	
		←SET TIME→	

INTERVAL

Figure RQ01

7. For the 8-pin timing relay shown in *Figure RQ01*, the delayed normally closed contacts are _____.

   a. pins 1–3 and 8–6
   b. pins 1–4 and 8–5
   c. pins 2–7
   d. pins 1–3 and 8–5

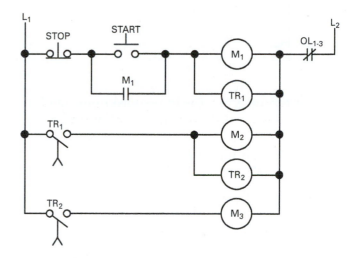

Figure RQ02

8. In *Figure RQ02*, the contacts of timing relays TR$_1$ and TR$_2$ are _____.

   a. normally open, timed-closed (NOTC) contacts
   b. normally closed, timed-open (NCTO) contacts
   c. normally open, timed-open (NOTO) contacts
   d. normally closed, timed-closed (NCTC) contacts

9. Which method of reduced-voltage starting provides for adjustment of the motor starting torque and current using taps?

   a. Solid state
   b. Wye-delta
   c. Resistive
   d. Autotransformer

10. The most common form of reduced-voltage soft-starting used with solid-state reduced-voltage controllers is _____.

   a. kick start
   b. current limit
   c. ramp start
   d. quick start

11. The NEMA classification for the form of adjustable-frequency drive with an AC input and an AC output that uses an intermediate DC conversion is _____.

    a. FA
    b. FB
    c. FC
    d. FD

12. Which of the following *best* describes the motor torque when the horsepower output of the motor varies directly with the speed of the motor?

    a. Constant torque
    b. Variable torque
    c. Constant horsepower
    d. Variable horsepower

13. In most applications, the motors used with AFDs are standard NEMA _____.

    a. Design A, Design B, or Design C
    b. Design A, Design B, or Design E
    c. Design B, Design D, or Design F
    d. Design C, Design D, or Design E

14. Parasitic voltage damage in motors powered by AFDs can be prevented using _____.

    a. nonmetallic shims
    b. AFD output filtering
    c. insulated motor housings
    d. electrically continuous shaft couplings

15. A method of braking in which a DC voltage is applied to the stator winding of an AC motor is called _____.

    a. regenerative braking
    b. friction braking
    c. dynamic braking
    d. DC injection braking

16. A method of braking an AC motor that involves the use of braking resistors is called _____.

    a. regenerative braking
    b. friction braking
    c. dynamic braking
    d. DC injection braking

17. To clean electronic circuit components, use _____.

    a. compressed air
    b. a vacuum
    c. a damp cloth
    d. spray cleaners

18. In any motor control circuit, the circuit load is the _____.

    a. motor
    b. relay contacts
    c. AFD
    d. control panel

19. Current imbalance in any leg of a three-phase power source must be within _____.

    a. 2%
    b. 5%
    c. 10%
    d. 25%

20. When testing a circuit breaker, the voltage readings on the line and load terminals with the breaker contacts closed should _____.

    a. be above 0 VAC
    b. read 120 VAC
    c. be the same
    d. have an imbalance of less than 2%

1. List four reasons for using solid-state relays (SSRs) instead of electromechanical relays.

   _____

2. An SSR that uses a light-emitting diode and photo detector to couple the switching command to the output circuit is called a(n) _____.

3. The inputs of two or more single-output SSRs can be connected _____ or _____ to obtain multiple-switched outputs.

4. True or False? Solid-state overload relays (SSOLRs) are designed to protect the motor and control devices against short circuits.

   _____

5. True or False? One reason for using reduced-voltage starting motor control is to reduce the current drawn from utility lines by the across-the-line start of large motors.

   _____

6. The autotransformer reduced-voltage starter motor control that will cause the least amount of electrical interference to the associated electrical distribution system uses the _____ method of voltage transition.

7. True or False? An open-circuit transition type of wye-delta reduced-voltage starting motor control uses a contactor to connect a resistor bank into the motor winding circuit during the start-to-run period.

   _____

8. Solid-state, reduced-voltage starting motor controllers are commonly called _____.

9. In a solid-state, reduced-voltage motor control, the effective voltage delivered to the motor is varied from zero to full voltage by controlling the precise time that the _____.

10. A metallized shielded storage bag should be used to protect microprocessors and other integrated circuit chips from _____.

11. True or False? Never clean motor control circuit components using compressed air.

    _____

12. An AFD controller converts three-phase input power to an adjustable _____ output for application in an AC motor.

13. List three common types of adjustable speed loads.

    _____

14. Name four different methods for motor braking.

    _____

15. True or False? AFD units and other microprocessor-controlled devices normally have built-in fault diagnostic circuits used to aid in troubleshooting.

    _____

Trade Terms Introduced in This Module

Base speed: The shaft speed at which the motor will develop rated horsepower at rated load and voltage, as stated on a motor nameplate.

Carrier frequency: In a pulse-width modulated (PWM) drive, the frequency at which the IGBT can be switched on or off to create an output waveform that mimics a sine wave of desired voltage and frequency. Carrier frequencies are seldom less than 500Hz nor more than 20,000Hz; but need to be at least 10 times the maximum drive output frequency.

Closed-circuit transition: A method of reduced-voltage starting in which the motor being controlled is never removed from the source of voltage while moving from one voltage level to another.

Dashpot: A device that uses a gas or fluid to absorb energy from or retard the movement of the moving parts of a relay, circuit breaker, or other electrical or mechanical device.

Fan Affinity Laws: The mathematical relationship between fan speed, airflow rate, power used, and torque. Airflow is directly proportional to speed. Torque is proportional to the square of speed. Power used is proportional to the cube of speed. For example, decreasing fan speed by 20% decreases the power used by about 50%.

Heat sink: A metal mounting base that dissipates the heat of solid-state components mounted on it by radiating the heat into the surrounding atmosphere.

Insulated gate bipolar transistors (IGBTs): A type of transistor that has low losses and low gate drive requirements. This allows it to be operated at higher switching frequencies.

Open-circuit transition: A method of reduced-voltage starting in which the motor being controlled may be disconnected temporarily from the source of voltage while moving from one voltage level to another.

Slip: The speed difference between the synchronous speed and the base speed. Induction motors develop their torque because of the slip. The amount of slip depends on motor design and motor load.

Solid-state overload relay (SSOLR): An electronic device that provides overload protection for electrical equipment.

Solid-state relay (SSR): A switching device that has no contacts and switches entirely by electronic means.

Thermal conductivity: The ability of a material or substance to transfer heat away from a given point of contact.

Thyristors: Bi-stable semiconductor devices that can be switched off and on. Thyristors turn on with a quick pulse of control current. They turn off only when the working current is interrupted elsewhere in the circuit. SCRs and triacs are forms of thyristors.

Additional Resources

This module presents thorough resources for task training. The following resource material is suggested for further study.

Application Guide for AC Adjustable Speed Drive Systems (PDF), Arlington, VA: National Electrical Manufacturers Association. Available for download as document NEMA ICS 7.2-2015 (or the most recent edition) from: **www.nema.org**.

Motor Control Technology for Industrial Maintenance, Thomas E. Kissell. New York, NY: Pearson Education, Inc.

National Electrical Code® Handbook, Latest Edition. Quincy, MA: National Fire Protection Association.

NFPA 70B, *Recommended Practice for Electrical Equipment Maintenance*. Quincy, MA: National Fire Protection Association.

Softstarter Handbook, publication 1SFC132060M0201. Available for download from ABB at: **https://library.e.abb.com/public/6b4e1a3530814df0c12579bb0030e58b/1SFC132060M0201.pdf**.

Figure Credits

Section Review Answer Key

SECTION 1.0.0

Answer	Section Reference	Objective
1. b	1.1.1	1a
2. a	1.2.0	1b

SECTION 2.0.0

Answer	Section Reference	Objective
1. d	2.1.1	2a
2. c	2.2.0	2b

SECTION 3.0.0

Answer	Section Reference	Objective
1. b	3.1.2	3a
2. d	3.2.3	3b

SECTION 4.0.0

Answer	Section Reference	Objective
1. b	4.1.0	4a
2. c	4.2.2	4b
3. d	4.3.0	4c

SECTION 5.0.0

Answer	Section Reference	Objective
1. b	5.1.0	5a
2. c	5.2.0	5b

SECTION 6.0.0

Answer	Section Reference	Objective
1. b	6.1.5	6a
2. a	6.2.1	6b

NCCER CURRICULA — USER UPDATE

NCCER makes every effort to keep its textbooks up-to-date and free of technical errors. We appreciate your help in this process. If you find an error, a typographical mistake, or an inaccuracy in NCCER's curricula, please fill out this form (or a photocopy), or complete the online form at **www.nccer.org/olf**. Be sure to include the exact module ID number, page number, a detailed description, and your recommended correction. Your input will be brought to the attention of the Authoring Team. Thank you for your assistance.

Instructors – If you have an idea for improving this textbook, or have found that additional materials were necessary to teach this module effectively, please let us know so that we may present your suggestions to the Authoring Team.

NCCER Product Development and Revision
13614 Progress Blvd., Alachua, FL 32615

Email: curriculum@nccer.org
Online: www.nccer.org/olf

❏ Trainee Guide   ❏ Lesson Plans   ❏ Exam   ❏ PowerPoints   Other _____

Craft / Level: _____ Copyright Date: _____

Module ID Number / Title: _____

Section Number(s): _____

Description: _____

_____

_____

_____

Recommended Correction: _____

_____

_____

_____

_____

Your Name: _____

Address: _____

_____

Email: _____ Phone: _____

This page is intentionally left blank.

HVAC Controls

OVERVIEW

Heating, ventilation, and air conditioning (HVAC) systems are among the electrically powered and controlled systems that electricians will encounter, especially in residential and commercial construction. During installation, electricians will be called upon to provide power and control connections to the various components of these systems. For that reason, it is important that electricians develop a basic understanding of HVAC systems and their components.

Module 26408-20

Objectives

When you have completed this module, you will be able to do the following:

1. Describe the operating principles and major components of HVAC systems.
   a. Describe the basic principles of heating and ventilation.
   b. Describe the basic principles and components of comfort air conditioning systems.
2. Identify the types of thermostats and their uses.
   a. State the operating principles of thermostats.
   b. Install different types of thermostats.
3. Identify and describe HVAC control systems and devices.
   a. Identify and describe controls used in comfort cooling systems.
   b. Identify and describe furnace controls.
   c. Identify and describe heat pump defrost controls.
   d. Describe digital controls used in HVAC systems.
   e. Analyze the operating sequences of various HVAC control systems.
   f. Describe troubleshooting methods used for HVAC systems.
4. Identify the *NEC*® requirements that apply to HVAC systems.
   a. Identify the *NEC*® requirements that apply to HVAC controls.
   b. Identify the *NEC*® requirements that apply to HVAC equipment.

Performance Tasks

Under the supervision of the instructor, you should be able to do the following:

1. Identify various types of thermostats and describe their operation and uses.
2. Install a thermostat and hook it up using the standard coding system for thermostat wiring.
3. Check and adjust a thermostat, including the heat anticipator setting and indicator adjustment.

Trade Terms

Analog-to-digital converter
Automatic-changeover thermostat
Bimetal
Compressor
Condenser
Cooling compensator
Deadband
Differential
Evaporator
Expansion device

Heat transfer
Invar®
Mechanical refrigeration
Refrigerant
Refrigeration cycle
Sub-base
Subcooling
Superheat
Thermostat

Industry Recognized Credentials

If you are training through an NCCER-accredited sponsor, you may be eligible for credentials from NCCER's Registry. The ID number for this module is 26408-20. Note that this module may have been used in other NCCER curricula and may apply to other level completions. Contact NCCER's Registry at 888.622.3720 or go to **www.nccer.org** for more information.

Contents

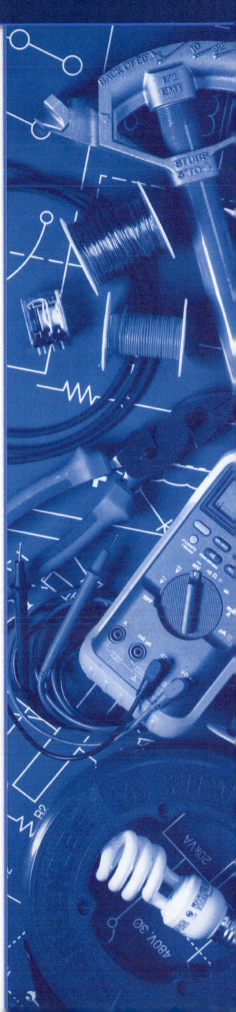

1.0.0 HVAC Operating Principles and Major Components 1

  1.1.0 Heating and Ventilation Systems 1

  1.2.0 Air Conditioning 4

    1.2.1 System Components 5

    1.2.2 Refrigeration Cycle 5

    1.2.3 Heat Pumps 8

2.0.0 Thermostats 10

  2.1.0 Thermostat Operating Principles 10

    2.1.1 Heating-Only Thermostats 11

    2.1.2 Cooling-Only Thermostats 12

    2.1.3 Heating-Cooling Thermostats 12

    2.1.4 Automatic-Changeover Thermostats 12

    2.1.5 Multistage Thermostats 13

    2.1.6 Programmable Thermostats 13

    2.1.7 Line-Voltage Thermostats 14

  2.2.0 Thermostat Installation 15

    2.2.1 Installation Guidelines 15

    2.2.2 Thermostat Wiring 16

    2.2.3 Checking Current Draw 16

    2.2.4 Adjusting Heat Anticipators 17

    2.2.5 Cycle Rate Settings 17

    2.2.6 Final Check 17

    2.2.7 Adjusting the Thermostat 18

3.0.0 HVAC Control Systems 19

  3.1.0 Comfort Cooling Controls 19

    3.1.1 Motor Speed Controls 19

    3.1.2 Lockout Control Circuit 20

    3.1.3 Time-Delay Relays 21

    3.1.4 Compressor Short-Cycle Timer 21

    3.1.5 Control Circuit Safety Switches 21

  3.2.0 Furnace Controls 22

    3.2.1 High-Temperature Limit Switch 22

    3.2.2 Rollout Switch 22

    3.2.3 Airflow Control 23

    3.2.4 Inducer Proving Switch 24

  3.3.0 Heat Pump Defrost Controls 24

  3.4.0 HVAC Digital Control Systems 25

    3.4.1 Controlling Devices 27

    3.4.2 Digital Control System Example 28

  3.5.0 Control Circuit Operating Sequence 29

  3.6.0 Troubleshooting 32

Contents (continued)

4.0.0 *NEC®* Requirements 39
  4.1.0 *NEC®* Requirements for HVAC Controls 39
  4.2.0 *NEC®* Requirements for HVAC Equipment 39
    4.2.1 Compressors...41
    4.2.2 Room Air Conditioners 41
    4.2.3 Electric Baseboard Heaters....................................43
    4.2.4 Electric Space-Heating Cable 43
    4.2.5 Energy Management Systems 45

Figures and Tables

Figure 1 Forced-air heating 2
Figure 2 High-efficiency condensing furnace 3
Figure 3 Humidifier and electronic air cleaner in a residential heating system 4
Figure 4 Basic refrigeration cycle 6
Figure 5 Heat pump—heating mode operation..............................8
Figure 6 Room thermostat 11
Figure 7 Bimetal sensing elements 11
Figure 8 Heating-only thermostat..12
Figure 9 Cooling-only thermostat...12
Figure 10 Cooling compensator...12
Figure 11 Heating-cooling thermostat...12
Figure 12 Automatic-changeover thermostat 13
Figure 13 Multistage thermostat hookup...14
Figure 14 Electronic programmable thermostat 14
Figure 15 Wi-Fi-enabled thermostat controlled using a smartphone app....15
Figure 16 Digital line-voltage thermostat 15
Figure 17 Thermostat wiring 16
Figure 18 Thermostat heat anticipator 17
Figure 19 Gas valve electric ratings...17
Figure 20 Motor speed control using triacs.....................................19
Figure 21 Electronic variable-speed furnace control 20
Figure 22 Lockout relay used in an HVAC control circuit...............21
Figure 23 Time-delay relay used to control a blower......................22
Figure 24 Freezestat...23
Figure 25 Example of a high-temperature limit switch...................23
Figure 26 Rollout switch 23
Figure 27 Sail switch 24
Figure 28 Combustion airflow pressure-sensing switch 24
Figure 29 Electronic defrost control 25
Figure 30 Defrost speed-up jumper..25
Figure 31 Centralized building management system 26
Figure 32 HVAC system control module..27
Figure 33 Changes are small and occur gradually.........................27

Figure 34 Typical constant-volume HVAC system 28
Figure 35 Room air conditioner circuit... 30
Figure 36 Typical cooling system control circuit.. 30
Figure 37 Circuit diagram of a cooling/gas heating system........................ 31
Figure 38 Summary of *NEC*® requirements for HVAC systems.................... 40
Figure 39 *NEC*® regulations governing motor compressors 43
Figure 40 Compressor branch and control circuits...................................... 44
Figure 41 *NEC*® requirements for lighting and switches for HVAC
          equipment... 45
Figure 42 *NEC*® requirements for locating 125V receptacles at HVAC
          equipment... 46
Figure 43 Branch circuits for room air conditioners 46
Figure 44 Room air conditioner connected to a dedicated branch circuit ... 47
Figure 45 *NEC*® installation guidelines for electric baseboard heaters 47
Figure 46 *NEC*® regulations for the installation of electric space-heating
          cable.. 48

Table 1A HVAC Troubleshooting Guide (1 of 3) 34
Table 1B HVAC Troubleshooting Guide (2 of 3) 35
Table 1C HVAC Troubleshooting Guide (3 of 3) 36
Table 2 Troubleshooting Guide for Electric Heat 37
Table 3 Summary of *NEC*® Requirements for Hermetically Sealed
        Compressors 42

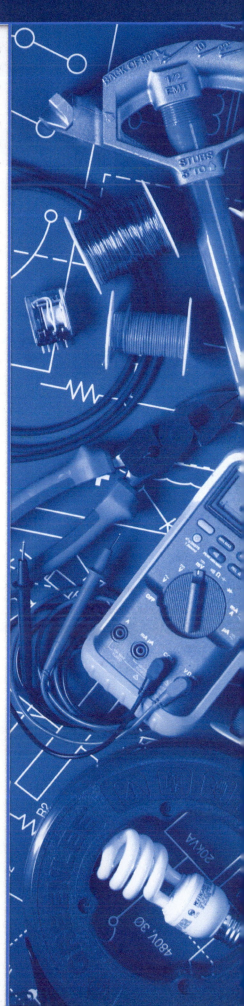

This page is intentionally left blank.

1.0.0 HVAC OPERATING PRINCIPLES AND MAJOR COMPONENTS

Objective

Describe the operating principles and major components of HVAC systems.

  a. Describe the basic principles of heating and ventilation.

  b. Describe the basic principles and components of comfort air conditioning systems.

Trade Terms

Compressor: A component of a refrigeration system that converts low-pressure, low-temperature refrigerant gas into high-temperature, high-pressure refrigerant gas through compression.

Condenser: A heat exchanger that transfers heat from the refrigerant flowing inside it to air or water flowing over it. Refrigerant enters as a hot, high-pressure vapor and is condensed to a liquid as the heat is removed.

Evaporator: A heat exchanger that transfers heat from the air or water flowing inside to the cooler refrigerant flowing through it. Refrigerant enters as a cold liquid/vapor mixture and changes to all vapor as heat is absorbed from the air or water.

Expansion device: A device that provides a pressure drop that converts the high-temperature, high-pressure liquid refrigerant from the condenser into the low-temperature, low-pressure liquid/vapor mixture entering the evaporator; also known as the *metering device*.

Heat transfer: The movement of heat energy from a warmer substance to a cooler substance.

Mechanical refrigeration: The use of machinery to provide a cooling effect.

Refrigerant: Typically, a fluid that picks up heat by evaporating at a low temperature and pressure and gives up heat by condensing at a higher temperature and pressure.

Refrigeration cycle: The process by which a circulating refrigerant absorbs heat from one location and transfers it to another location.

Subcooling: The measurable heat removed from a liquid after it has been cooled to its condensing temperature, has changed to a fully liquid state, and heat continues to be removed.

Superheat: The measurable heat added to a vapor after it has been heated to its boiling point, has changed to a fully vapor state, and heat continues to be added.

Heating, ventilating, and air conditioning (HVAC) systems provide the means to control the temperature, humidity, and even the cleanliness of the air in our homes, schools, offices, and factories. Electricians may be called upon to install the power distribution and control system wiring for this equipment and to troubleshoot the power distribution and control systems in the event of a malfunction. You may already be familiar with many types of control devices. You will now learn how these devices are used to control HVAC equipment.

This module begins with a review of the basic principles of heating and cooling systems and the equipment used in these systems.

1.1.0 Heating and Ventilation Systems

Early humans burned fuel as a source of heat. Today, a central heating source, such as a furnace or boiler, does the job using the principle of heat transfer; that is, heat is created in one place and carried to another place by means of air or water.

For example, in a common household furnace, fuel oil, natural gas, or propane gas is burned to create heat, which warms metal plates called *heat exchangers* (*Figure 1*). Air drawn from living spaces is circulated over the heat exchangers, where the heat is transferred to the air. The heated air is then returned to the living spaces. This type of system is known as a *forced-air system*, and it is the most common type of central heating system in the United States.

Figure 2 shows the interior of a high-efficiency gas furnace. This design has two heat exchangers to extract the maximum amount of heat from the burning fuel. High-efficiency furnaces and boilers are often eligible for energy-saving cash incentives from local power companies.

Water is used as a heat exchange medium as well. The water is heated in a boiler, then pumped through pipes to heat exchangers, where the heat it contains is transferred to the surrounding air. The heat exchangers are usually baseboard heating elements located in the space to be heated. This type of system is known as a *hydronic heating system*. It is more common in the Northeast and Midwest than in other parts of the country.

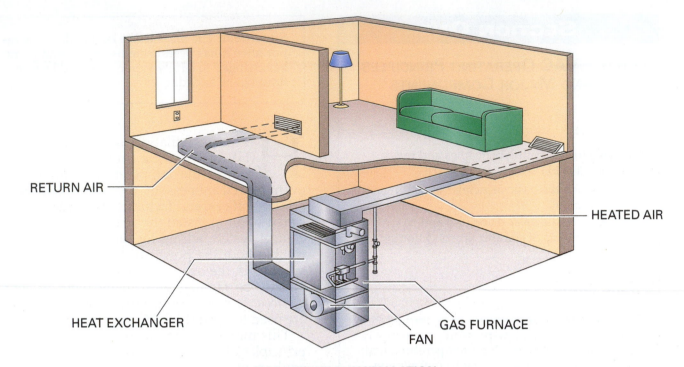

RETURN AIR

HEATED AIR

HEAT EXCHANGER

GAS FURNACE

FAN

(A) BASEMENT INSTALLATION

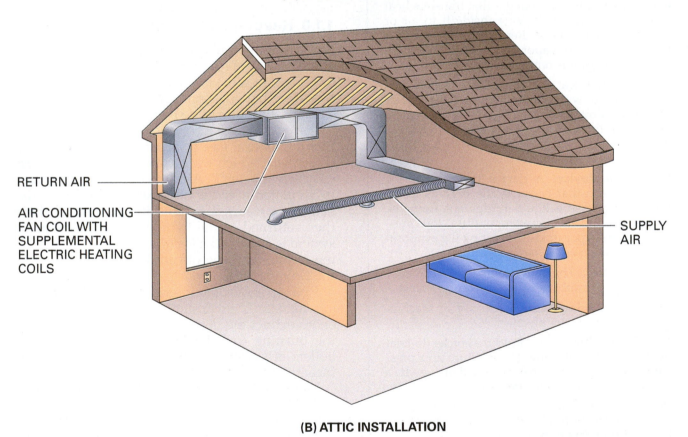

RETURN AIR

AIR CONDITIONING FAN COIL WITH SUPPLEMENTAL ELECTRIC HEATING COILS

SUPPLY AIR

(B) ATTIC INSTALLATION

Figure 1 Forced-air heating.

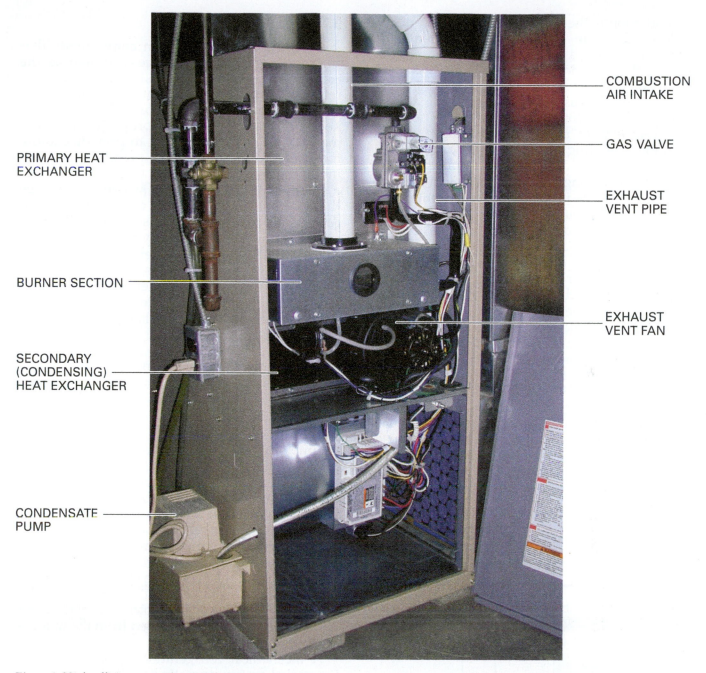

PRIMARY HEAT EXCHANGER

BURNER SECTION

SECONDARY (CONDENSING) HEAT EXCHANGER

CONDENSATE PUMP

COMBUSTION AIR INTAKE

GAS VALVE

EXHAUST VENT PIPE

EXHAUST VENT FAN

Figure 2 High-efficiency condensing furnace.

Natural gas and fuel oil are the most widely used heating fossil fuels. Of these two, natural gas is the most common. Oil heat is more prevalent in the Northeast and in rural areas. Propane gas is used in many parts of the country in place of oil or natural gas, also primarily in rural areas where natural gas pipelines are not available.

Electricity is another heat source. In an electric-resistance heating system, electricity flows through coils of resistive wire, causing the coils to become hot. Air from the conditioned space is passed over the coils, and the heat from the coils is transferred to the air. Because electricity is so expensive, electric-resistance heat is no longer common in cold climates. It is more likely to be used in warm climates where heat is seldom required.

In some areas of the country, a main or supplementary heating system may be fueled by wood or coal. Due to increases in the cost of fossil fuels (such as oil and gas) over the last several decades and because wood is a relatively cheap and renewable resource, wood-burning stoves and furnaces have remained quite popular.

Ventilation is the introduction of fresh air into a closed space in order to control air quality. Fresh air entering a building provides the oxygen we breathe. In addition to fresh air, we want clean air. The air in our homes, schools, and offices contains dust, pollen, and molds, as well as vapors and odors from a variety of sources. Relatively simple air circulation and filtration methods, including natural ventilation, are used to help keep the air in these environments clean and fresh. Among the common methods of improving the air in residential applications are the addition of humidifiers and electronic air cleaners to air handling systems, such as forced-air furnaces (*Figure 3*).

Many industrial environments require special ventilation and air management systems. Such systems are needed to eliminate noxious or toxic fumes and airborne particles that may be created by the processes and materials used at the facility.

The U.S. Government has strict regulations governing indoor air quality (IAQ) in industrial environments and the release of toxic materials to the outside air. Where noxious or toxic fumes may be present, the indoor air must be constantly replaced with fresh air. Fans and other ventilating devices are normally used for this purpose. Special filtering devices may also be required, not only to protect the health of building occupants, but also to prevent the release of toxic materials to the outside air.

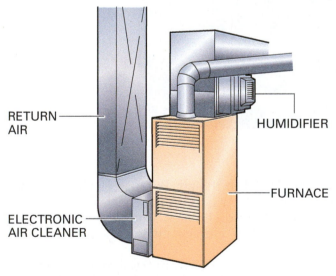

Figure 3 Humidifier and electronic air cleaner in a residential heating system.

1.2.0 Air Conditioning

All cooling systems contain components that function together in a process known as the refrigeration cycle. Understanding the basic refrigeration cycle will help you better understand how the various HVAC controls interact to keep the system operating properly. There are many types of systems used to provide cooling for personal comfort, food preservation, and industrial processes. Each of these systems uses a mechanical refrigeration system to produce the cooling. The operation of the mechanical refrigeration system is the same for all vapor compression systems. Varying from system to system are the type of refrigerant, the size and style of the components, and the installed locations of the basic components and refrigerant lines. We will review the basic refrigeration cycle before covering HVAC controls in more detail.

> **CAUTION**
>
> It is unlawful to release refrigerants into the atmosphere. All refrigerants must be recovered, recycled, and/or reclaimed.

Simply stated, the refrigeration cycle relies on the ability of chemical refrigerants to absorb heat. If a cold refrigerant is brought into contact with warm air, it will absorb heat from the air. Having given up heat to the refrigerant, the air occupying a space becomes cooler. The greater the difference in temperature between the air and the refrigerant, the more heat that will be transferred and the cooler the space will become. If the hot refrigerant is then exposed to cooler air—outdoor air for example—the refrigerant will give up the heat it absorbed from the indoors and become cool again.

A mechanical refrigeration system (*Figure 4*) is a sealed system operating under various pressures. The relationship between temperature and pressure is critical to mechanical refrigeration. The same refrigerant can be very cold at one point in the system (the evaporator inlet) and very hot at another (the condenser inlet). These two points are often not far apart. This is possible because of pressure changes caused by the compressor and expansion device. In addition to the circulation of refrigerant, air must circulate as well. Fans at the condenser and evaporator move air across the condenser and evaporator coils.

RETURN AIR

HUMIDIFIER

FURNACE

ELECTRONIC AIR CLEANER

1.2.1 System Components

There are many types of systems used to provide cooling for personal comfort, food preservation, and industrial processes. Each of these uses a mechanical refrigeration system with the following components:

- *Evaporator* – A heat exchanger where the heat from the area or item being cooled is transferred to the refrigerant.
- *Compressor* – A device that creates the pressure differences in the system needed to make the refrigerant flow and make the refrigeration cycle work.
- *Condenser* – A heat exchanger where the heat absorbed by the refrigerant is transferred to the cooler outdoor air or to another substance, such as water.
- *Expansion device* – A device that provides a pressure drop that lowers the boiling point of the refrigerant just before it enters the evaporator. It is also known as a *metering device*.

Also shown in *Figure 4* is the refrigerant piping used to connect the basic components and provide the path for refrigerant flow. The arrows show the direction of flow through the system. Together, the components and lines form a closed refrigeration system. The lines are designated as follows:

- *Suction line* – The tubing that carries heat-laden refrigerant gas from the evaporator to the compressor.
- *Hot gas line* (also called the *discharge line*) – The tubing that carries hot refrigerant gas from the compressor to the condenser.
- *Liquid line* – The tubing that carries the liquid refrigerant formed in the condenser to the expansion device.

The purpose of the refrigeration cycle is to move heat. The refrigerant is the medium by which heat can be moved into or from a space or substance. A refrigerant can be any liquid or gas that picks up heat by evaporating at a low temperature and pressure and gives up heat by condensing at a higher temperature and pressure. The events that take place within the system happen again and again in the same order.

Remember that the operation is basically the same for all mechanical refrigeration systems. Only the type of refrigerant used, the size and style of the components, and the installed locations of the components and the lines will change from system to system. Various accessories may be used in some systems to gain the desired cooling effect and to perform special functions.

1.2.2 Refrigeration Cycle

Although you are not required to be fully knowledgeable about all aspects of HVAC system operation, you need to have a basic understanding of how an HVAC system operates in order to perform your duties as an electrician. It is true that the majority of the problems in HVAC systems are electrical in nature, but it can be very difficult to pinpoint electrical failures unless you understand their corresponding impact on the refrigeration cycle. Therefore, it is important for you to understand the basic refrigeration cycle in order to trace mechanical failures back to their electrical sources.

The refrigeration cycle is based on two principles:

- As a liquid changes to a gas or vapor, it absorbs large quantities of heat. The reverse is also true: as a vapor condenses into a liquid, a great deal of heat must be expelled from it.
- The boiling and condensing temperatures of a liquid can be changed by altering the pressure exerted on the liquid.

The evaporator receives low-temperature, low-pressure liquid refrigerant from the expansion device. The evaporator is a series of tubing coils that expose the cool liquid refrigerant to the warmer air passing over the coils. Heat from the warm air is transferred through the tubing to the cooler refrigerant. This causes the refrigerant to boil and vaporize. It is important to realize that even though it has just boiled, it is still not considered "hot" because refrigerants boil at such low temperatures. So, it is a low-temperature, low-pressure refrigerant vapor that travels through the suction line to the compressor.

The compressor receives the low-temperature, low-pressure vapor and compresses it. It then becomes a high-temperature, high-pressure vapor. This vapor refrigerant travels to the condenser via the hot gas line.

The Mechanical Refrigeration Cycle

Many people think that an air conditioner adds cool air to an indoor space. In reality, the basic principle of air conditioning and the mechanical refrigeration cycle is that heat is extracted from the indoor air and transferred to another location (the outdoors) by the refrigerant that flows through the system.

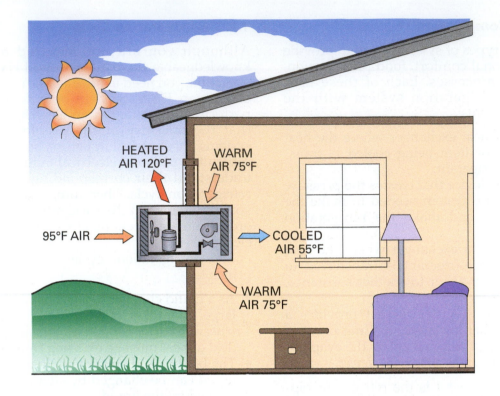

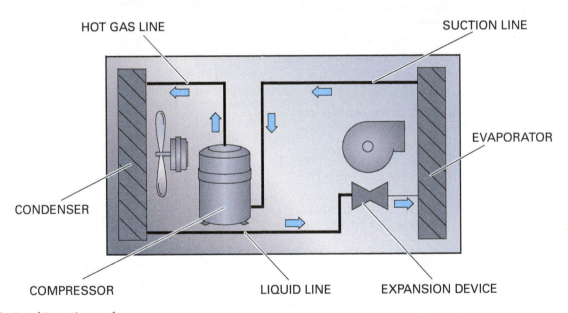

Figure 4 Basic refrigeration cycle.

The evaporator is designed so that the refrigerant is completely vaporized before it reaches the end of the evaporator. Thus the refrigerant absorbs more heat from the warm air flowing over the evaporator coils. Because the refrigerant is totally vaporized by this time, the additional heat it absorbs causes an increase in sensible (measurable) heat. This additional heat is known as superheat.

Like the evaporator, the condenser is a series of tubing coils through which the refrigerant flows.

As cooler air moves across the tubing, the hot refrigerant vapor gives up heat. As it continues to give up heat to the outside air, it cools to the condensing point, where it begins to change from a vapor into a liquid. As more cooling takes place, all of the refrigerant is eventually condensed to a liquid. Further cooling of the liquid, known as subcooling, takes place before the refrigerant leaves the condenser. The high-temperature, high-pressure liquid travels through the liquid line to the input of the expansion device.

Hermetic and Semi-Hermetic Compressors

Hermetic compressors are typically used in residential and light commercial air conditioners and heat pumps. Semi-hermetic compressors, also known as *serviceable compressors*, are used in large-capacity air conditioning, refrigeration, and chilled-water units. Semi-hermetic compressors can be partially disassembled for repair in the field. The term *hermetic* means to be sealed and air-tight. Thus, hermetic compressors are sealed and cannot be opened or repaired in the field.

HERMETIC RECIPROCATING COMPRESSOR

HERMETIC ROTARY COMPRESSOR

(A) HERMETIC COMPRESSORS

SEMI-HERMETIC (SERVICEABLE) COMPRESSOR

(B) SEMI-HERMETIC COMPRESSOR

The expansion device regulates the flow of refrigerant to the evaporator. It also decreases its pressure and temperature. It restricts the flow of refrigerant by forcing it through a tiny hole or orifice, converting the high-temperature, high-pressure liquid refrigerant from the condenser into the low-temperature, low-pressure liquid refrigerant needed to absorb heat in the evaporator.

1.2.3 Heat Pumps

A special type of air conditioner known as a *heat pump* is widely used in the warmer regions of the country to provide both cooling and heating. Heat pumps are extremely efficient; however, they are most effective in climates where the temperature does not generally fall below 25°F (–4°C) or 30°F (2°C). In colder parts of the country, heat pumps can be combined with supplemental electric, geothermal, or gas- or oil-fired, forced-air furnaces. In such arrangements, the furnace automatically supplements or takes over for the heat pump completely when the outdoor temperature falls below the heat pump's efficient range. Like high-efficiency furnaces and boilers, heat pumps are often eligible for energy-saving incentives.

Heat pumps operate by reversing the cooling cycle (*Figure 5*). For this reason, heat pumps are sometimes called *reverse-cycle air conditioners*.

They operate on the principle that there is always some extractable heat in the air, even though the air may seem very cold. In fact, the temperature would have to be –460°F (–273.15°C; referred to as *absolute zero*) for a total absence of heat to exist. In the heating mode, a special valve known as a *reversing valve* switches the compressor input and output so that the condenser operates as the evaporator and the evaporator becomes the condenser. Because of this role reversal, the coils in a heat pump are referred to as the *outdoor coil* and *indoor coil* instead of the condenser and evaporator.

Modern Air Conditioning

Dr. Willis Carrier, founder of Carrier Corporation, is credited with the invention of modern air conditioning. In 1902, he developed a system that could control both humidity and temperature using a non-toxic, non-flammable refrigerant. Air conditioning started out as a means of solving a problem in a printing facility where heat and humidity were causing paper shrinkage. Later systems served similar purposes in textile plants.

The concept was not applied to comfort air conditioning until about 20 years later when Carrier's centrifugal chillers began to be installed in department stores and movie theaters.

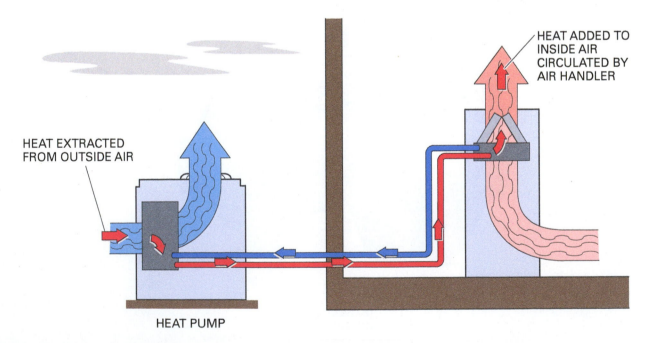

HEAT EXTRACTED FROM OUTSIDE AIR

HEAT ADDED TO INSIDE AIR CIRCULATED BY AIR HANDLER

HEAT PUMP

Figure 5 Heat pump—heating mode operation.

1.0.0 Section Review

1. In a hydronic heating system, the heat transfer medium is _____.
   a. air
   b. water
   c. natural gas
   d. ammonia

2. In mechanical refrigeration, the purpose of the evaporator is to _____.
   a. convert refrigerant from high pressure to low pressure
   b. transfer heat from the refrigerant to the outdoor air
   c. transfer heat from the indoor air to the refrigerant
   d. provide the pressure difference needed to circulate the refrigerant

2.0.0 THERMOSTATS

Objective

Identify the types of thermostats and their uses.

a. State the operating principles of thermostats.
b. Install different types of thermostats.

Performance Tasks

1. Identify various types of thermostats and describe their operation and uses.
2. Install a conventional 24V bimetal thermostat and hook it up using the standard coding system for thermostat wiring.
3. Check and adjust a thermostat, including the heat anticipator setting and indicator adjustment.

Trade Terms

Automatic-changeover thermostat: A thermostat that automatically selects heating or cooling based on the space temperature.

Bimetal: A control device made of two dissimilar metals that warp when exposed to heat, creating movement that can open or close a switch.

Cooling compensator: A fixed resistor installed in a thermostat to act as a cooling anticipator.

Deadband: A temperature band, often 3°F, that separates the heating mode and cooling mode in an automatic-changeover thermostat.

Differential: The difference between the cut-in and cut-out points of a thermostat.

Invar®: An alloy of steel containing 36% nickel. It is one of the two metals in a bimetal control device.

Sub-base: The portion of a two-part thermostat that contains the wiring terminals and control switches.

Thermostat: A device that is responsive to ambient temperature conditions.

The room **thermostat** (*Figure 6*) is the primary control in an HVAC system. It can be as simple as a single temperature-sensitive switch. It can also be a complex collection of sensing elements and switching devices that provides many levels of control. In this module, the term *thermostat* generally refers to the control devices that are mounted on a wall in a conditioned space.

Many residences and many small commercial businesses, such as retail stores and shops, have a single thermostat. Large office buildings, shopping malls, and factories are divided into cooling and heating zones and have several thermostats—one for each zone. Because each zone is independent of the others, each thermostat must control either a separate system or the airflow from a common system.

2.1.0 Thermostat Operating Principles

Programmable electronic thermostats using electronic sensing elements are increasingly common. However, thermostats with **bimetal** sensing elements are still being manufactured and are still widely used. Although not quite as precise as the electronic thermostat, bimetal devices are considerably less expensive and have proven to be effective and reliable. Most bimetal thermostats will maintain the space temperature within ±2° of the setpoint. A good electronic thermostat will maintain the temperature within ±1°.

A bimetal element (*Figure 7*) is composed of two different metals bonded together. One metal is usually copper or brass; the other is a metal called **Invar®**, which contains 36% nickel. When heated, the copper or brass has a more rapid expansion rate than the Invar® and changes the shape of the element. The movement that occurs when the bimetal changes shape is used to open or close switch contacts in the thermostat.

> **WARNING!**
>
> Older thermostats contain mercury. Mercury is toxic. Even short-term exposure may result in damage to the lungs and central nervous system. Mercury is also an environmental hazard. Do not dispose of thermostats containing mercury bulbs in regular trash. Contact your local waste management or environmental authority for disposal/recycling instructions.

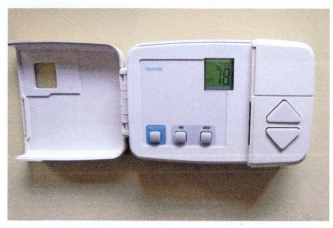

Figure 6 Room thermostat.

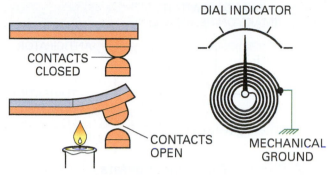

Figure 7 Bimetal sensing elements.

Most residential and small commercial thermostats are of the low voltage (24VAC) type. With low-voltage control circuits, there is less risk of electrical shock and less chance of fire from short circuits. Low-voltage components are also less expensive and less likely to produce arcing, coil burnout, and contact failure. Some self-generating systems use a millivolt power supply that generates about 750 millivolts (mV) to operate the thermostat circuit. These thermostats are very similar in construction and design to low-voltage thermostats; however, they are not interchangeable with 24V thermostats.

2.1.1 Heating-Only Thermostats

A wall-mounted heating-only thermostat typically contains a temperature-sensitive switch, as shown in *Figure 8*. In this arrangement, the heating device, such as a furnace, will not turn on unless the thermostatic switch is calling for heat.

When the temperature in the conditioned space reaches the thermostat setpoint, the thermostatic switch will open. Because of the residual heat in the heat exchangers and the continued rotation of the fan as it cools the heat exchanger before coming to a stop, the room temperature can overshoot the setpoint. To avoid occupant discomfort, an adjustable heat anticipator opens the thermostat before the temperature in the space reaches the setpoint. The heat anticipator is a small resistance heater in series with the switch contacts. The anticipator heats the bimetal strip, causing the contacts to open before the space becomes overheated.

Large Water-Cooled Chillers

GOING GREEN

As the name implies, chillers chill water to be used as a heat transfer medium. The chilled water is then circulated through the building in piping systems to a number of fan-coil or air handling units. Many chillers are air-cooled, but many are also water-cooled. Water-cooled chillers are generally combined with a cooling tower that serves as an additional heat exchanger. Water flowing through the chiller condenser is routed to the cooling tower, where the heat is then transferred to air before returning to the condenser. Water-cooled systems are more energy efficient than most air-cooled systems. The photo below shows a large commercial water-cooled chiller.

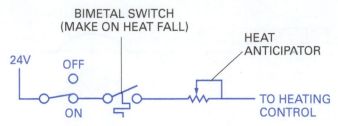

Figure 8 Heating-only thermostat.

2.1.2 Cooling-Only Thermostats

The cooling thermostat (*Figure 9*) works in reverse when compared to a heating thermostat. When the bimetal coil heats up, the switch closes its contacts and starts the cooling system compressor. When the cooling thermostat is turned down to make the conditioned space cooler, the bimetal element returns to its original position to turn the cooling system off.

Cooling thermostats contain a device called a **cooling compensator** to help improve indoor comfort. The cooling compensator (*Figure 10*) is a fixed resistance in parallel with the thermostatic switch. (Heating anticipators are normally adjustable.) No current flows through the compensator when cooling is on because it has a much higher resistance than the switch contacts. In this case, the contacts are essentially a short circuit. When the thermostat is open, however, a very small current can flow through the compensator and the contactor coil.

Because of the size of the compensator, the current is not enough to energize the contactor. The heat created by the current flowing through the compensator warms the thermostat slightly, making the thermostatic switch contacts close sooner than they would otherwise. In this way, the cooling compensator accounts for the lag between the call for cooling and the time when the system actually begins to cool the space.

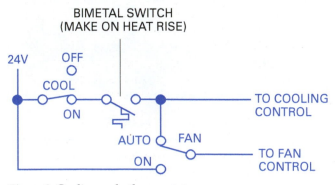

Figure 9 Cooling-only thermostat.

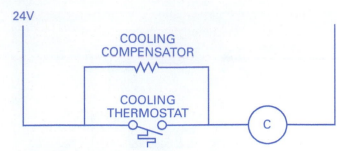

Figure 10 Cooling compensator.

2.1.3 Heating-Cooling Thermostats

When heating and cooling are combined for year-round comfort, it is impractical to use a separate thermostat for each mode. Therefore, the two are combined into one heating-cooling thermostat (*Figure 11*).

Unless a switch is provided in the heating-cooling thermostat, the thermostat will continuously switch back and forth from heating to cooling. In effect, heating and cooling will combat each other for control. A switch provides a means to direct the control to cooling, while disconnecting the heating control circuit. Likewise, when the switch is moved to heating, the switch connects the heating components, while electrically isolating the cooling circuit. When the switch is in the center or Off position, neither the heating nor cooling control circuit can be energized.

2.1.4 Automatic-Changeover Thermostats

The disadvantage of a common heating-cooling thermostat is that the building occupant must determine whether heating or cooling is needed at a particular time and set the thermostat switch accordingly. In some climates, that is impractical because the need could change several times a day.

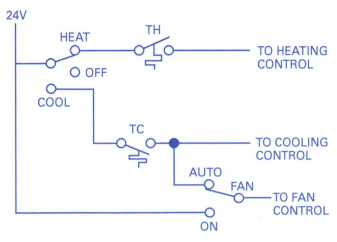

Figure 11 Heating-cooling thermostat.

The **automatic-changeover thermostat** automatically selects the mode, depending on the heating and cooling setpoints. The thermostat shown in *Figure 12* is essentially the same as the thermostat in *Figure 11*, with the exception of an Auto position on the main control switch. The occupant can still select either heating or cooling. When the switch is in the Auto position, however, the thermostat makes the selection. All that is necessary is for one of the thermostatic switches to close, indicating that the conditioned space is too warm or too cold.

The thermostat contains a built-in mechanical **differential**, which is the difference between the cut-in and cut-out points of a thermostat. The differential is normally 2°. For example, if the heating setpoint is 70°F (21°C), the furnace will turn on at 70°F (21°C) and run until the temperature is 72°F (22°C).

Automatic-changeover thermostats also have a minimum interlock setting, commonly known as the **deadband**. The deadband is a built-in feature that prevents the heating and cooling setpoints from being too close together. A typical deadband setting is 3°. The deadband keeps the system from cycling back and forth between heating and cooling.

2.1.5 Multistage Thermostats

Multistage thermostats (*Figure 13*) provide energy savings by cycling-on stages of heating or cooling based on demand. For example, a cooling system might use a two-speed compressor. If the demand for cooling is low, the first-stage cooling circuit of the room thermostat cycles the compressor on at the lower speed and capacity until the demand is met. If the demand for cooling is high, such as on a very hot day, the second-stage cooling circuit in the thermostat energizes the higher compressor speed. The stages of cooling can also be used to start and stop multiple compressors in larger systems.

Similarly, a two-stage heating thermostat could stage on burners in a gas furnace or electric elements in an electric furnace. Multistage room thermostats are commonly used with heat pumps. These thermostats come in a variety of configurations, such as single-stage heat with two stages of cooling; two stages of heat with one stage of cooling; or two stages of heating and two stages of cooling.

Heat pump thermostats usually have an emergency heat switch. If the heat pump compressor

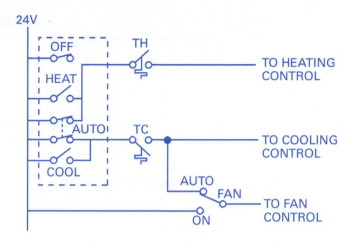

Figure 12 Automatic-changeover thermostat.

becomes inoperative, this switch locks out the normal heat pump operation and heats the area with auxiliary electric heat (or other connected heat source) until the problem can be corrected. An indicator light, usually red, is mounted on the thermostat. It will come on when the selector switch is in the emergency heat position. As soon as the unit has been repaired, the switch must be returned to the normal operating position.

2.1.6 Programmable Thermostats

Programmable thermostats are self-contained controls with the timer, temperature sensor, and switching devices all located in the unit mounted on the wall. Early programmable thermostats looked very much like conventional thermostats. This type of thermostat used a motor-driven time clock driving a wheel containing cams that were set to raise and lower the temperature settings at desired intervals.

Modern electronic programmable thermostats (*Figure 14*) use microprocessors and integrated circuits to provide a wide variety of control and energy-saving features. Their control panels use touch-screen technology and their displays are digital readouts. Different thermostats offer different features; the more sophisticated (and expensive) the thermostat, the more features it generally offers. Wireless models can be connected to the local network and accessed through various smartphone apps (*Figure 15*). These thermostats allow a user to change the setpoint without even being in the building.

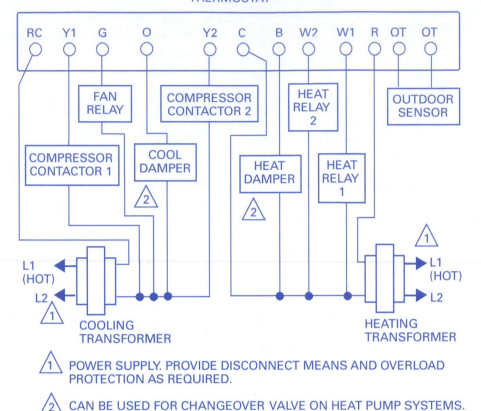

Figure 13 Multistage thermostat hookup.

Some of the features available on electronic thermostats are:

- *Override control* – This feature allows the occupant to override the program when desired. For example, you might override the night setback on Monday night so the thermostat isn't lowered at the normal time but is delayed until the football game is over.

Figure 14 Electronic programmable thermostat.

- *Multiple programs* – This feature allows the occupant to design and select different schedules for different conditions. For example, you could program a special schedule for a vacation away from home.
- *Battery backup* – This feature prevents program loss in the event of a power failure.
- *Staggered startup for multi-unit systems* – This feature avoids excessive current drain. This is an important feature in office buildings, shopping malls, and hotels.
- *Maintenance tracking* – This feature indicates when maintenance is to be performed (for example, when to replace filters).

The savings available from programmable thermostats are significant. For example, it is estimated that a setback of 10° for both daytime and nighttime can result in a 20% energy savings. A 5° setback will yield a 10% energy savings.

2.1.7 Line-Voltage Thermostats

Most of the thermostats you encounter will operate at low voltages. There are also thermostats that operate at line voltages (for example, 240VAC).

Figure 15 Wi-Fi-enabled thermostat controlled using a smartphone app.

Line-voltage thermostats, as their name implies, control heat by directly controlling the power supply. This is in contrast to low-voltage thermostats that indirectly control heat or cooling through a relay or contactor. Line-voltage thermostats are commonly used to control baseboard heaters and floor or ceiling radiant-heat grids. Conventional line-voltage thermostats use a bimetal sensing element that does not provide precise temperature control. This is because the bimetal elements and switch contacts must be heavier to carry the larger current. Because they are heavier, they do not respond as quickly or easily as the smaller and lighter contacts in 24VAC room thermostats. Digital line-voltage thermostats provide much more precise temperature control (*Figure 16*).

2.2.0 Thermostat Installation

Even the most sensitive thermostat cannot perform correctly if it is poorly installed. Selecting the proper location for the thermostat is the first step in any installation procedure.

2.2.1 Installation Guidelines

The thermostat should be installed in the space in which it will be called upon to control the temperature and other conditioning factors. The thermostat should be installed on a solid inside wall that is free from vibration that could affect operation by making the thermostat contacts chatter. For the same reason, it should not be located on a wall near slamming doors or near stairways. The following practices should be observed when installing a thermostat:

- The installer must be a trained, experienced technician.
- The manufacturer's instructions must be carefully read before installing the thermostat. Failure to follow them could lead to product damage or failure.
- The rating should be checked in the instructions and on the unit to make sure the thermostat is suitable for the particular application.
- When the installation is complete, the thermostat must be operationally checked and adjusted as indicated in the installation instructions.

> **CAUTION**
>
> Thermostats containing solid-state devices are sensitive to static electricity; therefore, you must discharge body static electricity before handling the instrument. This can be accomplished by touching a grounded metal pipe or conduit. Touch only the front cover when holding the device.

Figure 16 Digital line-voltage thermostat.

When unpacking the new thermostat and wallplate or sub-base, handle it with care. Rough handling can damage the thermostat. Save all instructional information and literature for future reference. Locate the thermostat and wallplate or sub-base about 5' (1.52 m) above the finished floor in an area with good air circulation at room temperature. Avoid locations that allow the following conditions:

- Drafts or dead air spots
- Hot or cold air from ducts or diffusers
- Radiant heat from direct sunlight or hidden heat from appliances
- Heat from concealed supply ducts and chimneys
- Unheated areas behind the thermostat, such as an outside wall or garage

2.2.2 Thermostat Wiring

The thermostat is connected through multiconductor thermostat wire to a terminal strip or junction box in the air conditioning unit. A standard coding method is used in the HVAC industry to designate wiring terminals and wire colors. *Figure 17* shows the coding method and illustrates how the terminals of the heatingcooling thermostat shown earlier would be designated. The terminal designation arrangement is fairly standard among thermostat and equipment manufacturers. It is not safe to assume, however, that the person who installed an existing system followed the color scheme when wiring the control circuits. There are additional codes for more complex thermostats; for example, the letter O designates orange and is connected to the reversing valve control in a heat pump. When there are multiple stages of heating or cooling, those terminals are designated with the appropriate letter plus a number.

Thermostat wiring should be done in accordance with national and local electrical codes. See **NEC Section 424.20** for limitations on using a thermostat as a disconnect device. All wiring connections should be tight. To avoid damaging the control wire conductor, always use a stripping tool designed to strip small-gauge wire. Color-coded wiring should be used where possible for easy reference should system troubleshooting be required at a later date.

Thermostat wires should not be spliced, but if splicing is absolutely necessary, soldered splices are recommended. If wires are stapled to prevent movement, take care to ensure that the staple does not go through the wire insulation. Seal the wire opening in the wall space behind the thermostat so it is not affected by drafts within the wall stud space.

2.2.3 Checking Current Draw

The next step is to determine the current draw. This may be done by locating the current draw of the primary control in the heating unit or the heat anticipator setting on the existing thermostat (*Figure 18*). The current draw is usually printed on the furnace nameplate and/or a primary control such as the gas valve (*Figure 19*), the relay, or the oil burner control. It may also be found in the manufacturer's installation and service literature.

> **WARNING!**
>
> To prevent electrical shock or equipment damage, make sure the power is off before connecting the wiring for the current draw adjustment.

THERMOSTAT WIRING CODES

TERMINAL DESIGNATION	WIRE COLOR	FUNCTION
R	RED	POWER
G	GREEN	FAN CONTROL
Y	YELLOW	COOLING CONTROL Y1 = STAGE 1 Y2 = STAGE 2
W	WHITE	HEATING CONTROL W1 = STAGE 1 W2 = STAGE 2
O	ORANGE	HEAT PUMP REVERSING VALVE CONTROL

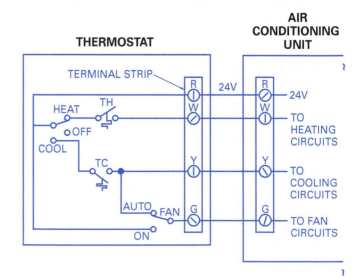

Figure 17 Thermostat wiring.

2.2.4 Adjusting Heat Anticipators

The heat anticipator on an electromechanical thermostat is adjustable and should be set at the amperage indicated on the primary control. Small variations from the required setting can be made to improve performance on individual jobs.

Changing the setting of the anticipator changes the resistance of the wire resistor. This shortens or lengthens the heating cycle. Some heat anticipators have arrows to indicate the heating cycle adjustment.

Some heat anticipators have a fixed resistance; others are equipped with an adjustable slide to change the resistance. The adjustable heat anticipator has a slide wire adjustment with the pointer scale marked in tenths of an ampere. This is used to set the anticipator to match the control amperage draw of the particular furnace. Each furnace is provided with an information sticker near the burner that states the amperage drawn by the control circuit of that particular furnace. This is the amperage at which the thermostat heat anticipator should be set under most conditions. The actual current in the circuit can also be measured for increased accuracy, because the stated current draw may be slightly inaccurate.

For example, if the amperage draw of a control circuit is shown as 0.45A, the installer should adjust the anticipator setting to 0.45A on the scale. The heat anticipator adjustment determines the length of the thermostat call-for-heat cycle by artificially heating the bimetal coil. As more heat is directed at the bimetal coil, a shorter heating cycle will occur. Conversely, as less heat is directed at the bimetal coil, a longer heating cycle will occur until the thermostat satisfies the call for heat.

When the control circuit amperage is high, less of the heater wire is needed; when the control amperage is low, more of the heater wire is needed. The control circuit amperage draw should be measured for each heating system as previously described.

2.2.5 Cycle Rate Settings

Electronic room thermostats do not contain a traditional heat anticipator. Instead, they contain a feature that allows the cycle rate (the number of times per hour the burner operates) to be adjusted, which in turn helps maintain indoor comfort. Most furnace manufacturers list the correct cycle rate for their product in the furnace installation literature. The room thermostat literature will contain directions for changing the cycle rate.

2.2.6 Final Check

The final step is to check the heating and/or cooling system when the thermostat is installed. With the thermostat in the heating mode, turn on the power, place the system switch at Heat, and leave the fan switch in the Auto position. Turn the setpoint dial to at least 5° above the room temperature. The burner should come on within 15 seconds. The fan will start after a short delay. Then turn the setpoint dial to 5° below the room temperature. The main burner should shut off within 15 seconds, but the blower may continue to run for several minutes. Next, set the thermostat to the cooling mode.

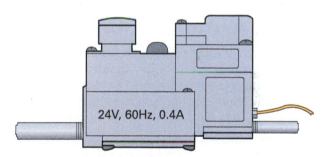

Figure 19 Gas valve electric ratings.

Thermostat Wiring

Thermostat wires come in a variety of conductor configurations. Simple two-conductor wire can be used in a heating-only installation. Thermostat wires with three, four, five, and eight conductors are readily available from any HVAC parts distributor for use in more complex installations. Normally, 18-gauge thermostat wire is adequate for most installations. However, long runs of thermostat wire can produce a voltage drop that might affect equipment operation. If you have a run of thermostat wire that seems excessively long, use 16-gauge or heavier wire to reduce the voltage drop.

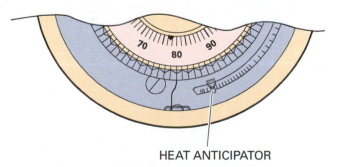

Figure 18 Thermostat heat anticipator.

If the outside temperature is at least 50°F (10°C), return power to the unit, set the thermostat to the Cool position, and set the fan switch to Auto. Leave the setpoint of the thermostat at 5° below the room temperature. The cooling system should come on either immediately or after any start delay, if the unit is so equipped. The indoor fan should come on immediately.

After the cooling system has come on, set the thermostat to at least 5° above the room temperature. The cooling equipment should turn off within 15 seconds. Place the system switch to Off. Move the setpoint dial to various positions. The system should not respond for heating or cooling.

2.2.7 Adjusting the Thermostat

Thermostats are calibrated or preset at the factory for accurate temperature response and normally will not need recalibration. If a thermostat seems out of adjustment, check and/or adjust the calibration as follows:

Step 1 Move the temperature setting lever to the lowest setting. Set the system switch to Heat and wait about 10 minutes.

Step 2 Remove the thermostat cover and move the thermostat temperature selector lever toward a higher temperature setting until the switch just makes contact.

Step 3 If the thermostat pointer and the setting lever read about the same at the instant the switch makes contact, no recalibration is needed. If recalibration is necessary, follow the manufacturer's instructions.

Heat Anticipators

Be very careful about adjusting a room thermostat heat anticipator, especially if there is no complaint of the system running very short periods of time repeatedly (short-cycling), or it running too long and overheating the space. This is a situation where it is better to do nothing than to run the risk of creating a comfort problem.

Compressor Short-Cycle Time Delay

Some systems are equipped with a compressor short-cycle time delay circuit to protect the compressor. This circuit, which is discussed later in this module, prevents the compressor from cycling back on for roughly five minutes after it has turned off. This allows the refrigerant pressures on the suction and discharge sides of the compressor to reach equilibrium.

2.0.0 Section Review

1. An auxiliary heat control is normally part of a _____.
   a. cooling-only thermostat
   b. heat pump thermostat
   c. bimetal thermostat
   d. heat-cool thermostat

2. The thermostat terminal normally assigned to cooling control is _____.
   a. Y
   b. G
   c. C
   d. R

3.0.0 HVAC CONTROL SYSTEMS

Objective

Describe HVAC control systems and devices.
   a. Identify and describe controls used in comfort cooling systems.
   b. Identify and describe furnace controls.
   c. Identify and describe heat pump defrost controls.
   d. Describe digital controls used in HVAC systems.
   e. Analyze the operating sequences of various HVAC control systems.
   f. Describe troubleshooting methods used for HVAC systems.

Trade Term

Analog-to-digital converter: A device designed to convert analog signals such as temperature and humidity to a digital form that can be processed by logic circuits.

Most HVAC control systems are designed to automatically maintain the desired heating, cooling, and ventilation conditions set into the system. The controls for a small system such as a window air conditioner are very simple—a couple of control switches and a thermostat. As the system gets larger and provides more features, the controls become more complicated. For example, add gas heating to a packaged cooling unit, and the size and complexity of the control circuits will more than double. Make it a heat pump instead, and the control complexity may triple.

3.1.0 Comfort Cooling Controls

Large commercial systems may use pneumatic or electronic controls in conjunction with conventional electrical controls. These systems may have 30 or 40 control devices, whereas a window air conditioner has just a handful.

The good news is that there are only a few different kinds of control devices. When you learn to recognize them and understand the role each plays, it won't matter how many are used to control a particular system.

All automatic control systems have the following basic characteristics in common:

- A sensing element (thermistor, thermostat, pressure switch, or humidistat) measures changes in temperature, pressure, and humidity.
- A control mechanism translates the changes into energy that can be used by devices such as motors and valves.
- The connecting wiring, pneumatic piping, and mechanical linkages transmit the energy to the motor, valve, or other devices that act at the point where the change is needed.
- The device then uses the energy to achieve some change. For example, motors operate compressors, fans, or dampers. Valves control the flow of gas to burners or refrigerant to cooling coils and permit the flow of air in pneumatic systems. Valves also control the flow of liquids in chilled-water systems.
- Sensing elements in the control detect the change in conditions and signal the control mechanism.
- The control stops the motor, closes the valves, or terminates the action of the component being used. As a result, the call for change is ended.

3.1.1 Motor Speed Controls

Greater efficiency can be achieved by varying the speed of compressors and fan motors to adapt to changing heating and cooling loads. In this way, the equipment consumes only as much power as is needed to meet the demand. For example, HVAC equipment commonly uses variable-speed blowers. In addition, two-speed compressors are used in small systems, while larger systems use unloaders or multiple compressors to adapt to changing loads.

The triac (*Figure 20*) is the most common motor control device. A knob on the control device is used to adjust a potentiometer; the setting of the potentiometer determines the motor speed. This circuit is similar to that of a dimmer switch.

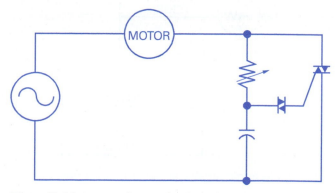

Figure 20 Motor speed control using triacs.

Modern electronic motor controls use microprocessor chips to achieve continuous control. The furnace control system shown in the simplified diagram in *Figure 21* controls the combustion system as well as the blower speed. The microprocessor monitors information, such as the length of the last heating cycle and how often the furnace is cycling on and off. It then optimizes the heating cycle by selecting the appropriate fan speed and adjusting the length of the low-fire and high-fire cycles to match the conditions in the space. It may also be able to sense any changes within the system, such as a dirty air filter or a closed zone valve, or vary the blower speed based on the demand for heating.

3.1.2 Lockout Control Circuit

The purpose of the lockout relay in a control circuit is to prevent the automatic restart of the HVAC equipment after a safety device has shut it down. If the lockout relay has been activated, the system may be reset only through interruption of the power supply to the control circuit by resetting the thermostat, turning the main power switch Off and then On again, or by other means.

In the circuit in *Figure 22*, the lockout relay coil, because of its high resistance, is not energized during normal operation. However, when any one of the safety controls opens the circuit to the compressor contactor coil, then full current flows through the lockout relay coil, causing it to become energized and to open its contacts. These contacts remain open, keeping the compressor contactor circuit open until the safety control has been reset (if it requires a manual reset) and power is interrupted. Performance of the lockout relay depends on the resistance of its coil being much greater than the resistance of the compressor contactor coil.

If the lockout relay becomes defective, it should be replaced with an exact duplicate to maintain the proper resistance balance. This type of relay is sometimes called an *impedance relay*.

It is permissible to add a control relay coil in parallel with the contactor coil when a system demands another control. The resistance of the contactor coil and the relay coil in parallel decreases the total resistance and does not affect the operation of the lockout relay.

> **CAUTION**
>
> Never place additional lockout relays, lights, or other load devices in parallel with the lockout relay coil. Doing so might prevent the lockout relay from operating, thus defeating its purpose.

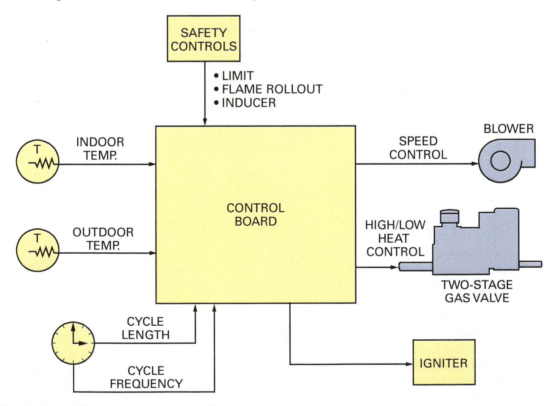

Figure 21 Electronic variable-speed furnace control.

3.1.3 Time-Delay Relays

The purpose of a time-delay relay is to delay the normal operation of a compressor or motor for a predetermined length of time after the control system has been energized. The length of the delay depends on the time built into the relay coil and may vary from a fraction of a second to several minutes. A common use for a time-delay relay is to delay the startup or shutdown of a furnace blower to improve heating efficiency (*Figure 23*).

Time-delay relays used to stage compressors in multiple-compressor systems have been replaced, for the most part, with electronic timers. The electronic timers can control conventional relays or may have relay contacts as an integral part of the timer. Advantages of the electronic timer include more accurate timing sequences and greater reliability.

A sequencer is a special type of time-delay relay used to energize resistance heaters in a heat pump air handler or electric furnace. The sequencer is basically multiple time-delay relays in a single enclosure, with contacts heavy enough to carry the current of an electric resistance heater. It is common for heat pump air handlers to be equipped with several banks of heaters, each controlled by its own sequencer. To avoid the current surge that would occur if all heaters were energized at once, the timing of the sequencers is selected to bring the various heaters on in stages.

3.1.4 Compressor Short-Cycle Timer

Attempting to start a compressor against an existing high pressure in the discharge side can damage the compressor. When a refrigerant system shuts down, it should not be restarted until the pressures in the high-pressure and low-pressure sides of the system have had time to equalize. Short-cycling can be caused by a momentary power interruption or by someone changing the thermostat setting.

A compressor short-cycle protection circuit contains a timing function that prevents the compressor contactor from re-energizing for a specified period of time after the compressor shuts off. Lockout periods typically range from 30 seconds to five minutes.

Short-cycle timers are available as self-contained modules that can be direct-wired into a unit. They are often sold as optional accessories.

3.1.5 Control Circuit Safety Switches

Refrigerant system compressor control circuits normally include several different types of safety switches. These include pressure switches, freezestats, and outdoor thermostats.

Pressure switches – Many systems use one or more pressure switches in the compressor control circuit. These are safety devices designed to protect the compressor. A pressure switch is normally closed and wired in series with the compressor contactor control circuit. The schematic diagram in *Figure 22* shows low-pressure cutout (LPCO) and high-pressure cutout (HPCO) switches used in circuits.

High-pressure switches are designed to open if the compressor head (discharge) pressure is too high. Low-pressure switches are designed to open if the suction pressure is too low. Pressure switches use a bellows mechanism that presses against switch contacts that have preset open and close settings. When the system pressure begins to rise above or drop below the normal operating pressure, the related high-pressure or low-pressure switch will open, causing the compressor contactor to de-energize.

Some manufacturers use a type of low-pressure switch called a *loss-of-charge switch* that removes power to the equipment if the refrigerant charge is too low.

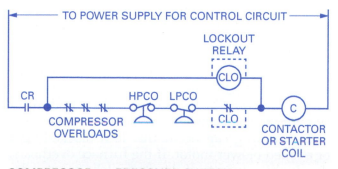

Figure 22 Lockout relay used in an HVAC control circuit.

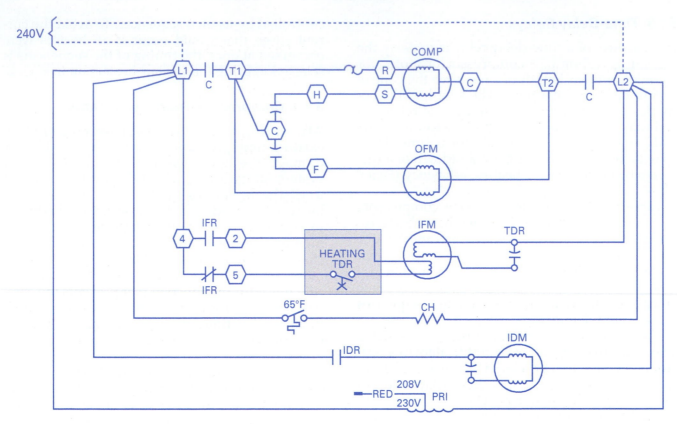

Figure 23 Time-delay relay used to control a blower.

Freezestat – Another type of safety switch is the freeze-protection thermostat (FPT), commonly called a *freezestat*. Its purpose is to prevent evaporator coil freeze-up. The freezestat switch (*Figure 24*) is a normally-closed switch that is usually attached to one of the end bells in the evaporator coil. It will open if the refrigerant temperature drops below a predetermined setpoint.

Outdoor thermostats – Some cooling equipment uses an outdoor thermostat to shut off the equipment when the ambient temperature drops below a predetermined outdoor temperature setpoint, typically between 55°F (13°C) and 65°F (18°C). This prevents equipment damage. Outdoor thermostats are often used in heat pump systems to enable or disable one or more stages of supplemental heat.

3.2.0 Furnace Controls

Gas furnaces are equipped with several devices that are designed to shut off the furnace if a hazardous condition is sensed. Common furnace controls include the high-temperature limit switch, rollout switch, airflow control, and inducer proving switch.

3.2.1 High-Temperature Limit Switch

Figure 25 shows an example of a high-temperature limit switch. This bimetal switch is thermally actuated and is usually mounted above or beside the heat exchangers. Temperature limit switches must be located where they can sense temperature accurately. Electrically, they are wired in series with the gas valve.

Overheating can occur due to a blocked vent or a failed blower motor. If the furnace overheats, the high-temperature limit switch opens, de-energizing the gas valve and shutting off the gas supply. These switches may reset automatically when the temperature drops below the setpoint.

Many manufacturers combine the limit switch with the fan switch. These switches are called fan/limit switches.

3.2.2 Rollout Switch

Another type of limit switch, known as a *rollout switch* (*Figure 26*), is also mounted near the heat exchanger—just outside the burner section in the vestibule—to stop the burner if an excessive temperature is reached. For example, if the air filter is badly clogged, there will not be enough airflow to transfer heat from the heat exchanger.

Figure 24 Freezestat.

It will overheat as a result, activating the high-temperature limit switch. If the heat exchanger itself is restricted, flames may roll out into the furnace vestibule, activating the rollout switch. It is usually a bimetal-actuated, normally-closed switch that must be manually reset before the furnace can be operated again.

3.2.3 Airflow Control

An airflow switch (*Figure 27*), also known as a *sail switch*, is often installed in ductwork as a safety device. It is used in electric heating systems to guarantee that air is circulating through the air distribution system before the heating elements are turned on, and in cooling systems to verify that there is air flowing across the condenser and evaporator before the compressor is turned on. Sail switches are also used to energize and control power to electronic air cleaners in the return air system.

Pressure Problems

High-pressure and low-pressure switches prevent system operation if the system pressure operates outside a preset range. Common causes of high head pressure are dirty condenser coils or a failed condenser fan motor. Common causes of low suction pressure include a loss of refrigerant charge and low evaporator airflow. A dual-stage pressure switch capable of monitoring both high and low pressures is shown here.

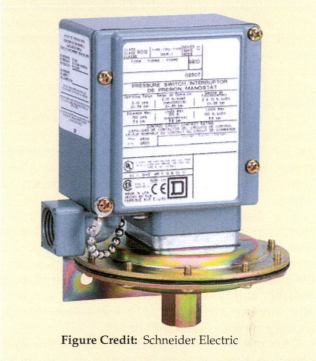

Figure Credit: Schneider Electric

Figure 25 Example of a high-temperature limit switch.

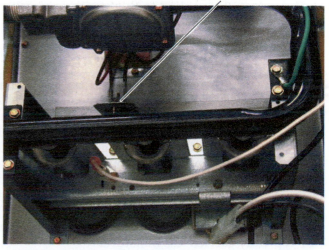

ROLLOUT SWITCH

Figure 26 Rollout switch.

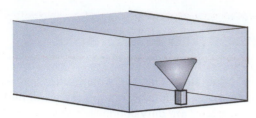

Figure 27 Sail switch.

3.2.4 Inducer Proving Switch

Modern gas furnaces use an inducer motor to induce (produce) a draft. These furnaces have either a centrifugal switch or a pressure-operated switch in the heating control circuit to monitor the operation of the inducer motor and verify that it is running. If the inducer motor fails to operate, the switch opens and prevents control voltage from being applied to the furnace gas valve.

The centrifugal switch has a set of contacts located near the inducer motor shaft. The centrifugal force created by the spinning shaft throws the contacts outward, closing a switch that allows control power to be applied to the furnace gas valve.

The disadvantage of a centrifugal switch is that it can stick in either the closed or open position. This can keep the unit from firing up, or in some cases, can allow the unit to continue firing without a demand from the thermostat. It is an indirect safety device, because it does not actually monitor airflow; it only senses whether the inducer motor is spinning or not.

The pressure switch (*Figure 28*) is the more common type of inducer switch. It has tubing connected to the housing, and in some cases, to the burner enclosure. When the inducer motor is operating, a negative pressure is created that causes the pressure switch to close, allowing control power to be applied to the furnace gas valve. This is a direct safety device.

3.3.0 Heat Pump Defrost Controls

When heat is transferred from the cold outdoor air to the refrigerant in the outdoor coil, moisture condenses on the coil. Because of the temperature of the outdoor coil surface, this moisture will freeze on the coil. Over time, the frost will build up on the coil, blocking airflow and preventing effective heat transfer. This is normal and expected, and it is more likely to be a problem at temperatures between 28°F (–2°C) and 40°F (4°C) than it is at lower outdoor temperatures. Colder air typically carries less moisture.

To eliminate ice buildup on the outdoor coil, most heat pumps have a defrost function.

Figure 28 Combustion airflow pressure-sensing switch.

In air-to-air heat pumps, the heating cycle is reversed, placing the unit back into the cooling mode. Hot refrigerant then flows through the outdoor coil to melt the frost buildup. This cycle usually lasts about 10 minutes. During that time, the heat pump is not providing any heat. In fact, cool air is blowing off the indoor coil into the conditioned space, because it is operating in the cooling mode. During the defrost cycle, the unit is essentially using heat from the indoor air to melt the ice on the outdoor coil. To prevent discomfort to building occupants from cold air blowing during the defrost cycle, a supplemental electric heater is usually energized to re-warm the air before it enters the room.

There are many different ways to control the defrost function. The most common method uses a combination of time and temperature. The logic behind this method is that defrost is needed only if the unit has been operating long enough for frost to form on the coil (for example, 90 minutes), and the temperature is low enough for moisture to freeze on the coil. If both conditions do not exist, defrost is unnecessary. Other defrost methods check for airflow across the frost-blocked coil or monitor outdoor coil temperature.

Modern heat pumps use electronic defrost controls in which all the defrost circuitry, including the relays, is mounted on a printed circuit board (*Figure 29*). The timer is electronic and therefore is less likely to fail. A temperature sensor, often a thermistor, sends temperature information from the outdoor coil to the board. An electronic clock in the timing logic receives a signal as long as the heating thermostat is closed. If defrost is interrupted by the opening of the heating thermostat, the timing logic will remember where it left off. If the coil temperature still indicates the presence of frost when the unit cycles on again, the unit will automatically go back into the defrost mode.

A jumper or other selection device on the board allows the installer to select the defrost frequency (for example, 30, 50, or 90 minutes) based on local conditions.

An important feature of many of these boards is a built-in defrost test cycle. It is activated by placing a jumper across two terminals on the board (*Figure 30*). The 90-minute defrost cycle is overridden, and the usual 10-minute running cycle can be reduced to just a few seconds, if desired. A major advantage of the electronic defrost control is that it is completely self-contained. However, external components such as temperature sensors can still fail.

Electronic defrost controls may be equipped with demand defrost, which combines time, temperature, and pressure information to determine if there is enough frost buildup to require defrost. It also terminates defrost when it senses that the coil is frost-free, regardless of whether the timed cycle is complete. Some electronic defrost controls contain a short-cycle timer.

3.4.0 HVAC Digital Control Systems

Modern buildings use computer technology to control all building functions, including the HVAC, lighting, fire protection, and building security systems. These control systems are known by many names, including *building management systems*, *building automation systems*, *facility management systems*, and *direct digital control (DDC) systems*.

Many building management systems are designed as part of a process called *integrated building design*. Integrated building design is a collaborative decision-making process that uses a project design team from a project's inception through its design and construction phases. The integrated building design process often incorporates green building design. A green building is a sustainable structure that is designed, built, and operated in a manner that efficiently uses resources while being friendly to the environment. The most common green building certification system is Leadership in Energy and Environmental Design (LEED) administered by the U.S. Green Building Council (USGBC). LEED provides a point system to score green building design and construction. The system is categorized into nine basic areas: Integrative Process, Location and Transportation, Sustainable Sites, Water Efficiency, Energy and Atmosphere, Materials and Resources, Indoor Environmental Quality, Innovation, and Regional Priority. Buildings are awarded points based on the number of sustainable strategies that are met in each category, and certifications are earned based on the total score. The certification levels are LEED Certified, Silver, Gold, and Platinum. HVAC systems provide energy management, indoor air quality, and other opportunities for points toward LEED certification.

In order to minimize resource use in many modern green building designs, the building management system must be able to integrate control strategies across multiple building systems. The basic building block of any building management system is a series of digital controllers that are used to monitor and control various building functions. *Figure 31* shows an example of building management system architecture.

Figure 29 Electronic defrost control.

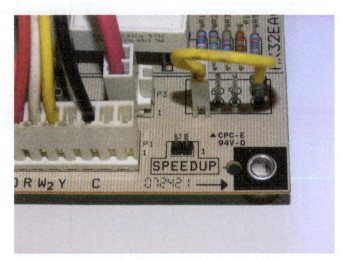

Figure 30 Defrost speed-up jumper.

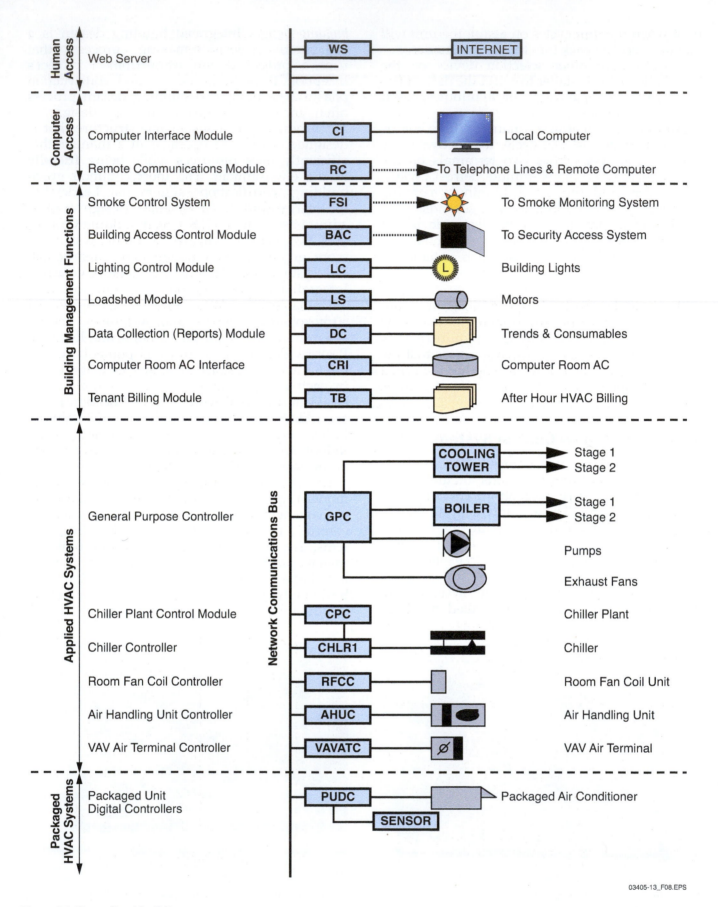

Figure 31 Centralized building management system.

03405-13_F08.EPS

The lower portion of *Figure 31* shows the controllers the BMS will interface with to control the individual pieces of HVAC equipment. The middle portion shows the next level of building control, the interface of the building's other system management functions. Building management systems have features and functions that go far beyond simply maintaining the desired conditions within a space. They include the following:

- Database management
- Alarm management
- Data collection
- Tenant billing
- Load shedding
- Lighting control
- Building access

The top portion of *Figure 31* shows the final level of building control—human access capability using local or remote computers, and/or the internet. The building's digital controllers are connected via the network communication bus. The communication network links all controllers in the building to a centralized computer for human access and integrated building system control strategies. BMS functions can be applied to installations ranging in size from a residence to a multi-building campus.

Figure 32 shows an example of a processor module used to control the interface between a building management system and a specific piece of HVAC equipment.

3.4.1 Controlling Devices

The microprocessor can execute instructions and obtain remarkable results, and it can do so at astounding speeds. If it has no way to communicate with other devices, however, it is useless. Communication is done through two categories of signals: digital and analog. These signals are referred to as *points* in the HVAC industry. The first group includes external digital devices. These might be relays, switches, lights, and other devices that can be operated in either a full on or full off (binary) condition.

The second group includes external analog devices. Devices in this group include thermistors, photocells, and DC motor controls. The problem of interfacing to an analog device is somewhat more complicated. In this case, a signal with an infinite number of values (analog) is converted to a form that can be represented and manipulated by a two-state or binary device (digital).

Most real-world processes are continuously changing (*Figure 33*). Physical quantities such as pressure, temperature, liquid levels, and fluid flow tend to change value rather gradually. Changes of this type produce a large number of discrete values before ever reaching a final state. The problem of converting from analog form to digital form requires the use of a circuit called an analog-to-digital converter.

The analog and digital signals can be further divided into inputs and outputs. Analog-in (AI) signals are obtained from sensors that represent characteristics such as temperature, pressure, and humidity. Analog-out (AO) signals are analog commands, such as the reset of system setpoints. Digital-in (DI) signals are contact closures or openings, showing status or alarm conditions in a two-position mode. Digital-out (DO) signals are two-position commands like start/stop or open/close states.

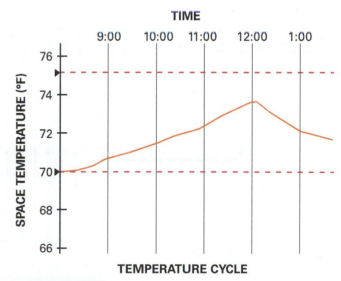

Figure 33 Changes are small and occur gradually.

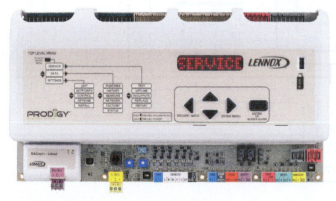

Figure 32 HVAC system control module.

3.4.2 Digital Control System Example

The key to a successful control system is the integration of all the sensors and unit controllers. We will look at the integration of a typical constant-volume HVAC system. A constant-volume system is one in which the volume of supply air remains constant and the temperature of the air is varied to achieve the desired comfort conditions. It is a single-zone system. If there is more than one zone, each must have its own dedicated unit. The constant-volume system contrasts with a variable air-volume system in which the supply air temperature is held constant and the volume of air changes to meet the changing demand. In a variable air-volume system, a single unit can serve several zones.

A schematic of a constant-volume system is shown in *Figure 34*. The control system is made up of several control loops: economizer control of mixed air, heating-cooling sequencing, and humidification-dehumidification sequencing.

When the unit fan is energized, as sensed by a static pressure sensor in the supply duct, the damper control system becomes activated. A mixed-air sensor maintains the mixed-air temperature by modulating the outdoor air, return air, and exhaust dampers. When the outdoor air temperature exceeds the setting of the outdoor air sensor, the outdoor and exhaust air dampers return to their minimum open position, as programmed, to provide required ventilation only. In large buildings, it is usually required that some outside ventilation air be provided at all times. Therefore, the damper has a minimum open position. The return air damper takes the corresponding open position.

A space temperature sensor, through the controller, maintains the space temperature by modulating the heating coil valve in sequence with the chilled water coil valve. A space humidity sensor, also through the controller, maintains the space humidity. Upon a drop in the relative

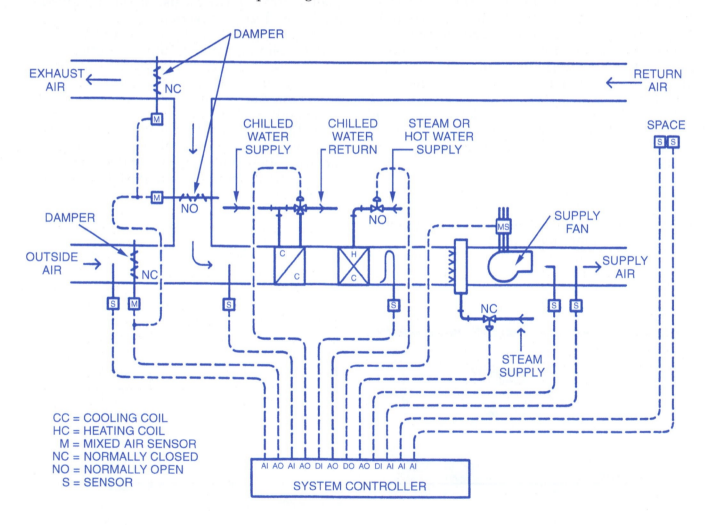

Figure 34 Typical constant-volume HVAC system.

humidity in the space, the humidifier steam valve modulates toward the open position, subject to a duct-mounted high-limit humidity sensor. With a rise in space relative humidity, the humidifier steam valve modulates to the closed position, followed by the opening of the chilled water coil valve to provide dehumidification. During the dehumidification cycle, the space temperature sensor modulates the heating coil valve to maintain space temperature conditions.

A low-temperature controller, with its capillary located on the discharge side of the heating coil, will de-energize the unit fan, close the outdoor and exhaust dampers, and open the return damper if the discharge air temperature drops below its setting. Whenever the unit fan is de-energized, as sensed by the supply duct static pressure sensor, the damper control system will be de-energized, closing the outdoor damper, exhaust damper, and humidifier steam control valve.

3.5.0 Control Circuit Operating Sequence

An automatic air conditioner control circuit cannot get much simpler than the one for a room air conditioner shown in *Figure 35*. It has a main control switch to enable the unit and a thermostat with a sensing bulb that responds to the temperature in the conditioned space. The key word is *automatic*; after the unit is plugged in and turned on, it will cycle on and off by itself based on the temperature at the sensing bulb. The machine would still operate without the thermostat, but the occupant would have to turn it on and off as the temperature varied; it would no longer be automatic.

Figure 36 shows a control circuit that is more typical of a basic cooling system. It has features that are common to most cooling control circuits. For example:

- The control devices (thermostat, compressor contactor, and fan relay) operate using 24VAC.
- The indoor fan has a separate control. By setting the fan switch to On, the occupant can use the fan for ventilation without operating the compressor. When the Fan switch is in the Auto position, the fan relay (IFR) coil will energize whenever the room thermostat (TC) closes.
- The outdoor fan motor (OFM) runs whenever the compressor is on. Follow this sequence: When the unit is on and TC closes, it completes the path to the coil of the compressor contactor (C). The normally open C contacts in the upper part of the circuit then close, completing the current path to the compressor and outdoor fan motor.
- The compressor will usually have some kind of overload protection. In this case, an automatic-reset thermal overload device (OL) is provided. It is embedded in the stator winding of the motor, or it can be externally mounted.

No matter how complex the control circuit appears to be, you will find the basic control arrangement shown in *Figure 36*—or something very much like it—at the heart of the circuit. Everything else will be special features to improve equipment safety or operating efficiency. This is illustrated in *Figure 37*. This diagram is for a combined cooling and gas heating unit. The circuit looks different, and there are several more components. However, if you trace the cooling control function, you will see that it is essentially the same as that previously shown. The differences are that the heating controls have been added near the bottom of the diagram, along with a heating-cooling thermostat. The cooling control has more extras (such as a compressor short-cycle protection circuit, a crankcase heater, and a current-sensitive overload device). Take away these components, and you have a circuit that is identical to the one in *Figure 36*.

Figure 37 appears different because instead of drawing L1 and L2 and the two sides of the transformer secondary as the verticals on a ladder, they are shown emanating from common terminals. This reflects the way the circuit is

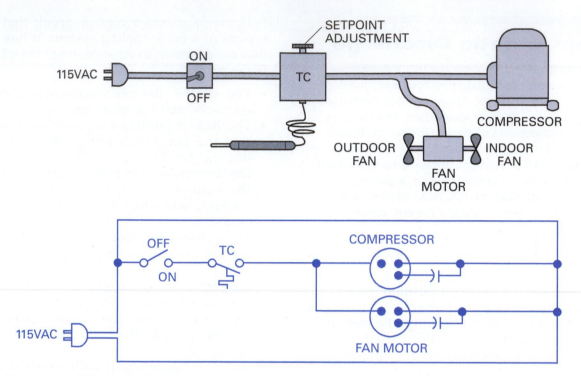

Figure 35 Room air conditioner circuit.

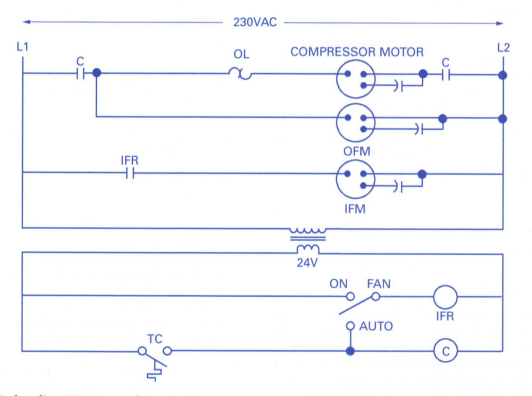

Figure 36 Typical cooling system control circuit.

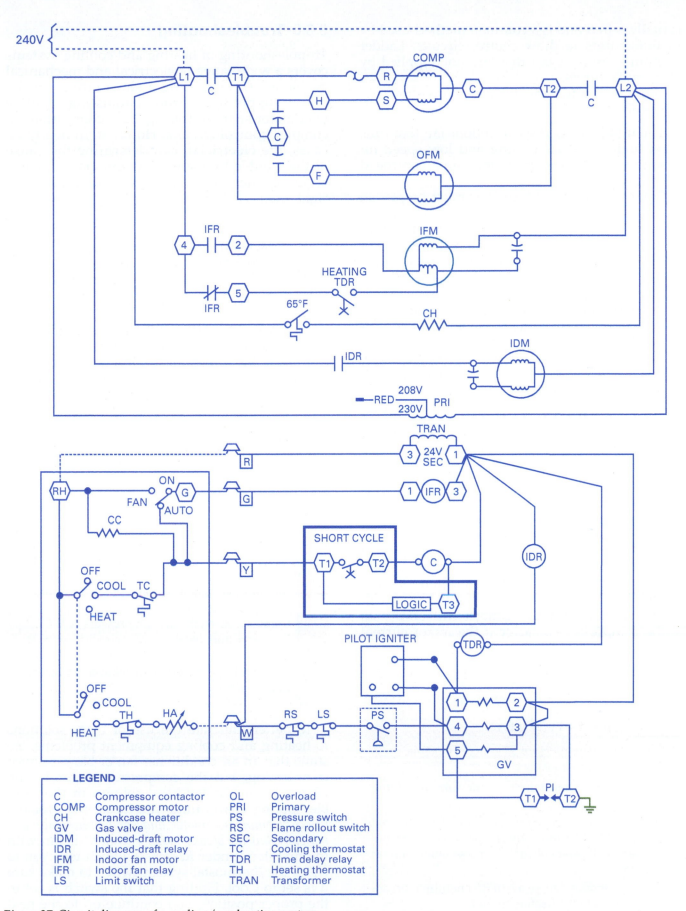

Figure 37 Circuit diagram of a cooling/gas heating system.

LEGEND

C	Compressor contactor	OL	Overload
COMP	Compressor motor	PRI	Primary
CH	Crankcase heater	PS	Pressure switch
GV	Gas valve	RS	Flame rollout switch
IDM	Induced-draft motor	SEC	Secondary
IDR	Induced-draft relay	TC	Cooling thermostat
IFM	Indoor fan motor	TDR	Time delay relay
IFR	Indoor fan relay	TH	Heating thermostat
LS	Limit switch	TRAN	Transformer

actually wired and is a common method used by manufacturers to draw control circuits. Ladder diagrams are nice, but they are not supplied by all manufacturers.

Some of the features of the circuit shown in *Figure 37* are:

- The unit has a two-speed indoor fan that runs on high speed for cooling and low speed for heating. In heating, the indoor fan is controlled by the time-delay relay.
- Operation of the inducer fan must be proven before the gas valve is turned on. The inducer pressure switch (PS), located in the draft hood, will close when the induced-draft motor (IDM) is running at the required speed. If the induced-draft fan stops, the stack pressure will drop and the switch will open, de-energizing the gas supply. The IDM is energized by the induced-draft relay (IDR) as soon as the heating thermostat closes.
- The heating section has two additional safety devices in series with the gas valve. A flame rollout switch (RS) will open if the burner flame escapes from the burner box. This usually indicates insufficient air for combustion or a leak in the heat exchanger. The limit switch (LS) is a heat-sensitive switch that extends into the air stream. If the heat is excessive, it will shut off the gas valve. One cause of excessive heat is insufficient air flowing over the heat exchangers. This could be caused by a blower failure or a dirty filter.
- This diagram has a pilot-duty compressor overload device in addition to the line-duty thermal overload included in *Figure 36*. The additional control is a current-sensitive device that will open the OL contacts in series with the compressor contactor if the motor current is excessive.

3.6.0 Troubleshooting

Troubleshooting of heating and cooling systems covers a wide range of electrical and mechanical problems, from finding a short circuit in the power supply line, through adjusting a pulley on a motor shaft, to tracing loose connections in complex control circuits. However, in nearly all cases, the electrician can determine the cause of the trouble by using a systematic approach, checking one part of the system at a time in the right order.

> **WARNING!**
> HVAC systems contain high voltages, rotating components, and hot surfaces. Be extremely careful when troubleshooting this equipment.

Table 1A, Table 1B, and *Table 1C* are arranged so that the problem is listed first. The possible causes of the problem are listed in the order in which they should be checked. Finally, solutions to the various problems are given, including step-by-step procedures where necessary. *Table 2* shows a troubleshooting guide for electric heat.

> **WARNING!**
> If damaged, the terminals of hermetic and semi-hermetic compressor motors in pressurized systems have been known to blow out. To avoid injury, do not disconnect or connect wiring at the compressor terminals. When testing compressors, do not place test probes on the compressor terminals. Instead, use test points downstream from the compressor. To be safe, measurements and connecting/disconnecting wiring should be done only at the compressor terminals when the system pressure is at 0 psig. The capacitors used in motor circuits can hold a high-voltage charge after the system power is turned off. Always discharge capacitors before touching them.

To better illustrate the use of these solutions to heating and cooling equipment problems, assume that an air conditioner fan or blower motor is operating, but the compressor motor is not. Glance down the left-hand column in the troubleshooting charts until you locate the problem titled *"Compressor motor and condenser motor will not start, but the blower motor operates."* Begin with the first item under *Responses*, which tells you to check the thermostat system switch to make sure it is set to Cool. Finding that the switch is set in the proper position, you continue on to the next item, which is to check the temperature setting.

Think About It

Circuit Analysis

Use the diagram in *Figure 37* to determine the answers to these questions.

1. Does the induced-draft motor turn off if the rollout switch opens?
2. If you connect a voltmeter from L2 to terminal 4 of IFR while the compressor contactor is de-energized, what voltage, if any, would the meter read?
3. Is the indoor fan relay (IFR) energized or de-energized in the heating mode?

Thermistors

Thermistors are temperature-sensitive semiconductor devices. Their resistance varies in a predictable way with variations in temperature. This allows them to be used in a variety of HVAC control applications. Thermistors have either a positive or negative coefficient of resistance. If the resistance increases as the temperature rises, it has a positive coefficient. If the resistance increases as the temperature drops, it has a negative coefficient. Thermistors come in different sizes and shapes to suit a variety of applications. In HVAC systems, they are used to sense the air temperature in room thermostats, the motor winding temperature when embedded in motor windings, the outdoor coil temperature as part of a defrost circuit, and countless other applications for which an accurate measure of temperature is needed.

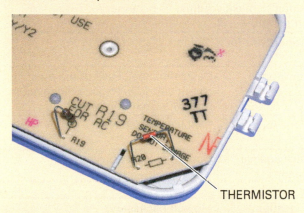

(A) THERMISTOR USED IN A ROOM THERMOSTAT

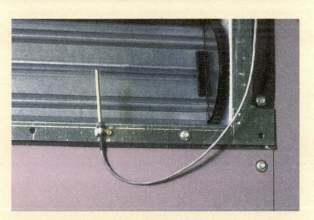

(B) THERMISTOR USED TO CONTROL OUTDOOR AIR DAMPER

Crankcase Heaters

Lubricating oil circulates with the refrigerant throughout the refrigerant circuit. The oil is used to lubricate the compressor, but the refrigerant flows through the crankcase, consistently picking up small amounts of oil and carrying it along. When a compressor sits in cool temperatures for a lengthy period of time, refrigerant vapors tend to migrate outdoors to the unit and condense inside the compressor crankcase. If the compressor is started with this liquid refrigerant in the crankcase, the rapid expansion of the refrigerant back to a vapor (a violent boiling action) can carry most or all of the oil away, leaving the compressor without lubrication. Crankcase heaters are used to keep refrigerant from condensing in the compressor crankcase when the unit is off.

Crankcase heaters come in a variety of sizes and shapes. One type, called a *bellyband heater*, clamps around the bottom of a welded hermetic compressor shell. Other heaters are inserted in a well in the base of the compressor shell, under the crankcase. Some equipment is designed so that power is applied to the crankcase heater at all times; others are controlled by thermostats that apply power to the heater only when the outdoor temperature drops below a certain point and the compressor is off.

Figure Credit: Pentair Thermal Management LLC

Table 1A HVAC Troubleshooting Guide (1 of 3)

Malfunction	Responses	Corrective Action
Compressor motor and condenser motor will not start, but the blower motor operates.	Check the thermostat system switch to ascertain that it is set to Cool.	Make necessary adjustments to settings.
	Check the thermostat to make sure that it is set below room temperature.	Make necessary adjustments.
	Check all low-voltage connections for tightness.	Tighten.
	Make a low-voltage check with a voltmeter on the condensate overflow switch; the condensate may not be draining.	The float switch is normally found in the condensate pump. Repair or replace.
	Make a low-voltage check of the pressure switches.	Replace if defective. Call a qualified HVAC technician to check the system charge.
Compressor, condenser, and blower motors will not start.	Check the thermostat system switch setting to ascertain that it is set to Cool.	Adjust as necessary.
	Check the thermostat setting to make sure it is below room temperature.	Adjust as necessary.
	Check all low-voltage connections for tightness.	Tighten.
	Check for a blown fuse or tripped circuit breaker.	Determine the cause of the open circuit and then replace the fuses or reset the circuit breaker.
	Make a voltage check of the low-voltage transformer.	Replace if defective.
	Check the electrical service against minimum requirements (correct voltage and amperage).	Update as necessary.
Condensing unit cycles too frequently.	Check room thermostat location.	Move if necessary.
	Check all low-voltage wiring connections for tightness.	Tighten.
	The equipment could be oversized.	Call qualified HVAC technician to check system capacity.
Inadequate cooling with condensing unit and blower running continuously.	Check for refrigerant leak or undercharge.	Call in a qualified HVAC technician to check for leaks in the refrigerant lines and/or to add charge.
	Check for undersized equipment.	Call qualified HVAC technician to check system capacity.

Malfunction	Responses	Corrective Action
Condensing unit cycles, but the evaporator blower motor does not run.	Check all low-voltage connections for tightness.	Tighten.
	Make a voltage check of the blower relay.	Replace if necessary.
	Check for correct voltage at motor terminals.	Replace if necessary.
	Make electrical and mechanical checks on the blower motor. Mechanical problems could be bad bearings or a loose blower wheel.	Repair or replace defective components.
Unit shows continuous short cycling of evaporator blower and provides Insufficient cooling.	Make electrical and mechanical checks.	Repair or replace motor if necessary.
Thermostat calls for heat, but the blower motor does not operate.	Check all low-voltage and line-voltage connections for tightness.	Tighten.
	Check for blown fuses or a tripped circuit breaker in the line.	Determine the reason for the open circuit and replace the fuses or reset the circuit breaker.
	Check the low-voltage transformer.	Replace if defective.
	Make a low-voltage check of the fan relay.	Repair or replace if necessary.
	Make electrical and mechanical checks on the blower motor.	Repair or replace the motor if defective.
Thermostat calls for heat; blower motor operates, but it delivers cold air.	Make a visual and electrical check on the heating elements.	If not operating, continue to the next check.
	Make an electrical check of the heater limit switch; begin by disconnecting all power to the unit, then use an ohmmeter to check for continuity between the two terminals of the switch.	If the limit switch is open, repair or replace.
	Make an electrical check of the sequencer. Most are rated at 24V and have one set of normally open auxiliary contacts for pilot duty.	If the sequencer coil is open or grounded, repair or replace.
	Check the electric service entrance and related circuits against the minimum recommendations.	Upgrade if necessary.

Malfunction	Responses	Corrective Action
Thermostat calls for heat and blower motor operates continuously; system delivers warm air, but the thermostat is not satisfied.	Check all joints in the ductwork for air leaks.	Make all defective joints tight.
	Check for a dirty air filter.	Clean or replace filter.
	Make a visual and electrical check of the electric heating element.	Repair or replace if necessary.
	Make an electrical check of the heater limit switch as described previously.	Repair or replace.
	Check for undersized equipment.	Call HVAC technician to check system capacity.
Electric heater cycles on limit switches, but the blower motor does not operate.	Make an electrical check of the fan relay.	Repair or replace if defective.
	Make electrical and mechanical checks on the blower motor.	Repair or replace if defective.
	Check the line connections for tightness.	Make any necessary changes.
There is excessive noise at the return air grille.	Check the return duct to make sure it has a 90° bend.	Correct if necessary.
	Make a visual check of the blower unit to ascertain that all shipping blocks and angles have been removed.	Remove if necessary.
	Check the blower motor assembly suspension and fasteners.	Remove if necessary.
There is excessive vibration at the blower unit.	Visually check for vibration isolators (which isolate the blower coil from the structure).	If missing, install as recommended by the manufacturer.
	Visually check to ascertain that all shipping blocks and angles have been removed from the blower unit.	Remove if necessary.
	Check the blower motor assembly suspension and fasteners.	Tighten if necessary.

Table 2 Troubleshooting Guide for Electric Heat

GENERALLY THE CAUSE Make these checks first.	OCCASIONALLY THE CAUSE Make these checks second.	RARELY THE CAUSE Make these checks last.
Problem: No heat—unit fails to operate		
Power failure	Control transformer	Faulty wiring
Blown fuses or tripped circuit breaker	Control relay or contactor	Loose terminals
Open disconnect switch or blown fuse	Bad thermal fuse	
Thermostat switch not in proper position		
Element open		
Problem: Not enough heat—unit cycles too often		
Limit controls	Control relay or contactor	Loose terminals
Low air volume	Fan control	Low voltage
Dirty blower wheel	Thermostat level	Faulty wiring
Dirty filters	Thermostat location	Loose blower wheel
Element fuse	Outdoor thermostat	Ductwork small or restricted
Heat sequencer	Blower bearings	
Heat relay	Blower motor	
Thermostat heat anticipator improperly set		
Problem: Too much heat—unit cycle is too long		
Thermostat heat anticipator incorrectly set	Thermostat level	Control relay or contactor
Thermostat out of calibration	Thermostat location	
Problem: No heat—unit runs continuously		
Thermostat faulty	Faulty wiring	Thermostat not level
Control relay or contactor		Thermostat location
Problem: Cost of operation too high		
Blower motor	Low air volume	Low voltage
Dirty filters	Insufficient building insulation	Fan control
Outdoor thermostat setting incorrect	Excessive building infiltration	Blower belt broken or slipping
		Dirty blower wheel
		Ductwork small or restricted
Problem: Mechanical noise		
Blower bearings	Blower motor	Low voltage
Blower out of balance	Cabinet	Control transformer
Blower belt slipping		
Problem: Air noise		
Blower	Ductwork or grilles restricted	Blower wheel
Cabinet		Dirty filters
Dirty coil		
Problem: Odor		
Foreign substances (dirt or dust) on heating elements	Low air volume	Faulty wiring
	Humidifier containing stagnantwater	Loose terminals
		Control transformer
	Water or moisture	Dirty filters
		Dirty or plugged heat exchanger

You may find that the temperature setting is above the room temperature, so the system is not calling for cooling. Set the thermostat below room temperature, and the cooling unit will function.

This example is, of course, very simple, but most of the heating and cooling problems can be just as simple if a systematic approach to troubleshooting is used.

Manufacturers of HVAC equipment also provide troubleshooting and maintenance manuals for their equipment. These manuals can be one of the most helpful tools for troubleshooting specific HVAC equipment. When unpacking equipment, controls, and other components for the system, always save any manuals or instructions that accompany the items. File them in a safe place so that you and other maintenance personnel can readily find them. Many electricians like to secure this type of manual on the inside of a cabinet door within the equipment. This way, it will always be available when needed.

Think About It

Baseboard Heat

What are the requirements for the disconnecting means for baseboard electric heaters?

3.0.0 Section Review

1. The purpose of the lockout relay in a comfort cooling control circuit is to _____.

   a. prevent the unit from operating on cool days
   b. keep unauthorized personnel from operating the equipment
   c. keep the equipment form operating during maintenance procedures
   d. prevent automatic restart of the HVAC equipment

2. The purpose of the inducer-proving control is to _____.

   a. monitor the operation of the inducer motor and verify that it is running
   b. prevent the blower fan from turning on if the gas valve is not operating
   c. prevent the heat exchanger temperature from exceeding a preset level
   d. make sure the gas valve is operating before turning on the gas

3. During a defrost cycle, heat pumps melt ice on the outside coil using _____.

   a. air supplied by another electric heater
   b. the compressor crankcase
   c. the thermostat heat anticipator
   d. warm indoor air

4. Which of the following would be considered a digital device?

   a. DC motor control
   b. relay
   c. thermistor
   d. photocell

5. Which of the following would be an optional feature in an air conditioning control system?

   a. Indoor fan motor
   b. Compressor contactor
   c. Crankcase heater
   d. Outdoor fan motor

6. Which of these causes could be immediately eliminated if a compressor and condenser fan will NOT start?

   a. Defective indoor fan motor
   b. Thermostat not level
   c. Defective pressure switch
   d. Thermostat not set below room temperature

4.0.0 *NEC*® REQUIREMENTS

Objective

Identify the *NEC*® requirements that apply to HVAC systems.

   a. Identify the *NEC*® requirements that apply to HVAC controls.

   b. Identify the *NEC*® requirements that apply to HVAC equipment.

This section summarizes the *NEC*® requirements for HVAC controls, compressors, room air conditioners, baseboard heaters, and electric space-heating cables.

4.1.0 *NEC*® Requirements for HVAC Controls

Several sections of the *NEC*® cover the requirements for HVAC controls. For example, *NEC Article 424, Part III* deals with the control and protection of fixed electric space heating equipment; *NEC Article 440, Part V* covers requirements for motor compressor controllers; and *NEC Article 440, Part VI* deals with motor compressor and branch circuit overload protection.

In general, a means must be provided to disconnect heating equipment, including motor controllers and supplementary overcurrent protective devices, from all ungrounded conductors. The disconnecting means may be a switch, circuit breaker, unit switch, or thermostatically controlled switching device. Both the selection and use of disconnecting devices are governed by the type of overcurrent protection and the rating of any motors that are part of the equipment.

In certain heating units, supplementary overcurrent protective devices other than the branch circuit overcurrent protection are required. These supplementary overcurrent devices are normally used when heating elements rated at more than 48A are supplied as a subdivided load. In this case, the disconnecting means must be on the supply side of the supplementary overcurrent protective device and within sight of it. This disconnecting means may also serve to disconnect the heater and any motor controllers, provided the means of disconnect is within sight of the controller and heater, or it can be locked in the open position. If the motor is rated over $\frac{1}{8}$hp, a disconnecting means must comply with the rules for motor disconnecting unless a unit switch is used to disconnect all ungrounded conductors.

A heater without supplementary overcurrent protection must have a disconnecting means that complies with rules similar to those for permanently connected appliances. A unit switch may be the disconnecting means in certain occupancies when other means of disconnection are provided as specified in the *NEC*®.

Figure 38 summarizes some of the *NEC*® requirements for HVAC controls, including thermostats, motor controllers, disconnects, and overcurrent protection. Additional *NEC*® regulations on motor controllers are covered in other modules.

You are also encouraged to study *NEC Articles 424 and 440*. Although most of this material has been covered in this module, interpreting these *NEC*® regulations is a good training exercise in itself.

4.2.0 *NEC*® Requirements for HVAC Equipment

The *NEC*® contains specific requirements for various types of HVAC equipment. These include compressors, room air conditioners, electric baseboard heaters, and energy management systems.

Heat Pump Installations

Most heat pump installations require that supplemental electric heaters be installed in the indoor air handling unit. They are required because the output of the heat pump declines as the outdoor temperature drops. The size of the electric heaters depends on a number of variables. Some codes require that the electric heaters be sized to satisfy the building's full heating load if the compressor is inoperative. In some colder climates, that could mean more than 20 kilowatts (kW) of electric heat. If a heat pump is being used to replace a fossil-fuel furnace in an existing structure and a large amount of electric resistance heat is required, carefully evaluate the existing electrical service to ensure that the service can handle the additional load. In some cases, a new electrical service will be required.

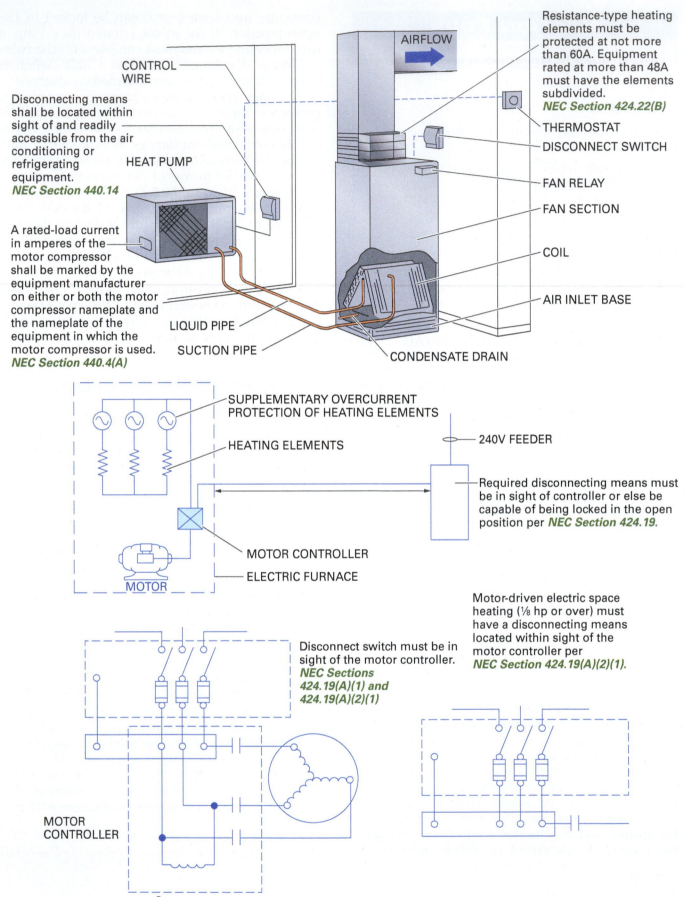

CONTROL
WIRE

Disconnecting means
shall be located within
sight of and readily
accessible from the air
conditioning or
refrigerating
equipment.
NEC Section 440.14

HEAT PUMP

A rated-load current
in amperes of the
motor compressor
shall be marked by the
equipment manufacturer
on either or both the motor
compressor nameplate and
the nameplate of the
equipment in which the
motor compressor is used.
NEC Section 440.4(A)

LIQUID PIPE

SUCTION PIPE

AIRFLOW

Resistance-type heating
elements must be
protected at not more
than 60A. Equipment
rated at more than 48A
must have the elements
subdivided.
NEC Section 424.22(B)

THERMOSTAT

DISCONNECT SWITCH

FAN RELAY

FAN SECTION

COIL

AIR INLET BASE

CONDENSATE DRAIN

SUPPLEMENTARY OVERCURRENT
PROTECTION OF HEATING ELEMENTS

HEATING ELEMENTS

240V FEEDER

Required disconnecting means must
be in sight of controller or else be
capable of being locked in the open
position per *NEC Section 424.19.*

MOTOR CONTROLLER

ELECTRIC FURNACE

MOTOR

Motor-driven electric space
heating (⅛ hp or over) must
have a disconnecting means
located within sight of the
motor controller per
NEC Section 424.19(A)(2)(1).

Disconnect switch must be in
sight of the motor controller.
*NEC Sections
424.19(A)(1) and
424.19(A)(2)(1)*

MOTOR
CONTROLLER

Figure 38 Summary of *NEC*® requirements for HVAC systems.

4.2.1 Compressors

NEC Article 440 contains provisions for motor-driven equipment, as well as for the branch circuits and controllers for the equipment. It also takes into account the special considerations involved with sealed (hermetic) motor compressors in which the motor operates using the cooling effect of the refrigerant as it passes through the crankcase.

It must be noted, however, that the rules of *NEC Article 440* are in addition to, or are amendments to, the rules given in *NEC Article 430*. The basic rules of *NEC Article 430* also apply to air conditioning and refrigerating equipment, unless exceptions are indicated in *NEC Article 440*. *NEC Article 440* further clarifies the application of *NEC®* rules to this type of equipment.

Where refrigeration compressors are driven by conventional motors (nonhermetic), the motors and controls are subject to *NEC Article 430*. They are not subject to *NEC Article 440*.

Other *NEC®* articles that will be covered in this module (besides *NEC Articles 430 and 440*) include:

- *NEC Article 422*, *Appliances*
- *NEC Article 424*, *Fixed Electric Space-Heating Equipment*

Room air conditioners are covered in *NEC Article 440, Part VII*. However, they must also comply with the rules of *NEC Article 422*.

Household refrigerators and freezers, drinking water coolers, and beverage dispensers are considered by the *NEC®* to be appliances, and their application must comply with *NEC Article 422* and must also satisfy the rules of *NEC Article 440*, because such devices contain sealed motor compressors.

Hermetic refrigerant motor compressors, circuits, controllers, and equipment must also comply with the applicable provisions of the following:

- *NEC Article 460*, *Capacitors*
- *NEC Article 470*, *Resistors and Reactors*
- *NEC Articles 500 through 503*, *Hazardous (Classified) Locations*
- *NEC Articles 511, 513 through 517, and all other articles covering special occupancies*

Table 3 summarizes the requirements of *NEC Article 440*. The figures that follow elaborate on many of these requirements. *Figure 39* provides detailed *NEC®* information about motor compressors; *Figure 40* illustrates compressor branch and control circuit requirements; *Figure 41* shows the requirements for HVAC-related lighting and switches; and *Figure 42* illustrates the requirements for 125VAC receptacles in relation to HVAC equipment locations.

4.2.2 Room Air Conditioners

There are millions of room air conditioners in use throughout the United States. Consequently, the *NEC®* deemed it necessary to provide *NEC Article 440, Part VII*, beginning with *NEC Section 440.60*, to ensure that such equipment would be installed so as not to provide a hazard to life or property. These *NEC®* requirements apply to electrically energized room air conditioners that control temperature and humidity. In general, this section of the *NEC®* considers a room air conditioner (with or without provisions for heating) to be an alternating current appliance of the air-cooled window, console, or through-wall type that is installed in a conditioned room or space and that incorporates one or more hermetic refrigerant motor compressor(s).

Furthermore, this *NEC®* provision covers only equipment rated at 250V or less, single-phase, and such equipment is permitted to be cord- and plug-connected.

Three-phase room air conditioners, or those rated at over 250V, are not covered under *NEC Article 440, Part VII*. This type of equipment must be directly connected using a wiring method as described in *NEC Chapter 3*.

The majority of room air conditioners covered under *NEC Article 440* are cord- and plug-connected to receptacle outlets of general-purpose branch circuits. The rating of any such unit must not exceed 80% of the branch circuit rating if connected to a 15A, 20A, or 30A general-purpose branch circuit. The rating of cord- and plug-connected room air conditioners must not exceed 50% of the branch circuit rating if lighting units and other appliances are also supplied.

Figure 43 depicts the *NEC®* application rules for room air conditioners on branch circuits that also serve other loads such as lighting. Note that the attachment plug and receptacle are allowed to serve as the disconnecting means. In some cases, the attachment plug and receptacle may also serve as the controller, or the controller may be a switch that is an integral part of the unit. The required overload protective device may be supplied as an integral part of the appliance and need not be included in the branch circuit calculations. *Figure 44* provides details related to room air conditioners that are on a dedicated circuit.

Table 3 Summary of *NEC®* Requirements for Hermetically Sealed Compressors

Application	*NEC®* Regulation	*NEC®* Reference
Marking on hermetic compressors	Hermetic compressors must be provided with a nameplate containing the manufacturer's name, trademark, or symbol, identifying designation, phase, voltage, frequency, rated-load current, locked-rotor current, and the words thermally protected, if appropriate. The fault current rating and date of calculation must also be made available.	*NEC Section 440.4(A)* *NEC Section 440.10(B)*
Marking on controllers	A controller serving hermetically sealed compressors must be marked with the maker's name, trademark, or symbol, identifying designation, voltage, phase, and full-load and locked-rotor currents (or hp) rating.	*NEC Section 440.5*
Ampacity and rating	Conductors for hermetically sealed compressors must be sized according to *NEC Tables 310.16 through 310.19* or calculated in accordance with *NEC Section 310.14*, as applicable.	*NEC Section 440.6*
Highest rated motor	The largest motor is the motor with the highest rated-load current.	*NEC Section 440.7*
Single machine	The entire HVAC system is considered one machine, regardless of the number of motors involved in the system.	*NEC Section 440.8*
Rating and interrupting capacity	The disconnecting means for hermetic compressors must be selected based on the nameplate rated-load current or branch circuit selection current, whichever is greater, and locked-rotor current, respectively.	*NEC Section 440.12(A)*
Cord-connected equipment	For cord-connected equipment, an attachment plug and receptacle are permitted to serve as the disconnecting means.	*NEC Section 440.13*
Location	A disconnecting means must be located within sight of the equipment. The disconnecting means may be mounted on or within the HVAC equipment.	*NEC Section 440.14*
Short circuit and ground fault protection	Amendments to *NEC Article 240* are provided in *NEC Article 440, Part III* for circuits supplying hermetically sealed compressors against overcurrent due to short circuits and grounds.	*NEC Section 440.21*
Rating of short circuit and ground fault protective device	The rating must not exceed 175% of the compressor rated-load current; if necessary for starting, the device may be increased to a maximum of 225%. [Do not exceed the manufacturer's value per *NEC Section 440.22(C)*].	*NEC Section 440.22(A)*
Compressor branch circuit conductors	Branch circuit conductors supplying a single compressor must have an ampacity of not less than 125% of either the motor compressor rated-load current or the branch circuit selection current, whichever is greater.	*NEC Section 440.32*
	Conductors supplying more than one compressor must be sized for the total load plus 25% of the largest motor's full-load amps.	*NEC Section 440.33*
Combination load	Conductors must be sufficiently sized for the other loads plus the required ampacity for the compressor as required in *NEC Sections 440.32 and 440.33*.	*NEC Section 440.34*
Multi-motor load equipment	Conductors must be sized to carry the circuit ampacity marked on the equipment as specified in *NEC Section 440.4(B)*.	*NEC Section 440.35*
Controller rating	Controllers must have both a continuous-duty full-load current rating and a locked-rotor current rating not less than the nameplate rated-load current or branch circuit selection current, whichever is greater, and locked-rotor current, respectively.	*NEC Section 440.41(A)*
Application and selection of controllers	Each motor compressor must be protected against overload and failure to start by one of the means specified in *NEC Sections 440.52(A)(1) through (4)*.	*NEC Section 440.52(A)*
Overload relays	Overload relays and other devices for motor overload protection that are not capable of opening short circuits must be protected by a suitable fuse or inverse time circuit breaker.	*NEC Section 440.53*
Equipment on 15A or 20A branch circuit; time delay required	Short circuit and ground fault protective devices protecting 15A or 20A branch circuits must have sufficient time delay to permit the motor compressor and other motors to start and accelerate their loads.	*NEC Section 440.54(B)*

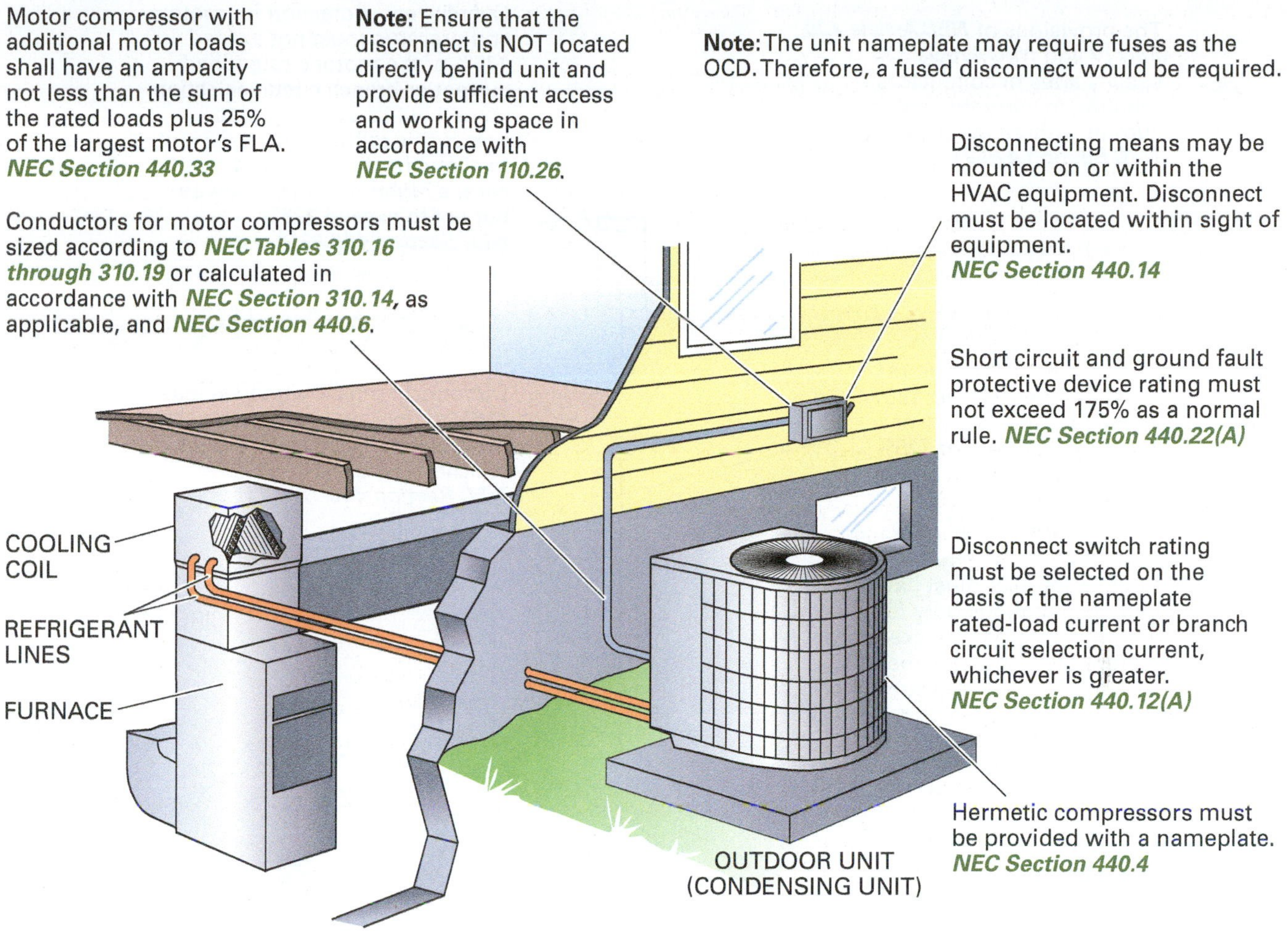

Motor compressor with additional motor loads shall have an ampacity not less than the sum of the rated loads plus 25% of the largest motor's FLA. *NEC Section 440.33*

Note: Ensure that the disconnect is NOT located directly behind unit and provide sufficient access and working space in accordance with *NEC Section 110.26*.

Note: The unit nameplate may require fuses as the OCD. Therefore, a fused disconnect would be required.

Conductors for motor compressors must be sized according to *NEC Tables 310.16 through 310.19* or calculated in accordance with *NEC Section 310.14*, as applicable, and *NEC Section 440.6*.

Disconnecting means may be mounted on or within the HVAC equipment. Disconnect must be located within sight of equipment. *NEC Section 440.14*

Short circuit and ground fault protective device rating must not exceed 175% as a normal rule. *NEC Section 440.22(A)*

Disconnect switch rating must be selected on the basis of the nameplate rated-load current or branch circuit selection current, whichever is greater. *NEC Section 440.12(A)*

Hermetic compressors must be provided with a nameplate. *NEC Section 440.4*

COOLING COIL

REFRIGERANT LINES

FURNACE

OUTDOOR UNIT (CONDENSING UNIT)

Figure 39 NEC® regulations governing motor compressors.

Equipment grounding, as required by *NEC Section 440.61*, may be handled by the grounded receptacle.

4.2.3 Electric Baseboard Heaters

All requirements of the *NEC®* apply for the installation of electric baseboard heaters, especially *NEC Article 424*, *Fixed Electric Space-Heating Equipment*. In general, electric baseboard heaters must not be used where they will be exposed to severe physical damage unless they are adequately protected from such damage. Heaters and related equipment installed in damp or wet locations must be approved for such locations and must be constructed and installed so that water cannot enter or accumulate in or on wired sections, electrical components, or ductwork.

Baseboard heaters must be installed to provide the required spacing between the equipment and adjacent combustible material, and each unit must be adequately grounded in accordance with *NEC Article 250*. *Figure 45* lists some of the *NEC®* regulations governing the installation of electric baseboard heaters.

4.2.4 Electric Space-Heating Cable

Always make certain that the heating cable is connected to the proper voltage. A 120V cable connected to a 240V circuit will melt the cable. A 240V heating cable connected to a 120V circuit will produce only 25% of the rated wattage of the cable. See *Figure 46* for a summary of *NEC®* requirements governing the installation of electric space-heating cable.

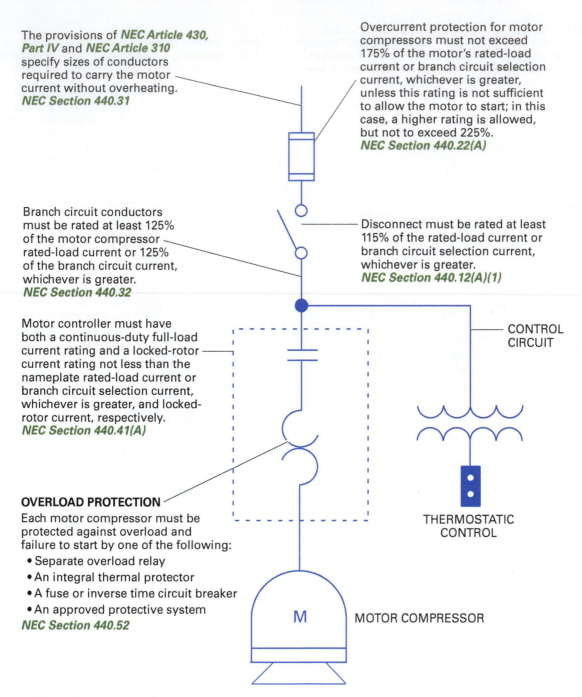

The provisions of *NEC Article 430, Part IV* and *NEC Article 310* specify sizes of conductors required to carry the motor current without overheating. *NEC Section 440.31*

Overcurrent protection for motor compressors must not exceed 175% of the motor's rated-load current or branch circuit selection current, whichever is greater, unless this rating is not sufficient to allow the motor to start; in this case, a higher rating is allowed, but not to exceed 225%. *NEC Section 440.22(A)*

Branch circuit conductors must be rated at least 125% of the motor compressor rated-load current or 125% of the branch circuit current, whichever is greater. *NEC Section 440.32*

Disconnect must be rated at least 115% of the rated-load current or branch circuit selection current, whichever is greater. *NEC Section 440.12(A)(1)*

CONTROL CIRCUIT

Motor controller must have both a continuous-duty full-load current rating and a locked-rotor current rating not less than the nameplate rated-load current or branch circuit selection current, whichever is greater, and locked-rotor current, respectively. *NEC Section 440.41(A)*

OVERLOAD PROTECTION
Each motor compressor must be protected against overload and failure to start by one of the following:
• Separate overload relay
• An integral thermal protector
• A fuse or inverse time circuit breaker
• An approved protective system
NEC Section 440.52

THERMOSTATIC CONTROL

M   MOTOR COMPRESSOR

Figure 40 Compressor branch and control circuits.

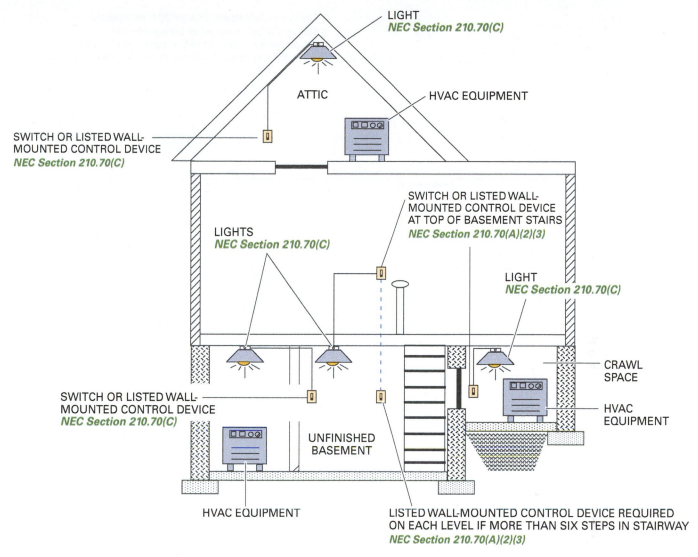

Figure 41 NEC® requirements for lighting and switches for HVAC equipment.

4.2.5 Energy Management Systems

An energy management system consists of monitors, communications equipment, controllers, timers, and other devices that monitor or control electrical loads, power production, or storage sources. *NEC Section 750.20* covers alternate power sources that can be controlled by the energy management system. *NEC Section 750.30(A)* covers the systems connected to load shedding controls that cannot be controlled by the energy management system. *NEC Article 705* covers interconnected electric power production sources, such as solar photovoltaic (PV) systems. *NEC Article 706* applies to permanently installed energy storage systems.

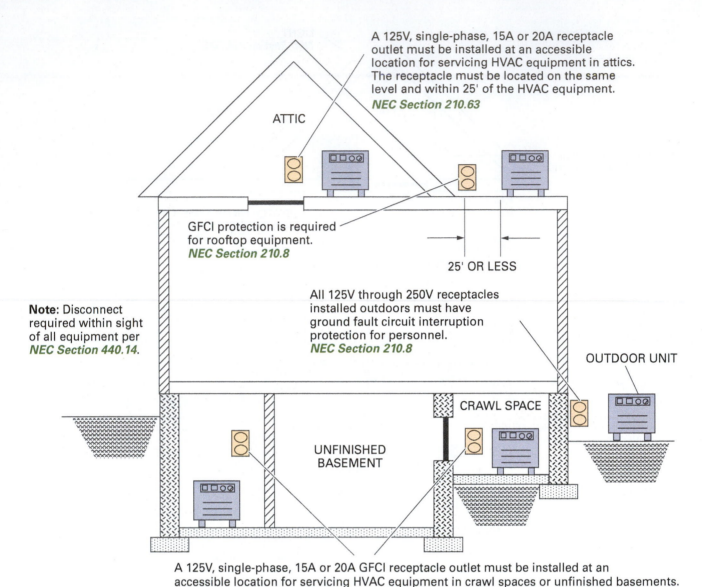

A 125V, single-phase, 15A or 20A receptacle outlet must be installed at an accessible location for servicing HVAC equipment in attics. The receptacle must be located on the same level and within 25' of the HVAC equipment. *NEC Section 210.63*

ATTIC

GFCI protection is required for rooftop equipment. *NEC Section 210.8*

25' OR LESS

Note: Disconnect required within sight of all equipment per *NEC Section 440.14*.

All 125V through 250V receptacles installed outdoors must have ground fault circuit interruption protection for personnel. *NEC Section 210.8*

OUTDOOR UNIT

CRAWL SPACE

UNFINISHED BASEMENT

A 125V, single-phase, 15A or 20A GFCI receptacle outlet must be installed at an accessible location for servicing HVAC equipment in crawl spaces or unfinished basements. The receptacle must be located on the same level and within 25' of the HVAC equipment. *NEC Sections 210.63 and 210.8*

Figure 42 NEC® requirements for locating 125V receptacles at HVAC equipment.

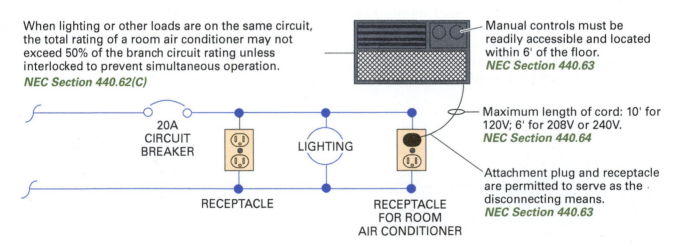

When lighting or other loads are on the same circuit, the total rating of a room air conditioner may not exceed 50% of the branch circuit rating unless interlocked to prevent simultaneous operation. *NEC Section 440.62(C)*

Manual controls must be readily accessible and located within 6' of the floor. *NEC Section 440.63*

20A CIRCUIT BREAKER

LIGHTING

Maximum length of cord: 10' for 120V; 6' for 208V or 240V. *NEC Section 440.64*

RECEPTACLE

RECEPTACLE FOR ROOM AIR CONDITIONER

Attachment plug and receptacle are permitted to serve as the disconnecting means. *NEC Section 440.63*

Figure 43 Branch circuits for room air conditioners.

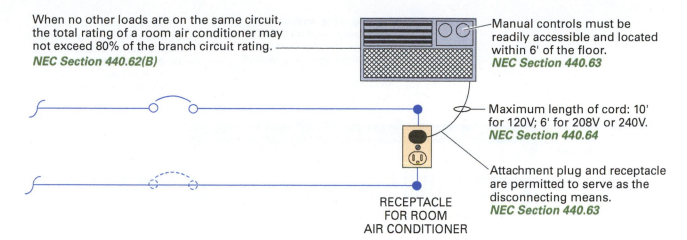

When no other loads are on the same circuit, the total rating of a room air conditioner may not exceed 80% of the branch circuit rating. *NEC Section 440.62(B)*

Manual controls must be readily accessible and located within 6' of the floor. *NEC Section 440.63*

Maximum length of cord: 10' for 120V; 6' for 208V or 240V. *NEC Section 440.64*

Attachment plug and receptacle are permitted to serve as the disconnecting means. *NEC Section 440.63*

RECEPTACLE FOR ROOM AIR CONDITIONER

Figure 44 Room air conditioner connected to a dedicated branch circuit.

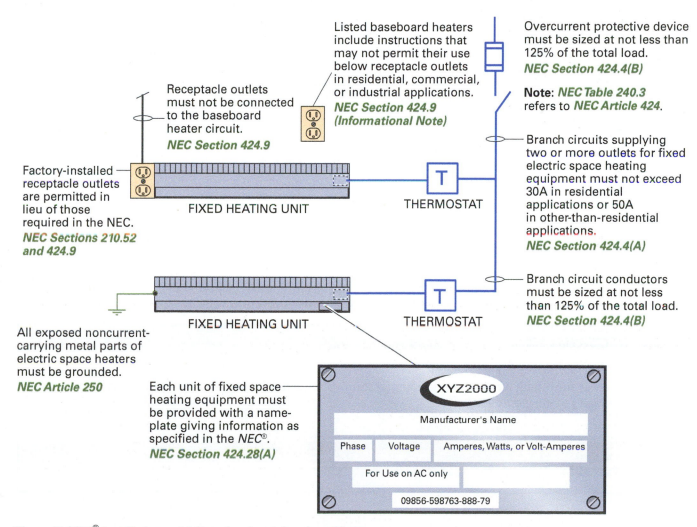

Listed baseboard heaters include instructions that may not permit their use below receptacle outlets in residential, commercial, or industrial applications. *NEC Section 424.9 (Informational Note)*

Overcurrent protective device must be sized at not less than 125% of the total load. *NEC Section 424.4(B)*

Note: *NEC Table 240.3* refers to *NEC Article 424*.

Receptacle outlets must not be connected to the baseboard heater circuit. *NEC Section 424.9*

Branch circuits supplying two or more outlets for fixed electric space heating equipment must not exceed 30A in residential applications or 50A in other-than-residential applications. *NEC Section 424.4(A)*

Factory-installed receptacle outlets are permitted in lieu of those required in the NEC. *NEC Sections 210.52 and 424.9*

FIXED HEATING UNIT

THERMOSTAT

Branch circuit conductors must be sized at not less than 125% of the total load. *NEC Section 424.4(B)*

FIXED HEATING UNIT

THERMOSTAT

All exposed noncurrent-carrying metal parts of electric space heaters must be grounded. *NEC Article 250*

Each unit of fixed space heating equipment must be provided with a name-plate giving information as specified in the *NEC®*. *NEC Section 424.28(A)*

XYZ2000

Manufacturer's Name

Phase | Voltage | Amperes, Watts, or Volt-Amperes

For Use on AC only

09856-598763-888-79

Figure 45 *NEC®* installation guidelines for electric baseboard heaters.

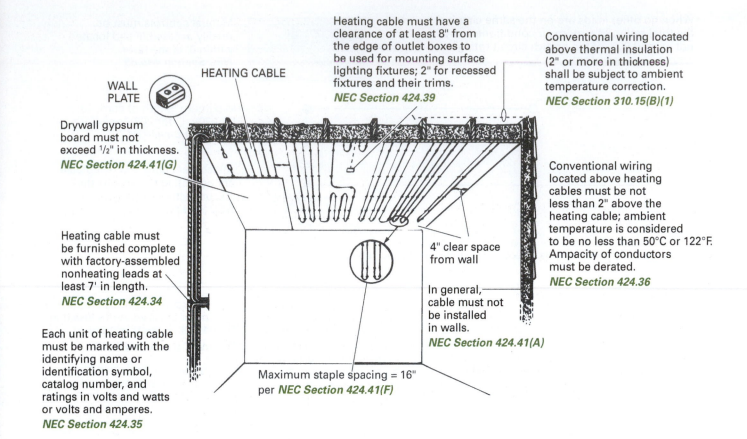

Drywall gypsum board must not exceed ½" in thickness. **NEC Section 424.41(G)**

WALL PLATE

HEATING CABLE

Heating cable must have a clearance of at least 8" from the edge of outlet boxes to be used for mounting surface lighting fixtures; 2" for recessed fixtures and their trims. **NEC Section 424.39**

Conventional wiring located above thermal insulation (2" or more in thickness) shall be subject to ambient temperature correction. **NEC Section 310.15(B)(1)**

Conventional wiring located above heating cables must be not less than 2" above the heating cable; ambient temperature is considered to be no less than 50°C or 122°F. Ampacity of conductors must be derated. **NEC Section 424.36**

Heating cable must be furnished complete with factory-assembled nonheating leads at least 7' in length. **NEC Section 424.34**

4" clear space from wall

In general, cable must not be installed in walls. **NEC Section 424.41(A)**

Each unit of heating cable must be marked with the identifying name or identification symbol, catalog number, and ratings in volts and watts or volts and amperes. **NEC Section 424.35**

Maximum staple spacing = 16" per **NEC Section 424.41(F)**

Figure 46 NEC® regulations for the installation of electric space-heating cable.

4.0.0 Section Review

1. The *NEC*® requirement for air conditioning disconnecting equipment is located in _____.

   a. *NEC Section 440.4*
   b. *NEC Section 440.14*
   c. *NEC Section 440.52*
   d. *NEC Section 440.65*

2. Conductors supplying more than one compressor must be sized for the total load plus which percentage of the largest motor's full-load amps?

   a. 10%
   b. 15%
   c. 25%
   d. 35%

1. In a boiler system, the heat exchange medium is _____.

   a. water
   b. refrigerant
   c. air
   d. oxygen

2. The component in an air conditioning or refrigeration system in which the heat from the conditioned space or item being cooled is transferred to the refrigerant is the _____.

   a. compressor
   b. condenser
   c. expansion device
   d. evaporator

3. The device that provides the pressure drop that lowers the boiling point of the refrigerant in a mechanical refrigeration system is the_____.

   a. compressor
   b. condenser
   c. expansion device
   d. evaporator

4. In the mechanical refrigeration cycle, the output of the evaporator is _____.

   a. high-pressure, high-temperature liquid refrigerant
   b. low-pressure, low-temperature refrigerant vapor
   c. high-pressure, high-temperature refrigerant vapor
   d. low-pressure, low-temperature liquid refrigerant

5. Low-voltage (24V) control circuits are used in HVAC systems because _____.

   a. step-down transformers produce only 24V output
   b. 24V power is safer
   c. it is mandated by law
   d. it is the same voltage used to operate the compressor and fan motors

6. A heat anticipator is used in a heating-only thermostat to _____.

   a. turn on the heat just before the temperature reaches the thermostat setpoint
   b. keep the thermostat warm on cold days
   c. turn on a light to let the occupants know when the furnace is about to come on
   d. open the thermostat just before the heat in the conditioned space reaches the thermostat setpoint

7. The differential in an auto-changeover thermostat is _____.

   a. the difference between the cut-in and cut-out points of the thermostat
   b. the difference between the cooling and heating setpoints
   c. normally at least 5°
   d. the difference between the settings of the heat anticipator and the cooling compensator

8. A good location for a thermostat is _____.

   a. anywhere that it will receive direct sunlight
   b. about 5' (1.52 m) above the floor on an outside wall
   c. about 5' (1.52 m) above the floor on an inside wall
   d. in a corner where there is minimum air circulation

9. In a standard thermostat wiring scheme, the R terminal would be connected to the _____.

   a. transformer common terminal
   b. secondary of the 24V control transformer
   c. reversing valve
   d. heating thermostat

10. In a standard thermostat wiring scheme, the G terminal would be connected to the _____.

   a. primary control
   b. cooling control
   c. fan control
   d. reversing valve

11. The function of a sequencer is to _____.
    a. make sure the condenser fan starts before the compressor
    b. prevent the compressor from restarting before system pressures can equalize
    c. stagger the start of electric heaters to prevent circuit overload
    d. run the compressor every five minutes to make sure it does not freeze up

12. The purpose of a compressor short-cycle timer is to _____.
    a. prevent the compressor from restarting before system pressures can equalize
    b. ensure the compressor runs for at least five minutes when it cycles on
    c. keep track of how often the compressor turns on
    d. run the compressor every five minutes to make sure it does not freeze up

13. When a heat pump is operating in the defrost mode, what usually happens in the conditioned space?
    a. The unit continues to provide compression heat.
    b. No heat is provided to the conditioned space.
    c. The furnace takes over.
    d. An electric heater is cycled on to provide heat.

14. The type of digital-control building system that is designed to serve multiple zones is the _____.
    a. constant-volume system
    b. variable temperature system
    c. variable air-volume system
    d. low-temperature system

15. In the electrical control circuit of a cooling-only system, the outdoor fan motor _____.
    a. is on continuously
    b. runs at low speed in the heating mode
    c. is on whenever the compressor is on
    d. is on whenever the compressor is off

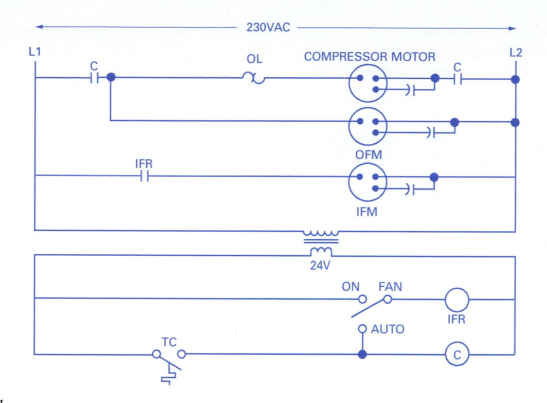

Figure RQ01

16. Referring to *Figure RQ01*, if the compressor motor, outdoor fan motor (OFM), and indoor fan motor (IFM) all fail to operate on a call from the thermostat (TC), the likely cause is _____.

    a. the outdoor fan motor
    b. a defective compressor
    c. the 24V transformer
    d. the disconnect is turned off

17. Referring to *Figure RQ01*, if the indoor fan relay (IFR) will run when the thermostat Fan switch is set to On, but does not run when the switch is set to Auto, the likely cause is a defective _____.

    a. fan relay
    b. thermostat
    c. fan motor
    d. compressor

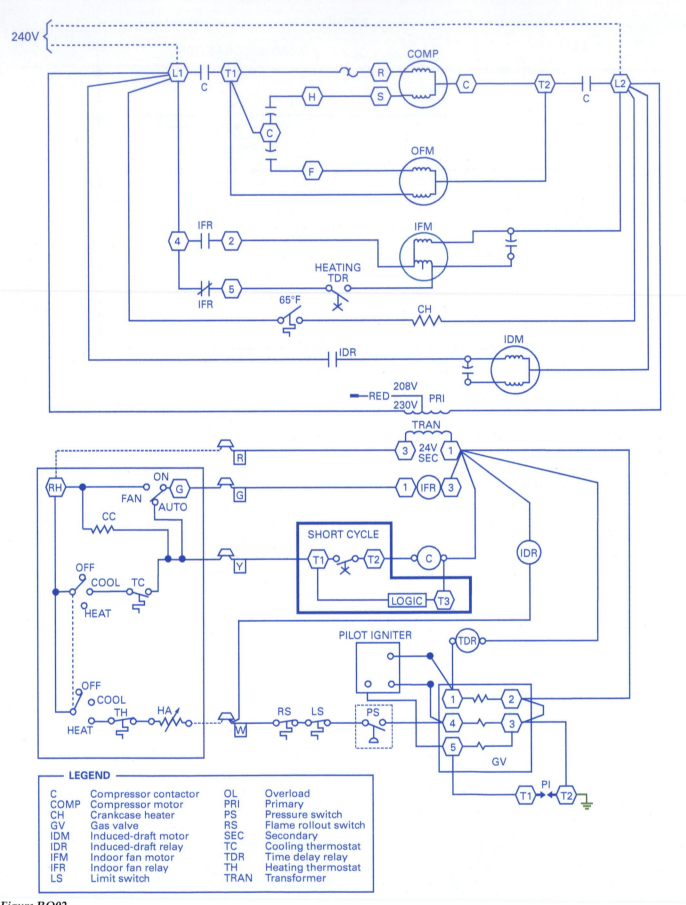

Figure RQ02

18. Referring to *Figure RQ02*, the yellow wire from the thermostat to the short cycle timer is connected to terminal _____.

    a. 1
    b. T3
    c. T2
    d. T1

19. According to *NEC Section 440.7*, the largest motor is considered to be the one that _____.

    a. weighs the most
    b. has the highest locked-rotor current
    c. has the highest rated load current
    d. has the highest input voltage

20. The provisions for room air conditioners are referenced in _____.

    a. *NEC Article 424, Part V*
    b. *NEC Article 427, Part IV*
    c. *NEC Article 430, Part XIII*
    d. *NEC Article 440, Part VII*

1. List the four mechanical refrigeration components used in a basic cooling/refrigeration system.
   _____.

2. Heat pumps provide heat by _____.

3. Why does a bimetal strip bend?
   _____.

4. What is a disadvantage of a bimetal thermostat when compared to an electronic thermostat?
   _____.

5. The heat anticipator in a heating-only thermostat is a(n) _____ in series with _____.

6. What prevents an automatic-changeover thermostat from cycling between heating and cooling?
   _____.

7. When you are installing a 24V thermostat, which pin of the thermostat terminal strip should you connect to the power supply, and which color wire would you use?
   _____.

8. Attempting to start a compressor against an _____ in the discharge side can damage the compressor.

9. A limit control is a heat-actuated switch positioned near the furnace _____.

10. The *NEC*® requirements for fixed electric space heating equipment are covered in _____.

11. Which two conditions are typical factors in initiating a defrost cycle in a heat pump?
    _____.

12. What is the difference between a digital signal and an analog signal?
    _____.

13. The key to a successful control system is the integration of the _____ and the _____.

14. True or False? Heating equipment must have a disconnect switch or other means of disconnecting it from all ungrounded conductors.
    _____.

15. Permanently installed energy storage systems are covered in the *NEC*® in _____.

Trade Terms Introduced in This Module

Analog-to-digital converter: A device designed to convert analog signals such as temperature and humidity to a digital form that can be processed by logic circuits.

Automatic-changeover thermostat: A thermostat that automatically selects heating or cooling based on the space temperature.

Bimetal: A control device made of two dissimilar metals that warp when exposed to heat, creating movement that can open or close a switch.

Compressor: A component of a refrigeration system that converts low-pressure, low-temperature refrigerant gas into high-temperature, high-pressure refrigerant gas through compression.

Condenser: A heat exchanger that transfers heat from the refrigerant flowing inside it to air or water flowing over it. Refrigerant enters as a hot, high-pressure vapor and is condensed to a liquid as the heat is removed.

Cooling compensator: A fixed resistor installed in a thermostat to act as a cooling anticipator.

Deadband: A temperature band, often 3°F, that separates the heating mode and cooling mode in an automatic-changeover thermostat.

Differential: The difference between the cut-in and cut-out points of a thermostat.

Evaporator: A heat exchanger that transfers heat from the air or water flowing inside to the cooler refrigerant flowing through it. Refrigerant enters as a cold liquid/vapor mixture and changes to all vapor as heat is absorbed from the air or water.

Expansion device: A device that provides a pressure drop that converts the high-temperature, high-pressure liquid refrigerant from the condenser into the low-temperature, low-pressure liquid/vapor mixture entering the evaporator; also known as the *metering device*.

Heat transfer: The movement of heat energy from a warmer substance to a cooler substance.

Invar®: An alloy of steel containing 36% nickel. It is one of the two metals in a bimetal control device.

Mechanical refrigeration: The use of machinery to provide a cooling effect.

Refrigerant: Typically, a fluid that picks up heat by evaporating at a low temperature and pressure and gives up heat by condensing at a higher temperature and pressure.

Refrigeration cycle: The process by which a circulating refrigerant absorbs heat from one location and transfers it to another location.

Sub-base: The portion of a two-part thermostat that contains the wiring terminals and control switches.

Subcooling: The measurable heat removed from a liquid after it has been cooled to its condensing temperature, has changed to a fully liquid state, and heat continues to be removed.

Superheat: The measurable heat added to a vapor after it has been heated to its boiling point, has changed to a fully vapor state, and heat continues to be added.

Thermostat: A device that is responsive to ambient temperature conditions.

Additional Resources

This module presents thorough resources for task training. The following resource material is suggested for further study.

ABC's of Air Conditioning. Syracuse, NY: Carrier Corporation.

NFPA 70®, *National Electrical Code (NEC®)*, Latest Edition. Quincy, MA: National Fire Protection Association (NFPA).

Refrigeration and Air Conditioning: An Introduction to HVAC/R. Larry Jeffus and David Fearnow. New York, NY: Pearson Education, Inc.

System Diagnostics and Troubleshooting Procedures. John Tomczyk. Mt. Prospect, IL: Esco Press.

Figure Credits

Section Review Answer Key

SECTION 1.0.0

Answer	Section Reference	Objective
1. b	1.1.0	1a
2. c	1.2.1	1b

SECTION 2.0.0

Answer	Section Reference	Objective
1. b	2.1.5	2a
2. a	2.2.2; *Figure 17*	2b

SECTION 3.0.0

Answer	Section Reference	Objective
1. d	3.1.2	3a
2. a	3.2.4	3b
3. d	3.3.0	3c
4. b	3.4.1	3d
5. c	3.5.0	3e
6. a	3.6.0; *Table 1*	3f

SECTION 4.0.0

Answer	Section Reference	Objective
1. b	4.1.0; *Figure 38*	4a
2. c	4.2.1; *Figure 39*	4b

This page is intentionally left blank.

NCCER CURRICULA — USER UPDATE

NCCER makes every effort to keep its textbooks up-to-date and free of technical errors. We appreciate your help in this process. If you find an error, a typographical mistake, or an inaccuracy in NCCER's curricula, please fill out this form (or a photocopy), or complete the online form at **www.nccer.org/olf**. Be sure to include the exact module ID number, page number, a detailed description, and your recommended correction. Your input will be brought to the attention of the Authoring Team. Thank you for your assistance.

Instructors – If you have an idea for improving this textbook, or have found that additional materials were necessary to teach this module effectively, please let us know so that we may present your suggestions to the Authoring Team.

NCCER Product Development and Revision

13614 Progress Blvd., Alachua, FL 32615

Email: curriculum@nccer.org
Online: www.nccer.org/olf

❏ Trainee Guide   ❏ Lesson Plans   ❏ Exam   ❏ PowerPoints   Other _____

Craft / Level: _____ Copyright Date: _____

Module ID Number / Title: _____

Section Number(s): _____

Description: _____

_____

_____

_____

Recommended Correction: _____

_____

_____

_____

Your Name: _____

Address: _____

_____

Email: _____ Phone: _____

This page is intentionally left blank.

Heat Tracing and Freeze Protection

OVERVIEW

Electrical heat-tracing systems keep piping systems at or above a given temperature for the purposes of freeze protection and temperature maintenance. In cold climates, freeze protection is needed because the thermal insulation installed around pipes is not enough to prevent the process fluids from freezing. This module presents heat-tracing and freeze-protection systems along with various applications and installation requirements.

Module 26409-20

Objectives

When you have completed this module, you will be able to do the following:

1. Describe heat-tracing applications, components, controls, and selection/installation considerations related to piping.
   a. Describe common pipeline heat-tracing applications, cables, and power distribution considerations.
   b. Describe methods of controlling and monitoring heat-tracing systems.
   c. Explain how typical heat-tracing systems operate.
   d. Explain how to select the equipment and components for a typical heat-tracing system.
   e. Explain how heat-tracing system components are installed and the related *NEC®* requirements.
2. Describe roof, gutter, and downspout de-icing systems and the relevant selection/installation considerations.
   a. Describe roof, gutter, and downspout de-icing systems.
   b. Explain how roof, gutter, and downspout de-icing system components are selected and installed.
3. Describe snow-melting and anti-icing systems and the relevant selection/installation considerations.
   a. Describe snow-melting and anti-icing system components.
   b. Explain how snow-melting and anti-icing system components are selected and installed.
4. Describe other electric heat-tracing and warming systems and the relevant selection/installation considerations.
   a. Describe domestic hot-water temperature maintenance systems and the relevant selection/installation considerations.
   b. Describe electric floor heating systems and the relevant selection/installation considerations.

Performance Task

Under the supervision of the instructor, you should be able to do the following:

1. Prepare and connect heat-tracing cable in a power connection box or splice box.

Trade Terms

Constant-wattage heating cable
Dead leg
Dew point
Heat source

Mineral-insulated (MI) heating cable
Nichrome
Parallel-resistance heating cable
Power-limiting heating cable

Self-regulating heating cable
Series-resistance heating cable
Startup temperature
Vessel

Industry Recognized Credentials

If you are training through an NCCER-accredited sponsor, you may be eligible for credentials from NCCER's Registry. The ID number for this module is 26409-20. Note that this module may have been used in other NCCER curricula and may apply to other level completions. Contact NCCER's Registry at 888.622.3720 or go to **www.nccer.org** for more information.

Contents

1.0.0 Introduction to Heat Tracing 1

  1.1.0 Piping Heat-Tracing Applications and Materials 2

    1.1.1 Self-Regulating Cables 2

    1.1.2 Power-Limiting Cables 3

    1.1.3 Mineral-Insulated Cables 4

    1.1.4 Long Cable Installations 5

    1.1.5 Heating Cable Components and Accessories 5

    1.1.6 Thermal Insulation Used with Heat-Tracing Systems 6

    1.1.7 Heat-Tracing System Power Distribution 7

  1.2.0 Heat-Tracing System Control 7

    1.2.1 Self-Regulating Control 8

    1.2.2 Ambient-Sensing Control 9

    1.2.3 Proportional Ambient-Sensing Control 9

    1.2.4 Line-Sensing Control 9

    1.2.5 Dead-Leg Control 9

    1.2.6 Ground-Fault Monitoring 10

    1.2.7 Continuity Monitoring 10

    1.2.8 Current Monitoring 11

    1.2.9 Temperature Monitoring 11

  1.3.0 Heat-Tracing System Operation 11

  1.4.0 Equipment Selection and Installation for Pipe Heat-Tracing
        Systems 13

  1.5.0 Installation Guidelines 14

    1.5.1 Installation of Heating Cable and Components 15

    1.5.2 Installation of Thermal Insulation 18

    1.5.3 *NEC*® Requirements 19

2.0.0 Roof, Gutter, and Downspout De-Icing Systems 21

  2.1.0 System Components and Control 21

  2.2.0 Component Selection and Installation 22

    2.2.1 Installation of Heating Cable and Components 23

    2.2.2 *NEC*® Requirements 24

3.0.0 Snow-Melting and Anti-Icing Systems 26

  3.1.0 Snow-Melting and Anti-Icing System Components 26

  3.2.0 Component Selection and Installation 28

    3.2.1 Installation of Heating Cable and Components 29

    3.2.2 *NEC*® Requirements 29

4.0.0 Other Tracing and Warming Systems 31

  4.1.0 Domestic Hot-Water Temperature Maintenance Systems 31

    4.1.1 Component Selection and Installation 31

    4.1.2 *NEC*® Requirements 32

  4.2.0 Floor Heating and Warming Systems 32

    4.2.1 Component Selection and Installation 33

    4.2.2 *NEC*® Requirements 34

Figures and Tables

Figure 1   Basic electric heat-tracing system 3

Figure 2   Typical construction of self-regulating heat-tracing cables 4

Figure 3   Resistance and power versus temperature for
self-regulating heat-tracing cable 4

Figure 4   Construction of a power-limiting heating cable........................... 4

Figure 5   Construction of mineral-insulated cables 4

Figure 6   Construction of a series-resistance heating cable...................... 5

Figure 7   Self-regulating/power-limiting heating system for Class I,
Division 2 6

Figure 8   Mineral-insulated cable system 7

Figure 9   Heat tracer/insulation configurations 7

Figure 10   Heat-tracing system power distribution.. 8

Figure 11   Ambient-sensing thermostat used to control
heat-tracing circuits... 9

Figure 12   Typical line-sensing thermostat used to control
heat-tracing circuits... 10

Figure 13   Simplified heat-tracing circuit with ground-fault monitoring 11

Figure 14   Simplified heat-tracing circuit with end-of-circuit continuity
monitoring 11

Figure 15   Simplified heat-tracing circuit with current monitoring.............. 12

Figure 16   Simplified heat-tracing circuit with temperature monitoring...... 12

Figure 17   Typical heat-tracing system circuit configuration..................... 13

Figure 18   Heat-tracing system using power-line carrier technology 14

Figure 19A   Installing a heating cable on common piping
components (Part 1 of 2).. 17

Figure 19B   Installing a heating cable on common piping
components (Part 2 of 2).. 18

Figure 20   Roof and gutter icing.. 21

Figure 21   Components of a typical roof, gutter, and downspout
de-icing system 22

Figure 22   Typical wiring for single- and multiple-circuit roof, gutter,
and downspout de-icing systems 23

Figure 23   Heating cable providing a continuous path for water to
flow off the roof.. 24

Figure 24   Bridge road surface protected by a snow-melting and
anti-icing system 26

Figure 25   Components of a typical snow-melting and anti-icing system ... 27

Figure 26   Typical wiring for a snow-melting and anti-icing system
controlled by an automatic snow controller 28

Figure 27   Heating cables in concrete sidewalk form................................ 29

Figure 28   Heat-traced domestic hot-water system.................................... 32

Figure 29   Basic concrete and masonry electric radiant
floor-heating system 33

Figure 30   Typical electric floor-heating mat installation 33

Table 1   Example of Data Used to Determine Maximum
Circuit Length 15

Table 2   Typical Lengths of Additional Cable Required to
Heat-Trace Piping Components.. 15

This page is intentionally left blank.

1.0.0 INTRODUCTION TO HEAT TRACING

Objective

Describe heat-tracing applications, components, controls, and selection/installation considerations related to piping.

a. Describe common pipeline heat-tracing applications, cables, and power distribution considerations.
b. Describe methods of controlling and monitoring heat-tracing systems.
c. Explain how typical heat-tracing systems operate.
d. Explain how to select the equipment and components for a typical heat-tracing system.
e. Explain how heat-tracing system components are installed and the related *NEC*® requirements.

Performance Task

1. Prepare and connect heat-tracing cable in a power connection box or splice box.

Trade Terms

Constant-wattage heating cable: A heating cable that has the same power output over a large temperature range. It is used in applications requiring a constant heat output regardless of varying outside temperatures. Zoned parallel-resistance, mineral-insulated, and series-resistance heating cables are examples of constant-wattage heating cables.

Dead leg: A segment of pipe designed to be in a permanent no-flow condition. This pipe section is often used as a control point for a larger system.

Dew point: The temperature at which a vapor begins to condense into its liquid state as the result of a cooling effect.

Heat source: Any component that adds heat to an object or system.

Mineral-insulated (MI) heating cable: A type of constant-wattage, series-resistance heating cable used in long line applications where high temperatures need to be maintained, high-temperature exposure exists, or high-power output is required.

Nichrome: An alloy based on nickel with chromium added (10–20 percent), and sometimes iron (up to 25 percent). Nichrome is often used for high-temperature applications such as in the fabrication of electric heating elements.

Parallel-resistance heating cable: An electric heating cable with parallel connections, either continuous or in zones. The watt density per lineal length is approximately equal along the length of the heating cable, allowing for a drop in voltage down the length of the heating cable.

Power-limiting heating cable: A type of heating cable that shows positive temperature coefficient (PTC) behavior based on the properties of a metallic heating element. The PTC behavior exhibited is much less (a smaller change in resistance in response to a change in temperature) than that shown by self-regulating heating cables.

Self-regulating heating cable: A type of heating cable that inversely varies its heat output in response to an increase or decrease in the ambient temperature.

Series-resistance heating cable: A type of heating cable in which the ohmic heating of the conductor provides the heat. The wattage output depends on the total circuit length and on the voltage applied.

Startup temperature: The lowest temperature at which a heat-tracing cable is energized.

Vessel: A container such as a barrel, drum, or tank used for holding fluids or other substances.

T his module focuses on the circuitry and components used for electric heat-tracing systems. It also introduces other widely used types of electric heating (heat-tracing) systems. The systems covered in this module include:

- Pipe heat-tracing systems
- Roof, gutter, and downspout de-icing systems
- Snow melting and anti-icing systems
- Domestic hot-water temperature maintenance systems
- Floor heating and warming systems

Applying heat to piping carrying process fluids is a common requirement in many industries including petroleum, plastics, chemicals, pharmaceuticals, power generation, and food processing. Electric-cable heating systems, called *heat-tracing systems*, maintain the piping systems at or above a specified temperature. This is done to prevent the systems from freezing, to maintain fluid temperature and consistent flow, and/or to prevent condensation in the system. Non-electrical steam and hot-liquid heat-tracing systems can also be used for these purposes. However, electric heating is often the most practical and least expensive method of protecting piping, especially when long pipe runs are involved.

1.1.0 Piping Heat-Tracing Applications and Materials

The main reasons for using electric heat tracing on piping are freeze protection and temperature maintenance of the process fluids. In cold climates, the thermal insulation installed around pipes is not enough to prevent the process fluids from freezing. Piping that carries water, fuel, or chemicals usually needs to be heat-traced to prevent freezing or thickening of the fluid at low temperatures. Heat tracing may also be applied to a tank or vessel to compensate for heat loss through the thermal insulation. Liquids that drop below their freezing point form a solid obstruction in the piping system, thereby stopping fluid flow. As the liquid turns into a solid, it may also expand, causing severe damage to the piping. Pipes installed in the ground below frost level are sometimes an exception and do not require heat tracing. This is because the temperature in this area remains relatively constant regardless of the outdoor air temperature.

Many industrial processes require temperature maintenance of piping to prevent the loss of heat in process fluids flowing through them or contained in tanks or vessels. Temperature maintenance can be divided into two main categories: broad temperature control and narrow temperature control. Broad temperature control is used to maintain the viscosity of process fluids in order to keep them flowing. Typical applications of broad temperature control are on fuel lines, cooking oil lines, and grease disposal lines. Narrow temperature control is used to keep process fluids within a narrow temperature range to maintain fluid viscosity and to prevent degradation of the fluid composition. Typical applications of narrow temperature control are on sulfur, acrylic acid, food syrup, and sugar solution lines.

Temperature maintenance is also used to prevent condensation from forming in piping systems. The dew point is the temperature at which a liquid first condenses from a vapor that is cooled. If an object's surface temperature drops below the dew point, condensation will occur. At night for example, humid air cools and often reaches its dew point, which results in the fallout of liquid water to the ground. This same event can occur inside some piping systems in the right conditions. The moisture that results from condensation in piping systems can cause operating difficulties, especially in gas burners where the control valve may become fouled by the liquid. In addition, natural-gas lines that contain condensed moisture may freeze at a variety of locations, stopping the flow. Moisture is especially harmful to compressors (since it cannot be compressed), and in the presence of hydrogen sulfide, it can produce harmfully corrosive sulfuric acid. Hydrogen sulfide is produced in some oil refineries during the crude oil distillation process.

A basic electric heat-tracing system consists of a power distribution system, electric heating cables, and interface components (*Figure 1*). Systems that are more complex may also incorporate a control and monitoring function with controllers, sensors, and alarms.

Several types of cables are used for heat tracing as well as other heating applications. These cables fall into three categories:

- Self-regulating heating cable
- Power-limiting heating cable
- Mineral-insulated (MI) heating cable

1.1.1 Self-Regulating Cables

Self-regulating cables are used for applications that require a heat output that varies with changes in the temperature of the air or other medium surrounding the heating cable. A self-regulating cable increases its heat output with a decrease in the ambient temperature and decreases its heat output with an increase in the ambient temperature. Typically, this type of cable is used for freeze protection and temperature maintenance on metal and plastic pipes. It is also widely used for other applications such as snow melting, anti-icing, and roof and gutter de-icing.

One type of self-regulating cable is made in the form of a heater strip consisting of two parallel bus wires embedded in a polymeric self-regulating conductive core heating element (*Figure 2*). The entire element is then surrounded by an insulating jacket, a braided tin-copper shield, and an outer jacket. Another type of self-regulating cable is similar in construction, but uses a polymeric self-regulating conductive fiber as the heating element instead of the core-type heating element. The self-regulating conductive core and conductive-fiber heating elements are made of a polymer mixed with conductive carbon. This composition creates multiple electrical paths for conducting current between the parallel bus wires along the entire length of the cable.

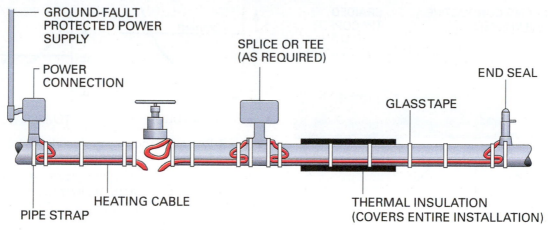

Figure 1 Basic electric heat-tracing system.

The core or fiber heating element of a self-regulating heating cable can be visualized as an array of parallel resistors connected between the bus wires, with each resistor having a value related to temperature. For this reason, a self-regulating cable is also referred to as a **parallel-resistance heating cable**.

Self-regulating heating cables produce heat from the flow of current through the polymeric core or fibers. In both types of cables, the number of electrical paths between the bus wires changes in response to temperature variations. As the temperature surrounding the heater cable gets colder, the conductive core or fibers contract microscopically. This decreases the electrical resistance and creates numerous parallel electrical paths between the bus wires. Increased current flows across these parallel paths, producing heat that warms the core or fiber along the length of the cable.

As the temperature surrounding the heater cable gets warmer, the conductive core or fibers expand microscopically. This increases the electrical resistance and decreases the number of electrical paths between the bus wires. The result is decreased current flow and a lower heat output in response to the rise in the ambient temperature. Through the process of contraction and expansion, the core or fiber heating element acts as a self-regulating thermostat protecting the cable from damage due to high temperatures. This feature allows this type of cable to be overlapped on pipes and valves during installation without causing damage to the cable from overheating. *Figure 3* shows a graph of resistance and power versus temperature for a self-regulating heat-tracing cable.

Because of their parallel construction, self-regulating heating cables can be cut to length and spliced in the field. Self-regulating cables with a core heating element, also called *monolithic cables*,

are used on metal and plastic pipes to maintain temperatures up to 225°F (107°C). Those with fiber heating elements are used on metal pipes for process temperature maintenance up to 250°F (121°C).

1.1.2 Power-Limiting Cables

Power-limiting heating cables are a type of parallel-resistance, self-regulating heating cable. They are used in process temperature-maintenance applications on metal pipes that require a high power output and/or high-temperature exposure.

The power-limiting heater cable (*Figure 4*) consists of a coiled-resistor alloy heating element wrapped around two parallel bus wires. It is connected alternately between the two bus wires at fixed intervals, usually every 2 or 3 feet (0.6 m to 0.9 m) along the cable length. The distance between the two contact points forms a heating zone. For this reason, power-limiting heating cables are also referred to as zoned heating cables. Power-limiting heating cables can be cut to length at the job site. However, the portion of cable between the cut and the next connection point of the heating and bus wires will not receive electricity and therefore will not provide any heating.

The resistor alloy heating element used in power-limiting heating cables has a positive temperature coefficient (PTC). This means the resistance of the heating element increases as its temperature increases and decreases as its temperature decreases. The result is an increase or decrease of power output as the temperature around the cable decreases or increases, respectively. This feature allows this type of cable to be overlapped on pipes and valves during installation without causing damage to the cable from overheating. This is because the temperature of the heating element is reduced at the points where the cable overlaps.

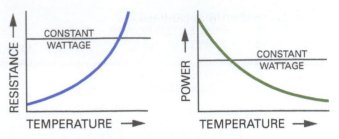

SELF-REGULATING CONDUCTIVE CORE HEATING ELEMENT

COPPER BUS WIRE

INSULATING JACKET

BRAIDED TIN-COPPER SHIELD AND OUTER JACKET

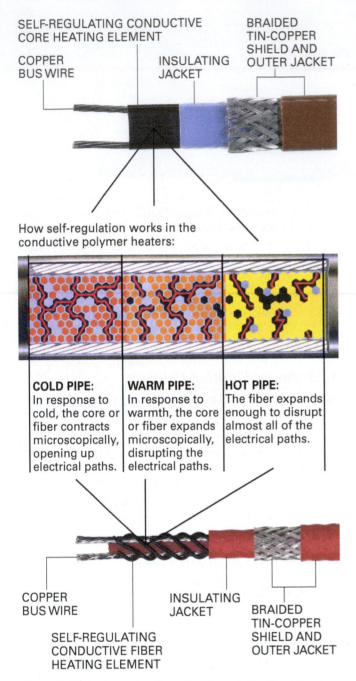

How self-regulation works in the conductive polymer heaters:

COLD PIPE: In response to cold, the core or fiber contracts microscopically, opening up electrical paths.

WARM PIPE: In response to warmth, the core or fiber expands microscopically, disrupting the electrical paths.

HOT PIPE: The fiber expands enough to disrupt almost all of the electrical paths.

COPPER BUS WIRE

INSULATING JACKET

SELF-REGULATING CONDUCTIVE FIBER HEATING ELEMENT

BRAIDED TIN-COPPER SHIELD AND OUTER JACKET

Figure 2 Typical construction of self-regulating heat-tracing cables.

1.1.3 Mineral-Insulated Cables

Mineral-insulated (MI) cables are a type of constant-wattage, series-resistance heating cable. They are widely used for tracing long pipes in applications where the temperature or power output requirements exceed the capabilities of self-regulating or power-limiting heating cables. MI cables (*Figure 5*) consist of one or two heating conductors made of nichrome, copper, or other metals insulated with magnesium oxide and encapsulated by a metal sheath made of copper, stainless steel, nickel (Alloy 825), or another

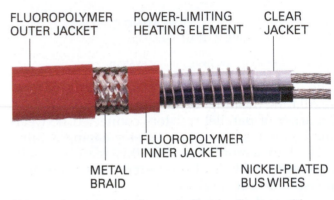

Figure 3 Resistance and power versus temperature for self-regulating heat-tracing cable.

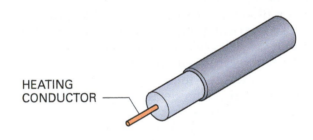

FLUOROPOLYMER OUTER JACKET

POWER-LIMITING HEATING ELEMENT

CLEAR JACKET

FLUOROPOLYMER INNER JACKET

METAL BRAID

NICKEL-PLATED BUS WIRES

Figure 4 Construction of a power-limiting heating cable.

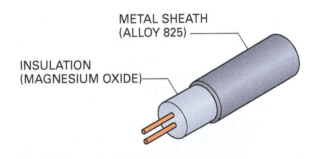

HEATING CONDUCTOR

SINGLE-CONDUCTOR CABLE

METAL SHEATH (ALLOY 825)

INSULATION (MAGNESIUM OXIDE)

DUAL-CONDUCTOR CABLE

Figure 5 Construction of mineral-insulated cables.

suitable metal. Unlike self-regulating and power-limiting cables, MI cables cannot be cut to length and terminated in the field. They must be ordered from the manufacturer and are cut to specified lengths and assembled with sealed connections at the factory.

1.1.4 Long Cable Installations

Heat tracing of long pipe runs is required in many applications, such as when transporting fluids between different processing stages in petrochemical facilities or to storage or transfer facilities, to tank farms, or to piers for ocean transport vessels. Such applications may involve cable lengths ranging between 1,250 and 5,000 feet (380 m to 1,500 m) that are powered from a single source. For many of these long runs, specially designed self-regulating and mineral-insulated heating cable systems are used.

For some long heat-tracing applications, specially designed series-resistance heating cables are used (*Figure 6*). They are generally used when the circuit length exceeds the ratings of conventional self-regulating heating cables. A series-resistance heating cable is a **constant-wattage heating cable** with one or two resistance conductors designed for use in single- or three-phase systems. Typically, the resistance conductors are electrically insulated and protected by a high-temperature, heavy-wall fluoropolymer jacket, fluoropolymer insulation, a metal grounding braid, and a fluoropolymer outer jacket. Resistance heating of the conductor provides the cable heat. The wattage output depends on the total circuit length and voltage applied.

Series-resistance heating cables are commonly used in circuit lengths of 5,000 feet (1,500 m) or more with a single power supply at voltages up to 600VAC.

1.1.5 Heating Cable Components and Accessories

Installation of a heat-tracing system requires the use of appropriate components and accessories to interface the cable with the power source, to properly support and terminate the cable(s), and to make splices in cable runs as needed to achieve a properly functioning system. The specific components and accessories must be designed for use with the particular type of cable being installed, its application, and its environment. Always refer to the manufacturer's catalogs and installation instructions to determine the proper components and accessories to use. Except for specific model numbers, the components and accessories used with self-regulating and power-limiting cable heating systems are similar in design. *Figure 7* shows a typical Class I, Division 2 self-regulating/power-limiting heating cable system. *Figure 8* shows a mineral-insulated cable system. The names shown in the figures for the various components and accessories also describe

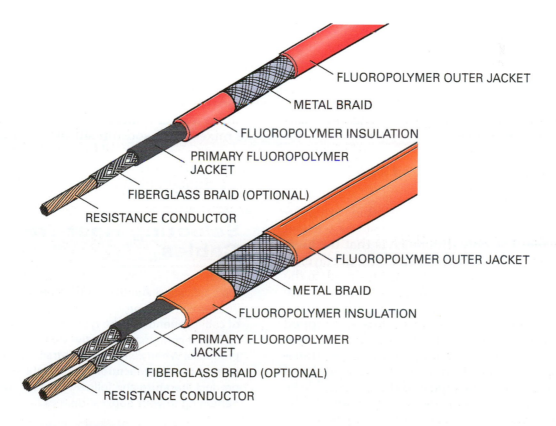

Figure 6 Construction of a series-resistance heating cable.

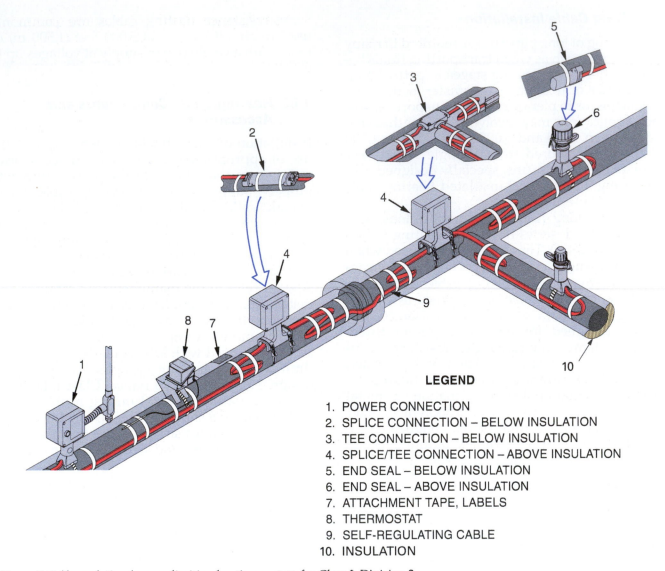

LEGEND

1. POWER CONNECTION
2. SPLICE CONNECTION – BELOW INSULATION
3. TEE CONNECTION – BELOW INSULATION
4. SPLICE/TEE CONNECTION – ABOVE INSULATION
5. END SEAL – BELOW INSULATION
6. END SEAL – ABOVE INSULATION
7. ATTACHMENT TAPE, LABELS
8. THERMOSTAT
9. SELF-REGULATING CABLE
10. INSULATION

Figure 7 Self-regulating/power-limiting heating system for Class I, Division 2.

the purpose for each item. Systems rated for Class I, Division 1 are available.

1.1.6 Thermal Insulation Used with Heat-Tracing Systems

Piping systems with heat-tracing cables are insulated in the same way as piping systems without heat tracing. The only difference is that the heat-tracing cables are installed first, under the layer of thermal insulation. It is important to insulate the heat-tracing cables and piping in order to reduce unnecessary heat loss to the surrounding atmosphere. In some cases, the traced line is wrapped with aluminum foil and then covered with insulation. The foil increases the radiation heat transfer. For proper transfer of heat, it is essential that the space between the pipe and the tracer cable be kept free of particles of insulating material. The transfer of the tracer heat may be improved by putting a layer of heat-conductive cement (graphite mixed with sodium silicate or other binders) between the tracer cable and the pipe. *Figure 9* shows various tracer and insulation configurations. Straight piping may be weather-protected

Selecting Heat-Tracing Cables

Select a self-regulating heat-tracing cable based on the application and the temperature requirements. The temperature produced by a specific type of cable depends on the type of polymeric core or fiber it contains and on the watts per foot it produces. When selecting heat-tracing cables, refer to the manufacturer's cable data sheets and product literature and follow the recommendations pertaining to your application.

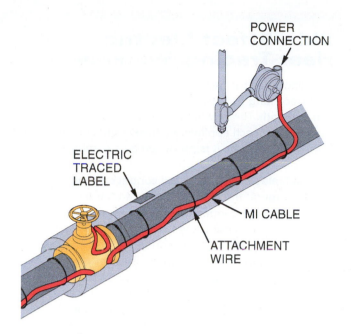

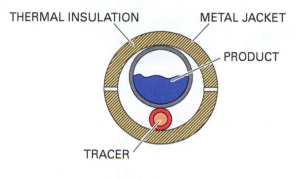

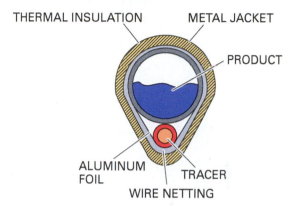

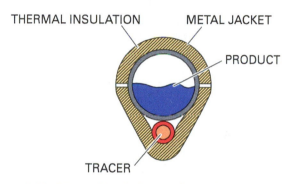

Figure 8 Mineral-insulated cable system.

with metal jacketing or a polymeric or mastic system. When metal jacketing is used, it should be sealed with closure bands and a sealant applied to the outer edge where the bands overlap. Some insulating materials commonly used with heat tracing are:

- Rigid cellular urethane
- Polyisocyanurate foam
- Fiberglass
- Calcium silicate
- Cellular glass

Some important factors to be considered when selecting an insulation material used with heat tracing are:

- Thermal characteristics
- Mechanical properties
- Chemical compatibility
- Moisture resistance
- Personal safety characteristics
- Fire resistance

1.1.7 Heat-Tracing System Power Distribution

A typical power distribution scheme used with heat-tracing systems is shown in *Figure 10*. As shown, three-phase utility power is applied to a dedicated step-down transformer for the heat-tracing system. The transformer supplies a reduced voltage to a distribution panel that contains the main circuit breaker and branch circuit breakers. The main circuit breaker is used to disconnect power from the panel board

Figure 9 Heat tracer/insulation configurations.

and to protect the panel buses, the transformer, and the wiring between the transformer and the distribution panel. The branch circuit breakers are ground-fault interrupting circuit breakers used to apply the operating voltages to the power connection boxes of the one or more heat-tracing circuits. Typical cable supply voltages are 120/277VAC and 208/240VAC.

1.2.0 Heat-Tracing System Control

Control of heat-tracing systems is used to adjust the output of the heating source to keep the pipes from freezing or to maintain process piping at correct temperatures. Temperature control of a heat-tracing system can be accomplished in several ways. The method selected depends on the application requirements and on how critical temperature maintenance is to the process.

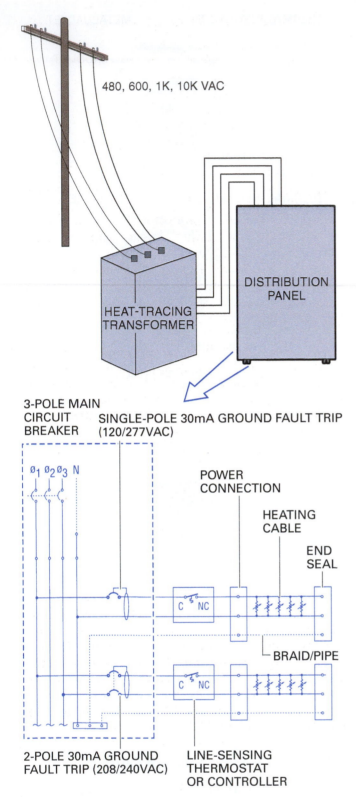

480, 600, 1K, 10K VAC

DISTRIBUTION PANEL

HEAT-TRACING TRANSFORMER

3-POLE MAIN CIRCUIT BREAKER

SINGLE-POLE 30mA GROUND FAULT TRIP (120/277VAC)

$\emptyset_1$ $\emptyset_2$ $\emptyset_3$ N

POWER CONNECTION

HEATING CABLE

END SEAL

C NC

BRAID/PIPE

C NC

2-POLE 30mA GROUND FAULT TRIP (208/240VAC)

LINE-SENSING THERMOSTAT OR CONTROLLER

Figure 10 Heat-tracing system power distribution.

The most widely used control methods are:

- Self-regulating control
- Ambient-sensing control
- Proportional ambient-sensing control

Skin-Effect Electric Heat-Tracing Systems

Skin-effect electric heat-tracing systems are used to heat-trace long transfer pipelines up to 12 miles (19 km) in length. These systems must be specifically designed for each application. In skin-effect heat-tracing systems, heat is produced by a combination of a ferromagnetic heat tube and a temperature-resistant, electrically insulated conductor that passes through the tube. At the far end, the two components are connected together; at the near end, the two are connected to an AC supply. Current flows through the insulated conductor and into the heat tube where, by skin and proximity effects, the return current passes through and is concentrated toward the inner surface of the heat tube, causing the outer surface of the heat tube to remain at ground potential. The current flowing in the heat tube is converted to heat that is supplemented by heat generated in the insulated conductor. The heat tube may be directly welded or cemented to the pipeline, allowing the heat to be transferred by conduction through the coupling media from the heat tube to the pipeline.

- Line-sensing control
- Dead-leg control

Beyond control, several schemes can be incorporated into heat-tracing systems to monitor their operation and detect any failures. The operating characteristic to be monitored depends on the application requirements and on how critical temperature maintenance is to the process. More than one characteristic can be monitored in a system if the application requires it. The most widely monitored system characteristics are:

- Ground fault
- Continuity
- Current
- Temperature

The temperature control methods and monitored characteristics identified here are described in greater detail in the sections that follow.

1.2.1 Self-Regulating Control

A self-regulating control uses the built-in properties of self-regulating heating cables to maintain the desired pipe temperature. It is used for basic freeze protection and broad temperature maintenance applications. Because no moving parts are involved, self-regulating control is the most economical and reliable method of control. However, self-regulating control should not be used for applications that require maintaining a narrow operating temperature range, involve temperature-sensitive fluids, or where energy consumption is a major concern.

1.2.2 Ambient-Sensing Control

An ambient-sensing control uses a mechanical thermostat or electronic controller to measure the ambient temperature and turn on the heat-tracing system only when the surrounding temperature drops below a predetermined setpoint. For this reason, it is more energy efficient than a self-regulating control and therefore is widely used in freeze protection systems. *Figure 11* shows an ambient-sensing thermostat typical of those used in nonhazardous areas to directly control a single heat-tracing circuit or as a pilot control of a contactor used to turn multiple heat-tracing circuits on and off.

Figure 11 Ambient-sensing thermostat used to control heat-tracing circuits.

1.2.3 Proportional Ambient-Sensing Control

A proportional ambient-sensing control (PASC) treats all pipes as if they contained stagnant fluid and provides power to the heat-tracing system based on the ambient temperature.

A PASC uses a programmable electronic controller connected to remote temperature-input modules and contactor control modules to regulate one or more heat-tracing cables. The PASC system monitors the ambient temperature and operates to continuously adjust the heat input to compensate for heat loss that occurs because of changing ambient temperature conditions. In order to maintain the desired temperature, a programmed algorithm calculates the cycle time during which the heating circuits are energized. The heat-tracing cable is energized more often as the temperature decreases and less often as the temperature increases. On cold days, the heat-tracing cable is turned on more often to maintain correct pipe temperatures. On warm days, it may not be turned on at all. A PASC is used when a self-regulating control is not precise enough. It is suitable for freeze protection, as well as all broad temperature-control and many narrow temperature-control applications. It is also one of the most energy-efficient forms of temperature control.

1.2.4 Line-Sensing Control

A line-sensing control uses a mechanical thermostat or electronic controller to measure the actual pipe temperature, not the ambient temperature surrounding the pipe. With this method of control, each heating circuit must have a separate control. Each heating circuit is turned on and off independently when signaled by its respective control device. Line-sensing controls are used in narrow temperature maintenance situations because they provide the highest degree of temperature control using the least amount of energy. Because this method requires the use of multiple control devices, it has the highest installation cost. *Figure 12* shows a line-sensing thermostat typical of those used in hazardous areas to directly control a single heat-tracing circuit or as a pilot control of a contactor used to turn multiple heat-tracing circuits on and off.

1.2.5 Dead-Leg Control

The dead-leg method of control uses a segment of inactive pipe (**dead leg**) in a piping system as a single control point for the larger system. The dead leg contains system fluid in a permanent no-flow condition. A thermostat measures the temperature of the fluid in the dead leg and uses it to represent the rest of the system. Dead-leg control provides much better control than ambient-

Figure 12 Typical line-sensing thermostat used to control heat-tracing circuits.

sensing control, at about the same cost. It is more economical than line-sensing control because it uses only one thermostat. However, temperature control is imprecise, especially in high-flow areas of the system. Today, the use of dead-leg control systems is on the decline. They are being replaced by electronic control systems such as PASC that duplicate the dead leg electronically.

1.2.6 Ground-Fault Monitoring

A ground-fault monitoring system monitors the current leakage from the system heating cable, power wiring, and components to ground using ground-fault (leakage) circuit breakers and/or current-sensing devices that measure current (*Figure 13*). NEC Section 427.22 requires the use of ground-fault equipment with heat-tracing circuits. The function of ground-fault circuit breakers is to protect the equipment, not personnel. The ground-fault circuit breakers used in heat-tracing circuits are designed to trip at 30mA, rather than at 5mA like the GFCIs used for personnel protection. Personnel protection GFCIs are not used in heat-tracing applications because they often experience nuisance trips in this application.

Ground-fault trips are usually caused by mechanical damage to the heating cable or power wiring, or by moisture in the junction boxes. When current leakage in the heat-tracing circuit exceeds the trip setting of the ground-fault circuit breaker or other protective device, it shuts off the circuit. If the protective device is a ground-fault circuit breaker, it may have an auxiliary contact used to trigger a remote trip alarm.

1.2.7 Continuity Monitoring

Continuity monitoring verifies that voltage is present at the end termination of a heat-tracing cable run. This can be done by using an end-of-circuit LED that remains on when the circuit is operative and shuts off if there is an open in the circuit (*Figure 14*). In some systems, an end-of-circuit transmitter is used. Installed at the end of the line, it communicates the status of the circuit (continuity or open) to a centralized receiver. This type of system has the capability to send a low-voltage monitoring signal down the heat-tracing circuit even when it is shut down.

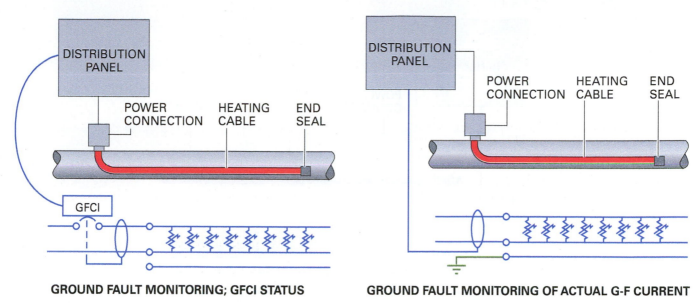

GROUND FAULT MONITORING; GFCI STATUS

GROUND FAULT MONITORING OF ACTUAL G-F CURRENT

Figure 13 Simplified heat-tracing circuit with ground-fault monitoring.

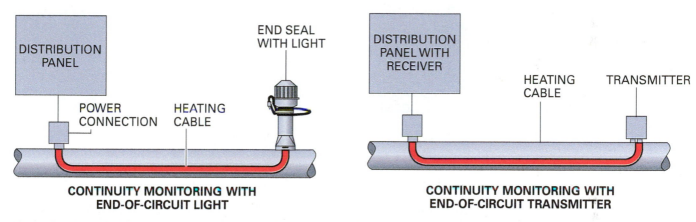

CONTINUITY MONITORING WITH
END-OF-CIRCUIT LIGHT

CONTINUITY MONITORING WITH
END-OF-CIRCUIT TRANSMITTER

Figure 14 Simplified heat-tracing circuit with end-of-circuit continuity monitoring.

1.2.8 Current Monitoring

Current monitoring uses a heat-tracing controller or current-monitoring relay to signal an alarm when the current in the heat-tracing circuit is too low or too high. This type of circuit will generate an alarm if there is a loss of power to the heating cable or if there is an open or damage to the heating cable bus wires or branch circuit wiring. *Figure 15* shows a simplified heat-tracing circuit with current monitoring.

1.2.9 Temperature Monitoring

Temperature monitoring continuously measures the temperature of the pipe with a digital controller connected to a resistance temperature detector (RTD) temperature sensor attached to the pipe (*Figure 16*). If the high and/or low temperature setpoints are exceeded, the controller generates an alarm. Low-temperature alarms can be generated by low power or loss of power to the

heating cable, wet or missing thermal insulation, heating cable damage, or controller failure. High-temperature alarms can be caused by fluid temperatures that exceed the preset threshold or by controller failure.

1.3.0 Heat-Tracing System Operation

Depending on the equipment manufacturer and the specific application, a wide variety of schemes can be used to control and monitor electric heat-tracing systems. This section briefly describes a circuit configuration (*Figure 17*) used by one heat-tracing equipment manufacturer to control and monitor a heat-tracing system consisting of multiple circuits. The system consists of a microprocessor-based controller that regulates the heating circuits by selecting different modes of control, including ambient control, PASC, or line-sensing control. The controller can be operated as a standalone unit, or it can be networked to one or more remote monitoring and control

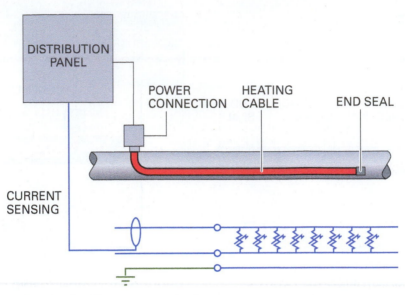

Figure 15 Simplified heat-tracing circuit with current monitoring.

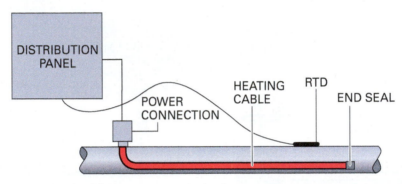

Figure 16 Simplified heat-tracing circuit with temperature monitoring.

modules (RMCs) and/or remote monitoring modules (RMMs). In response to signals from the controller, each RMC provides multiple relay-controlled outputs that cause the associated heating cable circuits to be switched on or off. Each RMM collects ambient temperature and/or pipe temperature inputs from RTD sensors located in the individual heater circuits. The RMM processes this information and forwards it to the controller. Based on the status inputs, the controller determines which circuits need to be turned on and which ones need to be turned off. It then signals the RMC to energize or de-energize the circuits, respectively, by means of the appropriate heating cable power contactors.

Data is communicated between the various devices in the monitoring and control network by means of twisted-pair RS-485 cables. This allows each module to be placed in the best location. RMCs are typically located in the heat-tracing system power distribution panel, whereas RMMs are placed at the local sensor location. A system's setup parameters, status, and alarm conditions can be viewed by the system operator at the controller. In some systems, this information is made available at a remote location by an RS-232/RS-485 link to a PC running heat-trace monitoring and control software.

In some systems, temperature-monitoring signals are communicated to the controller by means of power-line carrier technology. In such systems, a power-line interface unit (PLI) receives temperature and continuity inputs transmitted over the heat-tracing circuit power wires from special end-of-line heat cable circuit transmitters. Power-line carrier technology uses frequency-shift keying to encode digital data on the power line network. Digital ones and zeros are transmitted by coupling high-frequency signals onto the heat-tracing bus wires and the AC power line. The use of power-line carrier technology allows temperature and continuity data to be sent to the controller without running additional wiring to the sensor locations. *Figure 18* shows the wiring for a typical system using power-line carrier technology to transmit temperature and continuity status.

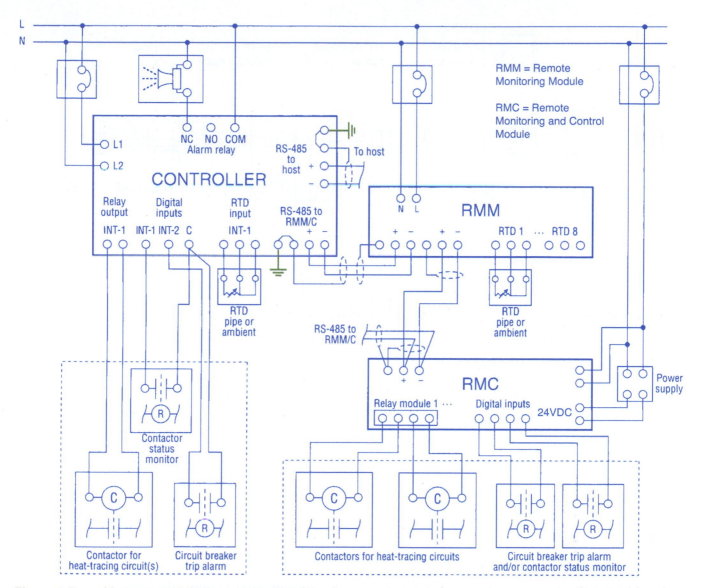

Figure 17 Typical heat-tracing system circuit configuration.

1.4.0 Equipment Selection and Installation for Pipe Heat-Tracing Systems

The design of a complex pipe heat-tracing system and the selection of its components requires extensive knowledge of the industrial processes and process fluids used in numerous industries. For this reason, the design and component selection for a pipe heat-tracing system is usually performed by a qualified master electrician, engineer, and/or the heat-tracing product manufacturer's engineering representative.

The component selection process involves the following stages and considerations:

Step 1 Determine the application:
- Freeze protection
- Temperature maintenance
- Both freeze protection and temperature maintenance

Step 2 Determine the application requirements based on:
- Pipe size and material (metal or PVC)
- Pipe location (above or below ground)
- Thermal insulation type and thickness
- Type of process fluid (water, fuel, oil, or grease)
- Ambient and maintenance temperatures
- Maximum fluid temperature

Step 3 Determine the appropriate type of heating cable based on:
- Power output
- Maximum exposure temperature
- Available voltages
- Outer protective jacket
- Cable length(s)

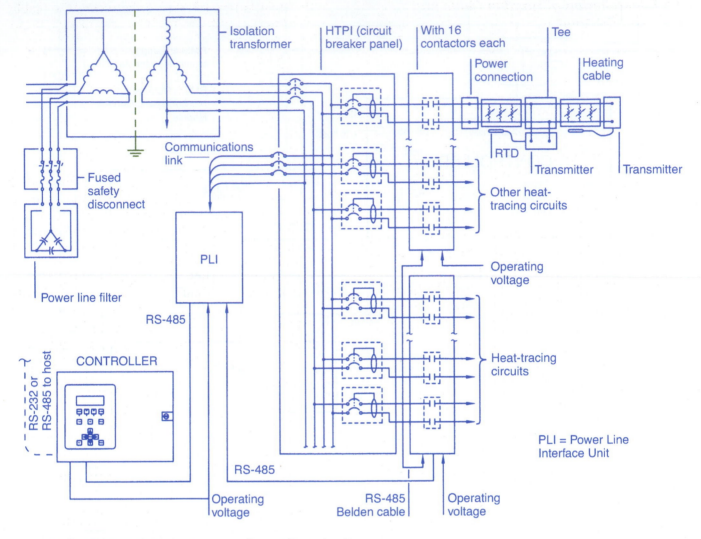

Figure 18 Heat-tracing system using power-line carrier technology.

Step 4 Determine the type of system temperature monitoring control based on the application requirements:
- Self-regulating control
- Ambient-sensing control
- Proportional ambient-sensing control
- Line-sensing control
- Dead-leg control

Step 5 Determine the number of electrical circuits and transformer load based on:
- Total heating cable length
- Supply voltage
- Minimum **startup temperature**

Step 6 Select the correct components and accessories.

The minimum number of circuits required for an installation is determined by dividing the total length of heating cable needed for the job by the maximum circuit length that can be used for that particular type of cable. The maximum circuit length is determined by the type of cable, supply voltage, startup temperature, and circuit breaker size. Manufacturers provide this information in their design and/or installation literature. *Table 1* shows some examples of data from one manufacturer's design guide used to determine the maximum circuit lengths of different cables.

1.5.0 Installation Guidelines

Always install the heat-tracing system and its components in strict accordance with the system design specifications, the manufacturer's instructions, and all prevailing local codes. *NEC Article 427* covers the requirements for fixed electric-heating equipment installed on pipes and vessels in unclassified areas. The *NEC®* requirements are covered later in this section.

1.5.1 Installation of Heating Cable and Components

Some general guidelines for the installation of heat-tracing cable and its components are listed here.

Prior to the installation:

- Inspect all the cables, components, and accessories to make sure they are the ones specified for the job and that they are not damaged. Make sure that the rating of the heating cable is suitable for the service voltage available.
- To verify that the heating cable has not been damaged during shipping, test it for continuity and electrical insulation resistance. Test the insulation resistance between the cable conductors and the metallic grounding braid in accordance with the cable manufacturer's instructions. When testing insulation, use a 2,500VDC megohmmeter. The minimum insulation resistance should meet the value specified by the cable manufacturer (typically greater than 20 megohms) regardless of length.
- Make sure there is enough heating cable available to complete the job. The minimum amount of heating cable required is equal to the total length of pipe plus the additional cable needed to make all the electrical connections and to trace each component such as valves, flanges, and pipe supports. Information on how much additional cable to provide for tracing each of these items can be found in the job specifications and/or in the manufacturer's installation instructions. Generally, 2 feet (600 mm) of cable is added for each electrical power connection, splice, tee, and end seal. *Table 2* lists some recommendations for the length of additional cable needed to trace valves, flanges, and supports.

Table 1 Example of Data Used to Determine Maximum Circuit Length

Heating Cable	Supply (VAC)	Startup Temperature	Circuit Breaker Size Maximum Circuit Length			Maximum Amp/Foot
			15A	20A	30A	
5XL1-CR or -CT	120	40°F	165	220	250	0.072
		0°F	110	145	220	0.108
8XL1-CR or -CT	120	40°F	120	160	190	0.100
		0°F	85	115	170	0.139
5XL2-CR or -CT	208/277	40°F	285	380	450	0.042
		0°F	190	255	385	0.042

Table 2 Typical Lengths of Additional Cable Required to Heat-Trace Piping Components

Pipe Size (Inches)	Screw Valves	Flange Valves	Butterfly Valves	Pipe Supports	Pipe Elbows	Pipe Flange
1	6''	6''	6''	6''	3''	3''
2	6''	1'	6''	1'	6''	6''
3	1'	2'	1'	1'	6''	6''
4	2'	3'	1'	1'	9''	6''
6	3'	4'	1'	2'	9''	1'
8	4'	5'	2'	2'	9''	1'
10	5'	7'	2'	3'	1'	1'
12	7'	8'	2'	3'	1'	1'
14	8'	10'	3'	4'	1'	2'

If the heating cable is spiraled (wrapped) around the pipe instead of being installed in a straight run, additional cable length must be allowed per foot of pipe. The amount of additional cable depends on the pipe size and the distance between the spirals. For a given pipe size, the amount of cable and the distance between the spirals, called the *pitch*, is determined by multiplying the total pipe length by an appropriate spiral factor. Equipment manufacturers normally provide tables in their installation instructions that give the spiral factors and cable lengths to add per foot for the different sizes of pipe.

- Check and/or test each control device before installation for the correct operating temperature range, proper span, and setpoints.
- Store the heating cable and its components in a clean, dry place where they will be protected from mechanical damage.

During installation:

- Handle heating cable in a way that avoids exposing the cable to sharp edges. Also avoid using excessive pulling force, kinking or crushing the cable, or walking on or driving over the cable.
- Heating cable must be installed on a clean, smooth portion of the pipe, avoiding any sharp bends or jagged edges. It should be positioned to avoid damage due to impact, abrasion, or vibration, while still maintaining proper heat transfer. Always install heat-tracing cable in accordance with the manufacturer's instructions. *Figure 19A* and *Figure 19B* show how a heating cable is installed on some common piping components.
- Heating cable must be installed in a manner that facilitates the removal of valves, small in-line devices, and instruments without completely removing the cable or excessive thermal insulation or cutting the heating cable.
- With the exception of self-regulating cable, avoid overlapping heating cable sections. This can cause excessive temperatures at overlap points.
- Heating-cable cold leads must be positioned to penetrate the thermal insulation in the lower 180-degree segment. This helps to minimize water entrance.

- Attach the heating cable to the pipes using fiberglass tape, aluminum tape, or plastic cable ties per the manufacturer's instructions. If using plastic cable ties, make sure they have a temperature rating that matches the system exposure rating.

Do not use metal attachments, vinyl electrical tape, or duct tape to fasten heating cable to the pipe. Do not twist the bus wires together in a self-regulating cable as this will cause a short circuit.

Heat-Tracing System Application and Design Guides

Most equipment manufacturers have application and design information to aid in the component selection of their various heat-tracing products and systems. This information is normally readily available at the manufacturer's website. It is also provided in product publications like the ones shown here. Even if you are not directly involved with the design process, reading these documents will provide you with a better understanding of these systems.

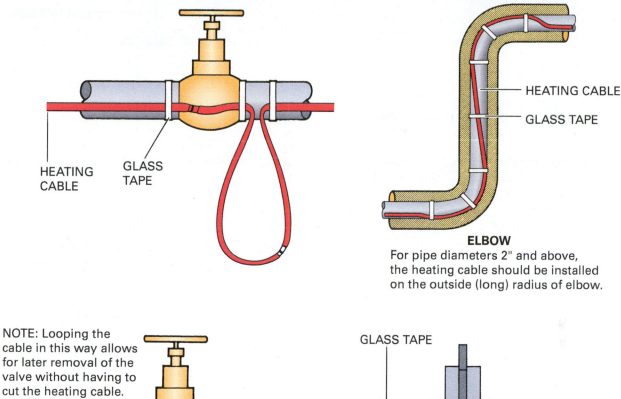

Figure 19A Installing a heating cable on common piping components (Part 1 of 2).

- Protect the heating-cable ends from moisture and mechanical damage if they are to be left exposed before connection.
- After the installation is complete, retest the cables electrically in the same manner as pre-installation testing. This test ensures that the integrity of the cable has not been compromised, and it ensures the integrity of splices and terminations.

The following conditions must be satisfied when installing temperature sensors:

- Sensors must be mounted to avoid the direct temperature effects of the heating cable.
- Sensors must be mounted 3 to 5 feet (0.9 m to 1.5 m) from any junction where two or more sections of cable meet or join.

- Sensors must be located 3 to 5 feet (0.9 m to 1.5 m) from any heat source or heat sink in the system.
- Where piping is routed through areas with different ambient conditions, such as inside and outside a heated building, separate sensors and associated controls are required for each area.

Pulling Heating Cable

When pulling heating cable, it helps to use a reel holder that plays the cable out smoothly with little tension. The heating cable should be strung loosely but close to the pipe being traced to avoid interference with supports and equipment.

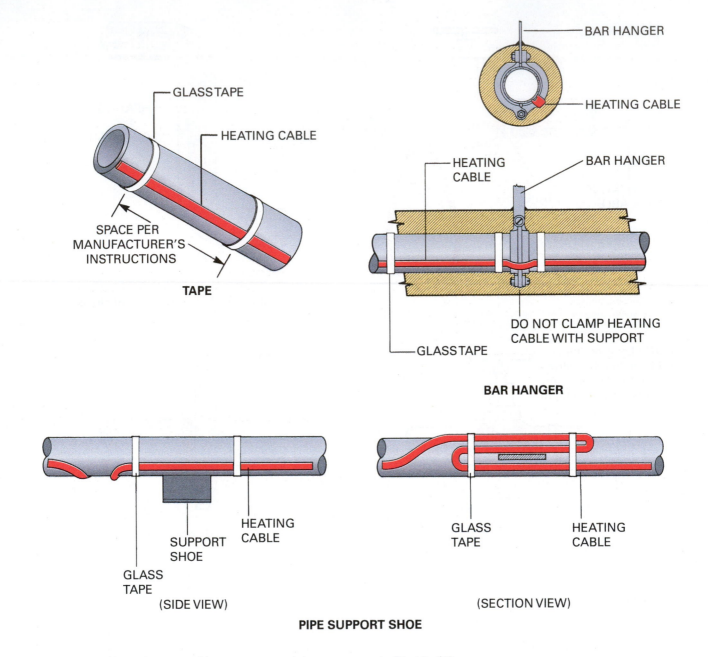

TAPE

BAR HANGER

PIPE SUPPORT SHOE

Figure 19B Installing a heating cable on common piping components (Part 2 of 2).

1.5.2 *Installation of Thermal Insulation*

Before installing thermal pipe insulation over the heat-tracing system components, the heat-tracing cable should be visually checked for signs of damage and electrically tested for continuity and electrical insulation resistance integrity. All testing should be done by properly trained personnel. Test the electrical insulation resistance between the cable conductors and the metallic grounding braid in accordance with the cable manufacturer's instructions. When testing insulation resistance, use a 2,500VDC megohmmeter. Regardless of the heating cable length, the resistance reading should meet the value specified by the cable manufacturer, usually 20 megohms or above. Record the original value of each circuit for future reference.

Correct temperature maintenance of a heat-traced pipe system requires that dry thermal insulation be properly installed. To reduce the chance of heating cable damage, install the thermal insulation as soon as possible after the heating cable and related devices have been attached. The insulation material and its thickness should be listed in the job specifications. If not, follow the recommendations of the heat-tracing equipment manufacturer. All pipe work, including fittings, wall penetrations, and other potential heat loss areas, must be completely insulated. The outside of the pipe insulation should be marked *Electric Traced* at appropriate intervals on alternate sides to warn people of the pipe tracing. Mark the location of all heating-cable system components on the outside of the insulation as well.

1.5.3 NEC® Requirements

NEC Article 427 covers the requirements for fixed electric-heating equipment installed on pipes and vessels in unclassified areas. If the equipment is installed in hazardous (classified) areas, *NEC Articles 500 through 516* also apply.

Some of the *NEC®* requirements for fixed electric-heating equipment installed on piping and vessels are summarized below. For complete *NEC®* requirements, refer to *NEC Article 427*.

- Branch circuits and overcurrent devices supplying the equipment shall be sized at not less than 125 percent of the total heater load per *NEC Sections 210.20(A) and Section 427.4*.
- Per *NEC Section 427.10*, the equipment must be identified as being suitable for the chemical, thermal, and physical environment, and be installed in accordance with the manufacturer's drawings and instructions.
- The external surfaces of any equipment that operates at temperatures above 140°F (60°C) shall be physically guarded, isolated, or thermally protected to prevent contact with personnel (*NEC Section 427.12*).
- The heating elements shall be attached to the pipe or vessel surface being heated with appropriate fasteners other than the thermal insulation. Where the heating element is not in direct contact with the surface, a means shall be provided to prevent overheating the heating element (*NEC Sections 427.14 and 427.15*).

- The presence of electric heating equipment shall be identified using permanent caution signs or markings at the intervals specified in *NEC Section 427.13*.
- Exposed heating equipment shall not bridge expansion joints unless provision is made for expansion/contraction (*NEC Section 427.16*).
- Ground-fault protection for equipment is required except where there is an alarm indication of ground faults and only qualified personnel maintain the system and/or continuous operation is necessary for safe operation of the equipment or process (*NEC Section 427.22*).
- Equipment must be provided with a readily accessible disconnecting means such as indicating switches or circuit breakers. A factory-installed cord and plug that is accessible to the equipment may be used if the equipment is rated at 20A or less and 150V or less (*NEC Section 427.55*).

Overcurrent Protective Devices

The overcurrent protective devices used in a heat-tracing system must be sized in accordance with the design specifications. Guidelines for correct sizing of overcurrent devices are normally given in the equipment manufacturer's design literature or computer selection programs. Ground-fault circuit breakers or other ground-fault protection devices with a nominal 30mA trip level (a common trip level for heat-tracing applications) are mandatory in most areas. See *NEC Section 427.22* for an exclusion related to industrial establishments. Because these devices need an effective ground path, the *NEC®* requires that a conductive covering of the heating cable be connected to an electrical ground.

1.0.0 Section Review

1. A mineral-insulated (MI) heating cable is a type of _____.

   a. self-regulating cable
   b. parallel-resistance heating cable
   c. power limiting cable
   d. constant-wattage, series-resistance heating cable

2. A line-sensing control uses a mechanical thermostat or electrical controller to measure the _____.

   a. actual pipe temperature
   b. ambient temperature around the pipe
   c. estimated pipe temperature
   d. heating cable temperature

3. The use of power-line carrier technology allows temperature and continuity data to be sent to the controller _____.

   a. using Bluetooth® technology
   b. without running additional wiring to the sensor locations
   c. as cellular data
   d. through twisted pair RS-485 cables

4. The design and component selection for a complex pipe heat-tracing system is usually performed by a master electrician, engineer, or the _____.

   a. electrician on site
   b. piping system designer
   c. heat-tracing product manufacturer's representative
   d. insulating contractor

5. When heating cable is wrapped in a spiral around the pipe instead of in a straight run along its length, the additional cable needed is determined by the _____.

   a. pipe size and the pitch
   b. pipe size and the cable wattage per foot
   c. pitch and the cable wattage per foot
   d. pitch and the cable type

SECTION TWO

2.0.0 ROOF, GUTTER, AND DOWNSPOUT DE-ICING SYSTEMS

Objective

Describe roof, gutter, and downspout de-icing systems and the relevant selection/installation considerations.

a. Describe roof, gutter, and downspout de-icing systems.
b. Explain how roof, gutter, and downspout de-icing system components are selected and installed.

Roof, gutter, and downspout de-icing systems are used to prevent ice dams and icicles from forming on roofs and in gutters and downspouts (*Figure 20*). Ice dams form because melting snow and ice freeze when they reach the colder roof edge. Once ice dams are formed, water tends to collect behind them. The water backs up and seeps under the roofing materials, where it can leak into the building and cause water damage. Another danger is icicles that form when water dripping off the roof flows over ice-filled gutters. When outside temperatures rise, these heavy icicles can become a safety hazard as they break away from the gutters and roof.

Figure 20 Roof and gutter icing.

2.1.0 System Components and Control

Roof, gutter, and downspout de-icing systems provide a continuous heated path for melting snow and ice. The heated path keeps the water above freezing temperatures so that it can drain from the roof to a safe discharge area. *Figure 21* shows the components used with a typical roof, gutter, and downspout de-icing system. In this system, the continuous heating path is created by self-regulating, low-wattage heating cables specifically designed for roof, gutter, and downspout de-icing applications. As current flows through the cable's self-regulating polymeric core, the heat output is adjusted in response to the outside temperature.

The method used to control the roof de-icing system determines its capability to operate when needed as well as the amount of power consumed. Control is usually exercised in one of three ways:

- Manual control
- Ambient thermostat control
- Automatic moisture/temperature control

Manual control involves turning a switch on and off to control the system power contactor. Someone must monitor the outside temperature and manually turn on the system whenever it drops below freezing. Less prone to human error is an automatic roof de-icing system control that uses an ambient-sensing thermostat. When it senses that the outside temperature is below freezing, it automatically turns on the roof and gutter system, energizing the heating cable(s).

The most energy-efficient method of roof de-icing system control is an automatic moisture/temperature controller. It consists of a control panel, one or more aerial-mounted snow sensors, and one or more ice sensors that are located in the gutters and at other critical areas that need to be

---

Electric Panel for Roof De-Icing

Some roof de-icing systems use mesh-type heating panels or mats instead of self-regulating heating cables to control ice and snow build-up and prevent ice dams from forming. For new roofing installations, the mats can be installed under the roofing material, making them invisible.

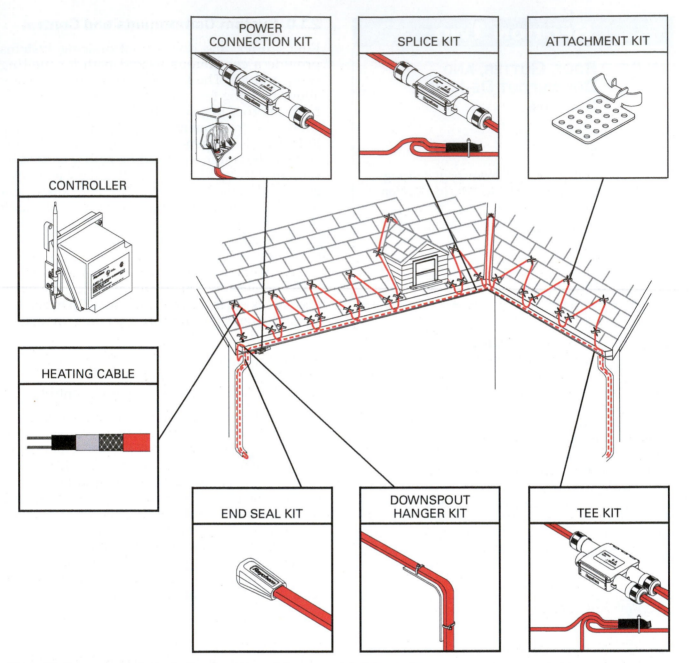

Figure 21 Components of a typical roof, gutter, and downspout de-icing system.

monitored. This arrangement turns on the roof, gutter, and downspout de-icing system when it begins to snow or when ice begins to form. *Figure 22* shows typical wiring for single- and multiple-circuit roof de-icing systems.

2.2.0 Component Selection and Installation

Each equipment manufacturer provides specific guidelines for the component selection and installation of its roof de-icing systems. Some general guidelines for the component selection process are described here.

Step 1 Determine the cable layout based on the roof construction:
- Type of construction (sloped roof, sloped roof with gutters, or flat roof)
- Location of roof/wall intersections
- Location of valleys, gutters, and downspouts

Step 2 Determine the method of cable attachment based on the roof construction materials and surface area:
- Roof, gutters, downspouts, and drip edges
- Mechanical or adhesive attachment clips

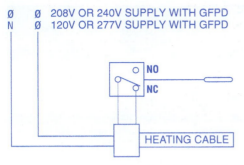

TYPICAL CONTROL WIRING – SINGLE CIRCUIT

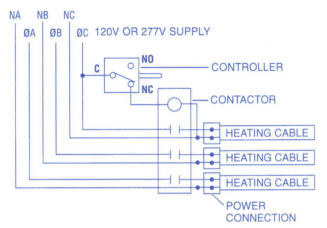

TYPICAL CONTROL WIRING – MULTIPLE CIRCUITS

Figure 22 Typical wiring for single- and multiple-circuit roof, gutter, and downspout de-icing systems.

Step 3 Determine the method of system temperature control based on dependability, precision, and energy consumption requirements:
- Manual control
- Ambient thermostat control
- Automatic moisture/temperature control

Step 4 Determine the number of electrical circuits and transformer load based on:
- Total heating cable length
- Supply voltage
- Minimum startup temperature

Step 5 Select the correct components and accessories.

Always install roof, gutter, and downspout de-icing systems in strict accordance with the system design specifications, the manufacturer's instructions, and all prevailing local codes. *NEC Article 426* covers the requirements for fixed outdoor electric de-icing and snow melting equipment. The *NEC*® requirements are covered later in this section.

2.2.1 Installation of Heating Cable and Components

Some general guidelines for the installation of roof, gutter, and downspout systems are listed here.

> **WARNING!**
> To prevent injury from falling off roofs when installing roof, gutter, and downspout systems, use appropriate fall protection equipment and follow all required safety procedures.

Prior to installation:

- Inspect all the cables, components, and accessories to make sure they are the ones specified for the job and that they are not damaged. Make sure that the rating of the heating cable is suitable for the service voltage available.
- To verify that the heating cable has not been damaged during shipping, test it for continuity and electrical insulation resistance. Test the insulation resistance between the heating cable conductors and the metallic grounding braid in accordance with the cable manufacturer's instructions. When testing insulation, use a 2,500VDC megohmmeter. The minimum insulation resistance should meet the value specified by the cable manufacturer, typically 1,000 megohms or more, regardless of length.
- Check and/or test each control device before installation for the correct operating temperature range, proper span, and setpoints.
- Store the heating cable and its components in a clean, dry place where they will be protected from mechanical damage.

As the installation progresses:

- Handle heating cable in a way that avoids exposing the cable to sharp edges. Also avoid using excessive pulling force, kinking or crushing the cable, walking or driving over the cable, or covering the heating cable with any roofing material unless it is a mat-type system designed for such use.
- Install the components and attachment accessories first in the locations indicated on the project drawings.
- Install the heating cable. Start at the end seal and work backwards. Make sure the heating cable provides a continuous path for water to flow off the roof (*Figure 23*). Leave drip loops where appropriate. Also, be sure to leave a drip loop at each component so that water will not track down the heating cable and into the component.

- Loop and secure the heating cable at the bottom of the downspouts in accordance with the manufacturer's instructions to prevent any ice buildup from the draining water and to protect the heating cable from mechanical damage.
- Use UV-resistant cable ties wherever you require two heating cables to stay together.
- Verify that the heating cable has not been damaged during installation by testing it for continuity and electrical insulation resistance. Test the insulation resistance between the heating cable conductors and the metallic grounding braid in accordance with the cable manufacturer's instructions. Use a 2,500VDC megger. If the heating cable is installed on a metal gutter, downspout, and/or metal roof, also measure the insulation resistance between the braid and metal surface. The minimum insulation resistance should meet the value specified by the cable manufacturer, typically 1,000 megohms or more, regardless of length. Record the original value for each circuit for future reference.

2.2.2 NEC® Requirements

NEC Article 426 covers the requirements for fixed outdoor electric de-icing and snow-melting equipment installed in unclassified areas. If the equipment is installed in hazardous (classified) areas, *NEC Articles 500 through 516* also apply.

The *NEC*® requirements for exposed fixed outdoor electric de-icing and snow-melting equipment are summarized below. For complete *NEC*® requirements, refer to *NEC Article 426*.

- Branch circuits and overcurrent devices supplying the equipment shall be sized at not less than 125 percent of the total heater load per *NEC Sections 210.20(A) and 426.4*.
- Per *NEC Section 426.10*, the equipment must be identified as being suitable for the environment, and be installed in accordance with the manufacturer's drawings and instructions.

Figure 23 Heating cable providing a continuous path for water to flow off the roof.

- The external surfaces of any equipment that operates at temperatures above 140°F (60°C) shall be physically guarded, isolated, or thermally protected to prevent contact with personnel (*NEC Section 426.12*).
- The presence of electric heating equipment shall be identified using clearly visible, permanent caution signs or markings (*NEC Section 426.13*).
- Exposed heating equipment shall not bridge expansion joints unless provision is made for expansion/contraction [*NEC Section 426.21(C)*].
- Ground-fault protection for equipment is required [*NEC Sections 210.8(A)(3) Exception and 426.28*].
- Equipment must be provided with a readily accessible disconnecting means such as branch circuit switches or circuit breakers per *NEC Section 426.50(A)*. A factory-installed cord and plug that is accessible to the equipment may be used if the equipment is rated at 20A or less and 150V or less, per *NEC Section 426.50(B)*.

2.0.0 Section Review

1. For roof and gutter de-icing systems, an automatic moisture/temperature controller is considered the _____.

   a. least expensive approach
   b. most energy-efficient method
   c. least reliable approach
   d. simplest method

2. On outdoor electric de-icing installations, prevent water from tracking into components by installing _____.

   a. collection containers
   b. the cable along the top edge of the gutter
   c. the cable in a serpentine pattern
   d. a drip loop at each component

3.0.0 SNOW-MELTING AND ANTI-ICING SYSTEMS

Objective

Describe snow-melting and anti-icing systems and the relevant selection/installation considerations.
  a. Describe snow-melting and anti-icing system components.
  b. Explain how snow-melting and anti-icing system components are selected and installed.

Figure 24 Bridge road surface protected by a snow-melting and anti-icing system.

Snow-melting and anti-icing systems are used to prevent the buildup of snow and ice on concrete surfaces of driveways, walkways, stairways, loading ramps, building entryways, and bridge road surfaces (*Figure 24*), where they can create hazards for people and/or vehicles.

3.1.0 Snow-Melting and Anti-Icing System Components

Figure 25 shows the components used with one manufacturer's snow-melting and anti-icing system. In this system, heating is provided by one or more self-regulating heating cables specifically designed for snow-melting and anti-icing applications. The cables are cut on the job. They are installed in a serpentine pattern with a 1-foot (30 cm) spacing between loops, securely fastened to rebar or wire mesh, and then completely encased in concrete. All electrical connections and terminations are made in externally accessible junction boxes. The capability of the heating cable to melt snow is proportional to its wattage rating. As current flows through the cable's self-regulating polymeric core, the heat output is adjusted in response to the ambient temperature of the surrounding concrete.

Some snow-melting and anti-icing systems use one or more factory-made electrical snow-melting mats instead of regular heating cable. The heating mats consist of self-regulating heating cable looped in a serpentine pattern within a screened mesh framework. Like the regular heating cable, the mats are embedded in concrete and electrically connected at externally accessible junction boxes.

The method used to control the snow-melting and anti-icing system determines its ability to operate when needed as well as the amount of power consumed. Control is done in one of three ways:

- Manual control
- Ambient thermostat control
- Automatic snow control

With the exception of specific models of equipment, the manual control and ambient thermostat control methods used with snow-melting and anti-icing systems are similar to those described earlier for roof de-icing systems. Ambient thermostat control is used to prevent the formation of ice on the pavement surface during all cold weather conditions. This control method keeps all points on the surface above 32°F (0°C) during freezing weather.

The most energy-efficient method of control is an automatic snow controller working in conjunction with a snow sensor. When the snow sensor detects rain or snow in the presence of low temperatures, it signals the snow controller to energize the contactor that supplies power to the snow-melting and anti-icing system. When the precipitation stops or the temperature rises above freezing, the controller de-energizes the system. An off-delay timing function incorporated into the controller can be used to extend the melting time. It keeps the melting equipment in operation

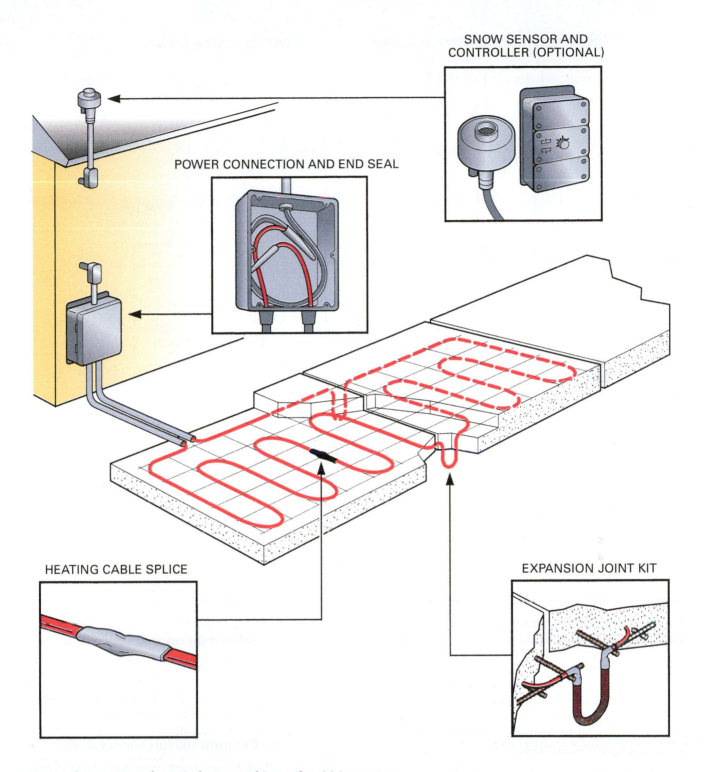

SNOW SENSOR AND
CONTROLLER (OPTIONAL)

POWER CONNECTION AND END SEAL

HEATING CABLE SPLICE

EXPANSION JOINT KIT

Figure 25 Components of a typical snow-melting and anti-icing system.

for a preset time after the sensor no longer detects precipitation accompanied by below-freezing
temperatures.

Figure 26 shows typical wiring for a snow-melting and anti-icing system under the control of an
automatic snow controller.

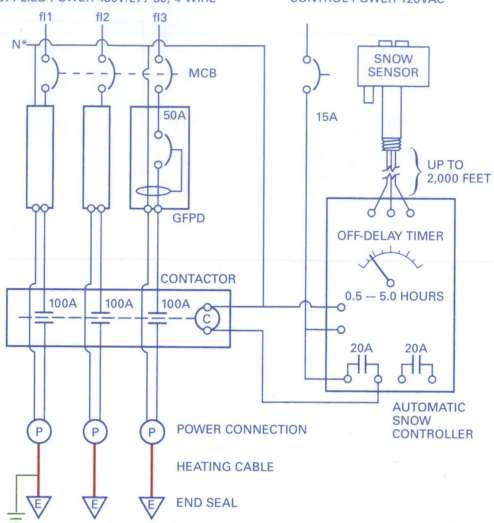

Figure 26 Typical wiring for a snow-melting and anti-icing system controlled by an automatic snow controller.

3.2.0 Component Selection and Installation

Each equipment manufacturer provides specific guidelines for the component selection and installation of its snow-melting and anti-icing systems. The component selection process for a regular heating cable (non-mat) system involves the following stages and considerations:

Step 1 Determine the spacing between the cables based on the ambient temperature and other weather characteristics of the region.

Step 2 Determine the cable layout and total length of cable based on the application and surface area to be heated.

Step 3 Determine the method of system temperature control based on dependability, precision, and energy consumption requirements:

- Manual control
- Ambient thermostat control
- Automatic snow control

Step 4 Determine the number of electrical circuits and transformer size based on:
- Total heating cable length
- Supply voltage
- Minimum startup temperature

Step 5 Select the correct components and accessories.

Always install snow-melting and anti-icing systems in strict accordance with the system design specifications, the manufacturer's instructions, and all prevailing local codes. *NEC Article 426* covers the requirements for fixed electric de-icing and snow-melting equipment installed in unclassified areas. The *NEC*® requirements are covered later in this section.

3.2.1 Installation of Heating Cable and Components

Before starting the installation, the cable and its components should be inventoried, tested, and stored as described earlier for the cable and components used in roof, gutter, and downspout systems. Make sure there is enough heating cable available to complete the job. The minimum amount of heating cable required can be calculated using the following formula. Typically, 3 feet (0.9 m) of additional cable should be allowed for making power connections, 1 foot (300 mm) for each splice, and $1\frac{1}{2}$ feet (450 mm) for each connection made at expansion joint kits.

Total heating cable length in feet =
[(heated area in feet^2 × 12) ÷ heated cable spacing] + end and component cable allowances

> **NOTE**
>
> Refer to the manufacturer's instructions for metric equivalents and equipment-specific spacing requirements.

The general guidelines for cable handling and installation described earlier also apply when installing heating cable for snow-melting and anti-icing systems. However, there are some cable and component installation guidelines unique to snow-melting and anti-icing systems installed in concrete. They are described as follows:

- Install the heating cable in a serpentine pattern covering the area to be heated. Space the cable in accordance with the job specifications, typically 1 foot (30 cm) on center. This distance should be maintained within 1 inch (25 mm). Some locations require closer spacing to provide more snow-melting or anti-icing capability.
- Do not route heating cable within 4 inches (100 mm) of the edges of the pavement, drains, anchors, or other objects in the concrete.
- Install the heating cable $1\frac{1}{2}$ to 2 inches (38 mm to 50 mm) below the finished surface and fasten it to the reinforcing bars or wire mesh (*Figure 27*).

Figure 27 Heating cables in concrete sidewalk form.

Cracks in Concrete

Heating cable should only be embedded in concrete pavement or slabs designed for that purpose. Concrete pavement or slabs with poor structural integrity can develop cracks that extend to the depth of the heating cable, possibly damaging the cable and causing a fire.

3.2.2 NEC® Requirements

NEC Article 426 covers the requirements for all fixed outdoor electric de-icing and snow-melting equipment. With the exception of the requirements that pertain only to exposed equipment, the *NEC Article 426* requirements described earlier for roof, gutter, and downspout de-icing systems also apply to snow-melting and anti-icing equipment. The requirements unique to equipment embedded in concrete or similar material are summarized as follows. For complete *NEC*® requirements, refer to *NEC Article 426*.

- Embedded panels or units shall not exceed 120W/ft² (1,300W/m²) [*NEC Section 426.20(A)*].
- Spacing between embedded adjacent cable runs depends on the cable rating, but shall not be less than 1 inch (25 mm) on center [*NEC Section 426.20(B)*].

- Embedded panels, units, or cables shall be placed on a masonry or asphalt base at least 2 inches (50 mm) thick, and there must be at least 1½ inches (38 mm) of masonry or asphalt applied over the finished surface [*NEC Section 426.20(C)(1)*].
- Embedded panels, units, or cables shall be fastened in place by approved means while the concrete, masonry, or asphalt finish is applied [*NEC Section 426.20(D)*].

3.0.0 Section Review

1. Snow-melting and anti-icing systems rely on cables installed in a serpentine pattern with a(n) _____.

    a. 8-inch (200 mm) spacing between the loops
    b. 1-foot (300 mm) spacing between the loops
    c. 18-inch (450 mm) spacing between the loops
    d. 3-foot (0.9 m) spacing between the loops

2. Snow-melting cable should not be routed within how many inches of the edges of the pavement, drains, anchors, or other objects in the concrete?

    a. 1 inch (25 mm)
    b. 2 inches (50 mm)
    c. 4 inches (100 mm)
    d. 6 inches (150 mm)

SECTION FOUR

4.0.0 OTHER TRACING AND WARMING SYSTEMS

Objective

Describe other electric heat-tracing and warming systems and the relevant selection/ installation considerations.

   a. Describe domestic hot-water temperature maintenance systems and the relevant selection/installation considerations.

   b. Describe electric floor heating systems and the relevant selection/installation considerations.

lectric heat-tracing cable and components can be used to maintain the temperature of domestic hot-water circulating loops. In addition, electric heating cable is often used for space heating by embedding or encasing heating cables in the floor. Both applications are presented in this section.

4.1.0 Domestic Hot-Water Temperature Maintenance Systems

Electrical heat tracing can be used for temperature maintenance of domestic hot water in lines wherever the hot-water circulating loop is expansive, such as hospitals, schools, prisons, and commercial high-rise office buildings. For these applications, specially designed self-regulating heating cable is installed on all the domestic hot-water pipe branches and risers within the building, underneath the standard pipe insulation (*Figure 28*). The heating cable adjusts its power output to compensate for supply-line heat losses that result from variations in water and ambient temperature. The power output of the cable is increased or decreased only at the points where the heat losses occur. Unlike conventional domestic hot-water systems, heat-traced hot-water systems do not require overheating the supply water to allow for cooling. System design is also simplified and costs are reduced because return water lines, circulating pumps, and balancing valves used in conventional domestic hot-water systems are not needed.

With the exception of the specific model, the components and accessories used in heat-traced domestic hot-water systems are the same as those used with other heat-tracing systems. These include power connections, heating cable, end seals, cable splices, and tees. The system control and monitoring methods are also similar.

4.1.1 Component Selection and Installation

Each equipment manufacturer provides specific guidelines for the component selection and installation of its domestic hot-water temperature maintenance systems. General guidelines for the component selection process involve the following stages and considerations:

Step 1 Review the plumbing design plan and determine its application.

Step 2 Determine the temperature requirements based on the application.

Step 3 Determine the type of self-regulating heating cable based on the application:
- Ambient temperature
- Nominal pipe temperature requirements

Step 4 Determine the maximum circuit cable length based on:
- Cable type
- Circuit breaker size
- Minimum water temperature at system startup

Step 5 Determine the number of electrical circuits based on:
- Total heating cable length
- Desired circuit layout for floors, risers, zones, and wings

Step 6 Select the correct components and accessories.

Installation and testing of heat-tracing cables used for domestic hot-water temperature maintenance is done in the same way as described earlier for heat tracing other types of pipe systems. The heating cables are typically installed on copper tubing beneath fiberglass insulation. At least 2 feet (0.6 m) of additional cable should be allowed for making power connections and splices, and 3 feet (0.9 m) should be allowed for heat tracing at pipe tees. Leave a service loop of extra heating cable at connections or at places in the piping where future service is expected.

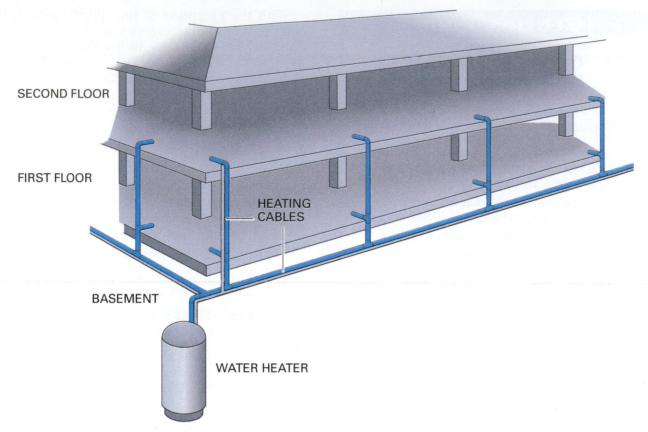

SECOND FLOOR

FIRST FLOOR

HEATING CABLES

BASEMENT

WATER HEATER

Figure 28 Heat-traced domestic hot-water system.

4.1.2 NEC® Requirements

NEC Article 427 covers the requirements for fixed electric-heating equipment installed on pipes and vessels, including domestic hot-water heat-tracing systems. Refer to the *NEC Article 427* requirements summarized earlier in the section on heat-tracing pipe systems. *NEC Article 517* contains specific requirements for health care facilities.

4.2.0 Floor Heating and Warming Systems

Floor heating and warming systems, also called radiant floor-heating systems, are used to heat or warm a floor mass that, in turn, radiates the heat to the cooler air in the room. Rooms with radiant heat in the floor tend to have a uniform air temperature from the ceiling to the floor, whereas a room heated with warm, forced air from a furnace will often be cold near the floor and hot near the ceiling. Floor heating systems are commonly used as either main or supplemental heating to heat or warm concrete floors, floors over unheated spaces, and tile and marble floors. Some floor heating systems can be used to warm hardwood and carpeted floors. To prevent frost-heave damage, floor systems are sometimes used to warm the soil beneath cold storage and large freezer floors.

Different types of heating elements are made for use in electric radiant floor-heating systems. One type uses continuous self-regulating or series-resistance heating cable specifically designed for floor warming applications (*Figure 29*). This heating cable is installed so that it snakes across the floor area in continuous loops. The closer together the loops, the more heat generated in that area. The heat from the cable is absorbed by the surrounding floor materials from which it is slowly radiated into the room. Heating cable can be incorporated into floors by embedding it in mortar; running it in conduits embedded in concrete, sand, or soil; or mounting it to the exposed bottom surface of concrete floors.

GOING GREEN

Electric Floor Heating

Often, the bathroom is the only room requiring additional warmth in the morning. Using a floor warming system in the bathroom allows the homeowner to keep the remainder of the home at a lower temperature, resulting in a significant energy savings over time.

Many floor warming systems use one or more factory-made electrical heating mats instead of a continuous cable. These are typically installed over a wood floor-joist system or concrete-slab floor system. Some heating mats consist of an electric cable fastened to a flexible mesh support structure. The mat is unrolled, cut to size, taped, or stapled in place, then embedded in thinset masonry beneath the finished floor material (*Figure 30*). Some systems use mesh mats that are installed on a plywood subfloor or concrete substrate over which carpet is installed.

Temperature control of a floor warming system is usually managed using an appropriate thermostatic control installed in accordance with the manufacturer's instructions. For systems using continuous cable, the sensing bulb of the thermostatic control is installed in the floor at the same depth as the heating cable, midway between adjacent runs of cable, and away from the room walls. The bulb should be installed inside a PVC conduit so that it can be easily removed for service, if necessary.

4.2.1 Component Selection and Installation

There is a wide diversity in floor heating and warming system applications, equipment, and related installation methods. Because of this diversity, it is impractical to give general guidelines for component selection and system installation. Each equipment manufacturer has readily available product literature that gives system design and component selection criteria for specific products and floor heating or warming applications. Floor heating and warming systems must be installed in strict accordance with the system design specifications, the manufacturer's instructions, and all prevailing local codes. When installing floor heating and warming systems, follow the general guidelines given earlier in this module for storing, handling, and testing the heating cable.

Radiant Floor-Heating Systems

Radiant floor-heating systems have been used for many centuries. Evidence shows that around 1300 BCE, a radiant floor heating system was used in a Turkish king's palace. Over a period ranging from about 80 BCE to 324 CE, the Romans used improved versions of these heating systems in upper-class houses and public baths throughout their empire. Those systems consisted of chambers built under the floor or tile flues built into a stone floor. Heated combustion gases, produced by a fire enclosed in a chamber located at one end of the floor, were routed through the floor chambers or flue tiles to exhaust vents placed in the outside walls at the other end of the floor.

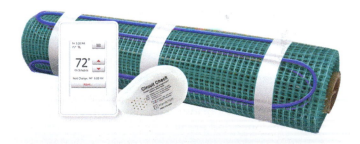

Figure 30 Typical electric floor-heating mat installation.

Figure 29 Basic concrete and masonry electric radiant floor-heating system.

4.2.2 NEC® Requirements

NEC Article 424 covers the requirements for fixed electric space-heating equipment. If the equipment is installed in hazardous (classified) areas, *NEC Articles 500 through 516* also apply. Refer to *NEC Article 424* for the requirements for heating cables installed in concrete or poured masonry floors. These requirements are summarized as follows:

- Heating cable shall be permitted to extend beyond the room or area where it originates [*NEC Section 424.38(A)*].
- Embedded cable shall be spliced only by using splices identified in the manufacturer's instructions (*NEC Section 424.40*).
- Adjacent runs of heating cable shall be installed in accordance with the manufacturer's instructions [*NEC Section 424.44(A)*].
- Embedded cable shall be fastened in place by approved nonmetallic means while the concrete is applied [*NEC Section 424.44(B)*].
- Non-heating cable leads exiting the floor shall be protected by installing them in conduit [*NEC Section 424.44(C)*].
- Ground-fault circuit interrupter protection for personnel shall be provided for cables installed in electrically heated floors of bathrooms and kitchens, and in hydromassage bathtub locations [*NEC Section 424.44(E)*].
- Cable installations shall be inspected and approved before cables are concealed (*NEC Section 424.46*).
- Heating panels or panel sets shall not be installed where they bridge expansion joints unless provision is made for expansion and contraction per *NEC Sections 424.98(B) and 424.99(B)(1)*.

4.0.0 Section Review

1. In hospitals, schools, and prisons, electrical heat-tracing systems are used for _____.

    a. temperature maintenance of domestic hot-water piping systems
    b. viscosity control on chemical piping
    c. adjusting the dew point
    d. moisture control inside air lines

2. On floor heating and warming systems, heating panels and/or panel sets are NOT installed without provisions for movement _____.

    a. near conduit
    b. in concrete slabs
    c. where they bridge expansion joints
    d. in concealed areas

---

1. As the ambient temperature surrounding a self-regulating heating cable gets colder, the _____.

   a. cable's heat output decreases
   b. cable's conductive polymeric core expands
   c. resistance of the bus wires decreases
   d. resistance of the polymeric core decreases

2. Which type of heating cable contains a resistor alloy heating element with a positive temperature coefficient?

   a. Power-limiting
   b. Self-regulating
   c. Mineral-insulated
   d. Series-resistance

3. Which of the following methods of heat-tracing system control uses a programmed algorithm to calculate the cycle time during which the heating circuits are energized?

   a. Line-sensing control
   b. Ambient-sensing control
   c. Proportional ambient-sensing control
   d. Dead-leg control

4. Which type of heat-tracing monitoring system measures current leakage from the system cables?

   a. Current monitoring
   b. Ground-fault monitoring
   c. Continuity monitoring
   d. Temperature monitoring

5. The minimum number of circuits required for a pipe heat-tracing installation is determined by dividing the total length of the heating cable needed _____.

   a. by seven
   b. by nine
   c. into sections with a demand of no more than 20A per section
   d. by the maximum circuit length that can be used for the specific cable in use

6. When determining the minimum amount of heating cable needed to heat-trace a run of pipe, add about _____.

   a. 1 foot (300 mm) of cable for each splice
   b. 2 feet (600 mm) of cable for each splice
   c. 3 feet (900 mm) of cable for each splice
   d. 4 feet (1.2 m) of cable for each splice

7. When installing temperature sensors in a pipe heat-tracing system, the sensors should be located _____.

   a. less than 1 foot (<300 mm) from any heat source or heat sink
   b. 1 to 3 feet (300 mm to 900 mm) from any heat source or heat sink
   c. 3 to 5 feet (0.9 to 1.5 m) from any heat source or heat sink
   d. 5 to 8 feet (1.5 to 2.4 m) from any heat source or heat sink

8. The continuity and electrical insulation resistance of heating cables should be tested _____.

   a. before installation
   b. during installation
   c. during and after installation
   d. before and after installation

9. Which of the following methods of control for a roof de-icing system is the most energy efficient?

   a. Manual control
   b. Ambient thermostat control
   c. Automatic moisture/temperature control
   d. Automatic snow control

10. What is the operating temperature of equipment above which de-icing or snow-melting cable must be guarded to prevent contact with personnel?

    a. 110°F (43°C)
    b. 120°F (49°C)
    c. 130°F (54°C)
    d. 140°F (60°C)

11. The most efficient method of snow-melting control is _____.

    a. manual control
    b. ambient thermostat control
    c. automatic snow control with a sensor
    d. end component control

12. The requirements for the installation of electrical snow-melting and de-icing systems are defined in _____.

    a. *NEC Article 424*
    b. *NEC Article 425*
    c. *NEC Article 426*
    d. *NEC Article 427*

13. The spacing between embedded adjacent cable runs for a snow-melting and de-icing system must not be less than _____.

    a. 1 inch (25 mm) on center
    b. 6 inches (150 mm) on center
    c. 10 inches (250 mm) on center
    d. 12 inches (300 mm) on center

14. Domestic hot-water temperature maintenance cable is typically _____.

    a. cut to the exact length required
    b. used with uninsulated systems
    c. installed on PVC pipe
    d. installed on copper tubing

15. While concrete is applied, embedded floor heating cable shall be fastened in place by _____.

    a. staples
    b. straps
    c. approved nonmetallic means
    d. wire ties

1. List five types of electric heat-tracing and freeze-protection systems.

   _____.

2. A self-regulating cable _____ its heat output when there is a decrease in the ambient temperature.

3. Power-limiting cables are a type of _____, _____ heating cable.

4. True or False? Mineral-insulated (MI) heating cables can be cut to length in the field as needed.

   _____.

5. The transfer of heat from a tracer heating cable to a pipe can be improved by putting a layer of _____ between the tracer cable and the pipe.

6. List four operating characteristics that may be monitored to detect failures in heat-tracing systems.

   _____.

7. When determining the amount of heating cable required for a job, you would typically add _____ of cable for each electrical power connection, splice, tee, and end seal.

8. Where in the *NEC*® would you find the requirements for roof, gutter, and downspout de-icing systems?

   _____.

9. The *NEC*® requirements for fixed electric-heating equipment installed on pipes and vessels also apply when heat tracing _____.
   a. snow melting and anti-icing systems
   b. domestic hot-water systems
   c. floor heating and warming systems
   d. roof, gutter, and downspout de-icing systems

10. Which of the following statements about the installation of the thermostatic control used to control the temperature in a floor warming system is NOT true?
    a. It should be placed in direct contact with the floor mortar or concrete.
    b. It should be placed midway between adjacent cable runs.
    c. It should be installed at the same depth as the heating cable.
    d. It should be installed away from the room walls.

Trade Terms Introduced in This Module

Constant-wattage heating cable: A heating cable that has the same power output over a large temperature range. It is used in applications requiring a constant heat output regardless of varying outside temperatures. Zoned parallel-resistance, mineral-insulated, and series-resistance heating cables are examples of constant-wattage heating cables.

Dead leg: A segment of pipe designed to be in a permanent no-flow condition. This pipe section is often used as a control point for a larger system.

Dew point: The temperature at which a vapor begins to condense into its liquid state as the result of a cooling effect.

Heat source: Any component that adds heat to an object or system.

Mineral-insulated (MI) heating cable: A type of constant-wattage, series-resistance heating cable used in long line applications where high temperatures need to be maintained, high-temperature exposure exists, or high-power output is required.

Nichrome: An alloy based on nickel with chromium added (10–20 percent), and sometimes iron (up to 25 percent). Nichrome is often used for high-temperature applications such as in the fabrication of electric heating elements.

Parallel-resistance heating cable: An electric heating cable with parallel connections, either continuous or in zones. The watt density per lineal length is approximately equal along the length of the heating cable, allowing for a drop in voltage down the length of the heating cable.

Power-limiting heating cable: A type of heating cable that shows positive temperature coefficient (PTC) behavior based on the properties of a metallic heating element. The PTC behavior exhibited is much less (a smaller change in resistance in response to a change in temperature) than that shown by self-regulating heating cables.

Self-regulating heating cable: A type of heating cable that inversely varies its heat output in response to an increase or decrease in the ambient temperature.

Series-resistance heating cable: A type of heating cable in which the ohmic heating of the conductor provides the heat. The wattage output depends on the total circuit length and on the voltage applied.

Startup temperature: The lowest temperature at which a heat-tracing cable is energized.

Vessel: A container such as a barrel, drum, or tank used for holding fluids or other substances.

Additional Resources

This module presents thorough resources for task training. The following resource material is suggested for further study.

IEEE Standard 515, IEEE Standard for the Testing, Design, Installation, and Maintenance of Electrical Resistance Trace Heating for Industrial Applications, Latest Edition. Piscataway, NJ: IEEE.

Industrial Heat-Tracing Systems Component Selection Guide. Menlo Park, CA: Raychem HTS.

National Electrical Code® Handbook, Latest Edition. Quincy, MA: National Fire Protection Association.

Selection Guide for Self-Regulating Heat-Tracing Systems. Houston, TX: Tyco Thermal Controls.

Figure Credits

Section Review Answer Key

SECTION 1.0.0

Answer	Section Reference	Objective
1. d	1.1.3	1a
2. a	1.2.4	1b
3. b	1.3.0	1c
4. c	1.4.0	1d
5. a	1.5.1	1e

SECTION 2.0.0

Answer	Section Reference	Objective
1. b	2.1.0	2a
2. d	2.2.1	2b

SECTION 3.0.0

Answer	Section Reference	Objective
1. b	3.1.0	3a
2. c	3.2.1	3b

SECTION 4.0.0

Answer	Section Reference	Objective
1. a	4.1.0	4a
2. c	4.2.2	4b

NCCER CURRICULA — USER UPDATE

NCCER makes every effort to keep its textbooks up-to-date and free of technical errors. We appreciate your help in this process. If you find an error, a typographical mistake, or an inaccuracy in NCCER's curricula, please fill out this form (or a photocopy), or complete the online form at **www.nccer.org/olf**. Be sure to include the exact module ID number, page number, a detailed description, and your recommended correction. Your input will be brought to the attention of the Authoring Team. Thank you for your assistance.

Instructors – If you have an idea for improving this textbook, or have found that additional materials were necessary to teach this module effectively, please let us know so that we may present your suggestions to the Authoring Team.

NCCER Product Development and Revision

13614 Progress Blvd., Alachua, FL 32615

Email: curriculum@nccer.org
Online: www.nccer.org/olf

❏ Trainee Guide   ❏ Lesson Plans   ❏ Exam   ❏ PowerPoints   Other _____

Craft / Level: _____   Copyright Date: _____

Module ID Number / Title: _____

Section Number(s): _____

Description: _____

_____

_____

_____

_____

Recommended Correction: _____

_____

_____

_____

_____

Your Name: _____

Address: _____

_____

Email: _____   Phone: _____

This page is intentionally left blank.

Motor Operation and Maintenance

OVERVIEW

For motors to operate at optimum levels over a long span of time, proper care and maintenance is required. This module covers motor care procedures, including cleaning, testing, and preventive maintenance. Basic troubleshooting procedures are also presented.

Module 26410-20

Trainees with successful module completions may be eligible for credentialing through the NCCER Registry. To learn more, go to **www.nccer.org** or contact us at 1.888.622.3720. Our website, **www.nccer.org**, has information on the latest product releases and training.

Your feedback is welcome. You may email your comments to **curriculum@nccer.org**, send general comments and inquiries to **info@nccer.org**, or fill in the User Update form at the back of this module.

This information is general in nature and intended for training purposes only. Actual performance of activities described in this manual requires compliance with all applicable operating, service, maintenance, and safety procedures under the direction of qualified personnel. References in this manual to patented or proprietary devices do not constitute a recommendation of their use.

26410-20 V10.0

Objectives

When you have completed this module, you will be able to do the following:

1. Identify the factors that affect motor reliability and lifespan.
   a. Identify the common causes of motor failure.
   b. Identify motor characteristics.
2. Describe maintenance and troubleshooting requirements for electric motors.
   a. Identify the tools and basic care and maintenance requirements for electric motors.
   b. Explain the requirements for maintaining motor bearings.
   c. Explain how to perform motor insulation testing.
   d. Explain how to troubleshoot an electric motor.
3. Describe the guidelines for installing and commissioning electric motors.
   a. Explain alignment and adjustment requirements.
   b. Describe startup procedures.

Performance Tasks

This is a knowledge-based module. There are no Performance Tasks.

Trade Terms

Generator
Insulation breakdown
Insulation classes
Leakage current

Megohmmeter
Soft foot
Totally enclosed motor

Industry Recognized Credentials

If you are training through an NCCER-accredited sponsor, you may be eligible for credentials from NCCER's Registry. The ID number for this module is 26410-20. Note that this module may have been used in other NCCER curricula and may apply to other level completions. Contact NCCER's Registry at 888.622.3720 or go to **www.nccer.org** for more information.

Contents

1.0.0   Motor Reliability and Lifespan 1
  1.1.0   Causes of Motor Failure 1
  1.2.0   Motor Characteristics 5
    1.2.1   Motor Service Conditions 5
    1.2.2   Starting Configurations 5
    1.2.3   Insulation Systems 6
2.0.0   Motor Maintenance 7
  2.1.0   Motor Maintenance Requirements 7
    2.1.1   Tools for Maintenance and Troubleshooting 8
    2.1.2   Basic Care and Maintenance 8
    2.1.3   Periodic Predictive Testing 9
    2.1.4   Receiving and Storing Motors 10
  2.2.0   Motor Bearing Maintenance 11
    2.2.1   Frequency of Lubrication 11
    2.2.2   Lubrication Procedure 11
    2.2.3   Checking Bearings 13
  2.3.0   Motor Insulation Testing 15
    2.3.1   Insulation Resistance Tests 15
    2.3.2   Determining the Polarization Index 22
    2.3.3   Insulation Testing Considerations 22
  2.4.0   Troubleshooting Motors 22
    2.4.1   Insulation Testing 23
    2.4.2   Grounded Coils 23
    2.4.3   Water-Damaged Motors 23
3.0.0   Motor Installation and Commissioning Guidelines 25
  3.1.0   Alignment and Adjustment 25
    3.1.1   Alignment 25
    3.1.2   Endplay Adjustment 25
    3.1.3   Doweling 27
  3.2.0   Startup 27
    3.2.1   First-Time Startup 27
    3.2.2   Coupled Startup 29

Figures and Tables

Figure 1   Squirrel-cage motor construction 1
Figure 2A   Stator winding failures and causes (Part 1 of 2) 3
Figure 2B   Stator winding failures and causes (Part 2 of 2) 4
Figure 3   TEFC explosion-proof motor 7
Figure 4   Infrared thermometer 9
Figure 5   Vibration tester 9
Figure 6   Online motor analysis instrument 10
Figure 7   Typical shaft rotation log tag 11
Figure 8   Typical equipment lubrication schedule and tag 12
Figure 9   Grease fitting and relief plug for a TEFC motor 13
Figure 10   Grease gun and meter 14
Figure 11   Checking for a bent shaft on a motor 14
Figure 12   Typical direct-drive configuration without coupling 14
Figure 13   Checking shaft alignment between a motor and its load 15
Figure 14   Typical dial indicator with various
                holders and stainless-steel slotted shims 16
Figure 15   Typical megohmmeter 17
Figure 16   Current curves related to insulation testing 18
Figure 17   Approximate temperature coefficient for insulation resistance 20
Figure 18   Chart of insulation resistance readings 21
Figure 19   Temperature-corrected resistance readings alone 21
Figure 20   Bench testing motor insulation 23
Figure 21   Typical vibration-reducing couplers 26
Figure 22   Typical coupling guard over rotating components 29

Table 1   Tools and Test Equipment for Electrical Maintenance 8

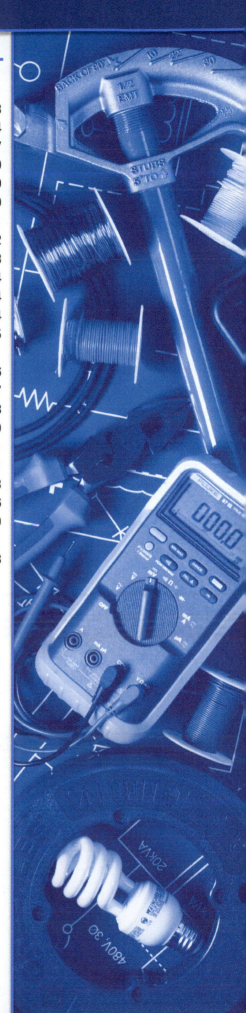

This page is intentionally left blank.

SECTION ONE

1.0.0 MOTOR RELIABILITY AND LIFESPAN

Objective

Identify the factors that affect motor reliability and lifespan.

  a. Identify the common causes of motor failure.
  b. Identify motor characteristics.

Trade Terms

Insulation classes: Categories of insulation based on the thermal endurance of the insulation system used in a motor. The insulation system is chosen to ensure that the motor will perform at the rated horsepower and service factor load.

Megohmmeter: An instrument or meter capable of measuring resistances in excess of 200MΩ It employs much higher test voltages than are used in ohmmeters, which measure up to 200MΩ. Commonly referred to as a megger.

Soft foot: When the feet of a motor frame are forced to anchor against an uneven surface, resulting in distortion of the motor frame. This can cause misalignment of bearings.

Totally enclosed motor: A motor that is encased to prevent the free exchange of air between the inside and outside of the case.

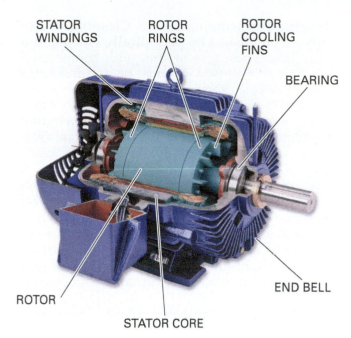

Figure 1 Squirrel-cage motor construction.

Three-phase, squirrel-cage induction motors (*Figure 1*) are the most common motors in heavy commercial and industrial use. The life expectancy of a squirrel-cage induction motor depends largely on the condition of its insulation, which should be suitable for the operating requirements, along with performance of any required maintenance. Electric motors are sometimes called *rotating electrical machinery, electric machinery,* or just *machines*.

1.1.0 Causes of Motor Failure

When a motor malfunctions, it is due to either mechanical or electrical causes. These failures can be traced to one or more of the following causes:

- *Worn or tight bearings* – Worn or tight bearings are usually caused by failing bearing lubrication or improper lubrication. If the bearings have failed or are worn to such an extent that they allow the rotor to drag on the stator, then the rotor, stator, and stator windings can be irreparably damaged. When bearings begin to fail, the condition can be detected by periodically checking the motor for excessive heat, noise, or vibration at the bearings before damage occurs. Heat and vibration can be measured using appropriate testers at the motor bearing housing. Bearing misalignment within the motor can also cause excessive wear or tightness. This is usually caused by a condition called **soft foot**. This occurs when the feet of a motor frame are forced to anchor against an uneven surface, resulting in distortion of the motor frame. This can cause misalignment of bearings. Proper mounting surfaces and shimming must be used to prevent soft foot when a motor is installed.

- *Vibration* – Motor vibration can be caused by a bent shaft, misalignment of shafts between a motor and a directly coupled load, or loose anchorage. Vibration may result in excessive bearing wear or failure. Dial indicators can be used to check for bent shafts or misalignment between the motor and a directly coupled load.

- *Moisture* – Motor insulation must be kept reasonably dry, although many applications make this practically impossible unless a **totally enclosed motor** is used. Infrequently used or stored motors must be kept dry to prevent premature insulation failure.

- *Dust and dirt* – Dust and dirt can restrict ventilation and increase a motor's operating temperature. Overheating can cause premature bearing and winding failure. Cleaning is usually accomplished by periodically blowing out ventilation passages or fins with compressed air. The compressed air must be dry and regulated to a safe pressure for use.

- *Insulation deterioration* – Assuming no other mechanical or electrical causes, the winding insulation of all motors will deteriorate over time and will eventually break down, causing arcing and/or burnout of the windings. A **megohmmeter**, commonly called a *megger*, is used to periodically check the insulation resistance of the windings to chart the progress of insulation deterioration. Megohmmeters are also used as a troubleshooting tool for motors. Some megohmmeters are also capable of performing other motor performance and evaluation tests.

- *Overloading* – Overloading a motor can cause overheating and eventual bearing failure or premature insulation failure of the windings. Overloading usually occurs because the driven load has been changed to accommodate a higher capacity, heavier capacity, or an increased duty cycle. When driven loads are altered for these reasons, existing motors must be evaluated to be certain that they can withstand the new load.

- *Single-phasing* – The loss of a single phase of a polyphase power source for a running motor can cause the failure of the remaining phase windings of the motor if the motor is not protected by the proper equipment. This can occur in older installations where failure of one voltage phase due to a partial primary power failure or a failed fuse does not cause removal of all voltage phases from the motor. These types of installations should be updated using proper motor controllers and/or loss-of-phase protection.

Severe electrical, mechanical, or environmental operating conditions can damage the stator winding of a three-phase, squirrel-cage motor. Such winding failures and their causes are shown in *Figure 2A* and *Figure 2B*. Motors that have suffered severe core damage because of overheating or mechanical damage from the rotor are usually discarded. This is because severe core damage generally results in abnormal core losses or loss of efficiency if the motor is rewound. In an emergency, some motors with severe damage can be rewound after some or most of the damaged core laminations are restacked. However, this will result in a shortened operating life and

these motors should be replaced as soon as possible. Depending on the type of facility and the number and size of the motors used within the facility, motors below some facility-established horsepower rating may not be salvaged when they fail, nor should they be included in periodic predictive testing programs. However, the cause of failure for any motor should be investigated and corrected to prevent rapid failures of any replacement motors.

Many times, a motor that was originally of adequate capacity for a given load is later found to be inadequate. The following are some reasons for this problem:

- More severe duty imposed on the machine tool, such as a different die job in a punch press
- Heavier material or material of different machining characteristics
- Changed machine operation

Connecting measuring instruments in the motor circuit may disclose the reason for motor overheating, failure to start the load, or other abnormal symptoms. It is frequently desirable to connect recording instruments in the motor circuit for other purposes, such as analyzing the output of the machine or to gauge processing operations.

Control circuits for many older induction motor installations were not provided with thermal overload relays or loss-of-phase protection. Single-phasing of polyphase induction motors on such circuits frequently caused motor winding failure, which often results from the blowing of one fuse while the motor is up to speed and under load. Under such conditions, the portion of the winding remaining in the circuit will attempt to carry the load until it fails due to overheating.

Increasing the load on the motor beyond its rated capacity increases the operating temperature, which shortens the life of the insulation. Momentary overloads usually do no damage; consequently, there is a tendency to use thermal overload protection in modern control systems for motors rated at 600V or less. Monitoring of the actual winding temperature is the ideal method of determining whether a motor is overloaded. Other methods, such as motor protection relays and electrical trip units for medium-voltage and 600V motors, are used to mimic thermal measurement and protect a motor from protracted overload conditions. These devices calculate (model) winding temperature by monitoring current and time using a microcomputer inside the device.

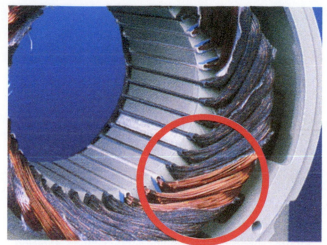

Single-phased undamaged winding (caused by a missing voltage phase in a wye-connected motor)

Single-phased damaged winding (caused by a missing voltage phase in a delta-connected motor)

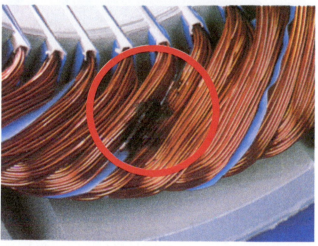

Phase-to-phase short (caused by contamination, abrasion, vibration, or voltage surge)

Turn-to-turn short (caused by contamination, abrasion, vibration, or voltage surge)

Shorted coil (caused by contamination, abrasion, vibration, or voltage surge)

Winding grounded at edge of slot (caused by contamination, abrasion, vibration, or voltage surge)

Figure 2A Stator winding failures and causes (Part 1 of 2).

Winding grounded in slot (caused by contamination, abrasion, vibration, or voltage surge)

Shorted connection (caused by contamination, abrasion, vibration, or voltage surge)

Unbalanced voltage damage (severe for one phase, slight for another phase, and caused by unbalanced power source loads or poor connections)

Overload damage to all windings
Note: Undervoltage and overvoltage (exceeding NEMA standards) will result in the same type of insulation deterioration.

Locked rotor overload damage to all windings (may also be caused by excessive starts or reversals)

Voltage surge damage (caused by lightning, capacitor discharges, voltage kickbacks caused by de-energizing large inductive loads, or voltage disturbances caused by solid-state devices)

Figure 2B Stator winding failures and causes (Part 2 of 2).

1.2.0 Motor Characteristics

Motors have specific service conditions and starting configurations in order to ensure optimum performance. In addition, motors are designed to operate within specific temperature ranges known as *insulation classes*.

1.2.1 Motor Service Conditions

The National Electrical Manufacturers Association (NEMA) has defined standards for usual service conditions for motors. When operated within the limits of the following NEMA usual service conditions, standard motors will perform in accordance with their ratings:

- *Ambient or room temperature not over 40°C (104°F)* – If the ambient temperature is over 40°C due to inadequate airflow in the motor location, the motor service factor must be reduced or a larger motor or motor with a higher insulation temperature must be used. The larger motor will be loaded below full capacity, so the temperature rise will be lower and the potential for overheating will be reduced.
- *Altitude does not exceed 3,300' (1,000 meters)* – Motors are typically designed to operate at altitudes below 3,300'. Motors having Class A or B insulation systems and lower temperature rises as defined by NEMA will operate satisfactorily at altitudes above 3,300' in locations where the decrease in ambient temperature compensates for the increase in temperature rise. Motors having a service factor of 1.15 or higher will operate satisfactorily at unity service factor and an ambient temperature of 40°C at altitudes above 3,300' up to 9,000' (2,743 meters).
- *Voltage variation of not more than 10% of nameplate voltage* – Operation outside these limits or a voltage imbalance over 5% can result in overheating or loss of torque and may require the use of a larger motor.
- *Frequency variation of not more than 5% of nameplate frequency* – Operation outside of these limits results in a substantial speed variation and causes both overheating and reduced torque.
- *Plugging, jogging, or multiple starts* – Unless rated for this use, standard motors should not be used.

Motors are often exposed to damaging atmospheres such as excessive moisture, steam, salt air, abrasive or conductive dust, lint, chemical fumes, and combustible or explosive dust or gases. To protect such motors, special enclosures or encapsulated windings and special bearing protection are required. The location of supplementary enclosures must not interfere with the ventilation of the motor, however.

Motors exposed to damaging mechanical or electrical loading such as unbalanced voltage conditions, abnormal shock or vibration, torsional impact loads, plugging, jogging, multiple starts, or excessive thrust or overhang loads may require special mountings or protection designed for the specific installation.

1.2.2 Starting Configurations

Squirrel-cage motors are usually designed for across-the-line starting, which means that they can be connected directly to the power source by means of a suitable contactor. For large motors and in some other types of motors, the starting currents are very high. Usually, the motor is built to withstand starting currents, but, because these currents are often six times or more the rated load current, there may be a large voltage drop in the power system. Some method of reducing the starting current must therefore be used to limit the system voltage drop to a tolerable value.

Reduced-voltage starting significantly reduces the torque and acceleration characteristics of any motor. Therefore, it is necessary to have the motor unloaded or nearly so when the motor is started. Reduction of starting voltage is the primary method of reducing motor starting current. Reduction of the starting current may be accomplished by the use of a variable-frequency motor controller or soft starter, or by any one of the following reduced-voltage starting methods:

- *Primary resistor or reactance* – Primary resistor or reactance starting uses series reactance or resistance to reduce the current on the first step. After a preset time interval, the motor is connected directly across the line. This method can be used with any standard motor.
- *Autotransformer* – Autotransformer starting uses autotransformers to reduce the voltage and current on the first step. After a preset time interval, the motor is connected directly across the line. It can be used with any standard motor.
- *Wye-delta* – Wye-delta starting impresses the voltage across the Y connection to reduce the current on the first step. After a preset time interval, the motor is connected in delta, permitting full-load operation. The motor must have a winding capable of wye-delta connection.

- *Part-winding*—Part-winding starting involves motors with two separate winding circuits. Upon starting, only one winding circuit is engaged and current is reduced. After a preset time interval, the full winding of the motor is put directly across the line. The motor must have two separate winding circuits. To avoid possible overheating and subsequent damage to the winding, the time between the connection of the first and second windings is limited to four seconds.

All of these starting methods require special starters designed for the particular method and controlled between the start and run functions by an adjustable timer or current relay.

1.2.3 Insulation Systems

An insulation system is an assembly of insulating materials in association with conductors and the supporting structural parts of a motor. Insulation systems are divided into classes according to the thermal endurance of the system for temperature rating purposes. Four classes of insulation systems are used in motors: A, B, F, and H. Do not confuse these insulation classes with motor designs, which are also designated by letter.

- *Class A* – Class A insulation systems have a suitable thermal endurance when operated at the limiting Class A temperature of 105°C (221°F). Typical materials used include cotton, paper, cellulose acetate films, enamel-coated wire, and similar organic materials impregnated with suitable substances.
- *Class B* – Class B insulation systems have suitable thermal endurance when operated at the limiting Class B temperature of 130°C (266°F). Typical materials include mica, glass fiber, and other materials (organic or inorganic), with compatible bonding substances having suitable thermal stability.
- *Class F* – Class F insulation systems have suitable thermal endurance when operated at the limiting Class F temperature of 155°C (311°F). Typical materials include mica, glass fiber, and other materials (organic or inorganic), with compatible bonding substances having suitable thermal stability.
- *Class H* – Class H insulation systems have suitable thermal endurance when operated at the limiting Class H temperature of 180°C (356°F). Typical materials used include mica, glass fiber, silicone elastomer, and other materials (organic or inorganic), with compatible bonding substances (for example, silicone resins) having suitable thermal stability.

1.0.0 Section Review

1. Excessive heat, noise, or vibration in a motor is an indication of _____.
   a. bad wiring
   b. worn bearings
   c. loose motor mount
   d. low input voltage

2. The NEMA standard for maximum voltage imbalance is _____.
   a. 5%
   b. 10%
   c. 20%
   d. 25%

2.0.0 MOTOR MAINTENANCE

Objective

Describe maintenance and troubleshooting requirements for electric motors.

  a. Identify the tools and basic care and maintenance requirements for electric motors.
  b. Explain the requirements for maintaining motor bearings.
  c. Explain how to perform motor insulation testing.
  d. Explain how to troubleshoot an electric motor.

Trade Terms

Generator: (1) A rotating machine that is used to convert mechanical energy to electrical energy. (2) General apparatus or equipment, that is used to convert or change energy from one form to another.

Insulation breakdown: The failure of insulation to prevent the flow of current, sometimes evidenced by arcing. If the voltage is gradually raised, breakdown will begin suddenly at a certain voltage level. Current flow is not directly proportional to voltage. When a breakdown current has flowed, especially for a period of time, the next gradual application of voltage will often show breakdown beginning at a lower voltage than initially.

Leakage current: AC or DC current flow through insulation and over its surfaces, and AC current flow through a capacitance. Current flow is directly proportional to voltage. The insulation and/or capacitance are thought of as a constant impedance unless breakdown occurs.

Motor failures account for a high percentage of industrial and heavy commercial electrical repair work. The care given to an electric motor while it is being stored or operated affects the life and usefulness of the motor. A motor that receives good maintenance will outlast a poorly treated motor many times over. Maintaining a motor in good operating condition requires periodic preventive maintenance actions along with periodic predictive testing. Predictive testing is required to determine if any faults exist or if faults are likely to occur in the near future. When a motor is determined to be in the process of failing, it should be removed and replaced. If a motor has been replaced before its core has been damaged, it can usually be sent to a motor rebuild facility for salvage as a spare.

2.1.0 Motor Maintenance Requirements

The frequency and thoroughness of the maintenance and testing of a motor depend on such factors as:

- Number of hours and days the motor operates
- Importance of the motor
- Nature of service
- Environmental conditions

Most heavy-duty commercial/industrial squirrel-cage motors are totally enclosed (TE) motors. TE motors are available in the following configurations:

- *TEFC motor* – A TEFC motor is a fan-cooled (FC) motor cooled by a fan attached to the rear shaft of the motor. The fan is covered by a shroud that directs the fan air through cooling passages and/or fins external to the motor. TEFC motors are the most common type of TE motor. They are also available in versions rated for severe service duty in corrosive environments, for wash-down duty in food service, and for explosion-proof duty in hazardous environments (*Figure 3*).

Figure 3 TEFC explosion-proof motor.

- *TEBC motor* – A TEBC motor is a blower-cooled (BC) motor cooled by a separately powered blower attached to the motor. TEBC motors are typically used in heavy-duty variable-speed motor applications.
- *TEAO motor* – A TEAO motor is a finned air-over (AO) motor that powers and is cooled by the slipstream from a large fan.
- *TENV motor* – A TENV motor is a nonventilated (NV) motor used in smaller-horsepower variable-speed applications.

> **WARNING!**
> TENV motors run at very hot temperatures. Do not touch these motors during or immediately after operation.

> **NOTE**
> Variable-speed motors are specified in the United States as inverter-grade motors that can withstand heat and surges up to 2,400V.

2.1.1 Tools for Maintenance and Troubleshooting

In addition to typical equipment such as voltmeters and ammeters, motor maintenance equipment should also include an insulation resistance tester, such as a multivoltage megohmmeter. *Table 1* provides a list of practical tools and test equipment for electrical maintenance of motors and other electrical apparatus.

2.1.2 Basic Care and Maintenance

Every motor in operation or storage should be observed regularly as recommended by the motor manufacturer to identify exposure to any adverse condition for which it is not rated. Operating motors should also be checked for excessive dust, debris, or lint on or about the motor. Ensure that airflow is not blocked by objects or room conditions and that the cooling fins are not dust-coated. Low-pressure compressed air may be used to blow clean the motor housing ventilation paths or cooling fins.

> **WARNING!**
> Before performing any motor maintenance procedures, other than external visual inspection or noise monitoring, always lock out and tag the equipment according to approved procedures. If guards covering rotating components must be removed to gain access to motor bearings, remove the guards in accordance with approved facility procedures and use great care near the rotating components. When using compressed air, exercise caution by wearing the appropriate personal protective equipment (PPE) and using an airflow tip with air pressure limited to 30 psi or less. Excess air pressure can result in injury.

Table 1 Tools and Test Equipment for Electrical Maintenance

Tools or Equipment	Application
Multimeters, voltmeters, ohmmeters, clamp-on ammeters, wattmeters, clamp-on power factor meter	Measure circuit voltage, resistance, current, and power. Useful for circuit tracing and troubleshooting.
Potential and current transformers, meter shunts	Increase range of test instruments to permit the reading of high-voltage and high-current circuits.
Motor rotation and phase tester	Determines the proper three-phase connections for the desired rotation direction of a motor prior to connecting the motor. Also checks the phasing of the power source so that the motor can be connected to the desired phases.
Tachometer	Checks rotating machinery speeds.
Recording meters	Provide a permanent record of voltage, current, power, temperature, and other data on charts for analytic study.
Insulation resistance tester, thermometer, psychrometer	Test and monitor insulation resistance; use a thermometer and psychrometer for temperature and humidity correction.
Vibration testers and transistorized stethoscope	Detect faulty rotating machinery bearings and leaky valves.
Milli or microhmmeter	Precise measurement of coil resistance.
Dial indicator with mounting devices	Used to check for bent motor shafts and performing shaft alignment for directly coupled motor loads.

Before using a vibration tester, make sure the equipment being checked is firmly anchored. When using the tester, make sure that its accelerometer is held firmly or secured against a bearing housing directly over the shaft center line of the bearing and as close to the middle of the actual bearing as possible. Magnetically mounted accelerometers can be used only on solid ferrous metal surfaces with no nonferrous metal in between. Make sure the magnet does not rock on the housing surface. For rounded housings, use a V-mount magnet. Place an identifying mark on the equipment where the test was made to ensure that subsequent vibration measurements are made in exactly the same spot.

Figure 4 Infrared thermometer.

Inspect the motor starter and verify that the motor reaches the proper speed each time it is started. Check other motor parts and accessories such as belts, gears, couplings, chains, and sprockets for excessive wear or misalignment. Periodically lubricate the motor as recommended by the manufacturer (discussed in detail later in this module). Be alert to any unusual noise, which may be caused by metal-to-metal contact, such as bad bearings. Note any abnormal odor, which might indicate scorched insulation varnish.

Use an infrared thermometer (*Figure 4*) to check the bearing housings for evidence of excess heat and the motor itself for excess temperature rise as listed on the motor nameplate. Compare to previously logged data. Use a vibration tester (*Figure 5*) to determine if excess vibration is occurring as compared to previously logged data. A stethoscope may be used to detect unusual bearing noise during rundown when the motor is powered down.

Vibration testers indicate the vibration velocity in inches per second (in/sec), millimeters/second (mm/sec), or centimeters/second (cm/sec), depending on the model selected. Other models are also available that display only the total peak value of the velocity amplitude in either in/sec or mm/sec. A printer or a data logger is used with some models to capture data for further analysis.

2.1.3 Periodic Predictive Testing

Periodic predictive testing can be conducted on two levels. Online (dynamic) tests can be conducted with the motor operating. Offline (static) tests are conducted with the motor stopped and the power locked out and tagged. Of the two types of testing, offline testing for insulation resistance and polarization index is the most common and

Figure 5 Vibration tester.

will be described in detail in a later section. Online motor analysis is a relatively new technology that evaluates a number of motor parameters on a noninterference basis while a motor is running. This type of testing uses computerized portable test equipment like that shown in *Figure 6*. This testing is normally done in facilities with many large motors and may be conducted more frequently than offline testing. This test equipment can be used for monitoring power circuit quality, overall motor condition, load conditions, and performance efficiency. It can provide data collection and trending information to aid in periodic maintenance evaluation, in addition to the data collected during offline testing. Some of these instruments are equipped with accelerometer probes that can provide sophisticated vibration

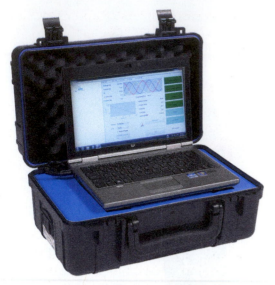

Figure 6 Online motor analysis instrument.

data in the form of time/amplitude velocity waveforms and/or Fast-Fourier Transform (FFT) velocity spectrum waveforms. The waveform data are usually analyzed by specially trained personnel and/or software to determine the probable causes of any abnormal vibration.

2.1.4 Receiving and Storing Motors

When an electric motor is received on a job site, always follow the manufacturer's recommendations for unloading, uncrating, and installing the motor. Failure to follow these recommendations can cause injury to personnel and possible damage to the motor. When the motor has been uncrated, check for damage that might have occurred during shipment. Check the motor shaft to verify that it turns freely. Clean the motor of any debris, dust, moisture, or foreign matter that might have accumulated during shipment. Make sure it is dry before putting it into storage or service.

> **WARNING!**
> Never attempt to start a wet or damp motor.

Motors may be put into storage when the project on which they are to be used is not complete or when they will be used as backups. The first consideration when storing motors for any length of time is the storage environment. A clean, dry, and warm location (one that does not undergo severe changes in temperature over a 24-hour period) is best. Ambient temperature changes cause condensation to form on and in the stored motor. Moisture in motor insulation can cause motors to fail on startup; therefore, guarding against moisture is vital when storing motors of any type. Some motors have internal heaters to protect against moisture buildup during storage. These heaters must be monitored while in use.

A means for transporting the motor from the place of storage is also important. Motors should not be lifted by their rotating shafts. Doing so can damage the alignment of the rotor in relationship to the stator. Only the eyebolts on motor frames are intended for lifting the motor. Some factory-mounted accessories or covers also have eyebolts, but they must not be used to lift the motor.

> **WARNING!**
> Never handle a motor by its shaft; motor shaft keyways have sharp edges that can cause severe cuts. The motor itself may also be damaged.

The following is a list of recommendations for proper motor storage:

- Keep the motors clean and dry.
- Supply supplemental heating in the storage area, if necessary, or connect and monitor the internal heaters if so equipped.
- Store motors in an orderly fashion (grouped by horsepower or other criteria).
- Rotate motor shafts periodically as specified by the motor manufacturer, so that they do not remain in the same position all the time. Maintain a shaft rotation log (*Figure 7*) for each motor.
- Motors should not need additional lubrication while in storage.

Ventilation and Temperature Control

Adequate ventilation and temperature control are essential for proper motor operation. Inspect the fan and auxiliary cooling system as part of the regular maintenance program. Many larger motors include resistance temperature detectors embedded in the windings and bearing housings. This provides for continuous monitoring of the winding and bearing temperatures to alert maintenance personnel to abnormal conditions.

- Protect shafts and keyways during storage and while transporting motors from one location to another.
- Test motor winding resistance upon receiving the unit and periodically during storage. Maintain a winding resistance test log for each motor.

2.2.0 Motor Bearing Maintenance

AC motors account for a large percentage of industrial and heavy commercial maintenance and repair, with many motor failures caused by faulty bearings. Consequently, most industrial facilities strongly emphasize proper care of motor bearings. Motor reliability is improved when a carefully planned lubrication schedule is followed.

If an AC motor failure does occur, the first step is to find out why. There are various causes of motor failures, including excessive load, binding or misalignment of motor drives, wet or dirty environments, and bearing failures. Motors equipped with sealed bearings are much less prone to bearing failure.

2.2.1 Frequency of Lubrication

The frequency of motor lubrication depends not only on the type of bearing, but also on the motor application and its service environment. Small- to medium-size motors equipped with ball bearings (except sealed bearings) should be greased every three to six years if the motor duty is normal. For larger motors or for severe applications (high temperature, wet or dirty locations, or corrosive atmospheres), lubrication may be required more often. In most cases, the lubrication schedule for a motor is specified in the manufacturer's operation and service literature.

Figure 8 shows a form that can be used to record annual equipment lubrication. The form can be developed based on the equipment manufacturer's recommendations and user experience. Usually, both motors and their driven equipment are included in the schedule. The specified type and quantity of lubricant can also be included. Multiple annual sheets can be set up for a series of years to accommodate equipment that requires lubrication at intervals greater than one year. Also shown is a typical lubrication tag that is replaced each time a motor is lubricated. The tag is used to indicate that the scheduled lubrication has occurred. Large facilities may use specialized software programs to schedule periodic maintenance tasks for various types of equipment. Typical tasks include lubrication, vibration testing, insulation testing, and temperature measurements. These programs issue daily, weekly, monthly, and annual task reminders as necessary for each piece of equipment.

2.2.2 Lubrication Procedure

Before lubricating a ball bearing motor, clean the bearing housing, grease gun, and fittings. Exercise care to keep out dirt and debris during lubrication. If multiple lubricants are allowed in accordance with the motor manufacturer's operation and service literature, use only one of them when periodically re-lubricating a motor. Do not mix lubricants.

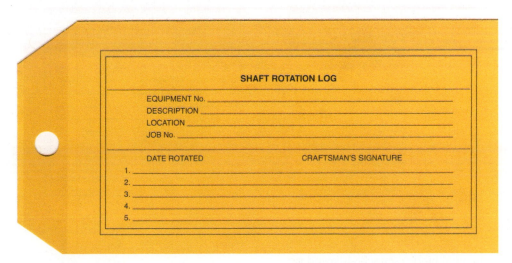

Figure 7 Typical shaft rotation log tag.

YEAR _____ EQUIPMENT LUBRICATION SCHEDULE

Equipment Location and/or Number	Lubricant	Jan	Feb	Mar	Apr	May	Jun	Jul	Aug	Sept	Oct	Nov	Dec

(A)

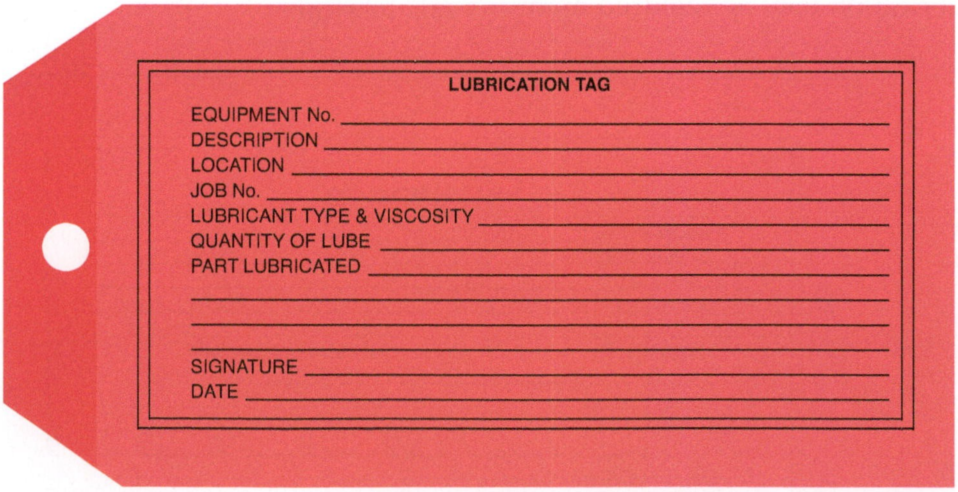

LUBRICATION TAG

EQUIPMENT No. _____

DESCRIPTION _____

LOCATION _____

JOB No. _____

LUBRICANT TYPE & VISCOSITY _____

QUANTITY OF LUBE _____

PART LUBRICATED _____

_____

_____

_____

SIGNATURE _____

DATE _____

(B)

Figure 8 Typical equipment lubrication schedule and tag.

GREASE FITTING

RELIEF PLUG

Figure 9 Grease fitting and relief plug for a TEFC motor.

To lubricate a motor, remove the relief plug from the bottom of the bearing housing (*Figure 9*). This prevents excessive pressure from building up inside the bearing housing during lubrication. If possible, run the motor and add grease to the grease fitting until it begins to flow from the relief hole. Some manufacturers specify that only a certain amount of lubricant be injected into the bearing housing. A grease meter can be used in conjunction with a grease gun to precisely measure the amount of grease injected (*Figure 10*).

After grease is injected, run the motor for 5 to 10 minutes to expel any excess grease, then reinstall the relief plug and clean the bearing housing. Avoid overlubrication. When excess grease is forced into a bearing, churning of the grease may occur, resulting in high bearing temperatures and eventual bearing failure.

2.2.3 Checking Bearings

Three simple methods are commonly used to check bearings during motor operation. These are: listening for unusual noises, performing temperature measurements, and performing vibration testing. If the bearing housing is unusually hot, or if a growling or grinding sound is being emitted from the area of the bearings, one of the bearings is probably nearing failure. Special stethoscopes can be used to listen to the bearings while the motor is running or while it runs down after being stopped. A "thunk" sound that occurs just before rotation stops is usually indicative of failure.

As an absolute minimum, bearing temperature and/or vibration testing should be done both before and after scheduled lubrication of the equipment. Enter the temperature and/or vibration data in a log and compare it to previously logged data. Note any increase in temperature and/or vibration. In some cases, it may be necessary to decouple the load to see if a defective load is causing the problem. Lockout/tagout procedures must be followed in such cases.

Motors should also be checked for endplay, which is the backward and forward movement in the shaft. Ball bearing motors typically will have $\frac{1}{32}$" to $\frac{1}{16}$" (0.79 mm to 1.59 mm) of end movement.

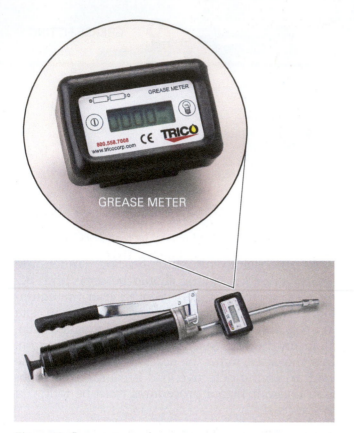

Figure 10 Grease gun and meter.

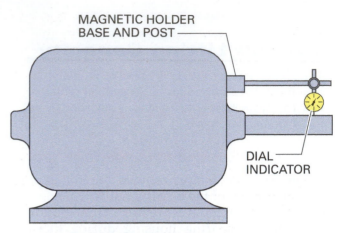

Figure 11 Checking for a bent shaft on a motor.

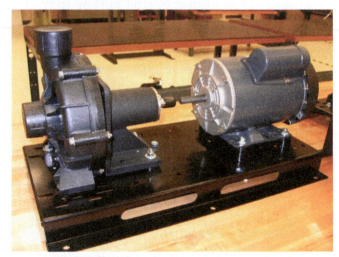

Figure 12 Typical direct-drive configuration without coupling.

> **WARNING!**
>
> For bent shaft or shaft alignment checks, the motor must be locked out and tagged before any guards are removed and any checks are performed.

Other inspections may include periodically checking for misalignment or bent shafts and for excessive belt tension. A magnetic base dial indicator can be used to check for a bent shaft on an uncoupled motor or driven load. The indicator is connected to either the motor or the load, or to a nearby solid ferrous metal surface, with the indicator feeler in contact with the end of the shaft (*Figure 11*). Rotate the shaft back and forth only between the edges of any keyway in the shaft. There should be no shaft misalignment (runout) detected for either the motor or the load; otherwise replace the motor or the load, whichever has the bent shaft.

Parallel or angular misalignment (runout) of an uncoupled motor shaft to a directly driven load shaft (*Figure 12*) can be checked using a magnetic V-base holder or clamp base holder with a dial indicator at the rear of one of the shafts (usually the motor shaft). Manually rotate the first shaft while the indicator feeler is in contact with the rear of the other shaft (usually the load shaft) (*Figure 13*). If the coupling halves are on the shafts and the outside diameter of the couplings is accurately machined in relation to the center of the coupling, then they can be used in the measurements; otherwise, the one on the shaft where the dial indicator will be used must be repositioned or removed. When positioning the dial indicator feeler against that shaft, make sure that it is first positioned at the rear of the shaft and is depressed enough to cause the indicator needle to rotate 180 degrees. Then rotate the dial markings so that the zero reading is at the needle position.

To make the alignment measurements, rotate the shaft with the holder base back and forth so that the indicator feeler does not enter any keyway on the other shaft. Note the locations of the minimum and maximum readings on the other shaft. Move the indicator to the front of that shaft and repeat the measurements. If the minimum and maximum readings are the same, there is a parallel shaft runout, and the load or motor will

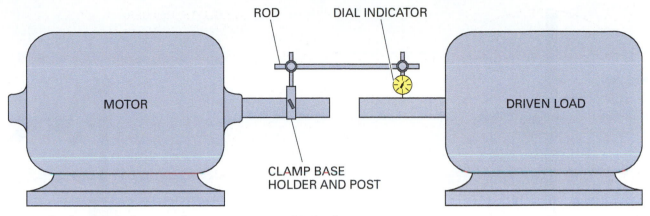

ROD DIAL INDICATOR

MOTOR

CLAMP BASE
HOLDER AND POST

DRIVEN LOAD

Figure 13 Checking shaft alignment between a motor and its load.

have to be moved or shimmed equally at both ends to retain the parallelism and correct the runout. If the readings are different, there is angular shaft runout, and the motor or load will have to be moved or shimmed unequally at each end to remove the angular runout. Then, any resulting parallel runout will have to be corrected.

By trial and error, loosen the anchoring, as necessary, and move or shim the motor/load in the direction necessary to equalize the maximum and minimum readings at both locations to correct any angular runout. Repeat the alignment measurements for any parallel runout until the difference between the minimum and maximum readings is essentially zero around the shaft at both positions of the dial indicator. Then anchor the motor/load and recheck for any resulting runout. If it is possible to rotate the other shaft, reverse the indicator setup and repeat the measurements as a double check. Before anchoring the motor/load, make sure that a soft foot condition was not caused by moving and/or shimming the motor/load.

Figure 14 shows a typical dial indicator with various holders, along with stainless-steel slotted shims. Stainless-steel shims are preferred in order to prevent rust deterioration over time. The shim assortment includes shims of various thicknesses. The thickness is marked on the shim so that records can be kept of the total thickness used under each foot of the equipment. This enables easier reinstallation of the same equipment if it must be removed.

> **NOTE**
>
> Other methods of shaft alignment are sometimes used. These include laser alignment or mechanically accurate adapter frames (bell housings) for some pumps/motors.

2.3.0 Motor Insulation Testing

An insulation resistance test indicates the condition of motor insulation and may indicate **insulation breakdown** or the presence of moisture, dirt, or other conductive material. Insulation resistance testing is high-energy testing and must be done in compliance with employer regulations, government regulations, and the tester manufacturer's instruction manuals.

2.3.1 Insulation Resistance Tests

Insulation resistance tests give an indication of the condition of insulation, particularly with regard to contamination by moisture and dirt. The actual value of the resistance will vary depending on the equipment type, size, voltage rating, and winding configuration.

The principal purpose of periodic insulation resistance tests is in comparing, or trending, relative values of insulation resistance of the same apparatus taken under similar conditions throughout its service life. Such tests usually reveal deterioration of insulation and may indicate approaching failure.

Measuring insulation resistance is straightforward. Identify any two points between which there is insulation, and make a connection across them using a megohmmeter. The measured value represents the equivalent resistance of all the insulation that exists between the two points and any component resistance present between the two points. Three-phase motors have three to twelve motor leads present in the motor terminal box, depending on the winding configuration and starting method. Most common are the nine leads present in the dual-voltage squirrel-cage motor.

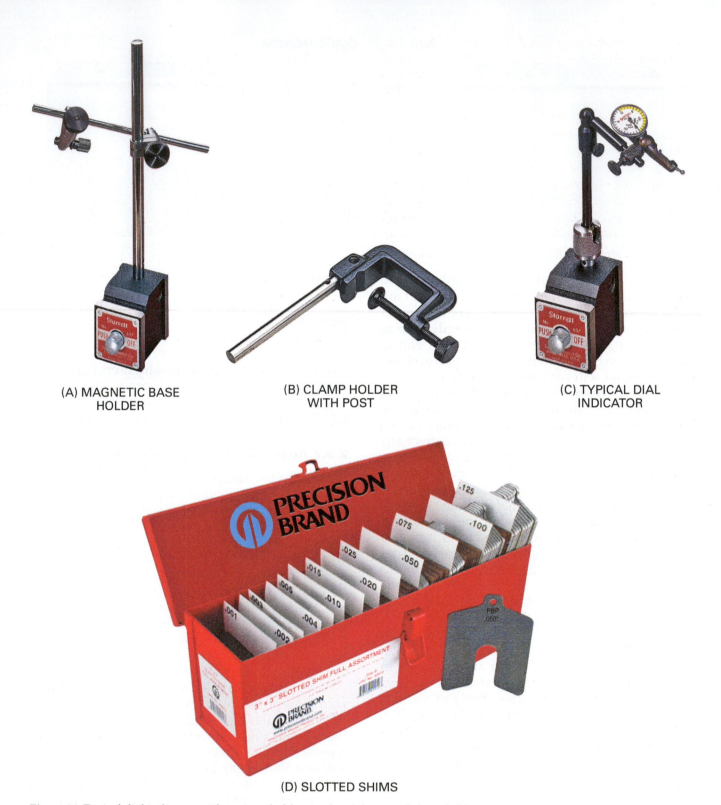

(A) MAGNETIC BASE HOLDER

(B) CLAMP HOLDER WITH POST

(C) TYPICAL DIAL INDICATOR

(D) SLOTTED SHIMS

Figure 14 Typical dial indicator with various holders and stainless-steel slotted shims.

Before a motor is placed in service, test the insulation resistance of each winding segment to ground and to all other winding segments connected to ground. A look at the connection diagrams on the motor nameplate will indicate point numbers for each test connection. The positive lead is connected to frame and ground along with all other motor leads except for the winding segment under test. The negative lead is connected to one end of the segment under test with the other end(s) isolated or insulated. The insulation resistance is tested from one segment to all other segments and to ground. The test is repeated until each winding segment has been individually tested.

This test is easily performed on a motor in storage or before connection to the electric supply. Some facilities require testing only the resistance of all windings to earth. The motor feeder conductors may be included in the test to avoid de-terminating and then re-terminating the motor connections. In this case, the test is often performed from the motor starter or local disconnect switch with one test lead connected to ground and the other to any T lead to the motor.

If the insulation resistance test indicates potential problems, isolate the feeders from the motor windings and test between individual phases or winding segments to identify the problem. Typical insulation resistance values used by one contractor for new equipment installations are shown in the *Appendix*. Other contractors may use different values. Specific values may also be obtained from the equipment manufacturer. In all cases in which comparison is used over time for trending data, the same test connections must be used.

Megohmmeters (*Figure 15*) are available in several varieties. Some are powered by a hand-cranked generator, others are battery-powered, and some have rechargeable batteries or a line-powered internal power supply. Some supply test voltages as low as 50V or as high as 10,000V or more, but the most common is 500V. In all cases, the test voltage is DC.

Before making resistance tests, you must first understand the concept of absorption current. The insulation between two connection points can be thought of as a dielectric, thus forming a capacitance. A phenomenon known as *dielectric absorption* occurs whereby the dielectric soaks up electrons and then releases them when the potential is removed. This is in addition to the current that charges the capacitance, and it occurs much more slowly. It is dependent on the nature of the dielectric. Two items for which this is of concern are capacitors and wound equipment. Such current is referred to as *absorption current*, or I_A in *Figure 16* (*A*). Absorption current consists of two components: polarization current and electron drift. Polarization current is due to the reorientation of molecules in the insulation impregnating

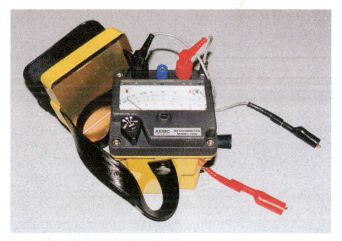

Figure 15 Typical megohmmeter.

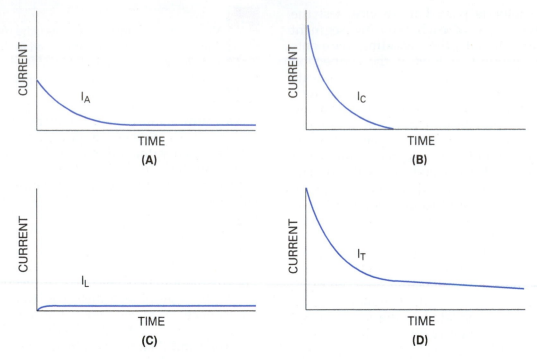

Figure 16 Current curves related to insulation testing.

materials. It can take several minutes for this to occur and for the polarization current to drop to zero. The second component is current caused by the gradual drift of electrons through most organic materials. This drift current is minor and occurs until the electrons and ions become trapped by the mica surfaces commonly found in insulation systems. For clean and dry motors, the insulation resistance between about 30 seconds and a few minutes is primarily determined by the absorption current.

This phenomenon may be demonstrated by taking a large capacitor, charging it to its rated voltage, and allowing it to remain at that voltage for some length of time. Then discharge the capacitor quickly and completely until a voltmeter placed across it reads zero. Remove the voltmeter and again allow the capacitor to sit for some length of time with its leads open-circuited. Take another voltmeter reading. Any voltage present is due to the dielectric absorption phenomenon. Some types of capacitors exhibit this phenomenon to a greater degree than other types. The larger the capacitor, the more apparent the effect.

The current required to charge whatever capacitance is present is known as *charging current*, or I_C in *Figure 16* (B). Like the dielectric absorption current, it decays to zero but more quickly. It is this current that typically determines how long it takes to make an accurate megohm measurement.

When the reading appears to stabilize, it means that the charging current has decayed to a point where it is negligible with respect to the **leakage current**.

The current that flows through the insulation is the leakage current, or I_L in *Figure 16* (C). The voltage across the insulation divided by the leakage current through it equals the insulation resistance. Thus, to accurately measure insulation resistance, the dielectric absorption current and the charging current must be allowed to decay to the point at which they are truly negligible with respect to the leakage current.

The total current that flows is the sum of the three components just mentioned, or I_T in *Figure 16* (D). It decays from an initial maximum and approaches a constant value, which is the leakage current alone. The megohm reading is dependent on the voltage across the insulation and the total current. It increases from an initial minimum and approaches a constant value, which is the true measured insulation resistance.

Motor manufacturers, installers, users, and repairers find megohm testing very useful in determining the quality of the insulation in a motor. A single insulation resistance measurement, along with experience and guidelines as to which reading to expect, can indicate whether a motor is fit for use. Information of real value is obtained from measurements made when a motor is new

Digital Surge/DC Hi-Pot/ Resistance Tester

The unit shown here combines automated standard temperature-compensated resistance tests and hi-pot/ surge tests in one portable field tester. It also provides a completely automated polarization index test and dielectric absorption test. Test results for up to 10 motors in the field can be stored, retrieved, printed, and uploaded to a desktop program for file management and analysis.

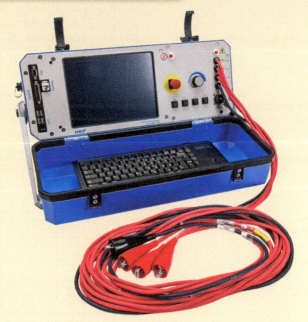

Figure Credit: SKF USA, INC.

and at least every year during service. Because the temperature of the windings has a great effect on the insulation resistance reading, the motor industry has standardized a winding temperature of 40°C (104°F) as a reference when specific recommendations can be made about the minimum acceptable megohm value for each type of motor. When measurements are made at a winding temperature other than 40°C, they can be corrected to reflect what the reading would have been at 40°C by using a temperature coefficient, K_t, obtained from a chart such as the one shown in *Figure 17*.

Figure 18 shows the actual readings obtained on a motor over several years (the dashed line) and the same readings corrected to a standard 40°C reference (the solid line). The temperature-corrected readings show a gradual decline characteristic of the motor's normal aging process until, in 1997, a sharp drop occurred. That drop signaled a problem with the insulation that would probably lead to an insulation failure, which would be costly in terms of repair and down time. Therefore, at the earliest practical opportunity, this motor was sent out for rewinding. The rewinding restored the high megohm readings, and the steady, gradual decline resumed. Today, a motor of this size (20hp) might simply be replaced instead of rewound, due to the high cost of shop labor. *Figure 19* shows just the temperature-corrected readings that are normally kept on motors.

Another factor that affects insulation resistance readings is moisture. Motors may have excessive moisture in and around the insulation from either high humidity or having been submerged. This is a temporary effect. By letting the motor dry out naturally, or accelerating the process by baking, the readings will increase. It is important not to start the motor until a satisfactory reading is obtained, because doing so might cause an insulation failure.

The tests used to evaluate the effect of absorption current on resistance are the dielectric absorption ratio (DAR), typically applied on smaller machines, and the polarization index tests.

The condition of insulation may be inferred by evaluating the effect of the dielectric absorption current on resistance over time. The DAR, also called a *60/30 test*, is determined by the following formula:

$$DAR = \frac{\text{resistance after 60 seconds}}{\text{resistance after 30 seconds}}$$

The resistance readings will gradually rise due to a decrease in dielectric absorption.

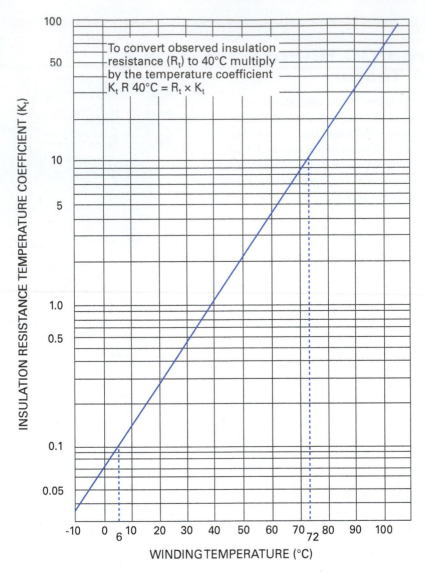

Figure 17 Approximate temperature coefficient for insulation resistance.

Causes of Invalid Test Data

Many times, valid data cannot be obtained for dielectric absorption ratios, polarization index ratios, and insulation resistance tests. For dielectric absorption ratios and polarization index ratios for small equipment, such as small motors, the capacitance is very low and the resistance readings will increase very rapidly. In this case, the test voltage must be reduced to spread the charging current over a longer period. For insulation resistance data, the insulation resistance of modern equipment is so effective and so high that the range of a megohmmeter may be exceeded when attempting to read valid resistance data. In these cases, insulation test equipment that has a gigaohm or teraohm range must be used to obtain valid data for calculating the polarization index or insulation resistance trend.

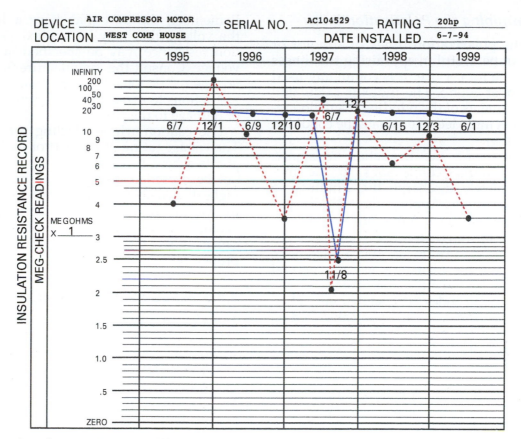

Figure 18 Chart of insulation resistance readings.

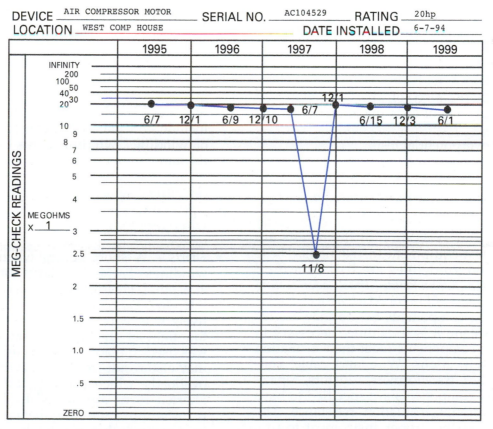

Figure 19 Temperature-corrected resistance readings alone.

2.3.2 Determining the Polarization Index

Knowing the polarization index of a motor or generator can be useful in appraising the fitness of the machine for service. It is commonly used for motors over 200hp. The index is calculated from measurements of the winding insulation resistance and is intended to evaluate the condition of the insulation.

Before measuring the insulation resistance, remove all external connections to the motor and completely discharge the windings to the grounded frame. Apply either 500VDC or 1,000VDC between the winding and ground using a direct-indicating, power-driven megohmmeter; use the higher value for motors rated 500V and above. Apply the voltage for ten minutes and keep it constant for the duration of the test.

The polarization index (PI), sometimes called the *polarization index ratio (PIR)*, is calculated as the ratio of the ten-minute value to the one-minute value of the insulation resistances measured consecutively:

$$PIR = \frac{\text{resistance after 10 minutes}}{\text{resistance after 1 minute}}$$

The recommended minimum value of polarization index for AC and DC motors and generators is 2.0. Machines having a PI under 2.0 are considered suspect. However, a PI over 4.0 may indicate dry, brittle insulation that is also suspect. The polarization index is useful in evaluating windings for the following:

- Buildup of dirt or moisture
- Gradual deterioration of the insulation (by comparing results of tests made earlier on the same motor)
- Fitness for overpotential tests
- Suitability for operation

2.3.3 Insulation Testing Considerations

The procedures for performing insulation resistance testing and evaluation of the results are found in the test equipment instructions. These procedures and guidelines are based on *IEEE Standard 43-2013, Recommended Practice for Testing Insulation Resistance of Electric Machinery.*

When testing components before they are installed in a system, it is sometimes necessary to verify that the components meet their insulation resistance specifications. Wire and cable, connectors, switches, transformers, resistors, capacitors, and other components are specified to have a certain minimum insulation resistance. The megohmmeter is the proper instrument to use.

Remember, the part may have a limitation on the voltage that may be applied, or the insulation resistance may be specified at a particular voltage. Consult the manufacturer's data for any voltage restrictions. Pay careful attention to these restrictions to avoid improper comparisons and/or damage to the part.

Use caution when selecting the test voltage. The voltage rating of the part or motor across the points where the measurement is made should not be exceeded. Many users prefer to use the highest available voltage that does not exceed the rating. In other cases, the customary test voltage is 500V, regardless of a higher rating, because higher voltage can make it harder to obtain proper data.

2.4.0 Troubleshooting Motors

To detect defects in electric motors, the windings are typically tested for ground faults, opens, shorts, and reverses. The methods used in performing these tests may depend on the type of motor being tested. If a defect is found, the motor must be removed and replaced. The possible defects include the following:

> **WARNING!**
>
> Before testing or troubleshooting a motor beyond a simple visual/auditory inspection, disconnect the power and follow the proper lockout/tagout procedures.

- *Grounded winding* – A winding becomes grounded when it makes electrical contact with the metal frame of the motor. Causes may include end plate bolts contacting conductors or windings, winding wires pressing against the laminations at the corners of slots damaged during rewinding, or conductive parts of the centrifugal switch contacting the end plate.
- *Open circuits* – Loose connections or broken wires represent an open circuit in an electric motor.
- *Shorts* – A short occurs when two or more windings of separate phases contact each other at a point, creating a path other than the one intended. This condition may develop in a relatively new winding if the winding is wound too tight at the factory and excessive force was used to place the wires in position. Shorts may also result from excessive heat created by overloads, which can degrade the insulation. Shorted turns are caused by loose or broken turns within a winding that contact each other in a manner that effectively causes the turns

within the winding to be reduced. This causes high current in the winding, which results in overheating and insulation damage.

• *Reverses* – Reverses occur when two phases of a three-phase power source are reversed. This causes high current, which results in overheating and insulation damage.

2.4.1 Insulation Testing

Before performing an insulation resistance test, first check to ensure that the circuits and equipment to be tested are rated at or above the output voltage of the megohmmeter and that all equipment scheduled for testing is disconnected and locked out according to facility or employer procedures.

Check the megohmmeter and other test equipment for proper operation in accordance with the manufacturer's instructions. Before conducting tests, temporarily ground all the leads of the equipment under test. Then conduct the insulation tests as prescribed in the manufacturer's instructions. For winding resistance tests, the test voltage is applied for one minute or until the reading is essentially stable. After testing, discharge the windings by placing a ground on the windings and maintaining it for four times the duration of the test.

Record the temperature and humidity immediately after the test. Also, note the condition of the environment and the running or idle time of the equipment.

After the readings have been recorded, correct them for the winding temperature using a temperature chart (supplied with most megohmmeters). For enclosed motors without internal winding temperature monitoring, the winding temperature must be estimated. The winding temperatures of motors that have not been operating for eight hours or more can be considered to be the same temperature as the outside motor skin temperature as measured with a thermometer (infrared or other). For motors that have been operating, the winding temperature can be estimated by adding the skin temperature to the ambient air temperature.

For new equipment, the corrected insulation resistance readings should be similar to those shown in the *Appendix* or as specified by the manufacturer. Elaborate motor insulation tests, such as time resistance and step voltage tests, use various megohmmeter reading ratios and indexes, such as absorption ratio and polarization index, to determine the effects of temperature and humidity on insulation breakdown (*Figure 20*).

Figure 20 Bench testing motor insulation.

2.4.2 Grounded Coils

The usual effect of one grounded coil in a winding is the repeated blowing of a fuse or tripping of the circuit breaker when the motor attempts to start, provided that the machine frame is grounded. Two or more grounds will produce the same result and will also short out part of the winding in that phase in which the grounds occur. A simple test to determine whether or not a ground exists in the winding can be made using a conventional continuity tester. Before testing with this instrument, first make certain that the motor disconnect switch is open, locked out, and tagged. Place one test lead on the frame of the motor and the other in turn on each of the line wires on the load side of the motor controller. If there is a grounded coil at any point in the winding, the lamp of the continuity tester will light or, in the case of a meter, the reading will show low or zero resistance.

2.4.3 Water-Damaged Motors

Motors that have been water damaged can sometimes be dried out and salvaged if rust is minimal. The NEMA guidelines for handling waterdamaged equipment recommend that salvage be done only by trained personnel in consultation with the motor manufacturer. This normally requires that these motors be sent to a manufacturer-authorized outside motor repair facility where they are disassembled, cleaned, and dried using ovens, steam heat, forced air, electric heat, or infrared heat in accordance with the motor manufacturer's instructions. When dried, they are subjected to extensive insulation testing and bearing replacement or lubrication, and then reassembled.

2.0.0 Section Review

1. A tachometer is used to _____.

   a. check winding resistance
   b. measure shaft rotation speed
   c. measure vibration
   d. record temperature variations

2. The back and forth movement of a motor shaft is known as _____.

   a. rotation
   b. end play
   c. vibration
   d. shaft roll

3. When using an insulation tester on a motor, the motor frame must be _____.

   a. at room temperature
   b. properly lubricated
   c. connected to ground
   d. suspended by the shaft

4. The condition in which two or more windings come into contact is a(n) _____.

   a. open circuit
   b. short circuit
   c. motor reversal
   d. grounded winding

3.0.0 MOTOR INSTALLATION AND COMMISSIONING GUIDELINES

Objective

Describe the guidelines for installing and commissioning electric motors.

a. Explain alignment and adjustment requirements.

b. Describe startup procedures.

A motor must be securely installed on a flat, rigid foundation or mounting surface in order to minimize vibration and maintain alignment between the motor and load shafts. Failure to provide a proper mounting surface may cause vibration, misalignment, and bearing damage.

Foundation caps and sole plates are designed to act as spacers for the equipment they support. If these devices are used, be sure that they are evenly supported by the foundation or mounting surface.

3.1.0 Alignment and Adjustment

Accurate alignment of the motor with the driven equipment is extremely important. Many different types of flexible couplings are available with a range of torque capabilities, bore sizes, and materials based on horsepower requirements.

3.1.1 Alignment

Two typical types of couplings used to reduce vibration between motors and their loads are shown in *Figure 21*. One is a jaw type that is sometimes called a *Lovejoy® coupling*. This type is secured to the shaft with a square key and setscrew. The flexible insert between the coupling halves is available in various materials with varying degrees of flexibility, depending on the application. They are also available with snap-in flexible inserts that can be externally removed and replaced without disturbing the coupling halves. The coupling halves are available in sintered iron (a magnetic alloy), cast iron, aluminum, or stainless steel. Couplings of this type are fail-safe because they will still work even if the insert fails. Installation of the coupling is easy, and alignment is not as critical as with other kinds of couplings.

The other type of coupling shown in *Figure 21* has internal splines in the coupling halves and the flexible insert between them. The coupling halves are secured to the shafts with square keys and setscrews. The couplings are available with different materials for the coupling halves and the flexible insert between them, depending on the application and torque requirements. Normally, precise alignment of these couplings is critical, but, depending on the flexibility of the insert material, some types can tolerate a certain amount of angular or parallel runout. Also note the balance compensator on the load shaft in this application. It is used to reduce vibration caused by the load.

If possible, use flexible couplings for direct-drive motors. Consult the drive or equipment manufacturer for more information. Mechanical vibration and roughness during operation may indicate poor alignment. Use dial indicators to check alignment as discussed previously. Maintain the manufacturer's recommended spacing between coupling hubs.

For belt-drive motors, make sure the pulley ratio recommended by the manufacturer is not exceeded. Align the sheaves carefully to minimize belt wear and axial bearing loads. Belt tension should be sufficient to prevent belt slippage at rated speed and load. However, belt slippage may occur during startup.

> **CAUTION**
>
> Do not over-tension belts. Over-tightening causes excessive bearing and pulley wear, leading to premature failure.

3.1.2 Endplay Adjustment

Except for motors designed for vertical mounting that have end thrust bearings, the axial position of a horizontal motor frame with respect to its load is extremely important. The motor bearings are not designed for excessive external axial thrust loads. Improper adjustment will cause failure. To establish the electrical center (live center) of the motor shaft, loosen the bearings (if applicable), and operate the motor with no load. Mark the shaft position, in relation to the end bell, on the shaft. Shim/reposition and tighten the bearings as necessary. After coupling the motor to the load, make sure to check that the marked position is maintained when the motor is operating.

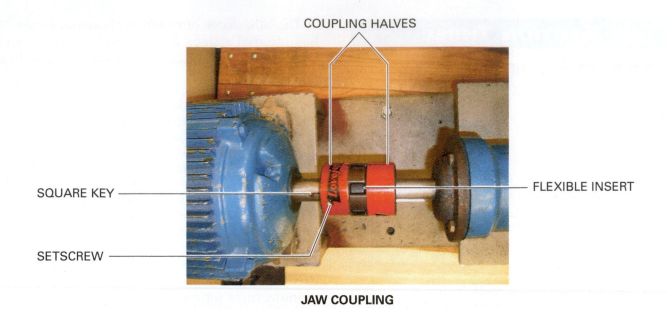

COUPLING HALVES

SQUARE KEY

FLEXIBLE INSERT

SETSCREW

JAW COUPLING

COUPLING HALVES

SQUARE KEY

BALANCE COMPENSATOR

SETSCREW

SPLINED FLEXIBLE INSERT

SPLINED COUPLING

Figure 21 Typical vibration-reducing couplers.

Endplay adjustment applies to large motors with adjustable bearings. It is almost never done on motors below 500hp, except for sleeve bearing direct-coupled motors.

3.1.3 Doweling

After proper alignment is verified, dowel pins are often inserted through the mounting feet of motors into their foundation. This will allow the motor to be returned to precisely the same position if removal for repairs is required. If doweling is desired or required, drill dowel holes through diagonally opposite motor feet and into the foundation or sole plate. Ream all holes to the proper size for the dowel pin, and then install the proper size dowels.

3.2.0 Startup

Before starting a motor, especially one that has been in storage, it is important to perform a prestart inspection and check procedure and to make sure the motor is properly lubricated. The initial checks are performed before the motor is connected to the load shaft.

3.2.1 First-Time Startup

Before making any prestart checks, make sure that all power to the motor and any accessories is locked out and tagged. Make sure the motor shaft is completely free of the load shaft and cannot cause it to rotate as you work with the motor. Then perform the following checks:

- Make sure that the motor installation and mounting is secure.
- If the motor has been in storage or idle for some time, check the insulation integrity using an insulation resistance tester.
- Verify the desired direction of the motor and corresponding power source phases using motor-rotation and phase-rotation testers.

In order to check for proper motor rotation, you must know the required rotation direction of the load. Depending upon the equipment being serviced, the proper rotation is often clearly marked on the motor or load housing. If not, a millwright or mechanic will likely be responsible for confirming proper rotation.

The One-Megohm Rule of Thumb

The one-megohm rule is an old minimum standard that is still used. It was originally developed for equipment manufactured 30 to 40 years ago. Newer equipment uses winding insulation that has much higher insulation resistance than earlier insulating material.

The old rule states that for equipment rated at 1,000V or less, the minimal acceptable rating is 1MΩ. For equipment rated over 1,000V, the minimal acceptable reading is 1MΩ minimum, plus 1MΩ per kilovolt rating. The rule only applies to winding temperatures corrected to 40°C (104°F). To correct for the ambient temperature, the reading must be adjusted up for winding temperatures above 40°C and down for winding temperatures below 40°C. The reasoning is that equipment insulation resistance decreases at higher temperatures and increases at lower temperatures. The readings are adjusted as follows:

- For every 10°C above 40°C, double the megohm reading.
- For every 10°C below 40°C, divide the megohm reading in half.

For example, assume a 2,300V, 600hp motor has a reading of 2.3MΩ. The motor has been operating, and the motor winding temperature is 60°C (140°F). The minimum acceptable reading is 3.3MΩ at 40°C. The temperature-corrected reading is as follows:

$$60°C - 40°C = 20°C$$
$$20°C \div 10°C = 2$$

Therefore, the insulation resistance reading must be doubled twice, because the reading must be doubled once for each 10°C above 40°C. This results in:

$$2.3MΩ \times 2 = 4.6MΩ$$
$$4.6MΩ \times 2 = 9.2MΩ$$

The temperature-corrected reading is 9.2MΩ. It is acceptable because it is greater than the 3.3MΩ minimum requirement.

- Lubricate the motor in accordance with the manufacturer's instructions.
- Connect the wiring and inspect all electrical connections for proper termination, clearance, mechanical strength, and electrical continuity.
- Perform a final insulation resistance test, including the wiring to the motor.
- Remove any protective covering from the motor shaft.
- Replace all panels and covers that were removed during installation.
- Apply power and momentarily start (bump) the motor to verify correct rotation. If wrong, again lock out and tag the motor and switch two phases. Then bump the motor again to check.
- Start the motor and ensure that operation is smooth without excessive vibration or noise. If so, run the motor for one hour with no load connected.
- After one hour of operation, again lock out and tag the motor. Connect the load to the motor shaft as previously described. For direct-drive couplings, make sure the coupling is aligned and is not binding.
- Verify that all coupling guards (*Figure 22*) and protective devices are installed and that the motor is properly ventilated.

Installation Checklist

Some contractors or facilities may use a motor installation checklist similar to the one shown here.

DESCRIPTION

EQUIP. #		
CIRCUIT #		
INITIAL MEGGER TEST @ _____ VOLTS	A to ⏚ B to ⏚ C to ⏚	A to B A to C B to C
ELECT. COMPLETE	INITIAL DATE	
FINAL ALIGN COMPLETE	INITIAL DATE	
INSTRUMENTATION COMPLETE	INITIAL DATE	
FINAL MEGGER TEST INCLUDES FEED WIRES	A to ⏚ B to ⏚ C to ⏚	A to B A to C B to C
ROTATION CHECK	INITIAL DATE	
COUPLING BELTS GUARD	INITIAL DATE	
TEST RUN	INITIAL DATE	
OPERATING AMPS	A B C	
ACCEPTED FOR SERVICE_____ DATE_____		

NW GRAPHICS FORM # EO-18 6/91

ALIGNMENT READINGS

I.D.

O.D.

PIPING COMPLETE	INITIAL DATE
FLUSH	INITIAL DATE
LUBE	INITIAL DATE
LEVEL ALIGN GROUT	INITIAL DATE
FINAL ALIGNMENT	INITIAL DATE
MECHANICAL OK TO RUN	INITIAL DATE
ACCEPTED FOR SERVICE_____ DATE_____	

Figure Credit: Jim Mitchem

Figure 22 Typical coupling guard over rotating components.

> **NOTE**
>
> This procedure assumes that the first-time startup procedure was performed and was successful.

3.2.2 Coupled Startup

The first coupled startup should be with no load applied to the driven equipment. Apply power and check the operating current on each phase of the motor. Verify that the driven equipment is not transmitting excessive vibration back to the motor though the coupling or the foundation. Vibration should be minimal. If not, balancing of the load, such as a large blower wheel, may be required. Pumps and similar loads may be suffering from bearing damage or failure, leading to vibration. Run the motor for approximately one hour with the driven equipment in an unloaded condition. The equipment can now be loaded and operated within specified limits. Do not exceed the nameplate ratings for amperes or for temperature rise for steady continuous loads.

Angular or Uneven Mounting Surfaces

For mounting surfaces that are angular or uneven under one or more equipment feet, the feet or mounting surface can be machined, or shims can be used. Shims are often required during the motor alignment process. Special shims can also be used, such as those shown here. These shims have an elastomeric coating that can be compressed to about half its thickness under final anchoring torque to fill an angular or uneven surface under the motor feet. After a short length of time, the coating hardens to prevent any change due to subsequent equipment operation.

Figure Credit: Precision Brand Products, Inc.

3.0.0 Section Review

1. The two types of couplings used to connect motor shafts to loads are _____.

   a. jaw and pulley
   b. spline and jaw
   c. spline and pulley
   d. bearing and jaw

2. "Bumping" is generally done in order to check _____.

   a. bearing wear
   b. excessive vibration
   c. shaft couplings
   d. direction of rotation

1. Single-phasing means that _____.
   a. a three-phase motor is attached to a single-phase power source
   b. two of the legs of a three-phase power source are lost
   c. one of the legs of a three-phase power source is lost
   d. phase rotation of the input voltage needs to be adjusted

2. If the load connected to a motor is increased beyond the motor's rating, which of the following is *most* likely to occur?
   a. Contact between the rotor and stator
   b. Decreased operating temperature
   c. Increased operating temperature and decreased motor life
   d. Loss of bearing lubrication

3. Motors with two separate winding circuits employ _____.
   a. primary resistor starting
   b. autotransformer starting
   c. wye-delta starting
   d. part-winding starting

4. An insulation system that, by experience or accepted test, can be shown to have a suitable thermal endurance when operated at a limiting temperature of 105°C is a _____.
   a. Class A insulation system
   b. Class B insulation system
   c. Class F insulation system
   d. Class H insulation system

5. Typical materials used in a Class H insulation system include _____.
   a. cotton
   b. paper
   c. cellulose acetate films
   d. silicone elastomer

6. Typical materials used in a Class F insulation system include _____.
   a. cotton
   b. paper
   c. cellulose acetate films
   d. mica

7. An insulation system that, by experience or accepted test, can be shown to have a suitable thermal endurance when operated at a limiting temperature of 130°C is a _____.
   a. Class A insulation system
   b. Class B insulation system
   c. Class F insulation system
   d. Class H insulation system

8. Which of the following may be eliminated as a concern in planning good motor maintenance?
   a. Number of hours and days the motor operates
   b. Environmental conditions
   c. Manufacturer of the motor
   d. Importance of the motor in the production scheme

9. Proper electrical connection of a motor is determined by _____.
   a. checking with a rotation tachometer
   b. using a rotation and phase tester
   c. using a recording meter
   d. observing the rotation of the motor

10. Which of the following statements concerning motors in storage is *true*?
    a. Shafts should be turned according to manufacturer instructions.
    b. Shafts should not be moved or turned until the motor is put in use.
    c. Shafts should be turned 180 degrees once a year to distribute bearing grease.
    d. Shafts should be turned once every three years to keep them from rusting.

11. Which of these is considered the *best* attachment point for lifting a motor?

    a. Eyebolt
    b. Shaft
    c. Base
    d. End bell

12. A type of motor bearing with a very low rate of failure is a _____.

    a. radial bearing
    b. pin bearing
    c. sealed bearing
    d. ball bearing

13. Which of the following is usually an indication of a bad motor bearing?

    a. Hot bearing housing
    b. Low current draw
    c. Low pull-in torque
    d. Sparking at the brushes

14. The polarization index (PI) is the ratio of the resistance after ten minutes to the resistance after _____.

    a. 60 minutes
    b. 30 minutes
    c. 5 minutes
    d. 1 minute

15. For motors with adjustable bearings, excessive endplay in a motor shaft is corrected by _____.

    a. shimming the bearings
    b. adjusting the coupling
    c. inserting dowel pins in the mounting feet
    d. repositioning the load

1. The life expectancy of a typical squirrel-cage induction motor depends largely upon the condition of its _____.

2. A(n) _____ is the tool used for an insulation resistance test.

3. The control circuits of many older induction motors were NOT provided with _____ or loss-of-phase protection.

4. With regard to NEMA usual service conditions, if the ambient temperature is over 40°C (104°F), the motor service factor must be _____ or a(n) _____ motor must be used.

5. The four classes of insulation systems used in motors are _____, _____, _____, and _____.

6. An electric motor winding that makes contact with the motor frame is described as being _____.

7. To detect defects in electric motors, the windings are normally tested for four conditions: _____, _____, _____, and _____.

8. Open circuits are often the result of _____ or _____.

9. When a motor suffers from shorted turns, it results in _____.

10. If you have a grounded winding on a motor, you would expect the reading on an ohmmeter to read a(n) _____.

Appendix

COMMON TEST VOLTAGES AND MINIMUM INSULATION RESISTANCE VALUES FOR VARIOUS DEVICES

Electrical Equipment and Wiring

Maximum Voltage Rating	Minimum DC Test Voltage	Normal DC Test Voltage for New Installations	Minimum Insulation Resistance at 20°C for New Installations
250V	250	500	25 megohms
600V	500	1,000	100 megohms
5,000V	2,500	2,500	1,000 megohms
8,000V	2,500	2,500	2,000 megohms
15,000V	2,500	5,000	5,000 megohms
25,000V	5,000	5,000	20,000 megohms

Transformer Windings

Transformer Winding Rating	Minimum DC Test Voltage	Normal DC Test Voltage for New Installations	*Minimum Insulation Resistance at 20°C for New Installations in Megohms	
			Oil-Filled	Dry-Type
0–600V	500	1,000	100	500
601–5,000V	1,000	2,500	1,000	5,000
>5,000–V	2,500	5,000	5,000	25,000

*Note: Refer to manufacturer's literature for specific acceptance criteria. Actual insulation resistance of transformer winding is dependent on transformer type, insulation, kVA rating, and winding temperature.

Motor Winding Insulation Resistance Testing

Motor Winding Rating	Minimum DC Test Voltage	Normal DC Test Voltage for New Installations	*Minimum Insulation Resistance at 20°C for New Installations
0–600V	500	1,000	15 megohms
601–5,000V	1,000	2,500	30 megohms
5,000–15,000V	2,500	5,000	50 megohms

*Note: Minimum insulation resistance for rotating equipment in megohms = equipment kV + 1

Trade Terms Introduced in This Module

Generator: (1) A rotating machine that is used to convert mechanical energy to electrical energy. (2) General apparatus or equipment that is used to convert or change energy from one form to another.

Insulation breakdown: The failure of insulation to prevent the flow of current, sometimes evidenced by arcing. If the voltage is gradually raised, breakdown will begin suddenly at a certain voltage level. Current flow is not directly proportional to voltage. When a breakdown current has flowed, especially for a period of time, the next gradual application of voltage will often show breakdown beginning at a lower voltage than initially.

Insulation classes: Categories of insulation based on the thermal endurance of the insulation system used in a motor. The insulation system is chosen to ensure that the motor will perform at the rated horsepower and service factor load.

Leakage current: AC or DC current flow through insulation and over its surfaces, and AC current flow through a capacitance. Current flow is directly proportional to voltage. The insulation and/or capacitance are thought of as a constant impedance unless breakdown occurs.

Megohmmeter: An instrument or meter capable of measuring resistances in excess of 200MΩ It employs much higher test voltages than are used in ohmmeters, which measure up to 200MΩ. Commonly referred to as a megger.

Soft foot: When the feet of a motor frame are forced to anchor against an uneven surface, resulting in distortion of the motor frame. This can cause misalignment of bearings.

Totally enclosed motor: A motor that is encased to prevent the free exchange of air between the inside and outside of the case.

Additional Resources

This module is intended as a thorough resource for task training. The following reference works are suggested for further study.

NCCER Module 15507, *Installing Electric Motors*.

Figure Credits

Section Review Answer Key

SECTION 1.0.0

Answer	Section Reference	Objective
1. b	1.1.0	1a
2. a	1.2.1	1b

SECTION 2.0.0

Answer	Section Reference	Objective
1. b	2.1.1; *Table 1*	2a
2. b	2.2.3	2b
3. c	2.3.1	2c
4. b	2.4.0	2d

SECTION 3.0.0

Answer	Section Reference	Objective
1. b	3.1.1	3a
2. d	3.2.1	3b

This page is intentionally left blank.

NCCER CURRICULA — USER UPDATE

NCCER makes every effort to keep its textbooks up-to-date and free of technical errors. We appreciate your help in this process. If you find an error, a typographical mistake, or an inaccuracy in NCCER's curricula, please fill out this form (or a photocopy), or complete the online form at **www.nccer.org/olf**. Be sure to include the exact module ID number, page number, a detailed description, and your recommended correction. Your input will be brought to the attention of the Authoring Team. Thank you for your assistance.

Instructors – If you have an idea for improving this textbook, or have found that additional materials were necessary to teach this module effectively, please let us know so that we may present your suggestions to the Authoring Team.

NCCER Product Development and Revision
13614 Progress Blvd., Alachua, FL 32615

Email: curriculum@nccer.org
Online: www.nccer.org/olf

❏ Trainee Guide   ❏ Lesson Plans   ❏ Exam   ❏ PowerPoints   Other _____

Craft / Level: _____ Copyright Date: _____

Module ID Number / Title: _____

Section Number(s): _____

Description: _____

_____

_____

_____

_____

Recommended Correction: _____

_____

_____

_____

_____

Your Name: _____

Address: _____

_____

Email: _____ Phone: _____

This page is intentionally left blank.

Medium-Voltage
Terminations/Splices

OVERVIEW

Both taping systems and manufactured slip-on kits are used to make splices and terminations. This module identifies types of medium-voltage cable and describes how to make various splices and terminations. It also covers hi-pot testing.

Module 26411-20

26411-20 V10.0

Objectives

When you have completed this module, you will be able to do the following:

1. Describe how to splice medium-voltage cable.
   a. Identify medium-voltage power cable configurations and components.
   b. Describe a typical procedure for making a straight splice.
   c. Describe a typical procedure for making an inline tape splice.
   d. Identify various manufactured termination and splice kits.
2. Describe termination classes and important considerations when creating terminations.
   a. Identify termination classes.
   b. Identify stress control methods.
3. Define high-potential testing and explain how such testing is conducted.
   a. Identify types of hi-pot tests.
   b. Explain how to make various test connections.
   c. Describe typical procedures for conducting high-potential tests.

Performance Task

Under the supervision of the instructor, you should be able to do the following:

1. Prepare a cable and complete a splice or stress cone.

Trade Terms

Armor
Dielectric constant (K)
High-potential (hi-pot) test
Insulation shielding
Pothead
Ribbon tape shield

Shield
Shielding braid
Splice
Strand shielding
Stress

Industry Recognized Credentials

If you are training through an NCCER-accredited sponsor, you may be eligible for credentials from NCCER's Registry. The ID number for this module is 26411-20. Note that this module may have been used in other NCCER curricula and may apply to other level completions. Contact NCCER's Registry at 888.622.3720 or go to **www.nccer.org** for more information.

> **NOTE**
>
> NFPA 70®, *National Electrical Code*® and *NEC*® are registered trademarks of the National Fire Protection Association, Quincy, MA.

Contents

1.0.0   Splicing Medium-Voltage Power Cable 1
   1.1.0   Medium-Voltage Cable Configurations and Components 1
     1.1.1   Strand Shielding.. 2
     1.1.2   Insulation... 2
     1.1.3   Insulation Shield System 3
     1.1.4   Jacket 4
   1.2.0   Making Straight Splices 4
     1.2.1   Preparing the Surface 5
     1.2.2   Joining the Conductors with Connectors 6
     1.2.3   Reinsulating 7
     1.2.4   Reshielding 8
     1.2.5   Rejacketing.. 8
   1.3.0   Making Inline Tape Splices 8
     1.3.1   Preparing the Cable 10
     1.3.2   Connecting the Conductors... 12
     1.3.3   Applying Primary Insulation.. 12
     1.3.4   Applying the Outer Sheath 14
     1.3.5   Tee Tape Splice 14
   1.4.0   Manufactured Termination and Splice Kits................................ 14
     1.4.1   Heat-Shrink Kits.. 14
     1.4.2   Cold-Shrink Kits 16
2.0.0   Terminating Medium-Voltage Cable 24
   2.1.0   Termination Classes.. 24
     2.1.1   Sealing to the External Environment.................................... 24
     2.1.2   Grounding Shield Ends 24
     2.1.3   External Leakage Insulation 25
   2.2.0   Stress Control Methods.. 26
     2.2.1   Geometric Stress Control... 26
     2.2.2   Capacitive Stress Control 26
3.0.0   High-Potential (Hi-Pot) Testing 29
   3.1.0   Types of Hi-Pot Tests 30
   3.2.0   Making Connections 33
     3.2.1   Selective Guard Service Connections 36
     3.2.2   Corona Guard Ring and Guard Shield 39
   3.3.0   Detailed Operating Procedure... 39
     3.3.1   Go/No-Go Testing 41
     3.3.2   Insulation Resistance Measurements 42

Figures and Tables ━━━━━━━━━━━━━━━━

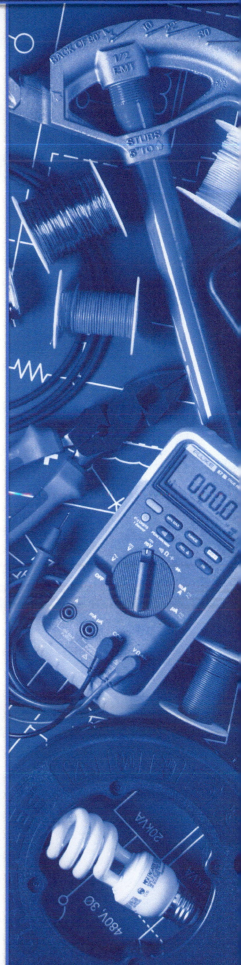

Figure 1   Basic types of medium-voltage cable...........................2
Figure 2   Basic cable configurations.......................................2
Figure 3   Basic components of medium-voltage cable................2
Figure 4   Strand shielding 3
Figure 5   Insulation shield system 4
Figure 6   Jacket and insulation 4
Figure 7   Cutaway view of a splice 4
Figure 8   Making a basic splice 6
Figure 9   High-voltage cable construction 10
Figure 10   Recommended cable preparation procedures for
            an inline splice..11
Figure 11   Wire shield procedure...11
Figure 12   Connecting conductors.. 13
Figure 13   Applying primary insulation................................... 13
Figure 14   Summary of an inline tape splice 15
Figure 15   Summary of a tee tape splice................................ 15
Figure 16   Preparing the cables 15
Figure 17   Placing the nested tubes on the cable.................... 17
Figure 18   Installing the connector 17
Figure 19   Installing the stress relief material......................... 17
Figure 20   Shrinking the stress control tube........................... 17
Figure 21   Shrinking the insulating tube 18
Figure 22   Applying sealant.. 18
Figure 23   Shrinking the insulating/conductive tube 18
Figure 24   Installing metallic shielding mesh 18
Figure 25   Shrinking the rejacketing tube and checking
            for adhesive flow.. 19
Figure 26   Cable preparation details...................................... 19
Figure 27   Components of a ribbon shielding cable splice 20
Figure 28   Installing PST cold-shrink insulator 20
Figure 29   Wrapping No 13 tape onto the cable...................... 21
Figure 30   Use bumps formed on splice ends as guides for centering 21
Figure 31   Wrap vinyl tape at each end of the splice body 22
Figure 32   Form the shield continuity strap over the splice shoulder
            on each side 22
Figure 33   Remove core by unwinding counterclockwise 22
Figure 34   A terminator may be considered an insulator 26
Figure 35   Field distribution over a radial cable section 27
Figure 36   Removing the shield 27
Figure 37   Geometric stress control 28
Figure 38   Equalizing electrical stresses................................ 28
Figure 39   Current-versus-voltage curve 31
Figure 40   Current-versus-time curve 32
Figure 41   Current-versus-voltage and current-versus-time curves
            at maximum voltage.. 32
Figure 42   Faulty insulation.. 33

Figures and Tables (continued)

Figure 43 Hi-pot connections for testing multiple-conductor shielded cable...35

Figure 44 Testing a high-voltage winding to a grounded core or case with a secondary winding to a bypass.........................36

Figure 45 Testing a high-voltage winding to a low-voltage winding with the ground and case to a bypass.........................36

Figure 46 Measuring bushing internal leakage.................................37

Figure 47 Measuring bushing surface leakage.................................37

Figure 48 Testing cable insulation between one conductor and the shield.................................37

Figure 49 Testing cable insulation between one conductor and all others and the shield.................................37

Figure 50 Testing cable insulation between conductors.................................37

Figure 51 Simplified hi-pot tester output circuit diagram.........................38

Figure 52 Corona shield.................................39

Table 1 Insulation Classes and BIL Ratings.................................25

1.0.0 SPLICING MEDIUM-VOLTAGE POWER CABLE

Objective

Describe how to splice medium-voltage cable.

a. Identify medium-voltage power cable configurations and components.
b. Describe a typical procedure for making a straight splice.
c. Describe a typical procedure for making an inline tape splice.
d. Identify various manufactured termination and splice kits.

Trade Terms

Armor: A mechanical protector for cables; usually a formed metal tube or helical winding of metal tape formed so that each convolution locks mechanically upon the previous one (interlocked armor).

Insulation shielding: An electrically conductive layer that provides a smooth surface in contact with the insulation outer surface; it is used to eliminate electrostatic charges external to the shield and to provide a fixed, known path to ground.

Ribbon tape shield: A helical (wound coil) strip under the cable jacket.

Shield: A conductive barrier against electromagnetic fields.

Shielding braid: A shield of small, interwoven wires.

Splice: Two or more conductors joined with a suitable connector and reinsulated, reshielded, and rejacketed with compatible materials applied over a properly prepared surface.

Strand shielding: The semiconductive layer between the conductor and insulation that compensates for air voids that exist between the conductor and insulation.

Stress: An internal force set up within a body to resist or hold it in equilibrium.

Medium-voltage cable, covered in *NEC Article 311*, is defined as a single or multiconductor solid dielectric insulated cable rated at 2,001V to 35,000V. There are many types of medium-voltage power cables in use today. Five of the most common are shown in *Figure 1* and listed as follows:

- Ribbon tape shield
- Drain wire shielded
- Cable UniShield®
- Concentric neutral (CN)
- Jacketed concentric neutral (JCN)

1.1.0 Medium-Voltage Cable Configurations and Components

Refer to *NEC Article 311* for requirements for medium-voltage cable. Conductors used with dielectric cables are available in four basic configurations (*Figure 2*). These configurations can be described as follows:

- *Concentric stranding* (*Class B*) – Not commonly used in modern shielded power cables because of the penetration of the extruded strand shielding between the conductor strands, making the strand shield difficult to remove during field cable preparation.
- *Compressed stranding* – Compressed to 97% of the concentric conductor diameters. This compression of the conductor strands blocks the penetration of an extruded strand shield, thereby making it easily removable in the field. For sizing lugs and connectors, sizes remain the same as with concentric stranding.
- *Compact stranding* – Compacted to 90% of the concentric conductor diameters. Although this conductor has full ampacity ratings, the general rule for sizing is to consider it one conductor size smaller than concentric or compressed cable. This reduced conductor size results in a proportional reduction of all cable layers, a consideration when sizing for molded rubber devices.
- *Solid wire* – This conductor is not commonly used in industrial shielded power cables.

Despite their differences, all medium-voltage power cables are essentially the same, consisting of the following:

- Conductor
- Strand shielding
- Insulation
- Insulation shielding system (semiconductive and metallic)
- Jacket

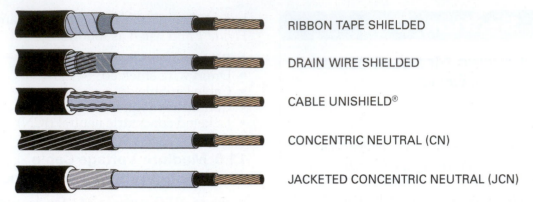

Figure 1 Basic types of medium-voltage cable.

RIBBON TAPE SHIELDED

DRAIN WIRE SHIELDED

CABLE UNISHIELD®

CONCENTRIC NEUTRAL (CN)

JACKETED CONCENTRIC NEUTRAL (JCN)

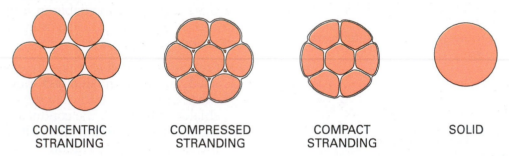

CONCENTRIC STRANDING

COMPRESSED STRANDING

COMPACT STRANDING

SOLID

Figure 2 Basic cable configurations.

Each component is vital to an optimal power cable performance and must be understood in order to make a dependable splice or termination (*Figure 3*).

1.1.1 Strand Shielding

Strand shielding is the semiconductive layer between the conductor and insulation that compensates for air voids between the materials.

Air is a poor insulator, having a nominal dielectric strength of only 76V/mil, whereas most cable insulations have dielectric strengths over 700V/mil. Without strand shielding, an electrical potential exists that will overstress these air voids.

As air breaks down or ionizes, it produces corona discharges. This forms ozone, which chemically deteriorates the cable insulation. The semiconductive strand shielding eliminates this potential by simply shorting out the air.

Modern cables are generally constructed with an extruded strand shield. Strand shielding is shown in *Figure 4*.

1.1.2 Insulation

Insulation consists of many different variations such as extruded solid dielectric or laminar (oil paper or varnish cambric). Its function is to contain the voltage within the cable system. The most common types of solid dielectric insulation in in-

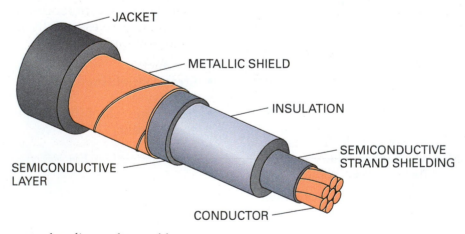

JACKET

METALLIC SHIELD

INSULATION

SEMICONDUCTIVE STRAND SHIELDING

SEMICONDUCTIVE LAYER

CONDUCTOR

Figure 3 Basic components of medium-voltage cable.

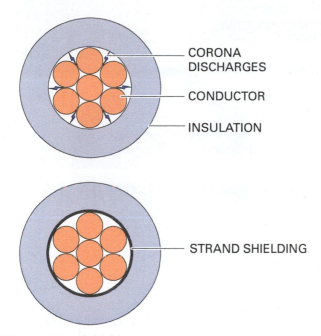

Figure 4 Strand shielding.

dustrial use today are polyethylene, cross-linked polyethylene (XLP), and ethylene propylene rubber (EPR). Each is preferred for different properties such as superior strength, flexibility, and temperature resistance, depending upon the cable characteristics required.

The selection of the cable insulation level to be used in a particular installation must be made based on the applicable phase-to-phase voltage and the general system category that follows:

- *100% level* – Cables in this category may be applied where the system is provided with relay protection such that ground faults will be cleared within one minute. Although these cables are applicable to the great majority of cable installations on grounded systems, they may also be used on other systems for which the application of cable is acceptable, provided the clearing requirement is met in completely de-energizing the faulted section.
- *133% level* – This insulation level corresponds to that formerly designated for ungrounded systems. Cables in this category may be applied in situations where the clearing time requirements of the 100% level category cannot be met, and yet there is adequate assurance that the faulted section will be de-energized in a period not exceeding one hour. In addition, they may be used when additional insulation strength over the 100% level category is desirable.

- *173% level* – Cables in this category should be applied on systems where the time required to de-energize a grounded section is indefinite. Their use is also recommended for resonant grounded systems. Consult the manufacturer for insulation thickness requirements.

The insulation level is determined by the thickness of the insulation. For example, 15kV cable at 100% has 175 mils of insulation, while the same type of cable at 133% has 220 mils. This variance holds true up to 1,000 kcmil.

1.1.3 Insulation Shield System

The outer shielding is composed of two conductive components: a semiconductive (semicon) layer under a metallic layer. The principal functions of a shield system (*Figure 5*) include:

- Confining the dielectric field within the cable
- Obtaining a symmetrical radial distribution of voltage stress within the dielectric
- Protecting the cable from induced potentials
- Limiting radio interference
- Reducing the hazard of shock
- Providing a ground path for leakage and fault currents

> **NOTE**
>
> The shield must be grounded for the cable to perform these functions.

The semiconductive component is available either as a tape or as an extruded layer. Some cables have an additional layer painted on between the semicon and the cable insulation. Its function is similar to strand shielding—to eliminate the problem of air voids between the insulation and the metallic component (in this case, the metallic shielding). In effect, it shorts out the air that underlies the metallic shield, preventing corona and its resultant ozone damage.

The metallic shield is the current-carrying component that allows the insulation shield system to perform the functions mentioned earlier. The various cable types differ most in this layer. Therefore, most cables are named after their metallic shield (for example, tape shielded, drain wire shielded, or UniShield®). The shield type (cable identification) thus becomes important information to know when selecting devices for splices and terminations.

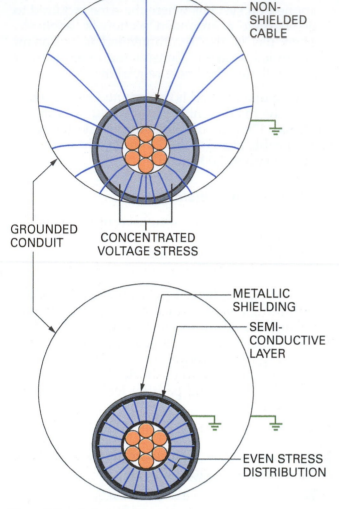

NON-SHIELDED CABLE

GROUNDED CONDUIT

CONCENTRATED VOLTAGE STRESS

METALLIC SHIELDING

SEMI-CONDUCTIVE LAYER

EVEN STRESS DISTRIBUTION

Figure 5 Insulation shield system.

1.1.4 *Jacket*

The jacket (*Figure 6*) is the tough outer covering that provides mechanical protection, as well as a moisture barrier. Often, the jacket serves as both an outer covering and the semiconductive component of the insulation shield system, combining two cable layers. However, because the jacket is external, it cannot serve as the semiconductive component of the complete system. Typical materials used for cable jackets are PVC, neoprene, and lead. Frequently, industrial three-conductor cables have additional protection in the form of an **armor** layer.

1.2.0 Making Straight Splices

Whenever possible, splicing should be avoided. However, splicing is often an economic necessity. There can be many reasons for making splices, including:

- Insufficient supplied length of cable due to reel limitations or cumbersome cable length

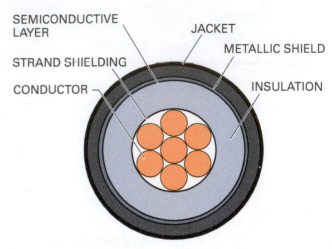

SEMICONDUCTIVE LAYER

STRAND SHIELDING

CONDUCTOR

JACKET

METALLIC SHIELD

INSULATION

Figure 6 Jacket and insulation.

- Cable failures
- Cable damaged after installation
- Insufficient room for cable training radius
- Excessive twisting of cable would otherwise result
- A tap into an existing cable (tee or wye splices)

For damaged cable and taps, the option is to either splice the cable or replace the entire length. In many cases, the economy of modern splicing products makes splicing an optimal choice (*Figure 7*).

Whatever the reason to splice, good practice dictates that splices have the same rating as the cable. In this way, the splice does not derate the cable and become the weak link in the system. Every splice must provide mechanical and electrical integrity at least equal to the original conductor insulation. Furthermore, all terminations and splices must be made under the best environmental conditions possible on the job site. The slightest amount of moisture, dirt, or foreign material can render the splice useless.

Splices can be completed by tapering the ends of the cables and taping with several layers of high-dielectric insulating tape covered with several layers of environmental jacketing tape. An alternative method is to use a stress control void-filling material and heat-shrink insulating and environmental tubing to rebuild the cable to its original insulation thickness. With this method, tapering, also known as *penciling*, is not required.

Figure 7 Cutaway view of a splice.

Medium-Voltage Cables

This is an example of a medium-voltage cable. It is typical of the cables you may have to splice and/or terminate using methods described in this module.

Figure Credit: General Cable

A typical straight splice for a single, unshielded conductor (ranging from 601V to 5,000V) is shown in *Figure 8*. In general, the conductor jacket and insulation should be removed for one-half of the connector length plus ¼" (6 mm). Then, a protective wrap of electrical tape is applied to the jacket ends to protect the jacket and insulation while cleaning the conductor and securing the connector, as shown in *Figure 8 (A)*.

After performing the preceding operations, remove the tape and also remove additional jacket and insulation in an amount equal to 25 times the thickness of the overall insulation, as shown in *Figure 8 (B)*, extending in both directions from the first insulation cut. For example, if the thickness of the insulation is ¼" (6 mm), the outside edge of the splice will be:

$$25 \times 0.25" = 6.25" \text{ or } 6\frac{1}{4}"$$

In metric:

$$25 \times 6 \text{ mm} = 150 \text{ mm}$$

Therefore, the outside edge of the splice will be 6¼" (150 mm) beyond the insulation cut. To continue, pencil the insulation with a sharp knife or insulation cutter, taking care not to nick the conductor.

Refer to *Figure 8 (C)*. Starting from A or B, wrap the splice area with high-voltage insulating rubber splicing tape, stretching the rubber tape about 25% and, if using lined tape, unwinding the interliner as you wrap. The finished splice should be built up to one and one-half times the factory insulation, if rubber (two times, if thermoplastic), as measured from the top of the connector. For example, if the rubber insulation is ¼" (6 mm), the finished splice would be ⅜" (10 mm) thick when measured from the top of the connector.

To finish off the splice and ensure complete protection, cover the entire splice with four layers of vinyl electrical tape, extending about 1½" (37 mm) beyond the edge of the splicing compound.

The five basic tasks in building a splice are:

- Preparing the surface
- Joining the conductors with connector(s)
- Reinsulating
- Reshielding
- Rejacketing

1.2.1 Preparing the Surface

Careful surface preparation is essential to making effective cable splices. It is necessary to begin with good cable ends. For this reason, it is a common practice to cut off a portion of the cable after pulling to ensure an undamaged end. Always use sharp, high-quality tools. When the various layers are removed, cuts should extend only partially through the layer. For example, when removing cable insulation, be careful not to cut completely

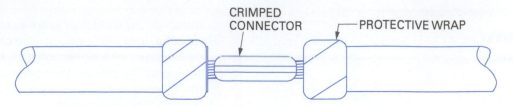

(A) APPLY A PROTECTIVE WRAP OF TAPE TO THE JACKET ENDS TO PROTECT THE JACKET AND INSULATION WHILE CLEANING AND SECURING THE CONNECTOR.

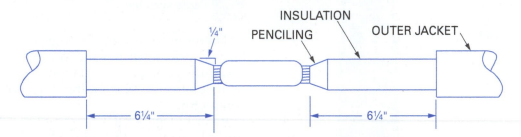

(B) REMOVE ADDITIONAL JACKET AND INSULATION BY AN AMOUNT EQUAL TO 25 TIMES THE THICKNESS OF THE OVERALL INSULATION.

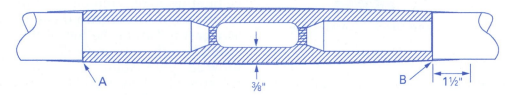

(C) INSULATE AND TAPE THE SPLICE.

Figure 8 Making a basic splice.

through and damage the conductor strands. Specialized tools are available to aid in the removal of the various cable layers. Another good technique for removing polyethylene cable insulation is to use a nylon string as the cutting tool.

When penciling is required (not normally necessary for molded rubber devices), a full, smooth taper is necessary to eliminate the possibility of air voids.

It is necessary to completely remove the semiconductive layer(s) and the resulting residue. Two methods are commonly used to remove the residue, including:

- *Abrasives* – Research has proven that a 120-grit abrasive cloth is fine enough to protect the medium-voltage interface and yet coarse enough to remove semicon residue without loading up the abrasive cloth. This abrasive must have nonconductive grit. Do not use emery cloth or any other abrasive that contains conductive particles because these could embed themselves into the cable insulation. When using an insulation diameter-dependent device (for example, molded rubber devices), take care not

to abrade the insulation below the minimums specified for the device.
- *Solvents* – Use the solvent recommended by the cable manufacturer and by the splicing kit manufacturer. Any solvent that leaves a residue should be avoided. Do not use excessive amounts of solvent because this can saturate the semicon layers and render them nonconductive. Always wipe the cable from the conductor back toward the jacket.

> **CAUTION**
>
> Always read the safety data sheet (SDS) before using any solvent. Avoid solvents that present health hazards, especially with long-term exposure. Choose a nontoxic solvent whenever possible.

1.2.2 Joining the Conductors with Connectors

After the cables are completely prepared, the rebuilding process begins. If a pre-molded or heat-shrink splice is being installed, the appropriate splice components must be slid onto the cable(s)

Making Terminations and Splices

Correct cable preparation, proper installation of all components, and good workmanship are all required to make splices and terminations like those shown here. Manufacturers have done much in the last few years to develop products that make splicing and terminating cables easier, but the expertise, skills, and care of the installer are still necessary to make dependable splices and terminations.

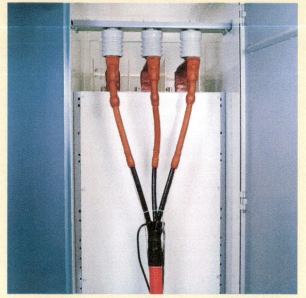

Figure Credit: TE Connectivity

before the connection is made. The first step is reconstructing the conductor with a suitable connector. Connector selection is based on the conductor material (copper or aluminum). A suitable connector for medium-voltage cable splices is a compression or crimp type. Do not use mechanical-type connectors, such as split-bolt connectors.

Aluminum conductors should be connected with an aluminum-bodied connector pre-loaded with contact aid (anti-oxide paste) to break down the insulating aluminum oxide coating on both the connector and conductor surfaces.

When the conductor is copper, connect it with either copper or aluminum-bodied connectors.

It is recommended that a UL-Listed connector be used that can be applied with any common crimping tool. This connector should be Listed for use at medium voltages. In this way, the choice of the connector is at the discretion of the user and is not limited by the tools available.

1.2.3 Reinsulating

There are several methods used to reinsulate splices:

- *Taping* – One method of reinsulating a splice is the tape method. This approach is not dependent upon cable types and dimensions. Tape has a history of dependable service and is generally available. However, wrapping tape on a medium-voltage cable can be time consuming and error prone, since the careful buildup of tape requires accurate half-lapping and constant tension in order to reduce air voids.

- *Heat-shrink tubing* – Heat-shrinkable products use cross-linked polymer tubing that shrinks when heated to its transition temperature. This provides a tight, even seal around the connection area and the cable components. These products can be installed using a flame torch or an electric hot air gun. Heat-shrink splices are range taking, which means that they can be used on a wide range of cable types. The only information required to select a heat-shrink splice is the conductor size and the voltage range of the cable. By using shim tubing, heat-shrink splice kits can be used as reducer splices over a very broad range of cable sizes. These products are also available with an adhesive coating on the inside that provides a superior water seal.

- *Molded rubber splices* – These factory-made splices are designed for the convenience of the installer. In many cases, these splices are also factory tested and designed to be installed without the use of special installation tools. Most molded rubber splices use EPR as the reinsulation material. The EPR must be cured during the molding process. Either a peroxide cure or a sulfur cure can be used.

1.2.4 Reshielding

The cable's two shielding systems (strand shield and insulation shield) must be rebuilt when constructing a splice. The same methods are used as outlined in the reinsulation process: tape, heat-shrink, and molded rubber.

For a tape splice, the cable strand shielding is replaced by a semiconductive tape. This tape is wrapped over the connector area to smooth the crimp indents and connector edges. The insulation shielding system is replaced by a combination of tapes. The semiconductive layer is replaced with the same semiconductive tape used to replace the strand shield.

The cable's metallic shield is generally replaced with a flexible woven mesh of tin-plated copper braid. This braid is for electrostatic shielding only and is not designed to carry shield currents. For conducting shield currents, a jumper braid is installed to connect the metallic shields of the two cable ends. This jumper must have an ampacity rating equal to that of the cable shields.

The insulation shielding of a heat-shrink splice contains a layer of conductive material bonded to the outside of the insulation tubing, similar to the way the cable is designed. The bonded conductive

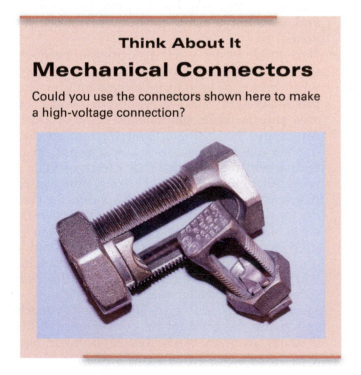

Think About It

Mechanical Connectors

Could you use the connectors shown here to make a high-voltage connection?

Extruded Semiconductors

Some extruded semiconductors peel more easily if they are heated slightly. A heat-shrink blowgun works well.

---

layer is continuous across the splice, so the placement of the tubing is not as critical as it is with a pre-molded splice. In addition, because the tube has the inner insulation and outer conductive layer bonded together in one dual-wall tube, it prevents air pockets between the insulation and semiconductive layers of the splice. The metallic shield is reinstated with ground braids and stainless steel spring clamps to ensure proper fault carrying capacity of the splice.

For a molded rubber splice, conductive rubber is used to replace the cable strand shielding and the semiconductive portion of the insulation shield system. Again, the metallic shield portion must be jumpered with a metallic component of equal ampacity.

A desirable design parameter of a molded rubber splice is that it be installable without special tools. To accomplish this, very short electrical interfaces are required. These interfaces are attained through proper design shapes of the conductive rubber electrodes.

1.2.5 Rejacketing

Rejacketing is accomplished in a tape splice by using rubber splicing tape overwrapped with vinyl tape.

In a heat-shrink splice, rejacketing is completed by using a heavy-wall, adhesive-coated tubing or, where space is limited, a wraparound sleeve.

In a molded rubber splice, rejacketing is accomplished by the proper design of the outer semiconductive rubber, effectively resulting in a semiconductive jacket.

When a molded rubber splice is used on internally shielded cable such as tape shield, drain wire shield, or UniShield® cables, a shield adapter is used to seal the opening that results between the splice and cable jacket.

1.3.0 Making Inline Tape Splices

This section provides an example of a tape splicing procedure using one manufacturer's splicing kit. The procedure applies to metallic and wire shield cable utilizing voltage ratings from 5kV to

Cable Preparation Tools

It is essential to use the proper tool when preparing conductors for termination or splicing. These photos show a cable preparation tool used for removing the outer cable jacket. It can also be used for stripping the insulation.

This semiconductor scoring tool has a blade that is adjustable for both cutting depth and angle. Once set to the proper depth, the blade is set perpendicular to the cable and a ring cut is made around the cable at the desired point. After the ring cut is made, the blade is angled so that each rotation around the cable results in a spiral cut.

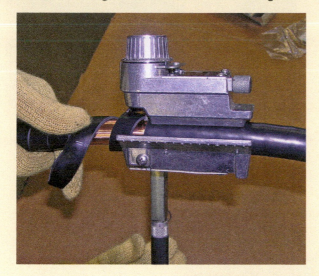 

Figure Credit: Jim Mitchem

Abrasives

Aluminum oxide (ALO) cloth is nonconductive and suitable for use in preparing cable splices. Most people assume that since it contains aluminum, this cloth is conductive, but the oxide of aluminum is actually an insulator.

Cable Preparation Kits

Cable preparation kits contain solvent-saturated cloths and 120-grit nonconductive abrasive cloth. Using these pre-moistened cloths saves time during surface preparation and prevents solvent spillage or storage issues on the job site.

Linerless Splicing Tapes

Linerless splicing tapes, such as the type shown here, reduce application time because there is no need to stop during taping to tear off the liner. This also allows the splicer to maintain a constant tape tension, thus reducing the possibility of taped-in voids.

Tape Splice Kits

Tape splice kits contain all the necessary tapes, along with detailed instructions. These versatile kits ensure that the proper materials are available at the job site and make an ideal emergency splice kit.

25kV, with a normal maximum temperature rating of 90°C and 130°C emergency operation. *Figure 9* shows the various cable sections.

1.3.1 Preparing the Cable

Prepare the cable as follows:

Step 1　Train the cable in position and cut it carefully so that the cable ends will butt squarely.

Step 2　Thoroughly scrape the jacket 3" (75 mm) beyond dimension A in *Figure 10* to remove all dirt, wax, and cable-pulling compound so the tape will bond, making a moisture-tight seal.

Step 3　Remove the cable jacket and nonmetallic filler tape from the ends to be spliced for distance A plus one-half the connector length.

> **CAUTION**
>
> Be careful not to cut into the metallic shielding or insulation when removing the jacket.

Step 4　Remove the cable metallic shielding to expose 1" (25 mm) beyond the cable jacket. If shielded wire cable is used, see *Figure 11*. Be careful not to nick the semiconductive material.

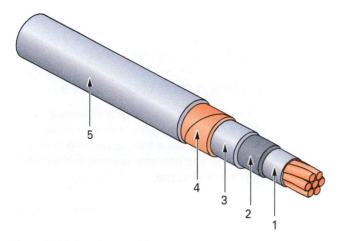

1. STRAND SHIELDING – Semi-conductive material

2. PRIMARY INSULATION – Ethylene propylene or cross-linked polyethylene

3. INSULATION SHIELD – Semiconductive material

4. CABLE METALLIC SHIELDING – Metallic or wire

5. JACKET – PVC, neoprene, polyethylene

Figure 9 High-voltage cable construction.

Rejacketing

This is an example of a heat-shrinkable wraparound jacket repair sleeve used to protect splices in cables subjected to mechanically abusive environments. It can also be used to provide a moisture barrier for repairing and rejacketing lead-covered shield and moisture-impervious cables.

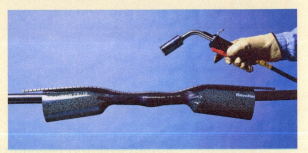

Figure Credit: TE Connectivity

Reshielding

This is an example of a partially completed splice of a shielded single-conductor power cable. The splice is being made using a heat-shrinkable splice kit for use with shielded power cables.

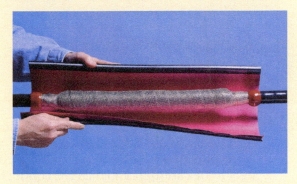

Figure Credit: TE Connectivity

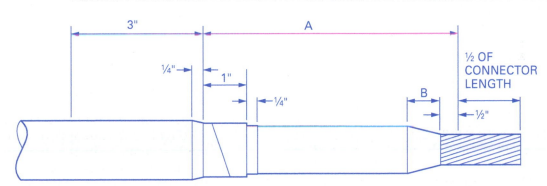

Figure 10 Recommended cable preparation procedures for an inline splice.

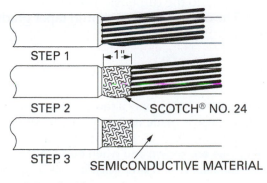

1. Remove cable jacket per previous instructions. Be careful not to cut any of the wires.

2. Wrap two unstretched layers of Scotch® No. 24 electrical shielding tape over shield wires for 1" beyond cable jacket. Tack in place with solder.

3. Cut shielding wires off flush with leading edge of tape.

Figure 11 Wire shield procedure.

Step 5 Tack the metallic shielding in place with solder.

> **CAUTION**
>
> Do not overheat the thermoplastic insulation.

Step 6 Remove the semiconductive material, leaving ¼" (6 mm) exposed beyond the cable metallic shielding. Be sure to remove all traces of this material from the exposed cable insulation by cleaning it with nonconductive abrasive cloth.

> **CAUTION**
>
> Do not use emery cloth, because it contains metal particles.

Step 7 Remove the cable insulation from the end of the conductor for a distance of ½" (13 mm) plus one-half the length of the connector. Be careful not to nick the conductor.

Step 8 Pencil the insulation at each end, using a penciling tool or a sharp knife, for a distance equal to B in *Figure 10*. Buff the tapers with an abrasive cloth so no voids will remain after the joint is insulated.

Step 9 Taper the cable jacket smoothly at each end for a distance equal to ¼" (6 mm). Buff the tapers with an abrasive cloth.

Step 10 Clean the entire area by wiping it with solvent-saturated cloth from the cable preparation kit. Solvent may be used on thermoplastic insulation, but the area must be absolutely dry and free of all solvent residue, especially in the conductor strands and underneath the cable shield, before proceeding to the next step.

1.3.2 *Connecting the Conductors*

Connect the conductors as follows:

Step 1 Join the conductors with an appropriate connector. Use a crimp connector to connect thermoplastic-insulated cables. Follow the connector manufacturer's directions. Use anti-oxidant paste for aluminum conductors. When using a solder-sweated connector, protect the cable insulation with temporary wraps of cotton tape. Be sure to remove all anti-oxidant paste from the connector area after crimping the aluminum connector.

> **CAUTION**
>
> Do not use acid core solder or acid flux.

Step 2 Fill the connector indents with Scotch® electrical semiconducting tape No. 13, as shown in *Figure 12*. Smoothly tape two half-lapped layers of highly elongated tape over the exposed conductor and connector area. Cover all threads of the semiconductive strand shielding and overlap the cable insulation with semiconductive tape. Stretching the tape increases its conductance and will not harm it in any way.

1.3.3 *Applying Primary Insulation*

Apply the primary insulation as follows:

Step 1 Apply Scotch® rubber splicing tape No. 23 (*Figure 13*) half-lapped over the hand-applied semiconductive tape and exposed insulation up to ¼" (6 mm) from the cable semiconductive material. Continue to apply the tape half-lapped in successive level, wound layers until the thickness is equal to the dimension shown in *Figure 13*. The outside surface should taper gradually

Selecting a Splice Method

For the versatility to handle practically any splicing emergency, for situations where only a few splices need to be made, or when little detail is known about the cable, the most effective splice is made with either a tape kit or a heat-shrink kit.

For those times when cable size, insulation diameter, and shielding type are known and when numerous splices will be made, use heat-shrink or molded rubber splices for dependability and simplicity, as well as quick application.

along distance C, reaching its maximum diameter over the penciled insulation.

Step 2 Wrap one half-lapped layer of Scotch® No. 13 tape over the splicing tape, extending over the cable semiconductive material and onto the cable metallic shielding ¼" (6 mm) at each end. Highly elongate this tape when taping over the edge of the cable metallic shielding.

Step 3 Wrap one half-lapped layer of Scotch® No. 24 electrical shielding tape over the semiconductive tape, overlapping ¼" (6 mm) onto the cable metallic shielding at each end. Solder the ends to the metallic-**shielding braid**.

Step 4 Wrap one half-lapped layer of Scotch® No. 33+ vinyl electrical tape, covering the entire area of the shielding braid. Stretch it tightly to flatten and confine the shielding braid.

Step 5 Attach Scotch® No. 25 ground braid as shown in *Figure 14*, soldering it to the cable metallic shielding at each end. Use a wire or strap having at least the same ampacity as the shield (No. 6 AWG is usually adequate).

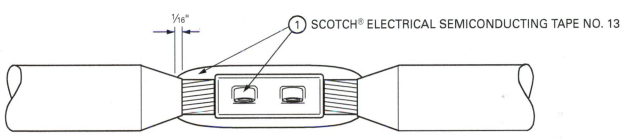

1/16"

① SCOTCH® ELECTRICAL SEMICONDUCTING TAPE NO. 13

Figure 12 Connecting conductors.

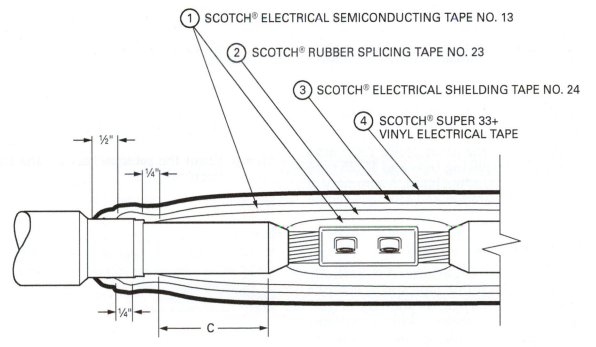

① SCOTCH® ELECTRICAL SEMICONDUCTING TAPE NO. 13

② SCOTCH® RUBBER SPLICING TAPE NO. 23

③ SCOTCH® ELECTRICAL SHIELDING TAPE NO. 24

④ SCOTCH® SUPER 33+ VINYL ELECTRICAL TAPE

½"

¼"

¼"

C

Figure 13 Applying primary insulation.

Step 6 If a splice is to be grounded, use the following procedure to construct a moisture seal at the ground braid:

- If a stranded ground is used, provide solder block to prevent moisture penetration into the splice.
- Wrap two half-lapped layers of Scotch® No. 23 tape, covering the last 2" (50 mm) of the cable jacket.
- Wrap two half-lapped layers of Scotch® No. 23 tape for 3" (75 mm) along the ground strap, beginning at a point where the ground is soldered to the shield.
- Lay the ground strap over the Scotch® No. 23 tape that was applied on the jacket for 1" (25 mm), and then bend the strap away from the cable. Press the strap hard against the cable jacket.

1.3.4 Applying the Outer Sheath

Apply the outer sheath as follows:

Step 1 Wrap four half-lapped layers of Scotch® No. 23 tape (or equivalent) over the entire splice 2" (50 mm) beyond the splicing tape. Highly stretch the tape to form a good moisture seal and eliminate voids. Make sure the tape covers the ground braid to complete the moisture seal.

Step 2 Wrap two half-lapped layers of Scotch® No. 33+ over the entire splice 1" (25 mm) beyond the splicing tape and 1" (25 mm) along the grounding braid. A summary of the end preparation and taping is shown in *Figure 14*.

1.3.5 Tee Tape Splice

A tee tape splice (tap) is performed in basically the same way as the inline tape splice, except a third conductor is attached, as shown in *Figure 15*. Tee splices require greater skill to construct because the tape must be highly elongated to prevent voids in the crotch area.

Heat-shrink splices are also available up to 25kV. These splices provide a reliable and efficient method of making tee splices.

1.4.0 Manufactured Termination and Splice Kits

Manufactured termination and splice kits consist of cable preparation materials, semiconductor material, stress control means, and materials for jacketing, sealing, and grounding. The stress control and jacket materials are applied using a shrink-to-fit method. Both heat-shrink and cold-shrink materials are available.

1.4.1 Heat-Shrink Kits

Heat-shrink splice kits have large use ranges and allow the connection of different conductor sizes, different insulation types (for example, paper insulation to polymeric insulation), and different cable constructions (for example, three-conductor cable spliced to single-conductor cable).

The following instructions are for a heat-shrink kit used on copper tape shielded cable. This same kit will splice UniShield®, wire shield, lead-sheathed cable, or any combination thereof.

To use a quick inline splicing kit, proceed as follows:

Step 1 Select the appropriate kit for the cable diameter using the selection charts provided by the kit manufacturer.

Step 2 Verify that the ground braid or bond wire has a cross-section equivalent to that of the cable's metallic shield.

Step 3 Cut back the cable to the dimensions specified in the kit manufacturer's instructions for the type of cable being spliced (*Figure 16*). If necessary, abrade the insulation to remove embedded semicon.

Step 4 Clean the cable jackets for the length of the tubes.

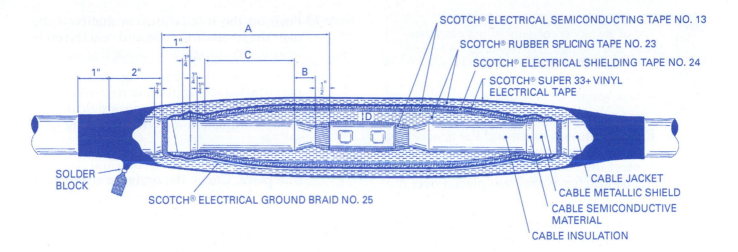

Figure 14 Summary of an inline tape splice.

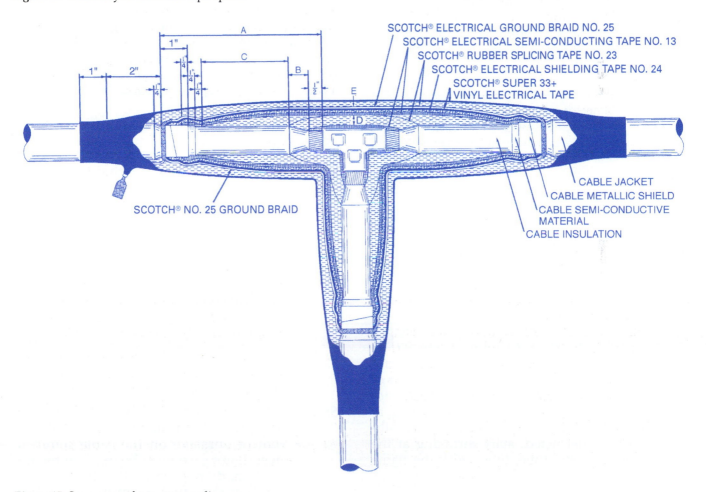

Figure 15 Summary of a tee tape splice.

Figure 16 Preparing the cables.

Step 5 Place the nested tubes and the rejacketing tube over the cable as shown in *Figure 17*. Protect the tubes from the sharp ends of the conductor as they are placed over the cable.

Step 6 Install the connector. Wipe the insulation down using an oil-free solvent (*Figure 18*).

Step 7 Tightly wrap the exposed conductor with the stress relief material (SRM) provided in the splicing kit. Be sure to fill all gaps and low spots around the connector. Also install SRM over the semicon cutback areas (*Figure 19*).

Step 8 Center the stress control tube over the splice and shrink it into place using a manufacturer-approved heat source (*Figure 20*). When shrinking the stress control or insulating tube, apply the outer tip of the flame using a smooth brushing motion. Be sure to keep the flame moving to avoid scorching the material. Unless otherwise instructed, start shrinking at the center of the tube, then work the flame outward on all sides to ensure uniform heat.

Step 9 Position the insulating tube over the stress control tube and shrink it into place (*Figure 21*).

Step 10 Apply sealant to the tube ends to provide a positive environmental seal. Build the sealant to the level of the insulating tube (*Figure 22*).

Step 11 Position the insulating/conductive tube over the insulating tube and seal the ends before shrinking it into place (*Figure 23*).

Step 12 Install metallic shielding mesh using manufacturer-supplied constant-tension clamps for metallic tape shielding or crimp connectors for wire shielding (*Figure 24*).

Step 13 Position the rejacketing tube and shrink it into place. Check the ends for correct adhesive flow (*Figure 25*).

1.4.2 Cold-Shrink Kits

The following instructions are for a cold-shrink kit used on ribbon shielding cable. Kits are also available for wire shield and UniShield®. Complete instructions are included in each kit, and once you understand the directions in this module for ribbon shielding cable, the directions for the other types should be readily understood as well. To use a cold-shrink kit, prepare the cables as follows:

Step 1 Clean each cable jacket by wiping it with a dry cloth about 2" (50 mm) from each end.

Step 2 Refer to *Figure 26* (note that the dimensions shown will vary with the type of kit used). Remove the jacket to distances B and C for cables X and Y, respectively.

Step 3 Remove the metallic shielding for distance A.

Step 4 Remove the cable semiconductive material (semicon), allowing it to extend $1/4$" (6 mm) beyond the metallic shielding (dimension A).

Step 5 Remove the insulation for dimension D and taper the edges $1/8$" (3 mm) at approximately 45°.

Step 6 Clean the exposed insulation using the cleaning pads in the kit. Do not use solvent or abrasive on the cable semicon layer. If abrasive must be used on the insulation, do not reduce the diameter below that specified for minimum splice application.

Step 7 Apply two highly elongated, half-lapped layers of Scotch® No. 13 semiconducting tape starting at 1" (25 mm) on the cable metallic shield and extending $1/2$" (13 mm) onto the cable insulation. Leave a smooth leading edge and tape back to the starting position (*Figure 27*).

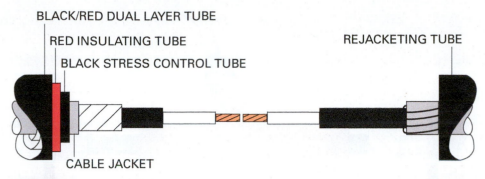

Figure 17 Placing the nested tubes on the cable.

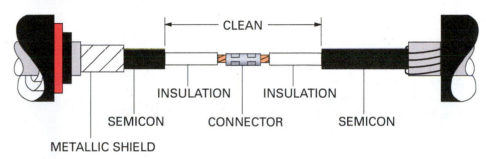

Figure 18 Installing the connector.

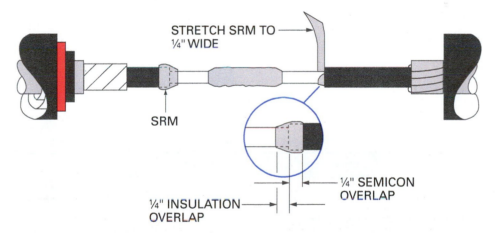

Figure 19 Installing the stress relief material.

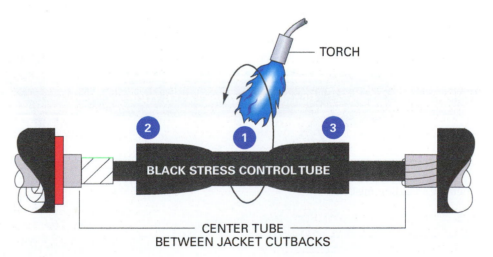

Figure 20 Shrinking the stress control tube.

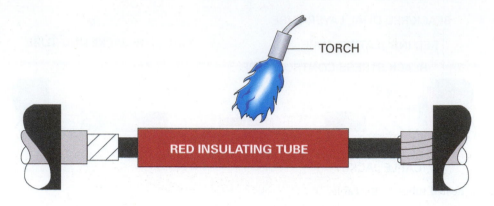

Figure 21 Shrinking the insulating tube.

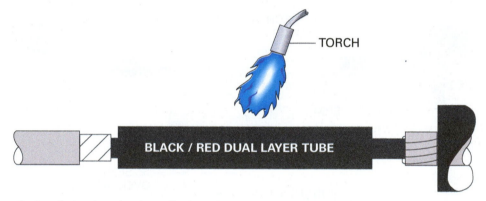

Figure 22 Applying sealant.

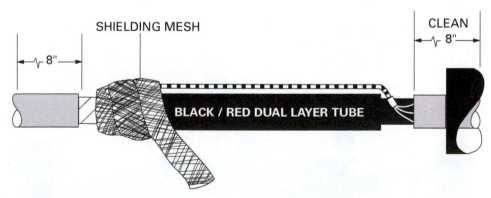

Figure 23 Shrinking the insulating/conductive tube.

Figure 24 Installing metallic shielding mesh.

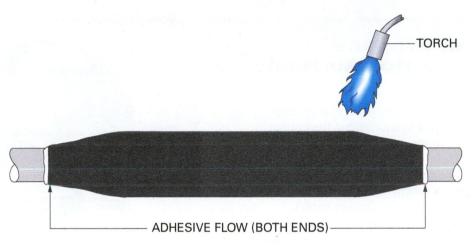

Figure 25 Shrinking the rejacketing tube and checking for adhesive flow.

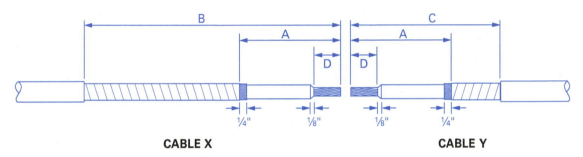

Figure 26 Cable preparation details.

Manufacturer's Instructions

High-quality products usually include detailed installation instructions. Always follow these instructions carefully. A suggested technique is to check off each step as it is completed. Some manufacturers also offer hands-on training programs designed to teach the proper installation of their products. It is highly recommended that inexperienced splice and termination installers take advantage of such programs.

Step 8 On cable X, apply one half-lapped layer (adhesive side out) of the orange vinyl tape supplied in the kit over the metallic shield, beginning approximately ¼" (6 mm) on the previously applied No. 13 tape and ending over the cable jacket.

Splice the installation as follows:

Step 1 Slide the longer PST cold-shrink insulator onto the X jacket and the shorter PST onto the Y jacket, directing the pull tabs away from the cable ends, as shown in *Figure 28*.

Step 2 Apply a few wraps of vinyl tape over cable X.

Step 3 Lubricate the exposed insulation, No. 13 tape, and orange vinyl tape of cables X and Y with silicone grease. Do not allow the grease to extend onto the metallic shield of cable Y.

Step 4 Lubricate the splice bore with silicone grease and install it onto cable X, leaving the conductor exposed for the connector, as shown in *Figure 29*.

NOTE

Rotating the splice while pulling will ease the installation. Make certain all splicing components (PSTs and splice body) are located properly on their respective cable ends before installing the connector.

Step 5 Remove the vinyl tape from the cable X conductor.

Step 6 Install a connector crimped per the manufacturer's instructions.

The Use of Propane Torches and Hot-Air Heat Guns

Making proper splices and terminations using heat-shrink tubing and materials requires that the proper heat level be applied to the heat-shrink material. One manufacturer recommends that a clean-burning propane torch, like the one shown here, be used for heating the heat-shrink tubing materials when making splices and terminations on high-voltage cables. The torch should have an output heat rating of about 90,000 Btuh. Heat-shrink splices and terminations on low-voltage cables and/or smaller high-voltage cables (up to 15kV) can be made using either a propane torch or a hot-air heat gun. The propane torch should have a heat output rating of about 20,000 Btuh; the hot-air heat gun should have a heat output rating of about 750°F to 1,000°F.

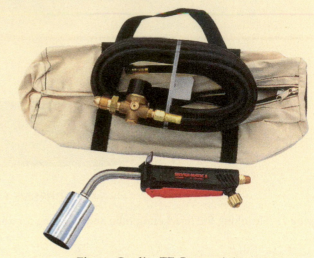

Figure Credit: TE Connectivity

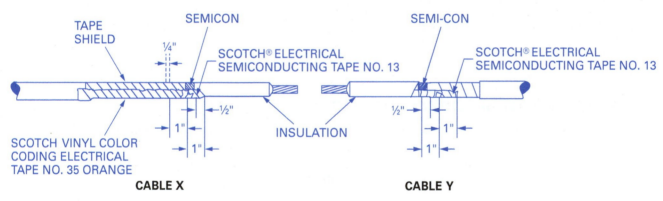

Figure 27 Components of a ribbon shielding cable splice.

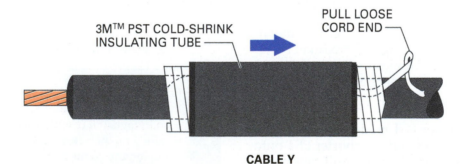

Figure 28 Installing PST cold-shrink insulator.

Step 7 Slide the splice body into its final position over the connector, using the bumps formed on the splice ends as guides for centering (*Figure 30*).

Step 8 Remove the orange vinyl tape and wipe off any remaining silicone grease.

Install the shield continuity assembly as follows:

Step 1 Position the shield continuity assembly over the splice body and hold it in place with a strap or two of vinyl tape at each end of the splice body (*Figure 31*). Form the shield continuity strap over the splice shoulder on each side (*Figure 32*).

> **NOTE**
>
> The assembly coils must be facing the cable and positioned so that they will only make contact with the metallic shielding when applied. Avoid positioning the flat strap over the splice grounding eyes.

Step 2 Unwrap the coil and straighten it for 1" to 2" (25 to 50 mm), then hold the coil and shield strap in place with your thumb. Pull to unwrap the coil around the cable and rewrap it around the cable metallic shield and itself. Cinch (tighten) the applied coil after the final wrap. Repeat for the other end of the splice.

Step 3 To seal the shield strap, cut the supplied strip of vinyl mastic into four equal lengths. After removing the liner, place one pad under the strap, with the mastic side toward the strap and located at the end of the splice body (*Figure 32*). Place a second pad over the strap and first pad and press together, mastic-to-mastic. This will be wrapped with No. 13 tape. Repeat for the other end of the splice body.

Step 4 Beginning just beyond the splice body, wrap No. 13 tape over the vinyl mastic pads extending onto the splice for approximately ¼" (6 mm) and returning to the starting point.

Install the cold-shrink insulators as follows:

Step 1 Position each PST so its leading edge will butt against the splice body grounding eye.

Step 2 Remove the core by unwinding it counter-clockwise and tugging (*Figure 33*).

If the installation calls for grounding the splice, fasten the alligator ground clamp (provided with most kits) to the exposed shield continuity strap approximately at the center of the splice. A ground strap conductor can then be fastened to the alligator clamp. Discard the black plastic shoe packaged with the ground clamp.

SCOTCH® ELECTRICAL SEMICONDUCTING TAPE NO. 13

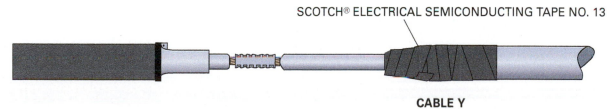

CABLE Y

Figure 29 Wrapping No. 13 tape onto the cable.

SCOTCH® VINYL COLOR CODING ELECTRICAL TAPE NO. 35 ORANGE

CENTER SPLICE BETWEEN BUMPS

CABLE METALLIC SHIELD

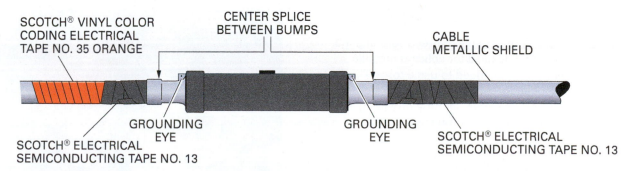

GROUNDING EYE

SCOTCH® ELECTRICAL SEMICONDUCTING TAPE NO. 13

GROUNDING EYE

SCOTCH® ELECTRICAL SEMICONDUCTING TAPE NO. 13

Figure 30 Use bumps formed on splice ends as guides for centering.

Figure 31 Wrap vinyl tape at each end of the splice body.

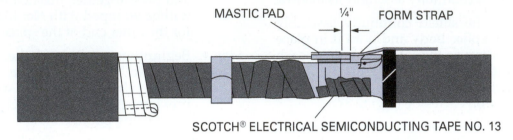

Figure 32 Form the shield continuity strap over the splice shoulder on each side.

Figure 33 Remove core by unwinding counterclockwise.

Installing Multiple Tubes

When installing multiple tubes, make sure that the surface of the last tube is still warm before positioning and shrinking the next tube. If the installed tube has cooled, reheat the entire surface.

Quick Inline Splicing Kits

This is an example of one manufacturer's cold-applied inline cable splice kit. It is used for splicing metallic shielded cables without the need to use a torch. The rejacketing component of the kit is a gel-wrap sleeve that seals the entire splice area, thereby providing mechanical and environmental protection for the splice.

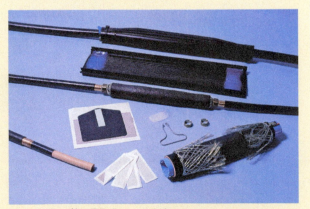

Figure Credit: TE Connectivity

1.0.0 Section Review

1. It is important to compensate for air voids in cable because air _____.

   a. is likely to contain contaminants
   b. is a poor insulator
   c. is an excellent resistor
   d. has a stronger dielectric value than insulation

2. When making a straight splice for a single, unshielded conductor, the conductor jacket and insulation should be removed for one-half of the connector length plus _____.

   a. ⅝" (16 mm)
   b. ½" (13 mm)
   c. ¼" (6 mm)
   d. ⅛" (3 mm)

3. To eliminate air voids in critical areas when making a splice, _____.

   a. loosely apply additional layers of tape
   b. stretch the tape nearly to its breaking point
   c. avoid overlapping the layers of tape
   d. apply the tape sticky side out

4. Before using a cold-shrink kit, clean the cable jacket using _____.

   a. de-ionized water
   b. a water-based solvent
   c. an oil-based solvent
   d. a dry cloth

2.0.0 TERMINATING MEDIUM-VOLTAGE CABLE

Objective

Describe termination classes and important considerations when creating terminations.
  a. Identify termination classes.
  b. Identify stress control methods.

Performance Task

1. Prepare a cable and complete a splice or stress cone.

Trade Terms

Dielectric constant (K): A measurement of the ability of a material to store a charge.

Pothead: A terminator for high-voltage circuit conductors that keeps moisture out of the insulation and protects the cable end, along with providing a suitable stress relief cone for shielded-type conductors.

According to *IEEE Standard 48-2020, Standard Test Procedures and Requirements for Alternating-Current Cable Terminations 2.5kV through 765kV*, a termination is a device used for terminating AC power cables having insulation rated 2.5kV and above. There are three main classes of terminations.

2.1.0 Termination Classes

A Class 1 cable termination (or more simply, a Class 1 termination) provides the following:

• Some form of electric stress control for the cable insulation shield terminus
• Complete external leakage insulation between the conductor(s) and ground
• A seal to prevent the entrance of the external environment into the cable and to maintain the pressure, if any, within the cable system

This classification encompasses the conventional potheads for which the original *IEEE Standard 48-1962* was written. With this new classification or designation, the term pothead is henceforth dropped from usage in favor of the term *Class 1 termination*.

A Class 2 termination is one that provides only some form of electric stress control for the cable insulation shield terminus and complete external leakage insulation, but no seal against external elements. Terminations falling into this classification would include stress cones with rain shields or special outdoor insulation added to give complete leakage insulation and the more recently introduced skip-on terminations for cables having extruded insulation when not providing a seal as in Class 1.

A Class 3 termination is one that provides only some form of electric stress control for the cable insulation shield terminus. This class of termination would be for use primarily indoors. Typically, this would include hand-wrapped stress cones (tapes or pennants), as well as slip-on stress cones.

Some Class 1 and Class 2 terminations have external leakage insulation made of polymeric material. It is recognized that there is some concern about the ability of such insulation to withstand weathering, ultraviolet radiation, contamination, and leakage currents and that a test capable of evaluating the various materials would be desirable. A number of test procedures are available today for this purpose; however, none of them has been recognized and adopted by the industry as a standard. Consequently, this module does not include such a test.

These IEEE classes make no distinction between indoor and outdoor environments. This is because contamination and moisture can be highly prevalent inside most industrial facilities such as paper plants, steel mills, and petrochemical plants.

As a general recommendation, if there are airborne contaminants or if fail-safe power requirements are critical, use a Class 1 termination.

2.1.1 Sealing to the External Environment

In order to qualify as a Class 1 termination, the termination must provide a seal to the external environment. Both the conductor/lug area and the shielding cutback area must be sealed. These seals keep moisture out of the cable to prevent degradation of the cable components. Several methods are used to make these seals, including factory-made seals, mastics used with heat-shrink kits, tape seals (silicone tape must be used for a top seal), and compound seals.

2.1.2 Grounding Shield Ends

All shield ends are normally grounded. However, individual circumstances may exist where

Lead Shield Repair Kits for PILC Cable

This is an example of a repair kit available from one manufacturer for the repair of a damaged lead sheath in paper-insulated, lead-covered (PILC) cable. The kit consists of a wraparound sleeve with an oil-resistant sealing mastic.

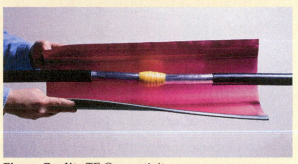

Figure Credit: TE Connectivity

the shield is grounded at only one end. The cable shield must be grounded somewhere in the system. When using solid dielectric cables, it is recommended that solderless ground connections be used, eliminating the danger of overheating cable insulation when soldering.

Terminations

This is an example of the types of high-voltage cable terminations encountered at a power substation.

Figure Credit: TE Connectivity

2.1.3 External Leakage Insulation

Insulation must provide for two functions: protection from flashover damage and protection from tracking damage.

Terminations can be subjected to flashovers such as lightning-induced surges or switching surges. A good termination must be designed to survive these surges. Terminations are assigned a basic lightning impulse insulation level (BIL) according to their insulation class. For example, a 15kV class termination has a BIL rating of 110kV crest (*Table 1*).

Terminations are also subjected to tracking. Tracking can be defined as the process that produces localized deterioration on the surface of the insulator, resulting in the loss of insulating function by the formation of a conductive path on the surface.

A termination can be considered an insulator having a voltage drop between the conductor and the shield. As such, a leakage current develops between these points. The magnitude of this leakage is inversely proportional to the resistance on the insulation surface (*Figure 34*).

Both contamination (dust, salt, and airborne particles) and moisture (humidity, condensation, and mist) will decrease this resistance. This results in surface discharges referred to as *tracking*. In an industrial environment, it is difficult to prevent these conditions. Therefore, a track-resistant insulator must be used to prevent failures.

Table 1 Insulation Classes and BIL Ratings

Insulation Class (kV)	BIL (kV Crest)
5.0	75
6.7	95
15.0	110
25.0	150
34.5	200
46.0	250
69.0	350

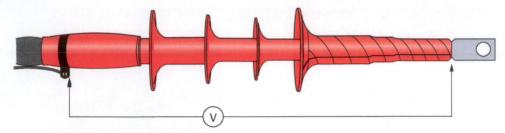

Figure 34 A terminator may be considered an insulator.

A high-performance Class 1 termination contains insulator skirts that are sized and shaped to form breaks in the moisture/contamination path, thus reducing the probability of tracking problems. In addition, these skirts can reduce the physical length of a termination by geometrically locating the creepage distance over the convolutions of the insulator, an important factor when space is a consideration.

Good termination design dictates that the insulators be efficiently applied under normal field conditions. This means the product should be applied by hand with a minimum of steps, parts, and pieces. Because they are inorganic materials, both porcelain and silicone rubbers provide excellent track resistance. Heat-shrink termination materials are the only non-tracking materials available and have shown superior long-term reliability.

2.2.0 Stress Control Methods

In a continuous shielded cable, the electric field is uniform along the cable axis and there is variation in the field only in a radial section. This is illustrated in *Figure 35*, which shows the field distribution over a radial section.

The spacing of the electric flux lines and corresponding equipotential lines is closer near the conductor than at the shield, indicating a higher electric stress on the insulation at the conductor. This stress increase or concentration is a direct result of the geometry of the conductor and shield in the cable section and is accommodated in practical cables by insulation thickness sufficient to keep the stress within acceptable values.

When terminating a shielded cable, it is necessary to remove the shield to a point some distance from the exposed conductor, as shown in *Figure 36 (A)*. This is to secure a sufficient length of insulation surface to prevent breakdown along the interface between the cable insulation and the insulating material to be applied in the termination. The particular length required is determined by the operating voltage and the properties of the insulating materials. This removal of a portion of the shield results in discontinuity of the electrical field. See *Figure 36 (B)* and *Figure 36 (C)*. The electric flux lines originating along the conductor are seen to converge on the end of the shield, with the attendant close spacing of the equipotential lines signifying the presence of high electric stresses in this area. This stress concentration is of much greater magnitude than that occurring near the conductor in the continuous cable. As a result, steps must be taken to reduce the stresses occurring near the end of the shield if cable insulation failure is to be avoided.

All terminations must at least provide stress control. This stress control may be accomplished by two commonly used methods:

- Geometric stress control
- Capacitive stress control

2.2.1 Geometric Stress Control

This method involves an extension of the shielding (*Figure 37*), which expands the diameter at which the terminating discontinuity occurs and thereby reduces the stress at the discontinuity. It also reduces stresses by enlarging the radius of the shield end at the discontinuity.

2.2.2 Capacitive Stress Control

This method consists of a material possessing a high **dielectric constant (K)**, generally around K30, and also a high dielectric strength. (The dielectric constant K is a measurement of the ability of a material to store a charge.)

The value of K is generally an order of magnitude higher than the cable insulation. Located at the end of the shield cutback, the material capacitively changes the voltage distribution in the electrical field surrounding the shield terminus. Lines of electrical flux are regulated to equalize the electrical stresses in a controlled manner along the entire area where the shielding has been removed, as shown in *Figure 38 (A)* and *Figure 38 (B)*.

By changing the electrical field surrounding the termination, the stress concentration is reduced from several hundred volts per mil to values found in continuous cable (usually less than 50V/mil at rated cable voltage).

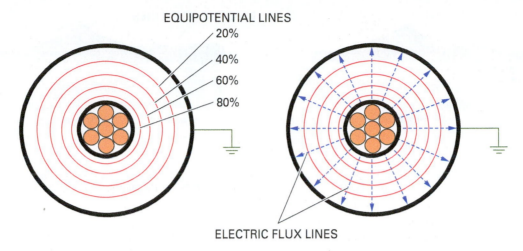

Figure 35 Field distribution over a radial cable section.

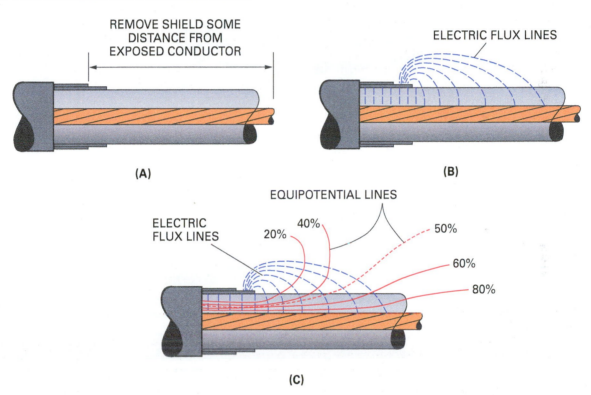

Figure 36 Removing the shield.

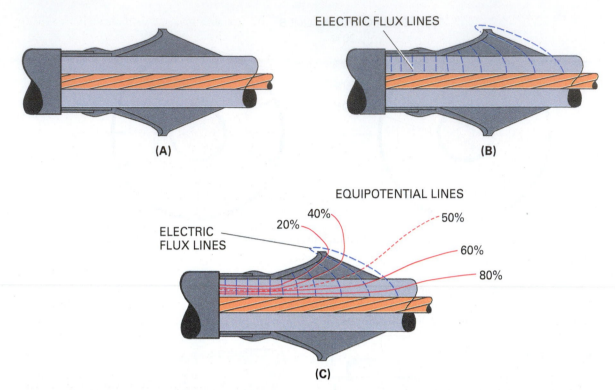

Figure 37 Geometric stress control.

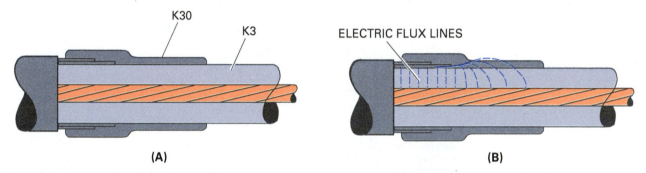

Figure 38 Equalizing electrical stresses.

2.0.0 Section Review

1. The BIL rating of a 15kV class termination is_____.
   a. 75kV crest
   b. 95kV crest
   c. 110kV crest
   d. 150kV crest

2. The dielectric strength of the material used for capacitive stress control is typically around _____.
   a. K10
   b. K20
   c. K30
   d. K40

SECTION THREE

3.0.0 HIGH-POTENTIAL (HI-POT) TESTING

Objective

Define high-potential testing and explain how such testing is conducted.

    a. Identify types of hi-pot tests.
    b. Explain how to make various test connections.
    c. Describe typical procedures for conducting high-potential tests.

Trade Term

High-potential (hi-pot) test: A high-potential test in which equipment insulation is subjected to a voltage level higher than that for which it is rated to find any weak spots or deficiencies in the insulation.

Suitable dielectric tests are commonly applied to determine the condition of insulation in cables, transformers, and rotating machinery. When properly conducted, a **high-potential (hi-pot) test** indicates such faults as cracks, discontinuity, thin spots, or voids in the insulation; excessive moisture or dirt; faulty splices; and faulty terminations. An experienced operator may not only frequently predict the expected breakdown voltage of the item under test, but can often make a good estimate of its future operating life.

The use of direct current (DC) has several important advantages over alternating current (AC). The test equipment itself may be much smaller, lighter in weight, and lower in price. When properly used and interpreted, DC tests give much more information than is obtainable with AC testing. There is far less chance of damage to equipment and less uncertainty in interpreting results. The high-capacitance current frequently associated with AC testing is not present to mask the true leakage current, nor is it necessary to actually break down the material being tested to obtain information regarding its condition. Although DC testing may not simulate the operating conditions as closely as AC testing, the many other advantages of using DC make it worthwhile.

The great majority of medium-voltage cable testing is done with direct current, primarily due to the smaller size of machines and the ability to easily provide quantifiable values of charging, leakage, and absorption currents. Although both AC and DC testers may damage a cable during tests, the damage in a DC test only occurs if the increasing current is not interrupted soon enough and the capacitance stored is discharged across the insulation.

Hi-pot testers for medium-voltage cable also have a microammeter and a range dial to allow measurement of up to 5mA, at which level the circuit breaker trips or the reactor collapses.

> **WARNING!**
> Extreme caution must be exercised when conducting hi-pot testing of high-voltage cable. The cable will recharge itself as the absorption current migrates after the test unless grounds are constantly maintained.

Hi-pot testing may be divided into the following broad categories:

- *Design tests* – Tests usually made in the laboratory to determine proper insulation levels prior to manufacturing.
- *Factory tests* – Tests made by the manufacturer to determine compliance with the design or production requirements.
- *Acceptance tests* – Tests made immediately after installation but prior to putting the equipment or cable into service.
- *Proof tests* – Tests made soon after the equipment has been put into service and during the guarantee period.
- *Maintenance tests* – Tests performed during normal maintenance operations or after servicing or repair of equipment or cable.
- *Fault locating tests* – Tests conducted to determine the location of a specific fault in a cable installation.

The maximum test voltage, testing techniques, and interpretation of the test results will vary somewhat, depending upon the particular type of test. Unfortunately, in many cases, the specifications do not spell out the test voltage, nor do they outline the test procedure to be followed. Therefore, it is necessary to apply a considerable amount of common sense and draw strongly upon experience when conducting these tests. There are certain generally accepted procedures, but the specific requirements of the organization requiring the tests are the ones that should govern. It is obvious that an acceptance test must be much more severe than a maintenance test, and a test made on equipment that is already faulted would be conducted in a manner different from a test being conducted on a piece of equipment in active service.

Special testing instruments are available for all of the tests listed here. However, testing instruments will be used mostly by electricians to perform routine production tests and acceptance tests for newly installed systems.

The general criteria for acceptance of a DC hi-pot test are a consistent leakage for each voltage step (linear) and a decrease in current over time at the final voltage. The level of charging current should also be consistent for each step.

3.1.0 Types of Hi-Pot Tests

DC hi-pot tests are performed primarily to determine the condition of the insulation. The extent of the highest breakdown voltage may or may not be of interest, but the relative condition of the insulation, in the neighborhood of and somewhat above the operating voltage, is certainly required. There are many ways of achieving these results. Two of the most popular are the leakage current-versus-voltage test and the current-versus-time test. The current-versus-time test is frequently made immediately after the current-versus-voltage test and yields the best results when performed in this manner.

Before performing a test, first determine what type of test will be performed and the appropriate test voltage. Refer to the cable and termination product supplier literature for test voltage recommendations and limitations. Next, the test configuration must be decided. When testing a circuit, the standard test setup is to test each conductor to its own shield and ground with all other conductors and their shields grounded also. The physical setup of the test equipment and circuit under test must be reviewed to ensure the safety of test participants and others. Barriers, flagging, or other means must be used to keep others away from the test area, including the other end of the cable under test. Testing cannot be done satisfactorily when the humidity is too high or in windy conditions. Also to be determined is the method of documenting test results. A test data sheet used by one company is shown in the *Appendix*.

After all safety precautions have been taken, disconnect all accessory equipment, connected loads, and potential transformers from the cable or rotating machine being tested. Make sure that the shield, frame, and unused phases are grounded, and the hot side of the hi-pot tester is connected to the conductor under test.

The voltage is slowly raised in discrete steps, allowing sufficient time at each step for the leakage current to stabilize. As the voltage is raised, the leakage current will first be relatively high and then will decrease with time until a steady minimum value is reached.

The initial high value of current is known as the *charging current*. It depends primarily on the capacitance of the item under test. The lower steady-state current consists of the actual leakage current and the dielectric absorption current (which is relatively minor as far as these tests are concerned). The value of the leakage current is noted at each step of voltage, and as the voltage is raised, a curve is plotted of leakage current versus voltage. As long as this curve is relatively flat (that is, equal increments of voltage giving equal increments of current, as shown in point A to point B in *Figure 39*), the item under test is considered to be in good condition. At some point, the current will start rising at a more rapid rate (point C). This will show up on the plot as a knee in the curve. It is very important that the voltage increments be small enough so that the starting point of this knee can be noted. If the test is carried

---

Hi-Pot Testing Overview

DC voltage hi-pot tests are made on cables by applying an appropriate, increasing level of DC voltage in steps and measuring the leakage current (microamperes) through the cable insulation at each step. The voltage level and measured current values are plotted on graph paper, and the shape of the resultant curve is used for determining the condition of the cable insulation being tested.

beyond the start of the knee, the current will increase at a much more rapid rate, and breakdown will soon occur (point D). Unless breakdown is actually desired, it is typical to halt a test as soon as the beginning of the knee is observed.

With a little experience, the operator can extrapolate the curve and estimate the point of breakdown voltage. This procedure enables the operator to anticipate the breakdown point without actually damaging the insulation, because if the test is stopped at the start of the knee, no harm is done.

Although the important thing to watch for is the rate of change of current as the voltage is raised, much information may be obtained from the value of the current. A comparison of the leakage current from phase to phase, a comparison of leakage current with values measured in other equipment under similar conditions, or a comparison with the leakage current values obtained on tests made of the same equipment on prior occasions will give an indication of the condition of the insulation. In general, higher leakage current is an indication of poorer insulation, but this is difficult to evaluate from a single test. As long as the leakage current-versus-voltage curve is of the general shape shown in *Figure 39* and the

knee occurs at a sufficiently high voltage (above the maximum test voltage required), the equipment being tested may be considered satisfactory.

On long cable runs or with rotating machinery having a high capacitance between windings or from winding to frame, the time for stabilization of the leakage current (decrease from E to F in *Figure 40*) may run as high as several minutes, several hours, or more. To minimize the testing time under these conditions, it is common to time the decay of current to where it definitely slows and to take a reading at each voltage step after the same fixed interval. For example, the voltage could be brought up to the first step, allowed to remain at that point for four minutes, and then the current reading taken. The voltage is next raised to the second step, the same four-minute wait is allowed, and the current reading is taken again. This is continued throughout the test. Although this method will not give the true leakage current at any point, it will still produce the proper curve and permit a comparative evaluation.

When the final test voltage is reached, the tester is left on and the current-versus-time curve (*Figure 41*) is plotted by recording the current at fixed intervals as it decays from the initial high

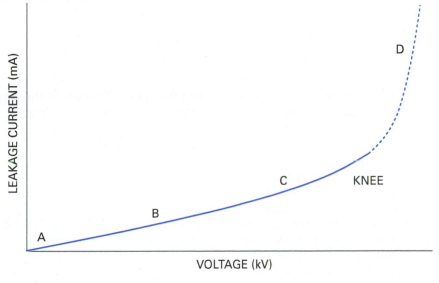

Figure 39 Current-versus-voltage curve.

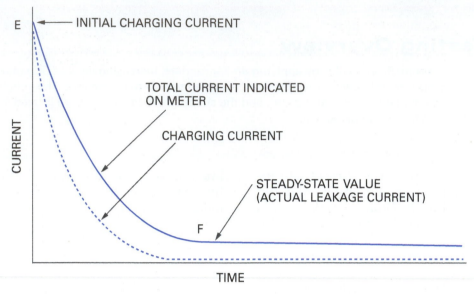

Figure 40 Current-versus-time curve.

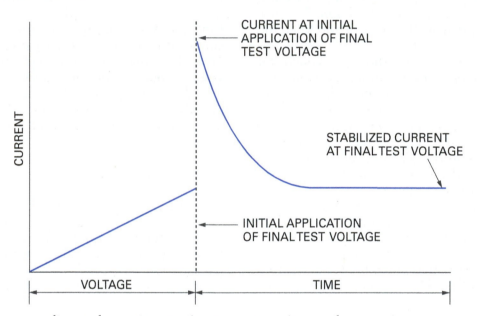

Figure 41 Current-versus-voltage and current-versus-time curves at maximum voltage.

charging value to the steady-state leakage value. This curve for good cable should indicate a continuous decrease in leakage current with time or a stabilization of the current without any increase during the test.

Any increase in current during this test would indicate a bad cable or machine. The test should be stopped immediately if such an indication occurs (*Figure 42*). After the current has stabilized and the last reading has been taken, shut off the high voltage. The kilovoltmeter on the tester will indicate the actual voltage present in the equipment under test as the tester's internal circuitry permits the charge to gradually leak off. After the voltage reaches zero, put an external ground on the cable or machine, and disconnect it from the tester. It is important to wear rubber gloves at this point,

since the effect of absorption currents may cause a buildup of voltage in the item under test until after it has been grounded for some time. The ground should be maintained until the cable is terminated and for at least five times the duration of the test.

When making the current-versus-voltage test, a rapid climb in current may occur not due to defects in the equipment itself, but because of surface leakage or corona at the connections, cable ends, or blades of connected switches. Corona discharge is usually not significant until about 20kV DC and is worse in windy and humid conditions. Bagging the cable ends in a plastic bag or slipping an empty plastic bottle over the cable end often satisfactorily controls corona. When corona cannot be controlled, then all DC testers incorporate internal circuitry (selective guard

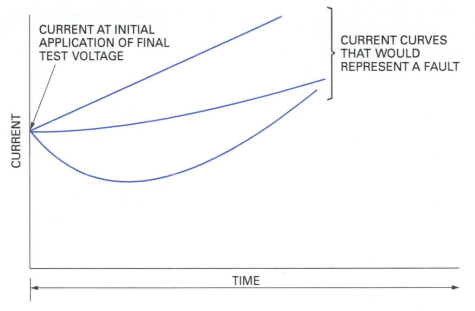

CURRENT AT INITIAL
APPLICATION OF FINAL
TEST VOLTAGE

CURRENT CURVES
THAT WOULD
REPRESENT A FAULT

CURRENT

TIME

Figure 42 Faulty insulation.

circuit) to bypass this stray current so that it does not interfere with the leakage measurements. By the proper use of the guard circuit and corona ring when necessary, good leakage current readings even below one microampere can be made. This technique consists of providing a separate path for these stray currents, bypassing them around the metering circuit, and only feeding the leakage current to be measured through the metering circuit. In order to accomplish this, DC testers have two return paths:

- The metered return binding post, which provides a return path through the microammeter circuit
- The bypass return binding post, which provides a return path bypassing the meter

Currents entering the first binding post will be measured, but currents entering the second binding post will not be measured. A panel switch is provided to connect the cabinet ground internally to either of these two binding posts (which enables ground currents to be measured or bypassed, depending on the test being performed).

3.2.0 Making Connections

Before performing any test, it is important that the test operator carefully plan the test procedure and connections so that the required measurements may be taken without misinterpretation or error.

Hi-Pot Testing Safety

All safety precautions must be taken when performing hi-pot testing of equipment or cables. This includes de-energizing and grounding the equipment to be tested to eliminate any charge. All circuit energizing switches must be opened, locked out, and tagged. Disconnect all jumpers running from terminations to feeders, potential transformers, and lightning arresters. Current transformers might be left in the circuit if their insulation level is sufficiently high.

After all safety precautions have been taken, make the hi-pot test connections as follows:

Step 1   On the tester, make sure the main On/Off switch is turned to the Off position, the high-voltage On switch is in the Off position, and the voltage control is turned fully counterclockwise to the zero voltage position.

Step 2   Connect a grounding cable from the safety ground stud of the tester (*Figure 43*) to a good electrical ground and make sure the connection is secure at both ends.

Step 3   Connect the return line from the item under test to the metered return or bypass return binding post of the tester, and move the grounding switch on the tester panel to the appropriate position. The return lines connected to either the bypass return or metered return binding post should be insulated for about 100V, so ordinary building grade wire may be used.

Step 4   Plug the shielded cable furnished with the tester into the receptacle at the top of the unit. If the tester has an oil-filled output receptacle, make sure that the oil level is at the mark indicated on the oil stick. If the oil level is low, add the proper amount of high-grade transformer or other insulating oil. After the cable is plugged into the receptacle, tighten the clamping nut to hold it securely in place.

Step 5   Connect the shield grounding strap from the cable to the shield ground stud on the tester adjacent to the receptacle.

Step 6   Connect the other end of the output cable to the item under test. This connection should be made mechanically secure, avoiding any sharp edges. It is usually advisable to tape the connection using a high-grade electrical tape to minimize corona at this point. If the voltages to be used are high or if for any other reason corona is anticipated, its effect may be minimized by using a corona ring or corona shield.

Step 7   Plug the line cord into a 120V, 60Hz, single-phase outlet and clip the ground wire coming from the line cord to the conduit receptacle mounting screw or another ground source.

If a bench-type tester or any other unit having an output bushing rather than a receptacle is used, the same procedure is followed except that the output cable is fastened to the bushing and should be supported along its length, preferably using nylon cord so that it stays well away from any grounded or conductive surface. The high-voltage connection between the tester and the item under test should be as short and direct as possible.

When shielded cable is used, it is important that the shield be trimmed back from each end for a distance of about an inch or more for every 10,000V. The shield at the tester end should be fastened to the ground stud on the tester. The shield at the test end should be taped and no connection made to it.

If the item under test is beyond the reach of the tester cable supplied, an extension running between the end of the tester cable and the item under test must be used. If this extension is made of shielded cable, it is necessary not only to splice the central conductor to the end of the tester cable, but also a jumper should connect the shields of the two cables together at the splice. Take care to run the jumper well away from the splice itself to avoid leakage or breakdown.

Hi-Pot Test Voltages

As a rough rule of thumb, acceptance tests are made at about 80% of the original factory test voltage. Proof tests are usually made at about 60% of the factory test voltage. The maximum voltage used in maintenance testing depends on the age, previous history, and condition of the equipment, but an acceptable value would be approximately 50% to 60% of the factory test voltage.

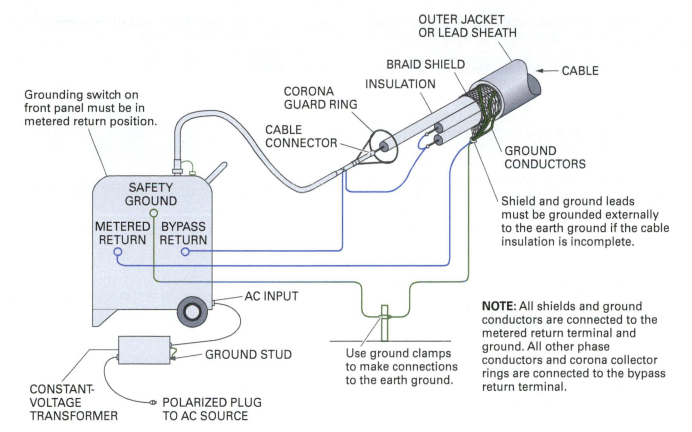

Figure 43 Hi-pot connections for testing multiple-conductor shielded cable.

The splice should be well taped. If no shielded cable is available, the single wire extension used should be supported well away from all other objects, as described previously. When operating at voltages within the corona levels, it is preferable that the tester be brought close enough to the item under test so that an extension cable is not needed. It should be noted that extension cable is a frequent cause of erratic readings, especially if the quality degrades at the high voltages used.

When line regulation is poor (because of location, heavy or intermittent equipment loads in the vicinity, or the use of portable generating equipment), it may be necessary to connect a voltage regulator (constant-voltage transformer) between the tester and the 120V supply. The use of a regulator is particularly desirable when using a tester with a sensitive vacuum tube current meter with ranges of less than 10µA full scale or when highly capacitive loads are being tested.

Because the DC output voltage of the tester is dependent on the AC line input voltage, an unstable line voltage causes fluctuation of the tester output voltage with corresponding fluctuation of the current meter. This is especially noticeable when measuring low current of a few microamperes or when testing loads having considerable capacitance. When the load is highly capacitive, it charges up to the test potential and tends to

maintain that potential because of the high leakage resistance. A sudden drop in line voltage causes the tester output voltage to also drop, so that the load capacitance is then charged to a higher value than the tester output. Since the load is at a higher voltage than the tester, the direction of current flow is reversed, and the current meter is driven down the scale. A sudden increase in line voltage causes the opposite effect, with the tester voltage suddenly rising above the voltage to which the load has been charged, causing a rapid increase in the current meter reading. As a result, the current meter fluctuates so that it might be difficult to obtain a meaningful leakage current reading.

The use of a line voltage regulator under such conditions greatly increases the usefulness and ease of operation of the equipment. In general, the power rating of the regulator should be based on the maximum power output capability of the tester multiplied by a factor of about 1.3. For example, a tester rated at 75kV at 2.5mA can deliver the following output power:

$$75kV \times 2.5mA = 187.5W$$
$$187.5W \times 1.3 = 243.75W$$

Consequently, for this example, a 250W, harmonic-free regulator should be used. This is preferable to the standard type of unit, although either may be used with satisfactory results. If

---

considerable testing at very light loads is done, it might also be desirable to have a smaller regulator to use under those conditions, since a small regulator operating at one-half to three-quarters of its capacity will give better results than a lightly loaded large regulator.

To use the voltage regulator, turn the microampere range switch to the highest range. On multiple voltage range units, turn the kilovolt range switch to the range that will allow testing to the maximum voltage required. The main power switch or circuit breaker may now be turned on and the test begun.

Figure 44 and *Figure 45* illustrate typical connection diagrams for testing high-voltage windings. Connections for monitoring the internal leakage at bushings are shown in *Figure 46* and *Figure 47*.

> **NOTE**
>
> *Figure 44* and *Figure 45* illustrate typical connection diagrams. The experienced operator will soon develop the technique of making the connections most suitable to the test being performed.

In the various connection diagrams given here, the guard circuit may be omitted by tying the connections shown going to the bypass return binding post to the metered return binding post instead. This is the difference shown by the connections in *Figure 48* and *Figure 49*. *Figure 50* shows a typical connection diagram for testing insulation between conductors.

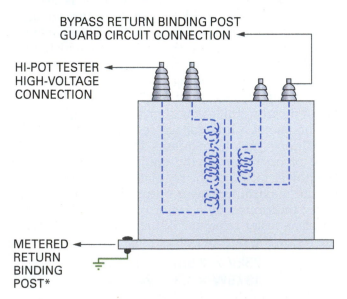

*Set panel grounding switch to metered return position.

Figure 44 Testing a high-voltage winding to a grounded core or case with a secondary winding to a bypass.

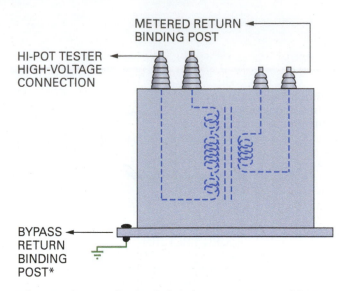

*Set panel grounding switch to bypass return position.

Figure 45 Testing a high-voltage winding to a low-voltage winding with the ground and case to a bypass.

If the return path is grounded, set the panel grounding switch to the metered return position. The total load current (leakage/corona) will be read on the meter.

If the return path is not grounded, set the panel grounding switch to the bypass return position, and the effect of corona currents on the meter reading will be minimized.

3.2.1 Selective Guard Service Connections

In the connection diagrams, frequent reference is made to the bypass return or metered return binding post. The simplified sketch of the tester output circuit shown in *Figure 51* will aid in understanding the use of these terminals.

Examination of the diagram shows that any current to be measured must be returned to the tester through the metered return binding post, and any current that is not to be measured, but is to be bypassed around the meter, should be returned to the bypass return binding post.

As corona currents flow from the high-voltage connection to ground, they may be bypassed around the milliammeter by grounding the bypass return binding post with the grounding switch on the control panel. This gives a direct return path back to the high-voltage supply for corona currents and bypasses them around the meter.

Assume a three-conductor shielded cable is to be tested for leakage current between one of the conductors and the outer shield. To make the proper tester connection, the test should be analyzed as follows.

The current that is to be measured must return to the low side of the tester through the metered return binding post; therefore, the high-voltage connection should be made to the one conductor in question and the shield should be connected to the metered return binding post.

Since the measurement of current flow between the high-voltage connection and the other two conductors in the cable is not desirable, these two conductors should be connected together and returned to the bypass return binding post.

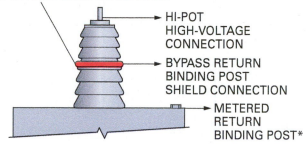

*Set panel grounding switch to metered return position.

Figure 46 Measuring bushing internal leakage.

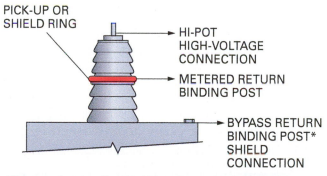

*Set panel grounding switch to bypass return position.

Figure 47 Measuring bushing surface leakage.

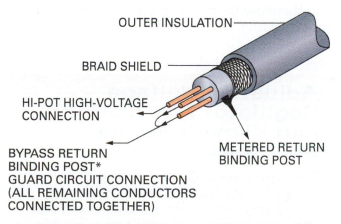

*Set panel grounding switch to metered return position.

Figure 48 Testing cable insulation between one conductor and the shield.

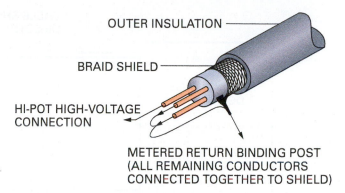

Figure 49 Testing cable insulation between one conductor and all others and the shield.

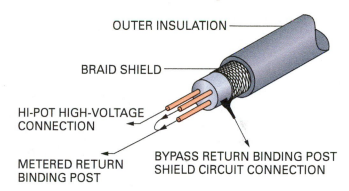

Figure 50 Testing cable insulation between conductors.

In order to minimize the measurement of the corona current from the high-voltage connection to ground, the bypass return binding post should be grounded by setting the grounding switch on the control panel to the bypass position. This will provide a return path for the corona current around the meter, and the corona current will not affect the meter readings.

To measure the leakage current between one conductor and the other two conductors (which are tied together), the return path for the two conductors should be through the metered return post, and the shield should then be connected to the bypass return post. Again, to minimize the measurement of the corona current, the grounding switch should be in the bypass return position.

The corona current alone may be measured by connecting the shield and the two conductors together and connecting them to the bypass return binding post. If the grounding switch is then placed in the metered return position, the return path for the corona current will be through the microammeter, and the corona current will be the only current measured.

The grounding switch on the control panel will either include or exclude the corona current in the meter reading, depending upon its position.

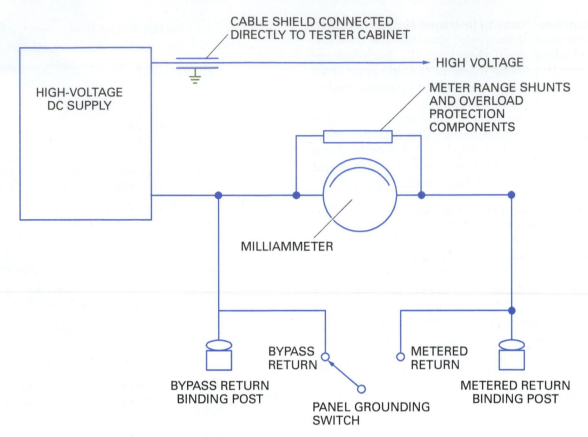

CABLE SHIELD CONNECTED
DIRECTLY TO TESTER CABINET

HIGH VOLTAGE

HIGH-VOLTAGE
DC SUPPLY

METER RANGE SHUNTS
AND OVERLOAD
PROTECTION
COMPONENTS

MILLIAMMETER

BYPASS
RETURN

METERED
RETURN

BYPASS RETURN
BINDING POST

METERED RETURN
BINDING POST

PANEL GROUNDING
SWITCH

Figure 51 Simplified hi-pot tester output circuit diagram.

It was previously assumed that there were no restrictions placed on the cable with respect to grounding. This was done to illustrate the general case. In many cases, however, the shield is permanently grounded, and therefore certain modifications in the test setup must be made in order to achieve corona-free stability and accuracy in the leakage current readings.

If the leakage current from one conductor to the shield is being measured, the shield should be connected to the metered return binding post as before. However, since the shield is now permanently grounded, the metered return binding post should also be grounded by placing the panel grounding switch in the metered return position. If the panel grounding switch were placed in the bypass return position, then the metered return post and the bypass return post would both be grounded, shorting out the microammeter. Therefore, the grounding switch must be put in the metered return position when one side of the test item is permanently grounded.

As outlined above, all unwanted currents that flow between the high-voltage point and any ungrounded elements in the cable may be returned through the bypass line and therefore be excluded from the leakage current reading. However, this does not include corona current because corona current always flows between the high-voltage point and ground. Since we cannot shunt the corona current around the microammeter by grounding the bypass return line as we did before, we must use some auxiliary method of intercepting the corona current before it reaches ground. This can be easily accomplished by the use of a corona or guard ring. A corona or guard ring is nothing more than a metal loop that encompasses the high-voltage connection and intercepts the corona current. The ring is connected to the tester through the bypass return binding post and therefore keeps the corona current out of the meter reading.

Adjusting Voltage Regulator Microampere and Kilovolt Range Switches

The microampere range may be changed while the test is underway, but the kilovolt range may not. Safety interlocks within the unit automatically shut off the high voltage as the voltage range switch is turned while the test is in progress.

3.2.2 Corona Guard Ring and Guard Shield

When using a corona ring, the diameter should be made as small as possible, without causing an arc to jump between the high-voltage connection and the ring. This will ensure interception of as much corona current as possible. A wire should be connected to the ring and the other end connected to the bypass return binding post. This will bypass the intercepted corona current around the micro-ammeter and allow essentially the same stability and accuracy that was obtainable previously.

Corona may also be minimized by electrically shielding or smoothing the connection. This may be done by smoothly wrapping the connection with metal foil or conductive tape or enclosing it in a round can or cover so that all jagged edges, wire points, and rough surfaces are within the metal enclosure. This corona shield (*Figure 52*) must be electrically connected to the high-voltage point and completely insulated from ground so that both the shield and high-voltage connection are at the same potential.

3.3.0 Detailed Operating Procedure

These instructions are meant to illustrate a typical operating procedure. An actual test must account for special conditions peculiar to the installation, ranges, and operating controls of the tester available, as well as the requirements of the operator. To test the system, proceed as follows:

Step 1 Assemble the necessary equipment and supplies at the test site. In addition to the tester, connecting cable, and so forth, bring a clipboard or other writing surface, pencils, colored pencils, scratch paper, graph paper (which is very convenient for plotting the test results while the test is underway), an electronic pocket calculator, and a watch.

Step 2 Determine the type of test to be made. For example, will the voltage be raised to breakdown, or will the voltage be raised to the specific test value only? If the item on the test is found to be defective, should the test stop before breakdown until the repairs can be made later, or should the test be continued to the preselected value regardless of breakdown so that repairs can be made immediately?

Step 3 Determine the maximum test voltage to use. Since the value chosen depends not only on the type of test being made, but also on the material to be tested, it is necessary that all factors involved be considered. As a rough rule of thumb, proof testing is usually conducted at 60% of the factory test voltage. Acceptance testing would call for a maximum value of about 80% of the factory test voltage. For maintenance testing or on older work, the minimum test voltage chosen should be at least 1.7 times the operating voltage up to a maximum value between 50% and 60% of the factory test voltage. In many cases, the value of the maximum test voltage to be used may be obtained from the cable manufacturer.

Step 4 Choose suitable increments so that the voltage may be raised to the final test value in about eight to ten steps (more or less to suit the individual conditions).

Step 5 Once the maximum test voltage and increments have been chosen, the graph paper may be prepared so that the test results can be plotted as the readings are taken. In this way, the condition of the cable is under constant surveillance and if testing short of breakdown is decided upon, the operator has the necessary control to stop the test at the start of the knee of the current-versus-voltage curve.

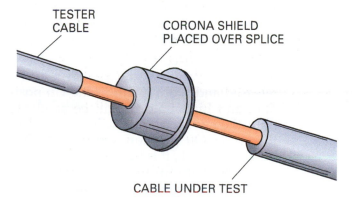

Figure 52 Corona shield.

To prepare the graph paper, mark the current scale on the shorter side of the paper and the voltage scale on the longer side. Suitable voltage increments should be chosen to produce a convenient plot. The current scale should include values well above the expected maximum leakage current. If the magnitude of leakage current is not known, it may be advisable to either make a low-voltage test run or to hold off plotting the first few points until the magnitude is seen.

Step 6 Connect the tester to the equipment under test, as described previously.

Step 7 Make sure the adjustable voltage control is turned to the Off position (fully counterclockwise), the high-voltage switch is in the center or Off position, and the power switch is in the Off position. Connect the input line cord to the 120V, 60Hz receptacle, and connect the alligator clip at the end of the line cord to ground.

Step 8 Turn the kilovoltmeter range switch and microammeter range switch to the appropriate positions.

Step 9 Turn the main On/Off switch or circuit breaker to the On position.

Step 10 Operate the high-voltage On switch. Note that the lever-type high-voltage switch has a spring return from the Down position but will remain in the Up position. If it is desired to keep the high voltage under strict manual control, use the Down position of the high-voltage switch. As soon as it is released to the Off position, the high voltage will be turned off, and it will become necessary to return the powerstat control to zero before the high voltage can be reapplied.

If the high-voltage switch is placed in the Up position, the switch may be released and will remain in the Up position, and high voltage will be available until the switch is manually returned to the center Off position. On units with a keyed interlock, it is necessary to turn the key to the right to enable the Up position of the high-voltage lever switch to be operated.

Step 11 Slowly rotate the variable voltage control from the extreme counterclockwise position (zero volts) to raise the voltage to the first increment, as previously deter-

mined. The voltage should be raised at a slow enough rate to avoid having the microammeter pointer go off the scale. The slower the voltage is raised, the lower the maximum charging current and microammeter reading will be. If the charging current is allowed to exceed the maximum current rating of the tester, the circuit breaker will trip. The circuit breaker will also trip if the unit under test fails.

Since most testers are equipped with zero return interlocks for operator safety, it is necessary to return the voltage control to the zero position before high voltage can be re-applied once it is shut off, whether due to circuit breaker tripping, kilovolt range selector switch rotation, or operation of the high-voltage On lever switch to the Off position.

Step 12 When the first increment of voltage is reached, watch the microammeter indicate gradually decreasing current as the cable or machine under test becomes charged. As the current drops to lower portions of the scale, switch the microammeter progressively to lower ranges until the current stabilizes. Note the length of time it takes for the current to stabilize before accurate results can be obtained. The same stabilization period is allowed at each voltage step. After the current is stabilized, whether it takes 10 seconds, a minute, or 10 minutes, record the value of the voltage and current.

Step 13 If the microampere scale for plotting the current-versus-voltage curve has been predetermined, the current reading at the first voltage step can now be plotted. If the scale has not yet been determined, record this value and after a few more increments, a scale can be determined and the plotting can take place.

Step 14 After the current is stabilized and a point is plotted, turn the microammeter range switch back to the higher range, and gradually raise the voltage to the next increment. Allow the current to stabilize for the same length of time as for the previous increment. Take a second reading and continue the test, gradually raising the voltage step-by-step and plotting the current versus voltage at each step. It is important that the voltage be raised at the same rate in each step.

Step 15 Closely watch the rate of change of current at each voltage increment. The curve should indicate a linear (even) rate of change up to the final test voltage. If this is the case, the material being tested is all right. Upon any indication of a rapid rise in current, stop the test immediately, unless, of course, you wish to test to breakdown. Even a slight knee in the curve may be an indication of imminent breakdown.

Step 16 This step-by-step raising of voltage and plotting of current should be continued until the maximum test voltage previously selected is reached. If a current-versus-time plot is also desired, the time should be noted immediately upon reaching the final voltage step. Then, as the current slowly decays, readings of current and time should be taken at suitable increments and the current-versus-time curve may be plotted. For a safe insulation system, this curve should indicate a continuous decrease in leakage current with time until the current stabilizes. At no time during this part of the test should there be any increase in current.

Step 17 At the end of the test, move the high-voltage lever switch to the Off position. Allow the material being tested to discharge either through the internal discharge circuit of the tester or by using a hot stick and rubber gloves and grounding the output.

Step 18 The kilovoltmeter on the tester will give a direct indication of the voltage at the output terminal regardless of whether the tester is On or Off. The cable or equipment being tested should not be touched until the kilovoltmeter reads zero.

Step 19 After the kilovoltmeter reads zero, disconnect all test leads, and immediately reconnect all cable conductors or equipment terminals to ground and leave grounded.

> **WARNING!**
> Rubber gloves must be worn at all times during this portion of the test. The dielectric absorption effect of cable in particular, and to some extent that of rotating equipment, will tend to restore a dangerous voltage on the disconnected item until it has been adequately grounded for a considerable length of time.

Step 20 The other phases of a multi-phase installation may be tested in a similar manner.

In many cases, once the fault occurs, its exact location may be determined without too much difficulty. However, in those cases where a fault occurs in a long run of cable, special fault-finding techniques incorporating special testing equipment could be of great assistance.

3.3.1 Go/No-Go Testing

The tests described previously provide a good indication of the condition of the equipment or cable, and with proper interpretation, may even allow prediction of the expected life of the device under test. However, to do the job properly takes time and a certain amount of experience and skill, both in conducting and evaluating the operation.

A much shorter and simpler test may be sufficient in those cases where the contractor, purchaser, or operator is only interested in knowing whether or not the installation meets a specific high-voltage breakdown requirement (that is, whether the equipment or cable will withstand X kilovolts without breakdown).

> **WARNING!**
> This test may only be performed by qualified individuals operating under the appropriate safe work plan or permit.

The procedure for conducting a go/no-go test is as follows:

Step 1 Set the microammeter range switch to the highest range.

Step 2 Gradually raise the test voltage at a rate that will keep the charging current below the microammeter's full-scale point. The voltage is raised at this steady rate until the required test value is reached.

Step 3 Maintain the voltage at this value for the length of time required by the specification, or in the absence of any specification, until after the current has stabilized for a minute or more.

Step 4 If the leakage current does not become excessive (remains below the maximum specified value or is low enough to avoid tripping the circuit breaker), the equipment has passed the test.

Step 5 Failure is indicated by a gradual or abrupt increase of current sufficient to trip the circuit breaker. This leakage current should not be confused with the charging current. The charging current may be minimized by raising the voltage at a slower rate.

Step 6 If the current seems to be rising too high, stop raising the voltage at that point and wait. If the current immediately starts decreasing, the indication is that of charging current, and the test may be continued.

Step 7 If the current holds steady or increases, it is due either to leakage current or to corona current. Methods of minimizing the effect of corona current were explained earlier.

Step 8 At the end of the test, reduce the voltage to zero, and follow the procedures recommended by the tester manufacturer.

> **NOTE**
>
> This type of test is not by any means a thorough analysis of cable condition. However, it is a sufficient test and is all that is necessary in a great number of cases. For example, the contractor installing the cable must immediately know if the installation is good. If it is defective, this test will reveal it in the shortest possible time. If the installation is faulty, the test cannot harm it any further, and the contractor may proceed to make the necessary repairs. The only information sought in this type of test is that the cable will not fail below a required test voltage level.

3.3.2 Insulation Resistance Measurements

The DC tester is an invaluable tool for making insulation resistance measurements at higher voltages. Using the same connections and following the same precautions given for high-potential testing, raise the voltage to the desired value and read the microammeter after the current has stabilized (no further decay). The insulation resistance may then be calculated using Ohm's law:

$$R = E \div I$$

Where:

R = insulation resistance (in megohms)
E = voltage (kV meter reading x 1,000)
I = current (in microamperes)

The insulation resistance may be calculated at each voltage step as the leakage current-versus-voltage test is made. Curves of insulation resistance versus voltage will yield much information to the experienced operator. It also helps to keep records of the insulation resistance at a specific voltage so that they can be compared from test to test or from year to year. This will give a good basis of comparison for proper insulation evaluation.

Insulation resistance may also be measured with a standard megger. These instruments will give a direct reading of insulation resistance at a 500V (or higher, depending on the model) test potential.

Insulation resistance measurements help in determining whether or not the items under test are suitable for hi-pot testing. If the insulation resistance is abnormally low, the material is defective and there is no need to make the hi-pot test. If the insulation resistance is high, then the hi-pot test may be made.

Instead of waiting for the current to stabilize before calculating the insulation resistance, readings of voltage and current can be taken at fixed time intervals during the leakage current-versus-time test, and calculations of insulation resistance as a function of time can be made. A plot of the insulation resistance versus time is known as the *absorption curve* and is often used when testing rotating machinery. In a good insulation system, this curve will rise rapidly at first (because of the decaying charging current) and gradually level off (an indication of stabilization of leakage current). The more pronounced the rise, the better the insulation. A relatively flat curve will indicate moist or dirty insulation.

The ratio of the insulation resistance at the end of 10 minutes to the insulation resistance at the end of one minute is called the *polarization index*. If the value of the polarization index is less than two, it usually indicates excessive moisture or contamination. Values of 10 or more are common when testing large motors or generators.

Caution should be exercised in evaluating the importance of insulation resistance measurements. The absolute value is not critical, but the relative order of magnitude or the change in value as the test progresses is significant. Such factors as temperature and humidity can have a very large effect on the reading.

Insulation Resistance Tests

The use of a megger for measuring insulation resistance should not be confused with a high-potential test, since insulation resistance, in general, has little or no direct relationship to dielectric or breakdown strength. For example, an actual void in the insulation may show an exceedingly high value of insulation resistance even though it would permit breakdown at a relatively low voltage.

3.0.0 Section Review

1. DC hi-pot tests are performed primarily to determine the _____.

   a. highest breakdown voltage
   b. condition of the insulation
   c. lowest breakdown voltage
   d. remaining service life

2. A hi-pot tester operates on a supply voltage of _____.

   a. 24V
   b. 60V
   c. 120V
   d. 240V

3. A simple test that may be used to determine whether or not an installation meets a specific high-voltage breakdown requirement is a _____.

   a. current-versus-voltage test
   b. current-versus-time test
   c. voltage-versus-time test
   d. go/no-go test

1. Medium-voltage cable is rated at _____.

    a. 480V to 1,200V
    b. 600V to 2,400V
    c. 1,000V to 6,000V
    d. 2,001V to 35,000V

2. Which of the following cable configurations has the smallest overall diameter for the same amount of current-carrying ability?

    a. Compact stranding
    b. Compressed stranding
    c. Eccentric stranding
    d. Concentric stranding

3. What are the two materials normally used to construct insulation shielding systems?

    a. Semicon and PVC plastic
    b. Semiconductive and metallic material
    c. Copper and aluminum
    d. Silver-plated aluminum (silver and aluminum)

4. When air breaks down or ionizes, it produces _____.

    a. moisture
    b. hydrogen sulfate
    c. photoreceptors
    d. corona discharges

5. Which of the following *best* describes the semiconductive layer between the conductor and the insulation that compensates for air voids that exist between the conductor and the insulation?

    a. Metallic shield
    b. Jacket
    c. Polyethylene
    d. Strand shielding

6. Resonant grounded systems typically require an insulation level of _____.

    a. 100 percent
    b. 133 percent
    c. 153 percent
    d. 173 percent

7. Which of the following can render a high-voltage splice useless?

    a. Movement
    b. Light
    c. Foreign matter
    d. Heat

8. What is normally the final step in a high-voltage cable splice?

    a. Rejacketing
    b. Reinsulating
    c. Reshielding
    d. Cleaning

9. When splicing aluminum conductors, which of the following should be applied to the contact areas of the connector?

    a. Anti-oxide paste
    b. Abrasives
    c. Solvents
    d. Vinyl insulating tape

10. The main difference between AC and DC current testing is that AC testing _____.

    a. uses smaller, less expensive test equipment
    b. results in less chance of equipment damage
    c. produces quantifiable test results
    d. provides true simulation of operating conditions

1. The five main components of a power cable include the _____, _____, _____, _____, and _____.

2. What are the three most common types of solid dielectric insulation?

   _____

3. What are the six main functions of a shield system?

   _____

   _____

   _____

   _____

   _____

   _____

4. List the five basic steps in building a splice.

   _____

   _____

   _____

   _____

   _____

5. When penciling is required, a full, smooth taper is necessary in order to _____.

6. True or False? Mechanical-type connectors are ideal for use in medium-voltage connections.

   _____

7. Contact aid (anti-oxide paste) is used on aluminum conductors and connectors to _____.

8. Name the two common methods for providing stress control.

   _____

9. True or False? All shield ends are normally grounded.

   _____

10. Name the six general categories of hi-pot tests.

    _____

    _____

    _____

    _____

    _____

    _____

Typical Test Data Sheet

CABLE DIELECTRIC TEST

PROJECT _____ DATE _____ 24-Aug-20

JOB # _____

CIRCUIT DESIGNATION CABLE P1022-P01C

TEST EQUIPMENT	MODEL #	SERIAL #	CAL DATE
Hypotronics	800PL	01911-08	28-Mar-20

Cable Nameplate Data 8kV 133% 115 mil EPR, TS, PVC, CT Rated

D-C TEST DATA			
	PHASE A	PHASE B	PHASE C
TIME STARTED	1:20	1:45	2:15
TEST VOLTAGE	CURRENT µA		
KV 7	1.5	1+	1+
KV 14	2+	1.6	2.0
KV 21	4.0	3.0	2.8
KV 28	5.0	4.0	4.3
KV 35	5.2	4.5	4.8

TIME AFTER 100% TEST VOLTAGE IS APPLIED			
0 SECONDS	5.2	4.6	4.9
15 SECONDS	4.8	4.4	4.5
30 SECONDS	4.0	3.9	4.1
45 SECONDS	3.7	3.5	3.6
1 MINUTE	3.5	2.9	3.3
2 MINUTES	3.2	2.6	3.4
3 MINUTES	3.0	2.8	3.0
4 MINUTES	3.2	2.5	3.0
5 MINUTES	3.1	2.6	2.9
10 MINUTES	3.1	2.5	2.7
15 MINUTES	3.1	2.4	2.5
KVDC AFTER 1 MIN DECAY	<1	0	<1
STABILIZATION TIME	20 SEC.		

Cable factory test voltage   45kV

Acceptance Test DC voltage   36kV

Cable is: NEW X   USED

Cable Size: 500KCM   Length: 750 FT

Operating System kV   4.16

System is: GROUNDED X   UNGROUNDED

Shield type: Copper Tape

Cable Mfr.   General

Temperature   82°F   Humidity   25%

Termination manufacturer / type   Raychem HVT-2

Splice type and locations   N/A

SAFETY CHECK: BY:

Test set grounded   MP

Personal safety equipment   MP

Personnel clearance   MP

Lockout in place   MP / KV

Barriers in place   MP

REMARKS & FAULT LOCATION   MP - Test normal

If leakage current rises with steady or decreasing voltage drop the voltage quickly to prevent insulation breakdown unless breakdown is desired to locate the fault.

TEST PERFORMED BY: Megan Paye   WITNESSED BY: Jordan Vidler

Trade Terms Introduced in This Module

Armor: A mechanical protector for cables; usually a formed metal tube or helical winding of metal tape formed so that each convolution locks mechanically upon the previous one (interlocked armor).

Dielectric constant (K): A measurement of the ability of a material to store a charge.

High-potential (hi-pot) test: A high-potential test in which equipment insulation is subjected to a voltage level higher than that for which it is rated to find any weak spots or deficiencies in the insulation.

Insulation shielding: An electrically conductive layer that provides a smooth surface in contact with the insulation outer surface; it is used to eliminate electrostatic charges external to the shield and to provide a fixed, known path to ground.

Pothead: A terminator for high-voltage circuit conductors that keeps moisture out of the insulation and protects the cable end, along with providing a suitable stress relief cone for shielded-type conductors.

Ribbon tape shield: A helical (wound coil) strip under the cable jacket.

Shield: A conductive barrier against electromagnetic fields.

Shielding braid: A shield of small, interwoven wires.

Splice: Two or more conductors joined with a suitable connector and reinsulated, reshielded, and rejacketed with compatible materials applied over a properly prepared surface.

Strand shielding: The semiconductive layer between the conductor and insulation that compensates for air voids that exist between the conductor and insulation.

Stress: An internal force set up within a body to resist or hold it in equilibrium.

Additional Resources

This module presents thorough resources for task training. The following resource material is suggested for further study.

3M™ Premium MV Outdoor Cold Shrink Termination - Fully Integrated (QTIII) video: **solutions.3m.com**.
Inline Crimp Connector Splice Installation video: **solutions.3m.com**.
National Electrical Code® Handbook, Latest Edition. Quincy, MA: National Fire Protection Association.

Figure Credits

Greenlee / A Textron Company, Module Opener
TE Connectivity, Figures 7, 16–25
3M, Figures 9–15, 26–33
Jim Mitchem, Figure 51

Section Review Answer Key

Section 1.0.0

Answer	Section Reference	Objective
1. b	1.1.1	1a
2. c	1.2.0	1b
3. b	1.3.3	1c
4. d	1.4.2	1d

Section 2.0.0

Answer	Section Reference	Objective
1. c	2.1.3; *Table 1*	2a
2. c	2.2.2	2b

Section 3.0.0

Answer	Section Reference	Objective
1. b	3.1.0	3a
2. c	3.2.0	3b
3. d	3.3.1	3c

This page is intentionally left blank.

NCCER CURRICULA — USER UPDATE

NCCER makes every effort to keep its textbooks up-to-date and free of technical errors. We appreciate your help in this process. If you find an error, a typographical mistake, or an inaccuracy in NCCER's curricula, please fill out this form (or a photocopy), or complete the online form at **www.nccer.org/olf**. Be sure to include the exact module ID number, page number, a detailed description, and your recommended correction. Your input will be brought to the attention of the Authoring Team. Thank you for your assistance.

Instructors – If you have an idea for improving this textbook, or have found that additional materials were necessary to teach this module effectively, please let us know so that we may present your suggestions to the Authoring Team.

NCCER Product Development and Revision
13614 Progress Blvd., Alachua, FL 32615

Email: curriculum@nccer.org
Online: www.nccer.org/olf

❏ Trainee Guide   ❏ Lesson Plans   ❏ Exam   ❏ PowerPoints   Other _____

Craft / Level: _____ Copyright Date: _____

Module ID Number / Title: _____

Section Number(s): _____

Description: _____

_____

_____

_____

_____

Recommended Correction: _____

_____

_____

_____

_____

Your Name: _____

Address: _____

_____

Email: _____ Phone: _____

This page is intentionally left blank.

Special Locations

Overview

As an electrician, it is your responsibility to familiarize yourself with all of the *NEC*® requirements for any electrical installation you undertake, as well as any local or regional codes that may also apply. This module describes the *NEC*® requirements for selecting and installing equipment, enclosures, and devices for special locations that require unique attention. These locations include places of public assembly, theaters, carnivals, agricultural and livestock facilities, marinas, swimming pools, and temporary facilities.

Module 26412-20

Trainees with successful module completions may be eligible for credentialing through the NCCER Registry. To learn more, go to **www.nccer.org** or contact us at 1.888.622.3720. Our website, **www.nccer.org**, has information on the latest product releases and training.

Your feedback is welcome. You may email your comments to **curriculum@nccer.org**, send general comments and inquiries to **info@nccer.org**, or fill in the User Update form at the back of this module.

This information is general in nature and intended for training purposes only. Actual performance of activities described in this manual requires compliance with all applicable operating, service, maintenance, and safety procedures under the direction of qualified personnel. References in this manual to patented or proprietary devices do not constitute a recommendation of their use.

26412-20 V10.0

Objectives

When you have completed this module, you will be able to do the following:

1. Identify and select equipment, components, and wiring methods for various special locations and applications.
   a. Identify and select equipment, components, and wiring methods for places of assembly.
   b. Identify and select equipment, components, and wiring methods for theaters and similar locations.
   c. Identify and select equipment, components, and wiring methods for carnivals, circuses, and fairs.
   d. Identify and select equipment, components, and wiring methods for agricultural buildings.
   e. Identify and select equipment, components, and wiring methods for temporary installations.
   f. Identify and select equipment, components, and wiring methods for wired office partitions.
2. Identify and select equipment, components, and wiring methods for marinas, boatyards, and bodies of water.
   a. Identify and select equipment, components, and wiring methods for marinas and boatyards.
   b. Identify and select equipment, components, and wiring methods for natural and man-made bodies of water.
3. Identify and select equipment, components, and wiring methods for pools, spas, tubs, and fountains.
   a. Identify general wiring requirements for pools, spas, tubs, and fountains.
   b. Identify and select equipment, components, and wiring methods for permanently installed pools.
   c. Identify and select equipment, components, and wiring methods for storable pools.
   d. Identify and select equipment, components, and wiring methods for spas, hot tubs, and therapeutic tubs.
   e. Identify and select equipment, components, and wiring methods for fountains.

Performance Tasks

This is a knowledge-based module. There are no Performance Tasks.

Trade Terms

Border light
Dead-front
Electrical datum plane
Equipotential plane
Extra-hard usage cable
Festoon lighting
Fire rating
Footlight

Landing stage
Portable switchboard
Positive locking device
Potting compound
Proscenium
Reactor-type dimmer
Resistance-type dimmer
Rheostat

Shunt trip
Site-isolating device
Solid-state phase-control dimmer
Solid-state sine-wave dimmer
Unbalanced load
Voltage gradients

Industry Recognized Credentials

If you are training through an NCCER-accredited sponsor, you may be eligible for credentials from NCCER's Registry. The ID number for this module is 26412-20. Note that this module may have been used in other NCCER curricula and may apply to other level completions. Contact NCCER's Registry at 888.622.3720 or go to **www.nccer.org** for more information.

Contents

1.0.0 Requirements for Special Locations and Applications 1

  1.1.0 Places of Assembly 2

    1.1.1 Wiring Methods in Assembly Occupancies 2

    1.1.2 Finish Ratings 3

  1.2.0 Theaters, Audience Areas, and Similar Locations 4

    1.2.1 Wiring Methods in Theaters, Audience Areas, and Similar Locations 4

    1.2.2 Fixed Stage Switchboards 6

    1.2.3 Wiring Methods for Fixed Equipment (Other than Switchboards) 8

    1.2.4 On-Stage Portable Switchboards 9

  1.3.0 Carnivals, Circuses, Fairs, and Similar Events............................ 12

    1.3.1 Overhead Conductor Clearances............................... 12

    1.3.2 Power Sources 13

    1.3.3 Wiring: Cords, Cables, and Connectors 13

    1.3.4 Wiring: Rides, Tents, and Concessions 14

    1.3.5 Grounding and Bonding...................................... 14

  1.4.0 Agricultural Buildings 14

    1.4.1 Wiring Methods 15

    1.4.2 Motors and Luminaires 16

    1.4.3 Electrical Supply from a Single Distribution Point...................... 16

    1.4.4 Equipotential Planes 17

  1.5.0 Temporary Installations 18

    1.5.1 Feeder and Branch Circuit Conductors 18

    1.5.2 Grounding 18

    1.5.3 Disconnects 19

    1.5.4 Protection and Support 19

    1.5.5 Receptacles... 19

    1.5.6 Temporary Lighting 19

    1.5.7 Wiring and Equipment Greater Than 600V 19

    1.5.8 Ground-Fault Protection 19

    1.5.9 Assured Equipment-Grounding Conductor Program 20

  1.6.0 Wired Partitions.. 20

2.0.0 Marinas, Boatyards, and Bodies of Water........................... 23

  2.1.0 Marinas and Boatyards...................................... 23

    2.1.1 General Requirements for Devices, Equipment, and Enclosures............................... 23

    2.1.2 Service and Feeder Conductor Load Calculations 24

    2.1.3 Wiring Methods 25

    2.1.4 Grounding 26

    2.1.5 Disconnecting Means for Shore Power........................ 26

    2.1.6 Receptacles.. 26

    2.1.7 Hazardous Locations in and Around Marinas and Boatyards..... 27

Contents (continued)

2.2.0   Natural and Manmade Bodies of Water 28
2.2.1   Electrical Datum Plane 28
2.2.2   Location of Equipment and Enclosures 29
2.2.3   GFCI Protection, Grounding, and Bonding 29
2.2.4   Equipotential Planes 29
3.0.0   Pools, Spas, Tubs, and Fountains 31
3.1.0   General Wiring Requirements 31
3.1.1   Grounding Requirements 31
3.1.2   Cord-and-Plug Connections 32
3.1.3   Clearances 32
3.1.4   Electric Pool Heaters 32
3.1.5   Underground Wiring 32
3.1.6   Disconnecting Means 33
3.2.0   Permanently Installed Pools 33
3.2.1   Motors 33
3.2.2   Receptacle and Area Lighting 34
3.2.3   Underwater Luminaires 35
3.2.4   Transformer or GFCI Junction Boxes and Enclosures 38
3.2.5   Feeders 39
3.2.6   Equipotential Bonding 39
3.2.7   Specialized Pool Equipment 40
3.3.0   Storable Pools 40
3.4.0   Spas, Hot Tubs, and Therapeutic Pools and Tubs 41
3.4.1   Therapeutic Pools and Tubs 42
3.4.2   Hydromassage Bathtubs 43
3.5.0   Fountains 43
3.5.1   Luminaires, Submersible Pumps, and Equipment 43
3.5.2   Junction Boxes and Other Enclosures 44
3.5.3   Bonding and Grounding 45
3.5.4   Cord-and-Plug Equipment 45
3.5.5   Signs Placed in Fountains 45

Figures and Tables

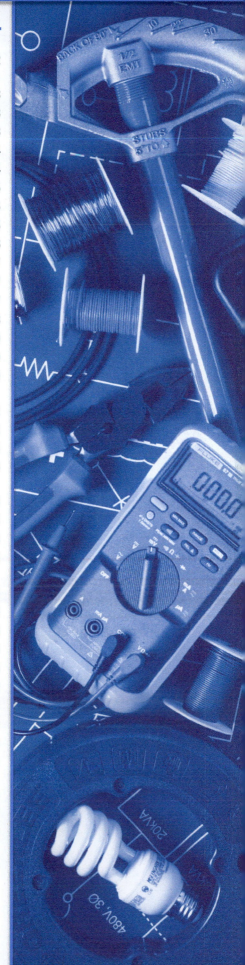

Figure 1 Oldest dance hall in Texas. 2
Figure 2 Combination civic center/chamber of commerce under
construction. 3
Figure 3 MC cable. 3
Figure 4 Dining facility in assembly occupancy. 3
Figure 5 Library. 4
Figure 6 Theater seating area. 4
Figure 7 Portable switchboard. 5
Figure 8 Gutter with more than 30 wires installed in a theater. 5
Figure 9 Switchboard panel with guard grating in place at the top. 5
Figure 10 Fixed stage switchboard. 6
Figure 11 Portable resistance-type dimming panel. 7
Figure 12 Fixed, solid-state programmable dimming system. 7
Figure 13 Manually controlled switchboard. 7
Figure 14 Strip lights. 8
Figure 15 Backstage lighting. 9
Figure 16 Lamp guard. 9
Figure 17 Properly marked grounded conductor. 10
Figure 18 Portable switchboard (dimmer) supplied by a multiconductor
cord. 12
Figure 19 Corridor pilot lights for a dressing room. 12
Figure 20 Carnival. 13
Figure 21 Tents. 13
Figure 22 Separately derived system (generator) powering portable
structures and rides. 13
Figure 23 Permanent wiring on amusement rides. 13
Figure 24 Disconnect switch for amusement ride. 14
Figure 25 Outlet box for concessions. 15
Figure 26 First means of disconnect supplied downstream of a carnival
generator. 15
Figure 27 Device box listed for wet location in agricultural building. 16
Figure 28 Totally enclosed motor. 16
Figure 29 Agricultural building distribution point. 16
Figure 30 Concrete area with accessible metal piping and animal contact. 18
Figure 31 Temporary wiring on construction site. 18
Figure 32 Holiday lighting. 18
Figure 33 Lamp guard in place on temporary lighting at construction site. 19
Figure 34 Receptacles on construction sites must be GFCI-protected. 20
Figure 35 Flexible assembly connecting two wired partitions. 21
Figure 36 Group of freestanding office partitions connected using a single
flexible cord. 21
Figure 37 Marina with floating docks. 23
Figure 38 Electrical distribution system at a marina. 23
Figure 39 Electrical connections within a box not rated for submersion
must be mounted at least 12" (305 mm) above the deck. 24

Figures and Tables (continued)

Figure 40 Accessible circuit breakers in water-tight enclosure. 24
Figure 41 Cable may be used on floating sections of piers. 25
Figure 42 Masts and lifts must clear overhead wiring. 25
Figure 43 Landing stage where people access or exit boats. 26
Figure 44 Water-tight shore power enclosure
          with accessible circuit breakers. 26
Figure 45 GFCI-protected non-shore power receptacle. 27
Figure 46 Class I, Division 1 area below fuel dispenser
          in the void between the pier and the water. 27
Figure 47 Docks or sections of pier that abut this fuel dispensing pier are
          classified as Class I, Division 2. 28
Figure 48 Natural body of water. 28
Figure 49 Manmade body of water. 28
Figure 50 Permanently installed pool. 31
Figure 51 Fountain. 32
Figure 52 Panel in equipment room of swimming pool. 32
Figure 53 Overhead conductors near a community swimming pool. 32
Figure 54 Service and disconnecting means for the pool equipment. 34
Figure 55 Pool recirculation motors connected with liquid-tight flexible
          metal conduit. 34
Figure 56 Pole-mounted area lighting around a pool. 34
Figure 57 Low-voltage area lighting around a pool. 35
Figure 58 Spa emergency shutoff switch and adjacent timer. 35
Figure 59 GFCI circuit breaker supplying the underwater luminaire. 35
Figure 60 Nonmetallic forming shell. 36
Figure 61 Sufficient cord must be provided to allow the luminaire to be
          relamped without entering the water. 37
Figure 62 Junction box for a wet-niche luminaire. 38
Figure 63 Pool perimeter conductive surfaces. 39
Figure 64 Pool heater bonded to the equipotential bonding system. 40
Figure 65 Storable pool. 41
Figure 66 Spa emergency shutoff switch located
          at least 5' (1.5 m) from the spa. 41
Figure 67 Spa handrails must be bonded. 42
Figure 68 Permanently installed fountain. 43
Figure 69 Fountain using pool water. 44
Figure 70 Luminaire and pump in fountains must have GFCI protection. 44

Table 1   Ampacities of Listed Extra-Hard Usage Cords
          and Cables [Data from *NEC Table 520.44(C)(2)*] 8
Table 2   Derating Factors for Extra-Hard Usage Cords
          and Cables [Data from *NEC Table 520.44(C)(3)*] 9
Table 3   Demand Factors (Data from *NEC Table 555.6*) 25
Table 4   Overhead Conductor Clearances [Data from *NEC Table 680.9(A)*] 33
Table 5   Minimum Cover Depths (Data from *NEC Table 300.5*) 33

1.0.0 REQUIREMENTS FOR SPECIAL LOCATIONS AND APPLICATIONS

Objective

Identify and select equipment, components, and wiring methods for various special locations and applications.

a. Identify and select equipment, components, and wiring methods for places of assembly.
b. Identify and select equipment, components, and wiring methods for theaters and similar locations.
c. Identify and select equipment, components, and wiring methods for carnivals, circuses, and fairs.
d. Identify and select equipment, components, and wiring methods for agricultural buildings.
e. Identify and select equipment, components, and wiring methods for temporary installations.
f. Identify and select equipment, components, and wiring methods for wired office partitions.

Trade Terms

Border light: A permanently installed overhead strip light.

Dead-front: Equipment or enclosures covered or protected in order to prevent exposure of energized parts on the operating side of the equipment.

Equipotential plane: An area in wire mesh or other conductive material embedded in or placed under concrete that may conduct current. It must be bonded to all metal structures and nonelectrical, fixed equipment that could become energized, and then it must be connected to the electrical grounding system.

Extra-hard usage cable: Cable designed with additional mechanical protection for use in areas where it may be subject to physical damage.

Festoon lighting: Outdoor lighting that is strung between supporting points.

Fire rating: Time in hours or minutes that a material or assembly can withstand direct exposure to fire without igniting and spreading the fire.

Footlight: A border light installed on or in a stage.

Portable switchboard: A movable lighting switchboard used for road shows and similar presentations, usually supplied by a flexible cord or cable.

Proscenium: The wall and arch that separate the stage from the auditorium (house).

Reactor-type dimmer: A dimming system that incorporates a number of windings or coils that are linked by a magnetic core, an inductive reactor that functions like a capacitor, and a relay. The reactor stores current from the source and reduces lamp power.

Resistance-type dimmer: A manual-type dimmer in which light intensity is controlled by increasing or decreasing the resistance in the lighting circuit.

Rheostat: A slide and contact device that increases or decreases the resistance in a circuit by changing the position of a movable contact on a coil of wire.

Shunt trip: A low-voltage device that remotely causes a higher voltage overcurrent protective device to trip or open the circuit.

Site-isolating device: A disconnecting means at the point of distribution that is used to isolate the service for the purpose of electrical maintenance, emergency disconnection, or the connection of an optional standby system.

Solid-state phase-control dimmer: A solid-state dimmer in which the wave shape of the steady-state current does not follow the wave shape of the applied voltage (that is, the wave shape is nonlinear).

Solid-state sine-wave dimmer: A solid-state dimmer in which the wave shape of the steady-state current follows the wave shape of the applied voltage (that is, the wave shape is linear).

Unbalanced load: Three-phase loads in which the loads are not evenly distributed between the phases.

Places of assembly, theaters, carnivals, agricultural buildings, temporary sites, and other special locations generally require more stringent wiring methods than traditional occupancies. These special wiring methods provide added protection against unique hazards that may exist, such as large crowds (places of assembly or theaters), exposed cable wiring (carnivals, fairs, and other temporary installations), and the presence of livestock (agricultural buildings).

Temporary wiring installations and wired partitions present their own unique electrical hazards. Temporary wiring does not provide long-term conductor or device protection, whereas wired partitions, such as those used in office buildings, are often constructed of sheet metal with sharp edges that create a potential for ground-fault conditions resulting from conductor insulation damage.

Electricians who install wiring in these and similar locations must refer to and comply with specific articles in the *NEC®*.

1.1.0 Places of Assembly

In order to qualify as an assembly occupancy per *NEC Article 518*, a building or structure must be designed to hold 100 or more people and be used for such purposes as meetings, worship, entertainment, eating and drinking, amusement, or awaiting transportation. Convention centers, churches, arenas, dance halls (*Figure 1*), auditoriums, cafeterias, and bus or train stations are some examples of places of assembly.

Often, an assembly occupancy is attached to other buildings, structures, or rooms, such as in the case of the civic center or chamber of commerce building shown in *Figure 2*. Only the portion of the building designed to assemble more than one hundred people is regulated by *NEC Article 518*. Other portions of the structure, such as offices and kitchens, are governed by different *NEC®* articles.

> **NOTE**
>
> If the assembly area contains a projection booth, stage platform, or accommodations for the presentation of theatrical or musical performances, it is governed by *NEC Article 520* (theaters or audience areas), not *NEC Article 518*.

Figure 1 Oldest dance hall in Texas.

Special Occupancies— Construction Job-Site Trailers

Wiring methods for construction job-site trailers, such as the one shown here, are regulated by *NEC Section 550.4(A)*.

1.1.1 Wiring Methods in Assembly Occupancies

For the most part, the building materials used in assembly occupancies are required to have a specific **fire rating** to retard the spread of fire through adjacent spaces. Approved wiring methods allowed in fire-rated construction include metal raceways; flexible metal raceways; nonmetallic raceways encased in at least 2" (50 mm) of concrete; mineral-insulated, metal-sheathed (MI) cable; metal-clad (MC) cable (*Figure 3*); or armored (AC) cable. Raceways or cable assemblies must either qualify as an approved equipment-grounding conductor per *NEC Section 250.118* or contain an insulated grounding conductor sized in accordance with *NEC Table 250.122*. In portions of assembly occupancies that are not required to be of fire-rated construction, wiring methods may include nonmetallic-sheathed (NM) cable, AC cable, electrical nonmetallic tubing, and rigid nonmetallic conduit (PVC). Wiring methods for assembly occupancies are covered in *NEC Section 518.4*.

Figure 2 Combination civic center/chamber of commerce under construction.

NOTE

Temporary wiring used to power display booths in trade shows and similar exhibitions commonly held in assembly occupancies is covered by *NEC Article 590, Temporary Installations.*

Figure 3 MC cable.

1.1.2 Finish Ratings

In addition to fire ratings, building codes often establish a finish rating for combustible (wood) supports when they are used in the construction of assembly buildings or structures. A finish rating for places of assembly is defined as the length of time it takes for a wood support exposed to fire to reach an average temperature rise of 250°F (121°C) or an individual temperature rise of 325°F (163°C) when the temperature is measured on the plane of the wood nearest the fire.

The use of electrical nonmetallic tubing and rigid nonmetallic conduit in places of assembly containing combustible supports is restricted to certain areas including club rooms, conference and meeting rooms in hotels or motels, courtrooms, dining facilities (*Figure 4*), restaurants, mortuary chapels, museums, libraries (*Figure 5*), and places of religious worship. In addition to location restrictions, the raceways must be installed

Figure 4 Dining facility in assembly occupancy.

and concealed in walls, floors, and ceilings that provide a thermal barrier of material with a finish rating of at least 15 minutes. See *NEC Section 518.4(C)(1)*.

NOTE

Check with the authority having jurisdiction to determine the fire and finish rating requirements for each area.

1.2.0 Theaters, Audience Areas, and Similar Locations

NEC Article 520 covers indoor or outdoor spaces used for dramatic or musical presentations, motion picture projection, or similar purposes. This article also covers specific audience seating areas within motion picture or television studios. A typical seating area is shown in *Figure 6*.

1.2.1 Wiring Methods in Theaters, Audience Areas, and Similar Locations

Wiring methods for theaters are covered in *NEC Section 520.5*. Fixed wiring methods include metal raceways, nonmetallic raceways encased in a minimum of 2" (50 mm) of concrete, MI cable, MC cable, or AC cable containing an insulated equipment grounding conductor sized in accordance with *NEC Table 250.122*. Areas in occupancies not required to be of fire-rated construction, as determined by applicable building codes, may contain wiring methods consisting of nonmetallic-sheathed cable, AC cable, electrical nonmetallic tubing, or rigid nonmetallic conduit.

Wiring for portable equipment, such as a portable switchboard (*Figure 7*), stage set lighting, and stage effects, may be in the form of approved flexible cords and cables as long as the cords or cables are not fastened or secured by uninsulated staples or nails. See *NEC Section 520.5(B)*.

Number of Conductors in a Raceway – The maximum number of conductors in metallic or non-metallic raceways as permitted by the *NEC®* for theaters or audience areas of motion picture and

Figure 5 Library.

Figure 6 Theater seating area.

Special Equipment—Electric Vehicle Charging Systems

GOING GREEN

As energy costs increase, hybrid and electric automobiles have become increasingly common. *NEC Article 625* covers wiring for electric vehicle charging systems. This photo shows an electrical charging system for wet-cell batteries. The code contains provisions for using a listed hybrid vehicle charging system as an emergency power backup in the event of a power failure (see *NEC Section 625.48*).

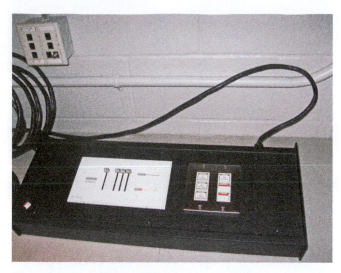

Figure 7 Portable switchboard.

television studios follows the requirements of *NEC Chapter 9, Table 1*: for one conductor, the raceway fill is limited to 53% of the raceway's interior cross section; for two conductors, it is 31%; and for more than two conductors, the limit is 40%.

When conductors are installed in a gutter or wireway in areas covered by *NEC Article 520*, the total cross-sectional area of all conductors cannot exceed more than 20% of the gutter or wireway cross-section in which they are installed. This complies with the *NEC®* cross-section requirements for most conductors installed in an auxiliary gutter (*NEC Section 366.22*) or wireway (*NEC Section 376.22*). However, if the number of current-carrying conductors in an auxiliary gutter or wireway exceeds 30, derating of the conductor ampacity (which is typically required in other locations) is not required in theaters or audience areas per *NEC Section 520.6* (*Figure 8*).

Protective Guarding, Branch Circuits, and Portable Equipment – Electrical circuits, switches, and equipment in theaters and similar areas covered by *NEC Article 520* are often operated by people with limited or no electrical safety training, so any energized part must be enclosed or guarded to prevent electrical shock. All switches must also be designed for external operation only. Every dimmer, rheostat, and similar device must be installed in cabinets that enclose all energized parts (*Figure 9*).

In addition to supplying receptacles, branch circuits in theaters and similar areas may also supply stage set lighting. Receptacles and conductors in these branch circuits must have an ampere rating equal to or greater than the branch circuit overcurrent protective device to which they are connected. As with any branch circuit receptacle, the voltage rating of the receptacle must not be less than the circuit voltage. See *NEC Section 520.9*.

Portable electrical equipment, such as stage or studio lighting, and power distribution equipment may be temporarily used outdoors as long as the use of such energized equipment is continuously supervised by qualified personnel and the equipment is barricaded to prevent public access.

Figure 8 Gutter with more than 30 wires installed in a theater.

Figure 9 Switchboard panel with guard grating in place at the top.

Fire Rating and Building Construction Standards

NFPA 703®, *Standard for Fire Retardant-Treated Wood and Fire-Retardant Coatings*, contains criteria for identifying and defining retardant-impregnated wood and building materials coated with fire retardant. This standard, along with NFPA 5000®, *Building Construction and Safety Code*, is applied in the design and construction of theaters and similar structures.

1.2.2 Fixed Stage Switchboards

The design and construction of switchboards installed in theaters and similar locations must be of the dead-front type (*Figure 10*) and shall be listed. Fixed stage switchboards are covered in *NEC Article 520, Part II*.

Per *NEC Section 520.21*, fixed stage switchboards:

- Shall be listed
- Shall be readily accessible but are not required to be located on or adjacent to the stage (multiple fixed stage switchboards are permitted at different locations)
- Shall contain overcurrent protective devices for all branch circuits supplied by that switchboard
- Are permitted to supply both stage and non-stage equipment

Dimmers – As stated in *NEC Section 520.25(A)*, dimmers used for stage and auditorium lighting may be designed to control voltage through either the grounded or ungrounded circuit conductor supplying the controlled lighting. In installations where the ungrounded conductor is connected through the dimmer, the dimmer must be protected by an overcurrent device with a rating no greater than 125% of the dimmer current rating. In addition, all ungrounded conductors connected to the line side of the dimmer must be opened when either the switch or circuit breaker supplying the dimmer is opened.

An older resistance-type dimmer or reactor-type dimmer may be installed in either the grounded or ungrounded lighting circuit conductor. *Figure 11* is an example of a portable resistance-type dimming panel. Newer solid-state dimmers may not be supplied with a voltage exceeding 150V between conductors unless the dimmer is rated for the voltage. *Figure 12* shows a solid-state programmable dimming system

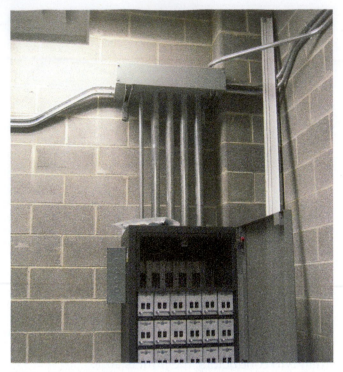

Figure 10 Fixed stage switchboard.

with the front cover removed. When a grounded conductor is connected to a solid-state dimmer, the conductor must be common to both the input and output circuits. The metal chassis of a solid-state dimmer must be connected to the equipment grounding conductor.

Types of Switchboard Control – Stage switchboard control may include manual control, remote control, intermediate control, constant power, or any combination of these types. See *NEC Section 520.26*. In a manually controlled stage switchboard, all dimmers and switches are manually operated by handles that are mechanically linked to the control devices. *Figure 13* shows two manually controlled, portable, solid-state stage switchboard lighting controls.

In a remotely controlled switchboard, dimmers and switches are electrically controlled from a pilot-type console or panel. These pilot panels may be part of the switchboard itself or located in a separate area.

Intermediate control incorporates a design in which primary and secondary switchboards are used. The secondary switchboard functions as a patch panel or secondary panelboard, containing lighting circuit interconnections and overcurrent protection devices. It is typically in a different location than the main or primary stage switchboard.

Figure 11 Portable resistance-type dimming panel.

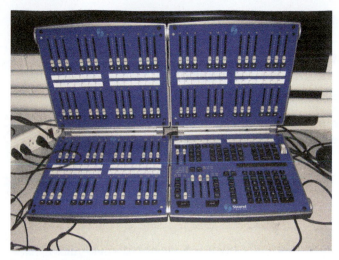

Figure 13 Manually controlled switchboard.

Figure 12 Fixed, solid-state programmable dimming system.

Constant-power stage switchboards contain only overcurrent protective devices and no control elements.

Stage Switchboard Feeders – As stated in *NEC Section 520.27(A)(1)*, a manual or remotely operated stage switchboard may be supplied by a single feeder that is disconnected by a single disconnecting device. In intermediate control, multiple feeders may supply the secondary switchboard (patch panel) as long as all feeders are part of the same electrical system. When neutral conductors are included in the same raceway with multiple feeders to a secondary switchboard, the neutral conductor must be of sufficient ampacity to carry the maximum unbalanced load of the secondary switchboard feeders but is not required to be of a greater ampacity than the neutral conductor supplying the primary switchboard.

Installations in which individual feeders supply a single primary switchboard shall have a disconnecting means for each feeder and have a labeling system that states the number and location of each disconnecting means. If the disconnecting means are located in more than one distribution panelboard, the primary stage switchboard must be designed so that barriers are installed to separate feeders from different disconnecting means locations, with labels indicating the disconnect locations.

In a three-phase, four-wire, solid-state phase-control dimmer, the feeder neutral conductor is considered a current-carrying conductor. However, the feeder neutral conductors in a three-phase, four-wire, solid-state sine-wave dimmer is not considered a current-carrying conductor. When a combination of phase-control and sine-wave control is used, the neutral conductor is considered a current-carrying conductor.

In order to calculate feeder sizes supplying stage switchboards, the total controlled load may be used as long as (1) all feeders supplying the switchboard are protected by an overcurrent device that is rated no greater than the feeder ampacity and (2) the opening of the overcurrent device does not affect the operation of the exit or emergency lighting systems.

1.2.3 Wiring Methods for Fixed Equipment (Other than Switchboards)

Wiring methods for fixed stage equipment, including the footlight units, border light units, proscenium sidelights, portable strip lights, backstage lamps, stage receptacles, connector strips, curtain machines, and smoke ventilators, are covered in *NEC Article 520, Part III*.

The maximum branch circuit rating for footlights, border lights, and proscenium sidelights is 20A unless heavy-duty lampholders are used. See *NEC Sections 520.41(A) and (B)*.

> **NOTE**
>
> *NEC Sections 210.23(B) and (C)* cover rules associated with circuits supplying heavy-duty lampholders.

Conductors that supply foot, border, proscenium, portable strip lights (*Figure 14*), and connector strips must have a minimum insulation temperature rating of 125°C (257°F); however, when calculating the conductor's ampacity using *NEC Table 310.16*, the 60°C (140°F) column must be used. See *NEC Section 520.42*. Cable or conductor drops from overhead connector strips must have an insulation rating of 90°C (194°F) and must also have their ampacity calculated using the 60°C (140°F) column of *NEC Table 310.16*. No more than 6" (150 mm) of conductor can extend into the connector strip. In any of these cases, the derating factor based on the number of conductors does not apply.

Footlights – As stated in *NEC Section 520.43(A)*, when footlights are enclosed in a metal trough, the trough shall be constructed of sheet metal that is at least 0.032" (0.81 mm) thick and treated to prevent rust. All lampholder terminals shall be spaced a minimum of ½" (13 mm) from any metal trough surface, and lampholder terminations must be made by a soldered connection.

In nonmetallic troughs, individual outlet boxes for lampholders must be connected with rigid metal conduit, intermediate metal conduit, flexible metal conduit, MC cable, or MI cable. Again, all terminations must be soldered connections.

> **CAUTION**
>
> Footlights that are designed to disappear into a storage recess when not in use must be designed so that the power supply is automatically disconnected when the lights are retracted into the storage recess.

Cords and Cables for Border Lights – Because border lights produce extreme heat, certain rules apply to the type of cord or cable that can be used to supply these lights. See *NEC Section 520.44(C)(1)*. The cord or cable must be listed as extra-hard usage cable, must be properly supported, and can only be used in applications that require flexible conductors. The ampacity of these cords or cables is governed by *NEC Section 400.5*.

If the cord or cable is not in direct contact with heat-producing lighting or equipment, the cable or cord ampacity may be calculated using *NEC Table 520.44(C)(2)*, as summarized in *Table 1*.

The values in this table reflect the ampacity for multiconductor cords and cables with no more than three current-carrying copper conductors

Table 1 Ampacities of Listed Extra-Hard Usage Cords and Cables [Data from *NEC Table 520.44(C)(2)*]

Temperature Rating of Cords and Cables			
Size AWG	75°C	90°C	Maximum Rating of Overcurrent Device
14	24	28	15
12	32	35	20
10	41	47	25
8	57	65	35
6	77	87	45
4	101	114	60
2	133	152	80

Reprinted with permission from NFPA 70®-2020, *National Electrical Code*®, Copyright © 2019, National Fire Protection Association, Quincy, MA. This reprinted material is not the complete and official position of the NFPA on the referenced subject, which is represented only by the standard in its entirety which may be obtained through the NFPA website at **www.nfpa.org**.

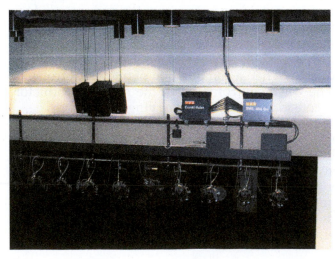

Figure 14 Strip lights.

and an ambient temperature of 86°F (30°C). The derating factors in *NEC Table 520.44(C)(3)* must be applied when more than three current-carrying conductors are in the cable or cord (*Table 2*).

Miscellaneous Fixed Equipment, Receptacles, and Outlets – Per **NEC Section 520.45**, any receptacle used to supply on-stage electrical equipment must be rated in amperes and the conductors supplying the receptacle must comply with **NEC Articles 310 and 400**. In addition, receptacles that supply portable stage-lighting equipment must be of the pendant type or installed in suitable pockets or enclosures (**NEC Section 520.46**).

Backstage lighting is usually utilitarian in design because it is out of view of the audience (*Figure 15*). When bare bulbs are used as backstage lighting, lamp guards must be installed (*Figure 16*) and a minimum space of 2" (50 mm) must be maintained between lamps and any combustible material (**NEC Section 520.47**). Electrical devices that are used to control the movement of curtains must be listed for that purpose (**NEC Section 520.48**).

Smoke Ventilators – Per **NEC Section 520.49**, electrical devices controlling stage smoke ventilators must be supplied by a normally closed circuit, rated for the full voltage of the circuit to which it is connected, located above the scenery in a loft area, and mounted in a metal enclosure having a tightly sealing self-closing door. In addition, the circuit supplying the device must be controlled by at least two externally operated switches, with one switch being located on stage at a readily accessible spot and the other switch location determined by the authority having jurisdiction for the electrical installation.

1.2.4 On-Stage Portable Switchboards

Portable stage switchboards are typically used for road-show presentations and other similar portable activities (**NEC Section 520.50**). Portable on-stage switchboards may be supplied with power from road-show connection panels, often referred to as *patch panels*, or from cord or cable power outlets. Portable on-stage switchboards are used to supply lighting and receptacles just like fixed on-stage switchboards.

Road-Show Connection Panel (Patch Panel) – A patch panel is often used to connect portable stage switchboards to fixed lighting outlets by way of permanently installed supplementary circuits. In these installations, the panel, supplementary

Table 2 Derating Factors for Extra-Hard Usage Cords and Cables [Data from *NEC Table 520.44(C)(3)*]

Number of Conductors	Percent of Ampacity Value in *NEC Table 520.44(C)(3)*
4–6	80%
7–24	70%
25–42	60%
43+	50%

Reprinted with permission from NFPA 70®-2020, *National Electrical Code®*, Copyright © 2019, National Fire Protection Association, Quincy, MA. This reprinted material is not the complete and official position of the NFPA on the referenced subject, which is represented only by the standard in its entirety which may be obtained through the NFPA website at **www.nfpa.org**.

Figure 15 Backstage lighting.

Figure 16 Lamp guard.

circuits, and outlets must follow certain rules as outlined in *NEC Section 520.50* and summarized here:

- Circuit conductors must terminate in a polarized, grounding-type inlet that is matched to the current and voltage rating of the fixed-load receptacle.
- When transfer switches are used between fixed and portable switchboards, all circuit conductors must transfer at the same time.
- Any exterior device that supplies current to supplementary circuits must be protected by overcurrent protective devices.
- Individual supplementary circuits within the road show patch panel and theater must be protected by overcurrent protective devices installed within the panel.

> **NOTE**
> Patch panel construction is regulated by *NEC Article 408*.

Power Outlets and Supply Conductors – Power outlets that supply on-stage portable switchboards shall be enclosed fused switches, fused disconnects, or circuit breakers that are externally operated with provisions for an equipment grounding conductor. See *NEC Section 520.51*. These supply devices shall be readily accessible from the stage floor, mounted on stage, or at the location of the permanent switchboard.

Extra-hard usage cords or cables shall be used to connect the power outlets to the switchboard. See *NEC Section 520.54(A)*. These cables or cords shall terminate totally within the enclosed power outlet and switchboard enclosure or in a connector assembly identified for that purpose.

A single-conductor current cable cannot be smaller than No. 2 AWG, and the equipment grounding conductor must be at least No. 6 AWG. See *NEC Section 520.54(C)*. If single conductors are installed in parallel to increase ampacity, the conductors must be the same size and length. Single-conductor supply cables may be grouped together but not bundled together.

Equipment grounding and grounded (neutral) conductors must be marked with approved marking methods in compliance with *NEC Sections 200.6 and 250.119*. One approved method for marking neutral conductors that supply portable switchboards is a white or gray marking 6" (150 mm) from each end of the conductor (*Figure 17*). Similarly, an approved marking method for

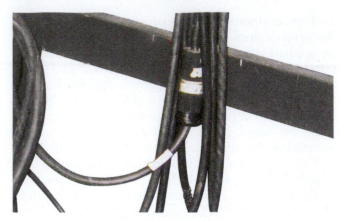

Figure 17 Properly marked grounded conductor.

equipment grounding conductors is green or green with yellow stripes 6" (150 mm) from each end of the conductor.

> **CAUTION**
> As with any electrical system, if more than one nominal voltage exists within the same premises, each ungrounded conductor must be identified according to the system from which it originates.

Portable on-stage switchboard supply conductors may be reduced in size based on the length of conductor required between the voltage supply and switchboard or between the voltage supply and an intermediate overcurrent protective device.

For supply conductors not longer than 10' (3 m), all of the conditions in *NEC Section 520.54(D)* must be met in order for the conductors to be reduced in size, as follows:

- Supply conductors shall be at least one-fourth the current rating of the associated overcurrent protective device.
- Supply conductors shall terminate at a single overcurrent protective device that limits the load to no greater than the supply conductor ampacity.
- Supply conductors cannot penetrate floors, walls, or ceilings or pass through doors or other traffic areas.
- Supply conductor terminations must be made using an approved method.
- Supply conductors cannot contain splices or connectors.
- Supply conductors must not be bundled together.
- Supply conductors must be supported above the floor in an approved manner.

If the supply conductors are over 10' (3 m) in length but not longer than 20' (6 m), all of the conditions in *NEC Section 520.54(E)* must be met in order for the conductors to be reduced in size:

- Supply conductors must have an ampacity that is at least one-half the current rating of the associated overcurrent protective device.
- Supply conductors must terminate at a single overcurrent protective device that limits the load to no greater than the supply conductor ampacity.
- Supply conductors cannot penetrate floors, walls, or ceilings or pass through doors or other traffic areas.
- Supply conductor terminations must be made using an approved method.
- Supply conductors must be supported at least 7' (2.1 m) above the floor except at termination points.
- Supply conductors must not be bundled together.
- Tap conductors must not contain splices or connectors (other than the tap itself).

Supply conductors that are not reduced in size based on the listed conditions may pass through holes in walls per *NEC Section 520.54(F)*. If the wall is fire-rated, the penetration must be in accordance with the rules of *NEC Section 300.21*.

All supply conductors must be protected against physical damage. See *NEC Section 520.54(G)*. When supply cables pass through enclosures, a protective bushing must be installed to protect the cable. In addition, the arrangement of the cables through and within the enclosure should be such that no tension is applied at the point of termination or connection.

For long lengths of supply conductors that require interconnecting connectors, *NEC Section 520.54(H)* states that no more than three mating connectors may be used where the total length of conductor from supply to switchboard does not exceed 100' (30 m). If the conductor exceeds 100' (30 m) in length, one additional mating connector is permitted for each additional 100' (30 m) of conductor.

Single-pole portable cable connectors must be of the locking type and, if installed in parallel as input devices, must be labeled with a warning that parallel connections are present. See *NEC Section 520.53(C)*. At least one of the following conditions must exist to permit the usage of single-pole separable connectors:

- Connectors can be connected or disconnected only if the supply is de-energized and the connectors are interlocked at the source.

- If sequential-interlocking line connectors are used, load connectors must be connected in the following sequence: equipment grounding conductor first, grounded conductor second, and ungrounded conductor(s) last, or a caution tag must be attached adjacent to the line connectors that lists the sequence of connection. Disconnection is in the reverse order.

The supply neutral terminal and busbar within a portable switchboard designed to operate from a three-phase, four-wire with ground supply must have a rated ampacity that is at least twice the ampacity of the largest ungrounded supply conductor terminal within the switchboard. See *NEC Section 520.53(B)*.

Neutral conductors supplying portable switchboards that incorporate solid-state phase-control dimmers are considered current-carrying conductors and must be sized accordingly. See *NEC Section 520.54(J)(2)*. However, neutral conductors supplying portable switchboards that incorporate solid-state sine-wave dimmers are not considered current-carrying conductors. When single conductors not installed in raceways are used as feeders for multiphase circuits powering portable switchboards containing solid-state phase-control dimmers, the neutral conductor must have a rated ampacity that is at least 130% of the ungrounded conductors. In cases in which the switchboard contains sine-wave dimmers, the neutral ampacity is required to be only at least 100% of the ungrounded conductors.

Only qualified personnel may install portable supply conductors, connect and disconnect connectors, or energize and de-energize supply circuits per *NEC Section 520.54(K)*. Portable switchboards must visibly display a label in a conspicuous place indicating this requirement. An exception to this rule states that a portable switchboard may be connected to a permanently installed supply receptacle by an unqualified person if the receptacle is protected by an overcurrent protective device that is not greater than 150A and if the switchboard complies with all of the following requirements:

- Listed multipole connectors are used for each supply interconnection.
- The general public may not have access to supply connections.
- Multiconductor cords and cables are listed for extra-hard usage, with a rated ampacity equal to or greater than the load and not less than the ampacity of the connectors used. *Figure 18* shows a small portable switchboard supplied by a multiconductor cord.

Portable Switchboard Construction, Circuits, Interior Wiring, and Devices – Portable switchboards must be installed within an enclosure that offers protection to the equipment and the general public. See *NEC Sections 520.53(A) through (E)*. There can be no exposed energized parts within the enclosure, and all switches and circuit breakers must be enclosed and externally operated. Each ungrounded conductor in every circuit supplied by the switchboard must have overcurrent protection. Dimmer terminals must not be exposed, and faceplates must be installed to prevent accidental contact. The short circuit current rating must be marked on the switchboard.

Portable on-stage switchboards must be equipped with a pilot light that is connected to the supply circuit of the switchboard. The power to the pilot light must not be interrupted by the opening of the master switch, and power must be supplied by an individual branch circuit rated at no more than 15A. Individual portable switchboards shall contain overcurrent protection for branch circuits (*NEC Section 520.52*).

Conductors within the switchboard enclosure (except busbars) must be stranded. Minimum insulation temperature ratings for dimmer conductors are dependent on the type of dimmer and must be at least equal to the operating temperature of the dimmer. Conductors connected to resistance-type dimmers must have a minimum insulation temperature rating of 200°C (392°F), while conductors connected to reactor-type, autotransformer, and solid-state dimmers must be rated for a minimum of 125°C (257°F).

Dressing Rooms – Pendant-type lampholders are not permitted in dressing or makeup rooms per *NEC Section 520.71*. Incandescent lamps installed less than 8' (2.5 m) from the floor must be equipped with open-ended guards that are riveted to the outlet box cover or locked in place in some other fashion. See *NEC Section 520.72*.

Wall switches must be installed in the dressing rooms to control lights or receptacles adjacent to mirrors and above the dressing table countertops (*NEC Section 520.73*). These switches must be equipped with a pilot light located adjacent to the door outside the dressing room to indicate when the receptacles are energized (*Figure 19*). Other general-use receptacles in dressing rooms do not require switch control.

1.3.0 Carnivals, Circuses, Fairs, and Similar Events

Carnivals, circuses, and fairs present several unique wiring and safety concerns. These concerns include large crowds, high power demands

Figure 18 Portable switchboard (dimmer) supplied by a multiconductor cord.

Figure 19 Corridor pilot lights for a dressing room.

for operating rides and concessions, outdoor installations, unpredictable animal behavior, and a need to rapidly set up and tear down equipment. *NEC Article 525* lists the requirements for portable wiring associated with equipment and portable structures used in carnivals (*Figure 20*), circuses, fairs, and similar functions. The wiring in permanent structures, such as audience areas and assembly halls, located within these activities are covered by *NEC Articles 518 and 520*. The wiring in permanent amusement attractions is covered by *NEC Article 522*.

1.3.1 Overhead Conductor Clearances

Tents and concessions must maintain vertical clearances from overhead conductors in accordance with *NEC Section 225.18*. For portable structures such as the one shown in *Figure 21*, a clearance of 15' (4.5 m) must be maintained in any direction from overhead conductors operating at

Figure 20 Carnival.

Figure 21 Tents.

600V or less, unless the conductors are supplying power to the portable structure per *NEC Section 525.5(B)(1)*. A portable structure may not be located under or within 15' (4.5 m) of conductors that have an energized voltage greater than 600V per *NEC Section 525.5(B)(2)*.

1.3.2 Power Sources

Areas that house service equipment must be lockable or located to prevent access by unqualified persons. The equipment must be mounted on a surface with a solid backing and protected from the weather or constructed with weather-proof enclosures and components. Refer to *NEC Section 525.10(B)*.

Portable structures at carnivals, circuses, or fairs often receive their power from separately derived systems or different services (*Figure 22*). In these situations, if the structures are less than 12' (3.7 m) apart, all equipment grounding conductors must be bonded together at portable

Figure 22 Separately derived system (generator) powering portable structures and rides.

structures per *NEC Section 525.11*. The bonding conductor must be sized in accordance with *NEC Table 250.122*, based on the largest overcurrent device supplying the portable structures and no smaller than No. 6 AWG.

1.3.3 Wiring: Cords, Cables, and Connectors

As stated in *NEC Section 525.20(A)*, flexible cords and cables exposed to potential physical damage—common for fairs and similar events—must be listed for extra-hard usage. Cords and cables not exposed to possible physical damage may be listed for hard usage. If the cords or cables are to be used outdoors, they must also be listed for wet locations and be sunlight resistant. Permanent wiring for portable amusement rides and other attractions, as opposed to temporary wiring, may be extra-hard usage flexible cords and cables as long as the cords or cables are not exposed to possible physical damage (*Figure 23*).

Figure 23 Permanent wiring on amusement rides.

Single-conductor cables must be a minimum of No. 2 AWG in size per *NEC Section 525.20(B)*. *NEC Section 525.20(C)* permits the use of open conductors only in festoon lighting, which is regulated by *NEC Article 225*. Flexible cords or cables may not contain any splices or taps between boxes or fittings.

Cord and cable connectors may not be laid on the ground unless the connector is rated for wet locations. Even then, connectors must not be located in any audience traffic area or other areas (unless guarded) that are accessible to the general public. Cords or cables may not be draped or hung on rides or similar structures unless designed for the purpose. If the cable or cord presents a tripping hazard, it may be covered with nonconductive matting as long as the matting does not present an additional tripping hazard. Cables or cords may also be buried to eliminate the tripping hazard.

1.3.4 Wiring: Rides, Tents, and Concessions

Per *NEC Section 525.21(A)*, all rides and similar portable structures must be equipped with a disconnect switch that is within 6' (1.8 m) of the operator and within the operator's sight. The disconnecting means must be readily available to the operator at all times, even when the ride is running (*Figure 24*). If the disconnecting means is accessible to the general public, it must also be lockable. A shunt trip in the operator's console may be used to open the fused disconnect or circuit breaker.

Lamps that are part of portable lighting inside tents and concessions must be guarded per *NEC Section 525.21(B)*. When outlet boxes or other device boxes are installed outdoors, the boxes must be weather-proof and mounted so the bottom of the box is at least 6" (150 mm) above the ground (*Figure 25*). See *NEC Section 525.22(A)*.

Figure 24 Disconnect switch for amusement ride.

NEC Section 525.23(A) states that ground-fault circuit interrupter (GFCI) protection is required on all 125V, single-phase, 15A and 20A non-locking-type receptacles that are used to assemble or disassemble rides, amusements, or concessions, or when those receptacles are exposed to the general public. Any equipment supplied by 125V, single-phase, 15A or 20A circuits and exposed to the general public must also be GFCI-protected. The GFCI protection may be an integral part of the attachment plug or installed directly in the power supply cord within 12" (300 mm) of the attachment plug. Locking-type receptacles are not required to be GFCI-protected per *NEC Section 525.23(B)*. Egress lighting shall not be GFCI protected per *NEC Section 525.23(C)*. Listed cord sets incorporating GFCI protection for personnel are permitted per *NEC Section 525.23(D)*.

1.3.5 Grounding and Bonding

All metal raceways, metal-sheathed cable, metal enclosures of electrical equipment, metal frames and metal parts of portable structures, trailers, trucks, or any other equipment that contains or supports electrical equipment connected to the same electrical source must be bonded together and connected to an equipment grounding conductor. See *NEC Section 525.30*. The equipment grounding conductor may serve as the bonding means.

The equipment grounding conductor must be connected to the system's grounded conductor at the service disconnect. In a separately derived system such as a generator, the equipment grounding conductor must be connected at the generator or in the first disconnecting means supplied by the generator (*Figure 26*). Refer to *NEC Section 525.31*.

> **CAUTION**
>
> Never connect the grounded circuit conductor to the equipment grounding conductor anywhere on the load side of the service disconnect or generator disconnect.

1.4.0 Agricultural Buildings

NEC Article 547 covers agricultural buildings in which excessive dust and dust with water may accumulate, including all areas of poultry, livestock, and fish confinement systems. It also contains provisions for agricultural buildings that contain a corrosive atmosphere that may be caused by poultry or animal excrement, corrosive particles mixed with water, and wet or damp corrosive areas caused by cleaning and sanitizing solutions.

Figure 25 Outlet box for concessions.

Figure 26 First means of disconnect supplied downstream of a carnival generator.

1.4.1 Wiring Methods

Wiring systems in agricultural buildings may contain jacketed MC cable, NMC (nonmetallic sheathed, corrosion resistant) cable, copper SE (service entrance) cable, UF (underground feeder) cable, rigid nonmetallic conduit, liquid-tight flexible nonmetallic conduit, and any other cable or raceway approved and suitable for the location, with approved fittings. See *NEC Section 547.5(A)*. Per *NEC Section 547.5(B)*, cables must be secured within 8" (200 mm) of any box, cabinet, or fitting to which they connect. If nonmetallic boxes are mounted directly to an agricultural building surface, the ¼" (6 mm) airspace requirement of *NEC Section 300.6(D)* does not apply.

Enclosures, boxes, conduit bodies, and fittings installed in dusty areas of agricultural buildings must be designed to minimize the entrance of dust, with no screw or mounting holes. In damp

What's wrong with this picture?

or wet locations, these enclosures, boxes, conduit bodies, and fittings must be listed for wet locations or mounted to prevent the entrance of moisture or water (*Figure 27*). Wiring methods used in corrosive atmospheres must provide corrosion resistance. Aluminum and magnetic ferrous (carbon steel) materials are not suitable for corrosive atmospheres.

> **NOTE**
>
> *NEC Table 110.28* lists properties for various types of electrical enclosures, including those that offer corrosion resistance.

Where flexible connections are required, such as supplying power to motors and other vibrating or moving equipment, *NEC Section 547.5(D)* requires the use of liquid-tight flexible metal or liquid-tight flexible nonmetallic conduit with dust-tight connectors, or flexible cord listed for hard usage.

Grounding conductors installed underground in agricultural buildings must be insulated. See *NEC Section 547.5(F)*.

Per *NEC Section 547.5(G)*, ground-fault protection must be provided for all 125V, single-phase, 15A and 20A receptacles installed in areas having an **equipotential plane**, outdoors, in damp or wet locations, and in dirt confinement areas for livestock.

Figure 27 Device box listed for wet location in agricultural building.

1.4.2 Motors and Luminaires

NEC Section 547.7 requires that all motors and rotating electrical machinery be totally enclosed or designed to reduce the entry of moisture, dust, or corrosive materials (*Figure 28*). Luminaires must also be designed to minimize the entry of dust, moisture, or corrosive particles. If luminaires are exposed to physical damage, guards must be installed. Luminaires exposed to water must be listed as suitable for use in wet locations. Refer to *NEC Sections 547.8(A) through (C)*.

1.4.3 Electrical Supply from a Single Distribution Point

NEC Section 547.9 states that any agricultural buildings or structures located on the same premises shall be electrically supplied from a distribution point (*Figure 29*). If two or more buildings are supplied overhead from a distribution point, a **site-isolating device** must be installed and the installation must qualify for exceptions or comply with the requirements listed in *NEC Sections 547.9(A)(1) through (10)*, summarized as follows:

- Pole-mounted
- Simultaneously open all ungrounded service conductors from the premises wiring

Figure 28 Totally enclosed motor.

Figure 29 Agricultural building distribution point.

- Mounted in an enclosure bonded to the grounded and grounding conductors
- System grounded conductors bonded to a grounding electrode system by means of a grounding electrode conductor at the site-isolating device
- Rated for the calculated load
- Not required to provide overcurrent protection
- If not readily accessible, capable of being remotely operated by a readily accessible operating handle not more than 6'7" (2 m) above grade at its highest position
- Permanently marked on or adjacent to the operating handle, identifying it as a site-isolating device

> **NOTE**
> If the electric utility provides a site-isolating device as part of their installation, no additional site-isolating device is required.

Service-disconnecting means and overcurrent protection for two or more buildings supplied from a distribution point may be located at the buildings or at the distribution point. If

Fans in Poultry Buildings

This chicken farm has an elaborate system of auxiliary power supplies in the form of generators and transfer switches. Failure of a livestock ventilation system can result in death by asphyxiation from a lack of oxygen and increased carbon dioxide, heat prostration, or poisoning from the buildup of gases such as ammonia and methane. The farmer who owned this building said that he would lose most of his chickens if the ventilator fans stopped operating for as little as 20 minutes.

located at the buildings, supply conductor sizes must comply with *NEC Article 220, Part V* and conductor installation must follow the requirements of *NEC Article 225, Part II*. Grounding and bonding of the supply conductors must be in accordance with *NEC Section 250.32*. If the equipment grounding conductor and the largest supply conductor are constructed from the same material, they must be the same size; if of different materials, the equipment grounding conductor must be adjusted according to *NEC Table 250.122*, as specified in *NEC Section 547.9(B)(3) (1)*.

When the service disconnecting means and overcurrent protection for two or more buildings are located at the distribution point, supply conductors to the buildings must meet the requirements of *NEC Section 250.32* and *NEC Article 225, Parts I and II*.

If more than one service supplies an agricultural site, each distribution point must have a permanent plaque or directory installed to indicate the location of each of the other distribution points and the buildings or structures served by each [*NEC Section 547.9(D)*].

1.4.4 Equipotential Planes

NEC Section 547.10(A)(1) requires that equipotential planes be installed in indoor and outdoor animal confinement areas having concrete floors or slabs with metallic equipment that may become energized and accessible to animals [*NEC Section 547.10(A)(2)*]. Refer to *Figure 30*. The plane must cover the area in which the animals may stand while accessing the metallic equipment.

Special Equipment— Electrified Truck Parking Spaces

NEC Article 626 covers the electrical conductors and equipment used to supply commercial power to tractor trailers, such as those with refrigeration units. A typical electrified truck parking lot is shown here.

Equipotential planes must be bonded to the grounding system with a bare or insulated solid copper conductor no smaller than No. 8 AWG [*NEC Section 547.10(B)*]. Bonding connections to wire mesh or other conductor elements may be made using approved pressure connectors or clamps of brass, copper, or copper alloy.

Figure 30 Concrete area with accessible metal piping and animal contact.

Figure 31 Temporary wiring on construction site.

1.5.0 Temporary Installations

Temporary wiring is covered in *NEC Article 590*. It includes wiring used to supply temporary power and lighting to construction sites (*Figure 31*) and for remodeling, maintenance, repair, demolition, holiday decorative lighting, electrical tests, emergencies, experiments, and developmental work. Temporary wiring, excluding holiday decorative lighting (*Figure 32*), must be removed immediately upon completion of the temporary need. Holiday lighting may remain up no longer than 90 days.

1.5.1 Feeder and Branch Circuit Conductors

Feeder and branch circuit conductors used in temporary wiring may be multiconductor cables that are identified for hard usage or extra-hard usage according to *NEC Table 400.4* and must originate from approved panelboards or power outlets. Per *NEC Sections 590.4(B)(1) and 590.4(C)(1)*, Type NM, Type NMC, and Type SE cable may be used for temporary wiring in any dwelling or other structure without consideration to height limitations or concealment within walls, floors, or ceilings. Single insulated feeder and branch circuit conductors may be used to supply power for temporary testing, emergencies, experiments, and developmental work where accessible to qualified persons only. Holiday decorative lighting may

Figure 32 Holiday lighting.

also be supplied by single insulated branch circuit conductors as long as the voltage to ground does not exceed 150V, the wiring is not subjected to physical damage, and the conductors are supported on insulators no more than 10' (3 m) apart. If the lighting is hung in festoon (swag) fashion, the conductors must be arranged so that no conductor strain is felt at the lampholder.

1.5.2 Grounding

Temporary branch circuits must include a separate equipment grounding conductor unless the branch circuit conductors are installed in a continuous metal raceway that meets the requirements of an equipment grounding conductor. Refer to *NEC Section 590.4(D)(1)*. Metal-clad cables that qualify as equipment grounding conductors may also be used.

1.5.3 Disconnects

Disconnecting switches or plug connectors must be installed to disconnect all ungrounded conductors supplying temporary circuits. See *NEC Section 590.4(E)*. Multiwire branch circuits must be provided with a disconnecting means that simultaneously disconnects all ungrounded conductors at the power outlet or panel from which they originate. Only identified handle ties may be used.

1.5.4 Protection and Support

All cables and conductors must be protected from pinch points, sharp corners, and other potentially damaging conditions. In addition, proper fittings must be installed where flexible cords and cables enter boxes. See *NEC Sections 590.4(H) and (I)*. Cable assemblies must also be adequately supported with staples, cable ties, straps, or other approved means at intervals to provide protection from physical damage [*NEC Section 590.4(J)*]. Vegetation may not be used to support overhead runs of branch circuit or feeder conductors, except for holiday lighting. In the latter case, the conductors or cables must be arranged and supported with strain relief or other devices to avoid conductor damage from the movement of limbs and branches.

1.5.5 Receptacles

Receptacles in temporary wiring must be of the grounding type and must be electrically connected to the equipment grounding conductor [*NEC Section 590.4(D)*].

> **WARNING!**
>
> Receptacles that supply temporary power on construction sites may not be installed on any branch circuit that supplies temporary lighting, nor may they be connected to the same ungrounded conductor of multiwire circuits supplying temporary lighting.

1.5.6 Temporary Lighting

General illumination luminaires must be designed to protect the lamp from contact or breakage per *NEC Section 590.4(F)*. All exposed lamps must be equipped with a lamp guard (*Figure 33*). Brass lamp shells, paper-lined sockets, or other metal-cased sockets may not be installed unless the lamp shell is grounded. All holiday or other decorative lighting must be listed per *NEC Section 590.5*.

1.5.7 Wiring and Equipment Greater Than 600V

If any wiring or equipment operates at greater than 600V at any temporary location, it must be guarded by a fence, barrier, or other effective means and be accessible only to authorized and qualified personnel familiar with the equipment or wiring. See *NEC Section 590.7*.

1.5.8 Ground-Fault Protection

The *NEC®* applies specific rules to power derived from an electric utility company or from an on-site generator (*NEC Section 590.6*). Ground-fault protection must be provided on all 125V, single-phase, 15A, 20A, or 30A temporary receptacles that are not part of the permanent wiring (*Figure 34*), unless temporary power is being supplied by the permanent wiring. If a receptacle that is part of the permanent wiring is used for temporary electric power, it, too, must be GFCI-protected.

Figure 33 Lamp guard in place on temporary lighting at construction site.

GFCI protection may also be incorporated into a cord set or device to comply with this rule.

The only exception to the GFCI rule is in industrial locations where a greater hazard would be introduced if power were interrupted or the system design is not compatible with GFCI protection [*NEC Section 590.6(A), Exception*]. In these locations, GFCI protection is not required as long as only qualified personnel are involved and an assured equipment-grounding conductor program is in place. However, if GFCI or special-purpose ground-fault circuit interrupter (SPGFCI) protection becomes available and is the appropriate voltage and amperage for the application, it must be used.

1.5.9 Assured Equipment-Grounding Conductor Program

If GFCI protection is not supplied, there must be a written assured equipment-grounding conductor program in place at each site requiring temporary wiring. See *NEC Section 590.6(B)(2)*. The program must be enforced by at least one designated person to ensure that all cord sets, receptacles (temporary), and cord-and-plug equipment are equipped with equipment grounding conductors and maintained in accordance with applicable requirements of the *NEC®*. At a minimum, the assured equipment-grounding conductor program must contain the following:

- The equipment grounding conductors must be electrically continuous and tested for continuity.
- Each receptacle and plug must be tested for correct grounding conductor termination. These tests must be performed before first use, when there is evidence of damage, after repairs and before next usage, and at intervals not exceeding three months.
- All test results must be recorded and available for review.

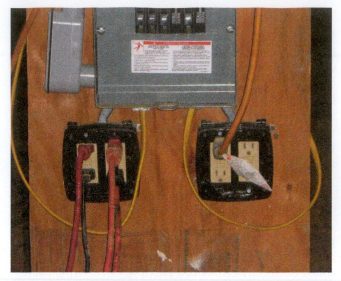

Figure 34 Receptacles on construction sites must be GFCI-protected.

1.6.0 Wired Partitions

Wired partitions allow for convenient rearrangement of office spaces. *NEC Article 605* provides the requirements for lighting accessories and wired office partitions. When installed in interconnecting office partitions, conductors and connections must be contained within wiring channels identified for such use (*NEC Section 605.4*). When two or more wired partitions are electrically connected together, the electrical connection between the partitions must be made with a flexible assembly specifically designed for such purposes (*Figure 35*) or with a flexible cord if all of the following conditions are met (*NEC Section 605.5*):

- The flexible cord must be rated for extra-hard usage, have No. 12 AWG or larger conductors, and contain an insulated grounding conductor.
- The partitions must be mechanically secured together (contiguous).
- The cord may be no longer than 2' (600 mm).

Figure 35 Flexible assembly connecting two wired partitions.

- The cord must be terminated at an attachment plug-and-cord connector equipped with a strain-relief device.

Per *NEC Section 605.7*, partitions that are fixed to the building surfaces must be permanently connected to the building's electrical system using the general wiring methods found in *NEC Chapter 3*. As with any multiwire branch circuit, a disconnecting means must be provided that simultaneously disconnects all ungrounded conductors.

Individual freestanding (not fixed) partitions, or groups of freestanding partitions that are mechanically secured together and do not exceed 30' (9 m) in length when connected, may be connected to the building's electrical system using a single flexible cord, provided all of the conditions in *NEC Sections 605.9(A) through (D)* are met:

- The flexible cord is rated for extra-hard usage, has No. 12 AWG or larger conductors, contains an insulated equipment grounding conductor, and is no longer than 2' (600 mm).
- The supply receptacle is on a separate circuit that serves only partitions and no other loads and is located no farther than 12" (300 mm) from the partition connected to it.
- Individual partitions or grouped partitions cannot contain more than thirteen 15A, 125V receptacles.
- Individual partitions or grouped partitions cannot contain multiwire circuits.

Figure 36 shows a group of freestanding partitions that are connected to the building's electrical system using a single flexible cord.

Figure 36 Group of freestanding office partitions connected using a single flexible cord.

1.0.0 Section Review

1. To accommodate fire-rated construction, wiring methods allowed in assembly occupancies include MI cable, MC cable, AC cable, metal raceways, flexible metal raceways, and nonmetallic raceways encased in _____.

   a. at least 18" (450 mm) of sand
   b. a metal cable tray
   c. at least 2" (50 mm) of concrete
   d. at least 6" (150 mm) of concrete

2. Wiring for portable equipment such as portable switchboards and stage effects in theaters and audience areas may be in the form of approved flexible cords and cables as long as they are _____.

   a. not fastened or secured by uninsulated staples or nails
   b. brightly colored for visibility
   c. contained within a metal or nonmetallic raceway
   d. rated for extra-hard usage

3. The wiring in permanent amusement attractions is covered by _____.

   a. *NEC Article 505*
   b. *NEC Article 511*
   c. *NEC Article 518*
   d. *NEC Article 522*

4. All 125V, single-phase, 15A and 20A receptacles mounted in an agricultural building having an equipotential plane are required to have _____.

   a. twist-load devices
   b. arc-fault protection
   c. tamper-resistant covers
   d. ground-fault protection

5. In temporary installations, which item or material below *cannot* be used for the support of overhead runs of branch circuit or feeder conductors unless it is holiday lighting?

   a. Vegetation
   b. Wooden structures
   c. Metal structures
   d. Concrete

6. When two or more wired partitions are electrically connected together, the connection may be made using a flexible cord as long as specific conditions are met, including that the cord is no longer than _____.

   a. 2' (600 mm) in length
   b. 4' (1.2 m) in length
   c. 6' (1.8 m) in length
   d. 8' (2.4 m) in length

2.0.0 MARINAS, BOATYARDS, AND BODIES OF WATER

Objective

Identify and select equipment, components, and wiring methods for marinas, boatyards, and bodies of water.

a. Identify and select equipment, components, and wiring methods for marinas and boatyards.
b. Identify and select equipment, components, and wiring methods for natural and man-made bodies of water.

Trade Terms

Electrical datum plane: Horizontal plane affecting electrical installations based on the rise and fall of water levels in land areas and floating piers and landing stages.

Landing stage: A floating platform used for the loading and unloading of passengers and cargo from boats or ships.

Figure 37 Marina with floating docks.

The permanent presence of water, for obvious reasons, creates additional electrical hazards that must be addressed using specific wiring components and techniques to achieve a safe installation. Marinas and boatyards in particular require significant attention to detail, as they often require power to be provided up to the water's edge and beyond.

2.1.0 Marinas and Boatyards

NEC Article 555 covers all commercial and dwelling unit docking facilities, boatyards, boat basins, boathouses, and yacht clubs used for the purpose of repairing, berthing, launching, storing, or fueling boats and/or the moorage of floating buildings (*Figure 37*).

2.1.1 General Requirements for Devices, Equipment, and Enclosures

A marine power outlet is any receptacle, circuit breaker, fused switch, meter, and any other similar device approved for marine use. Electrical distribution systems within boatyards and piers

(*Figure 38*) cannot exceed 250V phase-to-phase per *NEC Section 555.5*. Where qualified personnel service the equipment under engineering supervision, the system voltage is permitted to be up to 600V per *NEC Section 555.5*.

Ground-fault protection requirements are listed in *NEC Section 555.35(A)*. When mounting transformers in areas covered by *NEC Article 555*, the bottoms of transformers must be above the electrical datum plane as defined in *NEC Section 555.3*. Electrical service equipment cannot be installed on or in a floating structure but instead must be mounted adjacent to the structure per *NEC Section 555.4*.

Electrical connections within boxes or enclosures that are not rated for submersion must be located at least 12" (305 mm) above the deck of a floating pier as stated in *NEC Section 555.30* and shown in *Figure 39*. If the junction box is rated for submersion, it may be located below the electrical datum plane for floating piers but must be above

Figure 38 Electrical distribution system at a marina.

the water line. Any electrical connection on a fixed pier must be at least 12" (305 mm) above the deck of the fixed pier but never below the electrical datum plane.

> **NOTE**
>
> Metric conversions in the *NEC®* may vary depending on the type of installation and are often more specific in areas of increased hazard, such as near water.

Conduit may not be the sole support for electrical enclosures installed on piers above deck level. External box ears or lugs must be used to support these enclosures per *NEC Section 555.31(A)*. If internal screws are used to secure boxes or enclosures to the pier structure, the screw heads must be sealed to prevent water from entering through the mounting holes. The location of electrical equipment, boxes, or enclosures must not be in the way of mooring lines per *NEC Section 555.31(B)*.

Floating Buildings

Wiring methods for buildings that float on water, are permanently moored, and electrically supplied by a system not located on the premises, are covered by *NEC Article 555, Part III*.

NEC Section 555.32 states that a person must be able to operate circuit breakers or switches that have been installed in water-tight enclosures without removing the enclosure cover (*Figure 40*). However, water-tight enclosures housing such devices must be designed with a weep hole to discharge condensation.

2.1.2 Service and Feeder Conductor Load Calculations

All general lighting and receptacle loads for marinas and boatyards must be calculated using the standard general lighting load calculation procedures of *NEC Article 220, Part III*. When service or feeder circuit conductors are used to supply a number of receptacles for shore power to boats, the demand factors shown in *Table 3* may be applied to determine the circuit rating and conductor sizes.

Other rules may apply to shore power installations and are listed in the notes following *NEC Table 555.6*. One such rule is that if two separate receptacles having different voltages and amperage ratings are installed in a single boat slip, only the receptacle with the greater volt-ampere (VA) demand needs to be included in the total load calculation. For example, if one receptacle is rated 30A at 125V (3,750VA) and another is rated 50A at 250V (12,500VA) within the same slip, the total load calculation for these two receptacles would be based on the higher 12,500VA only.

Consideration must be given to ambient temperatures and types of equipment when applying demand factors. Extremely hot or cold temperatures may result in increased demand factors for circuits with loaded heating, air conditioning, or refrigeration equipment.

Figure 39 Electrical connections within a box not rated for submersion must be mounted at least 12" (305 mm) above the deck.

Figure 40 Accessible circuit breakers in water-tight enclosure.

Table 3 Demand Factors (Data from *NEC Table 555.6*)

Number of Shore Power Receptacles	Factor Applied to the Sum of the Rating of the Receptacles
1–4	100%
5–8	90%
9–14	80%
15–30	70%
31–40	60%
41–50	50%
51–70	40%
71 and up	30%

Reprinted with permission from NFPA 70®-2020, *National Electrical Code®*, Copyright © 2019, National Fire Protection Association, Quincy, MA. This reprinted material is not the complete and official position of the NFPA on the referenced subject, which is represented only by the standard in its entirety which may be obtained through the NFPA website at **www.nfpa.org**.

2.1.3 Wiring Methods

The standard wiring methods in *NEC Chapter 3* apply to locations covered by *NEC Article 555* as long as the methods are rated for wet locations. When portable power cables are used as permanent wiring on the underside of fixed or floating piers, or where flexibility is required on the floating sections of piers (*Figure 41*), the cable must meet the following criteria found in *NEC Sections 555.34(A)(2) and (B)(3)*:

- Rated for extra-hard usage
- Insulation rated for a minimum of 75°C (167°F) and 600V
- Listed for wet locations
- Sunlight resistant

- Manufactured with an outer jacket resistant to temperature extremes, oil, gasoline, ozone, abrasions, acids, and chemicals
- Properly supported by nonmetallic clips to pier structural members other than deck planking
- Located on the underside of piers where not subject to physical damage
- Protected by nonmetallic sleeves where penetrating structural members

In installations where cables are used to supply power to floating sections of a pier, either a listed marine power outlet with terminal blocks/bars or an approved, corrosion-resistant junction box containing terminal blocks must be installed on each pier section for conductor terminations. Any exposed metal boxes, covers, screws, and the like must be corrosion-resistant. See *NEC Section 555.34(B)(3)(b)*.

> **WARNING!**
> Temporary wiring such as extension cords cannot be used to supply power to boats, piers, or other locations covered by *NEC Article 555*, except as permitted in *NEC Article 590*. See *NEC Section 555.34(A)(3)*.

Overhead wiring in and around boatyards and marinas must be installed in a manner that avoids contact with masts and other boat parts, as well as portable lifts (*Figure 42*) moving about the yard. *NEC Section 555.34(B)(2)* requires a minimum overhead clearance of 18' (5.49 m). In addition, all electrical conductors and cables must maintain a minimum of 20' (6 m) from the outer perimeter or any section of the boatyard that may be used to move vessels or for stepping or

Figure 41 Cable may be used on floating sections of piers.

Figure 42 Masts and lifts must clear overhead wiring.

unstepping (raising or lowering) boat masts per *NEC Section 555.34(B)(1)*.

Per *NEC Section 555.34(B)(4)*, rigid metal, reinforced thermosetting resin conduit, or nonmetallic conduit must be installed above the deck of a pier and landing stage (*Figure 43*) and below the enclosure that the landing stage serves. Full-thread connections must be used to connect the conduit to the enclosure. When nonmetallic rigid conduit (PVC) is installed in these locations, nonmetallic fittings that provide a threaded connection must be used to connect the conduit to the enclosure. All installations must be approved for damp or wet locations.

2.1.4 Grounding

All metal boxes, metal cabinets, metal enclosures, metal frames of utilization equipment, and grounding terminals of grounding-type receptacles in locations covered by *NEC Article 555* must be connected to an equipment grounding conductor.

NEC Sections 555.37(B) and (C) require that the equipment grounding conductor must be copper, insulated, no smaller than No. 12 AWG, and green or green with one or more yellow stripes, unless the equipment grounding conductor is larger than No. 6 AWG. In the latter case, the equipment grounding conductor may be marked by other approved methods as described in *NEC Section 250.119*. For instances in which MI cable is used, the equipment grounding conductor may be identified with an approved means at each termination.

Branch circuit equipment-grounding conductors must terminate at a grounding terminal in a remote panelboard or on the grounding terminal in the main service equipment per *NEC Section 555.37(D)*. An insulated equipment-grounding conductor must interconnect the grounding

terminal in a remote panelboard to the grounding terminal in the service equipment. See *NEC Section 555.37(E)*.

2.1.5 Disconnecting Means for Shore Power

NEC Section 555.36 requires that a disconnecting means be provided to disconnect each individual shore power receptacle from the electrical supply. The disconnecting means may be a circuit breaker or switch that opens all ungrounded supply circuit conductors, and all disconnecting means must be properly marked to identify which receptacle they control. The disconnecting means must be located within 30" (762 mm) of the receptacle. Receptacles must be mounted at least 12" (305 mm) above the deck surface of the pier and not below the electrical datum plane on a fixed pier (*NEC Section 555.33*). *Figure 44* shows disconnects for receptacles in the form of circuit breakers.

2.1.6 Receptacles

The two types of receptacles in a marina are shore power receptacles and non-shore power receptacles. *NEC Section 555.33(A)(1)* states that shore power receptacles shall be listed as marina power outlets, listed for set locations or installed in en-

Figure 43 Landing stage where people access or exit boats.

Figure 44 Water-tight shore power enclosure with accessible circuit breakers.

closures that provide protection from the weather. Strain relief must be provided for the cable when it is plugged into the receptacle. Each shore power receptacle must be protected by an individual branch circuit that corresponds to the rating of the receptacle, which can be rated at no less than 30A. Shore power receptacles rated 30A and 50A must be of the locking and grounding type, while receptacles rated 60A or higher must be of the pin and sleeve type. See *NEC Section 555.33(A)(4)*.

GFCI protection must be provided on non-shore power receptacles, including all 15A and 20A, single-phase, 125V receptacles (*Figure 45*). See *NEC Section 555.33(B)(1)*.

Any non-shore power receptacle installed in a marine power outlet enclosure must be marked to indicate that the receptacle is not to be used as shore power for boats. See *NEC Section 555.33(B)(2)*.

2.1.7 *Hazardous Locations in and Around Marinas and Boatyards*

Marina or boatyard wiring that is located near fuel dispensing areas must comply with *NEC Article 514*. In general, all power and lighting wiring in marinas and boatyards must be located on the opposite side from fueling systems and piping on wharfs, piers, or docks.

NEC Tables 514.3(B)(1) and 514.3(B)(2) classify areas in and around fuel dispensing locations as either Class I, Division 1 or Class I, Division 2. When applying these tables to marina or boatyard fuel dispensing areas, you must first determine whether the area is considered closed construction or open construction.

Closed construction-type docks, piers, or wharfs are those areas in which no space exists between the bottom of the dock, pier, or wharf and the waterline. The class and division designations for closed construction are as follows:

- *Class I, Division 1* – Voids, pits, or similar areas below closed construction-type floating docks, piers, or wharfs that support fuel dispensers in which flammable liquid or vapor can accumulate. *Figure 46* shows a fuel dispenser mounted on a closed construction pier. Areas directly below the fuel dispenser are voids and are rated Class I, Division 1.

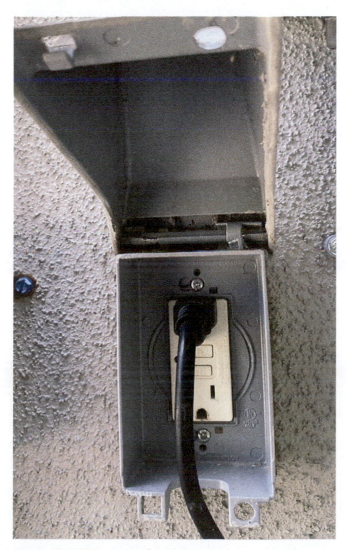

Figure 45 GFCI-protected non-shore power receptacle.

Figure 46 Class I, Division 1 area below fuel dispenser in the void between the pier and the water.

- *Class I, Division 2* – Spaces above the surface of closed-construction floating docks, piers, or wharfs that are within 18" (450 mm) horizontally in all directions extending to grade from the dispenser enclosure. Also sections of a dock, pier, or wharf that do not support fuel dispensers but directly abut and are 20' (6 m) or more from sections that do support fuel dispensers, with air spaces between them to allow for vapor dissipation. Those sections of pier that abut the fuel dispensing pier are classified as Class I, Division 2, as illustrated in *Figure 47*.

Open construction-type docks, piers, or wharfs are built on stringers that are supported by pilings, floats, pontoons, or similar devices in which space exists between the bottom of the dock, pier, or wharf and the waterline. The class and division designations for open construction are as follows:

- *Class I, Division 1* – Voids, pits, and similar areas within 20' (6 m) of the dispenser where flammable liquids or vapors may accumulate.
- *Class I, Division 2* – Areas 18" (450 mm) above the surface of open construction-type fuel dispensing docks, piers, or wharfs, extending 20' (6 m) horizontally in all directions from the outside edge of the dispenser, and down to the water level.

2.2.0 Natural and Manmade Bodies of Water

Natural bodies of water include lakes, streams, ponds, and rivers, all of which may naturally vary in depth. *Figure 48* is an example of a natural body of water, while *Figure 49* illustrates a manmade or artificial body of water. Manmade bodies of water refer to aeration ponds, fish-farm ponds, storm

Figure 47 Docks or sections of pier that abut this fuel dispensing pier are classified as Class I, Division 2.

Figure 48 Natural body of water.

Figure 49 Manmade body of water.

retention basins, water treatment ponds, and irrigation channels. Electrical installations in and around natural and manmade bodies of water are covered in *NEC Article 682*.

2.2.1 Electrical Datum Plane

The electrical datum plane is an imaginary horizontal plane based on changing water levels. Electrical datum planes are referred to extensively in *NEC Articles 555 and 682* because both articles deal with changing water levels that can affect electrical installations in and around natural and artificial bodies of water as well as marinas and boatyards. The definition of an electrical datum plane is provided in *NEC Section 682.5* and varies based on the following conditions:

- Where land areas are affected by tidal fluctuations, the electrical datum plane is 2' (600 mm) above the highest high tide for the area.

- Where land areas are not affected by tidal fluctuations, the electrical datum plane is 2' (600 mm) above the highest water level for the area under normal conditions.
- In areas subject to flooding, the electrical datum plane is 2' (600 mm) above the prevailing high water mark determined by the authority having jurisdiction.
- For floating structures that are installed to rise and fall in response to the water level without lateral movement, the electrical datum plane is 30" (750 mm) above the point at which the floating structure touches the water and a minimum of 12" (300 mm) above the deck level.

2.2.2 Location of Equipment and Enclosures

NEC Section 682.10 requires that all electrical equipment and enclosures installed in or near natural and man-made bodies of water be listed and approved for the location. Equipment or enclosures not identified for submersible use must not be located below the applicable electrical datum plane.

Per *NEC Section 682.11*, service equipment and the service disconnecting means that supply floating structures and submersible equipment must be located on land at least 5' (1.5 m) from the shoreline, within sight of the structure served, identified as to which structure or equipment it serves, and installed to allow a minimum of 12" (300 mm) between live parts and the electrical datum plane. In situations where the water level rises to the electrical datum plane, the service equipment must be designed to disconnect the electrical service when this occurs.

What's wrong with this picture?

Electrical connections that are not identified for submersible use must be located at least 12" (300 mm) above the deck of a floating or fixed structure but never below the established electrical datum plane (*NEC Section 682.12*).

2.2.3 GFCI Protection, Grounding, and Bonding

Outlets supplied by branch circuits not exceeding 150V to ground and 60A, single-phase, must be GFCI-protected per *NEC Section 682.15(A)*. The GFCI protection device must be located at least 12" (300 mm) above the established electrical datum plane.

Equipment grounding conductors must meet the requirements listed in *NEC Sections 682.31(A) through (D)* and summarized as follows:

- Equipment grounding conductors must be insulated copper conductors no smaller than 12 AWG.
- Feeders supplying a remote panelboard or other distribution equipment must include an insulated grounding conductor that connects from the service grounding terminal to a grounding terminal or busbar in the remote panelboard.
- Branch circuit equipment-grounding conductors must terminate on the grounding bus in the remote panelboard or the grounding terminal in the main service panel.
- Cord-and-plug equipment must be grounded by means of an equipment grounding conductor in the cord and connected to a grounding-type attachment plug at the supply end.

Metal piping, tanks, metal parts in contact with the water, and non-current-carrying metal parts that could become energized must be bonded to the grounding terminal in the distribution equipment per *NEC Section 682.32*.

2.2.4 Equipotential Planes

An equipotential plane bonds together the grounding grids, structural steel, conductive equipment enclosures, and all other metal parts in order to reduce or eliminate any gradient (variable) voltage potential between system components, structures, enclosures, and even the water itself. Per *NEC Section 682.33(A)*, equipotential planes must be installed in areas adjacent to all outdoor service equipment with metal enclosures or disconnecting means that are used to supply or control equipment in or on the water and are accessible to personnel. The plane must encompass the immediate area around and directly below the equipment and must extend outward at least 36" (900 mm) in all directions from which a person would be able to stand and reach the equipment.

Equipotential planes must be bonded to the electrical grounding system using an insulated, covered, or bare solid-copper conductor no smaller than No. 8 AWG [*NEC Section 682.33(C)*]. Connections of the bonding conductor to the grounding system must be made by exothermic welding or using listed stainless steel, brass, copper, or copper alloy pressure connectors or clamps.

2.0.0 Section Review

1. The factor applied to the sum of the rating for 35 shore power receptacles is _____.

   a. 40%
   b. 60%
   c. 70%
   d. 100%

2. The equipotential planes around natural and manmade bodies of water must be bonded to the electrical grounding system using a conductor no smaller than No. _____.

   a. 4 AWG
   b. 6 AWG
   c. 8 AWG
   d. 10 AWG

3.0.0 POOLS, SPAS, TUBS, AND FOUNTAINS

Objective

Identify and select equipment, components, and wiring methods for pools, spas, tubs, and fountains.

   a. Identify general wiring requirements for pools, spas, tubs, and fountains.

   b. Identify and select equipment, components, and wiring methods for permanently installed pools.

   c. Identify and select equipment, components, and wiring methods for storable pools.

   d. Identify and select equipment, components, and wiring methods for spas, hot tubs, and therapeutic tubs.

   e. Identify and select equipment, components, and wiring methods for fountains.

Trade Terms

Positive locking device: A screw, clip, or other mechanical locking device that requires a tool or key to lock or unlock it.

Potting compound: An approved and listed insulating compound, such as epoxy, which is mixed and installed in and around electrical connections to encapsulate and seal the connections to prevent the entry of water or moisture.

Voltage gradients: Fluctuating voltage potentials that can exist between two surfaces or components and may be caused by improper or insufficient bonding or grounding.

Figure 50 Permanently installed pool.

NEC Article 680 applies to electrical wiring and equipment associated with the construction and installation of permanently installed swimming pools (*Figure 50*), storable pools, spas and hot tubs, fountains (*Figure 51*), therapeutic tubs and tanks, and hydromassage bathtubs. The sections that follow summarize the requirements for each type of pool and tub, beginning with some general requirements.

3.1.0 General Wiring Requirements

NEC Article 680 applies to pools, spas, tubs, and fountains.

3.1.1 Grounding Requirements

Because of the proximity to water, all electrical equipment in a pool area must be grounded. The following equipment and devices must be grounded per *NEC Section 680.6*:

- Underwater and through-wall luminaires, except low-voltage lighting listed for usage without a grounding conductor
- Electrical equipment located within 5' (1.5 m) of the inside wall of the body of water
- All associated water-circulating equipment
- Junction boxes
- Transformer and power supply enclosures
- Ground-fault circuit interrupters
- Panelboards not associated with the service equipment but supplying electrical equipment associated with the body of water

Figure 52 shows a panel located within the equipment room of a permanently installed swimming pool.

Per the definition of corrosive environment in *NEC Section 680.2*, areas where pool sanitation chemicals are stored, as well as areas with circulation pumps, automatic chlorinators, filters, open areas under pool decks, and similar locations are considered corrosive environments. All wiring in these areas must be listed and identified for use in these locations.

Figure 51 Fountain.

Figure 52 Panel in equipment room of swimming pool.

3.1.2 Cord-and-Plug Connections

Flexible cords with a maximum length of 3' (900 mm) may be used to connect fixed or stationary equipment (except underwater luminaires) to their power supply for easy removal or disconnection for maintenance or repair (*NEC Section 680.8*). The cord must contain a copper equipment-grounding conductor no smaller than No. 12 AWG that terminates on the supply end to a grounding-type attachment plug and is connected to a fixed metal part of the equipment on the other end. All components of the removable equipment must be bonded together.

3.1.3 Clearances

Clearances between open overhead electrical conductors and the water's edge are set forth in *NEC Table 680.9(A)*. *Figure 53* shows overhead conductors in the voltage range of over 750V near a community swimming pool.

Table 4 lists overhead conductor clearances. The values in the table assume a maximum water level. The distances do not apply to communication, radio, and television coaxial cables. These cables must maintain a minimum height of 10' (3 m) above swimming and wading pools, diving boards and towers, or other observation platforms. However, the values in *Table 4* do apply to network-powered broadband communications system conductors operating at 0V to 750V to ground.

3.1.4 Electric Pool Heaters

Heater elements on electric pool-water heaters must be subdivided into loads not exceeding 48A each and must be protected by overcurrent protective devices not greater than 60A. The ampacity of branch circuit conductors and the overcurrent device rating may not exceed 125% of the total nameplate load. Refer to *NEC Section 680.10*.

3.1.5 Underground Wiring

Wiring may not be installed directly under a pool unless the wiring is needed to supply associated pool equipment [*NEC Section 680.11(B)*]. In these cases, the wiring must be installed in materials suitable for the conditions subject to that location, including rigid metal conduit, intermediate metal conduit, a nonmetallic raceway system (PVC), reinforced thermosetting resin conduit, or Type MC. Conduit installations must conform to the minimum cover requirements as outlined in *NEC Table 300.5* and summarized here in *Table 5*.

Figure 53 Overhead conductors near a community swimming pool.

Table 4 Overhead Conductor Clearances [Data from *NEC Table 680.9(A)*]

Parameters	Insulated Cables Supported By a Solidly Grounded Bare Messenger Wire or Grounded Neutral Conductor	Conductors Not Insulated or Supported by Solidly Grounded Bare Messenger Wire or Grounded Neutral Conductor	
	0–750V to Ground	0–15kV	Over 15KV–50kV
A. Clearance in any direction to the water level, edge of water surface, base of diving platform, or permanently anchored raft	22.5 feet	25 feet	27 feet
B. Clearance in any direction to observation stands, towers, or diving platform	14.5 feet	17 feet	18 feet
C. Horizontal limit measured from the inside wall of the pool	Must extend to the outer edge of all structures described in A and B but not less than 10 feet.		

Reprinted with permission from NFPA 70®-2020, *National Electrical Code*®, Copyright © 2019, National Fire Protection Association, Quincy, MA. This reprinted material is not the complete and official position of the NFPA on the referenced subject, which is represented only by the standard in its entirety which may be obtained through the NFPA website at **www.nfpa.org**.

3.1.6 Disconnecting Means

A disconnecting means that simultaneously disconnects all ungrounded conductors must be provided for all utilization equipment other than lighting (*NEC Section 680.13*). The disconnecting means must be installed in a readily accessible location at least 5' (1.5 m) horizontally from the inside wall of a pool, hot tub, spa, or fountain unless a barrier that requires a reach path of 5' (1.5 m) or more is permanently installed between the disconnect and pool's edge. *Figure 54* shows the meter socket and service panel for the pool, which houses the disconnecting means for all utilization equipment around the pool and in the adjacent equipment room. Note that the service equipment is greater than 5' (1.5 m) from the pool.

> **NOTE**
> The horizontal distance is the shortest path from the pool's edge to the disconnecting device.

3.2.0 Permanently Installed Pools

Permanently installed pools must follow specific wiring and equipment requirements as specified in *NEC Article 680, Parts I and II*. These include special requirements for motors and auxiliary equipment, receptacles, lighting, junction boxes, feeders, and bonding.

Table 5 Minimum Cover Depths (Data from *NEC Table 300.5*)

Wiring Method	Minimum Depth in Inches
Rigid Metal Conduit	6
Intermediate Metal Conduit	6
Nonmetallic Conduit Listed for Burial Under 4" Concrete Min.	6
Nonmetallic Direct Burial Conduit	18
Other Approved Raceways*	18

*Other approved raceways requiring concrete encasement for burial must have a concrete envelope no less than 2 inches thick.

Reprinted with permission from NFPA 70®-2020, the *National Electrical Code*®, Copyright © 2019, National Fire Protection Association, Quincy, MA. This reprinted material is not the complete and official position of the NFPA on the referenced subject, which is represented only by the standard in its entirety which may be obtained through the NFPA website at **www.nfpa.org**.

3.2.1 Motors

When installed in corrosive environments, as defined in *NEC Section 680.2*, wiring methods shall comply with *NEC Section 680.14* or be installed in Type MC cable listed for that location. Where installed in noncorrosive environments, branch circuits shall comply with the general requirements in *NEC Chapter 3*. All raceways or cables must contain an insulated copper equipment-grounding conductor sized in compliance with *NEC Section 250.122*, but no smaller than No. 12 AWG per *NEC Section 680.21(A)(1)*.

When flexibility is necessary due to motor vibration, *NEC Section 680.21(A)(2)* requires that the connection from the motor be made with liquid-tight flexible metal or liquid-tight flexible nonmetallic conduit (*Figure 55*). Cord-and-plug connections are also permitted as long as the cord is no longer than 3' (900 mm) and incorporates a copper equipment-grounding conductor sized in accordance with *NEC Section 250.122* but not smaller than a No. 12 AWG. The cord shall terminate in a grounding-type attachment plug.

Figure 54 Service and disconnecting means for the pool equipment.

Figure 55 Pool recirculation motors connected with liquid-tight flexible metal conduit.

3.2.2 Receptacle and Area Lighting

General-purpose receptacles must be located at least 6' (1.83 m) from the inside walls of a pool per *NEC Section 680.22(A)(1)*. Receptacles that are used to provide power to pump motors or other loads must be a grounding-type receptacle having GFCI protection per *NEC Section 680.22(A)(2)*. These types of receptacles must also be located at least 6' (1.83 m) from the wall of the pool.

Every pool installed at a dwelling unit must be provided with at least one 125V, 15A or 20A receptacle supplied by a general-purpose branch circuit. The receptacle must be located within 20' (6 m) of the pool's inside wall but no less than 6' (1.83 m) and at a height of no more than 6'6" (2 m) above the grade or deck level [*NEC Section 680.22(A)(1)*]. These receptacles must be GFCI-protected. All other receptacles rated 15A or 20A, 125V and located within 20' (6 m) of the inside walls of a pool must also have GFCI protection for personnel [*NEC Section 680.22(A)(4)*].

Area lighting includes all lighting around or above pool areas except those luminaires installed in or under water. *Figure 56* is an example of pole-mounted area lighting, while *Figure 57* shows low-voltage area lighting.

As stated in *NEC Section 680.22(B)*, the installation of new luminaires, lighting outlets, or paddle fans above outdoor pools or within 5' (1.5 m) horizontally from the inside walls of the pool must maintain a clearance of no less than 12' (3.7 m) above maximum water level of the pool. For new indoor installations, a reduced minimum clearance of 7'6" (2.3 m) is permitted if the branch circuit has ground-fault protection and the luminaires are totally enclosed or, in the case of fans, if the fan is identified for use beneath ceilings

Figure 56 Pole-mounted area lighting around a pool.

such as on porches or patios.

Figure 57 Low-voltage area lighting around a pool.

Figure 58 Spa emergency shutoff switch and adjacent timer.

Existing luminaires or lighting outlets (installed before the pool was built) that are located less than 5' (1.5 m) from the inside edge of the pool must also be at least 5' (1.5 m) above the maximum water level, rigidly attached to the existing structure, and ground-fault protected.

Luminaires, lighting outlets, and ceiling fans that are installed 5 to 10' (1.5 to 3 m) from the inside edge of a pool must be GFCI-protected, unless they are installed 5' (1.5 m) or more above the maximum water level and rigidly attached to the structure adjacent to or enclosing the pool. Luminaires within 16' (4.9 m) of the pool using cord-and-plug connections must comply with *NEC Section 680.22(B)(5)*. An exception is made for certain listed low-voltage luminaires in *NEC Section 680.22(B)(6)*.

Switches must be located at least 5' (1.5 m) from the pool's edge unless separated from the pool by a solid fence, wall, or other permanent barrier, unless the switch is listed for use within 5' (1.5 m) of a pool [*NEC Section 680.22(C)*]. *Figure 58* shows an emergency shutoff switch (red switch) for an attached spa. The switch on the right is an automatic timer switch for the spa that can be set for various time periods, with an automatic shutoff after the time expires. These switches must be readily accessible.

Outlets used for remote control operations, signaling, fire alarm, or communication circuits must be located a minimum of 10' (3 m) from the inside edge of the pool per *NEC Section 680.22(D)*.

3.2.3 Underwater Luminaires

NEC Section 680.23(A) states that the maximum voltage supplying underwater luminaires is limited to 150V between circuit conductors. Underwater luminaires may be directly supplied by a branch circuit or by way of a transformer or power supply and shall be listed for swimming pool and spa use, with an ungrounded secondary and a grounded metal barrier between the primary and secondary windings.

In order to reduce or prevent shock during re-lamping, *NEC Section 680.23(A)(3)* requires any underwater luminaire that operates at more than the low-voltage contact limit to have a ground-fault circuit interrupter installed in the branch circuit supplying the luminaire. *Figure 59* shows the GFCI circuit breaker in the panel serving

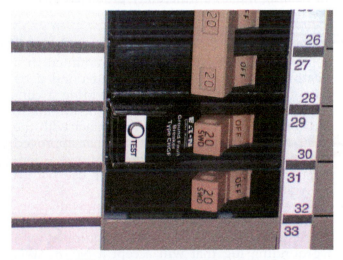

Figure 59 GFCI circuit breaker supplying the underwater luminaire.

Figure 60 Nonmetallic forming shell.

the pool lighting and equipment. The ground-fault circuit interrupter must be installed so that any shock hazard is eliminated should contact occur between a person and any ungrounded part of the branch circuit or the luminaire to ground. Underwater luminaires supplied by a transformer and operated at less than the low-voltage contact limit do not require ground-fault protection.

Wall-mounted underwater luminaires must be installed so that the top of the lens is at least 18" (450 mm) below the water level. Under no conditions may an underwater luminaire be installed less than 4" (100 mm) below the normal water level. If an underwater luminaire is installed at the bottom of the pool facing upward, it must either have the lens guarded to prevent contact with people, or be listed for use without a guard. Underwater luminaires that depend on water for cooling must be inherently protected against overheating when not submerged.

Wet-Niche Luminaires – NEC Section 680.23(B) governs wet-niche luminaires. A wet-niche luminaire in a pool or fountain is mounted in a forming shell that is completely surrounded by water, with the forming shell providing a conduit connection. All metal parts of the luminaire and forming shell that come in contact with the water must be made of brass or other approved corrosion-resistant metal. *Figure 60* shows a nonmetallic forming shell designed for use with a low-voltage LED luminaire.

Forming shells designed with nonmetallic conduit connections (other than low-voltage systems not requiring grounding) must be equipped with a grounding lug that will accept a No. 8 AWG copper conductor. Only rigid metal conduit, intermediate metal conduit, liquid-tight flexible nonmetallic conduit, or rigid nonmetallic (PVC)

conduit may connect to a wet-niche forming shell and must run directly from the shell to a junction box.

Conduit to shells may be metal or nonmetallic conduit that is approved for swimming pool installations. Metal conduit must be constructed of brass or other approved corrosion-resistant metal. Nonmetallic conduit must contain a No. 8 AWG insulated solid or stranded copper conductor to function as a bonding jumper between metal parts, unless a low-voltage luminaire not requiring grounding is installed. The bonding jumper must be terminated in the shell, junction box, transformer enclosure, or ground-fault circuit interrupter enclosure (if installed). After the grounding termination is made in the forming shell, it must be covered with a listed **potting compound** to protect it from deterioration.

Other than listed low-voltage lighting systems not requiring grounding, all wet-niche luminaires that are supplied by a flexible cord or cable must contain an insulated copper equipment-grounding conductor no smaller than No. 16 AWG that grounds all exposed non-current-carrying metal parts of the luminaire and shell. The grounding conductor must be terminated on a grounding terminal within the supply junction box or transformer enclosure and shell, as shown in *Figure 60*. The ends of the flexible cord jacket and the termination within the luminaire must be sealed with an approved potting compound to prevent water from entering the cable.

The wet-niche luminaire must be secured and bonded to the forming shell by a **positive locking device** (typically an Allen-head screw) that provides a low-resistance path, requiring a wrench or other tool to remove the luminaire from the forming shell. Wet-niche luminaires must be supplied with sufficient cable or cord length to allow the luminaire to be removed and placed on the deck or other dry location for servicing, without requiring the servicing technician to enter the water (*Figure 61*).

Other than Wet-Niche Luminaires – Other types of less common underwater luminaire installations include dry-niche, no-niche, and through-wall lighting assemblies. Refer to *NEC Sections 680.23(C) through (E)* when installing these devices.

Branch Circuit Wiring – Branch circuits supplying enclosures and junction boxes connected to wet-niche and no-niche luminaires and branch circuits supplying field wiring compartments of dry-niche luminaires in corrosive environments, as described in *NEC Section 680.2*, shall comply with requirements in *NEC Section 680.14* or shall be liquid-tight flexible nonmetallic conduit. Electrical metallic tubing (EMT) may be used if installed on or in a building. Electrical nonmetallic tubing, MC, or AC cable may be used only if installed inside a building.

Equipment Grounding – Other than listed low-voltage luminaires not requiring grounding, all through-the-wall lighting assemblies, wet-niche, dry-niche, and no-niche luminaires must all be connected to an insulated copper equipment-grounding conductor containing no joints or splices, sized in accordance with *NEC Table 250.122*, and must be no smaller than No. 12 AWG [*NEC Section 680.23(F)(2)*]. Two exceptions to an unbroken equipment-grounding conductor are as follows:

- Where more than one underwater luminaire is supplied by the same branch circuit, the equipment grounding conductor may be broken at the interconnection between the grounding terminals of junction boxes, transformer enclosures or other supply-circuit enclosures to wet-niche luminaires, or between the field-wiring compartments of dry-niche luminaires.

Figure 61 Sufficient cord must be provided to allow the luminaire to be relamped without entering the water.

- Underwater luminaires supplied from a transformer, GFCI, clock-operated switch, or manual snap switch installed between the panelboard and a junction box directly connected by conduit to the luminaire may have the circuit equipment-grounding conductor terminated on the enclosure holding the transformer, GFCI, clock-operated switch, or manual switch.

Conductors – Per *NEC Section 680.23(F)(3)*, load-side conductors of GFCIs or transformers supplying underwater luminaires operating at no more than the low-voltage contact limit must not be installed in raceways, boxes, or enclosures containing other conductors unless at least one of the following conditions are met:

- The other conductors are GFCI-protected.
- The other conductors are grounding and bonding conductors.
- The other conductors supply a feed-through GFCI.
- A GFCI shall be permitted to be installed in a panelboard containing other circuits protected by non-GFCI devices.

Matching Luminaire and Niche

Underwater luminaires and forming shells must be listed and matched. To ensure proper matching, listed luminaires are marked on the exterior of the luminaire to identify the compatible forming shell in which the luminaire may be installed. Likewise, forming shells are marked on the interior surfaces to identify the luminaires that may be installed inside the shell. This marking system is part of *Underwriters Laboratories UL® 676, Standard for Underwater Luminaires and Submersible Junction Boxes.*

3.2.4 Transformer or GFCI Junction Boxes and Enclosures

Junction boxes (*Figure 62*) that are used as terminal points for conduit running to wet-niche luminaire niches or the mounting brackets of no-niche luminaires must comply with the construction and installation requirements of *NEC Sections 680.24(A)(1) and (2)*:

- The junction box must be listed as a swimming pool junction box.
- It must be equipped with threaded hubs or a nonmetallic hub.
- The box must be made of copper, brass, plastic, or other approved corrosion-resistant material.
- The box must provide electrical continuity between all metal conduit connections and the grounding terminals by means of copper, brass, or other approved corrosion-resistant metal that is integrated into the box construction.
- If the luminaire operates at more than the low-voltage contact limit, the junction box must be mounted at least 4" (100 mm) above ground (measured from the bottom of the box) or at least 8" (200 mm) above the maximum water level of the pool, whichever provides the greater elevation, and at least 4' (1.2 m) from the inside wall of the pool unless separated by a fence or other permanent barrier.
- If the luminaire operates at or below the low-voltage contact limit, the junction box may be mounted as a flush deck box as long as the box is filled with an approved potting compound to prevent the entry of moisture and the flush box is located at least 4' (1.2 m) from the inside wall of the pool.

Enclosures used to house transformers, GFCIs, or similar devices that directly connect by conduit to a niche or no-niche luminaire mounting bracket must comply with the construction

Figure 62 Junction box for a wet-niche luminaire.

and installation requirements of *NEC Sections 680.24(B)(1) and (2)*. These requirements are the same as for wet-niche luminaires, except the use of a flush deck box is not permitted. Also, *NEC Section 680.24(B)(1)* requires the use of a duct seal or other approved seal at the conduit connection to prevent air circulation between the conduit and enclosure.

Regardless of type, every enclosure must meet the following requirements:

- Enclosures must be protected from physical damage, per *NEC Section 680.24(C)*.
- Enclosures must be provided with at least one grounding terminal more than the number of conduit connections provided by the enclosure, per *NEC Section 680.24(D)*.
- Terminations of flexible cords into enclosures must be provided with a strain relief to protect the cord, per *NEC Section 680.24(E)*.
- Grounding terminals of enclosures must be connected to the panelboard equipment grounding terminal, per *NEC Section 680.24(F)*.

Pool Electrocutions

The Consumer Product Safety Commission reported 14 electrocutions in addition to other electrical shock incidents involving circuits around swimming pools in the ten-year period between 2004 and 2014. Educating consumers about the hazards of electricity in pool areas and adhering to the electrical standards of the *NEC*® appears to be helping though—a similar number of electrocutions were reported in the six-year period between 1997 and 2003, so the average rate of incidence per year has dropped.

3.2.5 Feeders

Feeders that supply swimming pool branch-circuit panelboards, feeders on the load side of the service equipment associated with pools, or the source of a separately derived system supplying pools must follow specific wiring methods and grounding rules of *NEC Section 680.25*. Some of these requirements are as follows:

- Where feeders are installed in corrosive environments, as described in *NEC Section 680.2*, the wiring methods of that portion shall be as required in *NEC Section 680.14* or shall be liquid-tight flexible nonmetallic conduit.
- The grounding conductor must be sized according to *NEC Section 250.122* but never smaller than No. 12 AWG.
- Aluminum conduit may not be installed in the pool area where it is exposed to corrosion. See *NEC Section 680.25(B)*.

3.2.6 Equipotential Bonding

Equipotential bonding in pool or fountain systems can be described as the interconnection of all exposed and buried, non-current-carrying conductive parts and surfaces (not just metal), with an approved grounded bonding conductor to reduce any **voltage gradients** in the pool or fountain area.

Per *NEC Section 680.26(B)*, the bonding conductor used must be solid copper, insulated or bare, and no smaller than No. 8 AWG. Interconnections between boxes and enclosures made with noncorrosive rigid metal conduit such as brass may also be considered as bonded. However, bonding made with a solid No. 8 AWG copper is not required to be extended to remote panelboards, electrodes, or service equipment. All bonded connections must comply with *NEC Section 250.8*.

The following equipment, surfaces, and pool components must be included in the equipotential bonding system and comply with all rules of *NEC Articles 680 and 250*:

- All conductive pool shells including concrete or concrete block shells (not vinyl liners and fiberglass pools) must be bonded. Pool shells constructed using structural reinforcing steel that is not encapsulated in a nonconductive compound must be bonded together using steel tie wires per *NEC Section 680.26(B)(1)(a)*. Pool shells constructed with encapsulated structural reinforcing steel must have a copper conductive grid installed in accordance with *NEC Section 680.26(B)(1)(b)*.

- Pool perimeter surfaces including any paved, unpaved, poured concrete or other conductive surfaces that extend outward to a maximum of 3' (1 m) when measured horizontally from the inside walls of the pool require bonding per *NEC Section 680.26(B)(2)*. Refer to *Figure 63*.
- All metal components, metal fittings, and underwater lighting niches and mounting brackets must be included in the equipotential bonding system, with the exception of low-voltage luminaires in nonmetallic niches. See *NEC Sections 680.26(B)(3) through (5)*.
- Any metal part associated with the pool water circulating system or pool covers must be bonded, including pumps and motors [*NEC Section 680.26(B)(6)*].

> **NOTE**
>
> Even though double-insulated water pump motors are not required to be bonded, a solid No. 8 AWG copper bonding conductor of sufficient length must be provided in the area of the water pump motor in case the motor is replaced with other than a double-insulated motor. See *NEC Section 680.26(B)(6)(a)*.

- Follow the specific manufacturer's recommendations regarding grounding and bonding of pool water heaters. *Figure 64* shows the enclosure of a pool heater bonded to the equipotential bonding system.
- Metal raceways, metal-sheathed cables, metal piping, and any fixed metal parts that are within 5' (1.5 m) horizontally from the inside walls of the pool, or 12' (3.7 m) vertically from the maximum water level, tower, stand, platform, or diving structure, must also be bonded unless separated by permanent barrier per *NEC Section 680.26(B)(7)*.

Figure 63 Pool perimeter conductive surfaces.

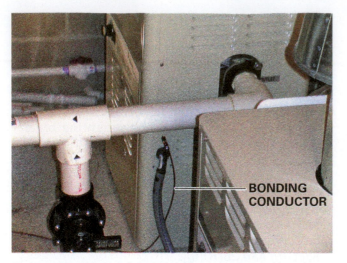

BONDING CONDUCTOR

Figure 64 Pool heater bonded to the equipotential bonding system.

- Per *NEC Section 680.26(C)*, the pool water itself must be bonded using a corrosion-resistant conductive surface with a minimum area of 9 in² (5,800 mm²) contacting the pool water.

3.2.7 Specialized Pool Equipment

Accessory equipment not directly related to the lighting and operation of the pool itself falls into the category of specialized pool equipment. Examples of specialized pool equipment include underwater audio equipment, electrically operated pool covers, and deck area heating.

Underwater Audio Equipment – Underwater speakers must be mounted in an approved metal forming shell that is recessed in the wall or floor of the pool and covered with a metal screen that is bonded to the forming shell [*NEC Section 680.27(A)(1)*]. The screen must be held in place by a positive locking device that requires a tool to open or remove the screen. Both the forming shell and screen must be constructed of brass or other approved corrosion-resistant metal per *NEC Section 680.27(A)(3)*. The forming shell must also be equipped with a bonding terminal that will accommodate a No. 8 AWG copper conductor.

Electrically Operated Pool Covers – All electrical equipment associated with the operation of an electric pool cover must be located at least 5' (1.5 m) from the inside wall of the pool unless a permanent barrier is in place between the pool and equipment. See *NEC Section 680.27(B)(1)*. Motors that are installed below grade level must be totally enclosed, and the device controlling the pool cover motor must be located so that the operator has a full view of the pool during cover operation. Both the motor and controller must be connected to a GFCI-protected circuit [*NEC Section 680.27(B)(2)*].

Deck Area Heating – Electrically operated unit heaters or permanently wired radiant heaters in-

> **WARNING!**
>
> Electrical (not steam) heating cables that are designed to be embedded in surfaces to produce heat transfer into the decking surface are not permitted in swimming pool decks.

stalled within 20' (6 m) from the inside wall of a pool must be rigidly mounted in place and guarded to prevent contact with the heating elements [*NEC Section 680.27(C)(1)*]. No type of heater may be installed over the pool or within 5' (1.5 m) from the pool's edge. Permanently wired radiant heaters must be installed at least 12' (3.7 m) above the pool decking per *NEC Section 680.27(C)(2)*.

3.3.0 Storable Pools

Cord-connected storable pool filter pumps must be double-insulated and have any internal, non-accessible, non-current-carrying metal parts grounded using an equipment grounding conductor that is part of the supply cord and terminated in a grounding-type attachment plug per *NEC Section 680.31*. The attachment plug must be designed with integral GFCI protection, or the power supply cord must incorporate GFCI protec-

Accessories Can Complete the Bonding Requirement

The bonding requirement of *NEC Section 680.26(C)* can be fulfilled by any of the other parts that are in direct contact with the water and already required to be bonded in other sections of *NEC Article 680*, such as shells, conduit, ladders, and other parts, as long as the total water contact area of the bonded metal parts measures a minimum of 9 in² (5,800 mm²).

tion within 12" (300 mm) of the attachment plug. *Figure 65* shows a type of storable pool.

The pool filter pump, while important, is not the only equipment requiring GFCI protection. *NEC Section 680.32* requires that all electrical equipment associated with storable pools and all 125V, 15A and 20A receptacles installed within 20' (6 m) of the inside walls of the pool have GFCI protection unless a permanent barrier is installed between the equipment or receptacle and the pool. In no case may a receptacle be installed within 6' (1.83 m) of the inside walls of the pool unless a permanent barrier is installed.

Underwater luminaires must be installed in or on the wall of a storable pool, spa, or hot tub. Refer to *NEC Section 680.33*. The lens, luminaire body, and transformer enclosure (if used) must be constructed of a polymeric impact-resistant material with no exposed metal parts. If the luminaire operates within the low-voltage contact limit or less, it must be a listed cord-and-plug assembly using a transformer listed for the application, with a primary voltage rating not to exceed 150V. Underwater luminaires operating at voltages greater than the low-voltage contact limit but less than 150V require GFCI protection.

3.4.0 Spas, Hot Tubs, and Therapeutic Pools and Tubs

Spas and hot tubs may be installed indoors or outdoors. Regardless of their location, all spas and hot tub electrical installations, except in single-family dwellings, must include a clearly labeled emergency shutoff or control switch that controls power to the recirculation and/or jet system motor (*Figure 66*). This emergency shutoff must be located at a readily accessible point for the user but no less than 5' (1.5 m) away from the spa or hot tub per *NEC Section 680.41*.

Outdoor Installations – Spas and hot tubs installed outdoors must follow the same rules as permanently installed pools installed outdoors, in addition to the requirements listed in *NEC Section 680.42* and summarized as follows:

- Packaged spas or hot tubs that are listed for outdoor installations or self-contained spas or hot tubs equipped with a factory-installed control panel or panelboard may be connected using a GFCI-protected cord and plug with a maximum cord length of 15' (4.6 m).
- Metal-to-metal bonding of common frame or base components is permitted.
- Metal parts used to secure wooden staves are not required to be bonded.
- If a self-contained or packaged outdoor spa/hot tub is supplied by the interior wiring of a single-family dwelling or associated building, the supply wiring method must comply with *NEC Chapter 3* and must include a copper equipment-grounding conductor (No. 12 AWG or larger) that is insulated or enclosed within the outer sheath of the wiring method.

Figure 65 Storable pool.

Figure 66 Spa emergency shutoff switch located at least 5' (1.5 m) from the spa.

Indoor Installations – Spas and hot tubs installed indoors must also follow the same rules for permanently installed pools and comply with the wiring methods of *NEC Chapter 3*. One exception is that indoor packaged spas or hot tubs rated 20A or less may be connected by cord and plug to allow for easy removal or disconnection during maintenance or repair. In addition, the rules of *NEC Section 680.43*, summarized below, also apply:

- At least one 125V, 15A, or 20A receptacle must be provided and must be located no less than 6' and no more than 10' (between 1.83 m and 3 m) from the inside wall of the unit. No receptacle may be located closer than 6' (1.83 m) from the inside walls of the spa or hot tub.
- Any 125V, 30A or less receptacle located within 10' (3 m) of the inside walls of an indoor spa or hot tub must be GFCI-protected.

Wall switches must be installed at least 5' (1.5 m) from the inside walls of the spa or hot tub [*NEC Section 680.43(C)*]. Luminaires, lighting outlets, and ceiling fans installed over indoor spas or hot tubs must be a minimum of 12' (3.7 m) above the maximum water level in installations without GFCI protection, per *NEC Section 680.43(B)(1)(a)*. For installations with GFCI protection, the minimum clearance is 7'6" (2.3 m) as specified in *NEC Section 680.43(B)(1)(b)*.

The following components associated with indoor spas and hot tubs must be bonded together per *NEC Section 680.43(D)*:

- All metal fittings within or attached to the spa or hot tub, including handrails (*Figure 67*)
- Non-current-carrying metal parts of electrical equipment, including circulating pump motors unless part of a listed self-contained spa or hot tub
- Any metal raceway, piping, or metal surface within 5' (1.5 m) of the inside walls of the spa or hot tub, unless separated by a permanent barrier
- Any electrical device or control not associated with the spa or hot tub, but located within 5' (1.5 m) of it

Bonding may be accomplished by the interconnection of threaded metal pipe and fittings, metal-to-metal mounting on a common frame or base, or by an insulated, covered, or bare No. 8 AWG or larger solid-copper bonding jumper. See *NEC Section 680.43(E)*. In addition to bonding, any electrical equipment that is located within 5' (1.5 m) of the inside wall of the spa or hot tub, or

Figure 67 Spa handrails must be bonded.

any electrical equipment associated with the circulating system, must also be grounded per *NEC Section 680.43(F)*. Receptacles that supply self-contained, packaged, or field-assembled spas and hot tubs must be GFCI-protected except as noted in *NEC Section 680.44*.

3.4.1 Therapeutic Pools and Tubs

NEC Article 680, Part VI covers pools and tubs used for therapeutic purposes in health care facilities, gymnasiums, athletic training rooms, and similar areas. Permanently installed therapeutic pools follow the same *NEC®* requirements as permanently installed swimming pools and must also comply with *NEC Article 680, Parts I and II*.

Electrical outlets that supply self-contained, packaged, or field-assembled therapeutic tubs or hydrotherapeutic tanks must be GFCI-protected unless the listed unit includes integral GCFI protection for all electrical parts within the unit or operates from a three-phase system greater than 250V or with a heater load of more than 50A [*NEC Section 680.62(A)(1) and (2)*]. Bonding requirements for therapeutic tubs or hydrotherapeutic tanks are listed in *NEC Section 680.62(B)* and summarized as follows:

- All metal fittings within or attached to the tub or tank must be bonded together.
- All non-current-carrying metal parts of the water circulation system (including pump motors) must be bonded.
- Any metal-sheathed cables, raceways, metal piping, and metal surfaces within 5' (1.5 m) of the inside walls of the tub or tank must be bonded unless separated by a permanent barrier.

- Any electrical device or control that is not associated with the operation of the tub or tank but is within 5' (1.5 m) of the tub or tank must be bonded.

This bonding requirement may be satisfied by the interconnection of threaded metal pipe and fittings, metal-to-metal mounting on a common frame, connections made using approved metal clamps, or by the installation of an insulated, covered, or bare solid-copper bonding jumper no smaller than No. 8 AWG.

All electrical equipment located within 5' (1.5 m) of the inside wall of the tub or tank and all electrical equipment associated with the tub or tank's water circulation system must be grounded to the equipment grounding conductor of the supply circuit as specified in *NEC Section 680.62(D)*. In addition, *NEC Section 680.62(E)* requires that any receptacle located within 6' (1.83 m) of a therapeutic tub or tank be GFCI-protected, and all luminaires installed in the area of therapeutic tubs or tanks must be totally enclosed [*NEC Section 680.62(F)*].

3.4.2 Hydromassage Bathtubs

Hydromassage bathtubs, also known as *whirlpool tubs*, are commonly installed in the master bathrooms of new residential construction and are often added when older bathrooms are renovated. A hydromassage bathtub is a permanently installed bathtub equipped with a recirculation pump, piping system, and jets. By design, the tub can accept, circulate, and discharge water in each usage. Hydromassage bathtubs must be installed in accordance with *NEC Article 680, Part VII*.

NEC Section 680.71 requires that hydromassage tubs be supplied by an individual branch circuit. This can be an issue when installing these tubs in retrofit installations, especially on the upper floors. In addition, a readily accessible GFCI must be provided between the supply and the tub, and all 125V receptacles located within 6' (1.83 m) of the inside wall of the tub must be GFCI-protected. The tub electrical equipment must be accessible without damaging the building structure or finish per *NEC Section 680.73*, which is often accomplished using a removable panel either on the tub itself or through an adjacent wall. Where the hydromassage tub is cord-and-plug connected with the supply receptacle accessible only through a service access opening, the receptacle shall be installed so that its face is within direct view and not more than 12" (300 mm) from the opening. Both metal piping

systems and grounded metal parts in contact with the water must be effectively bonded using an insulated, covered, or bare solid-copper bonding jumper no smaller than No. 8 AWG (*NEC Section 680.74*). The bonding jumper must be connected to a circulating pump motor terminal intended for this purpose. Double-insulated circulating pump motors do not require bonding. The bonding jumper is not required to be extended or attached to remote panelboards, service equipment, or any electrode.

3.5.0 Fountains

Fountains have become increasingly popular in residential and commercial landscapes. Permanently installed fountains (*Figure 68*), ornamental pools, display pools, and reflection pools must comply with *NEC Article 680, Parts I and V*. In addition, fountains that use common water from a swimming pool (*Figure 69*) must also comply with *NEC Article 680, Part II*.

> **NOTE**
> Self-contained portable fountains and water drinking fountains are not covered by *NEC Article 680*.

3.5.1 Luminaires, Submersible Pumps, and Equipment

Luminaires, submersible pumps, and other submersible equipment installed in fountains (*Figure 70*) must have GFCI protection unless they operate at the low-voltage contact limit or less and are supplied by a listed transformer intended for pool and fountain applications. See *NEC Section 680.51(A)*.

Figure 68 Permanently installed fountain.

Figure 69 Fountain using pool water.

Figure 70 Luminaire and pump in fountains must have GFCI protection.

Luminaires may not operate at voltages over 150V between conductors, and submersible equipment may not operate at voltages over 300V between conductors. See *NEC Section 680.51(B)*. Luminaires must also be installed so that the top of the luminaire is below the normal water level of the fountain, unless the luminaire is listed for out-of-water operation. Luminaires installed with the lens facing upward must have the lens guarded against contact by people or be listed for use without a guard per *NEC Section 680.51(C)*.

NEC Section 680.51(F) requires that luminaires and other submersible equipment be installed to allow easy removal from the fountain for relamping or other maintenance without having to reduce the water level. Luminaires and other submersible equipment may not be permanently embedded into the fountain's structure.

Hydromassage Bathtubs

When installing a hydromassage bathtub in an installation where the final finish will be a tiled surround, make sure the tub or surround design provides for permanent access to the electrical equipment.

Electrical submersible equipment that must remain submersed in water for proper operation and safety must be equipped with a low-water shutoff device per *NEC Section 680.51(D)*. Submersible equipment must also be equipped with threaded conduit entries or a factory- or field-installed flexible cord no longer than 10' (3 m) [*NEC Section 680.51(E)*]. If the cord extends beyond the perimeter of the fountain, the cord must be enclosed in an approved wiring enclosure. Any metal parts in contact with the water must be constructed of brass or other corrosion-resistant material.

3.5.2 Junction Boxes and Other Enclosures

Junction boxes and other enclosures associated with fountains must comply with *NEC Section 680.24*, which covers junction boxes and enclosures in permanently installed pools. In addition, underwater junction boxes and enclosures installed in fountains must also comply with the requirements listed in *NEC Section 680.52(B)*, summarized as follows:

- Designed with threaded conduit entries or compression-type sealing (grommet) fittings for cord entry
- Listed for submersible use and constructed of copper, brass, or other approved corrosion-resistant material
- Filled with an approved potting compound to prevent water entry
- Firmly supported and bonded
- Supported by conduit only if the conduit is copper, brass, stainless steel, or other approved metal conduit
- If supplied by nonmetallic conduit, provided with additional support using approved corrosion-resistant metal support devices or fasteners

3.5.3 Bonding and Grounding

Any metal piping associated with a fountain must be bonded to the branch circuit equipment-grounding conductor that supplies the fountain, per *NEC Section 680.54(B)*. Other than low-voltage luminaires not requiring grounding, all electrical equipment located within 5' (1.5 m) of the inside wall, any electrical equipment associated with the recirculation system, and any panelboards that are not part of the service equipment but supply electrical equipment associated with the fountain must be grounded per *NEC Section 680.54(A)*.

NEC Section 680.55(B) requires that any electrical equipment associated with a fountain that is supplied by a flexible cord have all of its exposed non-current-carrying metal parts grounded using an insulated copper equipment-grounding conductor that is part of the flexible cord. The other end of the equipment grounding conductor must be terminated to an equipment grounding terminal in the supply junction box or other supply enclosure. Any 15A or 20A, 125V through 250V receptacle installed within 20' (6 m) of a fountain must be GFCI-protected per *NEC Section 680.58*.

3.5.4 Cord-and-Plug Equipment

NEC Section 680.56(A) requires that any electrical equipment associated with a fountain and supplied by a cord and plug be GFCI-protected. In addition, the flexible cord must be rated for extra-hard usage and listed for wet usage per

---

Cord and Cable Types

NEC Table 400.4 lists cords and cables according to their trade names (for example, portable power cable or heater cord) and type letter (SE, SEW, SJ, or other designation). Cords or cables used in fountains must have a type letter that ends with W to indicate a cable or cord suitable for a wet location.

NEC Section 680.56(B). Ends and terminations of the flexible cord and the grounding connection within the equipment must be covered with or encapsulated in a suitable potting compound to prevent the entry of water into the cable or deterioration of the terminations [*NEC Section 680.56(C)*].

3.5.5 Signs Placed in Fountains

NEC Section 680.57(C)(1) requires that any fixed or stationary electric sign installed within a fountain be at least 5' (1.5 m) from the outside edge of the fountain. This places the sign out of reach of passersby or those sitting on the fountain edge. Fixed signs must also be equipped with a local disconnecting means in accordance with *NEC Sections 600.6 and 680.13* and must be grounded and bonded in accordance with *NEC Section 600.7*. Portable signs may not be installed in fountains or within 5' (1.5 m) from the inside walls of the fountain [*NEC Section 680.57(C)(2)*].

3.0.0 Section Review

1. The required clearance from overhead electrical conductors in any direction to swimming pool observation stands, towers, or diving platforms for insulated cables supported by a solidly grounded bare messenger wire or a grounded neutral conductor in a voltage range of 0–750V to ground is _____.

   a. 14.5' (4.4 m)
   b. 17' (5.2 m)
   c. 18' (5.5 m)
   d. 22.5' (6.9 m)

2. Raceways and cables serving motors associated with permanently installed pools *must* contain an insulated copper equipment-grounding conductor sized no smaller than No. _____.

   a. 6 AWG
   b. 8 AWG
   c. 10 AWG
   d. 12 AWG

3. Underwater luminaires installed in storable pools can be powered through a cord-and-plug arrangement using a transformer listed for the application if the transformer-primary power requirement does NOT exceed _____.

   a. the low-voltage contact limit
   b. 100V
   c. 150V
   d. 250V

4. Which of the following methods of bonding is acceptable for spas and hot tubs?

   a. No. 14 AWG solid-copper bonding jumper
   b. No. 10 AWG solid-copper bonding jumper
   c. Threaded metal pipe and fittings
   d. Plastic pipe and fittings

5. Junction boxes and other enclosures associated with fountains *must* comply with *NEC Section 680.24* and must be _____.

   a. designed to eliminate threaded conduit entries
   b. filled with an approved potting compound to prevent water entry
   c. supported only by nonmetallic conduit
   d. constructed of plastic

1. In order to qualify as an assembly occupancy, a structure must be designed to hold
   at least _____.

   a. 50 people
   b. 100 people
   c. 200 people
   d. 500 people

2. The use of electrical nonmetallic tubing and
   rigid nonmetallic conduit in spaces of assembly is restricted to certain areas such as club
   rooms if the building contains _____.

   a. combustible supports
   b. conduit supports
   c. conductive supports
   d. clamping supports

3. In a manually controlled stage switchboard,
   all dimmers and switches are operated by
   handles that are _____.

   a. automatically linked to the control
      devices
   b. mechanically linked to the control
      devices
   c. electrically linked to the control devices
   d. manually linked to the control devices

4. For portable structures such as tents at circuses and similar areas, what is the clearance
   required in any direction from overhead conductors operating at 600V or less?

   a. 10' (3 m)
   b. 15' (4.5 m)
   c. 18' (5.5 m)
   d. 20' (6 m)

5. The size of the bonding conductor for portable buildings at carnivals, circuses, or fairs
   must be accordance with *NEC Table 250.122*,
   based on the largest overcurrent device supplying the portable structures, but NEVER
   smaller than No. _____.

   a. 2 AWG
   b. 6 AWG
   c. 8 AWG
   d. 12 AWG

6. When outlet boxes or other device boxes
   are installed outdoors at circuses, fairs, and
   similar locations, the boxes must be weather-
   proof and mounted so the bottom of the box
   is *at least* _____.

   a. 6" (150 mm) off the ground
   b. 8" (200 mm) off the ground
   c. 12" (300 mm) off the ground
   d. 18" (450 mm) off the ground

7. If two or more agricultural buildings are
   supplied overhead from a distribution point,
   a _____.

   a. service-splitter device must be installed
   b. current-monitoring device must be
      installed
   c. motion-detecting device must be installed
   d. site-isolating device must be installed

8. Holiday lighting, considered temporary wiring, must be removed within _____.

   a. 30 days of the date of installation
   b. 60 days of the date of installation
   c. 90 days of the date of installation
   d. 180 days of the date of installation

9. It is NOT necessary to install a box for splices
   or connections on construction sites in nonmetallic multiwire cord or cable assemblies
   as long as the equipment grounding conductor's _____.

   a. termination is maintained
   b. insulation is maintained
   c. continuity is maintained
   d. location is maintained

10. Electrical distribution systems within boat-
    yards and piers CANNOT exceed _____.

    a. 250V phase-to-phase
    b. 480V phase-to-phase
    c. 600V phase-to-phase
    d. 1,000V phase-to-phase

11. Any electrical connection on a fixed pier must be at least 12" (305 mm) above the deck of the fixed pier, but NEVER below the _____.

    a. service height
    b. electrical datum plane
    c. seepage point
    d. fuel dispensing system

12. Where land areas are affected by tidal fluctuations, the electrical datum plane is 2' (600 mm) above the highest _____.

    a. water level for the area
    b. water mark for the area
    c. flood plane for the area
    d. high tide for the area

13. The required overhead clearance to the water's edge of a swimming pool for insulated cables supported by a solidly grounded, bare messenger wire or a grounded neutral conductor in the voltage range of 0–750V to ground is _____.

    a. 18' (5.5 m)
    b. 22.5' (6.9 m)
    c. 25' (7.5 m)
    d. 27' (8 m)

14. If a flexible cord or cable supplies a wet-niche luminaire, the cord or cable must contain an insulated copper equipment-grounding conductor no smaller than No. _____.

    a. 6 AWG
    b. 8 AWG
    c. 12 AWG
    d. 16 AWG

15. For spas and hot tubs, wall switches must be located away from the inside walls of the tub by *at least* _____.

    a. 2' (600 mm)
    b. 4' (1.2 m)
    c. 5' (1.5 m)
    d. 6' (1.8 m)

1. True or False? An office cafeteria with a capacity of 20 people qualifies as an assembly occupancy.

   _____

2. Raceways or cable assemblies must either qualify as an approved equipment-grounding conductor per *NEC Section 250.118*, or contain a(n) _____ sized in accordance with *NEC Table 250.122*.

3. Wiring methods for theatres are covered in _____.

4. The type of stage switchboard that contains only overcurrent protective devices and no control elements is known as a(n) _____ stage switchboard.

5. A portable structure, such as a carnival tent, may not be located under or within _____ of conductors that have an energized voltage greater than 600V.

6. True or False? Aluminum grounding conductors are permitted in agricultural buildings.

   _____

7. Equipotential planes are required in both indoor and outdoor animal confinement areas with concrete floors or slabs where _____.

8. Each individual shore power receptacle must be provided with a disconnecting means located within _____ of the receptacle.

9. In areas subject to flooding, the electrical datum plane is _____ above the prevailing high water mark as determined by the authority having jurisdiction.

10. Heater elements on electric pool water heaters must be subdivided into loads NOT exceeding _____ each.

Trade Terms Introduced in This Module

Border light: A permanently installed overhead strip light.

Dead-front: Equipment or enclosures covered or protected in order to prevent exposure of energized parts on the operating side of the equipment.

Electrical datum plane: Horizontal plane affecting electrical installations based on the rise and fall of water levels in land areas and floating piers and landing stages.

Equipotential plane: An area in wire mesh or other conductive material embedded in or placed under concrete that may conduct current. It must be bonded to all metal structures and nonelectrical, fixed equipment that could become energized, and then it must be connected to the electrical grounding system.

Extra-hard usage cable: Cable designed with additional mechanical protection for use in areas where it may be subject to physical damage.

Festoon lighting: Outdoor lighting that is strung between supporting points.

Fire rating: Time in hours or minutes that a material or assembly can withstand direct exposure to fire without igniting and spreading the fire.

Footlight: A border light installed on or in a stage.

Landing stage: A floating platform used for the loading and unloading of passengers and cargo from boats or ships.

Portable switchboard: A movable lighting switchboard used for road shows and similar presentations, usually supplied by a flexible cord or cable.

Positive locking device: A screw, clip, or other mechanical locking device that requires a tool or key to lock or unlock it.

Potting compound: An approved and listed insulating compound, such as epoxy, which is mixed and installed in and around electrical connections to encapsulate and seal the connections to prevent the entry of water or moisture.

Proscenium: The wall and arch that separate the stage from the auditorium (house).

Reactor-type dimmer: A dimming system that incorporates a number of windings or coils that are linked by a magnetic core, an inductive reactor that functions like a capacitor, and a relay. The reactor stores current from the source and reduces lamp power.

Resistance-type dimmer: A manual-type dimmer in which light intensity is controlled by increasing or decreasing the resistance in the lighting circuit.

Rheostat: A slide and contact device that increases or decreases the resistance in a circuit by changing the position of a movable contact on a coil of wire.

Shunt trip: A low-voltage device that remotely causes a higher voltage overcurrent protective device to trip or open the circuit.

Site-isolating device: A disconnecting means at the point of distribution that is used to isolate the service for the purpose of electrical maintenance, emergency disconnection, or the connection of an optional standby system.

Solid-state phase-control dimmer: A solid-state dimmer in which the wave shape of the steady-state current does not follow the wave shape of the applied voltage (that is, the wave shape is nonlinear).

Solid-state sine-wave dimmer: A solid-state dimmer in which the wave shape of the steady-state current follows the wave shape of the applied voltage (that is, the wave shape is linear).

Unbalanced load: Three-phase loads in which the loads are not evenly distributed between the phases.

Voltage gradients: Fluctuating voltage potentials that can exist between two surfaces or components and may be caused by improper or insufficient bonding or grounding.

Additional Resources

This module presents thorough resources for task training. The following resource material is suggested for further study.

National Electrical Code® Handbook. Latest Edition. Quincy, MA: National Fire Protection Association.

Figure Credits

Section Review Answer Key

SECTION 1.0.0

Answer	Section Reference	Objective
1. c	1.1.1	1a
2. a	1.2.1	1b
3. d	1.3.0	1c
4. d	1.4.1	1d
5. a	1.5.4	1e
6. a	1.6.0	1f

SECTION 2.0.0

Answer	Section Reference	Objective
1. b	2.1.2; *Table 3*	2a
2. c	2.2.4	2b

SECTION 3.0.0

Answer	Section Reference	Objective
1. a	3.1.3; *Table 4*	3a
2. d	3.2.1	3b
3. c	3.3.0	3c
4. c	3.4.0	3d
5. b	3.5.2	3e

NCCER CURRICULA — USER UPDATE

NCCER makes every effort to keep its textbooks up-to-date and free of technical errors. We appreciate your help in this process. If you find an error, a typographical mistake, or an inaccuracy in NCCER's curricula, please fill out this form (or a photocopy), or complete the online form at **www.nccer.org/olf**. Be sure to include the exact module ID number, page number, a detailed description, and your recommended correction. Your input will be brought to the attention of the Authoring Team. Thank you for your assistance.

Instructors – If you have an idea for improving this textbook, or have found that additional materials were necessary to teach this module effectively, please let us know so that we may present your suggestions to the Authoring Team.

NCCER Product Development and Revision
13614 Progress Blvd., Alachua, FL 32615

Email: curriculum@nccer.org
Online: www.nccer.org/olf

❏ Trainee Guide   ❏ Lesson Plans   ❏ Exam   ❏ PowerPoints   Other _____

Craft / Level: _____   Copyright Date: _____

Module ID Number / Title: _____

Section Number(s): _____

Description: _____

_____

_____

_____

Recommended Correction: _____

_____

_____

_____

Your Name: _____

Address: _____

_____

Email: _____   Phone: _____

This page is intentionally left blank.

Fundamentals of Crew Leadership

OVERVIEW

When a crew is assembled to complete a job, one person is appointed the leader. This person is usually an experienced craft professional who has demonstrated leadership qualities. While having natural leadership qualities helps in becoming an effective leader, it is more true that "leaders are made, not born." Whether you are a crew leader or want to become one, this module will help you learn more about the requirements and skills needed to succeed.

46101

NCCER

President: Don Whyte
Vice President: Steve Greene
Chief Operations Officer: Katrina Kersch
Fundamentals of Crew Leadership Project Manager: Mark Thomas
Senior Development Manager: Mark Thomas

Senior Production Manager: Tim Davis
Quality Assurance Coordinator: Karyn Payne
Desktop Publishing Coordinator: James McKay
Permissions Specialist: Adrienne Payne
Production Specialist: Adrienne Payne
Editor: Graham Hack

Writing and development services provided by Topaz Publications, Liverpool, NY

Lead Writer/Project Manager: Thomas Burke
Desktop Publisher: Joanne Hart
Art Director: Alison Richmond

Permissions Editor: Andrea LaBarge
Writer: Thomas Burke

Pearson

Director of Alliance/Partnership Management: Andrew Taylor
Editorial Assistant: Collin Lamothe
Program Manager: Alexandrina B. Wolf
Assistant Content Producer: Alma Dabral
Digital Content Producer: Jose Carchi
Director of Marketing: Leigh Ann Simms

Senior Marketing Manager: Brian Hoehl
Composition: NCCER
Printer/Binder: RR Donnelley
Cover Printer: RR Donnelley
Text Fonts: Palatino, Minion Pro, and Univers

Trainees with successful module completions may be eligible for credentialing through the NCCER Registry. To learn more, go to **www.nccer.org** or contact us at 1.888.622.3720. Our website has information on the latest product releases and training, as well as online versions of our *Cornerstone* magazine and Pearson's product catalog.

Your feedback is welcome. You may email your comments to **curriculum@nccer.org**, send general comments and inquiries to **info@nccer.org**, or fill in the User Update form at the back of this module.

This information is general in nature and intended for training purposes only. Actual performance of activities described in this manual requires compliance with all applicable operating, service, maintenance, and safety procedures under the direction of qualified personnel. References in this manual to patented or proprietary devices don't constitute a recommendation of their use.

46101 V3

From *Fundamentals of Crew Leadership, Trainee Guide.* NCCER.
Copyright © 2017 by NCCER. Published by Pearson Education. All rights reserved.

FUNDAMENTALS OF CREW LEADERSHIP

Objectives

When you have completed this module, you will be able to do the following:

1. Describe current issues and organizational structures in industry today.
   a. Describe the leadership issues facing the construction industry.
   b. Explain how gender and cultural issues affect the construction industry.
   c. Explain the organization of construction businesses and the need for policies and procedures.
2. Explain how to incorporate leadership skills into work habits, including communications, motivation, team-building, problem-solving, and decision-making skills.
   a. Describe the role of a leader on a construction crew.
   b. Explain the importance of written and oral communication skills.
   c. Describe methods for motivating team members.
   d. Explain the importance of teamwork to a construction project.
   e. Identify effective problem-solving and decision-making methods.
3. Identify a crew leader's typical safety responsibilities with respect to common safety issues, including awareness of safety regulations and the cost of accidents.
   a. Explain how a strong safety program can enhance a company's success.
   b. Explain the purpose of OSHA and describe the role of OSHA in administering worker safety.
   c. Describe the role of employers in establishing and administering safety programs.
   d. Explain how crew leaders are involved in administering safety policies and procedures.
4. Demonstrate a basic understanding of the planning process, scheduling, and cost and resource control.
   a. Describe how construction contracts are structured.
   b. Describe the project planning and scheduling processes.
   c. Explain how to implement cost controls on a construction project.
   d. Explain the crew leader's role in controlling project resources and productivity.

Performance Tasks

Under the supervision of your instructor, you should be able to do the following:

1. Develop and present a look-ahead schedule.
2. Develop an estimate for a given work activity.

Trade Terms

Autonomy
Bias
Cloud-based applications
Craft professionals
Crew leader
Critical path
Demographics
Ethics
Infer
Intangible
Job description
Job diary
Legend
Lethargy
Letter of instruction (LOI)
Local area networks (LAN)
Lockout/tagout (LOTO)

Look-ahead schedule
Negligence
Organizational chart
Paraphrase
Pragmatic
Proactive
Project manager
Return on investment (ROI)
Safety data sheets (SDS)
Sexual harassment
Smartphone
Superintendent
Synergy
Textspeak
Wide area networks (WAN)
Wi-Fi
Work breakdown structure (WBS)

Industry Recognized Credentials

If you are training through an NCCER-accredited sponsor, you may be eligible for credentials from NCCER's Registry. The ID number for this module is 46101. Note that this module may have been used in other NCCER curricula and may apply to other level completions. Contact NCCER's Registry at 888.622.3720 or go to **www.nccer.org** for more information.

Contents

1.0.0   Business Structures and Issues in the Industry 1
  1.1.0   Leadership Issues and Training Strategies 2
    1.1.1   Motivation 2
    1.1.2   Understanding Workers 2
    1.1.3   Craft Training 3
    1.1.4   Supervisory Training 3
    1.1.5   Impact of Technology 3
  1.2.0   Gender and Cultural Issues 4
    1.2.1   Communication Styles of Men and Women 5
    1.2.2   Language Barriers 5
    1.2.3   Cultural Differences 6
    1.2.4   Sexual Harassment 6
    1.2.5   Gender and Minority Discrimination 7
  1.3.0   Business Organization 8
    1.3.1   Division of Responsibility 8
    1.3.2   Responsibility, Authority, and Accountability 9
    1.3.3   Job Descriptions 9
    1.3.4   Policies and Procedures 10
2.0.0   Leadership Skills 13
  2.1.0   The Qualities and Role of a Leader 14
    2.1.1   Functions of a Leader 15
    2.1.2   Leadership Traits 15
    2.1.3   Expected Leadership Behavior 15
    2.1.4   Leadership Styles 15
    2.1.5   Ethics in Leadership 16
  2.2.0   Communication 17
    2.2.1   The Communication Process and Verbal Communication 17
    2.2.2   Written or Visual Communications 19
    2.2.3   Nonverbal Communication 19
    2.2.4   Communication Issues 20
  2.3.0   Motivation 20
    2.3.1   Accomplishment 20
    2.3.2   Change 22
    2.3.3   Job Importance 22
    2.3.4   Opportunity for Advancement 22
    2.3.5   Recognition 23
    2.3.6   Personal Growth 23
    2.3.7   Rewards 23
    2.3.8   Motivating Employees 24
  2.4.0   Team Building 24
    2.4.1   Successful Teams 24
    2.4.2   Building Successful Teams 24
    2.4.3   Delegating 25
    2.4.4   Implementing Policies and Procedures 26

2.5.0 Making Decisions and Solving Problems 27
  2.5.1 Decision Making Versus Problem Solving 27
  2.5.2 Types of Decisions 27
  2.5.3 Problem Solving 27
  2.5.4 Special Leadership Problems 28
3.0.0 Safety and Safety Leadership 32
  3.1.0 The Impact of Accidents 32
    3.1.1 Cost of Accidents 33
  3.2.0 OSHA 34
  3.3.0 Employer Safety Responsibilities 35
    3.3.1 Safety Program 35
    3.3.2 Safety Policies and Procedures 35
    3.3.3 Hazard Identification and Assessment 36
    3.3.4 Safety Information and Training 36
    3.3.5 Safety Record System 36
    3.3.6 Accident Investigation 37
  3.4.0 Leader Involvement in Safety 38
    3.4.1 Safety Training Sessions 38
    3.4.2 Inspections 39
    3.4.3 First Aid 39
    3.4.4 Fire Protection and Prevention 40
    3.4.5 Substance Abuse 40
    3.4.6 Job-Related Accident Investigations 41
    3.4.7 Promoting Safety 41
    3.4.8 Safety Contests 42
    3.4.9 Incentives and Rewards 42
    3.4.10 Publicity 43
4.0.0 Project Planning 46
  4.1.0 Construction Project Phases, Contracts, and Budgeting 46
    4.1.1 Development Phase 46
    4.1.2 Planning Phase 47
    4.1.3 Construction Phase 48
    4.1.4 Project-Delivery Systems 48
    4.1.5 Cost Estimating and Budgeting 48
  4.2.0 Planning 54
    4.2.1 Stages of Planning 57
    4.2.2 The Planning Process 58
    4.2.3 Planning Resources 60
    4.2.4 Scheduling 61
  4.3.0 Cost Control 65
    4.3.1 Assessing Cost Performance 65
    4.3.2 Field Reporting System 65
    4.3.3 Crew Leader's Role in Cost Control 67

4.4.0 Resource Control 68

    4.4.1 Ensuring On-Time Delivery 68

    4.4.2 Preventing Waste 68

    4.4.3 Verifying Material Delivery 69

    4.4.4 Controlling Delivery and Storage 69

    4.4.5 Preventing Theft and Vandalism 70

    4.4.6 Equipment Control 70

    4.4.7 Tool Control 70

    4.4.8 Labor Control 71

    4.4.9 Production and Productivity 71

Appendix OSHA Forms for Safety Records ....

Figures and Tables

Figure 1 Digital tablets are a convenient management and scheduling tool for supervisors 4
Figure 2 The modern workforce is diverse 5
Figure 3 Any unwelcome physical contact can be considered harassment 7
Figure 4 Sample organization chart for a construction company 9
Figure 5 Sample organization chart for an industrial company 10
Figure 6 A sample job description 11
Figure 7 Percentages of technical and supervisory work by role.............. 14
Figure 8 The authority-autonomy leadership style matrix 16
Figure 9 The communication process 17
Figure 10 Eye contact is important when communicating face-to-face 18
Figure 11 Avoid textspeak for on-the-job written communications............ 19
Figure 12 Tailor your message... 23
Figure 13 Falls are the leading cause of deaths and injuries in construction 33
Figure 14 Costs associated with accidents 34
Figure 15 Root causes of accidents 38
Figure 16 An impaired worker is a dangerous one 41
Figure 17 An example of a safety award.. 43
Figure 18 Architects and clients meet to refine plans 47
Figure 19 Project flow diagram 49
Figure 20 Project-delivery systems.. 50
Figure 21 Quantity takeoff sheet.. 52
Figure 22 Summary sheet 53
Figure 23 Proper prior planning prevents poor performance 54
Figure 24 Steps to effective planning... 59
Figure 25 Planning form 60
Figure 26 Sample page from a job diary... 60
Figure 27 Example of a bar-chart schedule 64
Figure 28 Example of a network schedule 65
Figure 29 Short-term schedule... 66
Figure 30 Idle workers cost money 67
Figure 31 Waste materials separated for recycling 69

Table 1 OSHA Penalties for violations established in 2016..................... 35

1.0.0 BUSINESS STRUCTURES AND ISSUES IN THE INDUSTRY

Objective

Describe current issues and organizational structures in industry today.

a. Describe the leadership issues facing the construction industry.
b. Explain how gender and cultural issues affect the construction industry.
c. Explain the organization of construction businesses and the need for policies and procedures.

Trade Terms

Cloud-based applications: Mobile and desktop digital programs that can connect to files and data located in distributed storage locations on the Internet ("the Cloud"). Such applications make it possible for authorized users to create, edit, and distribute content from any location where Internet access is possible.

Craft professionals: Workers who are properly trained and work in a particular construction trade or craft.

Crew leader: The immediate supervisor of a crew or team of craft professionals and other assigned persons.

Demographics: Social characteristics and other factors, such as language, economics, education, culture, and age, that define a statistical group of individuals. An individual can be a member of more than one demographic.

Job description: A description of the scope and responsibilities of a worker's job so that the individual and others understand what the job entails.

Letter of instruction (LOI): A written communication from a supervisor to a subordinate that informs the latter of some inadequacy in the individual's performance, and provides a list of actions that the individual must satisfactorily complete to remediate the problem. Usually a key step in a series of disciplinary actions.

Local area networks (LAN): Communication networks that link computers, printers, and servers within a small, defined location, such as a building or office, via hard-wired or wireless connections.

Organizational chart: A diagram that shows how the various management and operational responsibilities relate to each other within an organization. Named positions appear in ranked levels, top to bottom, from those functional units with the most and broadest authority to those with the least, with lines connecting positions indicating chains of authority and other relationships.

Paraphrase: Rewording a written or verbal statement in one's own words in such a way that the intent of the original statement is retained.

Pragmatic: Sensible, practical, and realistic.

Sexual harassment: Any unwelcome verbal or nonverbal form of communication or action construed by an individual to be of a sexual or gender-related nature.

Smartphone: The most common type of cellular telephone that combines many other digital functions within a single mobile device. Most important of these, besides the wireless telephone, are high-resolution still and video cameras, text and email messaging, and GPS-enabled features.

Synergy: Any type of cooperation between organizations, individuals, or other entities where the combined effect is greater than the sum of the individual efforts.

Wide area networks (WAN): Dedicated communication networks of computers and related hardware that serve a given geographic area, such as a work site, campus, city, or a larger but distinct area. Connectivity is by wired and wireless means, and may use the Internet as well.

Wi-Fi: The technology allowing communications via radio signals over a LAN or WAN equipped with a wireless access point or the Internet. (Wi-Fi stands for wireless fidelity.) Many types of mobile, portable, and desktop devices can communicate via Wi-Fi connections.

Today's managers, supervisors, and lower-level managers face challenges different from those of previous generations. To be a **crew leader** today, it is essential to be well prepared. Crew leaders must understand how to use various types of new technology. In addition, they must have the knowledge and skills needed to manage, train, and communicate with a culturally-diverse workforce whose attitudes toward work and job expectations may differ from those of earlier generations and cultures.

A summary of the changes in the workforce, in the work environment, and of industry needs include the following:

- A shrinking workforce
- The growth of construction, communication, and scheduling technology
- Changes in the attitudes and values of **craft professionals**
- The rapidly-changing **demographics** of gender, gender-identity, and foreign-born workers
- Increased emphasis on workplace safety and health
- Greater need for more education and training

1.1.0 Leadership Issues and Training Strategies

Effective craft training programs are necessary if the industry is to meet the forecasted worker demands. Many skilled, knowledgeable craft professionals, crew leaders, and managers have reached retirement age. In 2015, the generation of workers called *"Baby Boomers,"* who were born between 1946 and 1964, represented 50 percent of the workforce. Their departure creates a large demand for craftworkers across the industry. The US Department of Labor (DOL) concludes that the best way for industry to reduce shortages of skilled workers is to create more education and training opportunities. The DOL suggests that companies and community groups form partnerships and create apprenticeship programs. Such programs could provide younger workers, including women and minorities, with the opportunity to develop job skills by giving them hands-on experience.

When training workers, it is important to understand that people learn in different ways. Some people learn by doing, some people learn by watching or reading, and others need step-by-step instructions as they are shown the process. Most people learn best through a combination of styles. While you may have the tendency to teach in the style that you learn best, you must always stay in tune with what kind of learner you are teaching. Have you ever tried to teach somebody and failed, and then another person successfully teaches the same thing in a different way? A person who acts as a mentor or trainer needs to be able to determine what kind of learner they are addressing, and teach according to those needs.

"The mediocre teacher tells.
The good teacher explains.
The superior teacher demonstrates.
The great teacher inspires."
—*William Arthur Ward (1921)*

The need for training isn't limited to craft professionals. There must be supervisory training to ensure there are qualified leaders in the industry to supervise the craft professionals.

1.1.1 Motivation

As a supervisor or crew leader, it is important to understand what motivates your crew. Money is often considered a good motivator, but it is sometimes only a temporary solution. Once a person has reached a level of financial security, other factors come into play. Studies show that environment and conditions motivate many people. For those people, a great workplace may mean more to them than better pay.

If you give someone a raise, they tend to work harder for a period of time. Then the satisfaction wanes and they may want another raise. A sense of accomplishment is what motivates most people. That is why setting and working toward recognizable goals tends to make employees more productive. A person with a feeling of involvement or a sense of achievement is likely to be better motivated and help to motivate others.

1.1.2 Understanding Workers

Many older workers grew up in an environment where they learned to work hard with the same employer until retirement. They expected to stay with a company for a long time, and the structure of the companies created a family-type environment.

Times have changed. Younger workers have grown up in a highly mobile society and expect frequent rewards and rapid advancement. Some might perceive this generation of workers as lazy, narcissistic, and unmotivated, but in reality, they simply have a different perspective on life, work, and priorities. For such workers, it may be better to give them small projects or break up large projects into smaller pieces so that they are rewarded more often by successfully achieving short-term goals. The following strategies are important for keeping young workers motivated and engaged:

- *Goal setting* – Set short-term and long-term goals, including tasks to be done and expected time frames. Things can happen to change, delay, or upset the short-term goals. This is one reason to set long-term goals as well. Don't set workers up for failure, as this leads to frustration, and frustration can lead to reduced productivity.
- *Feedback* – Timely feedback is important. For example, telling someone they did a good job last year, or criticizing them for a job they did a month ago, is meaningless. Simple recognition isn't always enough. Some type of reward should accompany positive feedback, even if it is simply recognizing the employee in a public way. Constructive criticism or reprimands should always be given in private. You can also provide some positive action, such as one-on-one training, to correct a problem.

1.1.3 Craft Training

Craft training is often informal, taking place on the job site, outside of a traditional training classroom. According to the American Society for Training and Development (ASTD), a qualified co-worker or a supervisor conducts craft training generally through on-the-job instruction. The Society of Human Resources Management (SHRM) offers the following tips to supervisors in charge of training their employees:

- *Help crew members establish career goals* – Once career goals are established, you can readily identify the training required to meet the goals.
- *Determine what kind of training to give* – Training can be on the job under the supervision of a co-worker. It can be one-on-one with the supervisor. It can involve cross training to teach a new trade or skill, or it can involve delegating new or additional responsibilities.
- *Determine the trainee's preferred method of learning* – Some people learn best by watching, others from verbal instructions, and others by doing. These are categorized as visual learners, auditory learners, and tactile learners, respectively. When training more than one person at a time, try to use a mix of all three methods.

Communication is a critical component of training employees. SHRM advises that supervisors do the following when training their employees:

- Explain the task, why it is important, and how to do it. Confirm that the worker trainees understand these three areas by asking questions. Allow them to ask questions as well.
- Demonstrate the task. Break the task down into manageable parts and cover one part at a time.
- Ask your trainees to do the task while you observe them. Try not to interrupt them while they are doing the task unless they are doing something that is unsafe or potentially harmful.
- Give the trainees feedback. Be specific about what they did and mention any areas where they need to improve.

1.1.4 Supervisory Training

Because of the need for skilled craft professionals and qualified supervisory personnel, some companies offer training to their employees through in-house classes, or by subsidizing outside training programs. However, for a variety of reasons, many contractors don't offer training at all. Some common reasons include the following:

- Lack of money to train
- Lack of time to train
- Lack of knowledge about the benefits of training programs
- High rate of employee turnover
- Workforce is too small
- Past training involvement was ineffective
- The company hires only trained workers
- Lack of interest from workers

For craft professionals to move up into supervisory and managerial positions, they must continue their education and training. Those who are willing to acquire and develop new skills have the best chance of finding stable employment. This makes it critical for them to take advantage of training opportunities, and for companies to incorporate training into their business culture.

If your company has recognized the need for training, your participation in a leadership training program such as this will begin to fill the gap between craft and supervisory training.

1.1.5 Impact of Technology

Many industries, including the construction industry, have embraced technology to remain competitive. Benefits of technology include increased productivity and speed, improved quality of documents, greater access to common data, and better financial controls and communication. As technology becomes a greater part of supervision, crew leaders must be able to use it properly.

Cellphones (in particular the **smartphone**) have made it easy to keep in touch through numerous forms of media. As of 2016, 95 percent of all Americans own a cellphone of some kind and 77 percent own a smartphone, according to the Pew Research Center. They are particularly useful communication tools for contractors or crew leaders who are on a job site, away from their offices, or constantly on the go. Workers use smartphones at any time for phone calls, emails, text messages, and voicemail, as well as to share photos and videos. The hundreds of thousands of mobile computing apps available allow smartphones to perform numerous other functions on the go. However, the number of accidents due to inattention while focusing on a smartphone or other mobile device is also rapidly rising. Always check the company's policy regarding cell phone use on the job.

As a crew leader, you should be aware that smartphones and tablets (*Figure 1*) allow supervisors to plan their calendars, schedule meetings, manage projects, and access their company email from remote locations. These devices are far more powerful than the computers that took us to the moon, and they can hold years of information from various projects. Cloud-based applications now permit remote access to and updating of files, plans, and data from any place with wireless connectivity. In fact, it is becoming common for work sites to set up local area networks (LAN), wide area networks (WAN), and dedicated Wi-Fi services to support mobile communications among workers on the job.

In all forms of electronic communication (verbal, written, and visual), it is important to keep messages brief, factual, and legal. Text-based communications can be easily misunderstood because there are no visual or auditory cues to indicate the sender's intent. In other words, it is more difficult to tell if someone is just joking via email because you can't see the sender's expression or hear their tone of voice.

1.2.0 Gender and Cultural Issues

In the past several years, the construction industry in the United States has experienced a shift in worker expectations and diversity. These two issues are converging at a rapid pace.

The generation of learners is also a factor in the learning process and in the workplace. The various generations include Baby Boomers, Generation X, Millennials (Generation Y), and Generation Z. The ranges of birth years for these generations are as follows (note that there may be

Figure 1 Digital tablets are a convenient management and scheduling tool for supervisors.

some overlap, and opinions on the generational names as well as the range of years for each generation may vary):

- *Baby Boomers* – born 1946 through 1964
- *Generation X* – born 1965 through 1979
- *Millennials* (*Generation Y*) – born 1980 through 2000
- *Generation Z* – born 2000 through 2015

Each generation has been studied to some degree to determine their different interpretations of and approaches to learning and training. Family norms, religious values and morals, educational methods, music, movies, politics, and global events are all influential factors that define a generation. Remember that individuals within a generation can vary widely in their expectations and behaviors.

This trend, combined with industry diversity initiatives, has created a climate in which companies recognize the need to embrace a diverse workforce that crosses generational, gender, and ethnic boundaries (*Figure 2*). To do this effectively, they are using their own resources, as well as relying on associations with the government and trade organizations.

All current research indicates that industry will be more dependent on the critical skills of a diverse workforce—but a workforce that must be both culturally and ethnically fused. Across the United States, construction and other industries are aggressively seeking to bring new workers into their ranks, including women, racial, and ethnic minorities. Social and political issues are no longer the main factors driving workplace diversity, but by consumers and citizens who need more hospitals, malls, bridges, power plants, refineries, and many other commercial and residential structures. The construction industry needs more workers.

Figure 2 The modern workforce is diverse.

There are some potential issues relating to a diverse workforce that may be encountered on the job site. These issues include different communication styles of men and women, language barriers associated with cultural differences, **sexual harassment**, and gender or racial discrimination.

1.2.1 Communication Styles of Men and Women

As more and more women enter construction workforce, it becomes increasingly important to break down communication barriers between men and women and to understand differences in behaviors so that men and women can work together more effectively. The Jamestown, Area Labor Management Committee (JALMC) in New York offers the following explanations and tips (*emphasized text* is from *Exchange and Deception: A Feminist Perspective*, ed. by Caroline Gerschlager, Monika Mokre, 2002. Springer Science & Business Media):

- *Women tend to ask more questions than men do* – Men are more likely to proceed with a job and figure it out as they go along, while women are more likely to ask questions first.
- *Men tend to offer solutions before empathy; women tend to do the opposite* – Both men and women should say what they want up front, whether it's the solution to a problem, or simply a sympathetic ear. That way, both genders will feel understood and supported.
- *Women are more likely to ask for help when they need it* – Women are generally more **pragmatic** when it comes to completing a task. If they need help, they will ask for it. Men are more likely to attempt to complete a task by themselves, even when they need assistance.
- *Men tend to communicate more competitively, and women tend to communicate more cooperatively* – Both parties need to hear one another out without interruption.

This doesn't mean that any one method is better or worse than the other; it simply means that men and women inherently use different approaches to achieve the same result. It's either genetic or cultural, but awareness can overcome these tendencies.

1.2.2 Language Barriers

Language barriers are a real workplace challenge for crew leaders. Millions of American workers speak languages other than English. Bilingual job sites are increasingly common. As the makeup of the immigrant population continues to change with the influx of refugees from around the world, the number of non-English speakers in the American workforce is rising and diversifying dramatically.

In addition, there are many different dialects of the English language in America, which can present some communication challenges. For example, some workers may speak a dialect commonly called *Ebonics* (referred to by linguists as *African-American Vernacular English*). The dialect called *Rural White Southern English* can also be difficult to understand by those not born in the American South. Many find the American New England accent and dialect difficult to comprehend. Some sources list as many as 16 distinct American dialects.

Companies have the following options to overcome the language challenge, mainly for foreign or English-as-a-second-language (ESL) workers:

- Offer English classes either at the work site or through school districts and community colleges.
- Offer incentives for workers to learn English.

Communication with ESL workers, and varying dialects in general, becomes even more critical as the workforce grows more diverse. The following tips will help when communicating across language barriers:

- Be patient. Give workers time to process the information in a way they can comprehend.
- Avoid humor. Humor is easily misunderstood. The worker may misinterpret what you say as a joke at the worker's expense.
- Don't assume workers are unintelligent simply because they don't understand what you are saying. Explaining something in multiple ways, and having trainees paraphrase what they heard, is an excellent way to ensure mutual understanding and prevent miscommunication.
- If a worker is not fluent in English, ask the worker to demonstrate his or her understanding through other means.
- Speak slowly and clearly, and avoid the tendency to raise your voice.
- Use face-to-face communication whenever possible. Over-the-phone communication is often more difficult when a language barrier is involved.
- Use pictures or sketches to get your point across.

1.2.3 Cultural Differences

As workers from a multitude of backgrounds and cultures work together, there are bound to be differences and conflicts in the workplace. To overcome cultural conflicts, the SHRM suggests the following approach to resolving cultural conflicts between individuals:

- *Define the problem from both points of view* – How does each person involved view the conflict? What does each person think is wrong? This involves moving beyond traditional thought processes to consider alternate ways of thinking.
- *Uncover cultural interpretations* – What assumptions may the parties involved be making based on cultural programming? This is particularly true for certain gestures, symbols, and words that mean different things in different cultures. By doing this, the supervisor may realize what motivated an employee to act in a specific way.
- *Create cultural* synergy – Devise a solution that works for both parties involved. The purpose is to recognize and respect other's cultural values, and work out mutually acceptable alternatives.

1.2.4 Sexual Harassment

In today's business world, men and women are working side-by-side in careers of all kinds, creating increased opportunity for sexual harassment to occur. Sexual harassment can be defined as any unwelcome behavior that makes someone feel uncomfortable in the workplace by focusing attention on their gender or gender identity. Activities that might qualify as sexual harassment include, but are not limited to, the following:

- Telling an offensive, sexually-oriented joke
- Displaying a poster of a man or woman in a revealing swimsuit
- Wearing a patch or article of clothing blatantly promoting or degrading a specific gender
- Making verbal or physical advances (*Figure 3*)
- Speaking abusively about a specific gender

Historically, many have thought of sexual harassment as an act perpetrated by men against women, especially those in subordinate positions. However, the nature of sexual harassment cases over the years have shown that the perpetrator can be of any gender.

Figure 3 Any unwelcome physical contact can be considered harassment.

Sexual harassment can occur in various circumstances, including the following:

- The victim and harasser may be either male or female. The victim does not have to be of the opposite gender.
- The harasser can be the victim's supervisor, an agent of the employer, a supervisor in another area, a co-worker, a subordinate, or a nonemployee.
- Legally, the victim doesn't have to be the person harassed, but could be anyone offended by the relevant conduct.
- Unlawful sexual harassment may occur without economic injury to or the discharge of the victim.
- The harasser's conduct must be unwelcome.

The Equal Employment Opportunity Commission (EEOC) enforces sexual harassment laws within industries. When investigating allegations of sexual harassment, the EEOC looks at the whole record, including the circumstances and the context in which the alleged incidents occurred. A decision on the allegations is made from the facts on a case-by-case basis. Supervision can hold the crew leader responsible who is aware of sexual harassment and does nothing to stop it. The crew leader therefore should not only take action to stop sexual harassment, but should serve as a good example for the rest of the crew.

Prevention is the best tool to eliminate sexual harassment in the workplace. The EEOC encourages employers to take steps to prevent sexual harassment from occurring. Employers should clearly communicate to employees that they and their company won't tolerate sexual harassment. They do so by developing a policy on sexual harassment, establishing an effective complaint or grievance process, and taking immediate and appropriate action when an employee complains.

Controversial or sexually-related remarks, jokes, and swearing are not only offensive to co-workers, but also tarnish a worker's character. Crew leaders need to emphasize that abrasive or crude behavior may affect opportunities for advancement. If disciplinary action becomes necessary, written company policies should spell out the type of actions warranted by such behavior. A typical approach is a three-step process in which the perpetrator is first given a verbal reprimand. In the event of further violations, a written reprimand, often including a letter of instruction (LOI) and a warning are given. Dismissal typically accompanies any subsequent violations.

1.2.5 Gender and Minority Discrimination

Employers are giving more attention to fair recruitment, equal pay for equal work, and promotions for women and minorities in the workplace. Consequently, many businesses are analyzing their practices for equity, including the treatment of employees, the organization's hiring and promotional practices, and compensation.

The construction industry, which was once male-dominated, is moving away from this image and actively recruiting and training women, younger workers, people from other cultures, and disabled workers. This means that organizations hire the best person for the job, without regard for race, sex, religion, age, etc.

Did You Know?

Respectful Workplace Training

Some companies employ a tool called *sensitivity training* in cases where individuals or groups have trouble adapting to a multi-cultural, multi-gender workforce. Sensitivity training is a psychological technique using group discussion, role playing, and other methods to allow participants to develop an awareness of themselves and how they interact with others. Critics of earlier forms of sensitivity training noted that the methods used were not very different from brainwashing. Modern forms of cultural-awareness and gender-issue training are more ethical, and go by the term *respectful workplace training*.

To prevent discrimination cases, employers must have valid job-related criteria for hiring, compensation, and promotion. They must apply these measures consistently for every applicant interview, employee performance appraisal, and hiring or promotion decision. Therefore, employers must train all workers responsible for recruitment, selection, supervision of employees, and evaluating job performance on how to use the job-related criteria legally and effectively.

1.3.0 Business Organization

An organization is the relationship among the people within the company or project. The crew leader needs to be aware of two types of organizations: formal organizations and informal organizations.

A formal organization exists when the people within a work group led by someone direct their activities toward achieving a common goal. An example of a formal organization is a work crew consisting of four carpenters and two laborers led by a crew leader, all working together to accomplish a project.

An **organizational chart** typically illustrates a formal organization. It outlines all the positions that make up an organization and shows how those positions are related. Some organizational charts even identify the person within each position and the person to whom that person reports, as well as the people that the person supervises. *Figure 4* and *Figure 5* show examples of what an organization chart might look like for a construction company and an industrial company. Note that each organizational position represents an opportunity for advancement in the construction industry that a crew leader can eventually achieve.

An informal organization allows for communication among its members so they can perform as a group. It also establishes patterns of behavior that help them to work as a group, such as agreeing to use a specific training or certification program.

An example of an informal organization is a trade association such as Associated Builders and Contractors (ABC), Associated General Contractors (AGC), and the National Association of Women in Construction (NAWIC). Those, along with the thousands of other trade associations in the United States, provide forums in which members with common concerns can share information, work on issues, and develop standards for their industry.

Both formal and informal organizations establish the foundation for how communication flows. The formal structure is the means used to delegate authority and responsibility and to exchange information. The informal structure provides means for the exchange of information.

Members in an organization perform best when they can do the following:

- Know the job and how it will be done
- Communicate effectively with others in the group
- Understand their role in the organization
- Recognize who has the authority and responsibility

"Leadership is not about titles, positions, or flow charts. It is about one life influencing another."
—John C. Maxwell

1.3.1 Division of Responsibility

The conduct of a business involves certain functions. In a small organization, one or two people may share responsibilities. However, in a larger organization with many different and complex activities, distinct activity groups may lie under the responsibility of separate department managers. In either case, the following major department functions exist in most companies:

- *Executive* – This office represents top management. It is responsible for the success of the company through short-range and long-range planning.
- *Human Resources (HR)* – This office is responsible for recruiting and screening prospective employees, managing employee benefits programs, advising management on pay and benefits, and developing and enforcing procedures related to hiring practices.
- *Accounting* – This office is responsible for all recordkeeping and financial transactions, including payroll, taxes, insurance, and audits.
- *Contract Administration* – This office prepares and executes contractual documents with owners, subcontractors, and suppliers.
- *Purchasing* – This office obtains material prices and then issues purchase orders. The purchasing office also obtains rental and leasing rates on equipment and tools.
- *Estimating* – This office is responsible for recording the quantity of material on the jobs, the takeoff, pricing labor and material, analyzing subcontractor bids, and bidding on projects.
- *Operations* – This office plans, controls, and supervises all project-related activities.

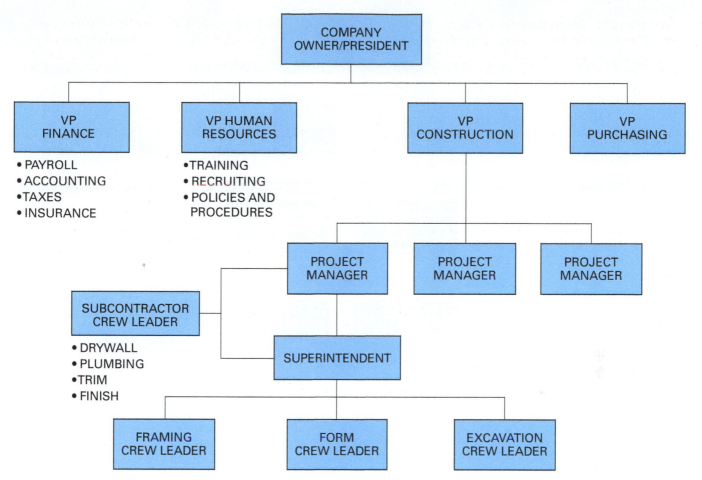

Figure 4 Sample organization chart for a construction company.

Other divisions of responsibility a company may create involve architectural and engineering design functions. These divisions usually become separate departments.

1.3.2 Responsibility, Authority, and Accountability

As an organization grows, the manager must ask others to perform many duties so that the manager can concentrate on management tasks. Managers typically assign (delegate) activities to their subordinates. When delegating activities, the crew leader assigns others the responsibility to perform the designated tasks.

Responsibility is the obligation to perform the duties. Along with responsibility comes authority. *Authority* is the right to act or make decisions in carrying out an assignment. The type and amount of authority a supervisor or worker has depends on the employee's company. There must be a balance between the authority and responsibility workers have so that they can carry out their assigned tasks. In addition, the crew leader must delegate sufficient authority to make a worker accountable to the crew leader for the results.

Accountability is holding an employee responsible for completing the assigned activities. Even though the crew leader may delegate authority and responsibility to crew members, the overall responsibility—the accountability—for the tasks assigned to the crew always rests with the crew leader.

1.3.3 Job Descriptions

Many companies furnish each employee with a written **job description** that explains the job in detail. Job descriptions set a standard for the employee. They make judging performance easier, clarify the tasks each person should handle, and simplify the training of new employees.

Each new employee should understand all the duties and responsibilities of the job after reviewing the job description. Thus, the time is shorter for the employee to make the transition from being a new and uninformed employee to a more experienced member of a crew.

A job description need not be long, but it should be detailed enough to ensure there is no misunderstanding of the duties and responsibilities of the position. The job description should

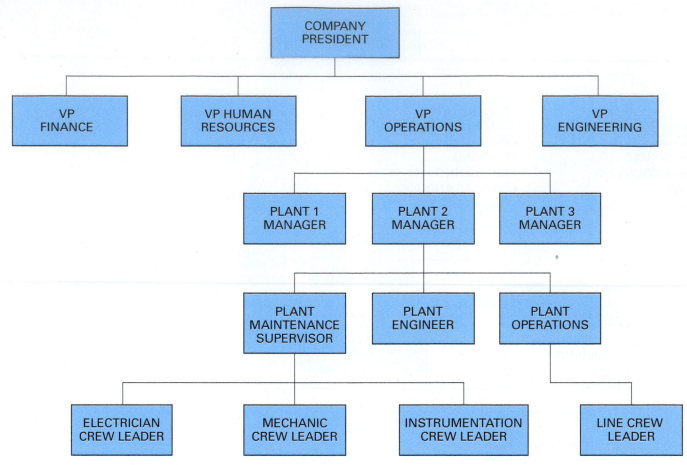

Figure 5 Sample organization chart for an industrial company.

contain all the information necessary to evaluate the employee's performance and hold the employee accountable.

A job description should contain, at minimum, the following:

- Job title
- General description of the position
- Minimum qualifications for the job
- Specific duties and responsibilities
- The supervisor to whom the position reports
- Other requirements, such as qualifications, certifications, and licenses

Figure 6 is an example of a job description.

1.3.4 Policies and Procedures

Most companies have formal policies and procedures established to help crew leaders carry out their duties. A policy is a general statement establishing guidelines for a specific activity. Examples include policies on vacations, breaks, workplace safety, and checking out tools. Procedures are formal instructions to carry out and meet policies. For example, a procedure written to implement a policy on workplace safety would include guidelines expected of all employees for reporting accidents and to follow general safety practices.

A crew leader must be familiar with the company policies and procedures, especially regarding safety practices. When OSHA inspectors visit a jobsite, they often question employees and crew leaders about the company policies related to safety. If they are investigating an accident, they will want to verify that the responsible crew leader knew the applicable company policy and followed it.

Position:
Crew Leader

General Summary:
First line of supervision on a construction crew installing concrete formwork.

Reports To:
Job Superintendent

Physical and Mental Responsibilities:
- Ability to stand for long periods
- Ability to solve basic math and geometry problems

Duties and Responsibilities:
- Oversee crew
- Provide instruction and training in construction tasks as needed
- Make sure proper materials and tools are on the site to accomplish tasks
- Keep project on schedule
- Enforce safety policies and procedures

Knowledge, Skills, and Experience Required:
- Extensive travel throughout the Eastern United States, home base in Atlanta
- Ability to operate a backhoe and trencher
- Valid commercial driver's license with no DUI violations
- Ability to work under deadlines with the knowledge and ability to foresee problem areas and develop a plan of action to solve the situation

Figure 6 A sample job description.

Did You Know?

The ADA

The Americans with Disabilities Act (ADA) is a law that prohibits employers from discriminating against disabled persons in their hiring practices. While the ADA doesn't require written job descriptions, investigators use existing job descriptions as evidence in determining whether discrimination occurred. This means that job descriptions must specifically define the job duties and the abilities needed to perform them. In making a determination of discrimination under the ADA, investigators give consideration to whether a reasonable accommodation could have qualified the disabled person to fill a job.

Additional Resources

Construction Workforce Development Professional, NCCER. 2016. New York, NY: Pearson Education, Inc.

Mentoring for Craft Professionals, NCCER. 2016. New York, NY: Pearson Education, Inc.

Generational Cohorts and their Attitudes Toward Work Related Issues in Central Kentucky, Frank Fletcher, et al. 2009. Midway College, Midway, KY. **www.kentucky.com**

The Young Person's Guide to Wisdom, Power, and Life Success: Making Smart Choices. Brian Gahran, PhD. 2014. San Diego, CA: Young Persons Press. **www.WPGBlog.com**

The following websites offer resources for products and training:

Aging Workforce News, **www.agingworkforcenews.com**

American Society for Training and Development (ASTD), **www.astd.org**

Equal Employment Opportunity Commission (EEOC), **www.eeoc.gov**

National Association of Women in Construction (NAWIC), **www.nawic.org**

Society for Human Resources Management (SHRM), **www.shrm.org**

United States Census Bureau, **www.census.gov**

United States Department of Labor, **www.dol.gov**

Wi-Fi® is a registered trademark of the Wi-Fi Alliance, **www.wi-fi.org**

1.0.0 Section Review

1. According to the US Department of Labor, the best way for the construction industry to reduce skilled-worker shortages is to _____.
   a. create training opportunities
   b. avoid discrimination lawsuits
   c. update the skills of older workers who are retiring at a later age than they previously did
   d. implement better policies and procedures

2. Which of the following is *not* helpful when dealing with language diversity in the workplace?
   a. Speaking slowly and clearly in an even tone
   b. Using humor when communicating
   c. Using sketches or diagrams to explain what needs to be done.
   d. Being patient and giving bilingual workers time to process your instructions.

3. Members tend to function best within an organization when they _____.
   a. are allowed to select their own uniform for each project
   b. understand their role within the organization
   c. don't disagree with the statements of other workers or supervisors
   d. are able to work without supervision

2.0.0 LEADERSHIP SKILLS

Objective

Explain how to incorporate leadership skills into work habits, including communications, motivation, team-building, problem-solving, and decision-making skills.

a. Describe the role of a leader on a construction crew.
b. Explain the importance of written and oral communication skills.
c. Describe methods for motivating team members.
d. Explain the importance of teamwork to a construction project.
e. Identify effective problem-solving and decision-making methods.

Trade Terms

Autonomy: The condition of having complete control over one's actions, and being free from the control of another. To be independent.

Bias: A preconceived inclination against or in favor of something.

Ethics: The moral principles that guides an individual's or organization's actions when dealing with others. Also refers to the study of moral principles.

Infer: To reach a conclusion using a method of reasoning that starts with an assumption and considers a set of logically-related events, conditions, or statements.

Legend: In maps, plans, and diagrams, an explanatory table defining all symbolic information contained in the document.

Lockout/tagout (LOTO): A system of safety procedures for securing electrical and mechanical equipment during repairs or construction, consisting of warning tags and physical locking or restraining devices applied to controls to prevent the accidental or purposeful operation of the equipment, endangering workers, equipment, or facilities.

Proactive: To anticipate and take action in the present to deal with potential future events or outcomes based on what one knows about current events or conditions.

Project manager: In construction, the individual who has overall responsibility for one or more construction projects; also called the general superintendent.

Superintendent: In construction, the individual who is the on-site supervisor in charge of a given construction project.

Textspeak: A form of written language characteristic of text messaging on mobile devices and text-based social media, which usually consists of acronyms, abbreviations, and minimal punctuation.

It is important to define some of the supervisory positions discussed throughout this module. You are already familiar with the roles of the craft professional and crew leader. A superintendent is essentially an on-site supervisor who is responsible for one or more crew leaders or front-line supervisors. A project manager or general superintendent may be responsible for managing one or more projects. This training will concentrate primarily on the supervisory role of the crew leader.

Craftworkers and crew leaders differ in that the crew leader manages the activities that the craft professionals perform. To manage a crew of craft professionals, a crew leader must have first-hand knowledge and experience in their activities. Additionally, the crew leader must be able to act directly in organizing and directing the activities of the various crew members.

This section explains the importance of developing leadership skills as a new crew leader. It will cover effective ways to communicate with co-workers and employees at all levels, build teams, motivate crew members, make decisions, and resolve problems.

Crew leaders are generally promoted up from a work crew. A worker's ability to accomplish tasks, get along with others, meet schedules, and stay within the budget have a significant influence on the selection process. The crew leader must lead the team to work safely and provide a quality product.

Making the transition from crew member to a crew leader can be difficult, especially when the new position involves overseeing a group of former peers. For example, some of the crew may try to take advantage of their friendship by seeking special favors. They may also want to be privy to supervisory information that is normally closely held. When you become a crew leader, you are no longer responsible for your work alone. Crew leaders are accountable for the work of an entire crew of people with varying skill levels, personalities, work styles, and cultural and educational backgrounds.

Crew leaders must learn to put personal relationships aside and work for the common goals of the entire crew. The crew leader can overcome these problems by working with the crew to set mutual performance goals and by freely communicating with them within permitted limits. Use their knowledge and strengths along with your own so that they feel like they are key players on the team.

As employees move from being a craftworker into the role of a crew leader and above, they will begin to spend more hours supervising the work of others (supervisory work) than practicing their own craft skills (technical work). *Figure 7* illustrates the relative amounts of time craft professionals, crew leaders, superintendents, and project managers spend on technical and supervisory work as their management responsibilities increase.

There are many ways to define a leader. One simple definition of a leader is a person who influences other people to achieve a goal. Some people may have innate leadership qualities that developed during their upbringing, or they may have worked to develop the traits that motivate others to follow and perform. Research shows that people who possess such talents are likely to succeed as leaders.

> *"Leadership is the art of getting someone else to do something you want done because he wants to do it."*
> —Dwight D. Eisenhower

2.1.0 The Qualities and Role of a Leader

Leadership traits are similar to the skills that a crew leader needs to be effective. Although the characteristics of leadership are many, there are some definite commonalities among effective leaders.

First and foremost, effective leaders lead by example. They work and live by the standards that they establish for their crew members or followers, making sure they set a positive example.

Effective leaders also tend to have a high level of drive and determination, as well as a persistent attitude. When faced with obstacles, effective leaders don't get discouraged. Instead, they identify the potential problems, make plans to overcome them, and work toward achieving the intended goal. In the event of failure, effective leaders learn from their mistakes and apply that knowledge to future situations. They also learn from their successes.

Effective leaders are typically good communicators who clearly lay out the goals of a project to their crew members. Accomplishing this may require that the leader overcome issues such as language barriers, gender bias, or differences in personalities to ensure that each member of the crew understands the established goals of the project.

Effective leaders can motivate their crew members to work to their full potential and become useful members of the team. Crew leaders try to develop crew-member skills and encourage them to improve and learn so they can contribute more to the team effort. Effective leaders strive for excellence from themselves and their team, so they work hard to provide the skills and leadership necessary to do so.

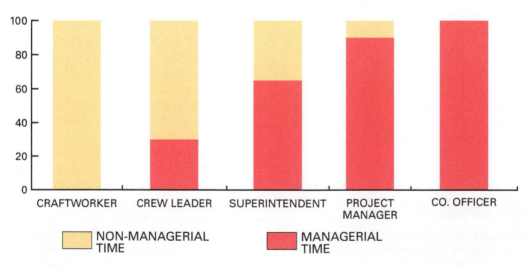

Figure 7 Percentages of technical and supervisory work by role.

In addition, effective leaders must possess organizational skills. They know what needs to be accomplished, and they use their resources to make it happen. Because they can't do it alone, leaders require the help of their team members to share in the workload. Effective leaders delegate work to their crew members, and they implement company policies and procedures to complete the work safely, effectively, and efficiently.

Finally, effective leaders have the experience, authority, and self-confidence that allows them to make decisions and solve problems. To accomplish their goals, leaders must be able to calculate risks, take in and interpret information, assess courses of action, make decisions, and assume the responsibility for those decisions.

2.1.1 Functions of a Leader

The functions of a leader will vary with the environment, the group of individuals they lead, and the tasks they must perform. However, there are certain functions common to all situations that the leader must fulfill. Some of the major functions are as follows:

- Accept responsibility for the successes and failures of the group's performance
- Be sensitive to the differences of a diverse workforce
- Ensure that all group members understand and abide by company policies and procedures
- Give group members the confidence to make decisions and take responsibility for their work
- Maintain a cohesive group by resolving tensions and differences among its members and between the group and those outside the group
- Organize, plan, staff, direct, and control work
- Represent the group

2.1.2 Leadership Traits

There are many traits and skills that help create an effective leader. Some of these include the following:

- Ability to advocate an idea
- Ability to motivate
- Ability to plan and organize
- Ability to teach others
- Enthusiasm
- Fairness
- Good communication skills
- Initiative
- Loyalty to their company and crew
- Willingness to learn from others

2.1.3 Expected Leadership Behavior

Followers have expectations of their leaders. They look to their leaders to do the following:

- Abide by company policies and procedures
- Be a loyal member of the team
- Communicate effectively
- Have the necessary technical knowledge
- Lead by example
- Make decisions and assume responsibility
- Plan and organize the work
- Suggest and direct
- Trust the team members

2.1.4 Leadership Styles

Through history, there have been many terms used to describe different leadership styles and the amount of autonomy they permit. The amount of crew autonomy directly relates to the leadership style used by the crew leader. *Figure 8* illustrates the linear relationship between leader authority and worker autonomy.

There are three main styles of leadership. At one extreme is the *controller style* of leadership, where the crew leader makes all of the decisions independently, without seeking the opinions of crew members. (Very little autonomy exists for workers under this type of leader.) At the other extreme is the *advisor style*, where the crew leader empowers the employees to make decisions (high autonomy). In between these extremes is the *directive style*, where the crew leader seeks crew member opinions and makes the appropriate decisions based on their input.

The following are some characteristics of each of the three leadership styles:

Controller style (high authority, low autonomy):

- Expect crew members to work without questioning procedures
- Seldom seek advice from crew members
- Insist on solving problems alone
- Seldom permit crew members to assist each other
- Praise and criticize on a personal basis
- Have no sincere interest in creatively improving methods of operation or production

Directive style (equal authority and autonomy):

- Discuss problems with their crew members
- Listen to suggestions from crew members
- Explain and instruct
- Give crew members a feeling of accomplishment by commending them when they do a job well

- Are friendly and available to discuss personal and job-related problems

Advisor style (low authority, high autonomy):

- Believe no supervision is best
- Rarely give orders
- Worry about whether they are liked by their crew members

Effective leadership takes many forms. The correct style for a particular situation or operation depends on the nature of the crew as well as the work it has to accomplish. For example, if the crew does not have enough experience for the job ahead, then a controller style may be appropriate. The controller style of leadership is also effective when jobs involve repetitive operations that require little decision-making.

However, if a worker's attitude is an issue, a directive style may be appropriate. In this case, providing the missing motivational factors may increase performance and result in the improvement of the worker's attitude. The directive style of leadership is also used when the work is of a creative nature, because brainstorming and exchanging ideas with such crew members can be beneficial.

The advisor style is effective with an experienced crew on a well-defined project. The company must give a crew leader sufficient authority to do the job. This authority must be commensurate with responsibility, and it must be made known to crew members when they are hired so that they understand who is in charge.

A crew leader must have an expert knowledge of the activities to be supervised in order to be effective. This is important because the crew members need to know that they have someone to turn to when they have a question or a problem, when they need some guidance, or when modifications or changes are warranted by the job.

Respect is probably the most useful element of authority. Respect usually derives from being fair to employees by listening to their complaints and suggestions, and by using incentives and rewards appropriately to motivate crew members. In addition, crew leaders who have a positive attitude and a favorable personality tend to gain the respect of their crew members as well as their peers. Along with respect comes a positive attitude from the crew members.

2.1.5 Ethics in Leadership

Crew leaders should maintain the highest standards of honesty and legality. Every day, the crew leader must make decisions that may have fair-

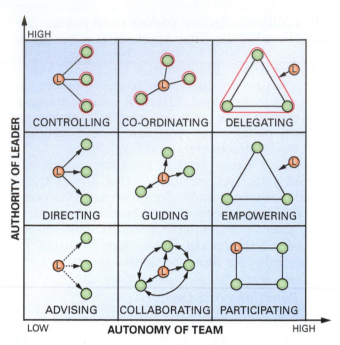

Figure 8 The authority-autonomy leadership style matrix.

ness and moral implications. When you make a dishonorable or unethical decision, it not only reflects on you, but also impacts other workers, peers, and the company as a whole.

There are three basic types of ethics:

- *Business or legal ethics* – Business, or legal, ethics concerns adhering to all laws and published regulations related to business relationships or activities.
- *Professional/balanced ethics* – Professional, or balanced, ethics relates to carrying out all activities in such a manner as to be honest and fair to everyone under one's authority.
- *Situational ethics* – Situational ethics pertains to specific activities or events that may initially appear to be a gray area. For example, you may ask yourself, "How will I feel about myself if my actions were going to be published in the newspaper or if I need to justify my actions to my family, friends, and colleagues? Would I still do the same thing?"

You will often find yourself in a situation where you will need to assess the ethical consequences of an impending decision. For example, if one of your crew members is showing symptoms of heat exhaustion, should you keep him working just because the superintendent says the project is behind schedule? If you are the only one aware that your crew did not properly erect the reinforcing steel, should you stop the pour and correct the situation? If supervision ever asks a crew leader to carry through on an unethical decision, it is up to that individual to inform the next-higher level of authority of the unethical nature of the issue. It is a sign of good character to refuse to carry out an unethical act.

"The right way is not always the popular and easy way. Standing for right when it is unpopular is a true test of character."
—Margaret Chase Smith

2.2.0 Communication

Successful crew leaders learn to communicate effectively with people at all levels of the organization. In doing so, they develop an understanding of human behavior and acquire communication skills that enable them to understand and influence others.

There are many definitions of communication. Communication is the act of accurately and effectively conveying to or exchanging facts, feelings, and opinions with another person. Simply stated, communication is the process of exchanging information and ideas. Just as there are many definitions of communication, it also comes in many forms, including verbal, written, and nonverbal.

2.2.1 The Communication Process and Verbal Communication

There are two basic steps to clear communication, as illustrated in *Figure 9*. First, a sender sends a message (either verbal or written) to a receiver. When the receiver gets the message, he or she figures out what it means by listening or reading carefully. If anything is unclear, the receiver gives

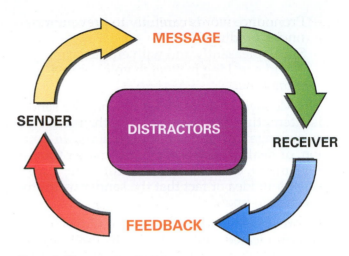

Figure 9 The communication process.

the sender feedback by asking the sender for more information.

This process is called *two-way communication*, and it is the most effective way to make sure that everyone understands what's going on. This process sounds simple, so you may ask why good communication is so hard to achieve? When we try to communicate, a lot of things can get in the way. These communication obstacles and distractors are called *noise*.

The communication process illustrated in *Figure 9* consists of the following major components:

- *The Sender* – The sender is the person who creates and transmits the message. In verbal communication, the sender speaks the message aloud to the person(s) receiving the message. The sender must be sure to speak in a clear and concise manner easily understood by others. This isn't a natural skill; it takes practice. Some basic speaking tips are as follows:

  - Avoid talking with anything in your mouth (food, gum, etc.).
  - Avoid swearing, or other crude language, and acronyms.
  - Don't speak too quickly or too slowly. In extreme cases, people tend to focus on the rate of speech rather than the words themselves.

Did You Know?

Supervisor's Communication Breakdown

No, this isn't about failure of communication, but how a supervisor's daily communications break down into the various types, and how much time the supervisor spends doing each. Research shows that about 80 percent of the typical supervisor's day is spent communicating through writing, speaking, listening, or using body language. Of that time, studies suggest that approximately 20 percent of communication is in the written form, and 80 percent involves speaking or listening.

- Pronounce words carefully to prevent mis-understandings.
- Speak pleasantly and with enthusiasm. Avoid speaking in a harsh voice or in a monotone.

- *The Message* – The message is what the sender is attempting to communicate to the receiver(s). A message can be a set of directions, an opinion, or dealing with a personnel matter (praise, reprimand, etc.). Whatever its function, a message is an idea or fact that the sender wants the receiver to know.

Before speaking, determine what must be communicated, and then take the time to organize what to say. Ensure that the message is logical and complete. Take the time to clarify your thoughts to avoid confusing the receiver. This also permits the sender to get to the point quickly.

In delivering the message, the sender should assess the audience. It is important not to talk down to them. Remember that everyone, whether in a senior or junior position, deserves respect and courtesy. Therefore, the sender should use words and phrases that the receiver can understand and avoid technical language or slang if the receiver is unfamiliar with such terms. In addition, the sender should use short sentences, which gives the audience time to understand and digest one point or fact at a time.

- *The Receiver* – The receiver is the person who takes in the message. For the verbal communication process to be successful, it is important that the receiver understands the message as the speaker intended. Therefore, the receiver must avoid things that interfere with the delivery of the message. There are many barriers to effective listening, particularly on a busy construction job site. Some of these obstacles include the following:

  - Noise, visitors, cell phones, or other distractions
  - Preoccupation, being under pressure, or daydreaming
  - Reacting emotionally to what is being communicated
  - Thinking about how to respond instead of listening
  - Giving an answer before the message is complete
  - Personal bias/prejudice against the sender or the sender's communication style
  - Finishing the sender's sentence

Some tips for overcoming these barriers include the following:

  - Take steps to minimize or remove distractions; learn to tune out your surroundings
  - Listen for key points
  - Take notes
  - Be aware of your personal biases, and try to stay open-minded; focus on the message, not the speaker
  - Allow yourself time to process your thoughts before responding
  - Let the sender communicate the message without interruption

There are many ways for a receiver to show that he or she is actively listening to the message. The receiver can even accomplish this without saying a word. Examples include maintaining eye contact (*Figure 10*), nodding your head, and taking notes. Feedback can also provide an important type of response to a message.

- *Feedback* – Feedback is the receiver's communication back to the sender in response to the message. Feedback is a very important part of the communication process because it shows the sender how the receiver interpreted the message and that the receiver understood it as intended. In other words, feedback is a checkpoint to make sure the receiver and sender are on the same page.

The receiver can paraphrase the message as a form of feedback to the sender. When paraphrasing, you use your own words to repeat the message. That way, you can show the sender that you interpreted the message correctly and could explain it to others if needed.

Figure 10 Eye contact is important when communicating face-to-face.

When providing feedback, the receiver can also request clarification or additional information, generally by asking questions.

One opportunity a crew leader can take to provide feedback is in the performance of crew evaluations. Many companies use formal evaluations on a yearly basis to assess workers' performance for pay increases and eligibility for advancement. These evaluations should not come as a once-a-year surprise to workers. An effective crew leader provides frequent performance feedback, and should ultimately summarize these communications in the annual performance evaluation. It is also important to stress the importance of self-evaluation with your crew.

2.2.2 Written or Visual Communications

Some communication must be written or visual. Written or visual communication includes messages or information documented on paper, or transmitted electronically using words, sketches, or other types of images.

Many messages on a job are in text form. Examples include weekly reports, requests for changes, purchase orders, and correspondence on a specific subject. These items are in writing because they can form a permanent record for business and historical purposes. Some communications on the job must be in the form of images. People are visual beings, and sketches, diagrams, graphs, and photos or videos can more effectively communicate some things. Examples include the plans or drawings used on a job, or a video of a work site accident scene.

"One picture is worth a thousand words."
—*San Antonio Express-News (1918)*

When writing or creating a visual message, it is best to assess the receiver (the reader) or the audience before beginning. The receiver must be able to read the message and understand the content; otherwise, the communication process will be unsuccessful. Therefore, the sender (the writer) should consider the actual meaning of words or diagrams and how others might interpret them. In addition, the writer should make sure that handwriting, if applicable, is legible.

Here are some basic tips for writing:

- Avoid emotion-packed words or phrases.
- Avoid making judgments unless asked to do so.
- Avoid using unfamiliar acronyms and text-speak (*Figure 11*).

Figure 11 Avoid textspeak for on-the-job written communications.

- Avoid using technical language or jargon unless the writer knows the receiver is familiar with the terms used.
- Be positive whenever possible.
- Be prepared to provide a verbal or visual explanation, if needed.
- Make sure that the document is legible.
- Present the information in a logical manner.
- Proofread your work; check for spelling, typographical, and grammatical errors—especially if relying on a spell-checker or an autocorrect typing feature.
- Provide an adequate level of detail.
- State the purpose of the message clearly.
- Stick to the facts.

Entire books are available that explain how to create effective diagrams and graphs. There are many factors that contribute to clear, concise, unambiguous images. The following are some basic tips for creating effective visual communications:

- Avoid making complex graphics; simplicity is better.
- Be prepared to provide a written or verbal explanation of the diagram, if needed.
- Ensure that the diagram is large enough to be useful.
- Graphs are better for showing numerical relationships than columns of numbers.
- Present the information in a logical order.
- Provide a legend if you include symbolic information.
- Provide an adequate level of detail and accuracy.

2.2.3 Nonverbal Communication

Unlike verbal or written communication, nonverbal communication doesn't involve spoken or written words, or images. Rather, nonverbal communication refers to things the speaker and the receiver see and hear in each other while communicating face to face. Examples include facial expressions, body movements, hand gestures, tone of voice, and eye contact.

Nonverbal communication can provide an external signal of an individual's inner emotions. It occurs at the same time as verbal communication. In most cases, the transmission of the sender and receiver nonverbal cues is totally unconscious.

People observe physical, nonverbal cues when communicating, making them just as important as words. Often, nonverbal signals influence people more than spoken words. Therefore, it's important to be conscious of nonverbal cues to avoid miscommunication based on your posture or expression. After all, these things may have nothing to do with the communication exchange; instead, they may be carrying over from something else going on in your day.

2.2.4 Communication Issues

It is important to note that everyone communicates differently; that is what makes us unique as individuals. As the diversity of the workforce changes, communication becomes even more challenging because the audience may include individuals from many different backgrounds. Therefore, it is necessary to assess the audience to determine how to communicate effectively with each individual.

The key to effective communication is to acknowledge that people are different, and to be able to adjust the communication style to meet the needs of the audience or the person on the receiving end of your message. This involves relaying the message in the simplest way possible, and avoiding the use of words that people may find confusing. Be aware of how you use technical language, slang, jargon, and words that have multiple meanings. Present the information in a clear, concise manner. Avoid rambling and always speak clearly, using good grammar.

> **NOTE**
> When working with bilingual workers, it is sometimes helpful to learn words and phrases commonly used in their language(s). They will respect you more for attempting to learn their language as they learn yours, and the effort will generate a closer comradery.

In addition, be prepared to communicate the message in multiple ways or adjust your level of detail or terminology to ensure that everyone understands the meaning as intended. For instance, a visual learner may need a map to comprehend directions. It may be necessary to overcome language barriers on the job site by using graphics or visual aids to relay the message. *Figure 12* shows how to tailor the message to the audience.

2.3.0 Motivation

The ability to motivate others is a key skill that leaders must develop. To motivate is to influence others to put forth more effort to accomplish something. For example, a crew member who skips breaks and lunch to complete a job on time is thought to be highly motivated, but crew members who do the bare minimum or just enough to keep their jobs are considered unmotivated.

A leader can **infer** an employee's motivation by observing various factors in the employee's performance. Examples of factors that may indicate a crew's motivation include the level or rate of unexcused absenteeism, the percentage of employee turnover, and the number of complaints, as well as the quality and quantity of work produced.

Different things motivate different people in different ways. Consequently, there is no single best approach to motivating crew members. It is important to recognize that what motivates one crew member may not motivate another. In addition, what works to motivate a crew member once may not motivate that same person again in the future.

Often, the needs and factors that motivate individuals are the same as those that create job satisfaction. These include the following:

- Accomplishment
- Change
- Job importance
- Opportunity for advancement
- Personal growth
- Recognition and praise
- Rewards

A crew leader's ability to satisfy these needs increases the likelihood of high morale within a crew. Morale refers to an individual's emotional outlook toward work and the level of satisfaction gained while performing the jobs assigned. High morale means that employees will be motivated to work hard, and they will have a positive attitude about coming to work and doing their jobs.

2.3.1 Accomplishment

Accomplishment refers to a worker's need to set challenging goals and achieve them. There is nothing quite like the feeling of achieving a goal, particularly a goal one never expected to accomplish in the first place.

Read the following verbal conversations, and identify any problems:

Conversation I:

Judy: Hey, José...

José: What's up?

Judy: Has the site been prepared for the job trailer yet?

José: Job trailer?

Judy: The job trailer—it's coming in today. What time will the job site be prepared?

José: The trailer will be here about 1:00 PM.

Judy: The job site! What time will the job site be prepared?

_____

_____

_____

_____

_____

Conversation II:

Jimar: Hey, Mike, I need your help.

Mike: What is it?

Jimar: You and Miguel go over and help Al's crew finish laying out the site.

Mike: Why me? I can't work with Miguel. He can't understand a word I say.

Jimar: Al's crew needs some help, and you and Miguel are the most qualified to do the job.

Mike: I told you, Jimar, I can't work with Miguel.

_____

_____

_____

_____

_____

Conversation III:

Hiro: Hey, Jill.

Jill: Sir?

Hiro: Have you received the latest DOL, EEO requirement to be sure the OFCP administrator finds our records up to date when he reviews them in August?

Jill: DOL, EEO, and OFCP?

Hiro: Oh, and don't forget the MSHA, OSHA, and EPA reports are due this afternoon.

Jill: MSHA, OSHA, and EPA?

_____

_____

_____

_____

_____

_____

_____

Conversation IV:

Lakeisha: Good morning, Roberto, would you do me a favor?

Roberto: Okay, Lakeisha. What is it?

Lakeisha: I was reading the concrete inspection report and found the concrete in Bays 4A, 3B, 6C, and 5D didn't meet the 3,000 psi strength requirements. Also, the concrete inspector on the job told me the two batches that came in today had to be refused because they didn't meet the slump requirements as noted on page 16 of the spec. I need to know if any placement problems happened on those bays, how long the ready mix trucks were waiting today, and what we plan to do to stop these problems in the future.

_____

_____

_____

_____

Read the following written memos, and identify any problems:

Memo I:

Let's start with the transformer vault $285.00 due. For what you ask? Answer: practically nothing I admit, but here is the story. Paul the superintendent decided it was not the way good ole Comm Ed wanted it, we took out the ladder and part of the grading (as Paul instructed us to do) we brought it back here to change it. When Comm Ed the architect or DOE found out that everything would still work the way it was, Paul instructed us to reinstall the work. That is the whole story there is please add the $285.00 to my next payout.

_____

_____

_____

_____

_____

_____

_____

Memo II:

Let's take rooms C 307-C-312 and C-313 we made the light track supports and took them to the job to erect them when we tried to put them in we found direct work in the way, my men spent all day trying to find out what to do so ask your Superintendent (Frank) he will verify seven hours pay for these men as he went back and forth while my men waited. Now the Architect has changed the system of hanging and has the gall to say that he has made my work easier, I can't see how. Anyway, we want an extra two (2) men for seven (7) hours for April 21 at $55.00 per hour or $385.00 on April 28th DOE Reference 197 finally resolved this problem. We will have no additional charges on DOE Reference 197, please note.

_____

_____

_____

_____

_____

Crew leaders can help their crew members attain a sense of accomplishment by encouraging them to develop performance plans, such as goals for the year, the attaining of which the crew leader will consider later in performance evaluations. In addition, crew leaders can provide the support and tools (such as training and coaching) necessary to help their crew members achieve these goals.

2.3.2 Change

Change refers to an employee's need to have variety in work assignments. Change can keep things interesting or challenging, and prevent the boredom that results from doing the same task day after day with no variety. However, frequent or significant changes in work can actually have a negative impact on morale, since people often prefer some consistency and predictability in their lives.

2.3.3 Job Importance

Job importance refers to an employee's need to feel that his or her skills and abilities are valued and make a difference. Employees who don't feel valued tend to have performance and attendance issues. Crew leaders should attempt to make every crew member feel like an important part of the team, as if the job wouldn't be possible without their help.

2.3.4 Opportunity for Advancement

Opportunity for advancement refers to an employee's need to gain additional responsibility and develop new skills and abilities. It is important that employees know that they aren't limited to their current jobs. Let them know that they have a chance to grow with the company and to be promoted as recognition for excelling in their work when such opportunities occur.

VERBAL INSTRUCTIONS (EXPERIENCED CREW)	VERBAL INSTRUCTIONS (INEXPERIENCED CREW)	WRITTEN INSTRUCTIONS	VISUAL INSTRUCTIONS (DIAGRAM/MAP)
"Please drive to the supply shop to pick up our order."	"Please drive to the supply shop. Turn right here and left at Route 1. It's at 75th Street and Route 1. Tell them the company name and that you're there to pick up our order."	1. Turn right at exit. 2. Drive 2 miles to Route 1. Turn LEFT. 3. Drive 1 mile (pass the tire shop) to 75th Street. 4. Look for supply store on right...	

Different people learn in different ways. Be sure to communicate so you can be understood.

Figure 12 Tailor your message.

Effective leaders encourage each of their crew members to work to his or her full potential. In addition, they share information and skills with their employees to help them advance within the organization.

2.3.5 Recognition

Recognition and praise refer to the need to have good work appreciated, applauded, and acknowledged by others. You can accomplish this by simply thanking employees for helping on a project, or it can entail more formal praise, such as an award for Employee of the Month, or tangible or monetary awards. Some tips for giving recognition and praise include the following:

- Be available on the job site so that you have the opportunity to witness good work.
- Look for and recognize good work, and look for ways to praise it.
- Give recognition and praise only when truly deserved; people can quickly recognize insincere or artificial praise, and you will lose respect as a result.
- Acknowledge satisfactory performance, and encourage improvement by showing confidence in the ability of the crew members to do above-average work.

2.3.6 Personal Growth

Personal growth refers to an employee's need to learn new skills, enhance abilities, and grow as a person. It can be very rewarding to master a new competency on the job. Personal growth prevents the boredom associated with contemplation of working the same job indefinitely with no increase of technical knowledge, skills, or responsibility.

Crew leaders should encourage the personal growth of their employees as well as themselves. Learning should be a two-way street on the job site; crew leaders should teach their crew members and learn from them as well. In addition, crew members should be encouraged to learn from each other

Personal growth also takes place outside the workplace. Crew leaders should encourage their employees to acquire formal education applicable to their vocation as well as in subjects that will expand their understanding of the larger world. Encourage them to develop outside interests and ways to express their innate creativity.

2.3.7 Rewards

Rewards are additional compensation for hard work. Rewards can include an increase in a crew member's wages, or go beyond that to include bo-

Did You Know?

The Importance of Continuing Education

Education doesn't stop the day a person receives a diploma or certificate. It is a lifelong activity. Employers have long recognized and promoted continuing education as a factor in advancement, but it is essential to simply remaining in place as well. Regardless of what you do, new construction materials, methods, standards, and processes are constantly emerging. Those who don't make the effort to keep up will fall behind.

Also, studies have shown that a worker's lifetime income improved by 7–10 percent per year of community college. Consider continuing and vocational technical education (CTE and VTE) through local colleges.

nuses or other incentives. They can be monetary in nature (salary raises, holiday bonuses, etc.), or they can be nonmonetary, such as free merchandise (shirts, coffee mugs, jackets, merchant gift cards, etc.), or other prizes. Attendance at costly certification or training courses can be another form of reward.

2.3.8 Motivating Employees

To increase motivation in the workplace, crew leaders must individualize how they motivate different crew members. It is important that crew leaders get to know their crew members and determine what motivates them as individuals. Once again, as diversity increases in the workforce, this becomes even more challenging; therefore, effective communication skills are essential. Some tips for motivating employees include the following:

- Keep jobs challenging and interesting.
- Communicate your expectations. People need clear goals in order to feel a sense of accomplishment when they achieve them.
- Involve the employees. Feeling that their opinions are valued leads to pride in ownership and active participation.
- Provide sufficient training. Give employees the skills and abilities they need to be motivated to perform.
- Mentor the employees. Coaching and supporting employees boosts their self-esteem, their self-confidence, and ultimately their motivation.
- Lead by example. Employees will be far more motivated to follow someone who is willing to do the same things they are required to do.
- Treat employees well. Be considerate, kind, caring, and respectful; treat employees the way that you want to be treated.
- Avoid using scare tactics. Approaching your leadership responsibilities by threatening employees with negative consequences can result in higher employee turnover instead of motivation.
- Reward your crew for doing their best by giving them easier tasks from time to time. It is tempting to give your best employees the hardest or dirtiest jobs because you know they will do the jobs correctly.
- Reward employees for a job well done.
- Maintain a sense of humor, especially toward your own failings as a human being. No one is perfect. Your employees will appreciate that and be more motivated to be an encouragement to you.

2.4.0 Team Building

Organizations are making the shift from the traditional boss-worker mentality to one that promotes teamwork. The manager becomes the team leader, and the workers become team members. They all work together to achieve the common goals of the team.

There are several benefits associated with teamwork. These include the ability to complete complex projects more quickly and effectively, higher employee satisfaction, and reduced turnover.

2.4.1 Successful Teams

Successful teams consist of individuals who are willing to share their time and talents to reach a common goal—the goal of the team. Members of successful teams possess an "Us" or "We" attitude rather than an "I" and "You" attitude; they consider what's best for the team and put their egos aside.

Some characteristics of successful teams include the following:

- Everyone participates and every team member counts.
- All team members understand the goals of the team's work and are committed to achieve those goals.
- There is a sense of mutual trust and interdependence.
- The organization gives team members the confidence and means to succeed.
- They communicate.
- They are creative and willing to take risks.
- The team leader has strong people skills and is committed to the team.

2.4.2 Building Successful Teams

To be successful in the team leadership role, the crew leader should contribute to a positive attitude within the team. There are several ways in which the team leader can accomplish this. First, he or she can work with the team members to create a vision or purpose of what the team is to achieve. It is important that every team member is committed to the purpose of the team, and the team leader is instrumental in making this happen.

Within the construction industry, the company typically assigns a crew to a crew leader. However, it can be beneficial for the team leader to be involved in selecting the team members. The willingness of people to work on the team and the resources that they bring should be the key factors for selecting them for the team.

You are the crew leader of a masonry crew. Sam Williams is the person whom the company holds responsible for ensuring that equipment is operable and distributed tothe jobs in a timely manner.

Occasionally, disagreements with Sam have resulted in tools and equipment arriving late. Sam, who has been with the company 15 years, resents having been placed in the job and feels that he outranks all the crew leaders.

Sam figured it was about time he talked with someone about the abuse certain tools and other items of equipment were receiving on some of the jobs. Saws were coming back with guards broken and blades chewed up, bits were being sheared in half, motor housings were bent or cracked, and a large number of tools were being returned covered with mud. Sam was out on your job when he observed a mason carrying a portable saw by the cord. As he watched, he saw the mason bump the swinging saw into a steel column. When the man arrived at his workstation, he dropped the saw into the mud.

You are the worker's crew leader. Sam approached as you were coming out of the work trailer. He described the incident. He insisted, as crew leader, you are responsible for both the work of the crew and how its members use company property. Sam concluded, "You'd better take care of this issue as soon as possible! The company is sick and tired of having your people mess up all the tools!"

You are aware that some members of your crew have been mistreating the company equipment.

1. How would you respond to Sam's accusations?

_____
_____
_____
_____
_____

2. What action would you take regarding the misuse of the tools?

_____
_____
_____
_____
_____

3. How can you motivate the crew to take better care of their tools? Explain.

_____
_____
_____
_____
_____

When forming a new team, team leaders should do the following:

- Explain the purpose of the team. Team members need to know what they will be doing, how long they will be doing it (if they are temporary or permanent), and why they are needed.
- Help the team establish goals or targets. Teams need a purpose, and they need to know what it is they are responsible for accomplishing.
- Define team member roles and expectations. Team members need to know how they fit into the team and what they can expect to accomplish as members of the team.
- Plan to transfer a sense of responsibility to the team as appropriate. Teams should feel responsible for their assigned tasks. However, never forget that management will still hold the crew leader responsible for the crew's work.

"My model for business is The Beatles. They were four guys who kept each other's kind of negative tendencies in check. They balanced each other and the total was greater than the sum of the parts. That's how I see business: Great things in business are never done by one person, they're done by a team of people.
—*Steve Jobs*

2.4.3 Delegating

Once the various activities that make up the job have been determined, the crew leader must identify the person or persons who will be responsible for completing each activity. This requires that the crew leader be aware of the skills and abilities of the people on the crew. Then, the crew leader must put this knowledge to work in matching the crew's skills and abilities to accomplish the specific tasks needed to complete the job.

After matching crew members to specific activities, the crew leader must then delegate the assignments to the responsible person(s). Generally, when delegating responsibilities, the crew leader verbally communicates directly with the person who will perform or complete the activity.

When delegating work, remember to do the following:

- Delegate work to a crew member who can do the job properly. If it becomes evident that the worker doesn't perform to the desired standard, either teach the crew member to do the work correctly or turn it over to someone else who can (without making a public spectacle of the transfer).
- Make sure crew members understand what to do and the level of responsibility. Be clear about the desired results, specify the boundaries and deadlines for accomplishing the results, and note the available resources.
- Identify the standards and methods of measurement for progress and accomplishment, along with the consequences of not achieving the desired results. Discuss the task with the crew member and check for understanding by asking questions. Allow the crew member to contribute feedback or make suggestions about how to perform the task in a safe and quality manner.
- Give the crew member the time and freedom to get started without feeling the pressure of too much supervision. When making the work assignment, be sure to tell the crew member how much time there is to complete it, and confirm that this time is consistent with the job schedule.
- Examine and evaluate the result once a task is complete. Then, give the crew member some feedback as to how well the worker did the task. Get the crew member's comments. The information obtained from this is valuable and will enable the crew leader to know what kind of work to assign that crew member in the future. It will also provide a means of measuring the crew leader's own effectiveness in delegating work.

Be aware that there may be times when someone else in the company will issue written or verbal instructions to a crew member or to the crew without going through the crew leader. This kind of situation requires an extra measure of maturity and discretion on the part of the crew leader to first understand the circumstances, then establish an understanding with the responsible individual, if possible, to avoid work orders that circumvent the crew leader in the future.

2.4.4 Implementing Policies and Procedures

Every company establishes policies and procedures that crew leaders are expected to implement, and employees are expected to follow. Company policies and procedures are essentially guidelines for how the organization does business. They can also reflect organizational philosophies, such as putting safety first or making the customer the top priority. Examples of policies and procedures include safety guidelines, credit standards, and billing processes.

The following tips can help you effectively implement policies and procedures:

- Learn the purpose of each policy. This will help you follow the policy and apply it appropriately and fairly.
- If you're not sure how to apply a company policy or procedure, check the company manual or ask your supervisor.

> **NOTE**
> Try to obtain a supervisor's policy interpretation in writing or print out the email response so that you can append the decision to your copy of the company manual for future reference.

- Always follow company policies and procedures. Remember that they combine what's best for the customer and the company. In addition, they provide direction on how to handle specific situations and answer questions.

Crew leaders may need to issue orders to their crew members. An order is a form of communication that initiates, changes, or stops an activity. Orders may be general or specific, written or oral, and formal or informal. The decision of how an order will be issued is up to the crew leader, but the policies and procedures of the company may govern the choice.

When issuing orders, do the following:

- Make them as specific as possible. Avoid being general or vague unless it is impossible to foresee all the circumstances that could occur in carrying out the order.
- Recognize that it isn't necessary to write orders for simple tasks unless the company requires that supervisors write all orders.
- Write orders for more complex tasks, tasks that will take considerable time to complete, or that are permanent (standing) orders.
- Consider what is being said, the audience to whom it applies, and the situation under which it will be implemented to determine the appropriate level of formality for the order.

2.5.0 Making Decisions and Solving Problems

Decision making and problem solving and are a large part of every crew leader's daily work. They are a part of life for all supervisors, especially in fast-paced, deadlineoriented industries.

2.5.1 Decision Making Versus Problem Solving

Sometimes, the difference between decision making and problem solving isn't clear. Decision making refers to simply initiating an action, stopping one, or choosing an alternative course, as appropriate for the situation. Problem solving involves recognizing the difference between the way things are and the way things should be, then taking action to move toward the desired condition. The two activities are related because, to make a decision, you may have to use problem-solving techniques, just as solving problems requires making decisions.

2.5.2 Types of Decisions

Some decisions are routine or simple, and can be made based on past experiences. An example would be deciding how to get to work. If you've worked at the same place for a long time, you are already aware of the options for traveling to work (take the bus, drive a car, carpool with a co-worker, take a taxi, etc.). Based on past experiences with the options identified, you can make a decision about how best to get to work.

Other decisions are more difficult to make. These decisions require more careful thought about how to carry out an activity by using problem-solving techniques. An example is planning a trip to a new vacation spot. If you're not sure how to get there, where to stay, what to see, etc., one option is to research the area to determine the possible routes, hotel accommodations, and attractions. Then, you can make a decision about which route to take, what hotel to choose, and what sites to visit, without the benefit of direct experience. The Internet makes these tasks much easier, but the research still requires prioritizing what is most important to you.

2.5.3 Problem Solving

The ability to solve problems is an important skill in any workplace. It's especially important for craft professionals, whose workday is often not predictable or routine. This section provides a five-step process for solving problems, which you can apply to both workplace and personal issues.

Review the following steps and then see how you can be apply them to a job-related problem. Keep in mind that you can't solve a problem until everyone involved acknowledges the problem.

Step 1 *Define the problem.* This isn't as easy as it sounds. Thinking through the problem often uncovers additional problems. Also, drilling down to the facts of the problem may mean setting aside your own biases or presumptions toward the situation or the individuals involved.

Step 2 *Think about different ways to solve the problem.* There is often more than one solution to a problem, so you must think through each possible solution and pick the best one. The best solution might be taking parts of two different solutions and combining them to create a new solution.

Step 3 *Choose the solution that seems best, and make an action plan.* It is best to receive input both from those most affected by the problem, those who must correct the problem, and from those who will be most affected by any potential solution.

Step 4 *Test the solution to determine whether it actually works.* Many solutions sound great in theory but in practice don't turn out to be effective. On the other hand, you might discover from trying to apply a solution that it is acceptable with a little modification. If a solution doesn't work, think about how you could improve it, and then test your new plan.

Step 5 *Evaluate the process.* Review the steps you took to discover and implement the solution. Could you have done anything better? If the solution turns out to be satisfactory, you can add the solution to your knowledge base.

These five steps can be applied in specific situations. Read the following example situation, and apply the five-step problem-solving process to come up with a solution.

Example:
You are part of a team of workers assigned to a new shopping mall project. The project will take about 18 months to complete. The only available parking is half a mile from the job site. The crew must carry heavy toolboxes and safety equipment from their cars and trucks to the work area at the start of the day, and then carry them back at the end of their shifts. The five-step problem-solving process can be applied as follows:

Step 1 *Define the problem.* Workers are wasting time and energy hauling all their equipment to and from the work site.

Step 2 *Think about different ways to solve the problem.* Several workers have proposed solutions:

- Install lockers for tools and equipment closer to the work site.
- Have workers drive up to the work site to drop off their tools and equipment before parking.
- Bring in another construction trailer where workers can store their tools and equipment for the duration of the project.
- Provide a round-trip shuttle service to ferry workers and their tools.

> **NOTE**
>
> Each solution will have pros and cons, so it's important to receive input from the workers affected by the problem. For example, workers will probably object to any plan (like the drop-off plan) that leaves their tools vulnerable to theft.

Step 3 *Choose the solution that seems best, and make an action plan.* The work site superintendent doesn't want an additional trailer in your crew's area. The workers decide that the shuttle service makes the most sense. It should solve the time and energy problem, and workers can keep their tools with them. To put the plan into effect, the project supervisor arranges for a large van and driver to provide the shuttle service.

Step 4 *Test the solution to determine whether it actually works.* The solution works, but there is another problem. The workers' schedule has them all starting and leaving at the same time. There isn't enough room in the van for all the workers and their equipment. To solve this problem, the supervisor schedules trips spaced 15 minutes apart. The supervisor also adjusts worker schedules to correspond with the trips. That way, all the workers won't try to get on the shuttle at the same time.

Step 5 *Evaluate the process.* This process gave both management and workers a chance to express an opinion and discuss the various solutions. Everyone feels pleased with the process and the solution.

2.5.4 Special Leadership Problems

Because they are responsible for leading others, it is inevitable that crew leaders will encounter problems and be forced to make decisions about how to respond to the problem. Some problems will be relatively simple to resolve, like covering for a sick crew member who has taken a day off from work. Other problems will be complex and much more difficult to handle.

Some complex problems that are relatively common include the following:

- Inability to work with others
- Absenteeism and turnover
- Failure to comply with company policies and procedures

Inability to Work with Others – Crew leaders will sometimes encounter situations where an employee has a difficult time working with others on the crew. This could be a result of personality differences, gender or gender-identity prejudices, an inability to communicate, or some other cause. Whatever the reason, the crew leader must address the issue and get the crew working as a team.

The best way to determine the reason for why individuals don't get along or work well together is to talk to the parties involved. The crew leader should speak openly with the employee, as well as the other individual(s) to uncover the source of the problem and discuss its resolution.

After uncovering the reason for the conflict, the crew leader can determine how to respond. There may be a way to resolve the problem and get the workers communicating and working as a team again. On the other hand, there may be nothing the crew leader can do that will lead to a harmonious solution. In this case, the crew leader would need to either transfer one of the involved employees to another crew or have the problem crew member terminated. Resorting to this latter option should be the last measure and taken after discussing the matter with one's immediate supervisor or the company's Human Resources Department.

Absenteeism and Turnover – Absenteeism and turnover in the industry can delay jobs and cause companies to lose money. Absenteeism refers to workers missing their scheduled work time on a job. It has many causes, some of which cannot be helped. Sickness, family emergencies, and funerals are examples of unavoidable causes of worker absence. However, there are some causes of unexcused absenteeism that crew leaders can prevent.

The most effective way to control absenteeism is to make the company's policy clear to all employees. The policy should be explained to all new employees. This explanation should include the number of absences allowed and acceptable reasons for taking sick or personal days. In addition, all workers should know how to inform their crew leaders when they miss work and understand the consequences of exceeding the number of sick or personal days allowed.

Once crew leaders explain the policy on absenteeism to employees, they must be sure to implement it consistently and fairly. This makes employees more likely to follow it. However, if enforcement of the policy is inconsistent and the crew leader gives some employees exceptions, it won't be effective. Thus, the rate of absenteeism is likely to increase.

Despite having a policy on absenteeism, there will always be employees who are chronically late or miss work. In cases where an employee abuses the absenteeism policy, the crew leader should discuss the situation directly with the employee, ensure that they understand the policy, and insist that they comply with it. If the employee's behavior does not improve, disciplinary action will be in order.

Turnover refers to the rate at which workers leave a company and are replaced by others. Like absenteeism, there are some causes of turnover that cannot be prevented and others that can. For instance, an employee may find a job elsewhere earning twice as much money. However, crew leaders can prevent some employee turnover situations. They can work to ensure safe working conditions for their crew, treat their workers fairly and consistently, and help promote good working conditions. The key is communication and promoting the motivational factors discussed earlier. Crew leaders need to know the problems if they are going to be able to successfully resolve them.

Some major causes of employee turnover include the following:

- Unfair/inconsistent treatment by the immediate supervisor
- Unsafe project sites
- Lack of job security or opportunities for advancement

For the most part, the actions described for absenteeism are also effective for reducing turnover. Past studies have shown that maintaining harmonious relationships on the job site goes a long way in reducing both turnover and absenteeism. This requires effective and proactive leadership on the part of the crew leader.

Failure to Comply with Company Policies and Procedures – Policies are rules that define the relationship between the company, its employees, its clients, and its subcontractors. Procedures include the instructions for carrying out the policies. Some companies have policies that dictate dress codes. The dress code may be designed partly to ensure safety, and partly to define the image a company wants to project to the outside world.

Companies develop procedures to ensure that everyone who performs a task does it safely and efficiently. Many procedures directly relate to safety. A lockout/tagout (LOTO) procedure is an example. In this procedure, the company defines who may perform a LOTO, how to properly complete and remove a LOTO, and who has the authority to remove it. Workers who fail to follow the procedure endanger themselves, as well as their co-workers.

Companies typically have a policy on disciplinary action, which defines steps to take if an employee violates company policies or procedures. The steps range from counseling by a supervisor for the first offense, to a written warning and/or LOI, to dismissal for repeat offenses. This will vary from one company to another. For example, some companies will fire an employee for the first violation of a safety procedure with potential for loss of life or serious injury.

The crew leader has the first-line responsibility for enforcing company policies and procedures. The crew leader should take the time with a new crew member to discuss the policies and procedures and show the crew member how to access them. If a crew member shows a tendency to neglect a policy or procedure, it is up to the crew leader to counsel that individual. If the crew member continues to violate a policy or procedure, the crew leader has no choice but to refer that individual to the appropriate authority within the company for disciplinary action.

> **NOTE**
>
> When a crew shows a pattern of consistent policy violations, management will scrutinize the crew leader and potentially take disciplinary action if the crew leader does not handle the violations appropriately. The crew leader is responsible for all aspects of the crew, not just its work accomplishment on the job.

Case I:

On the way over to the job trailer, you look up and see a piece of falling scrap heading for one of the laborers. Before you can say anything, the scrap material hits the ground about five feet in front of the worker. You notice the scrap is a piece of conduit. You quickly pick it up, assuring the worker you will take care of this matter.

Looking up, you see your crew on the third floor in the area from which the material fell. You decide to have a talk with them. Once on the deck, you ask the crew if any of them dropped the scrap. The men look over at Bob, one of the electricians in your crew. Bob replies, "I guess it was mine. It slipped out of my hand."

It is a known fact that the Occupational Safety and Health Administration (OSHA) regulations state that an enclosed chute of wood shall be used for material waste transportation from heights of 20 feet or more. It is also known that Bob and the laborer who was almost hit have been seen arguing lately.

1. Assuming Bob's action was deliberate, what action would you take?

   _____

   _____

   _____

2. Assuming the conduit accidentally slipped from Bob's hand, how can you motivate him to be more careful?

   _____

   _____

   _____

3. What follow-up actions, if any, should be taken relative to the laborer who was almost hit?

   _____

   _____

   _____

4. Should you discuss the apparent OSHA violation with the crew? Why or why not?

   _____

   _____

   _____

5. What acts of leadership would be effective in this case? To what leadership traits are they related?

   _____

   _____

   _____

Case II:

The company just appointed Antonio crew leader of a tile-setting crew. Before his promotion into management, he had been a tile setter for five years. His work had been consistently of superior quality.

Except for a little good-natured kidding, Antonio's co-workers had wished him well in his new job. During the first two weeks, most of them had been cooperative while Antonio was adjusting to his supervisory role.

At the end of the second week, a disturbing incident took place. Having just completed some of his duties, Antonio stopped by the job-site wash station. There he saw Steve and Ron, two of his old friends who were also in his crew, washing.

"Hey, Ron, Steve, you shouldn't be cleaning up this soon. It's at least another thirty minutes until quitting time," said Antonio. "Get back to your work station, and I'll forget I saw you here."

"Come off it, Antonio," said Steve. "You used to slip up here early on Fridays. Just because you have a little rank now, don't think you can get tough with us." To this Antonio replied, "Things are different now. Both of you get back to work, or I'll make trouble." Steve and Ron said nothing more, and they both returned to their work stations.

From that time on, Antonio began to have trouble as a crew leader. Steve and Ron gave him the silent treatment. Antonio's crew seemed to forget how to do the most basic activities. The amount of rework for the crew seemed to be increasing. By the end of the month, Antonio's crew was behind schedule.

1. How do you think Antonio should've handled the confrontation with Ron and Steve?

   _____

   _____

   _____

2. What do you suggest Antonio could do about the silent treatment he got from Steve and Ron?

   _____

   _____

3. If you were Antonio, what would you do to get your crew back on schedule?

   _____

   _____

4. What acts of leadership could be used to get the crew's willing cooperation?

   _____

   _____

5. To which leadership traits do they correspond?

   _____

   _____

Additional Resources

Construction Workforce Development Professional, NCCER. 2016. New York, NY: Pearson Education, Inc.

Mentoring for Craft Professionals, NCCER. 2016. New York, NY: Pearson Education, Inc.

It's Your Ship: Management Techniques from the Best Damn Ship in the Navy, Captain D. Michael Abrashoff, USN. 2012. New York City, NY: Grand Central Publishing.

Survival of the Fittest, Mark Breslin. 2005. McNally International Press.

The Definitive Book of Body Language: The Hidden Meaning Behind People's Gestures and Expressions, Barbara Pease and Allan Pease. 2006. New York City, NY: Random House / Bantam Books.

2.0.0   Section Review

1. A crew leader differs from a craftworker in that a crew leader _____.

    a. does not need direct experience in the job duties a craft professional typically performs
    b. can expect to oversee one or more workers in addition to performing typical craft duties
    c. is exclusively in charge of overseeing, since performing technical work isn't part of this role
    d. has no responsibility to be present on the job site

2. Feedback is important in verbal communication because it _____.

    a. requires the sender to repeat the message
    b. involves the receiver repeating back the message word for word
    c. informs the sender of how the message was received
    d. consists of a written analysis of the message

3. Once achieved, setting challenging goals for workers gives them a sense of _____.

    a. accomplishment
    b. entitlement
    c. persecution
    d. failure

4. Which of the following is *not* a characteristic of a successful team?

    a. Everyone participates and everyone counts.
    b. There is a sense of mutual trust.
    c. Members minimize communication with each other.
    d. The team leader is committed to the team.

5. Problem solving differs from decision making because it _____.

    a. involves finding an answer
    b. expresses an opinion
    c. involves starting or stopping an action
    d. separates facts from non-facts

SECTION THREE

3.0.0 SAFETY AND SAFETY LEADERSHIP

Objective

Identify a crew leader's typical safety responsibilities with respect to common safety issues, including awareness of safety regulations and the cost of accidents.

a. Explain how a strong safety program can enhance a company's success.
b. Explain the purpose of OSHA and describe the role of OSHA in administering worker safety.
c. Describe the role of employers in establishing and administering safety programs.
d. Explain how crew leaders are involved in administering safety policies and procedures.

Trade Terms

Intangible: Not touchable, material, or measureable; lacking a physical presence.

Lethargy: Sluggishness, slow motion, lack of activity, or a lack of enthusiasm.

Negligence: Lack of appropriate care when doing something, or the failure to do something, usually resulting in injury to an individual or damage to equipment.

Safety data sheets (SDS): Documents listing information about a material or substance that includes common and proper names, chemical composition, physical forms and properties, hazards, flammability, handling, and emergency response in accordance with national and international hazard communication standards. Also called material safety data sheets (MSDS).

Businesses lose millions of dollars every year because of on-the-job accidents. Work-related injuries, sickness, and fatalities have caused untold suffering for workers and their families. Resulting project delays and budget overruns can cause huge losses for employers, and work-site accidents damage the overall morale of the crew.

Craft professionals routinely face hazards. Examples of these hazards include falls from heights, working on scaffolds, using cranes in the presence of power lines, operating heavy machinery, and working on electrically-powered or pressurized equipment. Despite these hazards, experts believe that applying preventive safety measures could drastically reduce the number of accidents.

As a crew leader, one of your most important tasks is to enforce the company's safety program and make sure that all workers are performing their tasks safely. To be successful, the crew leader should do the following:

- Be aware of the human and monetary costs of accidents.
- Understand all federal, state, and local governmental safety regulations applicable to your work.
- Be the most visible example of the best safe work practices.
- Be involved in training workers in safe work methods.
- Conduct training sessions.
- Get involved in safety inspections, accident investigations, and fire protection and prevention.

"Example is not the main thing in influencing others. It is the only thing."
—*Albert Schweitzer*

Crew leaders are in the best position to ensure that their crew members perform all jobs safely. Providing employees with a safe working environment by preventing accidents and enforcing safety standards will go a long way towards maintaining the job schedule and enabling a job's completion on time and within budget.

3.1.0 The Impact of Accidents

Each day, workers in construction and industrial occupations face the risk of falls, machinery accidents, electrocutions, and other potentially fatal

---

Did You Know?

The Fatal Four

When OSHA inspects a job site, they focus on the types of safety hazards that are most likely to cause serious and fatal injuries. These hazards result in the following most common categories of injuries:

- Falls from elevations
- Struck-by hazards
- Caught-in/between hazards
- Electrical-shock hazards

hazards. The National Institute of Occupational Safety and Health (NIOSH) statistics show that roughly 1,000 construction workers are killed on the job each year, which is more than any other industry. Falls are the leading cause of deaths in the construction industry through accident or negligence, accounting for over 60 percent of the fatalities in recent years. Nearly half of the fatal falls occurred from roofs, scaffolds, or ladders. Roofers, structural metal workers, and painters are at the greatest risk of fall fatalities (*Figure 13*).

In addition to the number of fatalities that occur each year, there are a staggering number of work-related injuries. In 2015, for example, almost 200,000 job-related injuries occurred in the construction industry. NIOSH estimates that the total cost of fatal and non-fatal injuries in the construction industry represents about 15 percent of the costs for all private industry. The main causes of injuries on construction sites include falls, electrocution, fires, and mishandling of machinery or equipment. According to NIOSH, back injuries are the leading health-related problem in workplaces.

3.1.1 Cost of Accidents

Occupational accidents cost roughly $250 billion or more every year. These costs affect the individual employee, the company, and the construction industry as a whole.

Organizations encounter both direct and indirect costs associated with workplace accidents. Direct costs are the money companies must pay out to workers' compensation claims and sick pay; indirect costs are all the other tangible and intangible things and costs a company must account for as the result of a worker's injury or death. To compete and survive, companies must control these as well as all other employment-related costs. There are many costs involved with workplace accidents. A company can insure some of these costs, but not others.

Insured costs – Insured costs are those costs either paid directly or reimbursed by insurance carriers. Insured costs related to injuries or deaths include the following:

- Compensation for lost earnings (known as *worker's comp*)
- Funeral charges
- Medical and hospital costs
- Monetary awards for permanent disabilities
- Pensions for dependents
- Rehabilitation costs

Insurance premiums or charges related to property damages include the following:

- Fire or other safety-related peril
- Structural loss; material and equipment loss or damage
- Loss of business use and occupancy
- Public liability
- Replacement cost of equipment, material, and structures

Uninsured costs – The relative direct and indirect costs of accidents are comparable to the visible and hidden portions of an iceberg, as shown in *Figure 14*. The tip of the iceberg represents direct costs, which are the visible costs. Not all of these are covered by insurance. The more numerous indirect costs aren't readily measurable, but they can represent a greater financial burden than the direct costs.

Uninsured costs from injuries or deaths include the following:

- First aid expenses
- Transportation costs
- Costs of investigations
- Costs of processing reports
- Down time on the job site
- Costs to train replacement workers

Figure 13 Falls are the leading cause of deaths and injuries in construction.

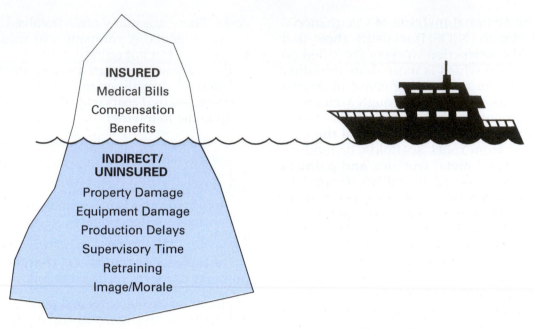

Figure 14 Costs associated with accidents.

Uninsured costs related to wage losses include the following:

- Idle time of workers whose work is interrupted
- Time spent cleaning the accident area
- Time spent repairing damaged equipment
- Time lost by workers receiving first aid
- Costs of training injured workers in a new career

Uninsured costs related to production losses include the following:

- Product spoiled by accident
- Loss of skill and experience; worker replacement
- Lowered production capacity
- Idle machine time due to lack of qualified operators

Associated costs may include the following:

- Difference between actual losses and amount recovered
- Costs of rental equipment used to replace damaged equipment
- Costs of inexperienced temp or permanent new workers used to replace injured workers
- Wages or other benefits paid to disabled workers
- Overhead costs while production is stopped
- Impact on schedule
- Loss of client bonus or payment of forfeiture for delays

Uninsured costs related to off-the-job activities include the following:

- Time spent on ensuring injured workers' welfare
- Loss of skill and experience of injured workers
- Costs of training replacement workers

Uninsured costs related to intangible factors include the following:

- Increased labor conflict
- Loss of bid opportunities because of poor safety records
- Loss of client goodwill
- Lowered employee morale
- Unfavorable public relations

3.2.0 OSHA

To reduce safety and health risks and the number of injuries and fatalities on the job, the federal government has enacted laws and regulations, including the Occupational Safety and Health Act of 1970 (OSH Act of 1970). This law created the Occupational Safety and Health Administration (OSHA), which is part of the US Department of Labor. OSHA also provides education and training for employers and workers. Through the administration of OSHA, the US Congress seeks "to assure so far as possible every working man and woman in the Nation safe and healthful working conditions and to preserve our human resources…" (OSH Act of 1970, Section 2[b]).

To promote a safe and healthy work environment, OSHA issues standards and rules for working conditions, facilities, equipment, tools, and work processes. It does extensive research into occupational accidents, illnesses, injuries, and deaths to reduce the number of occurrences and adverse effects. In addition, OSHA regulatory agencies conduct workplace inspections to ensure that companies follow the standards and rules.

To enforce OSHA regulations, the government has granted regulatory agencies the right to enter public and private properties to conduct workplace safety investigations. The agencies also have the right to take legal action if companies are not in compliance with the Act. These regulatory agencies employ OSHA Compliance Safety and Health Officers (CSHO), who are experts in the occupational safety and health field. The CSHOs are thoroughly familiar with OSHA standards and recognize safety and health hazards.

States with their own occupational safety and health programs conduct inspections by enlisting the services of qualified state CSHOs.

Companies are inspected for a multitude of reasons. They may be randomly selected, or they may be chosen as a result of employee complaints, a report of an imminent danger, or major accidents/fatalities that have occurred.

OSHA has established significant monetary fines for the violation of its regulations. *Table 1* lists the penalties as of 2016. In some cases, OSHA will hold a superintendents or crew leaders personally liable for repeat violations as well. In addition to the fines, there are possible criminal charges for willful violations resulting in death or serious injury. The attitude of the employer and their safety history can have a significant effect on the outcome of a case.

3.3.0 Employer Safety Responsibilities

Each employer must set up a safety and health program to manage workplace safety and health and to reduce work-related injuries, illnesses, and fatalities. The program must be appropriate for the conditions of the workplace. It should consider the number of workers employed and the hazards they face while at work.

To be successful, the safety and health program must have management, leadership, and employee participation. In addition, training and

Table 1 OSHA Penalties for violations established in 2016

Type of Violation	Maximum Penalty
Serious	
Other-Than-Serious	$12,471 per violation
Posting Requirements	
Failure to Abate	$12,471 per day
Willful or Repeated Violation	$124,709 per violation

informational meetings play an important part in effective programs. Being consistent with safety policies is the key. Regardless of the employer's responsibility, however, the individual worker is ultimately responsible for his or her own safety.

3.3.1 Safety Program

The crew leader plays a key role in the successful implementation of the safety program. The crew leader's attitude toward the program sets the standard for how crew members view safety. Therefore, the crew leader should follow all program guidelines and require crew members to do the same.

Safety programs should consist of the following:

- Safety policies and procedures
- Safety information and training
- Posting of safety notices
- Hazard identification, reporting, and assessment
- Safety record system
- Accident reporting and investigation procedures
- Appropriate discipline for not following safety procedures

3.3.2 Safety Policies and Procedures

Employers are responsible for following OSHA and state safety standards. Usually, they incorporate federal and state OSHA regulations into a safety policies and procedures manual. Employees receive such a manual when they are hired.

During orientation, appropriate company staff should guide the new employees through the general sections of the safety manual and the sections that have the greatest relevance to their job.

If the employee can't read, the employer should have someone read it to the employee and answer any questions that arise. The employee should then sign a form stating understanding of the information.

It isn't enough to tell employees about safety policies and procedures on the day they are hired and then never mention them again. Rather, crew leaders should constantly emphasize and reinforce the importance of adhering to all safety policies and procedures. In addition, employees should play an active role in determining job safety hazards and find ways to prevent and control hazards.

3.3.3 Hazard Identification and Assessment

Safety policies and procedures should be specific to the company. They should clearly present the hazards of the job and provide the means to report hazards to the proper level of management without prejudice to the individual doing the reporting. Crew leaders should also identify and assess hazards to which employees are exposed. They must also assess compliance with federal and state OSHA standards.

To identify and assess hazards, OSHA recommends that employers conduct periodic and random inspections of the workplace, monitor safety and health information logs, and evaluate new equipment, materials, and processes for potential hazards before they are used.

"You get what you inspect, not what you expect."
—Anonymous

Crew leaders and workers play important roles in identifying and reporting hazards. It is the crew leader's responsibility to determine what working conditions are unsafe and to inform employees of hazards and their locations. In addition, they should encourage their crew members to tell them about hazardous conditions. To accomplish this, crew leaders must be present and available on the job site.

The crew leader also needs to help the employee be aware of and avoid the built-in hazards to which craft professionals are exposed. Examples include working at elevations, working in confined spaces such as tunnels and underground vaults, on caissons, in excavations with earthen walls, and other naturally-dangerous projects. In addition, the crew leader can take safety measures, such as installing protective railings to prevent workers from falling from buildings, as well as scaffolds, platforms, and shoring.

3.3.4 Safety Information and Training

The employer must provide periodic information and training to new and long-term employees. This happens as often as necessary so that all employees receive adequate training. When safety and health information changes or workplace conditions create new hazards, the company must then provide special training and informational sessions. It is important to note that the company must present safety-related information in a manner that each employee will understand.

When a crew leader assigns an inexperienced employee a new task, the crew leader must ensure that the employee can do the work in a safe manner. The crew leader can accomplish this by providing safety information or training for groups or individuals.

When assigning an inexperienced employee a new task, do the following:

- Define the task.
- Explain how to do the task safely.
- Explain what tools and equipment to use and how to use them safely.
- Identify the necessary personal protective equipment and train the employee in its use.
- Explain the nature of the hazards in the work and how to recognize them.
- Stress the importance of personal safety and the safety of others.
- Hold regular safety training sessions with the crew's input.
- Review safety data sheets (SDS) that may be applicable.

3.3.5 Safety Record System

OSHA regulations (29 *CFR* 1904) require that employers keep records of hazards identified and document the severity of the hazard. The information should include the likelihood of employee exposure to the hazard, the seriousness of the harm associated with the hazard, and the number of exposed employees.

In addition, the employer must document the actions taken or plans for action to control the hazards. While it is best to take corrective action immediately, it is sometimes necessary to develop a plan to set priorities and deadlines and track progress in controlling hazards.

Employers who are subject to the recordkeeping requirements of the Occupational Safety and Health Act of 1970 must maintain records of all recordable occupational injuries and illnesses. The following are some OSHA forms that should be used for this recordkeeping:

- OSHA Form 300, *Log of Work-Related Injuries and Illnesses*
- OSHA Form 300A, *Summary of Work-Related Injuries and Illnesses*
- OSHA Form 301, *Injury and Illness Incident Report*

These three OSHA forms are included in the *Appendix* at the end of this module. Note that crew leaders directly handle the OSHA Form 301.

An SDS provides both workers and emergency personnel with the proper procedures for handling or working with a substance that may be dangerous. The document will include information such as physical data (melting point, boiling point, flash point, etc.), toxicity, health effects, first aid, reactivity, storage, disposal, protective equipment required for handling, and spill/leak procedures. These sheets are of particular use if a spill, fire, or other accident occurs.

Companies not exempted by OSHA must maintain required safety logs and retain them for 5 years following the end of the calendar year to which they relate. Logs must be available (normally at the company offices) for inspection and copying by representatives of the Department of Labor, the Department of Health and Human Services, or states given jurisdiction under the Act. Employees, former employees, and their representatives may also review these logs.

3.3.6 Accident Investigation

Employees must know from their training to immediately report any unusual event, accident, or injury. Policies and definitions of what these incidents consist of should be included in safety manuals. In the event of an accident, the employer is required to investigate the cause of the accident and determine how to avoid it in the future.

According to OSHA regulations, the employer must investigate each work-related death, serious injury or illness, or incident having the potential to cause death or serious physical harm. The employer should document any findings from the investigation, as well as the action plan to

Summaries of Work-Related Injuries and Illnesses

Most companies with 11 or more employees must post an OSHA Form 300A, *Summary of Work-Related Injuries and Illnesses*, between February 1 and April 30 of each year. Employees have the right to review this form. Check your company's policies regarding this and the related OSHA forms.

prevent future occurrences. The company should complete these actions immediately, with photos or video if possible. It's important that the investigation uncover the root cause of the accident to avoid similar incidents in the future. In many cases, the root cause was a flaw in the system that failed to recognize the unsafe condition or the potential for an unsafe act (*Figure 15*).

3.4.0 Leader Involvement in Safety

To be an effective, you must be actively involved in your company's safety program. Crew leader involvement includes conducting frequent safety training sessions and inspections, promoting first

aid, and fire protection and prevention, preventing substance abuse on the job, and investigating accidents. Most importantly, crew leaders must practice safety at all times.

3.4.1 Safety Training Sessions

A safety training session may be a brief, informal gathering of a few employees or a formal meeting with instructional videos and talks by guest speakers. The size of the audience and the topics addressed determine the format of the meeting. You should plan to conduct small, informal safety sessions weekly.

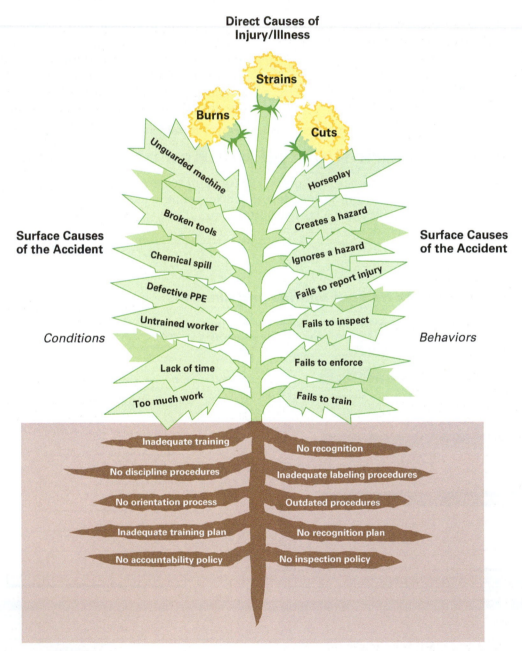

Figure 15 Root causes of accidents.

OSHA Accident Notification Requirements

There are urgent reporting requirements by the employer to OSHA for the following cases:

1. Within 8 hours: A work-related fatality.
2. Within 24 hours:
   - Work-related accident that resulted in an inpatient admission of one or more employees
   - Work-related amputation
   - Work-related loss of an eye

You should also plan safety training sessions in advance, and you should communicate the information to all affected employees. In addition, the topics covered in these training sessions should be timely and practical. Keep a log of each safety session signed by all attendees. It must be maintained as a record and available for inspection. It is advisable to attach a copy of a summary of the training session to the attendance record so you can keep track of what topics you covered and when.

3.4.2 Inspections

Crew leaders must make routine, frequent inspections to prevent accidents from happening. They must also take steps to avoid accidents. For that purpose, they need to inspect the job sites where their workers perform tasks. It's advisable to do these inspections before the start of work each day and during the day at random times.

Crew leaders must protect workers from existing or potential hazards in their work areas. Crews are sometimes required to work in areas controlled by other contractors. In these situations, the crew leader must maintain control over the safety exposure of the crew. If hazards exist, the crew leader should immediately bring the hazards to the attention of the contractor at fault, their superior, and the person responsible for the job site.

Crew leader inspections are only valuable if follow-up action corrects potential hazards. Therefore, crew leaders must be alert for unsafe and negligent acts on their work sites. When an employee performs an unsafe action, the crew leader must explain to the employee why the act was unsafe, tell that employee to not do it again, and request cooperation in promoting a safe working environment. The crew leader must document what happened and what the employee was told to do to correct the situation.

It is then important that crew leaders follow up to make certain the employee is complying with the safety procedures. Never allow a safety violation to go uncorrected. There are three courses of action that you, as a crew leader, can take in an unsafe work site situation:

- Get the appropriate party to correct the problem.
- Fix the problem yourself.
- Refuse to have the crew work in the area until the responsible party corrects the problem.

"As soon as you see a mistake and don't fix it, it becomes your mistake."
—Anonymous

3.4.3 First Aid

The primary purpose of first aid is to provide immediate and temporary medical care to employees involved in accidents, as well as employees experiencing non-work-related health emergencies, such as chest pain or breathing difficulty. To meet this objective, every crew leader should be aware of the location and contents of first aid kits available on the job site. Emergency numbers should be posted in the job trailer. In addition, OSHA requires that at least one person trained in first aid be present at the job site at all times. It's also advisable, but not required, that someone on site should have cardiopulmonary resuscitation (CPR) training.

> **NOTE**
> CPR certifications must be renewed every two years.

The victim of an accident or sudden illness at a job site may be harder to aid than elsewhere since the worker may be at a remote location. The site may be far from a rescue squad, fire department, or hospital, presenting a problem in the rescue and transportation of the victim to a hospital. The worker may also have received an injury from falling rock or other materials, so immediate special rescue equipment or first-aid techniques are often needed.

Employer benefits of having personnel trained in first aid at job sites include the following:

- The immediate and proper treatment of minor injuries may prevent them from developing into more serious conditions. Thus, these precautions can eliminate or reduce medical expenses, lost work time, and sick pay.

- It may be possible to determine if the injured person requires professional medical attention.
- Trained individuals can save valuable time preparing the inured for treatment for when professional medical care arrives. This service increases the likelihood of saving a life.

The American Red Cross, Medic First Aid, and the United States Bureau of Mines provide basic and advanced first aid courses at nominal costs. These courses include both first aid and CPR. The local area offices of these organizations can provide further details regarding the training available.

3.4.4 Fire Protection and Prevention

Fires and explosions kill and injure many workers each year, so it is important that crew leaders understand and practice fire-prevention techniques as required by company policy.

The need for protection and prevention is increasing as manufacturers introduce new building materials. Some building materials are highly flammable. They produce great amounts of smoke and gases, which cause difficulties for fire fighters, and can quickly overcome anyone present. Other materials melt when they burn and may puddle over floors, preventing fire-fighting personnel from entering areas where this occurs.

OSHA has specific standards for fire safety. Employers are required to provide proper exits, fire-fighting equipment, and employee training on fire prevention and safety. For more information, consult OSHA guidelines (available at **www.osha.gov**).

3.4.5 Substance Abuse

Substance abuse is a continuing problem in the workplace. Substance abuse is the inappropriate overuse of drugs and chemicals, whether they are legal or illegal. All substance abuse results in some form of mental, sensory, or physical impairment. Some people use illegal "street drugs", such as cocaine or crystal meth. Others use legal prescription drugs incorrectly by taking too many (or too few) pills, using other people's medications, or self-medicating. Alcohol can also be abused by consuming to the point of intoxication. Other substances that are legal in some states (e.g., marijuana) can cause prolonged impairment with heavy usage.

It is essential that crew leaders enforce company policies and procedures regarding substance abuse. Crew leaders must work with management to deal with suspected drug and alcohol abuse and should not deal with these situations themselves. The Human Resources department or a designated manager usually handles these cases.

There are legal consequences of substance abuse and the associated safety implications. If you observe an employee showing impaired behavior for any reason, immediately contact your supervisor and/or Human Resources department for assistance. You protect the business, and the employee's and other workers' safety by taking these actions. It is the crew leader's responsibility to maintain safe working conditions at all times. This may include removing workers from a work site where they may be endangering themselves or others.

For example, suppose several crew members go out and smoke marijuana or drink during lunch. Then, they return to work to erect scaffolding for a concrete pour in the afternoon. If you can smell marijuana on the crew member's clothing or alcohol on their breath, you must step in and take action. Otherwise, they might cause an accident that could delay the project, or cause serious injury or death to themselves or others.

> *"Concern for man himself and his safety must always form the chief interest of all technical endeavors."*
> —*Albert Einstein*

It is often difficult to detect drug and alcohol abuse because the effects can be subtle. The best way is to look for identifiable effects, such as those mentioned above, or sudden changes in behavior that aren't typical of the employee. Some examples of such behaviors include the following:

- Unscheduled absences; failure to report to work on time
- Significant changes in the quality of work
- Unusual activity or lethargy
- Sudden and irrational temper flare-ups
- Significant changes in personal appearance, cleanliness, or health

There are other more specific signs that should arouse suspicion, especially if more than one is visible:

- Slurring of speech or an inability to communicate effectively
- Shiftiness or sneaky behavior, such as an employee disappearing to wooded areas, storage areas, or other private locations
- Wearing sunglasses indoors or on overcast days to hide dilated or constricted pupils, conditions which impair vision (*Figure 16*)

Figure 16 An impaired worker is a dangerous one.

- Wearing long-sleeved garments, particularly on hot days, to cover marks from needles used to inject drugs
- Attempting to borrow money from co-workers
- The loss of an employee's tools or company equipment

3.4.6 Job-Related Accident Investigations

Crew leaders are sometimes involved with an accident investigation. When an accident, injury, or report of work-connected illness takes place. If present on site, the crew leader should proceed immediately to the accident location to ensure that the victim receives proper first aid. The crew leader will also want to make sure that responsible individuals take other safety and operational measures to prevent another incident.

If required by company policy, the crew leader will also need to make a formal investigation and submit a report after an incident (including the completion of an *OSHA Form 301* report). An investigation looks for the causes of the accident by examining the circumstances under which it occurred and talking to the people involved. Investigations are perhaps the most useful tool in the prevention of future accidents.

The following are four major parts to an accident investigation:

- Describing the accident and related events leading to the accident
- Determining the cause(s) of the accident
- Identifying the persons or things involved and the part played by each
- Determining how to prevent reoccurrences

3.4.7 Promoting Safety

The best way for crew leaders to encourage safety is through example. Crew leaders should be aware that their behavior sets standards for their crew members. If a crew leader cuts corners on safety, then the crew members may think that it is okay to do so as well.

> **CAUTION**
>
> Workers often "follow the leader" when it comes to unsafe work practices. It is common for supervisory personnel to engage in unsafe practices and take more risks because they are more experienced. However, inexperienced or careless workers who take the same risks won't be as successful avoiding injury. As a leader, you must follow all safety practices to encourage your crew to do the same.

"I cannot trust a man to control others who cannot control himself."
—Robert E. Lee

The key to effectively promote safety is good communication. It is important to plan and coordinate activities and to follow through with safety programs. The most successful safety promotions occur when employees actively participate in planning and carrying out activities.

Some activities used by organizations to help motivate employees on safety and help promote safety awareness include the following:

- Safety training sessions
- Contests
- Recognition and awards
- Publicity

Safety training sessions can help keep workers focused on safety and give them the opportunity to discuss safety concerns with the crew. A previous section addressed this topic.

> **Did You Know?**
>
> ## Substance Abuse
>
> An employee who is involved in an accident while under the influence of drugs or alcohol may be denied workers compensation insurance benefits.

Case Study

For years, a prominent safety engineer was confused as to why sheet-metal workers fractured their toes frequently. The crew leader had not performed thorough accident investigations, and the injured workers were embarrassed to admit how the accidents really occurred. Further investigation discovered they used the metal-reinforced cap on their safety shoes as a "third hand" to hold the sheet metal vertically in place when they fastened it. The rigid and heavy metal sheet was inclined to slip and fall behind the safety cap onto the toes, causing fractures. The crew leader could have prevented several injuries by performing a proper investigation after the first accident.

3.4.8 Safety Contests

Contests are a great way to promote safety in the workplace. Examples of safety-related contests include the following:

- Sponsoring housekeeping contests for the cleanest job site or work area
- Challenging employees to come up with a safety slogan for the company or department
- Having a poster contest that involves employees or their children creating safety-related posters
- Recording the number of accident-free workdays or worker-hours
- Giving safety awards (hats, T-shirts, other promotional items or prizes)

One of the positive aspects of safety contests is their ability to encourage employee participation. It is important, however, to ensure that the contest has a valid purpose. For example, workers can display the posters or slogans created in a poster contest throughout the organization as safety reminders.

CAUTION

One mistake that some companies make when offering safety contests is providing tangible or monetary awards to departments or teams specifically for the lowest number of reported accidents or accident-free work hours. While well-intentioned, this approach appeals to the tendency of people to inflate their performance to win. Consequently, history shows this has the negative effect of encouraging the underreporting of accidents and injuries, which defeats the purposes of safety contests.

3.4.9 Incentives and Rewards

Incentives and awards serve several purposes. Among them are acknowledging and encouraging good performance, building goodwill, reminding employees of safety issues, and publicizing the importance of practicing safety standards. There are countless ways to recognize and award safety. Examples include the following:

- Supplying food at the job site when a certain goal is achieved

- Providing a reserved parking space to acknowledge someone for a special achievement
- Giving gift items such as T-shirts or gift certificates to reward employees
- Giving awards to a department or an individual (*Figure 17*)
- Sending a letter of appreciation
- Publicly honoring a department or an individual for a job well done

You can use creativity to determine how to recognize and award good safety on the work site. The only precautionary measure is that the award should be meaningful and not perceived as a bribe. It should be representative of the accomplishment.

3.4.10 Publicity

Publicizing safety is the best way to get the message out to employees. An important aspect of publicity is to keep the message accurate and current. Safety posters that hang for years on end tend to lose effectiveness. It is important to keep ideas fresh.

Examples of promotional activities include posters or banners, advertisements or information on bulletin boards, payroll mailing stuffers, and employee newsletters. In addition, the company can purchase merchandise that promotes safety, including buttons, hats, T-shirts, and mugs.

Figure 17 An example of a safety award.

Participant Exercise D

Described here are three scenarios that reflect unsafe practices by craft workers. For each of these scenarios write down how you would deal with the situation, first as the crew leader of the craft worker, and then as the leader of another crew.

1. You observe a worker wearing his hard hat backwards and his safety glasses hanging around his neck. He is using a concrete saw.

_____

_____

_____

_____

_____

_____

2. As you are supervising your crew on the roof deck of a building under construction, you notice that a section of guard rail has been removed. Another contractor was responsible for installing the guard rail.

_____

_____

_____

_____

_____

_____

_____

3. Your crew is part of plant shutdown at a power station. You observe that a worker is welding without a welding screen in an area where there are other workers.

_____

_____

_____

_____

_____

_____

_____

NCCER – *Fundamentals of Crew Leadership*

Additional Resources

Construction Workforce Development Professional, NCCER. 2016. New York, NY: Pearson Education, Inc.

Mentoring for Craft Professionals, NCCER. 2016. New York, NY: Pearson Education, Inc.

The following websites offer resources for products and training:

National Census of Fatal Occupational Injuries (NCFOI), **www.bls.gov**

National Institute of Occupational Safety and Health (NIOSH), **www.cdc.gov/niosh**

National Safety Council, **www.nsc.org**.

Occupational Safety and Health Administration (OSHA), **www.osha.gov**

3.0.0 Section Review

1. One of a crew leader's most important responsibilities to the employer is to _____.

   a. enforce company safety policies
   b. estimate material costs for a project
   c. make recommendations for setting up a crew
   d. provide input for fixed-price contracts

2. What amount can OSHA fine a company for willfully committing a violation of an OSHA safety standard?

   a. $1,000
   b. $7,000
   c. $12,471
   d. $124,709

3. Who is ultimately responsible for a worker's safety?

   a. The individual worker
   b. The crew leader
   c. The project superintendent
   d. The company HR department head

4. A crew leader's safety responsibilities include all of the following *except* _____.

   a. conducting safety training sessions
   b. developing a company safety program
   c. performing safety inspections
   d. participating in accident investigations

4.0.0 PROJECT PLANNING

Objective

Demonstrate a basic understanding of the planning process, scheduling, and cost and resource control.

a. Describe how construction contracts are structured.
b. Describe the project planning and scheduling processes.
c. Explain how to implement cost controls on a construction project.
d. Explain the crew leader's role in controlling project resources and productivity.

Performance Tasks

1. Develop and present a look-ahead schedule.
2. Develop an estimate for a given work activity.

Trade Terms

Critical path: In manufacturing, construction, and other types of creative processes, the required sequence of tasks that directly controls the ultimate completion date of the project.

Job diary: A written record that a supervisor maintains periodically (usually daily) of the events, communications, observations, and decisions made during the course of a project.

Look-ahead schedule: A manual- or software-scheduling tool that looks several weeks into the future at the planned project events; used for anticipating material, labor, tool, and other resource requirements, as well as identifying potential schedule conflicts or other problems.

Return on investment (ROI): A measure of the gain or loss of money resulting from an investment; normally measured as a percentage of the original investment.

Work breakdown structure (WBS): A diagrammatic and conceptual method for subdividing a complex project, concept, or other thing into its various functional and organizational parts so that planners can analyze and plan for each part in detail.

This section describes methods of efficient project control. It examines estimating, planning and scheduling, and resource and cost control. All workers who participate in a job are responsible at some level for controlling cost and schedule performance, and for ensuring that they complete the project according to plans and specifications.

> **NOTE**
>
> This section mainly pertains to building-construction projects, but the project control principles described here apply generally to all types of projects.

The contractor, project manager, superintendent, and crew leader each have management responsibilities for their assigned jobs. For example, the contractor's responsibility begins with obtaining the contract, and it doesn't end until the client takes ownership of the project. The project manager is generally the person with overall responsibility for coordinating the project. Finally, the superintendent and crew leader are responsible for coordinating the work of one or more workers, one or more crews of workers within the company or, on occasion, one or more crews of subcontractors. The crew leader directs a crew in the performance of work tasks.

4.1.0 Construction Project Phases, Contracts, and Budgeting

Construction projects consist of three phases: the *development phase*, the *planning phase*, and the *construction phase*. Throughout these phases, the property owner must work directly with engineers and architects to gather data necessary for persuading a contractor to accept the job. Once a job is in progress, the crew leader fills an active role in maintaining the planned budget for the project.

4.1.1 Development Phase

A new building project begins when an owner has decided to build a new facility or add to an existing facility. The development process is the first stage of planning for a new building project. This process involves land research and feasibility studies to ensure that the project has merit. Architects or engineers develop the conceptual drawings that define the project graphically. They then provide the owner with sketches of room layouts and elevations and make suggestions about what construction materials to use.

During the development phase, architects, engineers, and/or the owner develop an estimate for the proposed project and establish a preliminary budget. Once that budget is established, the financing of the project with lending institutions begins. The development team begins preliminary reviews with government agencies. These reviews include zoning, building restrictions, landscape requirements, and environmental impact studies.

The owner must analyze the project's cost and potential **return on investment (ROI)** to ensure that its costs won't exceed its market value and that the project provides a reasonable profit during its existence. If the project passes this test, the responsible architects/engineers will proceed to the planning phase.

4.1.2 Planning Phase

When the architects/engineers begin to develop the project drawings and specifications, they consult with other design professionals such as structural, mechanical, and electrical engineers. They perform the calculations, make a detailed technical analysis, and check details of the project for accuracy.

The design professionals create drawings and specifications. They use these drawings and specifications to communicate the necessary information to the contractors, subcontractors, suppliers, and workers that contribute to a project.

During the planning phase, the owners hold many meetings (*Figure 18*) to refine estimates, adjust plans to conform to regulations, and secure a construction loan. If the project is a condominium, an office building, or a shopping center, then a marketing firm develops a marketing program. In such cases, the selling of the project often starts before actual construction begins.

Figure 18 Architects and clients meet to refine plans.

Next, the design team produce a complete set of drawings, specifications, and bid documents. Then the owner will select the method to obtain contractors. The owner may choose to negotiate with several contractors or select one through competitive bidding. Everyone concerned must also consider safety as part of the planning process. A safety crew leader may walk through the site as part of the pre-bid process.

Contracts for construction projects can take many forms. All types of contracts fall under three basic categories: *firm-fixed-price*, *cost reimbursable*, and *guaranteed maximum price*.

Firm-fixed-price – In this type of contract, the buyer generally provides detailed drawings and specifications, which the contractor uses to calculate the cost of materials and labor. To these costs, the contractor adds a percentage representing company overhead expenses such as office rent, insurance, and accounting/payroll costs. At the end, the contractor adds a profit factor.

When submitting the bid, the contractor will state very specifically the conditions and assumptions on which the company based the bid. These conditions and assumptions also form the basis from which parties can price allowable changes to the contract. Because contracting parties establish the price in advance, any changes in the job requirements once the job is started will impact the contractor's profit margin.

This is where the crew leader can play an important role by identifying problems that increase the amount of planned labor or materials. By passing this information up the chain of command, the crew leader allows the company to determine if the change is outside the scope of the bid. If so, they can submit a change order request to cover the added cost.

Cost reimbursable contract – In this type of contract, the buyer reimburses the contractor for labor, materials, and other costs encountered in the performance of the contract. Typically, the contractor and buyer agree in advance on hourly or daily labor rates for different categories of worker. These rates include an amount representing the contractor's overhead expense. The buyer also reimburses the contractor for the cost of materials and equipment used on the job.

The buyer and contractor also negotiate a profit margin. On this type of contract, the profit margin is likely to be lower than that of a fixed-price contract because of the significantly-reduced contractor's cost risk. The profit margin is often subject to incentive or penalty clauses that make the amount of profit awarded subject to performance by the contractor. The contract usually ties performance to project schedule milestones.

Guaranteed maximum price (GMP) contract – This form of contract, also called a *not-to-exceed contract*, is common on projects negotiated mainly with the owner. Owner's involvement in the process usually includes preconstruction, and the entire team develops the parameters that define the basis for the work.

With a GMP contract, the owner reimburses the contractor for the actual costs incurred by the contractor. The contract also includes a payment of a fixed fee up to the maximum price allowed in the contract. The contractor bears any cost overruns.

The advantages of the GMP contract vehicle may include the following:

- Reduced design time
- Allows for phased construction
- Uses a team approach to a project
- Reduction in changes related to incomplete drawings

4.1.3 Construction Phase

The designated contractor enlists the help of mechanical, electrical, elevator, and other specialty subcontractors to complete the construction phase. The contractor may perform one or more parts of the construction, and rely on subcontractors for the remainder of the work. However, the general contractor is responsible for managing all the trades necessary to complete the project. *Figure 19* shows the flow of a typical project from beginning to end.

As construction nears completion, the architect/engineer, owner, and government agencies start their final inspections and acceptance of the project. If the general contractor has managed the project, the subcontractors have performed their work, and the architects/engineers have regularly inspected the project to ensure it satisfied the local code, then the inspection process can finish up in a timely manner. This results in a satisfied client and a profitable project for all.

On the other hand, if the inspection reveals faulty workmanship, poor design, incorrect use of materials, or violation of codes, then the inspection and acceptance will become a lengthy battle and may result in a dissatisfied client and an unprofitable project.

The initial set of drawings for a construction project reflects the completed project as conceived by the architect and engineers. During construction, changes are usually necessary because of factors unforeseen during the design phase. For example, when electricians must reroute cabling or conduit, or the installed equipment location is different than shown on the original drawing, such changes must be marked on the drawings. Without this record, technicians called to perform maintenance or modify the equipment later will have trouble locating all the cabling and equipment.

Project supervision must document any changes made during construction or installation on the drawings as the changes occur. Architects usually note changes on hard-copy drawings using a colored pen or pencil, so users can readily spot the change. These marked-up versions are commonly called **redline drawings**. With mobile digital technology, architects and engineers can revise and promulgate the latest drawing versions almost instantly. After the drawings have been revised to reflect the redline changes, the final drawings are called **as-built drawings**, and are so marked. These become the drawings of record for the project.

4.1.4 Project-Delivery Systems

Project-delivery systems are processes for constructing projects, from development through construction. Project delivery systems focus on the following three primary systems, shown in *Figure 20*:

- *General contracting* – The traditional project delivery system uses a general contractor. In this type of project, the owner determines the design of the project, and then solicits proposals from general contractors. After selecting a general contractor, the owner contracts directly with that contractor, who builds the project as the prime, or controlling, contractor.
- *Design-build* – In the design-build system, a single entity manages both the design and construction of a project. Design-build delivery commonly use GMP contracts.
- *Construction management* – The construction management project delivery system uses a construction manager to facilitate the design and construction of a project. Construction managers are very involved in project control; their main concerns are controlling time, cost, and the quality of the project.

4.1.5 Cost Estimating and Budgeting

Before building a project, an estimate must be prepared. Estimating is the process of calculating the cost of a project. There are two types of costs to consider, including direct and indirect costs. Direct costs, also known as general conditions, are those that planners can clearly assigned to a

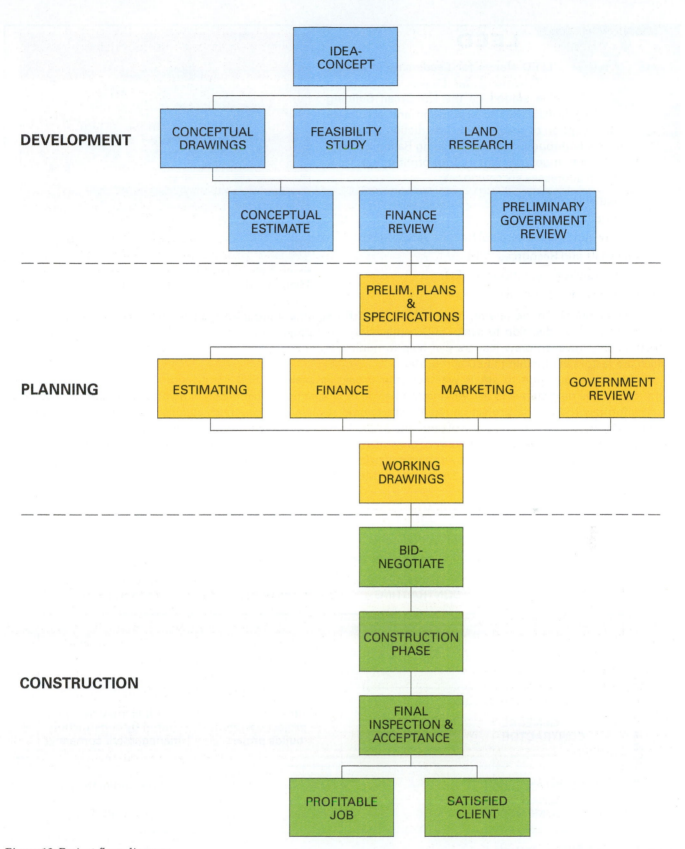

DEVELOPMENT

IDEA-CONCEPT

CONCEPTUAL DRAWINGS

FEASIBILITY STUDY

LAND RESEARCH

CONCEPTUAL ESTIMATE

FINANCE REVIEW

PRELIMINARY GOVERNMENT REVIEW

PLANNING

PRELIM. PLANS & SPECIFICATIONS

ESTIMATING

FINANCE

MARKETING

GOVERNMENT REVIEW

WORKING DRAWINGS

CONSTRUCTION

BID-NEGOTIATE

CONSTRUCTION PHASE

FINAL INSPECTION & ACCEPTANCE

PROFITABLE JOB

SATISFIED CLIENT

Figure 19 Project flow diagram.

LEED

LEED stands for *Leadership in Energy and Environmental Design*, which is an initiative started by the US Green Building Council (USGBC) to encourage and accelerate the adoption of sustainable construction standards worldwide through its Green Building Rating System. USGBC is a non-government, not-for-profit group. Their rating system addresses six categories:

1. Sustainable Sites (SS)
2. Water Efficiency (WE)
3. Energy and Atmosphere (EA)
4. Materials and Resources (MR)
5. Indoor Environmental Quality (EQ)
6. Innovation in Design (ID)

Figure Credit: © Linda Williams/Dreamstime.com

Building owners are the driving force behind the LEED voluntary program. Construction crew leaders may not have input into the decision to seek LEED certification for a project, or what materials are used in the project's construction. However, crew leaders can help to minimize material waste and support recycling efforts, both of which are factors in obtaining LEED certification.

An important question to ask is whether your project is seeking LEED certification. If the project is seeking certification, the next step is to ask what your role will be in getting the certification. If you are procuring materials, what information does the project need and who should receive it? What specifications and requirements do the materials need to meet? If you are working outside the building or inside in a protected area, what do you need to do to protect the work area? How does your crew manage waste? Are there any other special requirements that will be your responsibility? Do you see any opportunities for improvement? LEED principles are described in more detail in NCCER's *Your Role in the Green Environment* (Module ID 70101-15) and *Sustainable Construction Supervisor* (Module ID 70201-11).

DELIVERY SYSTEMS

	GENERAL CONTRACTING	DESIGN-BUILD	CONSTRUCTION MANAGEMENT
OWNER	Designs project (or hires architect)	Hires general contractor	Hires construction management company
GENERAL CONTRACTOR	Builds project (with owner's design)	Involved in project design, builds project	Builds, may design (hired by construction management company)
CONSTRUCTION MANAGEMENT COMPANY			Hires and manages general contractor and architect

(Left-side label: RESPONSIBLE PARTIES)

Figure 20 Project-delivery systems.

budget. Indirect costs are overhead costs shared by all projects. Planners generally calculate these costs as an overhead percentage to labor and material costs.

Direct costs include the following:

- Materials
- Labor
- Tools
- Equipment

Indirect costs refer to overhead items such as the following:

- Office rent
- Utilities
- Telecommunications
- Accounting
- Office supplies, signs

The bid price includes the estimated cost of the project as well as the profit. Profit refers to the amount of money that the contractor will make after paying all the direct and indirect costs. If the direct and indirect costs exceed the estimate for the job, the difference between the actual and estimated costs must come out of the company's profit. This reduces what the contractor makes on the job.

Profit is the fuel that powers a business. It allows the business to invest in new equipment and facilities, provide training, and to maintain a reserve fund for times when business is slow. In large companies, profitability attracts investors who provide the capital necessary for the business to grow. For these reasons, contractors can't afford to consistently lose money on projects. If they can't operate profitably, they are forced out of business. Crew leaders can help their companies remain profitable by managing budget, schedule, quality, and safety adhering to the drawings, specifications, and project schedule.

The cost estimate must consider many factors. Many companies employ professional cost estimators to do this work. They also maintain performance data for previous projects. They use this data as a guide in estimating new projects. Development of a complete estimate generally proceeds as follows:

Step 1 Using the drawings and specifications, an estimator records the quantity of the materials needed to construct the job. Construction companies call this step the *quantity takeoff*. The estimator enters the information on a hard-copy or digital takeoff sheet like the one shown in *Figure 21*.

Step 2 The estimator uses the company's productivity rates to calculate the amount of labor required to complete the project. Most companies keep records of these rates for the type and size of the jobs that they perform. The company's estimating department maintains and updates these records.

Step 3 The estimator calculates the labor hours required by dividing the estimated amount of work by the productivity rate.

- For example, if the productivity rate for concrete finishing is 40 square feet per hour, and there are 10,000 square feet of concrete to be finished, then 250 hours of concrete finishing labor is required (10,000 ÷ 4 = 250).
- The estimator multiplies this number by the hourly rate for concrete finishing to determine the cost of that labor category.
- If this work is subcontracted, then the estimator uses the subcontractor's cost estimate, raised by an overhead factor, in place of direct-labor cost.

Step 4 The estimator transfers the total material quantities from the quantity takeoff sheet to a summary or pricing sheet (*Figure 22*). The total cost of materials is calculated after obtaining material prices from local suppliers.

Step 5 Next, the estimator determines the cost of equipment needed for the project. This number could reflect rental cost or a factor applied by the company when they plan to use their own equipment.

Step 6 The estimator totals the cost of all resources on the summary sheet—materials, equipment, tools, and labor. The estimator can also calculate the material unit cost—the total cost divided by the total number of units of listed materials.

Step 7 The estimator adds the cost of taxes, bonds, insurance, subcontractor work, and other indirect costs to the direct costs of the materials, equipment, tools, and labor.

Step 8 A sum of direct and indirect costs yields the total project cost. The contractor adds the expected profit to that total.

WORKSHEET

Takeoff By: _____

Checked By: _____

PROJECT _____

ARCHITECT _____

DATE _____

SHEET ____ of ____

PAGE # _____

REF.	DESCRIPTION	DIMENSIONS				EXTENSION					TOTAL		REMARKS
		NO	LENGTH	WIDTH	HEIGHT		QUANTITY	UNIT	QUANTITY	UNIT	QUANTITY	UNIT	

Figure 21 Quantity takeoff sheet.

SUMMARY SHEET

By:

DATE ______ PROJECT ______

SHEET ______ of ______ TITLE: ______ WORK ORDER # ______

PAGE #

	DESCRIPTION	QUANTITY		MATERIAL COST		LABOR MAN HOURS FACTORS					LABOR COST		ITEM COST	
		TOTAL	UT	PER UNIT	TOTAL	CRAFT	PR UNIT	TOTAL	RATE	COST PR	PER	TOTAL	TOTAL	PER UNIT
					MATERIAL							LABOR	TOTAL	

Figure 22 Summary sheet.

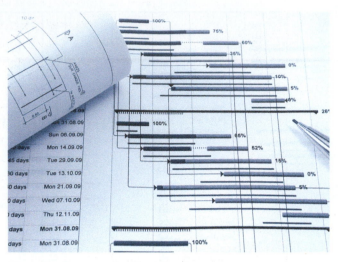

Figure 23 Proper prior planning prevents poor performance.

As a crew leader, you may be required to estimate quantities of materials. You will need a set of construction drawings and specifications to estimate the amount of a certain type of material required to perform a job. You should carefully review the appropriate section of the technical specifications and page(s) of drawings to determine the types and quantities of materials required. Then enter the quantities on the worksheet. For example, you should review the specification section on finished carpentry along with the appropriate pages of drawings before taking off the linear feet of door and window trim.

If insufficient materials are available to complete the job and an estimate is required, the estimator must determine how much more construction work is necessary. Knowing this, the crew leader can then determine the materials needed. You must also reference the construction drawings in this process.

4.2.0 Planning

One definition of planning is determining the methods used and their sequence to carry out the different tasks to complete a project (*Figure 23*). Planning involves the following:

- Determining the best method for performing the job
- Identifying the responsibilities of each person on the work crew
- Determining the duration and sequence of each activity
- Identifying the tools and equipment needed to complete a job
- Ensuring that the required materials are at the work site when needed
- Making sure that heavy construction equipment is available when required

- Working with other contractors in such a way as to avoid interruptions and delays

Some important reasons for crew-leader planning include the following:

- Controlling the job in a safe manner so that it is built on time and within cost
- Lowering job costs through improved productivity
- Preparing for bad weather or unexpected occurrences
- Promoting and maintaining favorable employee morale
- Determining the best and safest methods for performing the job

With a plan, a crew leader can direct work efforts efficiently and can use resources such as personnel, materials, tools, equipment, work area, and work methods to their full potential.

A proactive crew leader will always have a backup plan in case circumstances prevent the original plan from working. There are many circumstances that can cause a plan to go awry, including adverse weather, equipment failure, absent crew members, and schedule slippage by other crafts.

Project planners establish time and cost limits for the project; the crew leader's planning must fit within those constraints. Therefore, it is important to consider the following factors that may affect the outcome:

- Site and local conditions, such as soil types, accessibility, or available staging areas
- Climate conditions that should be anticipated during the project
- Timing of all phases of work
- Types of materials to be installed and their availability

Assume you are the leader of a crew building footing formwork for the construction shown in the figure below. You have used all of the materials provided for the job, yet you have not completed it. You study the drawings and see that the formwork consists of two side forms, each 12" high. The total length of footing for the entire project is 115'-0". You have completed 88'-0" to date; therefore, you have 27'-0" remaining (115' – 88' = 27').

Your task is to prepare an estimate of materials that you will need to complete the job. In this case, you will estimate only the side form materials (do not consider the miscellaneous materials here).

- Footing length to complete: 27'-0"
- Footing height: 1'-0"

Refer to the worksheet on the next page for a final tabulation of the side forms needed to complete the job.

1. Using the same footing as described in the example above, calculate the quantity (square feet) of formwork needed to finish 203 linear feet of the footing. Place this information directly on the worksheet.
2. You are the crew leader of a carpentry crew whose task is to side a warehouse with plywood sheathing. The wall height is 16', and there is a total of 480 linear feet of wall to side. You have done 360 linear feet of wall and have run out of materials. Calculate how many more feet of plywood you will need to complete the job. If you are using 4 × 8 plywood panels, how many will you need to order to cover the additional work? Write your estimate on the worksheet.

Show your calculations to the instructor.

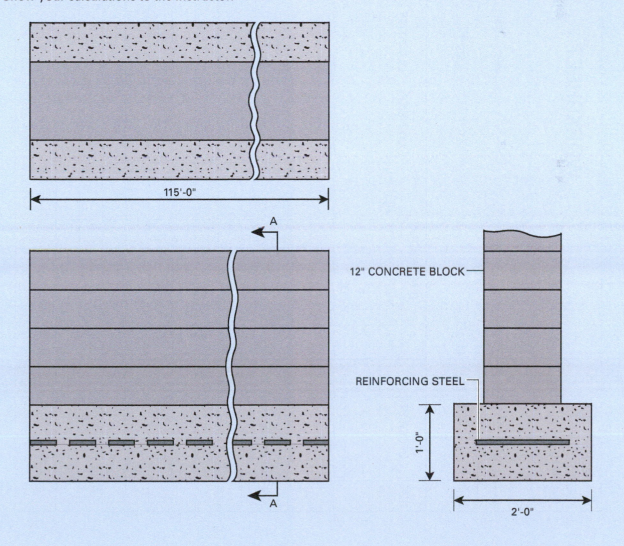

115'-0"

A

A

12" CONCRETE BLOCK

REINFORCING STEEL

1'-0"

2'-0"

Participant Exercise E Worksheet

WORKSHEET

Takeoff By: RWH

Checked By:

DATE 2/1/17

SHEET 01 of 01

PROJECT Sam's Diner

ARCHITECT 654b

PAGE #1

REF.	DESCRIPTION	DIMENSIONS				EXTENSION	QUANTITY	UNIT	TOTAL		REMARKS
		NO	LENGTH	WIDTH	HEIGHT				QUANTITY	UNIT	
	Footing Side Forms	2	27'0"		1'0"	2x27x1	54	SF	54	SF	

- Equipment and tools required and their availability
- Personnel requirements and availability
- Relationships with the other contractors and their representatives on the job

On a simple job, crews can handle these items almost automatically. However, larger or more complex jobs require the planner to give these factors more formal consideration and study.

4.2.1 Stages of Planning

Formal planning for a construction job occurs at specific times in the project process. The two most important stages of planning occur in the preconstruction phase and during the construction work.

Preconstruction planning – The preconstruction stage of planning occurs before the start of construction. The preconstruction planning process doesn't always involve the crew leader, but it's important to understand what it consists of.

There are two phases of preconstruction planning. The first is developing the proposal, bid, or negotiated price for the job. This is when the estimator, the project manager, and the field superintendent develop a preliminary plan for completing the work. They apply their experience and knowledge from previous projects to develop the plan. The process involves determining what methods, personnel, tools, and equipment the work will require and what level of productivity they can achieve.

The second phase of preconstruction planning occurs after the client awards the contract. This phase requires a thorough knowledge of all project drawings and specifications. During this process, planners select the actual work methods and resources needed to perform the work. Here, crew leaders might get involved, but their planning must adhere to work methods, production rates, and resources that fit within the estimate prepared during contract negotiations. If the project requires a method of construction different

---

Participant Exercise F

1. In your own words, define planning, and describe how a job can be done better if it is planned. Give an example.

   _____

   _____

   _____

   _____

2. Consider a job that you recently worked on to answer the following:
   a. List the material(s) used.
   b. List each member of the crew with whom you worked and what each person did.
   c. List the kinds of equipment used.

   _____

   _____

   _____

   _____

   _____

3. List some suggestions for how the job could have been done better, and describe how you would plan for each of the suggestions.

   _____

   _____

   _____

   _____

---

from what is normal, planners will usually inform the crew leader of what method to use.

Construction planning – During construction, the crew leader is directly involved in daily planning. This stage of planning consists of selecting methods for completing tasks before beginning work. Effective planning exposes likely difficulties, and enables the crew leader to minimize the unproductive use of personnel and equipment. Proper planning also provides a gauge to measure job progress.

Effective crew leaders develop a tool known as the look-ahead schedule. These schedules consider actual circumstances as well as projections two-to-three weeks into the future. Developing a look-ahead schedule helps ensure that all resources are available on the project when needed. Most scheduling apps and programs include this feature that automatically help you focus on this period in the job.

4.2.2 The Planning Process

The planning process consists of the following five steps:

Step 1 Establish a goal.

Step 2 Identify the completed work activities required to achieve the goal.

Step 3 Identify the required tasks to accomplish those activities.

Step 4 Communicate responsibilities.

Step 5 Follow up to verify the goal achievement.

Establishing a goal – The term *goal* has different meanings for different people. In general, a goal is a specific outcome that one works toward. For example, the project superintendent of a home construction project could establish the goal to have a house dried-in by a certain date. (The term *dried-in* means ready for the application of roofing and siding.) To meet that goal, the leader of the framing crew and the superintendent would need to agree to a goal to have the framing completed by a given date. The crew leader would then establish sub-goals (objectives) for the crew to complete each element of the framing (floors, walls, roof) by a set time. The superintendent would need to set similar goals with the crews that install sheathing, building wrap, windows, and exterior doors. However, if the framing crew doesn't meet its goal, that will delay the other crews.

Identifying the required work – The second step in planning is to identify the necessary work to achieve the goal as a series of activities in a certain sequence. You will learn how to break down a job into activities later in this section. At this point, the crew leader should know that, for each activity, one or more objectives must be set.

An objective is a statement of a condition the plan requires to exist or occur at a specific time. An objective must:

- Mean the same thing to everyone involved
- Be measurable, so that everyone knows when it has been reached
- Be achievable with the resources available
- Have everyone's full support

Examples of practical objectives include the following:

- "By 4:30 p.m. today, the crew will have completed installation of the floor joists."
- "By closing time Friday, the roof framing will be complete."

Notice that both examples meet the first three requirements of an objective. Planners assume that everyone involved in completing the task is committed to achieving the objective. The advantage in developing objectives for each work activity is that it allows the crew leader to determine if the crew is following the plan. In addition, objectives serve as sub-goals that are usually under the crew leader's control.

Some construction work activities, such as installing 12" footing forms, are done so often that they require little planning. However, other jobs, such as placing a new type of mechanical equipment, require substantial planning. This type of job demands that the crew leader set specific objectives.

Whenever faced with a new or complex activity, take the time to establish objectives that will serve as guides for accomplishing the job. You can use these guides in the current situation, as well as for similar work in the future.

Identifying the required tasks – To plan effectively, the crew leader must be able to break a work activity assignment down into smaller tasks. Large jobs include a greater number of tasks than small ones, but all jobs can be broken down into manageable components.

When breaking down an assignment into tasks, make each task identifiable and definable. A task is identifiable when one knows the types and amounts of resources it requires. A task is definable if it has a specific duration. For purposes of efficiency, the job breakdown should not be too detailed or complex, unless the job has never been done before or must be performed with strictest efficiency.

For example, a suitable breakdown for the work activity to install square vinyl floor tiles in a cafeteria might be the following:

Step 1 Prepare the floor.

Step 2 Stage the tiles.

Step 3 Spread the adhesive.

Step 4 Lay the tiles.

Step 5 Clean the tiles.

Step 6 Wax the floor.

The crew leader could create even more detail by breaking down any one of the tasks into subtasks. In this case, however, that much detail is unnecessary and wastes the crew leader's time and the project's money.

Planners can divide every work activity into three general parts:

- Preparing
- Performing
- Cleaning up

Some of the most frequent mistakes made in the planning process are forgetting to prepare and to clean up. The crew leader must not overlook preparation and cleanup.

After identifying the various tasks that make up the job and developing an objective for each task, the crew leader must determine what resources the job requires. Resources include labor, equipment, materials, and tools. In most jobs, the job estimate identifies these resources. The crew leader must make sure that these resources are available on the site when needed.

"By failing to prepare, you are preparing to fail."
—*Benjamin Franklin*

Communicating responsibilities – No supervisor can complete all the activities within a job alone. The crew leader must rely on other people to get everything done. Therefore, most jobs have a crew of people with various experiences and skill levels to assist in the work. The crew leader's job is to draw from this expertise to get the job done well and in a safe and timely manner.

Once the various activities that make up the job have been determined, the crew leader must identify the person or persons responsible for completing each activity. This requires that the crew leader be aware of the skills and abilities of the people on the crew. Then, the crew leader must put this knowledge to work in matching the crew's skills and abilities to specific tasks required to complete the job.

After matching crew members to specific activities, the crew leader must then assign work to the crew. The crew leader normally communicates responsibilities verbally; the crew leader often talks directly to the responsible person for the activity. There may be times when the crew leader assigns work through written instructions or indirectly through someone other than the crew leader. Either way, crew members should know what it is they are responsible for accomplishing on the job.

Following up – Once the crew leader has delegated the activities to the appropriate crew members, there must be follow-up to make sure that the crew has completed them correctly and efficiently. Task follow-up involves being present on the job site to make sure all the resources are available to complete the work, ensuring that the crew members are working on their assigned activities, answering any questions, and helping to resolve any problems that occur while the work is being done. In short, follow-up activity means that the crew leader is aware of what's going on at the job site and is doing whatever is necessary to make sure that the crew completes the work on schedule. *Figure 24* reviews the planning steps.

The crew leader should carry a small note pad or electronic device for planning and taking notes. That way, you can record thoughts about the project as they occur, and you won't forget pertinent details. The crew leader may also choose to use a manual planning form such as the one illustrated in *Figure 25*.

As the job progresses, refer to these resources to see that the tasks are being done according to plan. This is job analysis. Construction projects that don't proceed according to work plans usually end up costing more and taking longer. Therefore, it is important that crew leaders refer to the planning documents periodically.

The crew leader is involved with many activities on a day-to-day basis. Thus, it is easy to forget important events if they aren't recorded. To help keep track of events such as job changes, interruptions, and visits, the crew leader should keep a job diary. A job diary is a notebook in which the crew leader records activities or events that take place on the

Figure 24 Steps to effective planning.

DAILY WORK PLAN

"PLAN YOUR WORK AND WORK YOUR PLAN = EFFICIENCY"

Plan of _____   Date _____

PRIORITY	DESCRIPTION	✓ When Completed ✗ Carried Forward

Figure 25 Planning form.

job site that may be important later. When making entries in a job diary, make sure that the information is accurate, factual, complete, consistent, organized, and up-to-date. This is especially true if documenting personnel problems. If the company requires maintaining a job diary, follow company policy in determining which events and what details you should record. However, if there is a doubt about what to include, it is better to have too much information than too little. *Figure 26* shows a sample page from a job diary.

4.2.3 Planning Resources

Once a job has been broken down into its tasks or activities, the various resources needed to perform them must be determined and accounted for. Resource planning includes the following specific considerations:

- *Safety planning* – Using the company safety manual as a guide, the crew leader must assess the safety issues associated with the job and take necessary measures to minimize any risk to the crew. This may involve working with the company or site safety officer and may require a formal hazard analysis.
- *Materials planning* – Preconstruction planning identifies the materials required for the job and lists them on the job estimate. Companies usually order the materials from suppliers who have previously provided quality materials on schedule and within estimated cost.

The crew leader is usually not involved in the planning and selection of materials, which happens during the preconstruction phase. The crew leader does, however, have a major role to play in the receipt, storage, and control of the materials after they reach the job site.

- The crew leader is also involved in planning materials for tasks such as job-built formwork and scaffolding. In addition, the crew leader may run out of a specific material, such as fasteners, and need to order more. In such cases, be sure to consult the appropriate supervisor, since most companies have specific purchasing policies and procedures.

July 8, 2017

Weather: Hot and Humid

Project: Company XYZ Building

- The paving contractor crew arrived late (10 am).

- The owner representative inspected the footing foundation at approximately 1 pm.

- The concrete slump test did not pass. Two trucks had to be ordered to return to the plant, causing a delay.

- John Smith had an accident on the second floor. I sent him to the doctor for medical treatment. The cause of the accident is being investigated.

Figure 26 Sample page from a job diary.

- *Site planning* – There are many planning elements involved in site work. The following are some of the key elements:

  - Access roads
  - Emergency procedures
  - Material and equipment storage
  - Material staging
  - Parking
  - Sedimentation control
  - Site security
  - Storm water runoff

- *Equipment planning* – The preconstruction phase addresses much of the planning for use of construction equipment. This planning includes the types of equipment needed, the use of the equipment, and the length of time it will be on the site. The crew leader must work with the main office to make certain that the equipment reaches the job site on time. The crew leader must also ensure that crew equipment operators are properly trained.

  Coordinating the use of the equipment is also very important. Some equipment operates in combination with other equipment. For example, dump trucks are generally required when loaders and excavators are used. The crew leader should also coordinate equipment with other contractors on the job. Sharing equipment can save time and money and avoid duplication of effort.

  The crew leader must reserve time for equipment maintenance to prevent equipment failure. In the event of an equipment failure, the crew leader must know who to contact to resolve the problem. An alternate plan must be ready in case one piece of equipment breaks down, so that the other equipment doesn't sit idle. The crew leader should coordinate these contingency plans with the main office or the crew leader's immediate superior.

- *Tool planning* – A crew leader is responsible for planning tool usage for a job. This task includes the following:

  - Determining the tools required
  - Informing the workers who will provide the tools (company or worker)
  - Making sure the workers are qualified to use the tools safely and effectively
  - Determining what controls to establish for tools

- *Labor planning* – All jobs require some sort of labor because the crew leader can't complete all the work alone. When planning for labor, the crew leader must do the following:

  - Identify the skills needed to perform the work.
  - Determine the number of people having those specific skills that are required.
  - Decide who will be on the crew.

In many companies, the project manager or job superintendent determines the size and make-up of the crew. Then, supervision expects the crew leader to accomplish the goals and objectives with the crew provided. Even though the crew leader may not have any involvement in staffing the crew, the crew leader is responsible for training the crew members to ensure that they have the skills needed to do the job. In addition, the crew leader is responsible for keeping the crew adequately staffed at all times to avoid job delays. This involves dealing with absenteeism and turnover, two common problems discussed in earlier sections.

4.2.4 Scheduling

Planning and scheduling are closely related and are both very important to a successful job. Planning identifies the required activities and how and in what order to complete them. Scheduling involves establishing start and finish times/dates for each activity.

A schedule for a project typically shows the following:

- Operations listed in sequential order
- Units of construction
- Duration of activities
- Estimated date to start and complete each activity
- Quantity of materials to be installed

There are different types of schedules used today. They include the bar chart, the network schedule, also called the **critical path** method (CPM) or precedence diagram, and the short-term, or look-ahead schedule.

The following is a summary of the steps a crew leader must complete to develop a schedule.

Step 1   Make a list of all the activities that the plan requires to complete the job, including individual work activities and special tasks, such as inspections or the delivery of materials. At this point, the crew leader should just be concerned with generating a list, not with determining how to accomplish the activities, who will perform them, how long they will take, or the necessary sequence to complete them.

Step 2   Use the list of activities created in Step 1 to reorganize the work activities into a logical sequence. When doing this, keep in mind that certain steps can't happen before the completion of others. For example, footing excavation must occur before concrete emplacement.

Step 3 Assign a duration or length of time that it will take to complete each activity and determine the start time for each. Then place each activity into a schedule format. This step is important because it helps the crew leader compare the task time estimates to the scheduled completion date or time.

The crew leader must be able to read and interpret the job schedule. On some jobs, the form provides the beginning and expected end date for each activity, along with the expected crew or worker's production rate. The crew leader can use this information to plan work more effectively, set realistic goals, and compare the starts and completions of tasks to those on the schedule.

Before starting a job, the crew leader must do the following:

- Determine the materials, tools, equipment, and labor needed to complete the job.
- Determine when the various resources are needed.
- Follow up to ensure that the resources are available on the job site when needed.

The crew leader should verify the availability of needed resources three to four working days before the start of the job. This should occur even earlier for larger jobs. Advance preparation will help avoid situations that could potentially delay starting the job or cause it to fall behind schedule.

Supervisors can use bar chart schedules, also known as *Gantt charts*, for both short-term and long-term jobs. However, they are especially helpful for jobs of short duration.

Bar charts provide management with the following:

- A visual presentation of the overall time required to complete the job using a logical method rather than a calculated guess
- A means to review the start and duration of each part of the job
- Timely coordination requirements between crafts
- Alternative sequences of performing the work

A bar chart works as a control device to see whether the job is on schedule. If the job isn't on schedule, supervision can take immediate action in the office and the field to correct the problem and increases the likelihood of completing the activity on schedule. *Figure 27* illustrates a bar or Gantt chart.

Another type of schedule is the network schedule, which shows dependent (critical path) activities and other activities completed in parallel with but not part of the critical path. In *Figure 28*, for example, reinforcing steel can't be set until the concrete forms have been built and placed. Other activities are happening in parallel, but the forms are in the critical path.

When building a house, builders can't install and finish drywall until wiring, plumbing, and HVAC ductwork have been roughed-in. Because other activities, such as painting and trim work, depend on drywall completion, the drywall work is a critical-path operation. In other words, until it is complete, workers can't start the other tasks, and the project itself will likely experience delay by the amount of delay starting any dependent

activity. Likewise, workers can't even start the drywall installation until the rough-ins are complete. Therefore, the project superintendent is likely to focus on those activities when evaluating schedule performance.

The advantage of a network schedule is that it allows project leaders to see how a schedule change with one activity is likely to affect other activities and the project in general. Planners lay out a network schedule on a timeline and usually show the estimated duration for each activity. Planners use network schedules for complex jobs that take a long time to complete. The PERT (program evaluation and review technique) schedule is a form of network schedule.

Since the crew needs to hold to the job schedule, the crew leader needs to be able to plan daily production. As discussed earlier, short-term scheduling is a method used to do this. *Figure 29* displays an example of a short-term, look-ahead schedule.

The information to support short-term scheduling comes from the estimate or cost breakdown. The schedule helps to translate estimate data and the various job plans into a day-to-day schedule of events. The short-term schedule provides the crew leader with visibility over the immediate future of the project. If actual production begins to slip behind estimated production, the schedule will warn the crew leader that a problem lies ahead and that a schedule slippage is developing.

Crew leaders should use short-term scheduling to set production goals. Generally, workers can improve production when they:

- Know the amount of work to be accomplished
- Know the time available to complete the work
- Can provide input when setting goals

Example:

A carpentry crew on a retaining wall project is about to form and pour catch basins and put up wall forms. The crew has put in several catch basins, so the crew leader is sure that they can perform the work within the estimate. However, the crew leader is concerned about their production of the wall forms. The crew will work on both the basins and the wall forms at the same time. The scheduling process in this scenario could be handled as follows:

1. The crew leader notices the following in the estimate or cost breakdown:
   a. Production factor for wall forms = 16 worker-hours (w-h) per 100 ft^2
   b. Work to be done by measurement = 800 ft^2
   c. Total time: $(800 \text{ ft}^2 \times 16 \text{ w-h}/100 \text{ ft}^2) = 128$ w-h

2. The carpenter crew consists of the following:
   a. One carpenter crew leader
   b. Four carpenters
   c. One laborer

3. The crew leader determines the goal for the job should be set at 128 w-h (from the cost breakdown).

4. If the crew remains the same (six workers), the work should be completed in about 21 crew-hours (128 w-h ÷ 6 workers/crew = 21.3 crew-hours).

5. The crew leader then discusses the production goal (completing 800 ft^2 in 21 crew-hours) with the crew and encourages them to work together to meet the goal of getting the forms erected within the estimated time.

In this example, the crew leader used the short-term schedule to translate production into work-hours or crew-hours and to schedule work so that the crew can accomplish it within the estimate. In addition, setting production targets provides the motivation to produce more than the estimate requires.

No matter what type of schedule is used, supervision must keep it up to date to be useful to the crew leader. Inaccurate schedules are of no value. The person responsible for scheduling in the office handles the updates. This person uses information gathered from job field reports to do the updates.

The crew leader is usually not directly involved in updating schedules. However, completing field or progress reports used by the company may be a daily responsibility to keep the schedule up to date. It's critical that the crew leader fill out any required forms or reports completely and accurately.

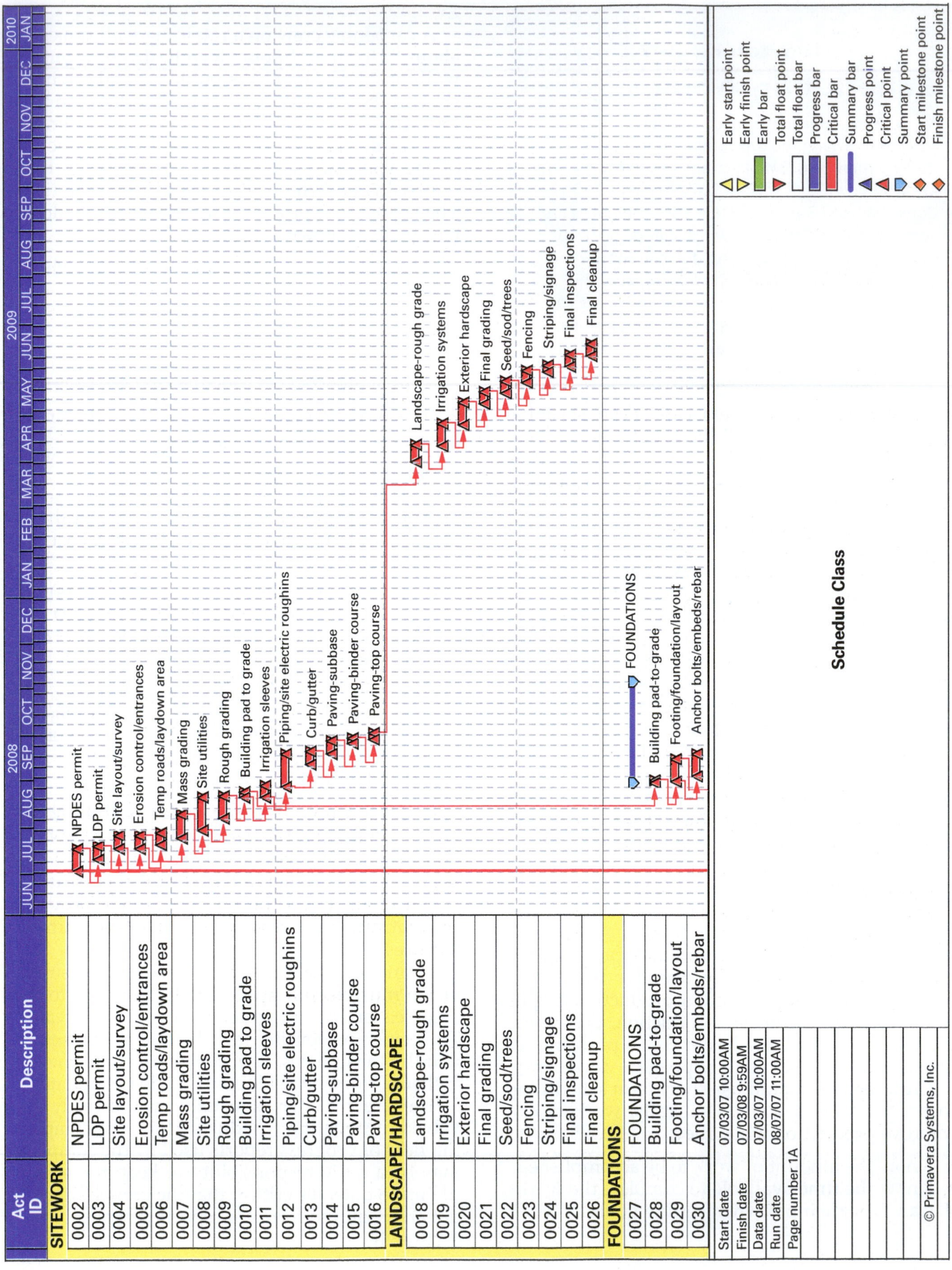

Figure 27 Example of a bar-chart schedule.

NCCER – *Fundamentals of Crew Leadership*

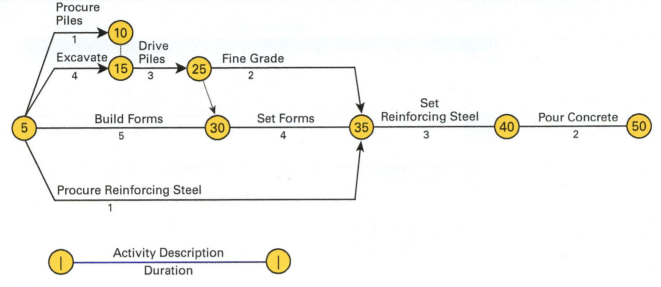

Figure 28 Example of a network schedule.

4.3.0 Cost Control

Being aware of costs and controlling them is the responsibility of every employee on the job. It's the crew leader's job to ensure that employees uphold this responsibility. Control refers to the comparison of estimated performance against actual performance and following up with any needed corrective action. Crew leaders who use cost-control practices are more valuable to the company than those who do not.

On a typical job, many activities are going on at the same time, even within a given crew. This can make it difficult to control the activities involved. The crew leader must be constantly aware of the costs of a project and effectively control the various resources used on the job.

When resources aren't controlled, the cost of the job increases. For example, a plumbing crew of four people is installing soil pipe and runs out of fittings. Three crew members wait (*Figure 30*) while one crew member goes to the plumbing-supply dealer for a part that costs only a few dollars. It takes the crew member an hour to get the part, so four worker-hours of production have been lost. In addition, the total cost of the delay must include the travel costs for retrieving the supplies.

4.3.1 Assessing Cost Performance

Cost performance on a project is determined by comparing actual costs to estimated costs. Regardless of whether the job is a contract-bid project or an in-house project, the company must first establish a budget. In the case of a contract bid, the budget is generally the cost estimate used to bid the job. For an in-house job, participants will submit labor and material forecasts, and someone in authority will authorize a project budget.

It is common to estimate cost by either breaking the job into funded tasks or by forecasting labor and materials expenditures on a timeline. Many companies create a **work breakdown structure (WBS)** for each project. Within the WBS, planners assign each major task a discrete charge number. Anyone working on that task charges that number on their time sheet, so that project managers can readily track cost performance. However, knowing how much money the company is spending doesn't necessarily determine cost performance.

Although financial reports can show that actual expenses are tracking with forecast expenses, they don't show if the work itself is occurring at the required rate. Thus, it is possible to have spent half the budget, but have less than half of the work compete. When the project is broken down into funded tasks related to schedule activities and events, there is far greater control over cost performance.

4.3.2 Field Reporting System

The total estimated cost comes from the job estimates, but managers obtain the actual cost of doing the work from an effective field-reporting system. A field-reporting system consists of a series of forms, which are completed by the crew leader and others. Each company has its own forms and methods for obtaining information. The following paragraphs describe the general information and the process of how they are used. First, records must document the number of hours each person worked on each task. This

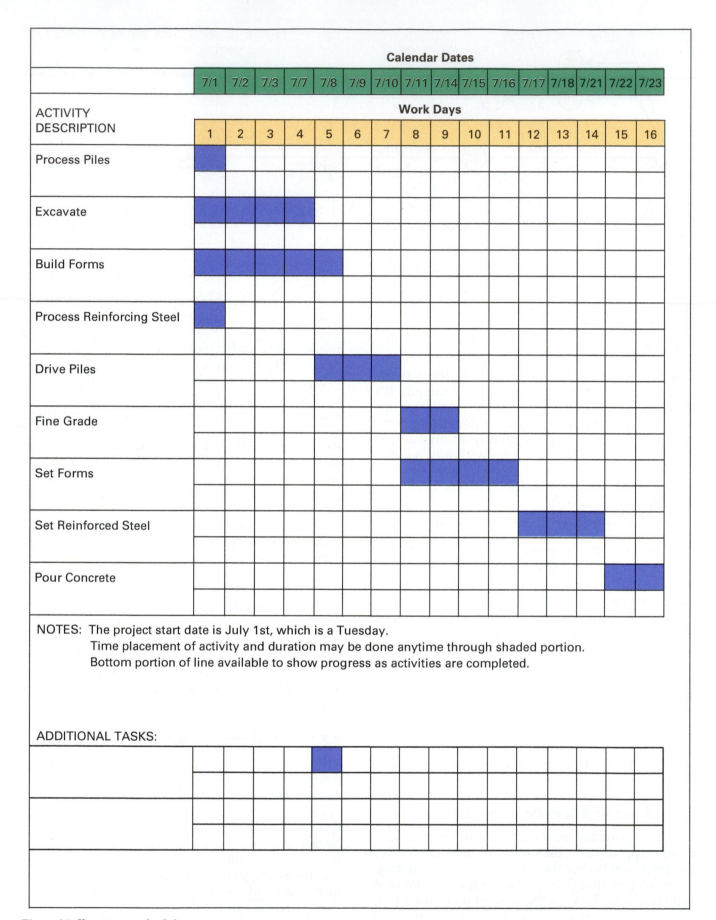

Calendar Dates

ACTIVITY DESCRIPTION	7/1	7/2	7/3	7/7	7/8	7/9	7/10	7/11	7/14	7/15	7/16	7/17	7/18	7/21	7/22	7/23
Work Days	1	2	3	4	5	6	7	8	9	10	11	12	13	14	15	16
Process Piles	■															
Excavate	■	■	■	■												
Build Forms	■	■	■	■	■											
Process Reinforcing Steel	■															
Drive Piles					■	■	■									
Fine Grade								■	■							
Set Forms								■	■	■	■					
Set Reinforced Steel												■	■	■		
Pour Concrete															■	■

NOTES: The project start date is July 1st, which is a Tuesday.
Time placement of activity and duration may be done anytime through shaded portion.
Bottom portion of line available to show progress as activities are completed.

ADDITIONAL TASKS:

				■											

Figure 29 Short-term schedule.

Figure 30 Idle workers cost money.

information comes from daily time cards. Once the accounting department knows how many hours each employee worked on an activity, it can calculate the total cost of the labor by multiplying the number of hours worked by the wage rate for each worker. Managers can then calculate the cost for the labor to do each task as the job progresses. They compare this cost and the estimated costs during the project and at its completion. Managers use a similar process to determine if the costs to operate equipment are comparable to the estimated equipment operating costs.

As the crew places materials, a designated person will measure the quantities from time to time, and send this information to the company office and, possibly, the crew leader. This information, along with the actual cost of the material and the number of hours it took the workers to install it, is compared to the estimated cost. If the cost is greater than the estimate, management and the crew leader must take action to reduce the cost.

For this comparison process to be of use, the information obtained from field personnel must be correct. It is important that the crew leader be accurate in reporting. This is another reason to maintain a daily job diary as discussed earlier.

In the event of a legal/contractual conflict with the client, courts can use such diaries as evidence in legal proceedings, and they can be helpful in reaching a settlement.

Example:

You are running a crew of five concrete finishers for a subcontractor. When you and your crew show up to finish a slab, the general contractor (GC) informs you, "We're a day behind on setting the forms, so I need you and your crew to stand down until tomorrow." What do you do?

Solution:

You should first call your office to let them know about the delay. Then, immediately record it in your job diary. A six-man crew for one day represents 48 worker-hours. If your company charges $30 an hour, that's a potential loss of $1,440, which the company would want to recover from the GC. If there is a dispute, your entry in the job diary could result in a favorable decision for your employer.

4.3.3 Crew Leader's Role in Cost Control

The crew leader is often the company's representative in the field, where the work takes place. Therefore, the crew leader contributes a great deal to determining job costs. When supervision assigns work to a crew, the crew leader should receive a budget and schedule for completing the job. It is then up to the crew leader to make sure the crew finishes the job on time and stays within budget. The crew leader achieves this by actively managing the use of labor, materials, tools, and equipment.

"Beware of little expenses;
a small leak will sink a great ship."
—*Benjamin Franklin*

If the actual costs are at or below the estimated costs, the job is progressing as planned and scheduled, then the company will realize the expected profit. However, if the actual costs exceed the estimated costs, one or more problems may result in the company losing its expected profit, and maybe more. No company can remain in business if it continually loses money.

One of the factors that can increase cost is client-related changes (often called scope creep). The crew leaders must be able to assess the potential impact of such changes and, if necessary, confer with the employer to determine the course of action. If contractor-related losses are occurring, the crew leader and superintendent will need to work together to get the costs back in line.

> *"There is no such thing as scope creep, only scope gallop."*
> —Cornelius Fitchner

The following are some causes that can make actual costs exceed estimated costs, along with some potential actions the crew leader can take to bring the costs under control:

There are many other methods to get the job done on time if it gets off schedule. Examples include working overtime, increasing the size of the crew, prefabricating assemblies, or working staggered shifts. However, these examples may increase the cost of the job, so the crew leader should not do them without the approval of the project manager.

4.4.0 Resource Control

The crew leader's job is to complete assigned tasks safely according to the plans and specifications, on schedule, and within the scope of the estimate. To accomplish this, the crew leader must closely control how resources of materials, equipment, tools, and labor are used. The crew must minimize waste whenever possible.

Control involves measuring performance and correcting departures from plans and specifications to accomplish objectives. Control anticipates predictable deviations from plans and specifications based on experience and takes measures to prevent them from occurring.

An effective control process can be broken down into the following steps:

Step 1 Establish standards and divide them into measurable units. For example, a baseline can be created using experience gained on a typical job, where 2,000 linear feet (LF) of $1\frac{1}{4}"$ copper water tube was installed in

five days. Thus, dividing the total LF installed by the number of installation days gives the average installation rate. In this case, for $1\frac{1}{4}"$ copper water tube, the standard rate was 400 LF/day.

Step 2 Measure performance against a standard. On another job, a crew placed 300 LF of the same tube during a given day. Thus, this actual production of 300 LF/day did not meet the standard rate of 400 LF/day.

Step 3 Adjust operations to meet the standard. In Step 2, if the plan scheduled the job for five days, the crew leader would have to take action to meet this goal. If 300 LF/day is the actual average daily production rate for that job, the crew must increase their placement rate by 100 LF/day to meet the standard. This increase may come by making the workers place piping faster, adding workers to the crew, or receiving authorization for overtime.

The crew leader's responsibility in materials control depends on the policies and procedures of the company. In general, the crew leader is responsible for ensuring on-time delivery, preventing waste, controlling delivery and storage, and preventing theft of materials.

4.4.1 Ensuring On-Time Delivery

It's essential that the materials required for each day's work be on the job site when needed. The crew leader should confirm in advance the placement of orders for all materials and that they will arrive on schedule. A week or so before the delivery date, follow up to make sure there will be no delayed deliveries or items on backorder.

If other people are responsible for providing the materials for a job, the crew leader must follow up to make sure that the materials are available when needed. Otherwise, delays occur as crew members stand around waiting for the delivery of materials.

4.4.2 Preventing Waste

Waste in construction can add up to loss of critical and costly materials and may result in job delays. The crew leader needs to ensure that every crew member knows how to use the materials efficiently. Crew leaders should monitor their crews to make certain that no materials are wasted.

An example of waste is a carpenter who saws off a piece of lumber from a full-sized piece, when workers could have found the length needed in the lumber scrap pile. Another example of waste involves rework due to bad technique or a conflict in craft scheduling requiring removal and reinstallation of finished work. The waste in time spent and replacement material costs occur when redoing the task a second time.

Under LEED, waste control is very important. Companies receive LEED credits for finding ways to reduce waste and for recycling construction waste. Workers should segregate waste materials by type for recycling, if feasible (*Figure 31*).

4.4.3 Verifying Material Delivery

A crew leader may be responsible for the receipt of materials delivered to the work site. When this happens, the crew leader should require a copy of the shipping invoice or similar document and check each item on the invoice against the actual materials to verify delivery of the correct amounts.

The crew leader should also check the condition of the materials to verify that nothing is defective before signing the invoice. This can be difficult and time consuming because it requires the crew to open cartons and examine their contents. However, this step cannot be overlooked, because a signed invoice indicates that the recipient accepted all listed materials and in an undamaged state. If the crew leader signs for the materials without checking them, and then finds damage, no one will be able to prove that the materials came to the site in that condition.

After checking and signing the shipping invoice, the crew leader should give the original or a copy to the superintendent or project manager. The company then files the invoice for future reference because it serves as the only record the company has to check bills received from the supplier company.

4.4.4 Controlling Delivery and Storage

Another essential element of materials control is the selection of where the materials will be stored on the job site. There are two factors in determining the appropriate storage location. The first is

Figure 31 Waste materials separated for recycling.

convenience. If possible, the materials should be stored near where the crew will use them. The time and effort saved by not having to carry the materials long distances will greatly reduce the installation costs.

Next, the materials must be stored in a secure area to avoid damage. It is important that the storage area suit the materials being stored. For instance, materials that are sensitive to temperature, such as chemicals or paints, should be stored in climate-controlled areas to prevent waste.

4.4.5 Preventing Theft and Vandalism

Theft and vandalism of construction materials increase costs. The direct and indirect costs of lost materials include obtaining replacement materials, production time lost while the needed materials are missing, and the costs of additional security precautions can add significantly to the overall cost to the company. In addition, the insurance that the contractor purchases will increase in cost as the theft and vandalism rate grows.

The best way to avoid theft and vandalism is a secure job site. At the end of each work day, store unused materials and tools in a secure location, such as a locked construction trailer. If the job site is fenced or the building has lockable accesses, the materials can be stored within. Many sites have security cameras and/or intrusion alarms to help minimize theft and vandalism.

4.4.6 Equipment Control

The crew leader may not be responsible for long-term equipment control. However, the equipment required for a specific job is often the crew leader's responsibility. The first step is to schedule when a crew member must transport the required equipment from the shop or rental yard. The crew leader is responsible for informing the shop of the location of its use and seeing a worker returns it to the shop when finished with it.

It is common for equipment to lay idle at a job site because of a lack of proper planning for the job and the equipment arrived early. For example, if wire-pulling equipment arrives at a job site before the conduit is in place, this equipment will be out of service while awaiting the conduit installation. In addition to the wasted rental cost, damage, loss, or theft can also occur with idle equipment while awaiting use.

The crew leader needs to control equipment use, ensure that the crew operates the equipment according to its instructions, and within time and cost guidelines. The crew leader must also assign or coordinate responsibility for maintaining and repairing equipment as indicated by the applicable preventive maintenance schedule. Delaying maintenance and repairs can lead to costly equipment failures. The crew leader must also ensure that the equipment operators have the necessary credentials to operate the equipment, including applicable licenses.

The crew leader is responsible for the proper operation of all other equipment resources, including cars and trucks. Reckless or unsafe operation of vehicles will likely result in damaged equipment, leading to disciplinary action in the case of the workers, fines and repair costs levied by the vehicle vendors, and a loss of confidence in the crew leader's ability to govern crew-members' actions.

The crew leader should also arrange to secure all equipment at the close of each day's work to prevent theft. If continued use of the equipment is necessary for the job, the crew leader should make sure to lock it in a safe place; otherwise, return it to the shop.

4.4.7 Tool Control

Among companies, various policies govern who provides hand and power tools to employees. Some companies provide all the tools, while others furnish only the larger power tools. The crew leader should be familiar with and enforce any company policies related to tools.

Tool control has two aspects. First, the crew leader must control the issue, use, and maintenance of all tools provided by the company. Second, the crew leader must control how crew members use the tools to do the job. This applies to tools supplied by the company as well as tools that belong to the workers.

Using the proper tools correctly saves time and energy. In addition, proper tool use reduces the chance of damage to the tool during use, as well as injury to the user and to nearby workers. This is especially true of edge tools and those with any form of fixed or rotating blade. A dull tool is far more dangerous than a sharp one.

Tools must be adequately maintained and properly stored. Making sure that tools are cleaned, dried, and lubricated prevents rust and ensures that the tools are in the proper working order. If a tool is damaged, it is essential to repair or replace it promptly. Otherwise, an accident or injury could result.

> **NOTE**
>
> Regardless of whether a worker or the company owns a tool, OSHA holds companies responsible for the consequences of using a tool on a job site. The company is accountable if an employee receives an injury from a defective tool. Therefore, the crew leader needs to be aware of any defects in the tools the crew members are using.

Users should take care of company-issued tools as if they were their own. Workers should not abuse tools simply because they don't their property.

One of the major sources of low productivity on a job is the time spent searching for a tool. To prevent this from occurring, supervision should establish a storage location for company-issued tools and equipment. The crew leader should make sure that crew members return all company-issued tools and equipment to this designated location after use. Similarly, workers should organize their personal toolboxes so that they can readily find the appropriate tools and return their tools to their toolboxes when they are finished using them. This is a matter of professionalism that crew leaders are in an excellent position to develop.

Studies have shown that some keys to an effective tool control system are:

- Limiting the number of people allowed access to stored tools
- Limiting the number of people held responsible for tools and holding them accountable
- Controlling the ways in which a tool can be returned to storage
- Making sure tools are available when needed

4.4.8 Labor Control

Labor typically represents more than half the cost of a project, and therefore has an enormous impact on profitability. For that reason, it's essential to manage a crew and their work environment in a way that maximizes their productivity. One of the ways to do that is to minimize the time spent on unproductive activities such as the following:

- Engaging in bull sessions
- Correcting drawing errors
- Retrieving tools, equipment, and materials
- Waiting for other workers to finish

If crew members are habitually goofing off, it is up to the crew leader to counsel those workers. The crew leader's daily diary should document the counseling. The crew leader will need to refer repeated violations to the attention of higher management as guided by company policy.

All sorts of errors will occur during a project that need correction. Some errors, such as mistakes on drawings, may be outside of the crew leader's control. However, before a drawing error results in performing an action that will need to be corrected later, a careful examination of the drawings by the crew leader may discover the error before work begins, saving time and materials. If crew members are making mistakes due to inexperience, the crew leader can help avoid these errors by providing on-the-spot training and by checking on inexperienced workers more often.

The availability and location of tools, equipment, and materials to a crew can have a profound effect on their productivity. As discussed in the previous section, the key to minimizing such problems is proactive management of these resources.

Delays caused by crews or contractors can be minimized or avoided by carefully tracking the project schedule. In doing so, crew leaders can anticipate delays that will affect the work and either take action to prevent the delay or redirect the crew to another task.

4.4.9 Production and Productivity

Production is the amount of construction materials put in place. It is the quantity of materials installed on a job, such as 1,000 linear feet of waste pipe installed for a task. On the other hand, productivity is a rate, and depends on the level of efficiency of the work. It is the amount of work done per hour or day by one worker or a crew.

Production levels are set during the estimating stage. The estimator determines the total amount of materials placed from the plans and specifications. After the job is complete, supervision can assess the actual amount of materials installed, and can compare the actual production to the estimated production.

Productivity relates to the amount of materials put in place by the crew over a certain period of time. The estimator uses company records during the estimating stage to determine how much time and labor it will take to place a certain quantity of materials. From this information, the estimator calculates the productivity necessary to complete the job on time.

Participant Exercise G

1. List the methods your company uses to minimize waste.

_____

_____

_____

_____

2. List the methods your company uses to control small tools on the job.

_____

_____

_____

_____

3. List five ways that you feel your company could control labor to maximize productivity.

_____

_____

_____

_____

For example, it might take a crew of two people ten days to paint 5,000 square feet. To calculate the required average productivity, divide 5,000 ft² by 10 days. The result is 500 ft²/day. The crew leader can compare the daily production of any crew of two painters doing similar work with this average, as discussed previously.

Planning is essential to productivity. The crew must be available to perform the work and have all the required materials, tools, and equipment in place when the job begins.

The time on the job should be for business, not for taking care of personal problems. Anything not work-related should be handled after hours, away from the job site, if possible. Even the planning of after-work activities, arranging social functions, or running personal errands should occur after work or during breaks. Only very limited and necessary exceptions to these rules should be permitted (e.g., making or meeting medical appointments), and these should be spelled out in company or crew policies.

Organizing field work can save time. The key to effectively using time is to work smarter, not necessarily harder. Working smart can save your crew from doing unnecessary work. For example, most construction projects require that the contractor submit a set of as-built plans at the completion of the work. These plans describe the actual structure and installation of materials. The best way to prepare these plans is to mark up a set of working plans as the work is in progress as described earlier. That way, workers and supervision won't forget pertinent details and waste time trying to reconstruct how the company did the work.

"Knowledge is power is time is money."
—*Robert Thier, Storm and Silence*

The amount of material used should not exceed the estimated amount. If it does, either the estimator has made a mistake, undocumented changes have occurred, or rework has caused the need for additional materials. Whatever the case, the crew leader should use effective control techniques to ensure the efficient use of materials.

When bidding a job, most companies calculate the cost per worker-hour. For example, a ten-day job might convert to 160 w-h (two painters for ten days at eight hours per day). If the company charges a labor rate of $30/hour, the labor cost would be $4,800. The estimator then adds the cost of materials, equipment, and tools, along with overhead costs and a profit factor, to determine the price of the job.

After completion of a job, information gathered through field reporting allows the company office to calculate actual productivity and compare it to the estimated figures. This helps to identify productivity issues and improves the accuracy of future estimates.

The following labor-related practices can help to ensure productivity:

- Ensure that all workers have the required resources when needed.
- Ensure that all personnel know where to go and what to do after each task is completed.
- Make reassignments as needed.
- Ensure that all workers have completed their work properly.

Additional Resources

Construction Workforce Development Professional, NCCER. 2016. New York, NY: Pearson Education, Inc.

Mentoring for Craft Professionals, NCCER. 2016. New York, NY: Pearson Education, Inc.

Blueprint Reading for Construction, James A. S. Fatzinger. 2003. New York, NY: Pearson Education, Inc.

Construction Leadership from A to Z: 26 Words to Lead By, Wally Adamchik. 2011. Live Oak Book Company.

The following websites offer resources for products and training:

Architecture, Engineering, and Construction Industry (AEC), **www.aecinfo.com**

US Green Building Council (USGBC), **www.usgbc.org/leed**

4.0.0 Section Review

1. Which of these activities occurs during the development phase of a project?

   a. Architect/engineer sketches are prepared and a preliminary budget is developed.
   b. Government agencies give a final inspection of the design, check for adherence to codes, and inspect materials used.
   c. Detailed project drawings and specifications are prepared.
   d. Contracts for the project are awarded.

2. A job diary should typically record any of the following *except* _____.

   a. items such as job interruptions and visits
   b. the exact times of scheduled lunch breaks
   c. changes needed to project drawings
   d. the actual time each job task related to a particular project took to complete

3. Which of the following is a correct statement regarding project cost?

   a. Cost is handled by the accounting department and isn't a concern of the crew leader.
   b. The difference between the estimated cost and the actual cost affects a company's profit.
   c. Wasted material is factored into the estimate and is never a concern.
   d. The contractor's overhead costs aren't included in the cost estimate.

4. To prevent job delays due to late delivery of materials, the crew leader should _____.

   a. demand a discount from the supplier to compensate for the delay
   b. tell a crew member to go look for the delivery truck
   c. refuse the late delivery and re-order the materials from another supplier
   d. check with the supplier in advance of the scheduled delivery

SUMMARY

In construction, the crew leader is the company's supervisor on the ground at the scene where the actual work takes place. The modern crew leader must be prepared to deal with a host of challenges ranging from differences in age, culture, race, gender, language, and educational backgrounds to attitudes regarding authority and personal integrity. Rather than being overwhelmed, the effective and successful crew leader will use these challenges to broaden his or her understanding of people and to grow in character.

Crew leaders must develop an understanding of the company, its policies, and how to become an effective and valuable member of the supervisory team. They must also be able to effectively communicate up and down the chain, to motivate the crew's workers, and develop a sense of ownership within the crew so that they have the confidence and will to do the jobs.

Safety is an overriding concern for the company, and crew leaders need to make safe work practices and conditions the number-one priority. Once a worker has lost an eye, a limb, or life itself, it's too late to reconsider how the task might have been done safer. Workers are a company's most valuable asset and the crew leader is in the position of being the best example of safety to the crew.

While crew leaders view their role as mainly people and work managers, these tasks gain a greater sense of importance when crew leaders understand their responsibilities to the company for whom they work. Effectively controlling worker productivity, resources, schedules, and other aspects of the project directly contribute to cost control, which reflects in greater profits for the company. Companies recognize effective crew leaders and they can expect long and successful careers as a result.

1. Younger-generation workers may have a better perception of success if _____.

   a. crew leaders assign their projects in smaller, well-defined units
   b. they receive closer supervision
   c. crew leaders leave them alone
   d. they are given the perception that they are in charge

2. Companies can help prevent sexual harassment in the workplace by _____.

   a. requiring employee training that avoids the potentially offensive subject of stereotypes
   b. developing a consistent policy with appropriate consequences for engaging in sexual harassment
   c. communicating to workers that, legally, the victim of sexual harassment is only the one being directly harassed
   d. establishing an effective complaint or grievance process for victims of harassment

3. Which of the following statements about a formal organization is *true*?

   a. It is a group of members created mainly to share professional information.
   b. It is a group established for developing industry standards.
   c. It is a group of individuals led by someone all working toward the same goal.
   d. It is a group that is identified by its overall function rather than by key positions.

4. Which of these departments in a large company would likely arrange equipment rentals?

   a. Accounting
   b. Estimating
   c. Human Resources
   d. Purchasing

5. When a person is promoted to crew leader, the amount of time they spend on craft work _____.

   a. increases
   b. decreases
   c. stays the same
   d. doubles

6. Which of the following is a *correct* statement regarding annual crew evaluations?

   a. Their results should never come as a surprise to the crew member.
   b. Formal evaluations should be conducted every three months.
   c. Comments on performance should be withheld until the formal evaluation.
   d. Advise the crew members to improve their performance just before the formal evaluation is scheduled.

7. As a crew leader, what should you avoid when giving recognition to crew members?

   a. Looking for good work and knowing what good work is
   b. Being available at the job site to witness good work
   c. Giving praise for poor work to make the worker feel better
   d. Encouraging work improvement through your confidence in a worker

8. Within the construction industry, who normally assigns a crew to the crew leader?

   a. The US Department of Labor
   b. The company
   c. The crew leader
   d. The crew members

9. Some decisions are difficult to make because you _____.

   a. have no experience dealing with the circumstances requiring a decision
   b. don't want to make a decision
   c. prefer to think it's someone else's responsibility to make a decision
   d. don't recognize the need for a decision

10. The first step in problem solving should be _____.

    a. choosing a solution to the problem
    b. evaluating the solution to the problem
    c. defining the problem
    d. testing the solution to the problem

11. The leading cause of fatalities in the construction industry is _____.
    a. electrocution
    b. asphyxiation
    c. struck by something
    d. falls

12. OSHA inspection of a business or job site _____.
    a. can be done only by invitation
    b. can be conducted at random
    c. is done only after an accident
    d. is conducted only if a safety violation occurs

13. To proactively identify and assess hazards, it is essential that employers _____.
    a. ensure that crew leaders are present and available on the job site
    b. conduct safety inspections of contractor offices
    c. design safety logs to document safety violations
    d. conduct accident investigations

14. Employers who are subject to OSHA record-keeping requirements must maintain a log of recordable occupational injuries and illnesses for _____.
    a. three years
    b. five years
    c. seven years
    d. as long as the company is in business

15. Which of the following statements regarding safety training sessions is *correct*?
    a. The project manager usually holds them.
    b. They are held only for new employees.
    c. They should be conducted frequently by the crew leader.
    d. They are required only after an accident has occurred.

16. The type of contract in which the client pays the contractor for their actual labor and material expenses they incur is known as a _____.
    a. firm fixed-price contract
    b. time-spent contract
    c. performance-based contract
    d. cost-reimbursable contract

17. On a design-build project, _____.
    a. the same contractor is responsible for both design and construction
    b. the owner is responsible for providing the design
    c. the architect does the design and the general contractor builds the project
    d. a construction manager is hired to oversee the project

18. When planning labor for onsite work, the crew leader must _____.
    a. factor in the absenteeism and turnover rates
    b. factor in time for training classes
    c. select the material to be used by each crew member
    d. identify the skills needed to perform the work

19. Which of the following is *not* a source of data in the field-reporting system?
    a. The cost-estimate for the project
    b. Crew leader's daily diary
    c. Onsite inventory of placed materials
    d. Daily time cards

20. The following duties are important in work site equipment control *except* _____.
    a. scheduling the pickup and return of a piece of equipment
    b. ensuring the equipment is operated and maintained correctly
    c. monitoring safe and responsible operation of vehicles
    d. selection and purchase of equipment to be stocked at the work site equipment room

Trade Terms Introduced in This Module

Autonomy: The condition of having complete control over one's actions, and being free from the control of another. To be independent.

Bias: A preconceived inclination against or in favor of something.

Cloud-based applications: Mobile and desktop digital programs that can connect to files and data located in distributed storage locations on the Internet ("the Cloud"). Such applications make it possible for authorized users to create, edit, and distribute content from any location where Internet access is possible.

Craft professionals: Workers who are properly trained and work in a particular construction trade or craft.

Crew leader: The immediate supervisor of a crew or team of craft professionals and other assigned persons.

Critical path: In manufacturing, construction, and other types of creative processes, the required sequence of tasks that directly controls the ultimate completion date of the project.

Demographics: Social characteristics and other factors, such as language, economics, education, culture, and age, that define a statistical group of individuals. An individual can be a member of more than one demographic.

Ethics: The moral principles that guides an individual's or organization's actions when dealing with others. Also refers to the study of moral principles.

Infer: To reach a conclusion using a method of reasoning that starts with an assumption and considers a set of logically-related events, conditions, or statements.

Intangible: Not touchable, material, or measureable; lacking a physical presence.

Job description: A description of the scope and responsibilities of a worker's job so that the individual and others understand what the job entails.

Job diary: A written record that a supervisor maintains periodically (usually daily) of the events, communications, observations, and decisions made during the course of a project.

Legend: In maps, plans, and diagrams, an explanatory table defining all symbolic information contained in the document.

Lethargy: Sluggishness, slow motion, lack of activity, or a lack of enthusiasm.

Letter of instruction (LOI): A written communication from a supervisor to a subordinate that informs the latter of some inadequacy in the individual's performance, and provides a list of actions that the individual must satisfactorily complete to remediate the problem. Usually a key step in a series of disciplinary actions.

Local area networks (LAN): Communication networks that link computers, printers, and servers within a small, defined location, such as a building or office, via hard-wired or wireless connections.

Lockout/tagout (LOTO): A system of safety procedures for securing electrical and mechanical equipment during repairs or construction, consisting of warning tags and physical locking or restraining devices applied to controls to prevent the accidental or purposeful operation of the equipment, endangering workers, equipment, or facilities.

Look-ahead schedule: A manual- or software-scheduling tool that looks several weeks into the future at the planned project events; used for anticipating material, labor, tool, and other resource requirements, as well as identifying potential schedule conflicts or other problems.

Negligence: Lack of appropriate care when doing something, or the failure to do something, usually resulting in injury to an individual or damage to equipment.

Organizational chart: A diagram that shows how the various management and operational responsibilities relate to each other within an organization. Named positions appear in ranked levels, top to bottom, from those functional units with the most and broadest authority to those with the least, with lines connecting positions indicating chains of authority and other relationships.

Paraphrase: Rewording a written or verbal statement in one's own words in such a way that the intent of the original statement is retained.

Pragmatic: Sensible, practical, and realistic.

Proactive: To anticipate and take action in the present to deal with potential future events or outcomes based on what one knows about current events or conditions.

Project manager: In construction, the individual who has overall responsibility for one or more construction projects; also called the general superintendent.

Return on investment (ROI): A measure of the gain or loss of money resulting from an investment; normally measured as a percentage of the original investment.

Safety data sheets (SDS): Documents listing information about a material or substance that includes common and proper names, chemical composition, physical forms and properties, hazards, flammability, handling, and emergency response in accordance with national and international hazard communication standards. Also called material safety data sheets (MSDS).

Sexual harassment: Any unwelcome verbal or nonverbal form of communication or action construed by an individual to be of a sexual or gender-related nature.

Smartphone: The most common type of cellular telephone that combines many other digital functions within a single mobile device. Most important of these, besides the wireless telephone, are high-resolution still and video cameras, text and email messaging, and GPS-enabled features.

Superintendent: In construction, the individual who is the on-site supervisor in charge of a given construction project.

Synergy: Any type of cooperation between organizations, individuals, or other entities where the combined effect is greater than the sum of the individual efforts.

Textspeak: A form of written language characteristic of text messaging on mobile devices and text-based social media, which usually consists of acronyms, abbreviations, and minimal punctuation.

Wide area networks (WAN): Dedicated communication networks of computers and related hardware that serve a given geographic area, such as a work site, campus, city, or a larger but distinct area. Connectivity is by wired and wireless means, and may use the Internet as well.

Wi-Fi: The technology allowing communications via radio signals over a LAN or WAN equipped with a wireless access point or the Internet. (Wi-Fi stands for wireless fidelity.) Many types of mobile, portable, and desktop devices can communicate via Wi-Fi connections.

Work breakdown structure (WBS): A diagrammatic and conceptual method for subdividing a complex project, concept, or other thing into its various functional and organizational parts so that planners can analyze and plan for each part in detail.

OSHA Forms for Safety Records

Employers who are subject to the recordkeeping requirements of the Occupational Safety and Health Act of 1970 must maintain records of all recordable occupational injuries and illnesses. The following are three important OSHA forms used for this recordkeeping (these forms can be accessed at **www.osha.gov**):

- OSHA Form 300, *Log of Work-Related Injuries and Illnesses*
- OSHA Form 300A, *Summary of Work-Related Injuries and Illnesses*
- OSHA Form 301, *Injury and Illness Incident Report*

OSHA's Form 300 (Rev. 01/2004)
Log of Work-Related Injuries and Illnesses

Attention: This form contains information relating to employee health and must be used in a manner that protects the confidentiality of employees to the extent possible while the information is being used for occupational safety and health purposes.

U.S. Department of Labor
Occupational Safety and Health Administration

Form approved OMB no. 1218-0176

Year _____

You must record information about every work-related injury or illness that involves loss of consciousness, restricted work activity or job transfer, days away from work, or medical treatment beyond first aid. You must also record significant work-related injuries and illnesses that are diagnosed by a physician or licensed health care professional. You must also record work-related injuries and illnesses that meet any of the specific recording criteria listed in 29 CFR 1904.8 through 1904.12. Feel free to use two lines for a single case if you need to. You must complete an injury and illness incident report (OSHA Form 301) or equivalent form for each injury or illness recorded on this form. If you're not sure whether a case is recordable, call your local OSHA office for help.

Establishment name _____

City _____ State _____

Identify the person				Describe the case		Classify the case							
(A) Case No.	(B) Employee's Name	(C) Job Title (e.g., Welder)	(D) Date of injury or onset of illness (mo./day)	(E) Where the event occurred (e.g. Loading dock north end)	(F) Describe injury or illness, parts of body affected, and object/substance that directly injured or made person ill (e.g. Second degree burns on right forearm from acetylene torch)	CHECK ONLY ONE box for each case based on the most serious outcome for that case:				Enter the number of days the injured or ill worker was:		Check the "Injury" column or choose one type of illness: (M)	
						Days away from work (H)	Remained at work		Other recordable cases (J)	Away From Work (days) (K)	On job transfer or restriction (days) (L)		
							Job transfer or restriction (I)	Other recordable cases (J)					

Classify the case column headers (M): (1) Injury (2) Skin Disorder (3) Respiratory Condition (4) Poisoning (5) Hearing Loss (6) All other illnesses

Page totals — 0 0 0 0 0 0 0 0 0 0 0

Be sure to transfer these totals to the Summary page (Form 300A) before you post it.

(1) Injury (2) Skin Disorder (3) Respiratory Condition (4) Poisoning (5) Hearing Loss (6) All other illnesses

Public reporting burden for this collection of information is estimated to average 14 minutes per response, including time to review the instruction, search and gather the data needed, and complete and review the collection of information. Persons are not required to respond to the collection of information unless it displays a currently valid OMB control number. If you have any comments about these estimates or any aspects of this data collection, contact: US Department of Labor, OSHA Office of Statistics, Room N-3644, 200 Constitution Ave, NW, Washington, DC 20210. Do not send the completed forms to this office.

Page _____ 1 of 1

Figure A01 OSHA Form 300

OSHA's Form 300A (Rev. 01/2004)
Summary of Work-Related Injuries and Illnesses

Year _____

U.S. Department of Labor
Occupational Safety and Health Administration

Form approved OMB no. 1218-0176

All establishments covered by Part 1904 must complete this Summary page, even if no injuries or illnesses occurred during the year. Remember to review the Log to verify that the entries are complete and accurate before you've added the entries from every page of the log. If you had no cases write "0."

Using the Log, count the individual entries you made for each category. Then write the totals below, making sure you've added the entries from every page of the log. If you had no cases write "0."

Employees former employees, and their representatives have the right to review the OSHA Form 300 in its entirety. They also have limited access to the OSHA Form 301 or its equivalent. See 29 CFR 1904.35, in OSHA's Recordkeeping rule, for further details on the access provisions for these forms.

Number of Cases

Total number of deaths	Total number of cases with days away from work	Total number of cases with job transfer or restriction	Total number of other recordable cases
0	0	0	0
(G)	(H)	(I)	(J)

Number of Days

Total number of days away from work	Total number of days of job transfer or restriction
0	0
(K)	(L)

Injury and Illness Types

Total number of...
(M)

(1) Injury	0	(4) Poisoning	0
(2) Skin Disorder	0	(5) Hearing Loss	0
(3) Respiratory Condition	0	(6) All Other Illnesses	0

Establishment Information

Your establishment name _____

Street _____

City _____ State _____ Zip _____

Industry description (e.g., Manufacture of motor truck trailers) _____

Standard Industrial Classification (SIC), if known (e.g., SIC 3715) _____

OR North American Industrial Classification (NAICS), if known (e.g., 336212) _____

Employment Information

Annual average number of employees _____

Total hours worked by all employees last year _____

Sign here

Knowingly falsifying this document may result in a fine.

I certify that I have examined this document and that to the best of my knowledge the entries are true, accurate, and complete.

_____ Company executive _____ Title _____

_____ Phone _____ Date _____

Post this Summary page from February 1 to April 30 of the year following the year covered by the form

Public reporting burden for this collection of information is estimated to average 58 minutes per response, including time to review the instruction, search and gather the data needed, and complete and review the collection of information. Persons are not required to respond to the collection of information unless it displays a currently valid OMB control number. If you have any comments about these estimates or any aspects of this data collection, contact: US Department of Labor, OSHA Office of Statistics, Room N-3644, 200 Constitution Ave. NW, Washington, DC 20210. Do not send the completed forms to this office.

Figure A02 OSHA Form 300A

OSHA's Form 301
Injuries and Illnesses Incident Report

U.S. Department of Labor

Occupational Safety and Health Administration

Form approved OMB no. 1218-0176

Attention: This form contains information relating to employee health and must be used in a manner that protects the confidentiality of employees to the extent possible while the information is being used for occupational safety and health purposes.

This *Injury and Illness Incident Report* is one of the first forms you must fill out when a recordable work-related injury or illness has occurred. Together with the *Log of Work-Related Injuries and Illnesses* and the accompanying *Summary*, these forms help the employer and OSHA develop a picture of the extent and severity of work-related incidents.

Within 7 calendar days after you receive information that a recordable work-related injury or illness has occurred, you must fill out this form or an equivalent. Some state workers' compensation, insurance, or other reports may be acceptable substitutes. To be considered an equivalent form, any substitute must contain all the information asked for on this form.

According to Public Law 91-596 and 29 CFR 1904, OSHA's recordkeeping rule, you must keep this form on file for 5 years following the year to which it pertains.

If you need additional copies of this form, you may photocopy and use as many as you need.

Completed by _____

Title _____

Phone _____ Date _____

Information about the employee

1) Full Name _____

2) Street _____

City _____ State _____ Zip _____

3) Date of birth _____

4) Date hired _____

5) ☐ Male   ☐ Female

Information about the physician or other health care professional

6) Name of physician or other health care professional

_____

7) If treatment was given away from the worksite, where was it given?

Facility _____

Street _____

City _____ State _____ Zip _____

8) Was employee treated in an emergency room?
☐ Yes
☐ No

9) Was employee hospitalized overnight as an in-patient?
☐ Yes
☐ No

Information about the case

10) Case number from the Log _____ *(Transfer the case number from the Log after you record the case.)*

11) Date of injury or illness _____

12) Time employee began work _____ AM/PM

13) Time of event _____ AM/PM   ☐ Check if time cannot be determined

14) What was the employee doing just before the incident occurred? Describe the activity, as well as the tools, equipment or material the employee was using. Be specific. Examples: "climbing a ladder while carrying roofing materials"; "spraying chlorine from hand sprayer"; "daily computer key-entry."

15) What happened? Tell us how the injury occurred. Examples: "When ladder slipped on wet floor, worker fell 20 feet"; "Worker was sprayed with chlorine when gasket broke during replacement"; "Worker developed soreness in wrist over time."

16) What was the injury or illness? Tell us the part of the body that was affected and how it was affected; be more specific than "hurt", "pain", or "sore." Examples: "strained back"; "chemical burn, hand"; "carpal tunnel syndrome."

17) What object or substance directly harmed the employee? Examples: "concrete floor"; "chlorine"; "radial arm saw." If this question does not apply to the incident, leave it blank.

18) If the employee died, when did death occur? Date of death _____

Public reporting burden for this collection of information is estimated to average 22 minutes per response, including time for reviewing instructions, searching existing data sources, gathering and maintaining the data needed, and completing and reviewing the collection of information. Persons are not required to respond to the collection of information unless it displays a current valid OMB control number. If you have any comments about this estimate or any other aspects of this data collection, including suggestions for reducing this burden, contact: US Department of Labor, OSHA Office of Statistics, Room N-3644, 200 Constitution Ave, NW, Washington, DC 20210. Do not send the completed forms to this office.

Additional Resources

This module presents thorough resources for task training. The following reference material is recommended for further study.

Construction Workforce Development Professional, NCCER. 2016. New York, NY: Pearson Education, Inc.

Mentoring for Craft Professionals, NCCER. 2016. New York, NY: Pearson Education, Inc.

Blueprint Reading for Construction, James A. S. Fatzinger. 2003. New York, NY: Pearson Education, Inc.

Construction Leadership from A to Z: 26 Words to Lead By, Wally Adamchik. 2011. Live Oak Book Company.

Generational Cohorts and their Attitudes Toward Work Related Issues in Central Kentucky, Frank Fletcher, et al. 2009. Midway College, Midway, KY. **www.kentucky.com.**

It's Your Ship: Management Techniques from the Best Damn Ship in the Navy, Captain D. Michael Abrashoff, USN. 2012. New York City, NY: Grand Central Publishing.

Survival of the Fittest, Mark Breslin. 2005. McNally International Press.

The Definitive Book of Body Language: The Hidden Meaning Behind People's Gestures and Expressions, Barbara Pease and Allan Pease. 2006. New York City, NY: Random House / Bantam Books.

The Young Person's Guide to Wisdom, Power, and Life Success: Making Smart Choices, Brian Gahran, PhD. 2014. San Diego, CA: Young Persons Press. **www.WPGBlog.com.**

The following websites offer resources for products and training:

Aging Workforce News, **www.agingworkforcenews.com.**

American Society for Training and Development (ASTD), **www.astd.org.**

Architecture, Engineering, and Construction Industry (AEC), **www.aecinfo.com.**

Equal Employment Opportunity Commission (EEOC), **www.eeoc.gov.**

National Association of Women in Construction (NAWIC), **www.nawic.org.**

National Census of Fatal Occupational Injuries (NCFOI), **www.bls.gov.**

National Institute of Occupational Safety and Health (NIOSH), **www.cdc.gov/niosh.**

National Safety Council, **www.nsc.org.**

Occupational Safety and Health Administration (OSHA), **www.osha.gov.**

Society for Human Resources Management (SHRM), **www.shrm.org.**

United States Census Bureau, **www.census.gov.**

United States Department of Labor, **www.dol.gov.**

US Green Building Council (USGBC), **www.usgbc.org/leed.**

Wi-Fi® is a registered trademark of the Wi-Fi Alliance, **www.wi-fi.org**

Figure Credits

© Photographerlondon/Dreamstime.com, Module Opener
© iStockphoto.com/shironosov, Figure 1
© iStockphoto.com/kali9, Figures 2, 10
© iStockphoto.com/nanmulti, Figure 3
© iStockphoto.com/Halfpoint, Figure 13
© iStock.com/vejaa, Figure 16
DIYawards, Figure 17
© iStockphoto.com/Tashi-Delek, Figure 18
© iStockphoto.com/kemaltaner, Figure 23
Sushil Shenoy/Virginia Tech, Figure 31
John Ambrosia, Figure 27
© iStockphoto.com/izustun, Figure 30

Section Review Answer Key

Answer	Section Reference	Objective
Section One		
1. a	1.1.0	1a
2. b	1.2.2	1b
3. b	1.3.0	1c
Section Two		
1. b	2.0.0	2a
2. c	2.2.1	2b
3. a	2.3.1	2c
4. c	2.4.1	2d
5. a	2.5.1	2e
Section Three		
1. a	3.0.0	3a
2. d	3.2.0; Table 1	3b
3. a	3.3.0	3c
4. b	3.4.0	3d
Section Four		
1. a	4.1.1	4a
2. b	4.2.2	4b
3. b	4.3.3	4c
4. d	4.4.1	4d

NCCER CURRICULA — USER UPDATE

NCCER makes every effort to keep its textbooks up-to-date and free of technical errors. We appreciate your help in this process. If you find an error, a typographical mistake, or an inaccuracy in NCCER's curricula, please fill out this form (or a photocopy), or complete the online form at **www.nccer.org/olf**. Be sure to include the exact module ID number, page number, a detailed description, and your recommended correction. Your input will be brought to the attention of the Authoring Team. Thank you for your assistance.

Instructors – If you have an idea for improving this textbook, or have found that additional materials were necessary to teach this module effectively, please let us know so that we may present your suggestions to the Authoring Team.

NCCER Product Development and Revision
13614 Progress Blvd., Alachua, FL 32615

Email: curriculum@nccer.org
Online: www.nccer.org/olf

❏ Trainee Guide ❏ Lesson Plans ❏ Exam ❏ PowerPoints Other _____

Craft / Level: _____ Copyright Date: _____

Module ID Number / Title: _____

Section Number(s): _____

Description:
_____
_____
_____
_____
_____

Recommended Correction:
_____
_____
_____
_____
_____

Your Name: _____

Address: _____
_____

Email: _____ Phone: _____

This page is intentionally left blank.

Glossary

Addressable device: A fire alarm system component with discrete identification that can have its status individually identified or that is used to individually control other functions.

Air sampling detector: A detector consisting of piping or tubing distribution from the detector unit to the area or areas to be protected. An air pump draws air from the protected area back to the detector through the air sampling ports and piping or tubing. At the detector, the air is analyzed for fire products.

Alarm: In fire systems, a warning of fire danger.

Alarm signal: A signal indicating an emergency requiring immediate action, such as an alarm for fire from a manual station, water flow device, or automatic fire alarm system.

Alarm verification: A feature of a fire control panel that allows for a delay in the activation of alarms upon receiving an initiating signal from one of its circuits. Alarm verification must not be longer than three minutes, but can be adjustable from 0 to 3 minutes to allow supervising personnel to check the alarm. Alarm verification is commonly used in hotels, motels, hospitals, and institutions with large numbers of smoke detectors.

Americans with Disabilities Act (ADA): An act of Congress intended to ensure civil rights for physically challenged people.

Ampere turns: The product of amperes times the number of turns in a coil.

Analog-to-digital converter: A device designed to convert analog signals such as temperature and humidity to a digital form that can be processed by logic circuits.

Approved: Acceptable to the authority having jurisdiction.

Armature: The assembly of windings and metal core laminations in which the output voltage is induced. It is the rotating part in a revolving field generator.

Armor: A mechanical protector for cables; usually a formed metal tube or helical winding of metal tape formed so that each convolution locks mechanically upon the previous one (interlocked armor).

Audible signal: An audible signal is the sound made by one or more audible indicating appliances, such as bells, chimes, horns, or speakers, in response to the operation of an initiating device.

Authority having jurisdiction (AHJ): The authority having jurisdiction is the organization, office, or individual responsible for approving equipment, installations, or procedures in a particular locality.

Automatic fire alarm system: A system in which all or some of the circuits are actuated by automatic devices, such as fire detectors, smoke detectors, heat detectors, and flame detectors.

Automatic-changeover thermostat: A thermostat that automatically selects heating or cooling based on the space temperature.

Autonomy: The condition of having complete control over one's actions, and being free from the control of another. To be independent.

Autotransformer: Any transformer in which the primary and secondary connections are made to a single winding. The application of an autotransformer is a good choice when a 480Y/277V or 208Y/120V, three-phase, four-wire distribution system is used.

Avalanche breakover: A form of electrical current multiplication that can occur in insulating and semiconductor materials. It could be considered a "jail-break" of electron flow, leading to large currents in materials that otherwise would allow little or no current flow.

Bank: An installed grouping of a number of units of the same type of electrical equipment, such as a bank of transformers, a bank of capacitors, or a meter bank.

Base speed: The shaft speed at which the motor will develop rated horsepower at rated load and voltage, as stated on a motor nameplate.

Bias: A preconceived inclination against or in favor of something.

Bimetal: A control device made of two dissimilar metals that warp when exposed to heat, creating movement that can open or close a switch.

Border light: A permanently installed overhead strip light.

Building: A structure that stands alone or that is separated from adjoining structures by fire-walls.

CABO: Council of American Building Officials.

Carrier frequency: In a pulse-width modulated (PWM) drive, the frequency at which the IGBT can be switched on or off to create an output waveform that mimics a sine wave of desired voltage and frequency. Carrier frequencies are seldom less than 500Hz nor more than 20,000Hz; but need to be at least 10 times the maximum drive output frequency.

Ceiling: The upper surface of a space, regardless of height. Areas with a suspended ceiling would have two ceilings: one visible from the floor, and one above the suspended ceiling.

Ceiling height: The height from the continuous floor of a room to the continuous ceiling of a room or space.

Certification: A systematic program using randomly selected follow-up inspections of the certified system installed under the program, which allows the listing organization to verify that a fire alarm system complies with all the requirements of the NFPA 72® code. A system installed under such a program is identified by the issuance of a certificate and is designated as a certificated system.

Chimes: A single-stroke or vibrating audible signal appliance that has a xylophone-type striking bar.

Circuit: The conductors or radio channel as well as the associated equipment used to perform a definite function in connection with an alarm system.

Class A circuit: Class A refers to an arrangement of supervised initiating devices, signaling line circuits, or indicating appliance circuits (IAC) that prevents a single open or ground on the installation wiring of these circuits from causing loss of the system's intended function. It is also commonly known as a *four-wire circuit*.

Class B circuit: Class B refers to an arrangement of initiating devices, signaling lines, or indicating appliance circuits that does not prevent a single open or ground on the installation wiring of these circuits from causing loss of the system's intended function. It is commonly known as a *two-wire circuit*.

Closed-circuit transition: A method of reduced-voltage starting in which the motor being controlled is never removed from the source of voltage while moving from one voltage level to another.

Cloud-based applications: Mobile and desktop digital programs that can connect to files and data located in distributed storage locations on the Internet ("the Cloud"). Such applications make it possible for authorized users to create, edit, and distribute content from any location where Internet access is possible.

Coded signal: A signal pulsed in a prescribed code for each round of transmission.

Compressor: A component of a refrigeration system that converts low-pressure, low-temperature refrigerant gas into high-temperature, high-pressure refrigerant gas through compression.

Condenser: A heat exchanger that transfers heat from the refrigerant flowing inside it to air or water flowing over it. Refrigerant enters as a hot, high-pressure vapor and is condensed to a liquid as the heat is removed.

Constant-wattage heating cable: A heating cable that has the same power output over a large temperature range. It is used in applications requiring a constant heat output regardless of varying outside temperatures. Zoned parallel-resistance, mineral-insulated, and series-resistance heating cables are examples of constant-wattage heating cables.

Control unit: A device with the control circuits necessary to furnish power to a fire alarm system, receive signals from alarm initiating devices (and transmit them to audible alarm indicating appliances and accessory equipment), and electrically supervise the system installation wiring and primary (main) power. The control unit can be contained in one or more cabinets in adjacent or remote locations.

Cooling compensator: A fixed resistor installed in a thermostat to act as a cooling anticipator.

Core loss: The electric loss that occurs in the core of an armature or transformer due to conditions such as the presence of eddy currents or hysteresis.

Craft professionals: Workers who are properly trained and work in a particular construction trade or craft.

Crew leader: The immediate supervisor of a crew or team of craft professionals and other assigned persons.

Critical path: In manufacturing, construction, and other types of creative processes, the required sequence of tasks that directly controls the ultimate completion date of the project.

Dashpot: A device that uses a gas or fluid to absorb energy from or retard the movement of the moving parts of a relay, circuit breaker, or other electrical or mechanical device.

Dead leg: A segment of pipe designed to be in a permanent no-flow condition. This pipe section is often used as a control point for a larger system.

Dead-front: Equipment or enclosures covered or protected in order to prevent exposure of energized parts on the operating side of the equipment.

Deadband: A temperature band, often 3°F, that separates the heating mode and cooling mode in an automatic-changeover thermostat.

Demographics: Social characteristics and other factors, such as language, economics, education, culture, and age, that define a statistical group of individuals. An individual can be a member of more than one demographic.

Dew point: The temperature at which a vapor begins to condense into its liquid state as the result of a cooling effect.

Diac: A three-layer diode designed for use as a trigger in AC power control circuits, such as those using triacs.

Dielectric constant (K): A measurement of the ability of a material to store a charge.

Differential: The difference between the cut-in and cut-out points of a thermostat.

Digital Alarm Communicator Receiver (DACR): A system component that will accept and display signals from digital alarm communicator transmitters (DACTs) sent over public switched telephone networks.

Digital Alarm Communicator System (DACS): A system in which signals are transmitted from a digital alarm communicator transmitter (DACT) located at the protected premises through the public switched telephone network to a digital alarm communicator receiver (DACR).

Digital Alarm Communicator Transmitter (DACT): A system component at the protected premises to which initiating devices or groups of devices are connected. The DACT will seize the connected telephone line, dial a preselected number to connect to a DACR, and transmit signals indicating a status change of the initiating device.

Dwelling unit: A single unit providing complete and independent living facilities for one or more persons, including permanent provisions for living, sleeping, cooking, and sanitation. A one-family dwelling consists solely of one dwelling unit. A two-family dwelling consists solely of two dwelling units. A multi-family dwelling is a building containing three or more dwelling units.

Eddy currents: The circulating currents that are induced in conductive materials by varying magnetic fields; they are usually considered undesirable because they represent a loss of energy and produce excess heat.

Electrical datum plane: Horizontal plane affecting electrical installations based on the rise and fall of water levels in land areas and floating piers and landing stages.

Electrolyte: A substance in which the conduction of electricity is accompanied by chemical action (for example, the paste that forms the conducting medium between the electrodes of a dry cell or storage cell).

End-of-line (EOL) device: A device used to terminate a supervised circuit. An EOL is normally a resistor or a diode placed at the end of a two-wire circuit to maintain supervision.

Equipotential ground plane: A mass of conducting material that, when bonded together, provides a low impedance to current flow over a wide range of frequencies.

Equipotential plane: An area in wire mesh or other conductive material embedded in or placed under concrete that may conduct current. It must be bonded to all metal structures and nonelectrical, fixed equipment that could become energized, and then it must be connected to the electrical grounding system.

Ethics: The moral principles that guides an individual's or organization's actions when dealing with others. Also refers to the study of moral principles.

Evaporator: A heat exchanger that transfers heat from the air or water flowing inside to the cooler refrigerant flowing through it. Refrigerant enters as a cold liquid/vapor mixture and changes to all vapor as heat is absorbed from the air or water.

Exciter: A device that supplies direct current (DC) to the field coils of a synchronous generator, producing the magnetic flux required for inducing output voltage in the stator coils.

Expansion device: A device that provides a pressure drop that converts the high-temperature, high-pressure liquid refrigerant from the condenser into the low-temperature, low-pressure liquid/vapor mixture entering the evaporator; also known as the metering device.

Extra-hard usage cable: Cable designed with additional mechanical protection for use in areas where it may be subject to physical damage.

Fan Affinity Laws: The mathematical relationship between fan speed, airflow rate, power used, and torque. Airflow is directly proportional to speed. Torque is proportional to the square of speed. Power used is proportional to the cube of speed. For example, decreasing fan speed by 20% decreases the power used by about 50%.

Fault: An open, ground, or short condition on any line(s) extending from a control unit, which could prevent normal operation.

Feeder: All circuit conductors between the service equipment, the source of a separately derived system, or other supply source and the final branch circuit overcurrent device.

Festoon lighting: Outdoor lighting that is strung between supporting points.

Field-effect transistor (FET): A transistor that controls the flow of current through it with an electric field.

Fire: A chemical reaction between oxygen and a combustible material where rapid oxidation may cause the release of heat, light, flame, and smoke.

Fire rating: Time in hours or minutes that a material or assembly can withstand direct exposure to fire without igniting and spreading the fire.

Flame detector: A device that detects the infrared, ultraviolet, or visible radiation produced by a fire. Some devices are also capable of detecting the flicker rate (frequency) of the flame.

Footlight: A border light installed on or in a stage.

Forward bias: Forward bias exists when voltage is applied to a solid-state device in such a way as to allow the device to conduct easily.

Garage: A building or portion of a building in which one or more self-propelled vehicles can be kept for use, sale, storage, rental, repair, exhibition, or demonstration purposes.

General alarm: A term usually applied to the simultaneous operation of all the audible alarm signals on a system, to indicate the need for evacuation of a building.

Generator: (1) A rotating machine that is used to convert mechanical energy to electrical energy. (2) General apparatus or equipment that is used to convert or change energy from one form to another.

Ground fault: A condition in which the resistance between a conductor and ground reaches an unacceptably low level.

Harmonic: An oscillation at a frequency that is an integral multiple of the fundamental frequency.

Heat detector: A device that detects abnormally high temperature or rate-of-temperature rise.

Heat sink: A metal mounting base that dissipates the heat of solid-state components mounted on it by radiating the heat into the surrounding atmosphere.

Heat source: Any component that adds heat to an object or system.

Heat transfer: The movement of heat energy from a warmer substance to a cooler substance.

High-potential (hi-pot) test: A high-potential test in which equipment insulation is subjected to a voltage level higher than that for which it is rated to find any weak spots or deficiencies in the insulation.

Horn: An audible signal appliance in which energy produces a sound by imparting motion to a flexible component that vibrates at some nominal frequency.

Hysteresis: The time lag exhibited by a body in reacting to changes in the forces affecting it; hysteresis is an internal friction.

Impedance: The opposition to current flow in an AC circuit; impedance includes resistance (R), capacitive reactance (X_C), and inductive reactance (X_L). It is measured in ohms (Ω).

Indicating device: Any audible or visible signal employed to indicate a fire, supervisory, or trouble condition. Examples of audible signal appliances are bells, horns, sirens, electronic horns, buzzers, and chimes. A visible indicator consists of an incandescent lamp, strobe lamp, mechanical target or flag, meter deflection, or the equivalent. Also called a *notification device (appliance)*.

Infer: To reach a conclusion using a method of reasoning that starts with an assumption and considers a set of logically-related events, conditions, or statements.

Initiating device circuit (IDC): A circuit to which automatic or manual signal-initiating devices such as fire alarm manual boxes (pull stations), heat and smoke detectors, and water flow alarm devices are connected.

Initiating device: A manually or automatically operated device, the normal intended operation of which results in a fire alarm or supervisory signal indication from the control unit. Examples of alarm signal initiating devices are thermostats, manual boxes (stations), smoke detectors, and water flow devices. Examples of supervisory signal initiating devices are water level indicators, sprinkler system valveposition switches, pressure supervisory switches, and water temperature switches.

Inpatient: A patient admitted into the hospital for an overnight stay.

Insulated gate bipolar transistor (IGBT): A type of transistor that has low losses and low gate drive requirements. This allows it to be operated at higher switching frequencies.

Insulation breakdown: The failure of insulation to prevent the flow of current, sometimes evidenced by arcing. If the voltage is gradually raised, breakdown will begin suddenly at a certain voltage level. Current flow is not directly proportional to voltage. When a breakdown current has flowed, especially for a period of time, the next gradual application of voltage will often show breakdown beginning at a lower voltage than initially.

Insulation classes: Categories of insulation based on the thermal endurance of the insulation system used in a motor. The insulation system is chosen to ensure that the motor will perform at the rated horsepower and service factor load.

Insulation shielding: An electrically conductive layer that provides a smooth surface in contact with the insulation outer surface; it is used to eliminate electrostatic charges external to the shield and to provide a fixed, known path to ground.

Intangible: Not touchable, material, or measureable; lacking a physical presence.

Invar®: An alloy of steel containing 36% nickel. It is one of the two metals in a bimetal control device.

Isolation transformer: A transformer that has no electrical metallic connection between the primary and secondary windings.

Job description: A description of the scope and responsibilities of a worker's job so that the individual and others understand what the job entails.

Job diary: A written record that a supervisor maintains periodically (usually daily) of the events, communications, observations, and decisions made during the course of a project.

Junction field-effect transistor (JFET): A field-effect transistor formed by combining layers of semiconductor material.

Labeled: In the context of fire alarm control panels, tags that identify various zones and sensors by descriptive names. Panels may include dual identifiers, one of which is meaningful to the occupants, while the other is helpful to firefighting personnel.

Landing stage: A floating platform used for the loading and unloading of passengers and cargo from boats or ships.

Latching: In a solid-state device, latching is the equivalent to the holding or seal-in circuit in relay logic. The current flowing through the relay maintains flow until the load is removed.

Leakage current: AC or DC current flow through insulation and over its surfaces, and AC current flow through a capacitance. Current flow is directly proportional to voltage. The insulation and/or capacitance are thought of as a constant impedance unless breakdown occurs.

Legend: In maps, plans, and diagrams, an explanatory table defining all symbolic information contained in the document.

Lethargy: Sluggishness, slow motion, lack of activity, or a lack of enthusiasm.

Letter of instruction (LOI): A written communication from a supervisor to a subordinate that informs the latter of some inadequacy in the individual's performance, and provides a list of actions that the individual must satisfactorily complete to remediate the problem. Usually a key step in a series of disciplinary actions.

Light scattering: The action of light being reflected or refracted off particles of combustion, for detection in a modern-day photoelectric smoke detector. This is called the Tyndall effect.

Listed: Equipment or materials included in a list published by an organization acceptable to the authority having jurisdiction that is concerned with product evaluation and whose listing states either that the equipment or materials meets appropriate standards or has been tested and found suitable for use in a specified manner.

Local area networks (LAN): Communication networks that link computers, printers, and servers within a small, defined location, such as a building or office, via hard-wired or wireless connections.

Lockout/tagout (LOTO): A system of safety procedures for securing electrical and mechanical equipment during repairs or construction, consisting of warning tags and physical locking or restraining devices applied to controls to prevent the accidental or purposeful operation of the equipment, endangering workers, equipment, or facilities.

Look-ahead schedule: A manual- or software-scheduling tool that looks several weeks into the future at the planned project events; used for anticipating material, labor, tool, and other resource requirements, as well as identifying potential schedule conflicts or other problems.

Maintenance: Repair service, including periodic inspections and tests, required to keep the protective signaling system and its component parts in an operative condition at all times. This is used in conjunction with replacement of the system and its components when for any reason they become undependable or inoperative.

Mechanical refrigeration: The use of machinery to provide a cooling effect.

Megohmmeter: An instrument or meter capable of measuring resistances in excess of 200 $M\Omega$ It employs much higher test voltages than are used in ohmmeters, which measure up to 200 $M\Omega$. Commonly referred to as a megger.

Mineral-insulated (MI) heating cable: A type of constant-wattage, series-resistance heating cable used in long line applications where high temperatures need to be maintained, high-temperature exposure exists, or high power output is required.

Multiplexing: A signaling method that uses wire path, cable carrier, radio, fiber optics, or a combination of these techniques, and characterized by the simultaneous or sequential (or both simultaneous and sequential) transmission and reception of multiple signals in a communication channel including means of positively identifying each signal.

N-type material: A material created by doping a region of a crystal with atoms from an element that has more electrons in its outer shell than the crystal.

National Fire Alarm Code®: This is the update of the NFPA standards book that contains the former NFPA 71, NFPA 72®, and NFPA 74 standards, as well as the NFPA 1221 standard. The NFAC was adopted and became effective May 1993.

National Fire Protection Association (NFPA): The NFPA administers the development and publishing of codes, standards, and other materials concerning all phases of fire safety.

Negligence: Lack of appropriate care when doing something, or the failure to do something, usually resulting in injury to an individual or damage to equipment.

Nichrome: An alloy based on nickel with chromium added (10–20 percent), and sometimes iron (up to 25 percent). Nichrome is often used for high-temperature applications such as in the fabrication of electric heating elements.

Noise: A term used in electronics to cover all types of unwanted electrical signals. Noise signals originate from numerous sources, such as fluorescent lamps, walkie-talkies, amateur and CB radios, machines being switched on and off, and power surges. Today, equipment must tolerate increasing amounts of electrical interference, and the quality of equipment depends on how much noise it can ignore and withstand.

Non-coded signal: A signal from any indicating appliance that is continuously energized.

Notification device (appliance): Any audible or visible signal employed to indicate a fire, supervisory, or trouble condition. Examples of audible signal appliances are bells, horns, sirens, electronic horns, buzzers, and chimes. A visible indicator consists of an incandescent lamp, strobe lamp, mechanical target or flag, meter deflection, or the equivalent. Also called an *indicating device*.

Obscuration: A reduction in the atmospheric transparency caused by smoke, usually expressed in percent per foot.

Open-circuit transition: A method of reduced-voltage starting in which the motor being controlled may be disconnected temporarily from the source of voltage while moving from one voltage level to another.

Organizational chart: A diagram that shows how the various management and operational responsibilities relate to each other within an organization. Named positions appear in ranked levels, top to bottom, from those functional units with the most and broadest authority to those with the least, with lines connecting positions indicating chains of authority and other relationships.

Outpatient: A patient who receives services at a hospital but is not admitted for an overnight stay.

Overhead service conductors: The overhead conductors between the service point and the first point of connection to the service-entrance conductors at the building or other structure.

P-type material: A material created when a crystal is doped with atoms from an element that has fewer electrons in its outer shell than the natural crystal. This combination creates empty spaces in the crystalline structure. The missing electrons in the crystal structure are called *holes* and are represented as positive charges.

Parallel-resistance heating cable: An electric heating cable with parallel connections, either continuous or in zones. The watt density per lineal length is approximately equal along the length of the heating cable, allowing for a drop in voltage down the length of the heating cable.

Paraphrase: Rewording a written or verbal statement in one's own words in such a way that the intent of the original statement is retained.

Path (pathway): Any conductor, optic fiber, radio carrier, or other means for transmitting fire alarm system information between two or more locations.

Photoelectric smoke detector: A detector employing the photoelectric principle of operation using either the obscuration effect or the light-scattering effect for detecting smoke in its chamber.

Portable switchboard: A movable lighting switchboard used for road shows and similar presentations, usually supplied by a flexible cord or cable.

Positive alarm sequence: An automatic sequence that results in an alarm signal, even when manually delayed for investigation, unless the system is reset.

Positive locking device: A screw, clip, or other mechanical locking device that requires a tool or key to lock or unlock it.

Pothead: A terminator for high-voltage circuit conductors that keeps moisture out of the insulation and protects the cable end, along with providing a suitable stress relief cone for shielded-type conductors.

Potting compound: An approved and listed insulating compound, such as epoxy, which is mixed and installed in and around electrical connections to encapsulate and seal the connections to prevent the entry of water or moisture.

Power supply: A source of electrical operating power, including the circuits and terminations connecting it to the dependent system components.

Power-limiting heating cable: A type of heating cable that shows positive temperature coefficient (PTC) behavior based on the properties of a metallic heating element. The PTC behavior exhibited is much less (a smaller change in resistance in response to a change in temperature) than that shown by self-regulating heating cables.

Pragmatic: Sensible, practical, and realistic.

Proactive: To anticipate and take action in the present to deal with potential future events or outcomes based on what one knows about current events or conditions.

Project manager: In construction, the individual who has overall responsibility for one or more construction projects; also called the general superintendent.

Projected beam smoke detector: A type of photoelectric light-obscuration smoke detector in which the beam spans the protected area.

Proscenium: The wall and arch that separate the stage from the auditorium (house).

Protected premises: The physical location protected by a fire alarm system.

Public Switched Telephone Network: An assembly of communications facilities and central office equipment, operated jointly by authorized common carriers, that provides the general public with the ability to establish communications channels via discrete dialing codes.

Rate compensation detector: A device that responds when the temperature of the air surrounding the device reaches a predetermined level, regardless of the rate of temperature rise.

Rate-of-rise detector: A device that responds when the temperature rises at a rate exceeding a predetermined value.

Reactance: The imaginary part of impedance; also, the opposition to alternating current due to capacitance (X_C) and/or inductance (X_L).

Reactor-type dimmer: A dimming system that incorporates a number of windings or coils that are linked by a magnetic core, an inductive reactor that functions like a capacitor, and a relay. The reactor stores current from the source and reduces lamp power.

Rectifier: A device or circuit commonly used to change AC voltage into DC voltage, or as a solid-state switch.

Refrigerant: Typically, a fluid that picks up heat by evaporating at a low temperature and pressure and gives up heat by condensing at a higher temperature and pressure.

Refrigeration cycle: The process by which a circulating refrigerant absorbs heat from one location and transfers it to another location.

Remote supervising station fire alarm system: A system installed in accordance with the applicable code to transmit alarm, supervisory, and trouble signals from one or more protected premises to a remote location where appropriate action is taken.

Reset: A control function that attempts to return a system or device to its normal, non-alarm state.

Resistance-type dimmer: A manual-type dimmer in which light intensity is controlled by increasing or decreasing the resistance in the lighting circuit.

Return on investment (ROI): A measure of the gain or loss of money resulting from an investment; normally measured as a percentage of the original investment.

Reverse bias: A condition that exists when voltage is applied to a device in such a way that it causes the device to act as an insulator.

Rheostat: A slide and contact device that increases or decreases the resistance in a circuit by changing the position of a movable contact on a coil of wire.

Ribbon tape shield: A helical (wound coil) strip under the cable jacket.

Safety data sheets (SDS): Documents listing information about a material or substance that includes common and proper names, chemical composition, physical forms and properties, hazards, flammability, handling, and emergency response in accordance with national and international hazard communication standards. Also called material safety data sheets (MSDS).

Self-regulating heating cable: A type of heating cable that inversely varies its heat output in response to an increase or decrease in the ambient temperature.

Semiconductors: Materials that are neither good insulators nor a good conductors. Such materials contain four valence electrons and are used in the production of solid-state devices.

Separately derived system: An electrical source, other than a service, having no direct connection(s) to circuit conductors of any other electrical source other than those established by grounding and bonding connections.

Series-resistance heating cable: A type of heating cable in which the ohmic heating of the conductor provides the heat. The wattage output depends on the total circuit length and on the voltage applied.

Service conductors: The conductors from the service point to the service disconnecting means.

Service: The conductors and equipment for delivering electric energy from the serving utility to the wiring system of the premises served.

Service drop: The overhead conductors between the utility electric supply system and the service point.

Service equipment: The necessary equipment, usually consisting of a circuit breaker or switch and fuses, and their accessories, connected to the load end of service conductors to a building or other structure, or an otherwise designated area, and intended to constitute the main control and means of cutoff of the supply.

Service lateral: The underground conductors between the utility electric supply system and the service point.

Service point: The point of connection between the facilities of the serving utility and the premises wiring.

Sexual harassment: Any unwelcome verbal or nonverbal form of communication or action construed by an individual to be of a sexual or gender-related nature.

Shield: A conductive barrier against electromagnetic fields.

Shielding braid: A shield of small, interwoven wires.

Shunt trip: A low-voltage device that remotely causes a higher voltage overcurrent protective device to trip or open the circuit.

Signal: A status indication communicated by electrical or other means.

Signaling line circuits (SLCs): A circuit or path between any combination of circuit interfaces, control units, or transmitters over which multiple system input signals or output signals (or both input signals and output signals) are carried.

Silicon-controlled rectifier (SCR): A device that is used mainly to convert AC voltage into DC voltage. To do so, however, the gate of the SCR must be triggered before the device will conduct current.

Site-isolating device: A disconnecting means at the point of distribution that is used to isolate the service for the purpose of electrical maintenance, emergency disconnection, or the connection of an optional standby system.

Slip: The speed difference between the synchronous speed and the base speed. Induction motors develop their torque because of the slip. The amount of slip depends on motor design and motor load.

Smartphone: The most common type of cellular telephone that combines many other digital functions within a single mobile device. Most important of these, besides the wireless telephone, are high-resolution still and video cameras, text and email messaging, and GPS-enabled features.

Smoke detector: A device that detects visible or invisible particles of combustion.

Soft foot: When the feet of a motor frame are forced to anchor against an uneven surface, resulting in distortion of the motor frame. This can cause misalignment of bearings.

Solid-state device: An electronic component constructed from semiconductor material. Such devices have all but replaced the vacuum tube in electronic circuits.

Solid-state overload relay (SSOLR): An electronic device that provides overload protection for electrical equipment.

Solid-state phase-control dimmer: A solid-state dimmer in which the wave shape of the steady-state current does not follow the wave shape of the applied voltage (that is, the wave shape is nonlinear).

Solid-state relay (SSR): A switching device that has no contacts and switches entirely by electronic means.

Solid-state sine-wave dimmer: A solid-state dimmer in which the wave shape of the steady-state current follows the wave shape of the applied voltage (that is, the wave shape is linear).

Spacing: A horizontally measured dimension related to the allowable coverage of fire detectors.

Splice: Two or more conductors joined with a suitable connector and reinsulated, reshielded, and rejacketed with compatible materials applied over a properly prepared surface.

Spot-type detector: A device in which the detecting element is concentrated at a particular location. Typical examples are bimetallic detectors, fusible alloy detectors, certain pneumatic rate-of-rise detectors, certain smoke detectors, and thermoelectric detectors.

Startup temperature: The lowest temperature at which a heat-tracing cable is energized.

Strand shielding: The semiconductive layer between the conductor and insulation that compensates for air voids that exist between the conductor and insulation.

Stratification: The phenomenon in which the upward movement of smoke and gases ceases due to a loss of buoyancy.

Stress: An internal force set up within a body to resist or hold it in equilibrium.

Sub-base: The portion of a two-part thermostat that contains the wiring terminals and control switches.

Subcooling: The measurable heat removed from a liquid after it has been cooled to its condensing temperature, has changed to a fully liquid state, and heat continues to be removed.

Superheat: The measureable heat added to a vapor after it has been heated to its boiling point, has changed to a fully vapor state, and heat continues to be added.

Superintendent: In construction, the individual who is the on-site supervisor in charge of a given construction project.

Supervisory signal: A signal indicating the need for action in connection with the supervision of guard tours, the fire suppression systems or equipment, or the maintenance features of related systems.

Switchboard: A large single panel, frame, or assembly of panels on which are mounted on the face, back, or both, switches, overcurrent and other protective devices, buses, and usually instruments. Switchboards are generally accessible from the rear as well as from the front and are not intended to be installed in cabinets.

Synchronous generator: An AC generator having a DC exciter. Synchronous generators are used as stand-alone generators for emergency and standby power systems and can also be paralleled with other synchronous generators and the utility system.

Synergy: Any type of cooperation between organizations, individuals, or other entities where the combined effect is greater than the sum of the individual efforts.

System unit: The active subassemblies at the central station used for signal receiving, processing, display, or recording of status change signals. The failure of one of these subassemblies causes the loss of a number of alarm signals by that unit.

Tap conductors: As defined in NEC Article 240, a tap conductor is a conductor, other than a service conductor, that has overcurrent protection ahead of its point of supply that exceeds the value permitted for similar conductors that are protected as described in NEC Section 240.4.

Textspeak: A form of written language characteristic of text messaging on mobile devices and text-based social media, which usually consists of acronyms, abbreviations, and minimal punctuation.

Thermal conductivity: The ability of a material or substance to transfer heat away from a given point of contact.

Thermostat: A device that is responsive to ambient temperature conditions.

Throwover: A transfer switch used for supplying temporary power.

Thyristor: A bi-stable semiconductor device that can be switched off and on. Thyristors turn on with a quick pulse of control current. They turn off only when the working current is interrupted elsewhere in the circuit. SCRs and triacs are forms of thyristors.

Totally enclosed motor: A motor that is encased to prevent the free exchange of air between the inside and outside of the case.

Transmitter: A system component that provides an interface between the transmission channel and signaling line circuits, initiating device circuits, or control units.

Triac: A bidirectional triode thyristor that functions as an electrically controlled switch for AC loads.

Trouble signal: A signal initiated by the fire alarm system or device that indicates a fault in a monitored circuit or component.

Unbalanced load: Three-phase loads in which the loads are not evenly distributed between the phases.

Underground service conductors: The underground conductors between the service point and the first point of connection to the service-entrance conductors in a terminal bus.

Vessel: A container such as a barrel, drum, or tank used for holding fluids or other substances.

Visible notification appliance: A notification appliance that alerts by the sense of sight.

Voltage gradients: Fluctuating voltage potentials that can exist between two surfaces or components and may be caused by improper or insufficient bonding or grounding.

Wavelength: The distance between peaks of a sinusoidal wave. All radiant energy can be described as a wave having a wavelength. Wavelength serves as the unit of measure for distinguishing between different parts of the spectrum. Wavelengths are measured in microns, nanometers, or angstroms.

Wi-Fi: The technology allowing communications via radio signals over a LAN or WAN equipped with a wireless access point or the Internet. (Wi-Fi stands for wireless fidelity.) Many types of mobile, portable, and desktop devices can communicate via Wi-Fi connections.

Wide area networks (WAN): Dedicated communication networks of computers and related hardware that serve a given geographic area, such as a work site, campus, city, or a larger but distinct area. Connectivity is by wired and wireless means, and may use the Internet as well.

Wide Area Telephone Service (WATS): Telephone company service that provides reduced costs for certain telephone call arrangements. In-WATS or 800-number service calls can be placed from anywhere in the continental United States to the called party at no cost to the calling party. Out-WATS is a service whereby, for a flat-rate charge, dependent on the total duration of all such calls, a subscriber can make an unlimited number of calls within a prescribed area from a particular telephone terminal without the registration of individual call charges.

Work breakdown structure (WBS): A diagrammatic and conceptual method for subdividing a complex project, concept, or other thing into its various functional and organizational parts so that planners can analyze and plan for each part in detail.

Zone: A defined area within the protected premises. A zone can define an area from which a signal can be received, an area to which a signal can be sent, or an area in which a form of control can be executed.

This page is intentionally left blank.

This page is intentionally left blank.

This page is intentionally left blank.

This page is intentionally left blank.

This page is intentionally left blank.

This page is intentionally left blank.

This page is intentionally left blank.